Emergence, Complexity and Computation

Volume 38

The Emergence, Complexity and Computation (ECC) series publishes new developments, advancements and selected topics in the fields of complexity, computation and emergence. The series focuses on all aspects of reality-based computation approaches from an interdisciplinary point of view especially from applied sciences, biology, physics, or chemistry. It presents new ideas and interdisciplinary insight on the mutual intersection of subareas of computation, complexity and emergence and its impact and limits to any computing based on physical limits (thermodynamic and quantum limits, Bremermann's limit, Seth Lloyd limits…) as well as algorithmic limits (Gödel's proof and its impact on calculation, algorithmic complexity, the Chaitin's Omega number and Kolmogorov complexity, non-traditional calculations like Turing machine process and its consequences,…) and limitations arising in artificial intelligence. The topics are (but not limited to) membrane computing, DNA computing, immune computing, quantum computing, swarm computing, analogic computing, chaos computing and computing on the edge of chaos, computational aspects of dynamics of complex systems (systems with self-organization, multiagent systems, cellular automata, artificial life,…), emergence of complex systems and its computational aspects, and agent based computation. The main aim of this series is to discuss the above mentioned topics from an interdisciplinary point of view and present new ideas coming from mutual intersection of classical as well as modern methods of computation. Within the scope of the series are monographs, lecture notes, selected contributions from specialized conferences and workshops, special contribution from international experts.

More information about this series at http://www.springer.com/series/10624

Nikolay Kuznetsov · Volker Reitmann

Attractor Dimension Estimates for Dynamical Systems: Theory and Computation

Dedicated to Gennady Leonov

Nikolay Kuznetsov
Department of Applied Cybernetics
Faculty of Mathematics and Mechanics
St. Petersburg State University
St. Petersburg, Russia

Faculty of Information Technology
University of Jyväskylä
Jyväskylä, Finland

Volker Reitmann
Department of Applied Cybernetics
Faculty of Mathematics and Mechanics
St. Petersburg State University
St. Petersburg, Russia

ISSN 2194-7287 ISSN 2194-7295 (electronic)
Emergence, Complexity and Computation
ISBN 978-3-030-50989-7 ISBN 978-3-030-50987-3 (eBook)
https://doi.org/10.1007/978-3-030-50987-3

This Springer imprint is published by the registered company Springer Nature Switzerland AG
The registered company address is: Gewerbestrasse 11, 6330 Cham, Switzerland

Preface

In this book, we continue the investigations of global attractors and invariant sets for dynamical systems by means of Lyapunov functions and adapted metrics. The effectiveness of such approaches for the approximation and localization of attractors for different classes of dynamical systems was already shown in Abramovich et al. [1, 2] and in [30, 32]. In particular, Lyapunov functions and adapted metrics were constructed for global stability problems and the existence of homoclinic orbits in the Lorenz system using frequency-domain methods and reduction principles.

In 1980, investigators of differential equations and general dynamical systems were greatly impressed by a paper about upper estimates of the Hausdorff dimension of flow and map invariant sets written by Douady and Oesterlé [12]. The Douady–Oesterlé approach, the significance of which can be compared with that of Liouville's theorem, has been developed and modificated in many papers for various types of dimension characteristics of attractors generated by dynamical systems: Ledrappier [22], Constantin et al. [11], Smith [42], Eden et al. [13], Chen [9], Hunt [17], Boichenko and Leonov [4].

After Ya. B. Pesin had worked out [39] a general scheme of introducing metric dimension characteristics, this method made it possible to define from a unique point of view various types of outer measures and dimensions, such as the Hausdorff dimension, the fractal dimension, the information dimension as well as the topological and metric entropies. The Pesin scheme naturally led to the characterization of a class of Carathéodory measures [25, 27], which are adapted to the specific character of attractors of autonomous differential equations, i.e. to the fact that these attractors consist wholly of trajectories. The neighborhoods of pieces of these trajectories form a covering of the attractor. It serves as the base for introducing the special outer Carathéodory measures. A number of effective tools for estimating these measures were developed within the theory of differential equations in Euclidean space and well-known results by Borg [8], Hartman and Olech [15], when analysing the orbital stability of solutions. The most important property, used in dimension theory, is the fact that the Carathéodory measures are majorants for the associated Hausdorff measures.

Early in the nineties of the last century, G. A. Leonov and his co-workers observed deep inner connections between the mentioned direct method of Lyapunov in stability theory and estimation technics for outer measures in dimension theory. Introducing the Lyapunov functions and varying Riemannian metrices (Leonov [24], Noack and Reitmann [38]) into upper estimates of dimension characteristics of invariant sets made it possible to generalize and improve [4, 6, 28, 29] some well-known results of R. A. Smith, P. Constantin, C. Foias, A. Eden, and R. Temam.

On the other hand, Pugh's closing lemma [40] and theorems about the spanning of two-dimensional surfaces on a given closed curve gave the opportunity to apply some theorems about the contraction of Hausdorff measures to global stability investigations of time-continuous dynamical systems (Smith [42], Leonov [24], Li and Muldowney [35]). In this book, the effectiveness of introducing the Lyapunov functions into dimensional characteristics is shown for a number of concrete dynamical systems: the Hénon map, the systems of Lorenz and Rössler as well as their generalizations for various physical systems and models (rotation of a rigid body in a resisting medium, convection of liquid in a rotating ellipsoid, interaction between waves in plasma, etc.).

Additionally to the derivation of upper Hausdorff dimension estimates, exact formulas for the Lyapunov dimensions for Lorenz type systems were shown [26, 28]. Many of these results were presented in [7].

In the following decade, the modified Douady–Oesterlé approach was also used for new classes of attractors [33].

It was also possible to get different versions of the Douady–Oesterlé theorem for piecewise continuous maps and differential equations [37, 41]. For cocycles generated by non-autonomous systems, the upper Hausdorff dimension estimates are derived in [31, 34]. Some of these results are included in the present book which provides a systematic presentation of research activities in the dimension theory of dynamical systems in finite-dimensional Euclidean spaces and manifolds. Let us briefly sketch the contents of the book.

In Part I, we consider the basic facts from attractor theory, exterior products and dimension theory. Chapter 1 is devoted to the investigation of various types of global attractors of dynamical systems in general metric spaces (global $\mathcal{B}$-attractors, minimal global $\mathcal{B}$-attractors and others). The theoretical results are applied to the generalized Lorenz system and dynamical systems on the flat cylinder. One section is concerned with the existence proof of a homoclinic orbit in the Lorenz system (Leonov [23], Hastings and Troy [16], Chen [10]).

In Chap. 2, some facts on singular values of matrices, the exterior calculus for spaces and matrices and the Lozinskii matrix norm, necessary for estimation techniques of outer measures, are presented. In addition to this, the Yakubovich–Kalman frequency theorem and the Kalman–Szegö theorem about the solvability of certain matrix inequalities are formulated and used for the estimation of singular values.

Chapter 3 is an introduction to dimension theory. It starts with the definition and the basic properties of the topological dimension in the spirit of Hurewicz and Wallman [18]. Next, the notions of Hausdorff measure, Hausdorff dimension and fractal dimension are introduced. After this, the topological entropy of a continuous map is discussed. The last part of the chapter deals with Pesin's scheme of introducing the Carathéodory dimension characteristics.

In Part II, we investigate dimension properties of dynamical systems in Euclidean spaces. It includes estimates of topological dimension of the Hausdorff and fractal dimensions for invariant sets of concrete physical systems and estimates of the Lyapunov dimension.

Chapter 4 is concerned with the investigation of dimension properties of almost periodic flows [3]. We thank M. M. Anikushin for helping us to prepare the first version of Chap. 4.

Chapter 5 begins with the so-called limit theorem about the evolution of Hausdorff measures under the action of smooth maps in Euclidean spaces. This theorem gives the opportunity to include into Hausdorff dimension and Hausdorff measure estimates, certain varying functions which turn over into Lyapunov functions when the theorem is applied to differential equations. The method of varying functions is used in estimations of fractal dimension and topological entropy. Simultaneously, with the introduction of Lyapunov functions in estimates for the Hausdorff measure, the logarithmic norms were used by Muldowney [36], for estimating the two-dimensional Riemannian volumes of compact sets shifted along the orbits of differential equations. One of the main goals of Chap. 5 is to combine the Lyapunov function and logarithmic norm approaches (Boichenko and Leonov [5]), in order to solve a number of problems in the qualitative theory of ordinary differential equations, such as the generalization of the Liouville formula and the Bendixson criterion.

Chapter 6 is devoted to finite-dimensional dynamical systems in Euclidean space and its aim is to explain, in a simple but rigorous way, the connection between the key works in the area: Kaplan and Yorke (the concept of Lyapunov dimension [19] 1979), Douady and Oesterlé (estimation of Hausdorff dimension via the Lyapunov dimension of maps [12]), Constantin, Eden, Foias, and Temam (estimation of Hausdorff dimension via the Lyapunov exponents and Lyapunov dimension of dynamical systems [13]), Leonov (estimation of the Lyapunov dimension via the direct Lyapunov method [20, 24]), and numerical methods for the computation of Lyapunov exponents and Lyapunov dimension [21]. We also concentrate in this chapter on the Kaplan–Yorke formula and the Lyapunov dimension formulas for the Lorenz and Hénon attractors (Leonov [26]).

In Part III, we consider dimension properties for dynamical systems on manifolds. Chapter 7 gives a presentation of the exterior calculus in general linear spaces. It contains also some results about orbital stability for vector fields on manifolds.

Chapter 8 is devoted to dimension estimates of invariant sets and attractors of dynamical systems on Riemannian manifolds. The Douady–Oesterlé theorem for the upper Hausdorff dimension estimates for invariant sets of smooth dynamical systems on Riemannian manifolds is proved. Chapter 8 contains also an important result on the estimation of the fractal dimension of an invariant set on an arbitrary finite-dimensional smooth manifold by the upper Lyapunov dimension, which goes back to Hunt [17], Gelfert [14]. Then we discuss the construction of the special Carathéodory measures for the estimation of Hausdorff measures connected with flow invariant sets on Riemannian manifolds.

In Chap. 9, we derive dimension and entropy estimates for invariant sets and global $\mathcal{B}$-attractors of cocycles. A version of the Douady–Oesterlé theorem (Leonov et al. [31]) is proved for local cocycles in a Euclidean space and for cocycles on Riemannian manifolds. As examples, we consider cocycles, generated by the Rössler system with variable coefficients. We also introduce time-discrete cocycles on fibered spaces and define the topological entropy of such cocycles. We thank A. O. Romanov for helping us to prepare this chapter.

In Chap. 10, we derive some versions of the Douady–Oesterlé theorem for systems with singularities. In the first part of this chapter, we consider a special class of non-injective maps, for which we introduce a factor describing the "degree of non-injectivity" (Boichenko et al. [6]). This factor can be included in the dimension estimates of Chap. 8 in order to weaken the condition to the singular value function. In the second part of Chap. 10, we derive the upper Hausdorff dimension estimates for invariant sets of a class of not necessarily invertible and piecewise smooth maps on manifolds with controllable preimages of the non-differentiability sets in terms of the singular values of the derivative of the smoothly extended map (Reitmann and Schnabel [41], Neunhäuserer [37]) These estimates generalize some Douady–Oesterlé type results for differentiable maps in a Euclidean space, derived in Chaps. 5 and 8.

In the last section of Chap. 10, we discuss some classes of functionals which are useful for the estimation of topological and metric dimensions.

St. Petersburg, Russia

Nikolay Kuznetsov
Volker Reitmann

References

1. Abramovich, S., Koryakin, Yu., Leonov, G., Reitmann, V.: Frequency-domain conditions for oscillations in discrete systems. I., Oscillations in the sense of Yakubovich in discrete systems. Wiss. Zeitschr. Techn. Univ. Dresden. **25**(5/6), 1153–1163 (1977) (German)
2. Abramovich, S., Koryakin, Yu., Leonov, G., Reitmann, V.: Frequency-domain conditions for oscillations in discrete systems. II., Oscillations in discrete phase systems. Wiss. Zeitschr. Techn. Univ. Dresden. **26**(1), 115–122 (1977) (German)
3. Anikushin, M.M.: Dimension theory approach to the complexity of almost periodic trajectories. Intern. J. Evol. Equ. **10**(3–4), 215–232 (2017)
4. Boichenko, V.A., Leonov, G.A.: Lyapunov's direct method in the estimation of the Hausdorff dimension of attractors. Acta Appl. Math. **26**, 1–60 (1992)

5. Boichenko, V.A., Leonov, G.A.: Lyapunov functions, Lozinskii norms, and the Hausdorff measure in the qualitative theory of differential equations. Amer. Math. Soc. Transl. **193**(2), 1–26 (1999)
6. Boichenko, V.A., Leonov, G.A., Franz, A., Reitmann,V.: Hausdorff and fractal dimension estimates of invariant sets of non-injective maps. Zeitschrift für Analysis und ihre Anwendungen (ZAA). **17**(1), 207–223 (1998)
7. Boichenko, V.A., Leonov, G.A., Reitmann, V.: Dimension Theory for Ordinary Differential Equations. Teubner, Stuttgart (2005)
8. Borg, G.: A condition for existence of orbitally stable solutions of dynamical systems. Kungl. Tekn. Högsk. Handl. Stockholm. **153**, 3–12 (1960)
9. Chen, Zhi-Min.: A note on Kaplan-Yorke-type estimates on the fractal dimension of chaotic attractors. Chaos, Solitons & Fractals **3**, 575–582 (1993)
10. Chen, X.: Lorenz equations, part I: existence and nonexistence of homoclinic orbits. SIAM J. Math. Anal. **27**(4), 1057–1069 (1996)
11. Constantin, P., Foias, C., Temam, R.: Attractors representing turbulent flows. Amer. Math. Soc. Memoirs., Providence, Rhode Island. **53**(314), (1985)
12. Douady, A., Oesterlé, J.: Dimension de Hausdorff des attracteurs. C. R. Acad. Sci. Paris, Ser. A. **290**, 1135–1138 (1980)
13. Eden, A., Foias, C., Temam, R.: Local and global Lyapunov exponents. J. Dynam. Diff. Equ. **3**, 133–177 (1991) [Preprint No. 8804, The Institute for Applied Mathematics and Scientific Computing, Indiana University, 1988]
14. Gelfert, K.: Maximum local Lyapunov dimension bounds the box dimension. Direct proof for invariant sets on Riemannian manifolds. Zeitschrift für Analysis und ihre Anwendungen (ZAA). **22**(3), 553–568 (2003)
15. Hartman, P., Olech, C.: On global asymptotic stability of solutions of ordinary differential equations. Trans. Amer. Math. Soc. **104**, 154–178 (1962)
16. Hastings, S.P., Troy, W.C.: A shooting approach to chaos in the Lorenz equations. J. Diff. Equ. **127**(1), 41–53 (1996)
17. Hunt, B.: Maximum local Lyapunov dimension bounds the box dimension of chaotic attractors. Nonlinearity. **9**, 845–852 (1996)
18. Hurewicz, W., Wallman, H.: Dimension Theory. Princeton Univ. Press, Princeton (1948)
19. Kaplan, J.L., Yorke, J.A.: Chaotic behavior of multidimensional difference equations. In: Functional Differential Equations and Approximations of Fixed Points, 204–227, Springer, Berlin (1979)
20. Kuznetsov, N.V.: The Lyapunov dimension and its estimation via the Leonov method. Physics Letters A, **380**(25–26), 2142–2149 (2016)
21. Kuznetsov, N.V., Leonov, G.A., Mokaev, T.N., Prasad, A., Shrimali, M.D.: Finite-time Lyapunov dimension and hidden attractor of the Rabinovich system. Nonlinear Dyn. **92** (2), 267–285 (2018)
22. Ledrappier, F.: Some relations between dimension and Lyapunov exponents. Commun. Math. Phys. **81**, 229–238 (1981)
23. Leonov, G.A.: On the estimation of the bifurcation parameter values of the Lorenz system. Uspekhi Mat. Nauk. **43**(3), 189–200 (1988) (Russian); English transl. Russian Math. Surveys. **43**(3), 216–217 (1988)
24. Leonov, G.A.: Estimation of the Hausdorff dimension of attractors of dynamical systems. Diff. Urav. **27**(5), 767–771 (1991) (Russian); English transl. Diff. Equations, **27**, 520–524 (1991)
25. Leonov, G.A.: Construction of a special outer Carathéodory measure for the estimation of the Hausdorff dimension of attractors. Vestn. S. Peterburg Gos. Univ. **1**(22), 24–31 (1995) (Russian); English transl. Vestn. St. Petersburg Univ. Math. Ser. 1, **28**(4), 24–30 (1995)
26. Leonov, G.A.: Lyapunov dimensions formulas for Hénon and Lorenz attractors. Alg. & Anal. **13**, 155–170 (2001) (Russian); English transl. St. Petersburg Math. J. **13**(3), 453–464 (2002)

27. Leonov, G.A., Gelfert, K., Reitmann, V.: Hausdorff dimension estimates by use of a tubular Carathéodory structure and their application to stability theory. Nonlinear Dyn. Syst. Theory, **1**(2), 169–192 (2001)
28. Leonov, G.A., Lyashko, S.: Eden's hypothesis for a Lorenz system. Vestn. S. Peterburg Gos. Univ., Matematika. **26**(3), 15–18 (1993) (Russian); English transl. Vestn. St. Petersburg Univ. Math. Ser. 1, **26**(3), 14–16 (1993)
29. Leonov, G.A., Ponomarenko, D.V., Smirnova, V.B.: Frequency-Domain Methods for Nonlinear Analysis. World Scientific, Singapore-New Jersey-London-Hong Kong (1996)
30. Leonov, G. A., Reitmann, V.: Localization of Attractors for Nonlinear Systems. Teubner-Texte zur Mathematik, Bd. 97, B. G. Teubner Verlagsgesellschaft, Leipzig, (1987) (German)
31. Leonov, G.A., Reitmann, V., Slepuchin, A.S.: Upper estimates for the Hausdorff dimension of negatively invariant sets of local cocycles. Dokl. Akad. Nauk, T. 439, No. 6 (2011) (Russian); English transl. Dokl. Mathematics. **84**(1), 551–554 (2011)
32. Leonov, G.A., Reitmann, V., Smirnova, V.B.: Non-local Methods for Pendulum-like Feedback Systems. Teubner-Texte zur Mathematik, Bd. 132, B. G. Teubner Stuttgart- Leipzig (1992)
33. Leonov, G.A., Kuznetsov, N.V., Mokaev T.N.: Homoclinic orbits, and self-excited and hidden attractors in a Lorenz-like system describing convective fluid motion. Eur. Phys. J. Special Topics. **224**(8), 1421–1458 (2015)
34. Maltseva, A.A., Reitmann, V.: Existence and dimension properties of a global B-pullback attractor for a cocycle generated by a discrete control system. J. Diff. Equ. **53**(13), 1703–1714 (2017)
35. Li, M.Y., Muldowney, J.S.: On Bendixson's criterion. J. Diff. Equ. **106**(1), 27–39 (1993)
36. Muldowney, J.S.: Compound matrices and ordinary differential equations. Rocky Mountain J. Math. **20**, 857–871 (1990)
37. Neunhäuserer, J.: A Douady-Oesterlé type estimate for the Hausdorff dimension of invariant sets of piecewise smooth maps. Preprint, University of Technology Dresden (2000)
38. Noack, A., Reitmann, V.: Hausdorff dimension estimates for invariant sets of time-dependent vector fields. Zeitschrift für Analysis und ihre Anwendungen (ZAA). **15**(2), 457–473 (1996)
39. Pesin, Ya. B.: Dimension type characteristics for invariant sets of dynamical systems. Uspekhi Mat. Nauk. **43**(4), 95–128 (1988) (Russian); English transl. Russian Math. Surveys. **43**(4), 111–151 (1988)
40. Pugh, C.C.: An improved closing lemma and a general density theorem. Amer. J. Math. **89**, 1010–1021 (1967)
41. Reitmann, V., Schnabel, U.: Hausdorff dimension estimates for invariant sets of piecewise smooth maps. ZAMM **80**(9), 623–632 (2000)
42. Smith, R.A.: Some applications of Hausdorff dimension inequalities for ordinary differential equations. Proc. Roy. Soc. Edinburgh. **104A**, 235–259 (1986)

Acknowledgements

While still working on this book, our coauthor Prof. G. A. Leonov, corresponding member of the Russian Academy of Science, died in 2018. This work is dedicated to his memory, with our deepest and most sincere admiration, gratitude, and love. He was an excellent mathematician with a sharp view on problems and a wonderful colleague and friend. He will stay forever in our mind.

The preparation of this book was carried out in 2017–2019 at the St. Petersburg State University, at the Institute for Problems in Mechanical Engineering of the Russian Academy of Science, and at the University of Jyväskylä within the framework of the Russian Science Foundation projects 14-21-00041 and 19-41-02002.

One of the authors (V.R.) was supported in 2017–2018 by the Johann Gottfried Herder Programme of the German Academic Exchange Service (DAAD).

The authors of the book are greatly indebted to Margitta Reitmann for her accurate typing of the manuscript in LaTeX.

St. Petersburg, Russia
February 2020

Nikolay Kuznetsov
Volker Reitmann

Contents

Part II Dimension Estimates for Almost Periodic Flows and Dynamical Systems in Euclidean Spaces

Part III Dimension Estimates on Riemannian Manifolds

Part I
Basic Elements of Attractor and Dimension Theories

Chapter 1
Attractors and Lyapunov Functions

Abstract The main tool in estimating dimensions of invariant sets and entropies of dynamical systems developed in this book is based on Lyapunov functions. In this chapter we introduce the basic concept of global attractors. The existence of a global attractor for a dynamical system follows from the dissipativity of the system. In order to show the last property we use Lyapunov functions. In this chapter we also consider some applications of Lyapunov functions to stability problems of the Lorenz system. A central result is the existence of homoclinic orbits in the Lorenz system for certain parameters.

1.1 Dynamical Systems, Limit Sets and Attractors

1.1.1 Dynamical Systems in Metric Spaces

Suppose that $(\mathcal{M}, \rho)$ is a complete metric space. Let $\mathbb{T}$ be one of the sets $\mathbb{R}$, $\mathbb{R}_+$, $\mathbb{Z}$ or $\mathbb{Z}_+$. A map $\varphi^{(\cdot)}(\cdot) : \mathbb{T} \times \mathcal{M} \to \mathcal{M}$ resp. a triple $(\{\varphi^t\}_{t\in\mathbb{T}}, \mathcal{M}, \rho)$ is called a *dynamical system on* $(\mathcal{M}, \rho)$ if the following conditions are satisfied ([2, 11]):

(1) $\varphi^0(u) = u\,, \quad \forall\, u \in \mathcal{M}\,;$
(2) $\varphi^{t+s}(u) = \varphi^t(\varphi^s(u))\,, \quad \forall\, t, s \in \mathbb{T}, \quad \forall\, u \in \mathcal{M}\,;$
(3) If $\mathbb{T} \in \{\mathbb{R}, \mathbb{R}_+\}$ the map $(t, u) \in \mathbb{T} \times \mathcal{M} \mapsto \varphi^t(u)$ is continuous;
if $\mathbb{T} \in \{\mathbb{Z}, \mathbb{Z}_+\}$ the map $u \in \mathcal{M} \mapsto \varphi^t(u)$ is continuous on $\mathcal{M}$ for any $t \in \mathbb{T}$.

If the metric space $(\mathcal{M}, \rho)$ is fixed we denote the dynamical system shortly by $\{\varphi^t\}_{t\in\mathbb{T}}$. The sets $\mathbb{T}$ and $\mathcal{M}$ are called *time sets* and *phase space*, respectively. The dynamical system $(\{\varphi^t\}_{t\in\mathbb{T}}, \mathcal{M}, \rho)$ forms a group, if $\mathbb{T} \in \{\mathbb{R}, \mathbb{Z}\}$, and a semi-group if $\mathbb{T} \in \{\mathbb{R}_+, \mathbb{Z}_+\}$. If $\mathbb{T} \in \{\mathbb{R}, \mathbb{R}_+\}$ we say that the dynamical system is with *continuous time*, if $\mathbb{T} \in \{\mathbb{Z}, \mathbb{Z}_+\}$ we say that the system is with *discrete time*. A dynamical system $(\{\varphi^t\}_{t\in\mathbb{T}}, \mathcal{M}, \rho)$ is called *flow* if $\mathbb{T} = \mathbb{R}$, *semi-flow* if $\mathbb{T} = \mathbb{R}_+$, and *cascade* if $\mathbb{T} = \mathbb{Z}$.

N. Kuznetsov and V. Reitmann, *Attractor Dimension Estimates for Dynamical Systems: Theory and Computation*, Emergence, Complexity and Computation 38,
https://doi.org/10.1007/978-3-030-50987-3_1

Example 1.1 Consider the autonomous differential equation

$$\dot{\varphi} = f(\varphi)\,, \tag{1.1}$$

where $f : \mathbb{R}^n \to \mathbb{R}^n$ is assumed to be locally Lipschitz. The Euclidean norm in $\mathbb{R}^n$ is denoted by $|\cdot|$. Suppose also that any maximal solution $\varphi(\cdot, u)$ of (1.1) starting in u at $t = 0$ exists for any $t \in \mathbb{R}$. Let now $\varphi^t(\cdot) \equiv \varphi(t, \cdot) : \mathbb{R}^n \to \mathbb{R}^n$ be the time t-map of (1.1). Clearly, that by the solution properties of (1.1) (uniqueness theorem and theorem of continuous dependence on initial condition, [12, 34]) the triple $(\{\varphi^t\}_{t\in\mathbb{R}}, \mathbb{R}^n, |\cdot|)$ defines a dynamical system with the additive group $\mathbb{T} = \mathbb{R}$, i.e. a flow.

Example 1.2 Assume that

$$\dot{\varphi} = f(t, \varphi) \tag{1.2}$$

is a non-autonomous differential equation with $f : \mathbb{R} \times \mathbb{R}^n \to \mathbb{R}^n$. Let us suppose that f is continuously differentiable, T-periodic in the first argument and that any solution exists on $\mathbb{R}$. Denote the solution of (1.2) starting in u at $t = t_0$ by $\varphi(\cdot, t_0, u)$. Again by the uniqueness theorem and the theorem of continuous dependence of solutions on initial conditions for ODE's it follows that the family of maps $\varphi^m(\cdot) \equiv \varphi(mT, 0, \cdot)$, $m \in \mathbb{Z}$, defines a dynamical system $(\{\varphi^m\}_{m\in\mathbb{Z}}, \mathbb{R}^n, |\cdot|)$ which is a cascade. Let us demonstrate this. Clearly, $\varphi^0(u) = \varphi(0, 0, u) = u, \ \forall u \in \mathbb{R}^n$. In order to show the property (2) of a dynamical system we consider arbitrary $m, k \in \mathbb{Z}$. Define the two functions $c_1(t) := \varphi(t, 0, \varphi(mT, 0, u))$ and $c_2(t) := \varphi(t + mT, 0, u)$. From the T-periodicity of f in the first variable it follows that c_2 is also a solution of (1.2) on $\mathbb{R}$, i.e.

$$\dot{c}_2(t) = f(t + mT, \varphi(t + mT, 0, u)) = f(t, c_2(t)).$$

Since $c_2(0) = \varphi(mT, 0, u)$, it follows by the uniqueness theorems that $c_1(t) = c_2(t), \ \forall t \in \mathbb{R}$.

If we put $t = kT$, the last property results in $c_1(kT) \equiv \varphi^k(\varphi^m(u)) = c_2(kT) \equiv \varphi^{k+m}(u)$.

Example 1.3 Suppose that

$$\varphi : \mathcal{M} \to \mathcal{M} \tag{1.3}$$

is a continuous invertible map on the complete metric space $(\mathcal{M}, \rho)$. Let us define the family of maps

$$\varphi^m := \begin{cases} \underbrace{\varphi \circ \varphi \circ \cdots \circ \varphi}_{m\text{-times}} & \text{for } \ m = 1, 2, \dots\,, \\ \mathrm{id}_{\mathcal{M}} & \text{for } \ m = 0\,, \\ \underbrace{\varphi^{-1} \circ \varphi^{-1} \circ \cdots \circ \varphi^{-1}}_{-m\text{-times}} & \text{for } \ m = -1, -2, \dots\,. \end{cases} \tag{1.4}$$

Using well-known properties of the composition of continuous invertible maps, i.e. of homeomorphisms, it is easy to show that the properties (1)–(3) of a dynamical system with the additive group $\mathbb{T} = \mathbb{Z}$ are satisfied. This means that (1.4) defines a cascade.

Example 1.4 Let $(\mathcal{M}, g)$ be a Riemannian n-dimensional C^k-manifold $(k \geq 2)$, $F : \mathcal{M} \to T\mathcal{M}$ a C^2-vector field (see Sect. A.6, Appendix A). Consider the corresponding differential equation

$$\dot{\varphi} = F(\varphi) . \tag{1.5}$$

Assume that any maximal integral curve $\varphi(\cdot, u)$ of (1.5) satisfying $\varphi(0, u) = u$ exists on $\mathbb{R}$. Define $\varphi^{(\cdot)}(u) := \varphi(\cdot, u)$ and denote with ρ the metric generated by the metric tensor g. Then $(\{\varphi^t\}_{t\in\mathbb{R}}, \mathcal{M}, \rho)$ is a flow defined by the vector field (1.5).

Instead of (1.5) we can consider a C^1-diffeomorphism

$$\varphi : \mathcal{M} \to \mathcal{M} . \tag{1.6}$$

It is clear that (1.6) generates the dynamical system $(\{\varphi^m\}_{m\in\mathbb{Z}}, \mathcal{M}, \rho)$.

Example 1.5 Suppose that $\Omega_2^+ := \{\omega = (\omega_0, \omega_1, \ldots) \mid \omega_i \in \{0, 1\}\}$ is the set of all one-sided infinite sequences of the symbols 0 and 1. The metric on Ω_2^+ is given by

$$\rho(\omega, \omega') := \sum_{i=0}^{\infty} 2^{-i} \mid \omega_i - \omega_i' \mid ,$$

where $\omega = (\omega_0, \omega_1, \ldots)$ and $\omega' = (\omega_0', \omega_1', \ldots)$ are from Ω_2^+. It is easy to see that ρ is really a metric and (Ω_2^+, ρ) is a complete metric space.

Define the (left) *shift map* $\vartheta : \Omega_2^+ \to \Omega_2^+$ by

$$\vartheta(\omega) = (\omega_1, \omega_2, \ldots) \quad \text{for} \quad \omega = (\omega_0, \omega_1, \ldots) \in \Omega_2^+ .$$

The map ϑ is continuous since for arbitrary $\omega, \omega' \in \Omega_2^+$ we have

$$\begin{aligned}\rho(\vartheta(\omega), \vartheta(\omega')) &= \sum_{i=0}^{\infty} 2^{-i} \mid \omega_{i+1} - \omega_{i+1}' \mid = 2 \sum_{i=0}^{\infty} \frac{1}{2^{i+1}} \mid \omega_{i+1} - \omega_{i+1}' \mid \\ &\leq 2 \sum_{i=-1}^{\infty} \frac{1}{2^{i+1}} \mid \omega_{i+1} - \omega_{i+1}' \mid = 2\, \rho(\omega, \omega') .\end{aligned}$$

It follows that $\left(\{\vartheta^m\}_{m\in\mathbb{Z}_+}, \Omega_2^+, \rho\right)$ is a dynamical system with discrete time.

Let us define now some properties of a dynamical system $(\{\varphi^t\}_{t\in\mathbb{T}}, \mathcal{M}, \rho)$. For an arbitrary fixed $u \in \mathcal{M}$, the map $t \mapsto \varphi^t(u)$, $t \in \mathbb{T}$ defines a *motion* of the dynamical system starting from u at time $t = 0$. For any $u \in \mathcal{M}$ the set $\gamma(u) := \bigcup_{t\in\mathbb{T}} \varphi^t(u)$

is the *orbit* through u. If $\mathbb{T} \in \{\mathbb{R}, \mathbb{Z}\}$ we consider also the *positive* and the *negative semi-orbit* through u defined by

$$\gamma^+(u) := \bigcup_{t \in \mathbb{T} \cap \mathbb{R}_+} \varphi^t(u) \quad \text{resp.} \quad \gamma^-(u) := \bigcup_{t \in \mathbb{T} \cap \mathbb{R}_-} \varphi^t(u) \ .$$

An orbit $\gamma(u)$ is called *stationary, critical* or an *equilibrium* if $\gamma(u) = \{u\}$. The orbit $\gamma(u)$ of a dynamical system is called T*-periodic* with *period* T if $T > 0$ is the smallest positive number in $\mathbb{T}$ such that $\varphi^t(u) = \varphi^{t+T}(u), \ \forall t \in \mathbb{T}$. A set $\mathcal{Z} \subset \mathcal{M}$ is said to be *positively invariant* if $\varphi^t(\mathcal{Z}) \subset \mathcal{Z}, \ \forall t \in \mathbb{T} \cap \mathbb{R}_+$, *invariant* if $\varphi^t(\mathcal{Z}) = \mathcal{Z}, \ \forall t \in \mathbb{T}$, and *negatively invariant*, if $\varphi^t(\mathcal{Z}) \supset \mathcal{Z}, \ \forall t \in \mathbb{T} \cap \mathbb{R}_+$. The positively invariant set $\mathcal{Z}$ of the dynamical system $(\{\varphi^t\}_{t\in\mathbb{T}}, \mathcal{M}, \rho)$ is said to be *stable* if in any neighborhood $\mathcal{U}$ of $\mathcal{Z}$ there exists a neighborhood $\mathcal{U}'$ such that $\varphi^t(\mathcal{U}') \subset \mathcal{U}, \forall t \in \mathbb{T}_+$. $\mathcal{Z}$ is called *asymptotically stable* if it is stable and $\varphi^t(u) \to \mathcal{Z}$ as $t \to +\infty$ for each $u \in \mathcal{U}'$.

The set $\mathcal{Z}$ is said to be *globally asymptotically stable* if $\mathcal{Z}$ is stable and $\varphi^t(u) \to \mathcal{Z}$ as $t \to +\infty$ for each $u \in \mathcal{M}$. The set $\mathcal{Z}$ is called *uniformly asymptotically stable* if it is stable and $\lim_{t\to+\infty}\{\text{dist}(\varphi^t(u), \mathcal{Z}) \mid u \in \mathcal{U}'\} = 0$.

For any $u \in \mathcal{M}$ the ω*-limit set* of u under $\{\varphi^t\}_{t\in\mathbb{T}}$ is the set

$$\omega(u) := \{v \in \mathcal{M} \mid \exists \{t_n\}_{n\in\mathbb{N}} \, , \ t_n \in \mathbb{T} \, , t_n \to +\infty \, , \varphi^{t_n}(u) \to v \quad \text{for} \quad n \to +\infty\} \ .$$

For a subset $\mathcal{Z} \subset \mathcal{M}$ we define its ω*-limit set* $\omega(\mathcal{Z})$ under $\{\varphi^t\}_{t\in\mathbb{T}}$ as the set of the limits of all converging sequences of the form $\varphi^{t_n}(u_n)$, where $u_n \in \mathcal{Z}, t_n \in \mathbb{T}$, and $t_n \to +\infty$.

Example 1.6 Consider a class of *modified horseshoe maps* φ which are defined on an open neighborhood $\mathcal{U} = (-\delta, 1+\delta) \times (-\delta, 1+\delta) \subset \mathbb{R}^2$ of the unit square $\mathcal{C} = [0, 1] \times [0, 1]$, where $\delta > 0$ is a sufficiently small number. The map is defined for points $(x, y) \in \mathcal{C}$ in such a way that it first contracts $\mathcal{C}$ horizontally with a factor $\alpha < \frac{1}{2}$ and stretches it vertically with a function f, then it is folded along a horizontal line such that the vertical edge of the resulting rectangle is greater than 2, and finally it is formed into an horseshoe (see Fig. 1.1) in such a way that the map can be continuously extended to $\mathcal{U}$ and is continuously differentiable on an open neighborhood $\widetilde{\mathcal{U}}$ of $\mathcal{K} = \bigcap_{i=-\infty}^{\infty} \varphi^i(\mathcal{C})$, where $\varphi^i(\cdot)$ for negative numbers i means the preimage under the map φ^{-i}.

For example we take $\alpha = \frac{1}{3}$ and let the function f stretch $\mathcal{C}$ with factor $\beta_1 = 3$, if $y \le \frac{51}{65} =: h$, with a factor β_2 between 3 and 5, if $\frac{51}{65} < y < \frac{4}{5}$ and with factor 5, if $y \ge \frac{4}{5}$, the resulting rectangle is folded on the image of the line $y = \frac{51}{65}$ and then it is formed to an horseshoe in such a way, that the map satisfies

$$\varphi(x, y) = \begin{cases} \left(\frac{1}{3}x, 3y - \frac{1}{13}\right) & \text{if} \ \ 0 \le y \le \frac{14}{39} \, , \\ \left(1 - \frac{1}{3}x, \frac{148}{65} - 3y\right) & \text{if} \ \ \frac{83}{195} \le y \le \frac{148}{195} \, , \\ \left(1 - \frac{1}{3}x, 5y - 4\right) & \text{if} \ \ \frac{4}{5} \le y \le 1 \, . \end{cases}$$

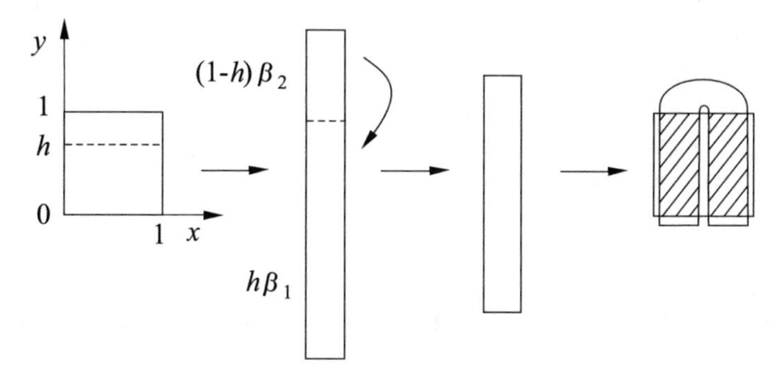

Fig. 1.1 Modified horseshoe map

The set $\mathcal{K} = \bigcap_{i=-\infty}^{\infty} \varphi^i(\mathcal{C})$ is invariant under the map. Therefore if we take $\widetilde{\mathcal{K}} = \mathcal{K}$, then $\varphi^j(\mathcal{K}) \subset \widetilde{\mathcal{K}}$ is satisfied for any $j = 1, 2, \ldots$. The set $\mathcal{K}$ can be constructed step by step starting with $\mathcal{K}^0 = \mathcal{C}$. At every step $i > 1$ we get $\mathcal{K}^i := \mathcal{K}^{i-1} \cap \varphi(\mathcal{K}^{i-1}) \cap \varphi^{-1}(\mathcal{K}^{i-1})$ and in the limit the invariant set $\mathcal{K}$ as $\mathcal{K} = \bigcap_{i=0}^{\infty} \mathcal{K}^i$. The set $\mathcal{K}^i$ consists of 6^i rectangles where the lengths of the edges are horizontally $\frac{1}{3^i}$ and vertically $\frac{1}{3^i}$ and $\frac{1}{5^i}$, respectively (see Fig. 1.1).

It is easy to verify that the following proposition is true ([4, 7, 38]).

Proposition 1.1 *For any subset $\mathcal{Z} \subset \mathcal{M}$ the ω-limit set under $\{\varphi^t\}_{t\in\mathbb{T}}$ is given by*

$$\omega(\mathcal{Z}) = \bigcap_{\substack{s\geq 0\\ s\in\mathbb{T}}} \overline{\bigcup_{\substack{t\geq s\\ t\in\mathbb{T}}} \varphi^t(\mathcal{Z})}\,.$$

Here for a set $\mathcal{Z} \subset \mathcal{M}$ we denote by $\overline{\mathcal{Z}}$ its closure in the topology of the metric space $(\mathcal{M}, \rho)$.

If $\mathbb{T} \in \{\mathbb{R}, \mathbb{Z}\}$ we also consider the α*-limit set* of a point $u \in \mathcal{M}$ under $\{\varphi^t\}_{t\in\mathbb{T}}$ defined by

$$\alpha(u) := \{v \in \mathcal{M} \mid \exists\{t_n\}_{n\in\mathbb{N}}\,,\ t_n \in \mathbb{T}, t_n \to -\infty, \varphi^{t_n}(u) \to v \quad \text{for} \quad n \to +\infty\}$$

and the α-limit set $\omega(\mathcal{Z})$ of a set $\mathcal{Z} \subset \mathcal{M}$ under $\{\varphi^t\}_{t\in\mathbb{T}}$ given as the set of the limits of all converging sequences of the form $\varphi^{t_n}(u_n)$, where $p_n \in \mathcal{Z}$, $t_n \in \mathbb{T}$, and $t_n \to -\infty$.

A set $\mathcal{Z}_{\min} \subset \mathcal{M}$ is called *minimal* for $(\{\varphi^t\}_{t\in\mathbb{T}}, \mathcal{M}, \rho)$ if it is closed, invariant, and does not have any proper subset with the same properties. The following proposition is taken from [33, 36].

Proposition 1.2 *Suppose that $\mathcal{Z} \subset \mathcal{M}$ is non-empty, compact and invariant for $(\{\varphi^t\}_{t\in\mathbb{T}}, \mathcal{M}, \rho)$. Then $\mathcal{Z}$ contains a minimal set $\mathcal{Z}_{\min}$.*

Proof (For the case that $\mathbb{T}$ is a semi-group). If $\mathcal{Z}$ has no proper subset, which is closed and invariant, then $\mathcal{Z}$ is minimal, and the proposition is proved.

Suppose that there exists $\mathcal{Z}_1 \subset \mathcal{Z}$, $\mathcal{Z}_1 \neq \mathcal{Z}$, such that $\mathcal{Z}_1$ is closed and invariant. If $\mathcal{Z}_1$ contains no proper subset being closed and invariant, then it is minimal.

Suppose again that there exists a closed invariant set $\mathcal{Z}_2 \subset \mathcal{Z}_1$ with $\mathcal{Z}_2 \neq \mathcal{Z}_1$. If we can continue this process and at any step we obtain a new minimal set $\mathcal{Z}_i$, we get the sequence of closed invariant sets

$$\mathcal{Z}_0 := \mathcal{Z} \supset \mathcal{Z}_1 \supset \mathcal{Z}_2 \supset \cdots .$$

The intersection of these sets $\mathcal{Z}_\omega := \bigcap_{i=0}^{\infty} \mathcal{Z}_i$ is non-empty and compact. Let us show that $\mathcal{Z}_\omega$ is invariant. Suppose that $u \in \mathcal{Z}_\omega$ is arbitrary. Then for any integer $k \geq 0$ we have $u \in \mathcal{Z}_k$. It follows that $\varphi^t(u) \in \mathcal{Z}_k$, $t \geq 0$, for any k. But this means that $\varphi^t(u) \in \mathcal{Z}_\omega$ and, consequently, $\mathcal{Z}_\omega \subset \varphi^t(\mathcal{Z}_\omega)$. The inverse inclusion can be proved analogously. Thus the set $\mathcal{Z}_\omega$ is invariant.

If the set $\mathcal{Z}_\omega$ is not minimal then there exists a closed and invariant set $\mathcal{Z}_{\omega+1} \subset \mathcal{Z}_\omega$ with $\mathcal{Z}_{\omega+1} \neq \mathcal{Z}_\omega$. If we can continue this process and if β is the transfinite limit number for which the sets $\mathcal{Z}_\alpha$ are constructed for all $\alpha < \beta$, we put

$$\mathcal{Z}_\beta := \bigcap_{\alpha<\beta} \mathcal{Z}_\alpha .$$

Clearly, the set $\mathcal{Z}_\beta$ is closed and invariant. Thus we get the transfinite sequence of sets

$$\mathcal{Z} \supset \mathcal{Z}_1 \supset \cdots \supset \mathcal{Z}_k \supset \cdots \supset \mathcal{Z}_\omega \supset \cdots \supset \mathcal{Z}_\beta \supset \cdots .$$

According to the Baire-Hausdorff theorem ([16], Theorem A.14.1) there exists a transfinite number β of the second class such that $\mathcal{Z}_\beta = \mathcal{Z}_{\beta+1}$, i.e. the set $\mathcal{Z}_\beta$ has no proper closed and invariant subset. Consequently, $\mathcal{Z}_\beta$ is a minimal set. □

The next result follows immediately from Proposition 1.2.

Proposition 1.3 *Suppose that the positive semi-orbit of* $(\{\varphi^t\}_{t\in\mathbb{T}}, \mathcal{M}, \rho)$, *starting in* p, *is relatively compact. Then* $\omega(p)$ *contains a minimal set.*

Some important properties of ω-limit sets are proved in the next proposition ([18]).

Proposition 1.4 *Let* $(\{\varphi^t\}_{t\in\mathbb{T}}, \mathcal{M}, \rho)$ *be a dynamical system with a semi-group as time set and let* $\mathcal{Z} \subset \mathcal{M}$ *be a non-empty set such that for some* $t_0 > 0$, $t_0 \in \mathbb{T}$, *the set* $\bigcup_{t\geq t_0, t\in\mathbb{T}} \varphi^t(\mathcal{Z})$ *is relatively compact.*

Then $\omega(\mathcal{Z})$ *is non-empty, compact and invariant. Furthermore,* $\omega(\mathcal{Z})$ *is a minimal closed set which attracts* $\mathcal{Z}$.

Proof Recall that relative compactness of a set means that the closure of this set is compact. Since $\mathcal{Z} \neq \emptyset$ the sets $\mathcal{B}_s = \bigcup_{t\geq s, t\in\mathbb{T}} \varphi^t(\mathcal{Z})$ are non-empty for all $s \geq 0$, $s \in \mathbb{T}$. Consequently, the sets $\overline{\mathcal{B}}_s$ are non-empty compact sets for $s \geq t_0$ and $\overline{\mathcal{B}}_{s_1} \subset \overline{\mathcal{B}}_{s_2}$

for all $s_1 \geq s_2$ in $\mathbb{T}$. Therefore $\omega(\mathcal{Z}) = \bigcap_{s \geq 0, s \in \mathbb{T}} \overline{\mathcal{B}}_s$ is a non-empty compact set and attracts $\mathcal{Z}$.

If $u \in \varphi^t(\omega(\mathcal{Z}))$ then for a certain $\upsilon \in \omega(\mathcal{Z})$ we have $u = \varphi^t(\upsilon)$. Hence there exists a sequence $u_m \in \mathcal{Z}$ and a sequence $\{t_m\}$, $t_m \in \mathbb{T}$, $t_m \to +\infty$ as $m \to +\infty$, such that $\lim_{m\to+\infty} \varphi^t(\varphi^{t_m}(u)) = \lim_{m\to+\infty} \varphi^{t+t_m}(u_m) = \varphi^t(\upsilon) = u$. But this means that $u \in \omega(\mathcal{Z})$.

Let us prove the reverse inclusion. If $\upsilon \in \omega(\mathcal{Z})$ a sequence $u_m \in \mathcal{Z}$ and a sequence $\{t_m\}$ from $\mathbb{T}$ exist such that $t_m \to +\infty$ as $m \to +\infty$, $1 + t_0 + t \leq t_1 < t_2 < \cdots$, and

$$\varphi^{t_m}(u_m) \to \upsilon \quad \text{for} \quad m \to +\infty \,. \tag{1.7}$$

For $t_m \geq t$ the sequence $\varphi^{t_m - t}(u_m)$ belongs to the relatively compact set $\bigcup_{t \geq t_0+1,\, t \in \mathbb{T}} \varphi^t(\mathcal{Z})$. Consequently, passing if necessary to a subsequence, we may suppose that there exists a point $u \in \mathcal{M}$ such that

$$\varphi^{t_m - t}(u_m) \to u \quad \text{as} \quad m \to +\infty \,.$$

But this means that $u \in \omega(\mathcal{Z})$. Since

$$\varphi^{t_m}(u_m) = \varphi^t(\varphi^{t_m - t}(u_m)) \to \varphi^t(u) \quad \text{as} \quad m \to +\infty \,,$$

by (1.7) we have $\upsilon = \varphi^t(u)$. Thus, $\upsilon \in \varphi^t(\omega(\mathcal{Z}))$.

It remains to show that $\omega(\mathcal{Z})$ is a minimal closed set which attracts $\mathcal{Z}$. Let us argue as in [18]. Suppose the contrary and let $\mathcal{C}$ be a proper closed subset of $\omega(\mathcal{Z})$ which attracts $\mathcal{Z}$. As $\omega(\mathcal{Z})$ is compact so is $\mathcal{C}$. Choose any $\upsilon \in \omega(\mathcal{Z}) \setminus \mathcal{C}$. For $\varepsilon > 0$ small enough the ε-neighborhoods $\mathcal{U}_\varepsilon(\upsilon)$ and $\mathcal{U}_\varepsilon(\mathcal{C})$ do not intersect. Since $\mathcal{C}$ attracts $\mathcal{Z}$ there is a $t = t(\varepsilon) \geq 0$ such that $\varphi^t(\mathcal{Z}) \subset \mathcal{U}_\varepsilon(\mathcal{C})$, $\forall t \geq t(\varepsilon)$. On the other hand since $\upsilon \in \omega(\mathcal{Z})$, $\upsilon = \lim_{k\to\infty} \varphi^{t_k}(u_k)$ for some $u_k \in \mathcal{Z}$ and $t_k \to +\infty$ is a sequence. Consequently, $\varphi^{t_k}(\mathcal{Z}) \cap \mathcal{U}_\varepsilon(\upsilon) \neq \emptyset$ for sufficiently large t_k. Hence $\mathcal{U}_\varepsilon(\upsilon) \cap \mathcal{U}_\varepsilon(\mathcal{C}) \neq \emptyset$, a contradiction. □

Assume that $(\{\varphi^t\}_{t\in\mathbb{T}}, \mathcal{M}, \rho)$ is a dynamical system with a group as time set and $u \in \mathcal{M}$ is an arbitrary point. The sets

$$W^s(u) := \{\upsilon \in \mathcal{M} | \lim_{t\to+\infty} \varphi^t(\upsilon) = u\} \qquad \text{and}$$

$$W^u(u) := \{\upsilon \in \mathcal{M} | \lim_{t\to-\infty} \varphi^t(\upsilon) = u\}$$

are called *stable* and *unstable manifold*, respectively, in u. Since the orbits are invariant, the sets $W^s(u)$ and $W^u(u)$ are also invariant. Suppose that u and υ are equilibria of the dynamical system. Then any *orbit* which is contained in $W^s(u) \cap W^u(\upsilon)$ is called *heteroclinic* if $u \neq \upsilon$, and *homoclinic* if $u = \upsilon$.

Let $\mathcal{Z} \subset \mathcal{M}$ be an arbitrary invariant subset. Then the *stable* and *unstable manifold* of $\mathcal{Z}$ are the sets

$$W^s(\mathcal{Z}) := \{v \in \mathcal{M} | \lim_{t \to +\infty} \operatorname{dist}(\varphi^t(v), \mathcal{Z}) = 0\} \qquad \text{and}$$
$$W^u(\mathcal{Z}) := \{v \in \mathcal{M} | \lim_{t \to -\infty} \operatorname{dist}(\varphi^t(v), \mathcal{Z}) = 0\},$$

respectively.

1.1.2 Minimal Global Attractors

Now we come to some of the basic definitions in our book. For arbitrary nonempty sets $\mathcal{Z}_1, \mathcal{Z}_2 \subset \mathcal{M}$ we define $\operatorname{dist}(\mathcal{Z}_1, \mathcal{Z}_2) := \sup_{u \in \mathcal{Z}_1} \inf_{v \in \mathcal{Z}_2} \rho(u, v)$. By $\mathcal{U}_\varepsilon(\mathcal{Z})$ we denote the ε-neighborhood of a set $\mathcal{Z}$, i.e. $\mathcal{U}_\varepsilon(\mathcal{Z}) := \{v \in \mathcal{M} \mid \operatorname{dist}(v, \mathcal{Z}) < \varepsilon\}$.

Definition 1.1 Suppose that $(\{\varphi^t\}_{t \in \mathbb{T}}, \mathcal{M}, \rho)$ is a dynamical system.

(1) We say that a set $\mathcal{Z}_0 \subset \mathcal{M}$ *attracts the set* $\mathcal{Z} \subset \mathcal{M}$ if for any $\varepsilon > 0$ there exists a $t_0 = t_0(\varepsilon, \mathcal{Z})$ such that for all $t \geq t_0$, $t \in \mathbb{T}$, we have $\varphi^t(\mathcal{Z}) \subset \mathcal{U}_\varepsilon(\mathcal{Z}_0)$.
(2) An *attractor* $\mathcal{A}$ for $(\{\varphi^t\}_{t \in \mathbb{T}}, \mathcal{M}, \rho)$ is a non-empty closed and invariant set which attracts all points from some set $\mathcal{Z}$ with a non-empty interior. The largest set with non-empty interior which is attracted by $\mathcal{A}$ is called the *domain of attraction*.
(3) A *global attractor* for $(\{\varphi^t\}_{t \in \mathbb{T}}, \mathcal{M}, \rho)$ is a non-empty, closed and invariant set which attracts all points of $\mathcal{M}$.
(4) A *global* $\mathcal{B}$*-attractor* is a non-empty, closed and invariant set which attracts any bounded set $\mathcal{B}$ of $\mathcal{M}$.
(5) A *minimal global attractor (minimal global* $\mathcal{B}$*-attractor)* is a global attractor (global $\mathcal{B}$-attractor) which is a minimal set among all global attractors (global $\mathcal{B}$-attractors).
(6) A set $\mathcal{Z}_0 \subset \mathcal{M}$ is said to be $\mathcal{B}$*-absorbing* for $(\{\varphi^t\}_{t \in \mathbb{T}}, \mathcal{M}, \rho)$ if for any bounded set $\mathcal{B}$ in $\mathcal{M}$ there exists a $t_0 = t_0(\mathcal{B})$ such that $\varphi^t(\mathcal{B}) \subset \mathcal{Z}_0$ for any $t \geq t_0$, $t \in \mathbb{T}$.
(7) A dynamical system is said to be *pointwise dissipative* ($\mathcal{B}$-dissipative) if it possesses a pointwise absorbing ($\mathcal{B}$-absorbing set) $\mathcal{B}_0$. The set $\mathcal{B}_0$ is called *region of pointwise dissipativity* (of $\mathcal{B}$*-dissipativity*).
(8) A set $\mathcal{Z}_0 \subset \mathcal{M}$ is said to be *pointwise absorbing* for $(\{\varphi^t\}_{t \in \mathbb{T}}, \mathcal{M}, \rho)$ if for any $u \in \mathcal{M}$ there exists a $t_0 = t_0(u)$ such that $\varphi^t(u) \subset \mathcal{Z}_0$ for any $t \geq t_0$, $t \in \mathbb{T}$.

Let us use the following abbreviations for the attractors of a dynamical system $(\{\varphi^t\}_{t \in \mathbb{T}}, \mathcal{M}, \rho)$: $\mathcal{A}$—an arbitrary attractor, $\mathcal{A}_\mathcal{M}$—a global $\mathcal{B}$-attractor, $\mathcal{A}_{\mathcal{M},\min}$—a minimal global $\mathcal{B}$-attractor, $\widehat{\mathcal{A}}_\mathcal{M}$—a global attractor, $\widehat{\mathcal{A}}_{\mathcal{M},\min}$—a minimal global attractor.

A direct consequence of Definition 1.1 is the following proposition.

Proposition 1.5 *Let $\mathcal{A}$ be a global $\mathcal{B}$-attractor and $\varepsilon > 0$ an arbitrary number. Then the ε-neighborhood of $\mathcal{A}$ is $\mathcal{B}$-absorbing for the dynamical system.*

Remark 1.1 Minimal global attractors and $\mathcal{B}$-attractors where introduced by O.A. Ladyzhenskaya in [18]. Our Definition 1.1 follows the representation given in [18]. Important properties of minimal global attractors are also derived in [6, 7, 15, 35, 37].

The existence of a global $\mathcal{B}$-attractor is shown in the next proposition ([17]).

Note that if a global $\mathcal{B}$-attractor exists, then it contains a minimal global $\mathcal{B}$-attractor.

Proposition 1.6 *Suppose that the dynamical system $(\{\varphi^t\}_{t\in\mathbb{T}}, \mathcal{M}, \rho)$ is $\mathcal{B}$-dissipative according to the bounded $\mathcal{B}$-absorbing set $\mathcal{B}_0$ and there exists a $t_0 > 0$ such that the set $\bigcup_{t\geq t_0, t\in\mathbb{T}} \varphi^t(\mathcal{B}_0)$ is relatively compact. Then*

$$\mathcal{A}_{\mathcal{M},\min} := \overline{\{\omega(\mathcal{B}) \mid \mathcal{B} \subset \mathcal{M}, \mathcal{B} \text{ bounded}\}}$$

is a minimal global $\mathcal{B}$-attractor and

$$\widehat{\mathcal{A}}_{\mathcal{M},\min} := \overline{\bigcup_{u\in\mathcal{M}} \omega(u)}$$

is a minimal global attractor of the dynamical system.

Proof Since by Proposition 1.4 every bounded set $\mathcal{B} \subset \mathcal{M}$ is attracted to its ω-limit set $\omega(\mathcal{B})$ and to $\mathcal{A}_{\mathcal{M},\min}$, it is attracted to $\omega(\mathcal{B}) \cap \mathcal{A}_{\mathcal{M},\min}$. Since $\omega(\mathcal{B})$ is minimal, it lies in $\mathcal{A}_{\mathcal{M},\min}$. The set $\mathcal{A}_{\mathcal{M},\min}$, is invariant and minimal. It follows that $\omega(\mathcal{A}_{\mathcal{M},\min}) = \mathcal{A}_{\mathcal{M},\min}$ and the representation for $\mathcal{A}_{\mathcal{M},\min}$, is shown.

The fact that $\overline{\bigcup_{u\in\mathcal{M}} \omega(u)}$ is a minimal global attractor follows from the properties of an ω-limit set. □

Remark 1.2 In contrast to the minimal global attractor given by Proposition 1.6 a minimal global attractor can be unbounded. The dynamical system generated by the ODE

$$\dot{x} = 0\,, \qquad \dot{y} = -ay\,, \qquad a > 0$$

has as a minimal global $\mathcal{B}$-attractor $\mathcal{A}_{\mathbb{R}^2,\min}$ the x-axis (see Sect. 2.1, Chap. 2)

Other examples of minimal global attractors and global $\mathcal{B}$-attractors will be considered in the sequel.

A dynamical system $(\{\varphi^t\}_{t\in\mathbb{T}}, \mathcal{M}, \rho)$ is called *locally completely continuous* if for any $u \in \mathcal{M}$ there exists a $\delta = \delta(u) > 0$ and an $l = l(u) > 0, l(u) \in \mathbb{T}_+$, such that $\varphi^l(\mathcal{B}_\delta(u))$ is relatively compact. It is clear that a dynamical system given in a locally compact space is locally completely continuous.

The next proposition is a result of [4].

Proposition 1.7 *For a locally completely continuous dynamical system pointwise dissipativity and $\mathcal{B}$-dissipativity are equivalent.*

Proof We have to show that a dynamical system which is pointwise dissipative is also $\mathcal{B}$-dissipative. From the pointwise dissipativity it follows that there exists a non-empty compact set $\widetilde{\mathcal{K}} \subset \mathcal{M}$ such that for each $\varepsilon > 0$ and $u \in \mathcal{M}$ there exists a $\delta(u) > 0$ and a $\tau(\varepsilon, u)$ such that

$$\operatorname{dist}(\varphi^t(v), \widetilde{\mathcal{K}}) < \varepsilon \tag{1.8}$$

for all $t \geq \tau(\varepsilon, u)$, $t \in \mathbb{T}$, and all $v \in \mathcal{B}_{\delta(u)}(u)$. Suppose $\mathcal{B}$ is an arbitrary bounded set which is contained in a compact set $\mathcal{K}$. Then for any $u \in \mathcal{K}$ there exists a $\delta = \delta(u) > 0$ and $\tau = \tau(\varepsilon, u) > 0$ such that (1.8) is satisfied. Consider an open cover $\{\mathcal{B}_{\delta(u)}(u)\}_{u\in\mathcal{K}}$ of $\mathcal{K}$. Since $\mathcal{K}$ is compact and $\mathcal{M}$ is a complete metric space there is a finite subcover $\{\mathcal{B}_{\delta(u_i)}(u_i)\}_{i=1}^m$ of $\mathcal{K}$.

Define $\tilde{l}(\varepsilon, \mathcal{K}) := \max\{\tau(\varepsilon, u_i) | i = 1, \ldots, m\}$. Then it follows from (1.8) that $\operatorname{dist}(\varphi^t(\mathcal{K}), \mathcal{K}) < \varepsilon, \forall t > \tilde{l}(\varepsilon, \mathcal{K})$. □

Since for dynamical systems in locally compact metric spaces pointwise dissipativity and $\mathcal{B}$-dissipativity are equivalent we call these properties shortly *dissipativity*.

The next proposition is proved in [7]. It shows that Lyapunov functions can give a good inside in the structure of an attractor.

In this book the term *Lyapunov function* for a dynamical system $(\{\varphi^t\}_{t\in\mathbb{T}}, \mathcal{M}, \rho)$ means a scalar valued continuous function V which is considered along the orbits and whose properties allow some conclusions about the qualitative behaviour of the dynamical system. If $\mathcal{M}$ is a manifold and V is differentiable the properties of V depend on the Lie derivative of V w.r.t. the dynamical system (see Subsect. 1.2.3).

Proposition 1.8 *Suppose for the dynamical system $(\{\varphi^t\}_{t\in\mathbb{T}}, \mathcal{M}, \rho)$ with a group as time set there exists a compact global $\mathcal{B}$-attractor $\mathcal{A}_\mathcal{M}$ and a continuous function $V : \mathcal{M} \to \mathbb{R}$ with the following properties:*

(1) For any $u \in \mathcal{M}$ the function $V(\varphi^t(u))$ is non-increasing with respect to $t \in \mathbb{T}_+$;

(2) If for some $t_0 > 0$, $t_0 \in \mathbb{T}_+$, the equation $V(u) = V(\varphi^{t_0}(u))$ holds, then u is an equilibrium of the dynamical system.

Then: (a) $\mathcal{A}_\mathcal{M} = W^u(\mathcal{C})$, where $\mathcal{C}$ is the set of equilibrium points of the dynamical system.

(b) The global minimal attractor $\widehat{\mathcal{A}}_{\mathcal{M},\min}$ of the dynamical system is $\mathcal{C}$.

Remark 1.3 Suppose $(\{\varphi^t\}_{t\in\mathbb{T}}, \mathcal{M}, \rho)$ is a dynamical system with a group as time set, which has a bounded minimal global $\mathcal{B}$-attractor $\mathcal{A}$. Then $W^u(\mathcal{C}) \subset \mathcal{A}$, where $\mathcal{C}$ is the set of equilibria of the dynamical system. For a proof see [7].

For dynamical systems $(\{\varphi^t\}_{t\in\mathbb{T}}, \mathcal{M}, \rho)$ given on a Riemannian smooth n-dimensional manifold $(\mathcal{M}, g)$ we define a *Milnor attractor* as a closed invariant set $\widetilde{\mathcal{A}}$ having the property $\lim_{t\to+\infty} \operatorname{dist}(\varphi^t(u), \mathcal{A}) = 0$ for each $u \in \mathcal{S}$, where $\mathcal{S}$ is a set of positive Lebesgue measure.

If the manifold is compact we define the *minimal global Milnor attractor* as a minimal closed invariant set $\mathcal{A}$ having the property $\lim_{t\to\infty} \operatorname{dist}(\varphi^t(u), \mathcal{A}) = 0$ for any $u \in \mathcal{S}$, where $\mathcal{S}$ is a Lebesgue measurable set with the full Lebesgue measure, i.e. $\mu_L(\mathcal{S}) = \mu_L(\mathcal{M})$.

In the following we denote a Milnor attractor by $\widetilde{\mathcal{A}}$ and a minimal global Milnor attractor by $\widetilde{\mathcal{A}}_{\min}$.

We will consider the minimal global Milnor attractor also for dynamical systems given in $\mathbb{R}^n$ and possessing a bounded open positively invariant absorbing set $\mathcal{B}_0$. In this case we can restrict our system on the positive semi-group $\mathbb{T}_+$ on $\mathcal{B}_0$, considering $\mathcal{B}_0$ with relative topology as compact manifold.

Example 1.7 Let us consider Van der Pol's equation

$$\ddot{x} + \varepsilon(x^2 - 1)\dot{x} + x = 0$$

where $\varepsilon > 0$ is a parameter. This equation can be written as planar system

$$\dot{x} = y, \quad \dot{y} = -\varepsilon(x^2 - 1)y - x. \tag{1.9}$$

It is well-known that (1.9) generates a semi-flow $(\{\varphi^t\}_{t\geq 0}, \mathbb{R}^2, |\cdot|)$ and the origin $(0, 0)$ is an unstable equilibrium of this semi-flow. Furthermore, there is a single orbitally stable periodic orbit (Fig. 1.2). Any orbit of the semi-flow, different from the equilibrium $(0, 0)$, tends for $t \to +\infty$ to this periodic orbit.

It is easy to see that the minimal global $\mathcal{B}$-attractor is $\mathcal{A}_{\mathbb{R}^2,\min}$ is the closed disk around the origin and bounded by the unit circle $\mathcal{S}^1 = \{(x, y) \mid x^2 + y^2 = 1\}$, the minimal global attractor is $\widehat{\mathcal{A}}_{\mathbb{R}^2,\min} = \mathcal{S}^1 \cup \{(0, 0)\}$, the minimal global Milnor attractor is $\widetilde{\mathcal{A}}_{\min} = \mathcal{S}^1$ and a non-global attractor is given by $\mathcal{A} = \mathcal{S}^1$.

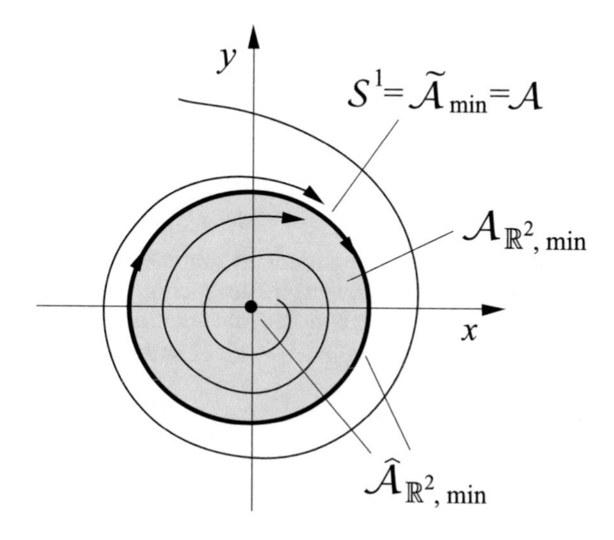

Fig. 1.2 Attractors of Van der Pol's system

Consider the dynamical system $(\{\varphi^t\}_{t\in\mathbb{T}}, (\mathcal{M}, \rho))$ on the metric space $(\mathcal{M}, \rho)$. Suppose $\mathfrak{B} = \mathfrak{B}(\mathcal{M})$ is the σ-algebra of Borel sets on $\mathcal{M}$ and μ is a finite Borel measure on $\mathfrak{B}$, i.e., $\mu(\mathcal{M}) < +\infty$. The bounded set $\widetilde{\mathcal{A}}_\mu(\mathcal{M}) \subset \mathcal{M}$ is called *global Milnor attractor* w.r.t. the dynamical system $(\{\varphi^t\}_{t\in\mathbb{T}}, (\mathcal{M}, \rho))$ and the measure μ if $\widetilde{\mathcal{A}}_\mu(\mathcal{M})$ is a minimal, closed and invariant set having the property $\lim_{t\to\infty} \text{dist}(\varphi^t(u), \widetilde{\mathcal{A}}_\mu(\mathcal{M}))$ for μ-a.e. point $u \in \mathcal{M}$. Sometimes the global Milnor attractor is called *stochastic attractor*.

If the metric space $(\mathcal{M}, \rho)$ is compact we define the *minimal global Milnor attractor* as a minimal closed invariant set $\widetilde{\mathcal{A}}_{\min,\mu}(\mathcal{M})$ having the property $\lim_{t\to\infty} \text{dist}(\varphi^t(u),$
$\widetilde{\mathcal{A}}_{\min,\mu}(\mathcal{M})) = 0$ for μ-a.e. point $u \in \mathcal{M}$.

Suppose that $\mathcal{M} = \mathbb{E}$ is a linear metric space and $\mathbb{E}^*$ is the dual to $\mathbb{E}$, i.e., the linear space of linear bounded functionals on $\mathbb{E}$. The sequence $\{u_n\}_{n=1}^\infty$ from $\mathbb{E}$ is called *weakly convergent* to $u \in \mathbb{E}$ if $\ell(u_n) \to \ell(u)$, $\forall \ell \in \mathbb{E}^*$. We denote this by $u_n \rightharpoonup u$ for $n \to \infty$. The set $\mathcal{Z} \subset \mathbb{E}$ is called *weakly closed* if it contains the weak limit u of arbitrary weakly convergent sequences $\{u_n\} \subset \mathcal{Z}$. In the *weak topology* the open sets are given by arbitrary unions of sets

$$\mathcal{O}(v; \ell_1, \ell_2, \dots, \ell_n; \varepsilon_1, \varepsilon_2, \dots, \varepsilon_n)$$
$$:= \Big\{u \in \mathbb{E} \;\Big|\; |\,\ell_1(u - v)\,| < \varepsilon_1\,,\ \ell_2(u - v) < \varepsilon_2, \dots,\ |\ell_n(u - v)\,| < \varepsilon_n\Big\}$$

where $v \in \mathbb{E}\,, \ell_i \in \mathbb{E}^*, \varepsilon_i > 0\ (i = 1, 2, \dots, n)$ are numbers.

By definition the empty set $\emptyset$ is open.

A non-empty set $\mathcal{O}$ which is open in the weak topology and which contains the set $\mathcal{Z} \subset \mathbb{E}$ is called *weak neighborhood* of $\mathcal{Z}$.

Suppose that $\mathcal{M} = \mathbb{E}$ is a linear metric space. The set $\mathcal{A}_w(\mathcal{M}) \subset \mathcal{M}$ is called *weak global $\mathcal{B}$-attractor* w.r.t. $(\{\varphi^t\}_{t\in\mathbb{T}}, (\mathbb{E}, \rho))$ if $\mathcal{A}_w(\mathcal{M})$ is a bounded and weakly closed invariant set such that for any weak neighborhood $\mathcal{O}$ of the set $\mathcal{A}_w(\mathcal{M})$ and any bounded set $\mathcal{B} \subset \mathcal{M}$ there exists a $t_0 = t_0(\mathcal{O}, \mathcal{B})$ such that $\varphi^t(\mathcal{B}) \subset \mathcal{O}$ for all $t \geq t_0$.

Let us note that if the linear space $\mathcal{M} = \mathbb{E}$ has finite dimension and for the dynamical system the global attractor $\mathcal{A}_\mathcal{M}$ exists and is weakly closed then also exists $\mathcal{A}_w(\mathcal{M})$ and $\mathcal{A}_w(\mathcal{M}) = \mathcal{A}_\mathcal{M}$.

In Table 1.1 we present the various types of attractors and their symbols.

Example 1.8 Let us consider as complete metric space $\mathcal{M}$ the *Hilbert space* $L^2(a, b)$ of quadratically integrable functions on (a, b). It follows from the Riesz theorem that the any linear bounded functional on $L^2(a, b)$ is given by $\ell(u) = \int_\Omega uv dx$, where $v \in L^2(a, b)$ and $\Omega = (a, b)$. Thus we have the properties $u_n \rightharpoonup u$ for $n \to \infty$ in $\mathbb{L}^2(a, b) \Leftrightarrow \int_\Omega u_n v dx \to \int_\Omega uv dx$ for $n \to \infty$ and any $v \in L^2(a, b)$. Assume that $\{e_i\}_{i=1}^\infty$ is an orthonormal basis of $L^2(a, b)$. Then any function $v \in L^2(a, b)$ can be represented as $v = \sum_{i>1}^\infty c_i e_i$, where $c_i = \int_\Omega v e_i dx, i = 1, 2, \dots,$ are the Fourier coefficients satisfying $\|v\|^2_{L^2(a,b)} = \sum_{i>1}^\infty c_i^2 < \infty$. Consider the functions $u_n = e_n\,,\ n = 1, 2, \dots$. Then for any $v \in L^2(a, b)$ we have

Table 1.1 Types of attractors and their symbols

Symbol	Type of attractor	Sections
$\mathcal{A}$	Arbitrary attractor	1.1.2
$\mathcal{A}_{\mathcal{M}}$	Global $\mathcal{B}$-attractor	1.1.2
$\mathcal{A}_{\mathcal{M},\min}$	Minimal global $\mathcal{B}$-attractor	1.1.2
$\widehat{\mathcal{A}}_{\mathcal{M}}$	Global attractor	1.1.2
$\widehat{\mathcal{A}}_{\mathcal{M},\min}$	Minimal global attractor	1.1.2
$\widetilde{\mathcal{A}}$	Milnor attractor	1.1.2
$\widetilde{\mathcal{A}}_{\min}$	Minimal global Milnor attractor	1.1.2
$\mathcal{A}_w(\mathcal{M})$	Weak global $\mathcal{B}$-attractor	1.1.2

$$\lim_{n\to\infty}\int_{\Omega} e_n v dx = \lim_{n\to\infty}\int_{\Omega} c_n e_n^2 dx = \lim_{n\to\infty} c_n = 0, \text{ i.e., } e_n \rightharpoonup 0 \text{ for } n\to\infty \text{ in } L^2(a,b).$$

From the other side we have $e_n \not\to 0$ as $n \to \infty$ in $L^2(a, b)$ since $\|e_n\|^2 = 1$, $n = 1, 2, \dots$.

1.2 Dissipativity

1.2.1 Dissipativity in the Sense of Levinson

This section is devoted to the concepts of dissipativity, region of dissipativity, and its estimation for autonomous differential equations. These notions arose for the first time in stability theory. Later they turned out to be very useful in the study of attractors since they give the possibility to localize attractors in the phase space.

Consider the dynamical system $(\{\varphi^t\}_{t\in\mathbb{T}}, \mathbb{R}^n, |\cdot|)$ which in the continuous-time case is given by the autonomous ODE

$$\dot{\varphi} = f(\varphi)\,, \tag{1.10}$$

where $f : \mathbb{R}^n \to \mathbb{R}^n$ is continuously differentiable, and in the discrete-time case is given by the continuous map

$$\varphi : \mathbb{R}^n \to \mathbb{R}^n\,. \tag{1.11}$$

Definition 1.2 The dynamical system $(\{\varphi^t\}_{t\in\mathbb{T}}, \mathbb{R}^n, |\cdot|)$ is called *dissipative in the sense of Levinson*, if there exists an $R > 0$ such that for any $u \in \mathbb{R}^n$

$$\limsup_{t\to+\infty} |\varphi^t(u)| < R\,.$$

Proposition 1.9 *The dynamical system* $(\{\varphi^t\}_{t\in\mathbb{T}}, \mathbb{R}^n, |\cdot|)$ *is dissipative in the sense of Levinson if and only if there exists a bounded set* $\mathcal{D} \subset \mathbb{R}^n$ *that attracts any point in* $\mathbb{R}^n$.

Proof Let the dynamical system be dissipative in the sense of Levinson. Choose as the set $\mathcal{D}$ a ball of radius R, where R is from Definition 1.2, and with center in the origin. It is obvious that such a $\mathcal{D}$ attracts every point in $\mathbb{R}^n$.

Conversely, let $\mathcal{D}$ be a set for the dynamical system which attracts every point of $\mathbb{R}^n$, and let $\varepsilon > 0$ be an arbitrary number. Choose R so large that the ball of radius R with center in the origin contains the ε-neighborhood $\mathcal{D}_\varepsilon$ of $\mathcal{D}$. It is clear that such an R satisfies Definition 1.2. □

It is easy to see that if the dynamical system is dissipative in the sense of Levinson with $\mathcal{D}$ as region of dissipativity, then any attractor $\mathcal{A}$ of the dynamical system satisfies the inclusion $\mathcal{A} \subset \mathcal{D}$.

1.2.2 Dissipativity and Completeness of The Lorenz System

Consider the Lorenz system ([28, 30, 39])

$$\dot{x} = -\sigma x + \sigma y\,, \quad \dot{y} = rx - y - xz\,, \quad \dot{z} = -bz + xy \tag{1.12}$$

where σ, r and b are positive parameters. Let us show that equation (1.12) is dissipative. Introduce the auxilary function $V : \mathbb{R}^3 \to \mathbb{R}_+$ given by

$$V(x, y, z) := \frac{1}{2}\left[x^2 + y^2 + (z - \sigma - r)^2\right]. \tag{1.13}$$

Direct computation along an arbitrary solution $u = (x, y, z)$ of (1.12) shows that the derivative of V along

$$\dot{V}(x, y, z) = -\sigma x^2 - y^2 - \frac{b}{2}(z - \sigma - r)^2 + \frac{b}{2}(\sigma + r)^2\,.$$

Thus in $\mathbb{R}^3$ we have

$$\dot{V} \le \frac{b}{2}(\sigma + r)^2\,. \tag{1.14}$$

On the set

$$\mathcal{E}_1 := \left\{(x, y, z)\,|\,\sigma x^2 + y^2 + \frac{b}{2}(z - \sigma - r)^2 \le \frac{b}{2}(\sigma + r)^2\right\}$$

the inequality $\dot{V} \geq 0$ is true and on the set $\mathbb{R}^3 \setminus \mathcal{E}_1$ we have $\dot{V} < 0$. For large R the ball $\mathcal{B}_R = \{(x, y, z) \mid V(x, y, z) < R\}$ contains the ellipsoid $\mathcal{E}_1$. On the boundary of $\mathcal{B}_R$, i.e. on the set $\mathcal{S}_R = \{(x, y, z) \mid V(x, y, z) = R\}$ the inequality $\dot{V} < 0$ is satisfied. It follows that $\mathcal{B}_R$ is a bounded absorbing set. If we put $\kappa = \min\{\sigma, 1, \frac{b}{2}\}$ we conclude that along the solution of (1.12)

$$\dot{V} \leq -2\kappa V + \frac{b}{2}(\sigma + r)^2 .$$

This means that any solution of (1.12) enters the ellipsoid

$$\mathcal{E}_2 := \left\{(x, y, z) \mid \frac{1}{2}\left[x^2 + y^2 + (z - \sigma - r)^2\right] \leq \frac{b}{4\kappa}(\sigma + r)^2\right\}$$

and remains there during the positive existence interval. From (1.14) we have

$$V(x(t), y(t), z(t)) \leq V(x(0), y(0), z(0)) + \frac{b}{2}(\sigma + r)^2 t$$

for $t \geq 0$. Since V cannot go to infinity in a finite positive time, each of $|x(t)|$, $|y(t)|$, and $|z(t)|$ cannot go to infinity in a finite positive time. Thus the Lorenz system is complete in positive time. Thus the Lorenz system defines a semi-flow in $\mathbb{R}^3$. From (1.13) it follows that for a sufficiently large $\kappa_1 > 0$ we have

$$\dot{V} + \kappa_1 V \geq \frac{b}{2}(\sigma + r)^2 =: c_1 \tag{1.15}$$

From (1.15) we get

$$\frac{d}{dt}(e^{\kappa_1 t} V) \geq c_1 e^{\kappa_1 t} .$$

Thus for $t \leq 0$ we have

$$V(x(0), y(0), z(0)) - e^{\kappa_1 t} V(x(t), y(t), z(t)) \geq \frac{c_1}{\kappa_1}[1 - e^{\kappa_1 t}]$$

$$\text{or} \quad V(x(t), y(t), z(t)) \leq e^{-\kappa_1 t} V(x(0), y(0), z(0)) + \frac{c_1}{\kappa_1}[1 - e^{-\kappa_1 t}] .$$

Thus $V(x(t), y(t), z(t))$ cannot go to infinity in finite negative time. Hence each of $|x(t)|$, $|y(t)|$ and $|z(t)|$ cannot got to infinity in finite negative time and the system is complete in negative time ([8]).

Let us obtain other estimates for the region of dissipativity for (1.12). Consider next the function

$$V_1(x, y, z) := \frac{1}{2}x^2 + \frac{1}{2}y^2 + \frac{1}{2}z^2 - (\sigma + r)z .$$

Let us show that for an arbitrary solution $u = (x, y, z)$ of (1.12) with $b > 1$ we have

$$\limsup_{t\to+\infty} V_1(x(t), y(t), z(t)) \le c_2 , \tag{1.16}$$

where $c_2 := \frac{(\sigma+r)^2(b-2)^2}{8(b-1)}$. Indeed, a calculation shows that

$$\begin{aligned}\dot{V}_1 + 2V_1 &= -(\sigma - 1)x^2 - (b-1)z^2 + (\sigma + r)(b - z)z \\ &\le -(b-1)z^2 + (\sigma + r)(b-2)z \le 2\,c_2 .\end{aligned}$$

Therefore we have

$$\frac{d}{dt}(V_1 - c_2) + 2\,(V_1 - c_2) \le 0 .$$

Multiplying the last inequality by e^{2t} we get for $t \ge 0$

$$\frac{d}{dt}\left[(V_1 - c_2)e^{2t}\right] \le 0 . \tag{1.17}$$

Integrating (1.17) on $[0, t]$, we obtain

$$V_1(x(t), y(t), z(t)) - c_2 \le [V_1(x(0), y(0), z(0)) - c_2]\,e^{-2t} ,$$

from which the inequality (1.16) results. From (1.16) it follows that the ellipsoid

$$\{(x, y, z) \in \mathbb{R}^3 \mid x^2 + \frac{1}{2}y^2 + \frac{1}{2}z^2 - (\sigma + r)z \le c_2\}$$

is a region of dissipativity for (1.12).

Let us now show that for an arbitrary solution $u = (x, y, z)$ of (1.12)

$$\limsup_{t\to+\infty}\,[y^2(t) + (z(t) - r)^2] \le l^2 r^2 \tag{1.18}$$

and, if $2\sigma - b \ge 0$,

$$\liminf_{t\to+\infty}\,[2\,\sigma\, z(t) - x^2(t)] \ge 0 . \tag{1.19}$$

The parameter l in (1.18) is defined by

$$l := \begin{cases} 1 , & \text{if } \ b \le 2 , \\ \frac{b}{2\sqrt{b-1}} , & \text{if } \ b \ge 2 . \end{cases} \tag{1.20}$$

In order to prove (1.18) we put for $(x, y, z) \in \mathbb{R}^3$

$$V_2(y, z) := \frac{1}{2}\left[y^2 + (z - r)^2\right] .$$

Suppose that $\kappa_0 := \min\{1, b\}$. Then for any $\kappa \in (0, \kappa_0)$ we have

$$\begin{aligned}\dot{V}_2 + 2\kappa V_2 &= (\kappa - 1)y^2 + (\kappa - b)z^2 - 2r\left(\kappa - \frac{b}{2}\right)z + \kappa r^2 \\ &\leq (\kappa - b)\left[z - \frac{r(\kappa - b/2)}{\kappa - b}\right]^2 - \frac{r^2(\kappa - b/2)^2}{\kappa - b} + \kappa r^2 \\ &\leq \left[\kappa - \frac{(\kappa - b/2)^2}{\kappa - b}\right]r^2 = \frac{b^2 r^2}{4(b - \kappa)} .\end{aligned}$$

It follows that

$$\limsup_{t \to +\infty} V_2(y(t), z(t)) \leq \frac{b^2 r^2}{8(b - \kappa)\kappa} . \tag{1.21}$$

Minimizing the right-hand side of (1.21) over $\kappa \in (0, \kappa_0)$ we obtain (1.18). To prove (1.19) we put

$$V_3(x, z) := \sigma z - \frac{1}{2}x^2 .$$

The direct computation shows that

$$\dot{V}_3 = -b\left[\sigma z - \frac{2\sigma}{b}\frac{1}{2}x^2\right] \geq -b V_3 .$$

The last relation implies (1.19).

From (1.18) it follows that the region of dissipativity $\mathcal{D}$ satisfies the inclusion

$$\mathcal{D} \subset \overline{\mathcal{D}}_1 ,$$

where $\mathcal{D}_1 := \{(x, y, z) \mid y^2 + (z - r)^2 < l^2 r^2\}$ is a cylinder in $\mathbb{R}^3$.

Under the condition $2\sigma - b \geq 0$ it follows from (1.18) and (1.19) that

$$\mathcal{D} \subset \overline{\mathcal{D}}_1 \cap \{(x, y, z) \mid z \geq 0\} .$$

Remark 1.4 Using the Lyapunov function (1.13) and Proposition 1.7 one sees that the Lorenz system has a compact attracting set which attracts bounded sets. It follows that the Lorenz system is $\mathcal{B}$-dissipative and has a minimal $\mathcal{B}$-attractor $\mathcal{A}_{\mathbb{R}^3,\min}$ which satisfies $\mathcal{A}_{\mathbb{R}^3,\min} \subset \mathcal{D}$. Note that any other attractor of (1.12) also belongs to $\mathcal{D}$.

Remark 1.5 Let us consider the system

$$\dot{x} = -\sigma x + \sigma y , \quad \dot{y} = rx - y + xz , \quad \dot{z} = -bz + xy \tag{1.22}$$

with positive parameters σ, r and b. This system differs from the Lorenz system (1.12) only in the sign of the nonlinearity xz in the second equation and the divergence of the right-hand side of (1.22) is $-(\sigma + 1 + b) < 0$, i.e. the same as in the Lorenz system. However, it was shown in [14] that system (1.22) for any positive parameters has solutions converging to infinity for $t \to +\infty$. But this means that system (1.22) is not dissipative. We need certain extra conditions on the right-hand side in order to guarantee dissipativity.

1.2.3 Lyapunov-Type Results for Dissipativity

Let us consider the dynamical system $(\{\varphi^t\}_{t\in\mathbb{T}}, \mathcal{M}, \rho)$ on the Riemannian n-dimensional C^k-manifold $(\mathcal{M}, g)$ which is, for continuous time, given by (1.5) and for discrete time by (1.6). Suppose that there exists a scalar valued function $V : \mathcal{M} \to \mathbb{R}$ which is C^1 in the continuous-time case and C^0 in the discrete-time case. Define the *Lie derivative* $\dot{V}(u)$ *w.r.t. the dynamical system* in the continuous-time case by

$$\dot{V}(u) := \frac{d}{dt} V(\varphi^t(u))|_{t=0} = (F(u), \operatorname{grad} V(u)) \tag{1.23}$$

and in the discrete-time case by

$$\dot{V}(u) := V(\varphi(u)) - V(u). \tag{1.24}$$

Let us establish the following theorem which is a generalization of a result from [40, 41], obtained for differential equations in $\mathbb{R}^n$.

Proposition 1.10 *Suppose that there exists a function as introduced above and such that the following conditions are satisfied:*

(1) V is proper for $\mathcal{M}$, i.e. for any compact set $\mathcal{K} \subset \mathbb{R}$ the set $V^{-1}(\mathcal{K}) \subset \mathcal{M}$ is compact and V is bounded from below on $\mathcal{M}$;

(2) There exists an $r > 0$ such that $\dot{V}(u) \le 0$ for $u \notin \overline{\mathcal{B}_r(0)}$;

(3) The dynamical system does not have a motion $\varphi^{(\cdot)}(v)$ with $\varphi^t(v) \notin \overline{\mathcal{B}_r(0)}$ and $\dot{V}(\varphi^t(v)) \equiv 0$ for $t \ge t_0$.

Then the dynamical system $(\{\varphi^t\}_{t\in\mathbb{T}}, \mathcal{M}, \rho)$ is dissipative.

Proof Let us put $\eta := \max\limits_{u\in\overline{\mathcal{B}_r(0)}} V(u)$ and consider the set $\mathcal{D} := \{u \in \mathcal{M} \,|\, V(x) \le \eta\}$. In the discrete-time case we choose η so large that additionally $\mathcal{D} \supset \varphi^1(\overline{\mathcal{B}_r(0)})$. By assumption (1) we can write $\mathcal{D} = \{u \in \mathcal{M} \,|\, \theta \le V(x) \le \eta\}$, where $\theta := \inf_{u\in\mathcal{M}} V(u) > -\infty$. Since $\mathcal{K} := [\theta, \eta]$ is compact again by assumption (1) we conclude that $\mathcal{D}$ is bounded. It follows from the definition of $\mathcal{D}$ that $\varphi^t(\mathcal{D}) \subset \mathcal{D}$ for all $t \in \mathbb{T}_+$, proposed that $u \in \mathcal{B}_r(0)$.

Let us show this. Assume to the contrary that there is a $u \in \mathcal{D}$ and a time $t_1 \in \mathbb{T}_+$ such $\varphi^{t_1}(u) \notin \mathcal{D}$.

Consider at first the continuous-time case. Here exists a maximum time t' such that $0 < t' < t_1$ and $\rho(\varphi^{t'}(u), 0) = r$ or put $t' := 0$.

It follows that $V(u) \le \eta$ and on the interval (t', t_1) we have $\rho(\varphi^t(u), 0) > r$. Now we conclude by continuity that $V(\varphi^{t_1}(u)) \le V(u) \le \eta$, a contradiction. In the discrete-time case there must exist a time $t_2 \in (0, t_1) \cap \mathbb{Z}_+$ such that $\varphi^{t_2}(u) \in \mathcal{D} \cap \mathcal{B}_r(0)$, but $\varphi^{t_2+1}(u) \notin \mathcal{D}$. But this is impossible by the choice of $\mathcal{D}$ in the discrete-time case.

Let us show now that for any point $u \in \mathcal{M}$ with $u \notin \mathcal{B}_r(0)$ there exists a time $t_1 > 0$ such that $\varphi^{t_1}(u) \in \overline{\mathcal{B}_r(0)}$. Suppose the opposite, i.e. $\varphi^t(u) \notin \overline{\mathcal{B}_r(0)}$ for all $t \in \mathbb{T}_+$. Then the positive semi-orbit of the motion $\varphi^{(\cdot)}(u)$ is bounded. Indeed, by condition (2) we have $V(\varphi^t(u)) \leq V(u)$ for all $t \in \mathbb{T}_+$. From this and assumption (1) we obtain the boundedness of the semi-orbit. In virtue of this boundedness according to Proposition 1.4, the ω-limit set of the semi-orbit of $\varphi^{(\cdot)}(u)$ is non-empty. Let $\upsilon \in \omega(u)$ be an arbitrary point. According to our assumption we have $\upsilon \notin \overline{\mathcal{B}_r(0)}$. The function $V(\varphi^t(u))$ is bounded on $\mathbb{T}_+$ and does not increase. Therefore, there exists the limit

$$\lim_{t \to +\infty} V(\varphi^t(u)) = V(\upsilon) \,. \tag{1.25}$$

Consider the motion $\varphi^{(\cdot)}(\upsilon)$. By Proposition 1.4 we have $\varphi^t(\upsilon) \in \omega(u)$ for all $t \in \mathbb{T}_+$. It follows that for any $t \in \mathbb{T}_+$ there exists a sequence $t_m \to +\infty$ as $m \to +\infty$ such that $\varphi^{t_m}(u) \to \varphi^t(\upsilon)$ as $m \to +\infty$. Hence by (1.25) we have

$$V(\varphi^t(\upsilon)) \equiv V(\upsilon) \,,$$

which contradicts the assumption (3). Thus, for arbitrary $u \in \mathbb{R}^n$ with $u \notin \overline{\mathcal{B}_r(0)}$ there exists a time $t_1 \in \mathbb{T}_+$ such that $\varphi^{t_1}(u) \in \overline{\mathcal{B}_r(0)}$. □

Example 1.9 Consider the equation of a pendulum

$$\ddot{x} + \varepsilon \dot{x} + \sin x = 0 \,,$$

where $\varepsilon > 0$ is a parameter. This equation is equivalent to the planar system

$$\dot{x} = y \,, \quad \dot{y} = -\varepsilon y - \sin x \,. \tag{1.26}$$

Since the right-hand side of (1.26) is globally Lipschitz we have global existence and uniqueness of all solutions. Let us denote the dynamical system generated by (1.26) by $(\{\varphi^t\}_{t \geq 0}, \mathbb{R}^2, |\cdot|)$. It is well-known that any semi-orbit of this system tends to an equilibrium for $t \to +\infty$. The phase portrait is shown in Fig. 1.3

It follows that the minimal global attractor $\widehat{\mathcal{A}}_{\mathbb{R}^2,\min}$ of (1.26) is the stationary set, i.e. the set of all equilibria $\mathcal{C}$.

Consider now a ball $\mathcal{B}_\delta$ with small radius $\delta > 0$ and center at a point $u_0 = (x_0, y_0)$ on the stable manifold of a saddle (Figs. 1.3 and 1.4).

It is obvious that $\varphi^t(\mathcal{B}_\delta)$ converges for $t \to +\infty$ to a set consisting of a saddle point and of two heteroclinic orbits coming from this saddle point and going to stable equilibria. It follows that the minimal global $\mathcal{B}$-attractor is the union of the stationary set $\mathcal{C}$ and the heteroclinic orbits (Fig. 1.5).

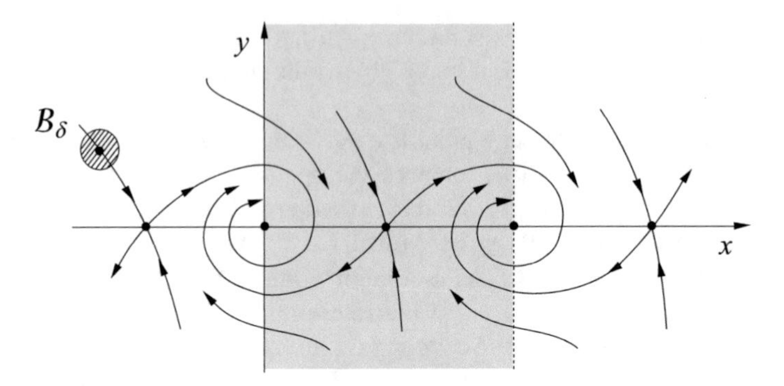

Fig. 1.3 Minimal global attractor of (1.26)

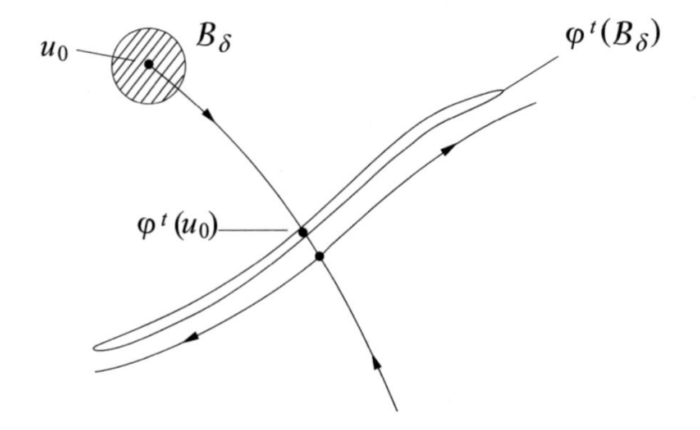

Fig. 1.4 Deformation of a small ball under the flow of (1.26)

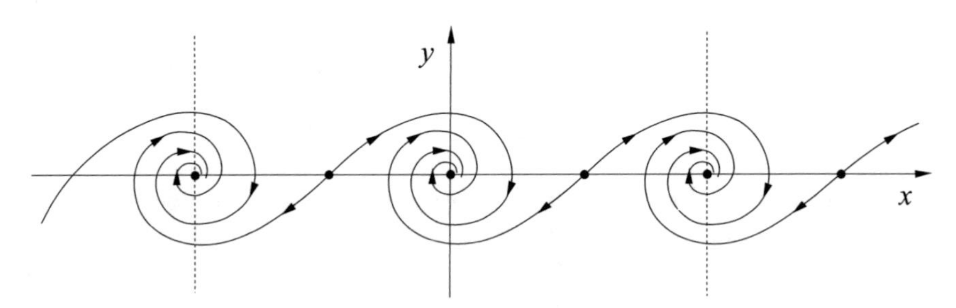

Fig. 1.5 Minimal global $\mathcal{B}$-attractor of (1.26)

In order to apply Proposition 1.10 one has to construct a Lyapunov-type function V satisfying the assumptions (1)–(3). Very often this is a sufficiently difficult problem. In some cases one can avoid this as the next proposition ([32]) shows.

Consider the dynamical system $(\{\varphi^t\}_{t\in\mathbb{T}}, \mathbb{R}^n, |\cdot|)$ which is given for continuous time by the ODE

$$\dot{\varphi} = A\varphi + g(\varphi)\ , \tag{1.27}$$

and for discrete time by the map

$$u \mapsto Au + g(u)\,, u \in \mathbb{R}^n\,. \tag{1.28}$$

In both cases A is an $n \times n$ matrix and $g : \mathbb{R}^n \to \mathbb{R}^n$ is continuous. The matrix A is assumed to be stable, i.e. all eigenvalues of A have negative real part in the continuous-time case, and all eigenvalues have moduli smaller one in the discrete-time case.

Proposition 1.11 *Suppose that the dynamical system is given by (1.29) resp. (1.30) and g is a bounded map. Then the dynamical system is dissipative.*

Proof Suppose that $|g(u)| \le c_0$ in $\mathbb{R}^n$ with some constant $c_0 > 0$. Any motion $\varphi^{(\cdot)}(u)$ of the dynamical system can be written as

$$\varphi^t(u) = e^{At}u + \int_0^t e^{A(t-\tau)} g(\varphi^\tau(u))\, d\tau\ , \quad t \ge 0\,, \tag{1.29}$$

in the continuous-time case, and as

$$\varphi^t(u) = A^t u + \sum_{\tau=0}^{t-1} A^{t-\tau-1} g(\varphi^\tau(u))\,, \quad t = 1, 2, \dots\,, \tag{1.30}$$

in the discrete-time case.

From (1.29), and the stability of A it follows that there exist constants $\gamma > 0$ and $c_1 > 0$ such that

$$|\varphi^t(u)| \le c_1(e^{-\gamma t}|u| + c_0 \int_0^\infty e^{\gamma(t-\tau)}\, d\tau), \quad t \ge 0\,. \tag{1.31}$$

From (1.30) and the stability of A we get with some constants $\delta \in (0, 1)$ and $c_2 > 0$ the representation

$$|\varphi^t(u)| \le c_2 \left(\delta^t |u| + c_0 \sum_{\tau=0}^{\infty} \delta^{t+\tau+1}\right), \quad t = 1, 2, \dots\,. \tag{1.32}$$

From (1.31) and (1.32) the assertion follows immediately. □

Definition 1.3 The equilibrium p of the dynamical system $(\{\varphi^t\}_{t\in\mathbb{T}}, \mathcal{M}, \rho)$ is said to be *globally asymptotically stable* if p is asymptotically Lyapunov stable and for any $q \in \mathcal{M}$ we have $\varphi^t(q) \to p$ for $t \to +\infty$.

The next theorem was proved by E. A. Barbashin and N. N. Krasovskii in [1] for continuous time. For discrete time it was shown in [29].

Theorem 1.1 *Suppose that p is an equilibrium of the dynamical system $(\{\varphi^t\}_{t\in\mathbb{T}}, \mathbb{R}^n, |\cdot|)$ and there exists a function $V : \mathbb{R}^n \to \mathbb{R}$ (C^1 in the continuous-time case and C^0 in the discrete-time case) such that the following conditions are satisfied:*
(1) $V(p) = 0$ and $V(u) > 0$ for all $u \in \mathbb{R}^n \setminus \{p\}$;
(2) $\dot{V}(p) = 0$ and $\dot{V}(u) < 0$ for all $u \in \mathbb{R}^n \setminus \{p\}$;
(3) $V(u) \to +\infty$ for $|u| \to +\infty$.
Then the equilibrium p is globally asymptotically stable.

Proof Let for simplicity be $p = 0$. It follows from the Lyapunov theorem that $p = 0$ is asymptotically Lyapunov stable. Suppose that $\varphi^{(\cdot)}(q)$ is an arbitrary motion of the dynamical system.

Using assumption (3) we choose r so large that

$$q \in \mathcal{B}_r(0) \quad \text{and}$$

$$V(u) > V(q) \quad \text{for all} \quad |u| \geq r \,. \tag{1.33}$$

From assumption (2) we conclude that

$$V(\varphi^t(q)) \leq V(q) \quad \text{for all} \quad t \in \mathbb{T}_+ \,. \tag{1.34}$$

Thus, if we take into consideration (1.33) we have

$$|\varphi^t(q)| < r \quad \text{for all} \quad t \in \mathbb{T}_+ \,.$$

Let us put $c = \lim_{t\to+\infty} V(\varphi^t(q))$ and show that $c = 0$. If we assume that $c > 0$ there exists a number $r_1 \in (0, r)$ such that $|\varphi^t(q)| \geq r_1$ for all $t \in \mathbb{T}_+$. It follows that $r_1 \leq |\varphi^t(q)| < r$ for $t \in \mathbb{T}_+$. The proof is complete if we argue as in the Lyapunov theorem. □

Example 1.10 Let us show that the equilibrium $u_1 = (0, 0, 0)$ of the Lorenz system (1.12) is globally asymptotically stable. Take the function

$$V(x, y, z) := \frac{1}{2}(x^2 + \sigma y^2 + \sigma z^2) \,.$$

A direct computation shoes that

$$\begin{aligned} \dot{V}(x, y, z) &= -\sigma \left[x^2 - (1 + r)xy + y^2 + bz^2\right] \\ &= -\sigma \left[\frac{1 - r}{2}(x^2 + y^2) + bz^2 + \frac{1 + r}{2}(x - y)^2\right] \\ &\leq -\sigma \left[\frac{1 - r}{2}(x^2 + y^2) + bz^2\right] . \end{aligned}$$

Thus by the continuous-time version of Theorem 1.1 we conclude that $u_1 = (0, 0, 0)$ is globally asymptotically stable if $0 < r < 1$.

1.3 Existence of a Homoclinic Orbit in the Lorenz System

1.3.1 Introduction

In this section we consider again the Lorenz system. We give estimates for the shape of a global $\mathcal{B}$-attractor and prove the existence of homoclinic orbits for certain parameter values. It will be shown that in certain cases these estimates are asymptotically exact. Since all estimates are uniform with respect to the parameters, it becomes possible to prove the existence of a homoclinic orbit using the formulae of asymptotic integration. The Lorenz system

$$\dot{x} = -\sigma(x - y), \quad \dot{y} = rx - y - xz, \quad \dot{z} = -bz + xy, \tag{1.35}$$

which is a three-mode approximation of a two-dimensional thermal convection, is now one of the classical models for the transition from global stability to chaotic behaviour and to the generation of attractors with non-integer Hausdorff dimension. Sometimes the phrase "homoclinic explosion" is used to refer to the appearance of various types of chaotic behaviour when parameters are perturbed from the bifurcation parameter of a homoclinic orbit. In such a process the role of homoclinic orbits, which appear for bifurcation values of parameters is very important. These orbits and the attractors of the Lorenz system are located in certain domains of the phase space which can be estimated. We shall suppose further that σ, r, b are positive numbers. Let, in addition, $r > 1$ and $2\sigma > b$. Note that if one of these restrictions is violated then system (1.35) is convergent, i.e. any its orbit tends to a certain equilibrium when $t \to +\infty$ (Example 1.10). Along with system (1.35) we consider the equivalent system

$$\dot{\xi} = \eta, \quad \dot{\eta} = -\mu\eta - \zeta\xi - \varphi(\xi), \quad \dot{\zeta} = -A\zeta - B\xi\eta, \tag{1.36}$$

where $\varphi(\xi) = -\xi + \gamma\xi^3$, $\xi = \varepsilon x/\sqrt{2\sigma}$, $\eta = \varepsilon^2(y - x)/\sqrt{2}$, $\zeta = \varepsilon^2(z - x^2/b)$, $t = t_1\sqrt{\sigma}/\varepsilon$, $\mu = \varepsilon(\sigma + 1)/\sqrt{\sigma}$, $A = \varepsilon b/\sqrt{\sigma}$, $\varepsilon = (r - 1)^{-1/2}$, $B = 2(2\sigma - b)/b$, $\gamma = 2\sigma/b$.

1.3.2 Estimates for the Shape of Global Attractors

In this section we shall obtain estimates which are for $b \le 2$ and great r asymptotically exact for a global $\mathcal{B}$-attractor with respect to the coordinates ξ and η. From these estimates if follows that a global $\mathcal{B}$-attractor of system (1.36) is located in domains which are uniformly bounded with respect to parameter $r \in (1, +\infty)$. This fact will be used for the demonstration of the existence of homoclinic orbits.

It was shown in Sect. 1.2 that the surfaces

$$\mathcal{S}_1 := \{(r-z)^2 + y^2 = l^2 + \varsigma\} \quad \text{and} \quad \mathcal{S}_2 := \{z - x^2/(2\sigma) = -\varsigma\},$$

where $\varsigma > 0$ and l is given by (1.20), are transversal ("contact-free") for the solutions of system (1.35). Hence the following inequalities hold on a global attractor of system (1.35):

$$(r-z)^2 + y^2 \le l^2, \quad z \ge x^2/(2\sigma)\,. \tag{1.37}$$

Hence it follows that on a global attractor of system (1.36)

$$-\frac{l}{\sqrt{2}\,(r-1)} - \frac{\sqrt{\sigma}\,\xi}{\sqrt{r-1}} \le \eta \le \frac{l}{\sqrt{2}\,(r-1)} - \frac{\sqrt{\sigma}\,\xi}{\sqrt{r-1}}\,, \tag{1.38}$$

$$\zeta > -B\xi^2/2, \quad \forall\, \xi \ne 0\,. \tag{1.39}$$

Using estimate (1.37), we introduce the comparison system ([19, 20])

$$\dot{\xi} = \eta, \qquad \dot{\eta} = -\mu\eta + \xi - \xi^3, \tag{1.40}$$

which is equivalent to the first-order equation

$$P\,\frac{dP}{d\xi} + \mu P - \xi + \xi^3 = 0. \tag{1.41}$$

Let us consider positive solutions $P_1(\xi)$ of (1.41) on the set $[0, \xi_0)$ with initial condition $P_1(\xi_0) = 0$. They define for system (1.36) in the half-space $\{\xi \ge 0\}$ the contact-free surfaces

$$\big\{\eta = P_1(\xi), \quad \eta > 0, \quad \xi \in [0, \xi_0]\big\}, \quad \{\eta < 0, \quad \xi = \xi_0\}. \tag{1.42}$$

Negative solutions $P_2(\xi)$ of (1.41) on $(-\xi_0, 0]$ with the initial condition $P_2(-\xi_0) = 0$ define for system (1.36) in the half-space $\{\xi \le 0\}$ the contact-free surfaces

$$\big\{\eta = P_2(\xi), \quad \eta < 0, \quad \xi \in [-\xi_0, 0]\big\}, \quad \{\eta > 0, \quad \xi = -\xi_0\}. \tag{1.43}$$

From this and from estimate (1.38) it follows that if the graph of the function $\eta = P_1(\xi)$ intersects the graph of the straight line

$$\eta = \frac{l}{\sqrt{2}\,(r-1)} - \frac{\sqrt{\sigma}}{\sqrt{r-1}}\,\xi$$

in a certain point ξ_1 on the interval $(0, \xi_0)$, then the inequalities

$$\xi < \xi_0, \qquad \eta < P_1(\xi) \quad \text{for} \quad \xi \in [\xi_1, \xi_0] \tag{1.44}$$

hold on a global attractor $\widehat{\mathcal{A}}$ of system (1.36). Similarly, if the graph of the function $\eta = P_2(\xi)$ intersects the graph of the straight line

$$\eta = -\frac{l}{\sqrt{2}\,(r-1)} - \frac{\sqrt{\sigma}\,\xi}{\sqrt{r-1}}$$

in a certain point ξ_2 on the interval $(-\xi_0, 0)$ then the inequalities

$$\xi > -\xi_0, \quad \eta > P_2(\xi) \quad \text{for} \quad [-\xi_0, \xi_2] \tag{1.45}$$

are true on the global attractor $\widehat{\mathcal{A}}$ of system (1.36). Note that the surfaces $\{\zeta = C - B\xi^2/2,\ C > B\xi_0^2/2\}$ are contact-free for system (1.36) in the strip $\{|\xi| \leq \xi_0\}$. Hence the estimate

$$\zeta \leq B(\xi_0^2 - \xi^2)/2 \tag{1.46}$$

holds on a global attractor of system (1.36). We have thus proved the following result ([24]).

Theorem 1.2 *Estimates (1.37)–(1.38), (1.44)–(1.46) hold on a global attractor of system (1.35).*

Let us give now a simple estimate of ξ_0. To do this we note that the inequalities (1.44) hold if the graph of $\eta = P_1(\xi)$ intersects the graph of the straight line $\eta = l/(\sqrt{2}\,(r-1))$. From the positiveness of μ in equation (1.41) it follows that

$$P_1(\xi)^2 > (\xi^2 - \xi_0^2) - \frac{1}{2}\,(\xi^4 - \xi_0^4).$$

Therefore, a sufficient condition for the above intersection to take place is that

$$(1 - \xi_0^2) - \frac{1}{2}\,(1 - \xi_0^4) = \frac{l^2}{2(r-1)^2}\,.$$

This inequality implies that

$$\xi_0 = \sqrt{1 + \frac{l}{r-1}}\,. \tag{1.47}$$

Similar reasoning may also be applied to estimate (1.45). It follows from relation (1.47) that any global attractor of (1.36) lies in a domain which is bounded uniformly with respect to the parameter $r \in (1, +\infty)$. For a global $\mathcal{B}$-attractor in the case $b \leq 2$, the estimates (1.37)–(1.38) are asymptotically the best possible as $r \to +\infty$. Indeed, in this case, as $r \to +\infty$ the following inequalities hold on the global $\mathcal{B}$-attractor:

$$|\eta| \le 1/\sqrt{2}, \qquad |\xi| \le \sqrt{2}.$$

We recall that a part of a global $\mathcal{B}$-attractor consists of the unstable manifold of the zero equilibrium, which may be represented approximately (for small ε) by the formulae

$$\left\{\zeta = -B\xi^2/2, \quad \eta^2 = \xi^2 - \xi^4/2\right\}.$$

So for large r the global $\mathcal{B}$-attractor has points close to the planes $\{|\xi| = \sqrt{2}\}$, $\{|\eta| = 1/\sqrt{2}\}$.

1.3.3 The Existence of Homoclinic Orbits

Let ξ^+, η^+, ζ^+ denote a solution of (1.36) associated with the positive branch of the unstable manifold of the saddle point (0, 0, 0), that goes into the half-plane $\{\xi > 0\}$, that is, a solution of (1.36) such that

$$\lim_{t\to-\infty} (\xi^+(t), \eta^+(t), \zeta^+(t)) = (0, 0, 0)$$

and $\xi^+(t) > 0$ for $t \in (-\infty, T)$. Here T is a certain number or $+\infty$. It is well-known ([20, 25, 27]) that if the values of σ and b and the value of r are close enough to 1, then $T = +\infty$.

Let us consider a smooth path $s \in [0, 1] \mapsto (b(s)\,,\ \sigma(s)\,,\ r(s))$ in the parameter space $\{b, \sigma, r\}$. The main result of this section is the following theorem ([24]).

Theorem 1.3 *Suppose that for system (1.36) with parameters $b(0)$, $\sigma(0)$, $r(0)$ there exist numbers $T > \tau$ such that the relations*

$$\xi^+(T) = \eta^+(\tau) = 0, \qquad \xi^+(t) > 0, \quad \forall\, t < T, \tag{1.48}$$

$$\eta^+(t) \neq 0, \qquad \forall\, t < T, \quad t \neq \tau \tag{1.49}$$

hold. Suppose also that for system (1.36) with parameters $b(1)$, $\sigma(1)$, $r(1)$ the inequality

$$\xi^+(t) > 0, \quad \forall\, t \in \mathbb{R} \tag{1.50}$$

is true. Then there exists a number $s_0 \in [0, 1]$ such that system (1.36) with parameters $b(s_0)$, $\sigma(s_0)$, $r(s_0)$ has a solution (ξ^+, η^+, ζ^+) corresponding to a homoclinic orbit.

In order to prove this assertion we shall need the following lemmas.

Lemma 1.1 *If the conditions*

$$\eta^+(\tau) = 0, \quad \eta^+(t) > 0, \quad \forall\, t \in (-\infty, \tau),$$

hold for system (1.36), then $\dot{\eta}^+(\tau) < 0$.

Proof Suppose the contrary, i.e. $\dot{\eta}^+(\tau) = 0$. Then we derive from the two last equations of system (1.36) that

$$\ddot{\eta}^+(\tau) = A\zeta^+(\tau)\,\xi^+(\tau)\,. \tag{1.51}$$

It follows from the relations $\eta^+(t) > 0, \xi^+(t) > 0, \forall\, t \in (-\infty, \tau)$ and from the last equation of (1.36) that $\zeta^+(t) < 0$, $\forall\, t \in (-\infty, \tau]$. This inequality and (1.51) imply the inequality $\ddot{\eta}^+(\tau) < 0$ follows. But this contradicts the assumption $\dot{\eta}^+(\tau) = 0$ and the conditions of the lemma. This contradiction proves Lemma 1.1. □

Lemma 1.2 *Consider system (1.36). Suppose that the relations (1.48), (1.49) and the inequalities*

$$\eta^+(t) > 0, \quad \forall\, t \in (-\infty, \tau), \qquad \eta^+(t) \leq 0, \quad \forall\, t \in (\tau, T) \tag{1.52}$$

are true. Then inequality (1.49) also holds.

Proof Suppose the contrary. Then we conclude that a number $\varsigma \in (\tau, T)$, exists such that

$$\eta^+(\varsigma) = \dot{\eta}^+(\varsigma) = 0, \quad \ddot{\eta}^+(\varsigma) = A\xi^+(\varsigma)\zeta^+(\varsigma) < 0, \quad \eta^+(t) < 0, \quad \forall\, t \in (\varsigma, T)$$

are valid. Note that the orbit corresponding to the solution $(\xi(t), \eta(t), \zeta(t)) = (0, 0, \zeta(0)\exp(-At))$ belongs to the stable manifold of the saddle point $(0, 0, 0)$. Hence, from the conditions (1.48), (1.49) and from the above relations it follows that the positive branch of the unstable manifold corresponding to the solution (ξ^+, η^+, ζ^+) and the stable manifold intersect. Then the positive branch of the unstable manifold belongs completely to the stable manifold of the saddle, and the relation $\xi^+(t) > 0, \forall\, t \geq \varsigma$ is valid. The latter relation contradicts the hypothesis (1.48). This contradiction proves Lemma 1.2. □

Remark 1.6 It is possible to give the following geometrical interpretation of this proof in the phase space with the coordinates ξ, η, ζ. A piece of the stable manifold of the saddle $\xi = \eta = \zeta = 0$ is situated "under" the set $\{\xi > 0,\ \eta = 0,\ \zeta \leq 1 - \gamma\xi^2\}$. This property does not allow the trajectory with the initial data from the set to reach the plane $\xi = 0$ if it remains in the quadrant $\{\xi \geq 0,\ \eta \leq 0\}$.

Let us consider the polynomial

$$\lambda^3 + a\lambda^2 + b\lambda + c, \tag{1.53}$$

where a, b, c are positive numbers.

Lemma 1.3 *Either all zeros of the polynomial (1.53) have negative real parts, or two of them have non-zero imaginary parts.*

Proof It is well-known ([9]) that all the zeros of (1.53) have negative real parts if and only if $ab > c$. If $ab = c$ the polynomial (1.53) has two pure imaginary zeros. Suppose now that for certain a, b, c with $ab < c$ the polynomial (1.53) has only real zeros. Since the coefficients are positive it follows that these zeros are negative. This leads to the inequality $ab > c$ which contradicts our assumption. □

Proof of Theorem 1.3 It is well-known ([12]) that the semi-orbit of system (1.36) $\{(\xi^+(t), \eta^+(t), \zeta^+(t)) \mid t \in (-\infty, t_0)\}$ depends continuously on parameter s. Here t_0 is an arbitrary fixed number. It follows from this and from Lemma 1.1 that, if conditions (1.48)–(1.49) hold for system (1.36) with parameters $b(s_1), \sigma(s_1), r(s_1)$ then these conditions also hold for $b(s), \sigma(s), r(s)$ provided that $s \in (s_1 - \delta, s_1 + \delta)$. Here δ is a certain sufficiently small number and the numbers τ and T depend on parameter s. It follows from the above reasoning that the relations (1.48)–(1.49) are valid for a certain interval $(0, s_0)$. Further we shall assume that $(0, s_0)$ is the maximal interval where the relations (1.48)–(1.49) are valid. Let us demonstrate that there exists a certain homoclinic orbit which corresponds to the values $b(s_0), \sigma(s_0), r(s_0)$. We first note that for these parameters and some value τ

$$\begin{aligned} &\eta^+(t) > 0, \ \forall t < \tau, \quad \eta^+(t) \le 0, \ \forall t \ge \tau, \\ &\xi^+(t) > 0, \ \forall t \in (-\infty, +\infty). \end{aligned} \tag{1.54}$$

Indeed, if there exist numbers $T_2 > T_1 > \tau$, for which

$$\begin{aligned} &\xi^+(t) > 0, \ \forall t \in (-\infty, T_2), \qquad && \xi^+(T_2) = 0, \quad \eta^+(T_1) > 0, \\ &\eta^+(t) > 0, \ \forall t < \tau, && \eta^+(\tau) = 0, \quad \dot{\eta}^+(\tau) < 0 \end{aligned}$$

are true then for s sufficiently close to s_0 and such that $s < s_0$ the inequality $\eta^+(T_1) > 0$ still holds. This contradicts the definition of the number s_0. If there exist numbers $T_1 > \tau$, for which $\eta^+(T_1) > 0$, $\eta^+(t) > 0, \forall t < \tau$, $\eta^+(\tau) = 0$, $\dot{\eta}^+(\tau) < 0$, and $\xi^+(t) > 0, \forall t \in (-\infty, +\infty)$, then again for s sufficiently close to s_0 and such that $s < s_0$ the inequality $\eta^+(T_1) > 0$ remains true and again we have a contradiction with the definition of the number s_0. If there exist numbers $T > \tau$, for which $\xi^+(t) > 0, \forall t < T$, $\xi^+(T) = 0$, $\eta^+(t) > 0, \forall t < \tau$, $\eta^+(t) \le 0, \forall t \in [\tau, T]$, then the inequality (1.49) is true according Lemma 1.2. Consequently for $s = s_0$ relations (1.48)–(1.49) are fulfilled and then $(0, s_0)$ is not the maximal interval, for which these relations are true. By these contradictions inequalities (1.54) are proven. It follows from (1.54) that for $s = s_0$ only an equilibrium can be the ω-limit set of the orbit of (ξ^+, η^+, ζ^+). Let us demonstrate that the equilibrium $(\xi, \eta, \zeta) = (1/\sqrt{\gamma}, 0, 0)$ can not be an ω-limit point of this orbit. The linearization in the neighborhood of this equilibrium gives the characteristic polynomial

$$\lambda^3 + (A + \mu)\lambda^2 + (A\mu + 2/\gamma)\lambda + 2A.$$

Suppose that for $s = s_0$ the ω-limit set of the positive branch of the unstable manifold corresponding to the solution ξ^+, η^+, ζ^+ includes the point $(\xi, \eta, \zeta) = (1/\sqrt{\gamma}, 0, 0)$. By Lemma 1.3 and by the fact that the semi-orbits $\{\xi^+(t), \eta^+(t), \zeta^+(t) \mid t \in (-\infty, t_0)\}$ depend continuously on the parameter s we obtain the following assertion. For the values s which are sufficiently close to s_0 the positive branch of the unstable manifold corresponding to (ξ^+, η^+, ζ^+) either tend to an equilibrium state $(\xi, \eta, \zeta) = (1/\sqrt{\gamma}, 0, 0)$ as $t \to +\infty$, or oscillate in some time-interval with changing sign of the coordinate η. Both of these possibilities contradict properties (1.48)–(1.49). Hence, for system (1.36) with parameters $b(s_0)$, $\sigma(s_0)$, $r(s_0)$ the orbit of (ξ^+, η^+, ζ^+) tends to the trivial equilibrium state as $t \to +\infty$. □

Remark 1.7 Note that the proof of Theorem 1.3 actually yields a stronger result, which may be formulated as follows. If relations (1.48)–(1.49) hold for $s \in [0, s_0)$, but not for $s = s_0$, then system (1.36) with parameters $b(s_0)$, $\sigma(s_0)$, $r(s_0)$ has a homoclinic orbit.

Let us apply Theorem 1.3 in various specific cases. Fix the numbers b and σ. It is well-known ([20, 27]) that inequality (1.50) is true for r sufficiently close to 1. We will show that if

$$3\sigma - 2b > 1 \tag{1.55}$$

and r is sufficiently large, then relations (1.48)–(1.49) will hold. Indeed, consider the system

$$Q\frac{dQ}{d\xi} = -\mu Q - P\xi - \varphi(\xi), \qquad Q\frac{dP}{d\xi} = -AP - BQ\xi, \tag{1.56}$$

which is equivalent to system (1.36) in the sets $\{\xi \geq 0,\ \eta > 0\}$ and $\{\xi \geq 0,\ \eta < 0\}$, where P and Q are solutions of (1.56) which are functions of ξ. Since Theorem 1.2 implies that the quantities $(\xi^+(t), \eta^+(t), \zeta^+(t))$ are bounded uniformly with respect to the parameter r, we can carry out an asymptotic integration of the solutions of system (1.56) with a small parameter ε that correspond to the branch of the unstable manifold under consideration. In the first approximation these solutions may be written in the form

$$Q_1(\xi)^2 = \xi^2 - \frac{\xi^4}{2} - 2\mu \int_0^{\xi} \xi\sqrt{1-\xi^2/2}\ d\xi - 2AB \int_0^{\xi} \xi\left(1 - \sqrt{1-\xi^2/2}\right) d\xi,$$

$$Q_1(\xi) \geq 0, \qquad P_1(\xi) = -\left(\frac{\beta}{2}\right)\xi^2 + AB\left(1 - \sqrt{1-\xi^2/2}\right),$$

$$Q_2(\xi)^2 =$$

$$\xi^2 - \frac{\xi^4}{2} - 2\mu \int_{\xi}^{\sqrt{2}} \xi \sqrt{1-\xi^2/2}\, d\xi - \frac{4}{3}\mu + 2AB \int_{\xi}^{\sqrt{2}} \xi \left(1 + \sqrt{1-\xi^2/2}\right) d\xi - \frac{2}{3} AB,$$

$$Q_2(\xi) \le 0, \quad P_2(\xi) = -\left(\frac{B}{2}\right)\xi^2 + AB\left(1 + \sqrt{1-\xi^2/2}\right).$$

It follows from these formulae that if inequality (1.56) holds, then for some $T > \tau$ relations (1.48)–(1.49) will also hold and at the same time

$$\zeta^+(T) = P_2(0) = 2AB,$$
$$\eta^+(T) = Q_2(0) = -\sqrt{8(AB-\mu)/3} = -\sqrt{8\varepsilon(3\sigma - 2b - 1)/(3\sqrt{\sigma})}.$$

Thus, all the conditions of Theorem 1.3 hold for the path $s \mapsto (b(s), \sigma(s), r(s))$ with $b(s) \equiv b, \sigma(s) = \sigma, r(0) = r_1, r(1) = r_2$, where r_1 is sufficiently large and r_2 is sufficiently close to 1. We may therefore formulate the following result.

Corollary 1.1 *For any positive numbers b and σ satisfying the inequality (1.55) a number $r \in (1, +\infty)$ exists, such that system (1.36) with these parameters b, σ and r has a solution (ξ^+, η^+, ζ^+) corresponding to a homoclinic orbit.*

Remark 1.8 Corollary 1.1 was first obtained in [21, 22] and discussed later in [5, 13, 23].

Now fix $\sigma = 10$ and $r = 28$, and consider the parameter $b \in (0, +\infty)$. It is well-known ([5]) that for

$$b > \frac{3\sigma - 1}{2}$$

condition (1.50) is fulfilled. To analyse system (1.36) for small b, we reduce it to the form

$$\dot{\xi} = \eta, \qquad \dot{\eta} = -\mu\eta - u\xi + \xi - \xi^3, \qquad \dot{u} = -Au + \frac{\varepsilon(2\sigma - b)}{\sqrt{\sigma}}\xi^2, \tag{1.57}$$

where $u = \zeta + B\xi^2/2$. Since the semi-orbit $\{(\xi^+(t), \eta^+(t), \zeta^+(t)) | t \in (-\infty, t_0]\}$ depends continuously on the parameter b, it follows that, when b is small, the system (1.57) may be replaced by the system

$$\dot{\xi} = \eta, \qquad \dot{\eta} = -\frac{\varepsilon(\sigma+1)}{\sqrt{\sigma}}\eta - u\xi + \xi - \xi^3, \qquad \dot{u} = 2\varepsilon\sqrt{\sigma}\,\xi^2. \tag{1.58}$$

Numerical integration of the solution (ξ^+, η^+, ζ^+) of system (1.58) for $\sigma = 10$, $r = 28$ shows that conditions (1.48)–(1.49) are satisfied. Hence, the above arguments, using Theorem 1.3, yield the following.

Corollary 1.2 *Let $\sigma = 10$ and $r = 28$. Then there exists a positive number b_0 such that system (1.36) with parameters $b = b_0$, $\sigma = 10$ and $r = 28$ has a solution (ξ^+, η^+, ζ^+) corresponding to a homoclinic orbit.*

1.4 The Generalized Lorenz System

1.4.1 Definition of the System

To obtain examples for the illustration of the results proved above, we consider a differential equation in $\mathbb{R}^3$ with four parameters ([10, 31]). Since this system includes as a special case the well-known Lorenz system, it is called by us "generalized Lorenz system". Many systems which appear in physics can be reduced in a certain sense to this system.

In this chapter the basic properties of the system are considered. This concerns the existence of equilibrium states, conditions for global stability, dissipativity, and estimates of the dissipativity region. At the end of the section we prove a theorem on the convergence behaviour of the generalized Lorenz system.

Thus, consider the *generalized Lorenz system*

$$\dot{x} = -\sigma x + \sigma y - ayz, \quad \dot{y} = rx - y - xz, \quad \dot{z} = -bz + xy, \tag{1.59}$$

where σ, b, r are positive parameters, a is a real parameter. In the case $a = 0$ this system coincides with the Lorenz system, which, as it was noted in Sect. 1.3, for certain values of parameters has a strange attractor . System (1.59) in the form given above, or in a form very close to it, was studied by numerical methods in many physical papers. In these papers it was shown numerically that (1.59) may posses a strange attractor also for $a \neq 0$. Besides interest connected with the existence of strange attractors, the Lorenz system also attracts the attention of scientists because it has appeared as a model of convection in the atmosphere, and was used for the description of other physical processes. The utility of reducing systems describing different phenomena to the Lorenz system is conditioned by the fact that in previous years the Lorenz model was studied intensively both by numerical and analytical methods. Consequently, many results obtained can be applied to the original systems. Similar to system (1.12), it encloses a large family of concrete systems. Several of them will be considered below.

Let us pass to the investigation of the simplest properties of the generalized Lorenz system. We shall define its equilibrium states in depending on the parameters and shall show that in the case of unique equilibrium state the system (1.59) is globally stable. In addition to this we shall prove the dissipativity of the system and find some estimates for its dissipativity region.

1.4.2 Equilibrium States

In Subsect. 1.4.1 we have investigated the equilibrium states of system (1.59) for $a = 0$. The following theorem ([26]) describes the equilibrium states of the system for $a \neq 0$.

Theorem 1.4 *Suppose that* $a \neq 0$.

(1) If $\sigma + a > 0$, $r < 1$ *or*

(2) $\sigma + a < 0$, $r < \sigma/a + 2\sqrt{-\sigma/a}$, *then system (1.59) has the unique equilibrium state* $(0, 0, 0)$.

(3) If $r > 1$, *then system (1.59) has exactly the three equilibrium states* $(0, 0, 0)$ *and* $(\pm x_1, \pm y_1, z_1)$.

(4) If $\sigma + a < 0$, $\sigma/a + 2\sqrt{-\sigma/a} < r < 1$, *then (1.59) has exactly the five equilibrium states* $(0, 0, 0)$, $(\pm x_1, \pm y_1, z_1)$ *and* $(\pm x_2, \pm y_2, z_2)$.

Here

$$x_k = \frac{\sigma b \sqrt{\zeta_k}}{\sigma b + a\zeta_k}, \quad y_k = \sqrt{\zeta_k}, \quad z_k = \frac{\sigma \zeta_k}{\sigma b + a\zeta_k} \qquad (k = 1,\ 2)$$

and the numbers ζ_1 *and* ζ_2 *are given by*

$$\zeta_{1,2} = \frac{\sigma b}{2a^2}\big[a(r-2) - \sigma \pm \sqrt{(ar-\sigma)^2 + 4a\sigma}\,\big].$$

Proof The number of equilibrium states of system (1.59) is defined by the number of positive roots of the quadratic equation

$$\zeta^2 - \frac{\sigma b}{a^2}\big[a(r-2) - \sigma\big]\zeta - \frac{\sigma^2 b^2}{a^2}(r-1) = 0. \tag{1.60}$$

If positive roots are missing, then there is only one equilibrium state; if there exists exactly one positive root, then there are three equilibrium states; if both roots are positive and different, then there are five equilibrium states. The above-mentioned numbers ζ_1 and ζ_2 are the roots of the equation (1.60). Let us write them in the form

$$\zeta_{1,2} = \frac{\sigma b}{2a^2}\big[a(r-2) - \sigma \pm \sqrt{\big(a(r-2)-\sigma\big)^2 + 4a^2(r-1)}\,\big]. \tag{1.61}$$

From (1.61) the validity of the theorem follows directly under condition (3).

Denote by Δ the expression under the square root in equality (1.61). Then we get

$$\Delta = a^2\big[(r - \sigma/a)^2 + 4\sigma/a\big].$$

It is easy to see that in the case when $a > 0$ or $a < 0$ and $r > \sigma/a + 2\sqrt{-\sigma/a}$ then $\Delta > 0$; if $a < 0$ and $r < \sigma/a + \sqrt{-\sigma/a}$ then we have $\Delta < 0$.

From the last relation the validity of the theorem follows under condition (2).

Let condition (1) be satisfied. If $a > 0$, then $a(r-2) - \sigma < 0$ and $\zeta_1 \leq 0$. If

$a < 0$, then $1 \leq -\sqrt{-\sigma/a}$. Therefore if $r < 2 + \sigma/a$, then $r < \sigma/a + 2\sqrt{-\sigma/a}$ and, consequently, $\Delta < 0$, and if $r > 2 + \sigma/a$, then $a(r-2) - \sigma < 0$ and $\zeta_1 < 0$. Thus, the theorem holds if condition (1) is satisfied.

Let condition (4) be true. Then $\Delta > 0$, $a(r-2) - \sigma = a(r-1) - (a+\sigma) > 0$, and the assertion of the theorem follows from (1.61). □

1.4.3 Global Asymptotic Stability

It is easy to check that for $r < 1$ and any a the unique equilibrium state $(0, 0, 0)$ in (1.59) is asymptotically Lyapunov stable.

The following theorem on the *global asymptotic stability* of the equilibrium state $(0, 0, 0)$ generalizes the result of Example 1.10 in the case of arbitrary a ([3, 30, 39]).

Theorem 1.5 *The equilibrium state* $(0, 0, 0)$ *of system (1.59) is global asymptotically stable if one of two conditions is satisfied:*

(1) $\sigma + a > 0$ *and* $r < 1$;

(2) $\sigma + a < 0$ *and* $r < \sigma/a + 2\sqrt{-\sigma/a}$.

Proof Suppose that condition (1) is satisfied and we put

$$V_1(x, y, z) := \frac{1}{2}[x^2 + \sigma y^2 + (\sigma + a)z^2].$$

Then

$$\dot{V}_1(x, y, z) = -b(\sigma + a)z^2 - \sigma x^2 + \sigma(r+1)xy - \sigma y^2 < 0$$

for all $(x, y, z) \neq (0, 0, 0)$, because the quadratic form $x^2 - (r+1)xy + y^2$ is positive definite for $r < 1$.

Suppose that condition (2) is satisfied. In this case the following inequality $r^2 - 2\frac{\sigma}{a}r + \frac{\sigma^2}{a^2} + 4\frac{\sigma}{a} < 0$ holds. Consequently, one can find a small ε such that the inequality

$$r^2 + 2\frac{\sigma}{\tilde{a}}r + \frac{\sigma^2}{\tilde{a}^2} - 4\frac{\sigma}{\tilde{a}} < 0, \tag{1.62}$$

is true, where $\tilde{a} = -a + \varepsilon$. Let us put

$$V_2(x, y, z) := \frac{1}{2}(x^2 + \tilde{a}y^2 + \varepsilon z^2).$$

Then

$$\dot{V}_2(x, y, z) = -\varepsilon b z^2 - \sigma x^2 + (\sigma + \tilde{a}r)xy - \tilde{a}y^2 < 0$$

for all $(x, y, z) \neq 0$, since the quadratic form $\sigma x^2 - (\sigma + \tilde{a}r)xy + \tilde{a}y^2$ is positive definite in virtue of inequality (1.62). □

Now the statement of the present theorem follows from Theorem 1.1.

Remark 1.9 From Theorem 1.5 it follows in particular that chaotic behaviour of system (1.59), as in the case of the Lorenz system, is possible only if the system has several equilibrium states.

1.4.4 Dissipativity

Let us show that system (1.59) is $\mathcal{B}$-dissipative and let us obtain some estimates of the dissipativity region.

Take arbitrary numbers κ, κ_1 and a number ς, such that

$$\kappa < \min(1, b, \sigma), \qquad \kappa_1 + a > 0, \qquad \varsigma_- \leq \varsigma \leq \varsigma_+,$$

where

$$\varsigma_\pm := \sigma + \kappa_1 r \pm 2\sqrt{\kappa_1(\sigma - \kappa)(1 - \kappa)},$$

and put for arbitrary $(x, y, z) \in \mathbb{R}^3$

$$W(x, y, z) := \frac{1}{2}x^2 + \frac{1}{2}\kappa_1 y^2 + \frac{1}{2}(\kappa_1 + a)z^2 - \varsigma z\,.$$

Lemma 1.4 *Let $u = (x, y, z)$ be an arbitrary solution of system (1.59). Then*

$$\limsup_{t \to +\infty} W\big(x(t), y(t), z(t)\big) \leq R, \tag{1.63}$$

where

$$R := \frac{\varsigma^2 (b - 2\kappa)^2}{8\kappa(\kappa_1 + a)(b - \kappa)}.$$

Proof We have

$$\begin{aligned} \dot{W}(x, y, z) + 2\kappa W(x, y, z) = &- (\sigma - \kappa)x^2 - \kappa_1(1 - \kappa)y^2 - (\kappa_1 + a)(b - \kappa)z^2 \\ &+ (\sigma + \kappa_1 r - \varsigma)xy + \varsigma(b - 2\kappa)z \\ \leq -(\gamma\kappa_1 + a)(b - \kappa)z^2 &+ \varsigma(b - 2\kappa)z \leq 2\kappa R, \quad \forall t \geq 0. \end{aligned}$$

Therefore along an arbitrary solution $u = (x, y, z)$ of (1.59) we have

$$\frac{d}{dt}(W(x(t), y(t), z(t)) - R) + 2\kappa(W(x(t), y(t), z(t)) - R) \leq 0, \quad \forall t \geq 0.$$

From this, after multiplication with $e^{2\kappa t}$, we get for all $t \geq 0$

$$\frac{d}{dt}\big[(W(x(t), y(t), z(t)) - R)e^{2\kappa t}\big] \le 0.$$

Integrating the last inequality from 0 to t, we obtain

$$W\big(x(t), y(t), z(t)\big) - R \le \big[W\big(x(0), y(0), z(0)\big) - R\big]e^{-2\kappa t}, \quad \forall t \ge 0.$$

whence it follows that (1.63) holds. □

Thus, all orbits of system (1.59) enter the ellipsoid

$$\mathcal{E} := \big\{(x, y, z) \in \mathbb{R}^3 \mid x^2 + \kappa_1 y^2 + (\kappa_1 + a)z^2 - 2\varsigma z \le 2R\big\}$$

and remain in it.

Let us obtain two other estimates for the dissipativity region of the generalized Lorenz system ([26]).

Lemma 1.5 *Let $u = (x, y, z)$ be an arbitrary solution of system (1.59). Then*

$$\limsup_{t\to+\infty}\big[y(t)^2 + (z(t) - r)^2\big] \le l^2 r^2,$$

where the number l is defined by (1.20).

Proof Let us put for arbitrary $y, z \in \mathbb{R}$

$$W(y, z) := \frac{1}{2}[y^2 + (z - r)^2].$$

Then for any $\kappa \in (0, \kappa_0)$, where $\kappa_0 = \min(1, b)$, we have

$$\begin{aligned}
\dot{W}(x, y, z) + 2\kappa W(x, y, z) &= (\kappa - 1)y^2 + (\kappa - b)z^2 - 2r(\kappa - \frac{b}{2})z + \kappa r^2 \\
&\le (\kappa - b)\left[z - \frac{r(\kappa - b/2)}{\kappa - b}\right]^2 - \frac{r^2(\kappa - b/2)^2}{\kappa - b} + \kappa r^2 \\
&\le \left[\kappa - \frac{(\kappa - b/2)^2}{\kappa - b}\right] r^2 = \frac{b^2 r^2}{4(b - \kappa)}, \quad \forall (x, y, z) \in \mathbb{R}^3.
\end{aligned}$$

From this it follows that along an arbitrary solution $u = (x, y, z)$ of (1.59)

$$\limsup_{t\to+\infty} W(y(t), z(t)) \le \frac{b^2 r^2}{8(b - \kappa)\kappa}.$$

Minimizing by κ the right-hand side of the last inequality, we obtain the claimed estimate.

□

It follows from Lemma 1.5 that system (1.59) has a dissipativity region $\mathcal{D}$ satisfying the inclusion

$$\mathcal{D} \subset \mathbb{R} \times \overline{\mathcal{D}}_1,$$

where $\mathcal{D}_1 = \{(y, z) \in \mathbb{R}^2 \mid y^2 + (z - r)^2 < l^2 r^2\}$.

Lemma 1.6 *Suppose that $2\sigma - b \geq 0$ and $a(b-2) \geq 0$. Let $u = (x, y, z)$ be an arbitrary solution of system (1.59). Then the estimate*

$$\liminf_{t \to +\infty} \big[2(\sigma - ar)z(t) - x(t)^2 + ay(t)^2\big] \geq 0 \tag{1.64}$$

is valid.

Proof If we introduce in $\mathbb{R}^3$ the function

$$W(x, y, z) = (\sigma - ar)z - \frac{1}{2}x^2 + \frac{a}{2}y^2,$$

then the derivative of W with respect to (1.59) is given by

$$\dot{W}(x, y, z) = -b\big[(\sigma - ar)z - \frac{2\sigma}{b}\frac{1}{2}x^2 + \frac{2}{b}\frac{a}{2}y^2\big] \geq -bW.$$

Using this we obtain inequality (1.64). □

References

1. Barbashin, E.A., Krasovskii, N.N.: On global stability of motion. Dokl. Akad. Nauk, SSSR, **86**, 453–456 (1952) (Russian)
2. Birkhoff, G.D.: Dynamical Systems. Amer. Math. Soc. Colloquium Publications, New York (1927)
3. Boichenko, V.A., Leonov, G.A.: On estimates of attractors dimension and global stability of generalized Lorenz equations. Vestn. Leningrad Gos. Univ. Ser. 1, Matematika, **2**, 7–13 (1990) (Russian); English transl. Vestn. Leningrad Univ. Math., **23**(2), 6–12 (1990)
4. Cheban, D.N., Fakeeh, D.S.: Global Attractors of Dynamical Systems without Uniqueness. Sigma, Kishinev (1994). (Russian)
5. Chen, X.: Lorenz equations, part I: existence and nonexistence of homoclinic orbits. SIAM J. Math. Anal. **27**(4), 1057–1069 (1996)
6. Chueshov, I.D.: Global attractors for nonlinear problems of mathematical physics. Uspekhi Mat. Nauk, **48** (3), 135–162 (1993) (Russian); English transl. Russian Math. Surveys, **48**(3), 133–161 (1993)
7. Chueshov, I.D.: Introduction to the Theory of Infinite-Dimensional Dissipative Systems. ACTA Scientific Publishing House, Kharkov. Electron. library of mathematics. ACTA (2002)
8. Coomes, B.A.: The Lorenz system does not have a polynomial flow. J. Diff. Equ. **82**, 386–407 (1989)
9. Demidovich, B.P.: Lectures on Mathematical Stability Theory. Nauka, Moscow (1967). (Russian)

10. Glukhovsky, A.B., Dolzhanskii, F.V: Three-component geostrophic model of convection in rotating fluid. Izv. Akad. Nauk SSSR, Fiz. Atmos. i Okeana, **16**, 451–462 (1980) (Russian)
11. Gottschalk, W.H., Hedlund, G.A.: Topological Dynamics, vol. 36. Amer. Math. Soc. Colloquium Publications (1955)
12. Hartman, P.: Ordinary Differential Equations. John Wiley & Sons, New York (1964)
13. Hastings, S.P., Troy, W.C.: A shooting approach to chaos in the Lorenz equations. J. Diff. Equ. **127**(1), 41–53 (1996)
14. Hirsch, M.W.: A note on the differential equations of Gleick-Lorenz. Proc. Amer. Math. Soc. **105**, 961–962 (1989)
15. Kapitansky, L.V., Kostin, I.N.: Attractors of nonlinear evolution equations and their approximations. Leningrad Math. J. **2**, 97–117 (1991)
16. Kuratowski, K., Mostowski, A.: Set Theory. North-Holland Publishing Company, Amsterdam (1967)
17. Ladyzhenskaya, O.A.: On finding the minimal global attractors for the Navier-Stokes equations and other partial differential equations. Uspekhi Mat. Nauk., **42**, 25–60 (1987) (Russian); English transl. Russian Math. Surveys, **42**, 27–73 (1987)
18. Ladyzhenskaya, O.A.: Attractors for Semigroups and Evolution Equations. Cambridge Univ. Press, Cambridge (1991)
19. Leonov, G.A.: Global stability of the Lorenz system. Prikl. Mat. Mekh., **47**(5), 869–871 (1983) (Russian)
20. Leonov, G.A.: On a method for constructing of positive invariant sets for the Lorenz system. Prikl. Mat. Mekh., **49** (5), 860–863 (1985) (Russian); English transl. J. Appl. Math. Mech., **49**, 660–663 (1986)
21. Leonov, G.A.: On the estimation of the bifurcation parameter values of the Lorenz system. Uspekhi Mat. Nauk., **43** (3), 189–200 (1988) (Russian); English transl. Russian Math. Surveys, **43**(3), 216–217 (1988)
22. Leonov, G.A.: Estimation of loop-bifurcation parameters for a saddle-point separatrix of the Lorenz system. Diff. Urav., **26** (6), 972–977 (1988) (Russian); English transl. J. Diff. Equ., **24**(6), 634–638 (1988)
23. Leonov, G. A.: Existence of homoclinic trajectories in the Lorenz system. Vestn. S. Peterburg Gos. Univ. Ser. 1, Matematika, **32**, 13–15 (1999) (Russian); English transl. Vestn. St. Petersburg Univ. Math., **32**(1), 13–15 (1999)
24. Leonov, G.A.: Bounds for attractors and the existence of homoclinic orbits in the Lorenz system. Prikl. Mat. Mekh., **65** (1), 21–35 (2001) (Russian); English transl. J. Appl. Math. Mech., **65**(1), 19–32 (2001)
25. Leonov, G.A.: General existence conditions of homoclinic trajectories in dissipative systems. Lorenz, Shimizu-Morioka, Lu and Chen systems. Phys Lett. A. **376**(45), 3045–3050 (2012)
26. Leonov, G.A., Boichenko, V.A.: Lyapunov's direct method in the estimation of the Hausdorff dimension of attractors. Acta Appl. Math. **26**, 1–60 (1992)
27. Leonov, G.A., Reitmann, V.: Lokalisierung der Lösung diskreter Systeme mit instationärer periodischer Nichtlinearität. ZAMM **66**(2), 103–111 (1986)
28. Leonov, G.A., Reitmann, V.: Localization of Attractors for Nonlinear Systems. Teubner-Texte zur Mathematik, Bd. 97, B. G. Teubner Verlagsgesellschaft, Leipzig, (1987) (German)
29. Leonov, G.A., Reitmann, V., Smirnova, V.B.: Non-local Methods for Pendulum-like Feedback Systems. Teubner-Texte zur Mathematik, Bd. 132, B. G. Teubner Stuttgart-Leipzig (1992)
30. Lorenz, E.N.: Deterministic nonperiodic flow. J. Atmos. Sci. **20**, 130–141 (1963)
31. Pikovsky, A.S., Rabinovich, M.I., Trakhtengerts, VYu.: Appearance of stochasticity on decay confinement of parametric instability. JTEF **74**, 1366–1374 (1978). (Russian)
32. Pliss, V.A.: Nonlocal Problems in the Theory of Oscillations. Academic Press, New York (1966)
33. Pliss, V.A.: Integral Sets of Periodic Systems of Differential Equations. Nauka, Moscow (1977). (Russian)
34. Reitmann, V.: Dynamical Systems. St. Petersburg State University Press, St. Petersburg, Attractors and their Dimension Estimates (2013). (Russian)

35. Reitmann, V.: Dimension estimates for invariant sets of dynamical systems. In: Fiedler, B. (ed.) Ergodic Theory, Analysis, and Efficient Simulation of Dynamical Systems, pp. 585–615. Springer, New York-Berlin (2001)
36. Robinson, J.C.: Global attractors: topology and finite-dimensional dynamics. J. Dynam. Diff. Equ. **11**, 557–581 (1999)
37. Shestakov, A.A.: Generalized Lyapunov's Direct Method in Distributed-Parameter Systems. Nauka, Moscow (1990). (Russian)
38. Sibirsky, K.S.: Introduction to Topological Dynamics. Nordhoff Intern. Publishing, Leyda (1975)
39. Sparrow, C.: The Lorenz Equations, Bifurcations, Chaos, and Strange Attractors. Springer, New York (1982)
40. Yakubovich, V.A.: The frequency theorem in control theory. Sibirsk. Mat. Zh. **14**(2), 384–420 (1973). (Russian); English transl. Siberian Math. J., **14**(2), 265–289 (1973)
41. Yakubovich, V.A., Leonov, G.A., Gelig, AKh: Stability of Stationary Sets in Control Systems with Discontinuous Nonlinearities. World Scientific, Singapore (2004)

Chapter 2
Singular Values, Exterior Calculus and Logarithmic Norms

Abstract Global stability and dimension properties of nonlinear differential equations essentially depend on the contraction properties of k-parallelopipeds or k-ellipsoids under the flow of the associated variational equations. The goal of this second chapter is to develop some elements of multilinear algebra for the investigation of linear differential equations. This includes the discussion of singular value inequalities for linear operators in finite-dimensional spaces, the Fischer-Courant theorem as an extremal principle for eigenvalues of Hermitian matrices, exterior powers of operators and spaces, the logarithmic norm calculation and the use of the Kalman-Yakubovich frequency theorem for the effective estimation of time-dependent singular values of the solution operator to linear differential equations. The Kalman-Yakubovich frequency theorem is also used to get sufficient conditions for convergence in dynamical systems.

2.1 Singular Values and Covering of Ellipsoids

2.1.1 Introduction

Dimension theory for *linear* differential equations is mainly connected with dimensions of linear spaces $\mathbb{V}$ over a field $\mathbb{K} \in \{\mathbb{R}, \mathbb{C}\}$. If $\mathbb{V}$ has a basis with a finite number of vectors, then $\mathbb{V}$ is called *finite-dimensional*. Any basis of a finite-dimensional linear space $\mathbb{V}$ has the same number of vectors. This number is said to be the *geometric dimension* of $\mathbb{V}$ and denoted by $\dim \mathbb{V}$. If there is no finite basis for $\mathbb{V}$ we say that the space is *infinite dimensional* and write $\dim \mathbb{V} = \infty$. The geometric dimension of the linear space consisting only of the null vector is 0. A linear space $\mathbb{V}$ has $\dim \mathbb{V} = 0$ if and only if $\mathbb{V}$ consists only of the null vector.

Suppose that $(\mathbb{V}, (\cdot, \cdot))$ is a real or complex Euclidean space given by the n-dimensional linear space $\mathbb{V}$ over $\mathbb{K}$ and the scalar product $(\cdot, \cdot)$. Using the standard norm $|z| := (z, z)^{1/2}$, $z \in \mathbb{V}$, we can, in addition to the linear structure, consider the topology generated by this norm. It will be shown in Chap.3 that other dimensions

N. Kuznetsov and V. Reitmann, *Attractor Dimension Estimates for Dynamical Systems: Theory and Computation*, Emergence, Complexity and Computation 38,
https://doi.org/10.1007/978-3-030-50987-3_2

of $\mathbb{V}$ such as the topological dimension, Hausdorff dimension or Fractal dimension (for bounded subsets) coincide with the geometrical dimension of $\mathbb{V}$.

Let us briefly consider one of the main topics of this book restricted to linear differential equations with constant coefficients in the plane. Assume the differential equation in $\mathbb{R}^2$

$$\dot{x} = -x \,, \quad \dot{y} = -y \,. \tag{2.1}$$

The solution of (2.1) starting for $t = 0$ at (x_0, y_0) has the form $\varphi^t((x_0, y_0)) := (e^{-t}x_0, e^{-t}y_0)\,, t \in \mathbb{R}$. It follows that for each $(x_0, y_0) \in \mathbb{R}^2$ we have $\lim_{t\to+\infty} \varphi^t((x_0, y_0)) = (0, 0)$ and, according to Definition 1.1, Chap. 1, $\mathcal{A}_{\mathbb{R}^2,\min} := \{(0, 0)\}$ is the minimal global $\mathcal{B}$-attractor of (2.1), i.e. $\mathcal{A}_{\mathbb{R}^2,\min}$ is a minimal closed and invariant set (satisfying $\varphi^t((0, 0)) \equiv (0, 0)$, $t \in \mathbb{R}$) which attracts all bounded sets in $\mathbb{R}^2$. To see this last property it is sufficient to consider arbitrary parallelopipeds $\mathcal{P} := [a, b] \times [c, d] \subset \mathbb{R}^2$ with numbers $a < b$, $c < d$. Since $\varphi^t(\mathcal{P}) = [e^{-t}a, e^{-t}b] \times [e^{-t}c, e^{-t}d]$ we have $\lim_{t\to+\infty} \varphi^t(\mathcal{P}) = (0, 0)$. Furthermore the flow $\varphi^{(\cdot)}(\cdot)$ of (2.1) shrinks arbitrary 2-dimensional and 1-dimensional volumes vol_2 and vol_1, respectively, of parallelopipeds. This follows from the fact that

$$\begin{aligned}
&\mathrm{vol}_2\, \varphi^t(\mathcal{P}) = e^{-2t}(b-a)\,(d-c) \to 0 \quad \text{as} \quad t \to +\infty, \\
&\mathrm{vol}_1\,(\varphi^t(\mathcal{P}) \cap (\mathbb{R} \times \{0\})) = e^{-t}(b-a) \to 0 \quad \text{as} \quad t \to +\infty \qquad \text{and} \\
&\mathrm{vol}_1\,(\varphi^t(\mathcal{P}) \cap (\{0\} \times \mathbb{R})) = e^{-t}(d-c) \to 0 \quad \text{as} \quad t \to +\infty\,.
\end{aligned}$$

These properties imply, as it will be shown, that for any considered dimension $\dim \mathcal{A}_{\mathbb{R}^2,\min} = 0$.

Since in the example $\mathcal{A}_{\mathbb{R}^2,\min} = \{(0, 0)\}$ is known, this last property is evident.

Consider now a second linear equation in the plane given by

$$\dot{x} = 0 \,, \quad \dot{y} = -y \,. \tag{2.2}$$

The solution of (2.2) starting at $t = 0$ in $(x_0, y_0) \in \mathbb{R}^2$ has the form $\varphi^t((x_0, y_0)) := (x_0, e^{-t}y_0)$. It follows that $\lim_{t\to+\infty} \varphi^t((x_0, y_0)) = (x_0, 0)$ for each $(x_0, y_0) \in \mathbb{R}^2$. This implies that $\mathcal{A}_{\mathbb{R}^2,\min} := \mathbb{R} \times \{0\}$ is the minimal global $\mathcal{B}$-attractor of (2.2), i.e. $\mathcal{A}_{\mathbb{R}^2,\min}$ is a minimal closed and invariant set (satisfying $\varphi^t(\mathbb{R} \times \{0\}) = \mathbb{R} \times \{0\}$, $t \in \mathbb{R}$) which attracts all bounded sets. To see the last property we consider parallelopipeds of the type $\mathcal{P} := [a, b] \times [c, d]$. Since $\varphi^t(\mathcal{P}) = [a, b] \times [e^{-t}c, e^{-t}d]$ it follows that $\mathrm{dist}(\varphi^t(\mathcal{P}), \mathcal{A}_{\mathbb{R}^2,\min}) \to 0$ as $t \to +\infty$. Again 2-dimensional volumes of parallelopipeds are shrinking since $\mathrm{vol}_2\varphi^t(\mathcal{P}) = e^{-t}(b-a)(d-c) \to 0$ as $t \to +\infty$. But arbitrary 1-dimensional volumes are not shrinking. Consider for example $\mathrm{vol}_1(\varphi^t(\mathcal{P}) \cap \mathbb{R} \times \{0\}) = (b-a) \nrightarrow 0$ as $t \to +\infty$. It will be shown in the following that this property is one of the reasons that $\dim \mathcal{A}_{\mathbb{R}^2,\min} = 1$ in this case.

2.1.2 Definition of Singular Values

In this subsection some important properties of matrices which are useful for dimension estimates are considered. If $\mathbb{K} = \mathbb{R}$ or $\mathbb{K} = \mathbb{C}$ we denote by $\mathbb{K}^n$ the n dimensional vector space over $\mathbb{K}$. As elements of the vector space we write $u \in \mathbb{K}^n$ as a row, meanwhile in the matrix calculus u is written as column. Let $M_{m,n}(\mathbb{K})$ and $M_n(\mathbb{K})$ denote the $m \times n$ resp. $n \times n$ matrices over $\mathbb{K} \in \{\mathbb{R}, \mathbb{C}\}$. The *transpose* of a matrix $A \in M_{m,n}(\mathbb{K})$ is denoted by A^T. The *adjoint matrix* A^* is defined as $A^* := (\overline{A})^T$ where $\overline{A}$ denotes the matrix A consisting of conjugate complex elements. For $A \in M_{m,n}(\mathbb{R})$ we use both notations A^T and A^* (which coincide in this case) for the transpose of A. We define the scalar product in $\mathbb{K}^n = \mathbb{C}^n$ by $(u, v) := \sum_{i=1}^n \xi_i \overline{\eta}_i$ and in $\mathbb{K}^n = \mathbb{R}^n$ by $(u, v) := \sum_{i=1}^n \xi_i \eta_i$ for all $u = (\xi_1, \ldots, \xi_n)$ and $v = (\eta_1, \ldots, \eta_n)$ from $\mathbb{K}^n$. The Euclidean norm for $u \in \mathbb{K}^n$ is $|u| := \sqrt{(u, u)}$. If it is necessary to distinguish between the scalar product in $\mathbb{C}^n(\mathbb{R}^n)$ and a second space $\mathbb{C}^m(\mathbb{R}^m)$, we write $(\cdot, \cdot)_n$ for the scalar product in $\mathbb{C}^n(\mathbb{R}^n)$ and $(\cdot, \cdot)_m$ for the scalar product in $\mathbb{C}^m(\mathbb{R}^m)$, respectively.

Let us recall some well-known definitions related to matrices. A matrix $A \in M_n(\mathbb{C})$ ($A \in M_n(\mathbb{R})$) is said to be *Hermitian* (*symmetric*) if $A^* = A$ ($A^T = A$). The Hermitian (symmetric) matrix $A \in M_n(\mathbb{C})$ ($A \in M_n(\mathbb{R})$) is called *positive semi-definite* if $(Au, u) \geq 0$ for all $u \in \mathbb{C}^n$ ($u \in \mathbb{R}^n$) and *positive definite* if $(Au, u) > 0$ for all $u \in \mathbb{C}^n, u \neq 0$ ($u \in \mathbb{R}^n, u \neq 0$).

For a given positive semi-definite matrix $A \in M_n(\mathbb{K})$ any positive semi-definite matrix $B \in M_n(\mathbb{K})$ satisfying the relation $B^2 = A$ is called (positive semi-definite) *square root of the matrix A*.

A positive semi-definite root of A is denoted by $\sqrt{A}$ or $A^{\frac{1}{2}}$. It is well-known that the square root of a positive semi-definite matrix exists and can be uniquely determined.

A non-singular matrix $A \in M_n(\mathbb{C})$ ($A \in M_n(\mathbb{R})$) is called *unitary (orthogonal)* if $A^{-1} = A^*$ ($A^{-1} = A^T$).

From the last definition one immediately concludes that for a unitary (orthogonal) matrix $A \in M_n(\mathbb{C})$ ($A \in M_n(\mathbb{R})$)

$$(Au, Av) = (u, v), \quad |Au| = |u|, \quad \forall\, u, v \in \mathbb{C}^n(\mathbb{R}^n).$$

Consequently the transformation given by a unitary (orthogonal) matrix transforms an *orthonormal* basis of $\mathbb{C}^n(\mathbb{R}^n)$ onto itself.

Let us introduce the basic definition of singular values [24, 25, 38, 44, 52].

Definition 2.1 For a matrix $A \in M_n(\mathbb{K})$ the *singular values* are the non-negative square roots of the eigenvalues of either A^*A or AA^*.

The singular values of a matrix $A \in M_n(\mathbb{K})$ are denoted by $\alpha_i(A)$ (or shortly α_i) and are arranged in a non-increasing order $\alpha_1(A) \geq \alpha_2(A) \geq \cdots \geq \alpha_n(A)$.

Note that the number of positive singular values of A is equal to the rank of A.

Proposition 2.1 *Suppose that $A \in M_n(\mathbb{K})$ is non-singular and $\alpha_i > 0, i = 1, 2, \ldots, n$ are its singular values. Then $\alpha_i^{-1}, i = 1, 2, \ldots, n$ are the singular values of A^{-1}.*

Proof It is sufficient to prove that the matrices $(A^{-1})^* A^{-1}$ and $(A^* A)^{-1} = A^{-1} (A^*)^{-1}$ have the same eigenvalues. But this follows immediately from the equation

$$\det((A^{-1})^* A^{-1} - \lambda I) = \det\big(A^{-1}((A^{-1})^* A^{-1} - \lambda I)A\big) = \det(A^{-1}(A^*)^{-1} - \lambda I)\,, \quad \forall \lambda \in \mathbb{C}\,.$$

□

Let us demonstrate some geometric properties of the singular values of a matrix $A \in M_n(\mathbb{R})$ viewing A as a linear map on $\mathbb{R}^n$. Since $(A^* A)^{1/2}$ is self-adjoint and non-negative there exists an orthonormal basis of $\mathbb{R}^n$, $e_1, \ldots, e_n$, which consists of eigenvectors of $(A^* A)^{1/2}$ associated with the eigenvalues $\alpha_1 \geq \alpha_2 \geq \cdots \geq \alpha_n \geq 0$, (the singular values of A) such that $(A^* A)^{1/2} e_i = \alpha_i e_i$, $i = 1, 2, \ldots, n$. We observe that $(Ae_i, Ae_j) = (A^* Ae_i, e_j) = \alpha_i^2 (e_i, e_j)$, so that the vectors Ae_i are orthogonal and $|Ae_i| = \alpha_i$, $Ae_i \neq 0$, if and only if $\alpha_i > 0$. So we obtain an orthogonal decomposition of $\mathbb{R}^n$ into the space $\mathbb{R}^n_0$, the nullspace of A, and the space $\mathbb{R}^n_1$, the space spanned by the vectors $Ae_i \neq 0$, i.e. with $\alpha_i > 0$. Any $u \in \mathbb{R}^n$ can be written as $u = \sum_{j=1}^n \xi_j e_j$ with $\xi_j \in \mathbb{R}$. It follows that $Au = \sum_{j=1}^n \xi_j Ae_j = \sum_{\alpha_j > 0} \xi_j \alpha_j \frac{Ae_j}{\alpha_j}$, i.e. the image of the closed unit ball in $\mathbb{R}^n$ under the map $u \mapsto Au$ is the set

$$\left\{ \sum_{\alpha_j > 0} \eta_j \frac{Ae_j}{\alpha_j} \;\middle|\; \sum_{\alpha_j > 0} \left(\frac{\eta_j}{\alpha_j} \right)^2 \leq 1 \right\},$$

an ellipsoid[1] in $\mathbb{R}^n_1$ with semi-axes Ae_j and the length of these semi-axes equal to $\alpha_j, \alpha_j > 0$. So we have observed the following proposition [52].

Proposition 2.2 *Let $A \in M_n(\mathbb{R})$ be an arbitrary matrix with singular values $\alpha_1 \geq \alpha_2 \geq \cdots \geq \alpha_n \geq 0$ and let $\mathcal{B}_r(0)$ be an arbitrary closed ball in $\mathbb{R}^n$ of radius $r > 0$ and with the center in 0. Then the image of $\mathcal{B}_r(0)$ under the map $u \mapsto Au, u \in \mathbb{R}^n$, is an ellipsoid $\mathcal{E}$ in the subspace $\mathbb{R}^n_1$ whose semi-axes are the vectors Ae_i (e_i the eigenvectors of $(A^* A)^{1/2}$ associated with eigenvalues $\alpha_i > 0$); the length of the semi-axes are the numbers $\alpha_j r, \alpha_j > 0$.*

Further, we use the following notation. Let $\alpha_1(A) \geq \alpha_2(A) \geq \cdots \geq \alpha_n(A)$ be the singular values of $A \in M_n(\mathbb{K})$. For any $k \in \{0, 1, 2, \ldots, n\}$ we put

---

[1] Suppose $(\mathbb{E}, (\cdot, \cdot)_{\mathbb{E}})$ is an m-dimensional Euclidean space and $\{u_i\}_{i=1}^m$ is an orthonormal basis of $\mathbb{E}$. If $a_1 \geq a_2 \geq \cdots a_m > 0$ are arbitrary positive numbers the set $\mathcal{E} := \{\sum_{i=1}^m \xi_i u_i \mid \sum_{i=1}^m (\frac{\xi_i}{a_i})^2 \leq 1\}$ is called *(non-degenerated) ellipsoid* in $\mathbb{E}$ with *semi-axes* $u_1, \ldots, u_m$ and *length* of semi-axes $a_1, \ldots, a_m$.

$$\omega_k(A) := \begin{cases} \alpha_1(A)\alpha_2(A)\ldots\alpha_k(A)\,, & \text{for } k > 0\,, \\ 1, & \text{for } k = 0\,. \end{cases} \tag{2.3}$$

Suppose $d \in [0,\ n]$ is an arbitrary number. Clearly, it can be represented as $d = d_0 + s$, where $d_0 \in \{0, 1, \ldots, n-1\}$ and $s \in (0,\ 1]$. Now we put

$$\omega_d(A) := \begin{cases} \omega_{d_0}(A)^{1-s}\omega_{d_0+1}(A)^s\,, & \text{for } d \in (0, n]\,, \\ 1, & \text{for } d = 0\,, \end{cases} \tag{2.4}$$

and call $\omega_d(A)$ the *singular value function* of A of order d.

2.1.3 Lemmas on Covering of Ellipsoids

For the proof of the limit theorem in Chap. 5 we need some lemmas on the covering of ellipsoids by balls. The substantial role of ellipsoids and singular values of linear maps for the estimates of the Hausdorff measure and dimension can be understood by the following. The definition of the Hausdorff measure of a set (see Chap. 3) involves its coverings by balls. Considering some differentiable transformation of this set, one can replace the image of each ball entering into the covering by the image of the differential of this map. But under a linear map a ball of radius r transforms into an ellipsoid with semi-axes $\{r\alpha_i\}$, where α_i are the singular values of the linear map (see Proposition 2.2).

Let $\mathcal{E}$ be an ellipsoid in $\mathbb{R}^n$, $a_k(\mathcal{E})$ be the lengths of its semi-axes, $a_1(\mathcal{E}) \geq a_2(\mathcal{E}) \geq \cdots \geq a_n(\mathcal{E})$ ordered with respect to its size. For an arbitrary number $d \in [0, n]$, which we represent in the form $d = d_0 + s$ with $d_0 \in \{0, 1, \ldots, n-1\}$ and $s \in (0, 1]$, let us introduce the *d-dimensional ellipsoid measure* by

$$\omega_d(\mathcal{E}) := \begin{cases} a_1(\mathcal{E})\cdots a_{d_0}(\mathcal{E})a_{d_0+1}(\mathcal{E})^s\,, & \text{for } d \in (0, n] \\ 1\,, & \text{for } d = 0\,. \end{cases} \tag{2.5}$$

As before we denote by $\mathcal{B}_r(u)$ a ball of radius r with center in u. The next three lemmas have been obtained in [13, 26, 28, 52].

Lemma 2.1 *Let $\mathcal{E} \subset \mathbb{R}^n$ be an ellipsoid such that $a_1(\mathcal{E}) \leq \delta$, $\omega_d(\mathcal{E}) \leq \kappa$ and $0 < \kappa \leq \delta^d$. Then for any $\eta > 0$ the set $\mathcal{E} + \mathcal{B}_\eta(0)$ is contained in the ellipsoid $\mathcal{E}'$ such that*

$$\omega_d(\mathcal{E}') \leq (1 + c\eta)^d \kappa,$$

where $c := (\delta^{d_0}/\kappa)^{1/s}$.

Proof Denote $\varsigma := a_{d_0+1}(\mathcal{E})$. Without loss of generality we can assume that

$$\omega_d(\mathcal{E}) = \kappa, \qquad a_{d_0+1}(\mathcal{E}) = \cdots = a_n(\mathcal{E}) = \varsigma.$$

Then $\kappa \leq \delta^{d_0} \varsigma^s$ and, consequently,

$$\varsigma \geq (\kappa/\delta^{d_0})^{1/s}. \tag{2.6}$$

It is clear that $\mathcal{B}_\varsigma(0) \subset \mathcal{E}$. Therefore

$$\mathcal{E} + \mathcal{B}_\eta(0) = \mathcal{E} + \frac{\eta}{\varsigma}\mathcal{B}_\varsigma(0) \subset \mathcal{E} + \frac{\eta}{\varsigma}\mathcal{E} = \Big(1 + \frac{\eta}{\varsigma}\Big)\mathcal{E}. \tag{2.7}$$

Put $\mathcal{E}' := (1 + \eta/\varsigma)\mathcal{E}$. We have

$$\omega_d(\mathcal{E}') = (1 + \eta/\varsigma)^d \omega_d(\mathcal{E}) \leq (1 + \eta/\varsigma)^d \kappa \tag{2.8}$$

and the assertion of the lemma follows from (2.6)–(2.8). □

Lemma 2.2 *Let $\mathcal{E} \subset \mathbb{R}^n$ be an ellipsoid with center in 0. Then for any $\eta > 0$ the ellipsoid $\mathcal{E}' := \big(1 + \eta/a_n(\mathcal{E})\big)\mathcal{E}$ contains the set $\mathcal{E} + \mathcal{B}_\eta(0)$.*

Proof Let $\delta := a_n(\mathcal{E})$. Since $\mathcal{B}_\delta(0) \subset \mathcal{E}$, we have

$$\mathcal{E} + \mathcal{B}_\eta(0) = \mathcal{E} + \eta/\delta\, \mathcal{B}_\delta(0) \subset \mathcal{E} + \eta/\delta\, \mathcal{E} = (1 + \eta/\delta)\, \mathcal{E}\,.$$

□

For an arbitrary bounded set $\mathcal{A} \subset \mathbb{R}^n$ we shall denote by $N_{\mathcal{A}}(r)$ the minimal number of balls of radius r necessary for covering of $\mathcal{A}$.

Lemma 2.3 *Let $\mathcal{E} \subset \mathbb{R}^n$ be an ellipsoid and $0 < r \leq a_n(\mathcal{E})$. Then*

$$N_{\mathcal{E}}(\sqrt{n}\, r) \leq 2^n \omega_n(\mathcal{E})/r^n.$$

Proof The ellipsoid $\mathcal{E}$ can be inscribed into a parallelepiped with edges of length $2a_1(\mathcal{E}), \ldots, 2a_n(\mathcal{E})$ that adopts a covering by N cubes with edges of length $2r$, where

$$N = \prod_{j=1}^{n} \left(\left[\frac{a_j(\mathcal{E})}{r} \right] + 1 \right) \leq 2^n \prod_{j=1}^{n} \frac{a_j(\mathcal{E})}{r}.$$

To finish the proof, remark that in $\mathbb{R}^n$ a cube with edge of length $2r$ can be inscribed into a ball of radius $\sqrt{n}\, r$. □

2.2 Singular Value Inequalities

2.2.1 The Fischer-Courant Theorem

In the following we need a classical theorem from linear algebra which is due to Courant and Fischer. Let us formulate this theorem for completeness (see [18]).

Theorem 2.1 (Fischer-Courant) *Suppose that $A \in M_n(\mathbb{K})$ is Hermitian ($\mathbb{K} = \mathbb{C}$) or symmetric ($\mathbb{K} = \mathbb{R}$), and $\lambda_1 \geq \lambda_2 \geq \cdots \geq \lambda_n$ are its eigenvalues. Then we have*

$$\lambda_1 = \max_{u \neq 0} (Au, u)/|u|^2 , \tag{2.9}$$

$$\lambda_k = \min_{\substack{|v_j|=1 \\ j=1,2,\ldots,k-1}} \max_{\substack{u \neq 0, \\ (u,v_j)=0,}} (Au, u)/|u|^2, \quad k = 2, \ldots, n . \tag{2.10}$$

From the Fischer-Courant theorem we directly derive the following corollaries.

Corollary 2.1 *Let $A, B \in M_n(\mathbb{K})$ be two Hermitian ($\mathbb{K} = \mathbb{C}$) or symmetric ($\mathbb{K} = \mathbb{R}$) matrices. Suppose that $\lambda_1 \geq \lambda_2 \geq \cdots \geq \lambda_n$ and $\nu_1 \geq \nu_2 \geq \cdots \geq \nu_n$ are the eigenvalues of A and $A + B$, respectively. Then, if B is positive semi-definite (positive definite) we have for $j = 1, 2, \ldots, n$ the inequalities $\lambda_j \leq \nu_j$ ($\lambda_j < \nu_j$ respectively).*

Recall now that for an arbitrary $A \in M_{n,m}(\mathbb{K})$ the *operator norm* $|A|$ is defined as

$$|A| := \sup_{\substack{|u|=1, \\ u \in \mathbb{K}^m}} |Au| .$$

Corollary 2.2 *Let the matrix $A \in M_n(\mathbb{K})$ be arbitrary with the singular values $\alpha_1 \geq \alpha_2 \geq \cdots \geq \alpha_n$. Then we have*

$$|A| = \alpha_1 .$$

Proof From the Fischer-Courant theorem it follows that

$$\alpha_1^2 = \sup_{|u|=1} (A^* Au, u) = \sup_{|u|=1} |Au|^2 = |A|^2.$$

□

In the next corollary denote by $\mathbb{L}_k \subset \mathbb{R}^n$ an arbitrary linear subspace of dimension $k \leq n$ and by $\gamma > 0$ an arbitrary number.

Corollary 2.3 *Suppose $A \in M_n(\mathbb{R})$ and $|Au| \leq \gamma |u|$, $\forall u \in \mathbb{L}_k$. Then we have*

$$\alpha_{n-k+1} \leq \gamma. \tag{2.11}$$

If the matrix A is non-singular and $|Au| \geq \gamma |u|$, $\forall\, u \in \mathbb{L}_k$, then

$$\alpha_k \geq \gamma\ . \tag{2.12}$$

Proof Let us prove (2.11). Suppose that $m \in \{1, 2, \dots, n\}$ is arbitrary and $\mathbb{P}_{n-m+1}$ is an arbitrary linear subspace of $\mathbb{R}^n$ of dimension $n - m + 1$. By the Fischer-Courant theorem we have

$$\alpha_m^2 \leq \max_{\substack{u \in \mathbb{P}_{n-m+1},\\ u \neq 0}} (A^* Au, u)/|u|^2.$$

Putting $k = n - m + 1$ and taking $\mathbb{L}_k$ in the role of the subspace $\mathbb{P}_{n-m+1}$, we obtain from above that

$$\alpha_{n-k+1}^2 \leq \max_{\substack{u \in \mathbb{L}_k,\\ u \neq 0}} |Au|^2/|u|^2 \leq \gamma^2\ .$$

The last inequality is equivalent to (2.11). Let us now prove (2.12). Since A is non-singular, the subspace $\mathbb{P}_k := \{Au \mid u \in \mathbb{L}_k\}$ of $\mathbb{R}^n$ has the dimension k. Let $u \in \mathbb{P}_k$ be arbitrary. Then $A^{-1}u \in \mathbb{L}_k$ and by assumption, $|u| = |AA^{-1}u| \geq \gamma |A^{-1}u|$, i.e. $|A^{-1}u| \leq \gamma^{-1}|u|$. From Proposition 2.1 it follows that $1/\alpha_k$ is the $(n - k + 1)$th singular value of A^{-1}. Thus by (2.10) we have $\frac{1}{\alpha_k} \leq \gamma^{-1}$ and the corollary is proved. □

2.2.2 *The Binet–Cauchy Theorem*

In this subsection we discuss the Binet-Cauchy formula [2, 6] for matrices, and derive from this formula Horn's inequality for the singular value function. Our representation is based on the book [18].

Proposition 2.3 (Binet-Cauchy formula) *Let $A = \left(a_{ij}\right) \in M_{m,n}(\mathbb{K})$ and $B = \left(b_{ij}\right) \in M_{n,m}(\mathbb{K})$ with $m \leq n$ be two arbitrary matrices and $C := AB$. Then we have*

$$\det C = \sum_{1 \leq r_1 < \cdots < r_m \leq n} \det \begin{pmatrix} a_{1r_1} & \dots & a_{1r_m} \\ \dots & \dots & \dots \\ a_{mr_1} & \dots & a_{mr_m} \end{pmatrix} \det \begin{pmatrix} b_{1r_1} & \dots & b_{1r_m} \\ \dots & \dots & \dots \\ b_{mr_1} & \dots & b_{mr_m} \end{pmatrix}.$$

The Binet–Cauchy formula allows us to express arbitrary minors of the product of two rectangular matrices by minors of the two cofactors. Let two matrices $A = \left(a_{ij}\right) \in M_{n,l}(\mathbb{K})$ and $B = \left(b_{ij}\right) \in M_{l,m}(\mathbb{K})$ be given, and define $C = (c_{ij}) = AB$. Consider an arbitrary minor $C[\alpha \mid \beta]$ of C with $\alpha = (i_1, \dots, i_k)$ and $\beta = (j_1, \dots, j_k)$, i.e.

$$C[\alpha \mid \beta] := \det \begin{pmatrix} c_{i_1 j_1} & \dots & c_{i_1 j_k} \\ \dots & \dots & \dots \\ c_{i_k j_1} & \dots & c_{i_k j_k} \end{pmatrix}.$$

We assume here that $1 \le i_1 < i_2 < \cdots < i_k \le n,\ 1 \le j_1 < j_2 < \cdots < j_k \le m$, and $k \le \min\{n, m, l\}$. The matrix composed of elements of this minor is the product of the two rectangular matrices

$$\begin{pmatrix} a_{i_1 1} \dots a_{i_1 l} \\ \dots\dots\dots\dots \\ a_{i_k 1} \dots a_{i_k l} \end{pmatrix} \quad \text{and} \quad \begin{pmatrix} b_{1 j_1} \dots b_{1 j_k} \\ \dots\dots\dots\dots \\ b_{l j_1} \dots b_{l j_k} \end{pmatrix}.$$

Therefore, by the Binet–Cauchy formula we get for the minor $C[\alpha \mid \beta]$ the useful representation

$$C[\alpha \mid \beta] = \sum_{1 \le r_1 < \cdots < r_k \le l} \det \begin{pmatrix} a_{i_1 r_1} \dots a_{i_1 r_k} \\ \dots\dots\dots\dots \\ a_{i_k r_1} \dots a_{i_k r_k} \end{pmatrix} \det \begin{pmatrix} b_{r_1 j_1} \dots b_{r_1 j_k} \\ \dots\dots\dots\dots \\ b_{r_k j_1} \dots b_{r_k j_k} \end{pmatrix}. \tag{2.13}$$

Let us now derive a further corollary from the Binet–Cauchy formula. Suppose that $v_1, v_2, \dots, v_k$ are vectors from the space $\mathbb{K}^n$, given in the canonical basis of $\mathbb{K}^n$ by $v_j = (v_{j1}, v_{j2}, \dots, v_{jn}), j = 1, 2, \dots, k$. Then it holds the *generalized Lagrange identity*

$$\det \begin{pmatrix} (v_1, v_1) \dots (v_1, v_k) \\ \dots\dots\dots\dots\dots\dots \\ (v_k, v_1) \dots (v_k, v_k) \end{pmatrix} = \sum_{1 \le r_1 < \cdots < r_k \le n} \det^2 \begin{pmatrix} v_{1 r_1} \dots v_{k r_1} \\ \dots\dots\dots\dots \\ v_{1 r_k} \dots v_{k r_k} \end{pmatrix}. \tag{2.14}$$

Note that the validity of (2.14) results from the Binet–Cauchy formula since in the left-hand side of equation (2.14) there is the determinant of the matrix A^*A, where A is an $n \times k$ matrix whose columns are the vectors $v_1, v_2, \dots, v_k$.

For $k = 2$ equality (2.14) gives the well-known *Lagrange identity*

$$\sum_{1 \le i < j \le n} \left(v_{1i} v_{2j} - v_{1j} v_{2i}\right)^2 = |v_1|^2 |v_2|^2 - (v_1, v_2)^2.$$

The Binet–Cauchy formula can be used to prove the following lemma which we need in the sequel for singular value estimations. Let $k \in \{1, 2, \dots, n\}$ be a natural number and consider in $\mathbb{K}^n$ the two arbitrary sets of vectors $v_1, v_2, \dots, v_k$ and $w_1, w_2, \dots, w_k$. Let us introduce for them a determinant of the form

$$[(v_i, w_j)] := \det \begin{pmatrix} (v_1, w_1) \ (v_1, w_2) \dots (v_1, w_k) \\ (v_2, w_1) \ (v_2, w_2) \dots (v_2, w_k) \\ \dots\dots\dots\dots\dots\dots\dots\dots \\ (v_k, w_1) \ (v_k, w_2) \dots (v_k, w_k) \end{pmatrix}.$$

The following lemma is due to Horn [24].

Lemma 2.4 *Let $P \in M_n(\mathbb{K})$ be a positive semi-definite matrix, $k \in \{1, 2, \dots, n\}$ be a natural number and $u_1, u_2, \dots, u_k$ be an arbitrary set of vectors from $\mathbb{K}^n$. Then*

$$[(Pu_i, u_j)] \leq \omega_k(P)^2[(u_i, u_j)] \,.$$

2.2.3 The Inequalities of Horn, Weyl and Fan

Let us consider for arbitrary $n \times n$ matrices A and B the singular value function $\omega_d(AB)$ which was introduced in Sect. 2.1.2. Our aim is to estimate this function by $\omega_d(A)$ and $\omega_d(B)$.

The next proposition which gives this estimate and which also goes back to Horn and Johnson [25] is a direct consequence of Lemma 2.4.

Proposition 2.4 (Horn's inequality) *Let $A, B \in M_n(\mathbb{K})$ and $d \in [0, n]$ be arbitrary. Then we have*

$$\omega_d(AB) \leq \omega_d(A)\omega_d(B). \tag{2.15}$$

The next proposition is a direct corollary of the previous one.

Proposition 2.5 *Suppose that $A, S \in M_n(\mathbb{K})$, S is non-singular and $k \in \{1, 2, \dots, n\}$ is arbitrary. Then we have*

$$\omega_k\big(S^{-1}AS\big) \leq |S|^k|S^{-1}|^k\omega_k(A), \tag{2.16}$$

where $|\cdot|$ denotes the operator norm of a matrix.

Proof By Proposition 2.4 we have

$$\omega_k\big(S^{-1}AS\big) \leq \omega_k(S)\omega_k(S^{-1})\omega_k(A). \tag{2.17}$$

Denote by $s_1 \geq s_2 \geq \cdots \geq s_n > 0$ the singular values of the matrix S. Using Proposition 2.1 and Corollary 2.2 from the Fischer-Courant theorem (Theorem 2.1) we get

$$\omega_k(S)\omega_k(S^{-1}) = \frac{s_1}{s_n}\frac{s_2}{s_{n-1}}\cdots\frac{s_k}{s_{n-k+1}} \leq \left(\frac{s_1}{s_n}\right)^k = |S|^k|S^{-1}|^k. \tag{2.18}$$

Thus (2.16) follows from (2.17) and (2.18). □

The inequality to be proved in the following proposition goes back to Weyl [53]. The proof is borrowed from [18].

Proposition 2.6 (Weyl's inequality) *Let $\lambda_1, \lambda_2, \dots, \lambda_n$ be the eigenvalues of the matrix $A \in M_n(\mathbb{K})$, ordered as $|\lambda_1| \geq |\lambda_2| \geq \cdots \geq |\lambda_n|$, and let $k \in \{1, 2, \dots, n\}$ be arbitrary. Then*

$$|\lambda_1||\lambda_2|\cdots|\lambda_k| \le \omega_k(A).$$

Proof Assume $\mathbb{K} = \mathbb{C}$. Let us consider the matrix A as operator in the space $\mathbb{C}^n$ and let $e_1, e_2, \ldots, e_n$ be a basis of $\mathbb{C}^n$ in which $A = (a_{ij})$ has a triangular form. The latter is possible on the basis of Schur's lemma [18, 44].

For $i, j \in \{1, 2, \ldots, k\}$ we have

$$(Ae_i, Ae_j) = \left(\sum_{r=1}^{i} a_{ri}e_r, \sum_{s=1}^{j} a_{sj}e_s\right) = \sum_{r,s=1}^{q} a_{ri}\overline{a}_{sj}(e_r, e_s), \text{ where } q = \min\{i, j\}.$$

Therefore,

$$\left[(Ae_i, Ae_j)\right] = \det A_k^T \left[(e_i, e_j)\right] \det \overline{A}_k, \tag{2.19}$$

where A_k is the $k \times k$ matrix, consisting of the first k rows and columns of the matrix A, A_k^T is the transposed of the matrix A_k, and $\overline{A}_k$ is the matrix, consisting of the complex conjugated elements of A_k. The matrix A is triangular, whose main diagonal consists of eigenvalues. Without loss of generality it can be assumed that they are ordered as $\lambda_1, \lambda_2, \ldots, \lambda_n$. Consequently, $\det A_k = \lambda_1 \ldots \lambda_k$ and

$$\left[(Ae_i, Ae_j)\right] = |\lambda_1|^2 \ldots |\lambda_k|^2 \left[(e_i, e_j)\right]. \tag{2.20}$$

On the other hand, by Horn's lemma (Lemma 2.4) we have

$$\left[(Ae_i, Ae_j)\right] = \left[(A^*Ae_i, e_j)\right] \le \omega_k^2(A)\left[(e_i, e_j)\right]. \tag{2.21}$$

The validity of Proposition 2.6 follows now from (2.20) and (2.21). □

The next proposition was shown in [14] (see also [37, 39]).

Proposition 2.7 (Fan's inequality) *If $\alpha_1, \alpha_2, \ldots, \alpha_n$ is the complete system of eigenvalues of the matrix $A \in M_n(\mathbb{K})$ ordered so that*

$$\operatorname{Re} \alpha_1 \ge \operatorname{Re} \alpha_2 \ge \cdots \ge \operatorname{Re} \alpha_n,$$

$\lambda_1 \ge \lambda_2 \ge \cdots \ge \lambda_n$ is the complete system of eigenvalues of the matrix $(A^* + A)/2$, and $k \in \{1, 2, \ldots, n\}$ is arbitrary, then

$$\operatorname{Re}\,(\alpha_1 + \alpha_2 + \cdots + \alpha_k) \le \lambda_1 + \lambda_2 + \cdots + \lambda_k\,.$$

Proof Let us assume that $\varepsilon > 0$ is arbitrary and $s_1 \ge s_2 \ge \cdots \ge s_n$ are the singular values of the matrix $I + \varepsilon A$. By the Weyl inequality (Proposition 2.6) we have for any $k \in \{1, 2, \ldots, n\}$

$$|1 + \varepsilon\,\alpha_1|^2\,|1 + \varepsilon\,\alpha_2|^2 \ldots |1 + \varepsilon\,\alpha_k|^2 \le s_1^2 s_2^2 \ldots s_k^2\,. \tag{2.22}$$

Here the numbers $s_1^2, s_2^2, \ldots, s_n^2$ are the eigenvalues of the matrix

$$(I + \varepsilon A^*)(I + \varepsilon A) = I + \varepsilon(A^* + A) + \varepsilon^2 A^* A.$$

For small ε this gives for any $j = 1, 2, \ldots, n$

$$s_j^2 = 1 + 2\,\varepsilon\lambda_j + O(\varepsilon^2). \tag{2.23}$$

It follows then by (2.22) and (2.23) that

$$1 + 2\,\varepsilon \mathrm{Re}\,(\alpha_1 + \alpha_2 + \cdots + \alpha_k) + O(\varepsilon^2) \le 1 + 2\,\varepsilon(\lambda_1 + \lambda_2 + \cdots + \lambda_k) + O(\varepsilon^2).$$

The last inequality completes the proof of Proposition 2.7. □

2.3 Compound Matrices

2.3.1 Multiplicative Compound Matrices

Let $n \in \mathbb{N}$ and $k \in \{1, \ldots, n\}$ be arbitrary numbers. For any $i \in \{1, \ldots, \binom{n}{k}\}$ let $(i) := (i_1, i_2, \ldots, i_k)$ be the ith k-tupel of natural numbers $1 \le i_1 < i_2 < \cdots < i_k \le n$ with respect to the lexicographic ordering, i.e. $(i) = (i_1, \ldots, i_k) < (j) = (j_1, \ldots, j_k)$ if and only if there exists a $t \in \{1, \ldots, k\}$, such that $i_t < j_t$ and $i_s = j_s$ for all $s \in \{1, \ldots, t-1\}$. Denote by $Q_{k,n}$ the lexicographically ordered set of such k-tupels.

Example 2.1 Suppose that $n = 3$ and $k = 2$. Then we can write
$Q_{2,3} = \{(1), (2), (3)\}$ with $(1) = (1, 2)$, $(2) = (1, 3)$ and $(3) = (2, 3)$.

Suppose that $A = (a_{ij}) \in M_{n,m}(\mathbb{K})$ and $k \in \{1, \ldots, \min(n, m)\}$. For any $(i) \in Q_{k,n}$ and $(j) \in Q_{k,m}$ let $A[(i), (j)]$ denote the $k \times k$ submatrix of A the (s, t)th element ($s, t \in \{1, \ldots, k\}$) of which is $a_{i_s j_t}$. Now we can give the following definition [8, 16, 18, 37, 44].

Definition 2.2 Let $A \in M_{n,m}(\mathbb{K})$ and $k \in \{1, \ldots, \min(n, m)\}$ be arbitrary. The *kth multiplicative compound matrix* of A is the $\binom{n}{k} \times \binom{m}{k}$ matrix $A^{(k)}$ the (i, j)th element of which is $\det A[(i), (j)]$ with $(i) \in Q_{k,n}$ and $(j) \in Q_{k,m}$.

Example 2.2 Assume that $n = 2$, $m = 3$ and $k = 2$. Then $Q_{2,2} = \{(1, 2)\}$, $Q_{2,3} = \{(1, 2), (1, 3), (2, 3)\}$. If $A = (a_{ij})$ is a given 2×3 matrix then $A^{(2)}$ is the $1 \times \binom{3}{2} = 1 \times 3$ matrix given by $A^{(2)} = [\det A[(1, 2)|(1, 2)], \det A[(1, 2)|(1, 3)], \det A[(1, 2)|(2, 3)]]$.

Using now the language of compound matrices we can state the following second version of the Binet-Cauchy theorem. We omit a direct proof. The result can be immediately deduced from Proposition 7.10, Chap. 7.

Proposition 2.8 *Let* $A \in M_{m,n}(\mathbb{K})$, $B \in M_{n,p}(\mathbb{K})$ *and the number* $k \in \{1, \ldots, \min(m, n, p)\}$ *be arbitrary. Then we have*

$$(AB)^{(k)} = A^{(k)} B^{(k)}.$$

In the next proposition [19, 29, 37, 39, 44] we shall give an overview over the most important properties of multiplicative compound matrices.

Proposition 2.9 *Let* $A \in M_n(\mathbb{K})$ *and* $k \in \{1, \ldots, n\}$ *be arbitrary. Then it holds:*

(a) $A^{(1)} = A$; $A^{(n)} = \det A$;
(b) $I_n^{(k)} = I_{\binom{n}{k}}$; $(\lambda A)^{(k)} = \lambda^k A^{(k)} \quad \forall \lambda \in \mathbb{K}$;
(c) If A is non-singular then $A^{(k)}$ *is non-singular and* $(A^{(k)})^{-1} = (A^{-1})^{(k)}$;
(d) $(A^T)^{(k)} = (A^{(k)})^T$; $(A^*)^{(k)} = (A^{(k)})^*$;
(e) If A is unitary or symmetric, then also $A^{(k)}$;
(f) If A is a lower (upper) triangular matrix, then $A^{(k)}$ *is also a lower (upper) triangular matrix;*
(g) The complete system of eigenvalues of $A^{(k)}$ *consists of all possible* $\binom{n}{k}$ *products of the form* $\lambda_{i_1} \cdot \lambda_{i_2} \cdots \lambda_{i_k}$ *with* $1 \le i_1 < i_2 < \cdots < i_k \le n$ *where* $\lambda_1, \ldots, \lambda_n$ *is the complete system of eigenvalues of A (Kronecker's theorem);*
(h) $\det A^{(k)} = (\det A)^{\binom{n-1}{k-1}}$, $\operatorname{tr} A^{(k)} = \sum_{1 \le i_1 < i_2 < \cdots < i_k \le n} A\begin{pmatrix} i_1, \ldots, i_k \\ i_1, \ldots, i_k \end{pmatrix}$.

Proof We restrict ourselves to the proof of (c), (e), (g) and (f).

(c) From $AA^{-1} = I_n$ it follows with Proposition 2.8 and the part (b) of this proposition that

$$(AA^{-1})^{(k)} = A^{(k)}(A^{-1})^{(k)} = I_{\binom{n}{k}}.$$

(e) With $AA^* = I_n$, by Proposition 2.8 and (d) we conclude that

$$(AA^*)^{(k)} = A^{(k)}(A^*)^{(k)} = A^{(k)}(A^{(k)})^* = I_{\binom{n}{k}}.$$

(g) From Schur's lemma it follows that there exists an upper triangular matrix $T \in M_n(\mathbb{C})$ and a unitary matrix $U \in M_n(\mathbb{C})$ such that $A = U\,T\,U^*$ and the main diagonal elements of T are the eigenvalues $\lambda_1, \ldots, \lambda_n$ of A. Hence from Proposition 2.8 and (d) it follows that $A^{(k)} = (U\,T\,U^*)^{(k)} = U^{(k)}T^{(k)}(U^*)^{(k)} = U^{(k)}T^{(k)}(U^{(k)})^*$.
From (f) we see, that $T^{(k)}$ is upper triangular and the main diagonal elements are the numbers $\lambda_{i_1} \cdot \lambda_{i_2} \cdots \lambda_{i_k}$ with arbitrary indices $1 \le i_1 < i_2 < \cdots < i_k \le n$. Since $U^{(k)}$ is unitary (see (e)), the eigenvalues of $A^{(k)}$ are the diagonal elements of $T^{(k)}$.
(h) From (g) it follows that $\det A^{(k)} = \prod_{1 \le i_1 < \cdots < i_k \le n} \lambda_{i_1} \cdot \lambda_{i_2} \cdots \lambda_{i_k} = \left(\prod_{i=1}^n \lambda_i\right)^{\binom{n-1}{k-1}} = (\det A)^{\binom{n-1}{k-1}}$, since $\binom{n-1}{k-1}$ gives the number of combinations to the $(k-1)$th class of $n-1$ elements without repeating.
The second claim of (h) follows immediately from the definition. □

From Proposition 2.9 we can deduce the following corollaries.

Corollary 2.4 *If $A \in M_n(\mathbb{K})$ is positive definite (positive semi-definite) then $A^{(k)}$ is also positive definite (positive semi-definite).*

Corollary 2.5 *Suppose $A \in M_n(\mathbb{K})$ is a positive semi-definite matrix and $\lambda_1(A) \geq \cdots \geq \lambda_n(A) \geq 0$ are the ordered eigenvalues of A. If $k \in \{1, \ldots, n\}$ is arbitrary and $\lambda_1(A^{(k)}) \geq \cdots \geq \lambda_{\binom{n}{k}}(A^{(k)})$ denotes the ordered eigenvalues of $A^{(k)}$ then*

$$\lambda_1(A^{(k)}) = \lambda_1(A) \cdot \lambda_2(A) \cdots \lambda_k(A)$$

and

$$\lambda_{\binom{n}{k}}(A^{(k)}) = \lambda_{n-k+1}(A) \cdots \lambda_n(A).$$

Definition 2.3 The *exterior product* $u_1 \wedge u_2 \wedge \cdots \wedge u_k$ of the vectors $u_1, \ldots, u_k \in \mathbb{K}^n$ $(k \leq n)$ is the $\binom{n}{k}$-vector $A^{(k)}$ where the $n \times k$ matrix A has as columns the vectors $u_1, \ldots, u_k$.

Let us consider the most important properties of exterior products [15, 16, 29, 37, 44].

Proposition 2.10 *(a) The exterior product is multilinear, i.e. linear in any argument if the remaining arguments are fixed;*

(b) $\operatorname{span}\{u_1 \wedge u_2 \wedge \cdots \wedge u_k,\ u_i \in \mathbb{K}^n\} = \mathbb{K}^{\binom{n}{k}}$;

(c) The vectors $u_1, u_2, \ldots, u_k \in \mathbb{K}^n$ are linearly independent if and only if $u_1 \wedge u_2 \wedge \cdots \wedge u_k \neq 0$;

(d) If π is an arbitrary permutation of $\{1, 2, \ldots, k\}$ then

$$u_{\pi(1)} \wedge \cdots \wedge u_{\pi(k)} = (\operatorname{sign} \pi) u_1 \wedge \cdots \wedge u_k\,;$$

(e) For arbitrary $A \in M_n(\mathbb{K})$, $k \in \{1, \ldots, n\}$ and $u_1, \ldots, u_k \in \mathbb{K}^n$ we have

$$A^{(k)}(u_1 \wedge \cdots \wedge u_k) = Au_1 \wedge \cdots \wedge Au_k\,;$$

(f) For arbitrary vectors $u_1, \ldots, u_k$ and $v_1, \ldots, v_k$ from $\mathbb{K}^n$ $(k \leq n)$ we have

$$(u_1 \wedge \cdots \wedge u_k, v_1 \wedge \cdots \wedge v_k)_{\mathbb{K}^{\binom{n}{k}}} = \det[(u_s, v_t)_{\mathbb{K}^n}|_{s,t=1}^k]\,.$$

Proof We shall restrict ourselves to the proof of (e) and (f).

(e) Let $u_1, \ldots, u_k \in \mathbb{K}^n$ be arbitrary column vectors. By definition of the exterior product and by Proposition 2.8 we have

$$\begin{aligned} A^{(k)}(u_1 \wedge \cdots \wedge u_k) &= A^{(k)}[u_1, \ldots, u_k]^{(k)} = (A[u_1, \ldots, u_k])^{(k)} \\ &= [Au_1, \ldots, Au_k]^{(k)} = Au_1 \wedge \cdots \wedge Au_k\,. \end{aligned}$$

(f) Denote by C and D the $n \times k$ matrices whose columns consist of $u_1, \dots, u_k$ and $\upsilon_1, \dots, \upsilon_k$, respectively. By definition of the exterior product, the scalar product in $\mathbb{K}^{\binom{n}{k}}$, and by Proposition 2.8 we conclude that

$$\begin{aligned}(u_1 \wedge \cdots \wedge u_k, \upsilon_1 \wedge \cdots \wedge \upsilon_k)_{\mathbb{K}^{\binom{n}{k}}} \\ &= \sum_{\gamma \in Q_{k,n}} \det C\,[\gamma|1,\dots,k] \cdot \overline{\det D[\gamma|1,\dots,k]} \\ &= \sum_{\gamma \in Q_{k,n}} \det C^T\,[1,\dots,k|\gamma] \cdot \overline{\det D^T[1,\dots,k|\gamma]} \\ &= \det(C^T\overline{D}) = \det[(u_s, \upsilon_t)_{\mathbb{K}^n}|_{s,t=1}^k]\,.\end{aligned}$$

□

Corollary 2.6 *If $A \in M_n(\mathbb{K})$ and $u_1, \dots, u_k, \upsilon_1, \dots, \upsilon_k \in \mathbb{K}^n$ are arbitrary, then*

$$(A^{(k)}(u_1 \wedge \cdots \wedge u_k), \upsilon_1 \wedge \cdots \wedge \upsilon_k)_{\mathbb{K}^{\binom{n}{k}}} = \det\left[(Au_i, \upsilon_j)_{\mathbb{K}^n}|_{i,j=1}^k\right]\,.$$

Corollary 2.7 *For arbitrary vectors $u_1, \dots, u_k \in \mathbb{K}^n$ and a positive semi-definite matrix $A \in M_n(\mathbb{K})$ with the complete system of real eigenvalues $\lambda_1(A) \geq \lambda_2(A) \geq \cdots \geq \lambda_n(A)$ we have*

$$\begin{aligned}|u_1 \wedge \cdots \wedge u_k|^2_{\mathbb{K}^{\binom{n}{k}}} \prod_{i=1}^k \lambda_{n-i+1}(A) &\leq (A^{(k)}(u_1 \wedge \cdots \wedge u_k), u_1 \wedge \cdots \wedge u_k)_{\mathbb{K}^{\binom{n}{k}}} \\ &\leq |u_1 \wedge \cdots \wedge u_k|^2_{\mathbb{K}^{\binom{n}{k}}} \prod_{i=1}^k \lambda_i(A)\,.\end{aligned}$$

Remark 2.1 (a) For arbitrary $u_1, \dots, u_n \in \mathbb{K}^n$ and $A \in M_n(\mathbb{K})$ by Propositions 2.9(a) and 2.10(e) it follows that

$$Au_1 \wedge \cdots \wedge Au_n = A^{(n)}(u_1 \wedge \cdots \wedge u_n) = \det A(u_1 \wedge \cdots \wedge u_n).$$

This means that $\det A$ is the *distortion factor* of the exterior product of $u_1, \dots, u_k$ w.r.t. A.

(b) For arbitrary $u_1, \dots, u_k \in \mathbb{K}^n (k \leq n)$ we have by Proposition 2.10(f)

$$|u_1 \wedge \cdots \wedge u_k|_{\mathbb{K}^{\binom{n}{k}}} = \sqrt{\det[(u_s, u_t)|_{s,t=1}^k]} = \sqrt{\mathrm{Gr}\{u_1, \dots, u_k\}}\,, \qquad (2.24)$$

where $\mathrm{Gr}\{u_1, \dots, u_k\}$ denotes *Gram's determinant* of the vectors $u_1, \dots, u_k$.

2.3.2 Additive Compound Matrices

We define the additive compound matrix in the following way [37, 39, 46]:

Definition 2.4 Let $A \in M_n(\mathbb{K})$ and $k \in \{1, \ldots, n\}$ be arbitrary. Then we call

$$A^{[k]} := \frac{d}{dh}(I + hA)^{(k)}|_{h=0}$$

the *kth additive compound matrix of* A.

We shall present some of the most important properties of additive compound matrices in the next proposition [37, 39, 44].

Proposition 2.11 *Let* $A, B \in M_n(\mathbb{K})$ *and* $k \in \{1, \ldots, n\}$ *be arbitrary. Then we have:*

(a) $A^{[k]}(u_1 \wedge \cdots \wedge u_k) = Au_1 \wedge \cdots \wedge u_k + u_1 \wedge Au_2 \wedge \cdots \wedge u_k + \cdots + u_1 \wedge \cdots \wedge Au_k , \quad \forall\, u_1, \ldots, u_k \in \mathbb{K}^n$;
(b) $(\beta A + \beta' B)^{[k]} = \beta A^{[k]} + \beta' B^{[k]}, \quad \forall\, \beta, \beta' \in \mathbb{K}$;
(c) The complete system of eigenvalues of $A^{[k]}$ *consists of all possible* $\binom{n}{k}$ *sums of the form* $\lambda_{i_1} + \cdots + \lambda_{i_k}$ *, where* $\lambda_1, \ldots, \lambda_n$ *is the complete system of eigenvalues of the matrix* A *and* $1 \le i_1 < \cdots < i_k \le n$;
(d) $A^{[1]} = A$; $A^{[n]} = \operatorname{tr} A$.

Proof (a) By definition, from Proposition 2.9 and the product rule for exterior products it follows that for arbitrary $u_1, \ldots, u_k \in \mathbb{K}^n$

$$\begin{aligned} A^{[k]}(u_1 \wedge \cdots \wedge u_k) &= \frac{d}{dh}(I + hA)^{(k)}(u_1 \wedge \cdots \wedge u_k)|_{h=0} \\ &= \frac{d}{dh}[(I + hA)u_1 \wedge \cdots \wedge (I + hA)u_k]\,|_{h=0} \\ &= Au_1 \wedge \cdots \wedge u_k + u_1 \wedge Au_2 \wedge \cdots \wedge u_k + \cdots + u_1 \wedge \cdots \wedge Au_k . \end{aligned}$$

(b) From Proposition 2.9 we have for arbitrary $h \in \mathbb{R}, \beta, \beta' \in \mathbb{K}$,

$$\begin{aligned} (I + h\,\beta A)^{(k)}(I + h\,\beta' B)^{(k)} &= ((I + h\,\beta A)(I + h\,\beta' B))^{(k)} \\ &= (I + h(\beta A + \beta' B) + h^2 \beta\beta' AB)^{(k)} , \end{aligned}$$

which proves the second assertion.

(c) As a conclusion of Schur's lemma [44] there exists an upper triangular matrix $T \in M_n(\mathbb{C})$ with elements λ_i on the main diagonal and a unitary matrix $U \in M_n(\mathbb{C})$, so that $A = UTU^*$. Hence by definition and by Propositions 2.8 and 2.9 one gets for arbitrary $h \in \mathbb{R}$

$$(I_n + hA)^{(k)} = (I_n + h\,UTU^*)^{(k)}$$
$$= (U(I_n + h\,T)U^*)^{(k)} = U^{(k)}(I_n + h\,T)^{(k)}(U^{(k)})^*.$$

Obviously $I_n + h\,T$ is an upper triangular matrix with main diagonal elements of the form $1 + h\,\lambda_i$ $(i = 1, \ldots, n)$. Hence by Proposition 2.9(f, g) it follows, that $(I_n + h\,T)^{(k)}$ is also an upper triangular matrix with main diagonal elements of the form $\prod_{j=1}^{k}(1 + h\lambda_{i_j})$, where $1 \le i_1 < \cdots < i_k \le n$ are arbitrary elements from $Q_{k,n}$. Since U is unitary, then by Proposition 2.9(e) the matrix $U^{(k)}$ has the same property and the eigenvalues of $A^{[k]}$ are the eigenvalues of $\frac{d}{dh}(I_n + h\,T)^{(k)}|_{h=0}$, i.e. the complete system of eigenvalues of $A^{[k]}$ consists of all possible values

$$\frac{d}{dh}\prod_{j=1}^{k}(1 + h\lambda_{i_j})|_{h=0} = \lambda_{i_1} + \cdots + \lambda_{i_k}$$

with $1 \le i_1 < i_2 < \cdots < i_k \le n$.

(d) The first property follows immediately from definition, the second follows from the relation

$$A^{[n]} = \frac{d}{dh}\,(I + hA)^{(n)}_{|_{h=0}} = \frac{d}{dh}\,\det\,(I + hA)_{|_{h=0}} = \operatorname{tr} A\ .$$

□

Let us express explicitly the elements of $A^{[k]}$ through the elements of A. We give the result in form of a proposition [16, 37, 46].

Proposition 2.12 *Suppose $A = (a_{ij}) \in M_n(\mathbb{K})$ and $k \in \{1, \ldots, n\}$ are arbitrary. Then the $\binom{n}{k} \times \binom{n}{k}$ matrix $A^{[k]} = (a_{ij}^{[k]})$ can be computed as follows:*

$$a_{ij}^{[k]} = \begin{cases} \sum_{s=1}^{k} a_{i_s i_s}\ , & \text{if } (i) = (j) \in Q_{k,n}\ , \\ (-1)^{r+s} a_{i_r j_s}, & \text{if exactly one entry } i_r \text{ in } (i) \text{ does not occur in} \\ & (j) \text{ and } j_s \text{ does not occur in } (i), \\ 0\ , & \text{if } (i) \text{ differs from } (j) \text{ in two or more entries.} \end{cases}$$

Proof Consider for simplicity $k = 2$. By definition the element of $(I + hA)^{(2)}$ which belongs to the ith row and the jth column with $(i) = (i_1, i_2)$ and $(j) = (j_1, j_2)$ is

$$(\delta_{i_1 j_1} + ha_{i_1 j_1})(\delta_{i_2 j_2} + ha_{i_2 j_2}) - (\delta_{i_2 j_1} + ha_{i_2 j_1})(\delta_{i_1 j_2} + ha_{i_1 j_2})\ ,$$

where δ_{kl} is the Kronecker symbol. It follows immediately that

$$a_{ij}^{[2]} = \begin{cases} a_{i_1 i_1} + a_{i_2 i_2}, & \text{if } (i) = (j) , \\ (-1)^{r+s} a_{i_r j_s}, & \text{if exactly one } i_r \text{ does not occur in} \\ & (j) \text{ and } j_s \text{ does not occur in } (i), \\ 0, & \text{if none of numbers from } (i) \text{ occurs in } (j). \end{cases}$$

□

Example 2.3 Suppose $n = 3$ and

$$A = \begin{pmatrix} a_{11} & a_{12} & a_{13} \\ a_{21} & a_{22} & a_{23} \\ a_{31} & a_{32} & a_{33} \end{pmatrix} \quad \text{is a given matrix.}$$

Let us consider the cases $k = 1, 2, 3$.
For $k = 1$ we have $Q_{1,3} = \{1, 2, 3\}$ and

$$a_{ij}^{[1]} = \begin{cases} a_{ii} , & \text{if } i = j , \\ (-1)^{1+1} a_{ij} , & \text{if } i \neq j . \end{cases}$$

It follows that $A^{[1]} = A$.

For $k = 2$ we have $Q_{2,3} = \{(1), (2), (3)\}$ with $(1) = (1, 2), (2) = (1, 3)$ and $(3) = (2, 3)$. Using the above formula we get

$$A^{[2]} = \begin{pmatrix} a_{11} + a_{22} & a_{23} & -a_{13} \\ a_{32} & a_{11} + a_{33} & a_{12} \\ -a_{31} & a_{21} & a_{22} + a_{33} \end{pmatrix} .$$

For $k = 3$ we have $Q_{3,3} = \{(1, 2, 3)\}$ and $A^{[3]} = \sum_{i=1}^{3} a_{ii} = \text{tr}\, A$.

2.3.3 Applications to Stability Theory

Let us demonstrate how to use the properties of compound matrices for the investigation of linear differential equations. Suppose that $A(\cdot)$ is a continuous $n \times n$ matrix valued function and consider the equation

$$\frac{dv}{dt} = A(t)v . \tag{2.25}$$

Proposition 2.13 *Let $\Phi(\cdot)$ be the fundamental matrix of (2.25) and let the number $k \in \{1, \ldots, n\}$ be arbitrary. Then it holds:*
(a) $\Phi(\cdot)^{(k)}$ is the fundamental matrix of the system

$$\frac{dw}{dt} = A(t)^{[k]}w\,; \tag{2.26}$$

(b) If $\upsilon_1(\cdot), \upsilon_2(\cdot), \ldots, \upsilon_k(\cdot)$ are arbitrary solutions of system (2.25) then $w(\cdot) = \upsilon_1(\cdot) \wedge \upsilon_2(\cdot) \wedge \cdots \wedge \upsilon_k(\cdot)$ is a solution of (2.26).

Proof (a) Since $\Phi(\cdot)$ is the fundamental matrix of (2.25) we have for small h and arbitrary fixed $t \in \mathbb{R}$

$$\Phi(t+h) = \big(I + h\,A(t)\big)\,\Phi(t) + o\,(h)\,.$$

Thus, by Proposition 2.8

$$\Phi(t+h)^{(k)} = (I + hA(t))^{(k)}\,\Phi(t)^{(k)} + o\,(h)\,.$$

From this representation it follows that $\Phi(\cdot)^{(k)}$ is a solution of (2.26). It is clear that for any t the matrix $\Phi(t)^{(k)}$ is non-singular. Thus the statement (a) is proved.

Let us prove (b). For this we put $U(t) := (\upsilon_1(t), \upsilon_2(t), \ldots, \upsilon_k(t))$. Let us take a constant $n \times k$ matrix C such that $U(t) = \Phi(t)C$ where $\Phi(\cdot)$ is the fundamental matrix of (2.25). By Proposition 2.8 we have $U(t)^{(k)} = \Phi(t)^{(k)}C^{(k)}$, i.e. by (a) the function $U(t)^{(k)} = \upsilon_1(t) \wedge \cdots \wedge \upsilon_k(t)$ is a solution of system (2.26). □

Let us show now that the set of all solutions of (2.25) that converge to zero for $t \to +\infty$ can be characterized by the geometric dimension of a solution subspace or by the asymptotic stability of a compound equation (2.26). In Chap. 5 linear equations (2.25) arise as variational equations with respect to solutions of nonlinear differential equations. It will be shown there that, under certain conditions, the geometric dimension properties of the variational equations are connected with Hausdorff and fractal dimension bounds of invariant sets of the nonlinear equation.

Let us recall some standard stability properties for the trivial solution $\upsilon(t) \equiv 0$ of (2.25). We consider this solution for $t \geq t_0$.

Definition 2.5 We say that the trivial solution of (2.25) is:

(a) *Lyapunov stable*, if for every $\varepsilon > 0$ and every $t_1 \geq t_0$ there exists $\delta > 0$ such that any solution $\upsilon(\cdot)$ of (2.25) with $|\upsilon(t_1)| < \delta$ exists on $[t_1, \infty)$ and satisfies $|\upsilon(t)| < \varepsilon$ for all $t \geq t_1$;

(b) *asymptotically Lyapunov stable*, if it is Lyapunov stable, and if for each $t_1 \geq t_0$ there exists an $\eta > 0$ such that for any solution $\upsilon(\cdot)$ of (2.25) with $|\upsilon(t_1)| < \eta$ we have $\lim_{t\to+\infty} |\upsilon(t)| = 0$;

(c) *globally asymptotically Lyapunov stable*, if it is Lyapunov stable, and if for each $t_1 \geq t_0$ and each solution $\upsilon(\cdot)$ of (2.25) on $[t_1, +\infty)$ we have $\lim_{t\to+\infty} |\upsilon(t)| = 0$;

(d) *uniformly Lyapunov stable*, if for every $\varepsilon > 0$ there exists $\delta > 0$ such that for any $t_1 \geq t_0$ each solution $\upsilon(\cdot)$ of (2.25) with $|\upsilon(t_1)| < \delta$ exists on $[t_1, \infty)$ and satisfies $|\upsilon(t)| < \varepsilon$ for all $t \geq t_1$.

(e) *exponentially asymptotically stable*, if there exists a $\lambda > 0$ and, given any $\varepsilon > 0$, there exists a $\delta = \delta(\varepsilon) > 0$ such that for any $t_1 \geq t_0$ each solution $\upsilon(\cdot)$ of (2.25) with $|\upsilon(t_1)| < \delta$ exists on $[t_1, \infty)$ and satisfies $|\upsilon(t)| \leq \varepsilon e^{-\lambda(t-t_1)}$ for all $t \geq t_1$.

It is easy to see that the trivial solution of (2.25) is asymptotically Lyapunov stable if and only if it is globally asymptotically Lyapunov stable. Note that the stability property of any solution of the linear system, (2.25) is equivalent to the stability of the trivial solution. Thus one may associate that property to the Eq. (2.25).

Now we come to a proposition from [46] for the dimension-like characterization of the set of solutions of (2.25) which converge to zero.

Proposition 2.14 *Suppose that (2.25) is uniformly Lyapunov stable. Then (2.25) has an $(n-k+1)$-dimensional subspace with $k \in \{1, \dots, n\}$ of solutions $v(\cdot)$ satisfying $\lim_{t\to+\infty} v(t) = 0$ if and only if the kth compound equation (2.26) is asymptotically Lyapunov stable.*

Proof (a) Suppose (2.25) has an $(n-k+1)$-dimensional subspace of solutions $v(\cdot)$ satisfying $\lim_{t\to+\infty} v(t) = 0$. We can assume that there are $(n-k+1)$-linearly independent solutions $v_1, \dots, v_{n-k+1}$ having the property that $\lim_{t\to+\infty} v_i(t) = 0,\ i = 1, \dots, n-k+1$. Let us complete this system to a basis $\{v_1, \dots, v_n\}$ of the n-dimensional solution space of (2.25). It is clear from Proposition 2.10 that the solution space of the kth compound equation (2.26) is $\binom{n}{k}$-dimensional and has a basis consisting of all exterior k-products $v_{i_1} \wedge \dots \wedge v_{i_k}$ with $1 \le i_1 < \dots < i_k \le n$. It follows that any solution of (2.26) can be written as linear combination of such k-products. But in any k-products $v_{i_1} \wedge \dots \wedge v_{i_k}$ is at least one solution from the set $\{v_1, \dots, v_{n-k+1}\}$ which converges to zero as $t \to +\infty$. Using the uniform Lyapunov stability of (2.25) it follows now that (2.26) is Lyapunov stable and that all solutions v_i are bounded on $[t_0, \infty)$. Thus the k-products $v_{i_1}(t) \wedge \dots \wedge v_{i_k}(t)$ converge to zero as $t \to +\infty$.

(b) Let us assume now that (2.26) is globally asymptotically Lyapunov stable. We have to show that (2.25) has an $(n-k+1)$-dimensional subspace of solutions $v(\cdot)$ which satisfy $\lim_{t\to+\infty} v(t) = 0$.

Suppose that $v_1, \dots, v_n$ are n linearly independent solutions of (2.25) on $[t_0, \infty)$ and that there are $m \ge k$ of them, say $v_1, \dots, v_m$, having the property that $v_i(t) \nrightarrow 0$ for $t \to +\infty, i = 1, \dots, m$. Consider any k-product of solutions $v_{i_1} \wedge \dots \wedge v_{i_k}$, $1 \le i_1 < \dots < i_k \le m$. Since this is a solution of (2.26) we have $v_{i_1}(t) \wedge \dots \wedge v_{i_k}(t) \to 0$ as $t \to +\infty$. But this implies that $v_{i_1}(t_0) \wedge \dots \wedge v_{i_k}(t_0) = 0$, a contradiction to the fact that the vectors $v_{i_1}(t_0), \dots, v_{i_k}(t_0)$ are linearly independent. The contradiction shows that $m < k$ and that there are at least $n-k+1$ solutions of the basis set $\{v_1, \dots, v_n\}$ which converge to zero as $t \to +\infty$. □

2.4 Logarithmic Matrix Norms

2.4.1 Lozinskii's Theorem

For estimating the growth of solutions of differential equations, the well-known Wåzewski inequality can be used. But a more flexible result gives an estimate of Lozinskii, which is based on the use of logarithmic matrix norms [17, 20]. In this subsection we shall give a short introduction to such norms which is based on [9, 10, 41]. Let $\|\cdot\|$ be some vector norm in $\mathbb{R}^n$. Any such vector norm induces for $A \in M_n(\mathbb{R})$ an operator matrix norm which is given by the equality

$$\|A\| := \max_{\substack{\|\xi\|=1, \\ \xi \in \mathbb{R}^n}} \|A\xi\|.$$

Further, some such matrix norms will be used under the assumption that they are induced by means of the corresponding vector norms. The operator matrix norm of A induced by the Euclidean vector norm $|\cdot|$ is denoted by $|A|$.

Definition 2.6 The *logarithmic norm* or *Lozinskii norm* of a square matrix $A \in M_n(\mathbb{R})$ is the number $\Lambda(A)$ given by

$$\Lambda(A) := \lim_{h \to 0+} \frac{\|I + hA\| - 1}{h}.$$

The number $\Lambda(\cdot)$ depends on the choice of the matrix norm. Note that the logarithmic norm does not possess all the properties of a usual norm. The following example shows that it may be negative.

Example 2.4 Suppose that the matrix norm of $A = (a_{ij}) \in M_2(\mathbb{R})$ is given by $\|A\| = \max_i \sum_j |a_{ij}|$. Consider the matrix

$$A = \begin{pmatrix} -5 & 1 \\ 2 & -3 \end{pmatrix}.$$

Then $\Lambda(A) = -1$.

Table 2.1 shows the matrix norm and other notions.

Proposition 2.15 *Definition 2.6 is correct, i.e. for any $A \in M_n(\mathbb{R})$ the number $\Lambda(A)$ exists.*

Table 2.1 Frequently used notions and their symbols

Symbol	Notion	Sections
$A^{(k)}$	kth multiplicative compound matrix	2.3.1
$u_1 \wedge \cdots \wedge u_k$	Exterior product of vectors	2.3.1
$A^{[k]}$	kth additive compound matrix	2.3.2
$\Lambda(A)$	Logarithmic norm or Lozinskii norm	2.4.1

Proof Consider for $h > 0$ the function $f(h) := \frac{\|I+hA\|-1}{h}$. Let us show that $f(h)$ is decreasing for $h \to 0+$ and bounded from below. Suppose $k \in (0, 1)$ is arbitrary. Then we have by the triangle inequality $khf(kh) = \|I + khA\| - 1 = \|k(I + hA) + (1-k)I\| - 1 \leq k\|I + hA\| + 1 - k - 1 = kh\frac{(\|I+hA\|-1)}{h} = khf(h)$.

It follows that $f(kh) \leq f(h) \quad \forall k \in (0, 1)$, i.e. $f(\theta)$ is decreasing if θ is decreasing. The boundedness from below follows immediately from the inequality

$$f(h) \geq \frac{\big|1 - h\|A\|\big| - 1}{h} = -\|A\| ,$$

which is satisfied for small $h > 0$. □

The logarithmic matrix norm has several basic properties, which follow directly from its definition:

Proposition 2.16 *Suppose that A and B are real $n \times n$ matrices, $\|\cdot\|$ is the considered matrix norm and $\beta \in \mathbb{R}_+$ is a number. Then it holds:*

(1) $\Lambda(A + B) \leq \Lambda(A) + \Lambda(B)$;
(2) $\Lambda(\beta A) = \beta\Lambda(A)$;
(3) $\Lambda(A) \leq \|A\|$.

Proof In order to prove the assertion (1) we can write for any $h > 0$

$$\begin{aligned}\|I + h(A + B)\| - 1 &= \frac{1}{2}(\|I + 2hA + I + 2hB\| - 2) \\ &\leq \frac{1}{2}(\|I + 2hA\| - 1) + \frac{1}{2}(\|I + 2hB\| - 1) .\end{aligned}$$

From this it follows that

$$\begin{aligned}\lim_{h\to 0+} \frac{\|I + h(A + B)\| - 1}{h} &\leq \lim_{h\to 0+} \frac{\|I + 2hA\| - 1}{2h} \\ &+ \lim_{h\to 0+} \frac{\|I + 2hB\| - 1}{2h} = \Lambda(A) + \Lambda(B) .\end{aligned}$$

Let us show (2). If $\beta = 0$ then $\Lambda(0) = 0 = 0 \cdot \Lambda(0)$. Suppose $\beta > 0$. We have by definition

$$\Lambda(\beta A) = \lim_{h\to 0+} \frac{\|I + \beta hA\| - 1}{h} = \beta \lim_{\beta h\to 0+} \frac{\|I + \beta hA\| - 1}{\beta h} = \beta\Lambda(A).$$

The last assertion follows from the inequality

$$\Lambda(A) \leq \lim_{h\to 0+} \frac{1 + h\|A\| - 1}{h} = \|A\| .$$

□

We now consider the system

$$\frac{dv}{dt} = A(t)v, \tag{2.27}$$

where $A(\cdot)$ is a continuous $n \times n$ matrix function on $\mathbb{R}_+$ and $\sup_{t\geq 0} |A(t)| < \infty$. Let $v(\cdot)$ be an arbitrary non-trivial solution of system (2.27) and $\lambda_1(t)$ be the greatest eigenvalue of the matrix $\big(A(t) + A(t)^*\big)/2$. By the Fischer-Courant theorem (Theorem 2.1) we have

$$\frac{d}{dt}|v(t)|^2 = \big(A(t)v, v\big) + \big(v, A(t)v\big) = \Big(\big(A(t) + A(t)^*\big)v, v\Big) \leq 2\lambda_1|v(t)|^2, \ \forall t \geq 0.$$

It follows that

$$|v(t)| \leq |v(0)|e^{\int_0^t \lambda_1(\tau)\, d\tau}, \quad \forall t \geq 0.$$

This estimate is called the *Wåzewski inequality.*

Denote by $\Phi(t)$ the *Cauchy matrix* of system (2.27), i.e. the fundamental matrix, satisfying the initial condition $\Phi(0) = I$. Then the following theorem holds [12, 56].

Theorem 2.2 (Lozinskii estimate) *Suppose that* $\|\cdot\|$ *is an arbitrary operator norm generated by the vector norm* $\|\cdot\|$ *and* Λ *is the corresponding logarithmic norm. Then*

$$\big\|\Phi(t)\big\| \leq e^{\int_0^t \Lambda(A(\tau))\, d\tau}, \quad \forall t \geq 0. \tag{2.28}$$

Proof Assume that $v(t) = \Phi(t)v_0$ is an arbitrary solution of (2.27). Consider for $t \geq 0$ the function $n(t) := \|v(t)\|$ and denote by $D^+n(t)$ the right derivative of $n(\cdot)$ at t. By definition we have

$$D^+n(t) = \lim_{h\to 0^+} \frac{n(t+h) - n(t)}{h}$$
$$= \lim_{h\to 0^+} \frac{\|v(t) + hA(t)v(t)\| - \|v(t)\|}{h}. \tag{2.29}$$

If we use the inequality

$$\|v(t) + hA(t)v(t)\| \leq \|I + hA(t)\|\, \|v(t)\|$$

in (2.29) we conclude that

$$D^+n(t) \leq \Lambda(A(t))n(t). \tag{2.30}$$

Since $n(t) > 0$ for $t \geq 0$, we get from (2.30)

$$\frac{D^+ n(t)}{n(t)} \leq \Lambda(A(t)), \quad \forall t \geq 0. \tag{2.31}$$

The integration of inequality (2.31) on $[0, t]$ gives

$$\|\upsilon(t)\| = \|\Phi(t)\upsilon_0\| \leq \|\upsilon_0\| e^{\int_0^t \Lambda(A(\tau))d\tau}. \tag{2.32}$$

Since $\upsilon_0 \neq 0$ is arbitrary (2.32) implies (2.28). □

The Lozinskii estimate (2.28) can be used to derive sufficient conditions for the stability of (2.27). This is shown in the next corollary [46] where $t_0 \geq 0$ is an arbitrary number.

Corollary 2.8 *The linear system (2.27) is:*

(a) Lyapunov stable, if $\limsup_{t\to+\infty} \displaystyle\int_{t_0}^{t} \Lambda(A(\tau))d\tau < \infty$;

(b) asymptotically Lyapunov stable, if $\lim_{t\to+\infty} \displaystyle\int_{t_0}^{t} \Lambda(A(\tau))d\tau = -\infty$;

(c) uniformly Lyapunov stable, if $\displaystyle\int_{s}^{t} \Lambda(A(\tau))d\tau \leq M,\ t_0 \leq s \leq t < \infty$, *$M$ independent of s and t.*

If we make use of logarithmic norms (Lozinskii's estimate, in particular), then the following proposition may be useful.

Proposition 2.17 *Let the logarithmic norm Λ for $n \times n$ matrices be defined through the Euclidean vector norm, let $\lambda_1 \geq \lambda_2 \geq \cdots \geq \lambda_n$ be the eigenvalues of the matrix $(A + A^*)/2$ and let $k \in \{1, 2, \ldots, n\}$. Then*

$$\Lambda(A^{[k]}) = \lambda_1 + \lambda_2 + \cdots + \lambda_k.$$

Proof Firstly, we prove that

$$\Lambda(A) = \lambda_1. \tag{2.33}$$

For any $h \in \mathbb{R}$ a vector ξ_h, $|\xi_h| = 1$, can be found such that

$$|I + hA|^2 = \big((I + hA)\xi_h, (I + hA)\xi_h\big) = \big((I + h(A^* + A))\xi_h, \xi_h\big) + o(h).$$

Consequently,

$$|I + hA| - 1 = \\ \frac{h}{2}\big((A^* + A)\xi_h, \xi_h\big) + o(h) \leq \frac{h}{2} \sup_{|\xi|=1} \big((A^* + A)\xi, \xi\big) + o(h) = h\lambda_1 + o(h).$$

Therefore, $\Lambda(A) \leq \lambda_1$. Let us show the opposite inequality. Choose a vector η, $|\eta| = 1$, such that $\big((A^* + A)\eta, \eta\big) = 2\lambda_1$. Then in the same way as above we get

$$\big|I + hA\big| - 1 \geq \frac{h}{2}\big((A^* + A)\eta, \eta\big) + o(h) = h\lambda_1 + o(h).$$

It follows that $\Lambda(A) \geq \lambda_1$ and (2.33) is proved.

Since the square matrix A from (2.33) is arbitrary, we can replace it by $A^{[k]}$. Then

$$\Lambda(A^{[k]}) = \lambda,$$

where λ is the maximal eigenvalue of the matrix $\big(A^{[k]} + A^{[k]*}\big)/2$. Since

$$A^{[k]} + A^{[k]*} = \big(A + A^*\big)^{[k]}$$

(see (2.25) and (2.26)), by Proposition 2.12 we get

$$\lambda = \lambda_1 + \lambda_2 + \cdots + \lambda_k.$$

□

Remark 2.2 Often used vector norms for $u = (u_1, \ldots, u_n) \in \mathbb{R}^n$, different from the standard Euclidean norm, are

$$\|u\|_1 = \max_{i=1,\ldots,n} |u_i| \quad \text{and} \quad \|u\|_2 = \sum_i |u_i|.$$

By using Definition 2.6 it is easy to check that for $A = (a_{ij}) \in M_n(\mathbb{R})$

$$\Lambda_1(A) = \max_i \Big\{a_{ii} + \sum_{j\neq i} |a_{ij}|\Big\}, \quad \Lambda_2(A) = \max_j \Big\{a_{jj} + \sum_{i\neq j} |a_{ij}|\Big\}.$$

Writing the elements of the matrix $A^{[k]}$ through the elements of the matrix A, we obtain the value $\Lambda(A^{[k]})$. In the case $k = 2$ which is of great interest (as it will be seen further) we have

$$\Lambda_1(A^{[2]}) = \max_{(i)} \Big\{a_{i_1i_1} + a_{i_2i_2} + \sum_{j\notin(i)} \big(|a_{i_1j}| + |a_{i_2j}|\big)\Big\}, \tag{2.34}$$

$$\Lambda_2(A^{[2]}) = \max_{(j)} \Big\{a_{j_1j_1} + a_{j_2j_2} + \sum_{i\notin(j)} \big(|a_{ij_1}| + |a_{ij_2}|\big)\Big\}. \tag{2.35}$$

Remark 2.3 Generally speaking, the Lozinskii estimate (2.28) is non-invariant with respect to a Lyapunov transformation $\widetilde{v} = Q(t)v$, where Q is a Lyapunov matrix, i.e. a continuously differentiable square matrix, satisfying the conditions $|\det Q(t)| \geq \text{const} > 0, \forall t \geq 0$, $\sup_{t\geq 0} |Q(t)| < \infty$, and $\max_{t\geq 0} |\dot{Q}(t)| < \infty$. The non-invariance takes place, for example, for the norms $\|\cdot\|_1$ and $\|\cdot\|_2$. If $\widetilde{\Phi}(t)$ is a Cauchy matrix for the transformed system, then by Theorem 2.2 we have for $\tau \geq 0$

$$\|\widetilde{\Phi}(\tau)\| \leq \|Q(\tau)^{-1}\| \, \|Q(0)\| \exp \int_0^\tau \Lambda(\dot{Q}(t)\, Q(t)^{-1} + Q(t)\, A(t)\, Q(t)^{-1})\, dt.$$

Sometimes this inequality may give an estimate which is better than (2.28).

2.4.2 Generalization of the Liouville Equation

It is well-known that the Cauchy matrix $\Phi(t)$ of system (2.27) satisfies the Liouville equation

$$\det \Phi(t) = \exp \int_0^t \operatorname{tr} A(\tau)\, d\tau, \quad \forall\, t \in \mathbb{R}_+ . \tag{2.36}$$

Now we find some inequalities for singular values of the Cauchy matrix, which generalize formula (2.36). Let $\alpha_1(t) \geq \alpha_2(t) \geq \cdots \geq \alpha_n(t)$ be the singular values of the matrix $\Phi(t)$, $\Lambda(\cdot)$ be an arbitrary logarithmic norm and $\|\cdot\|$ be the vector norm being used for the definition of the norm Λ. The next proposition goes back to [5].

Proposition 2.18 *For $k = 1, 2, \ldots, n$ and $t \geq 0$ the following inequalities are true:*

$$\alpha_1(t)\alpha_2(t)\cdots\alpha_k(t) \leq \beta_1(t;k) \exp \int_0^t \Lambda(A(\tau)^{[k]})\, d\tau, \tag{2.37}$$

$$\alpha_n(t)\alpha_{n-1}(t)\cdots\alpha_{n-k+1}(t) \geq \beta_2(t;k) \exp \left\{ -\int_0^t \Lambda(-[A(\tau)^*]^{[k]})\, d\tau \right\} . \tag{2.38}$$

Here $\beta_1(\cdot\,;k), \beta_2(\cdot\,;k) \in C(\mathbb{R}_+)$ are two functions, depending on k and the norm $\|\cdot\|$ such that there exist positive constants c_{i1}, c_{i2}, satisfying the inequalities $c_{i1} \leq \beta_i(t;k) \leq c_{i2}$. In addition, $\beta_i(0;k) = 1$ and, if the norm $\|\cdot\|$ coincides with the Euclidean norm, $\beta_i(t;k) \equiv 1 (i = 1, 2;\ k = 1, 2, \ldots, n)$.

To prove Proposition 2.18 we need the following lemma. Let us for the numbers $k = 1, 2, \ldots, n$ consider the kth compound system of (2.27), i.e.

$$\frac{dw}{dt} = A(t)^{[k]} w. \tag{2.39}$$

Lemma 2.5 *Suppose that $k = 1, 2, \ldots, n$, Ψ is the Cauchy matrix of (2.39) and $|\cdot|$ is the operator matrix norm based on the Euclidean vector norm. Then*

$$\alpha_1(t)\alpha_2(t)\cdots\alpha_k(t) \leq \left|\Psi(t)\right|, \quad \forall\, t \geq 0.$$

Proof Let us fix $t \geq 0$. An orthogonal matrix T can be found such that for the Cauchy matrix $\Phi(\cdot)$ of (2.36)

$$T^*\Phi(t)^*\Phi(t)T = \operatorname{diag}\big(\alpha_1(t)^2, \alpha_2(t)^2, \ldots, \alpha_n(t)^2\big).$$

Denote by $\widehat{T}$ the matrix, consisting of the first k columns of T. Then

$$\widehat{T}^*\Phi(t)^*Y\Phi(t)\widehat{T} = \operatorname{diag}\big(\alpha_1(t)^2, \alpha_2(t)^2, \ldots, \alpha_k(t)^2\big).$$

The matrix $\Phi(t)\widehat{T}$ takes the form

$$\Phi(t)\widehat{T} = \big(\upsilon_1(t), \upsilon_2(t), \ldots, \upsilon_k(t)\big),$$

where $\upsilon_i(t)$ are some solutions of system (2.27). Therefore,

$$\alpha_1(t)^2 \cdots \alpha_k(t)^2 = \det \begin{pmatrix} (\upsilon_1, \upsilon_1) & (\upsilon_2, \upsilon_1) & \ldots & (\upsilon_k, \upsilon_1) \\ (\upsilon_1, \upsilon_2) & (\upsilon_2, \upsilon_2) & \ldots & (\upsilon_k, \upsilon_2) \\ \cdots & \cdots & \cdots & \cdots \\ (\upsilon_1, \upsilon_k) & (\upsilon_2, \upsilon_k) & \ldots & (\upsilon_k, v_k) \end{pmatrix}.$$

Let us introduce now the notation $U(t) = \big(\upsilon_1(t), \upsilon_2(t), \ldots, \upsilon_k(t)\big)$. By (2.24)

$$\alpha_1(t)^2 \cdots \alpha_k(t)^2 = |\upsilon_1(t) \wedge \cdots \wedge \upsilon_k(t)|^2 = \big|U^{(k)}(t)\big|^2.$$

Taking into account that $\Phi(0) = I$ and T is an orthogonal matrix, we obtain $\big|U^{(k)}(0)\big| = 1$. But by Proposition 2.13 $\upsilon_1(t) \wedge \cdots \wedge \upsilon_k(t)$ is a solution of system (2.39). Hence

$$\alpha_1(t) \cdots \alpha_k(t) \leq \sup_{\substack{w_0 \in \mathbb{R}^n, \\ |w_0|=1}} |w(t, w_0)|,$$

where $w(t, w_0)$ is the solution of (2.39) satisfying for $t = 0$ the initial condition $w(0, w_0) = w_0$. □

Proof of Proposition 2.18. By Lemma 2.5 we have for $t \geq 0$

$$\alpha_1(t) \cdots \alpha_k(t) \leq \big|\Psi(t)\big|,$$

where $\Psi(t)$ is the Cauchy matrix of system (2.39). By Lozinskii's estimate (2.28) we get

$$\big\|\Psi(t)\big\| \leq \exp \int_0^t \Lambda(A(\tau)^{[k]})\, d\tau.$$

Estimate (2.37) follows now from the last two inequalities, if we put $\beta_1(t;k) = |\Psi(t)|/\|\Psi(t)\|$. Note that the existence of two constants, bounding β_1, is a consequence of the equivalence of any two norms in the finite-dimensional space.

In order to prove the lower estimate (2.38), consider the Cauchy matrix W for the following system which is the *adjoint* of (2.27):

$$\frac{dw}{dt} = -A(t)^* w. \tag{2.40}$$

Since the identity $\Phi(t)^* W(t) \equiv I$ holds, by Proposition 2.1 we see that $\alpha_n(t)^{-1} \geq \cdots \geq \alpha_1(t)^{-1}$ are the singular values of the matrix $W(t)$. Applying to W the upper estimate, proved above, we obtain

$$\frac{1}{\alpha_n(t)\alpha_{n-1}(t)\cdots\alpha_{n-k+1}(t)} \leq \beta(t;k) \exp \int_0^t \Lambda\big[(-A(\tau)^*)^{[k]}\big]\, d\tau.$$

From this inequality, putting $\beta_2(t;k) = 1/\beta(t;k)$, we get (2.38). □

Let $\lambda_1(t) \geq \lambda_2(t) \geq \cdots \geq \lambda_n(t)$ be the complete system of eigenvalues of the matrix $\big(A(t) + A(t)^*\big)/2$.

Choosing in Proposition 2.18 the logarithmic norm, defined by the Euclidean vector norm, and taking into account Proposition 2.17, we obtain the following inequality from [51]:

Corollary 2.9 *For $k = 1, 2, \ldots, n$ and $t \geq 0$ the inequalities*

$$\alpha_1(t)\alpha_2(t)\cdots\alpha_k(t) \leq \exp \int_0^t \big(\lambda_1(\tau) + \lambda_2(\tau) + \cdots + \lambda_k(\tau)\big)\, d\tau, \tag{2.41}$$

$$\alpha_n(t)\alpha_{n-1}(t)\cdots\alpha_{n-k+1}(t) \geq \exp \int_0^t \big(\lambda_n(\tau) + \lambda_{n-1}(\tau) + \cdots + \lambda_{n-k+1}(\tau)\big)\, d\tau \tag{2.42}$$

are true.

For $k = n$ Corollary 2.9 results in the Liouville equation (2.36).

The next corollary is based on the paper [51].

Corollary 2.10 *Suppose that there exists a constant real symmetric positive-definite $n \times n$ matrix Q and a real valued function Θ continuous on $[0, t]$ such that*

$$A(\tau)^* Q + QA(\tau) + 2\Theta(\tau)Q \geq 0\,, \quad \forall \tau \in [0, t]\,. \tag{2.43}$$

If $1 \leq k \leq n$, then

$$\alpha_1(\tau)\alpha_2(\tau)\dots\alpha_k(\tau)$$
$$\leq \lambda_1(Q)^{k/2}\lambda_1(Q^{-1})^{k/2} \exp \int_0^\tau \big[(n-k)\Theta(s) + \operatorname{tr} A(s)\big] ds, \quad \forall\, \tau \in [0,t], \tag{2.44}$$

where $\lambda_1(Q)$ and $\lambda_1(Q^{-1})$ denote the largest eigenvalues of Q and Q^{-1}, respectively.

Proof Let the symmetric matrix R be the positive-definite square root of the positive-definite matrix Q. Then $Q = R^2$, $R = R^*$ and $\lambda_1(R) = \lambda_1(Q)^{1/2}$. When (2.43) is multiplied on both sides by R^{-1} it reduces to

$$M(\tau)^* + M(t) + 2\Theta(t)I_n \geq 0,$$

where $M(\tau) := R\dot{A}(\tau)R^{-1}$, $\tau \in (0,t]$. This shows that $\widetilde{\lambda}_i(\tau) \geq -\Theta(\tau)$ for $i = 1, 2, \dots, n$, where $\widetilde{\lambda}_1(\tau) \geq \dots \geq \widetilde{\lambda}_n(\tau)$ are the eigenvalues of the symmetric matrix $\frac{1}{2}\big[M(\tau)^* + M(\tau)\big]$. Hence for any $\tau \in [0,t]$

$$\widetilde{\lambda}_1(\tau) + \widetilde{\lambda}_2(\tau) + \dots + \widetilde{\lambda}_k(\tau) + (n-k)\Theta(\tau)$$
$$\leq \widetilde{\lambda}_1(\tau) + \dots + \widetilde{\lambda}_n(\tau) = \operatorname{tr} M(\tau).$$

If $S(\tau) := RA(\tau)R^{-1}$, then $S(0) = I_n$ and $\dot{S}(\tau) = M(\tau)S$ by (2.43). Applying (2.41) to the differential equation $\dot{S} = M(\tau)S$, we obtain

$$\widetilde{\alpha}_1(\tau)\widetilde{\alpha}_2(\tau)\dots\widetilde{\alpha}_k(\tau) \leq \exp \int_0^\tau \big[\widetilde{\lambda}_1(s) + \widetilde{\lambda}_2(s) + \dots + \widetilde{\lambda}_k(s)\big] ds$$
$$\leq \exp \int_0^\tau \big[(n-k)\Theta(s) + \operatorname{tr} M(s)\big]\, ds,$$

where $\widetilde{\alpha}_1(\tau), \dots, \widetilde{\alpha}_n(\tau)$ denote the singular values of $S(\tau)$ arranged so that $\widetilde{\alpha}_1(\tau) \geq \widetilde{\alpha}_2(\tau) \geq \dots \geq \widetilde{\alpha}_n(\tau) > 0$. It follows from Proposition 2.5 that the singular values $\widetilde{\alpha}_i(\tau)$ of the matrix $S(\tau)$ are connected with the singular values $\alpha_i(\tau)$ of the matrix $X(\tau) = R^{-1}S(\tau)R$ by the relation $\alpha_i(\tau) \leq \lambda_1(R^{-1})\widetilde{\alpha}_i(\tau)\lambda_1(R)$. Consequently

$$\alpha_1(\tau)\alpha_2(\tau)\dots\alpha_k(\tau) \leq \lambda_1(R)^k\lambda_1(R^{-1})^k \exp \int_0^\tau \big[(n-k)\Theta(s) + \operatorname{tr} M(s)\big]\, ds.$$

Since $\operatorname{tr} M(\tau) = \operatorname{tr} A(\tau)$ and $\lambda_1(R)^k\lambda_1(R^{-1})^k = \lambda_1(Q)^{k/2}\lambda_1(Q^{-1})^{k/2}$, (2.44) is proved. □

2.4.3 Applications to Orbital Stability

In this subsection we consider only the continuous time case. Suppose that $(\{\varphi^t\}_{t\in\mathbb{R}_+}, \mathbb{R}^n, |\cdot|)$ is a semi-flow generated by the equation

$$\dot{u} = f(u) \tag{2.45}$$

with the C^1-vector field $f : \mathbb{R}^n \to \mathbb{R}^n$. A solution of (2.45) starting in p for $t = 0$ is denoted by $u(\cdot, p)$. Clearly, then $\varphi^{(\cdot)}(\cdot) \equiv u(\cdot, \cdot)$.

We introduce some notations [7, 11, 43].

Definition 2.7

(1) A solution $u(\cdot, p)$ of (2.45) is called *orbitally stable* or *Poincaré stable* if for any $\varepsilon > 0$ there exists $\delta > 0$ such that for any $q \in \mathcal{B}_\delta(p)$ the inequality

$$\operatorname{dist}(u(t, q), \gamma^+(p)) < \varepsilon , \quad \forall t \geq 0$$

is satisfied.

(2) If in addition to (1) there exists an $\eta > 0$ such that for each $q \in \mathcal{B}_\eta(p)$ the relation

$$\operatorname{dist}(u(t, q), \gamma^+(p)) \to 0 \text{ as } t \to +\infty$$

holds, then the solution $u(\cdot, p)$ is said to be *asymptotically orbitally stable* or *asymptotically Poincaré stable*.

(3) We say that the solution $u(\cdot, p)$ of (2.45) has an *asymptotic phase* if there exists an $\eta_1 > 0$ such that for any solution $u(\cdot, q)$ of (2.45) with $\operatorname{dist}(q, \gamma^+(p)) < \eta_1$ a constant Δ can be found such that

$$|u(t + \Delta, q) - u(t, p)| \to 0 \text{ as } t \to +\infty . \tag{2.46}$$

Proposition 2.19 *If an orbitally stable solution $u(\cdot, p)$ of (2.45) has an asymptotic phase, then it is asymptotically orbitally stable.*

Proof For $t \geq |\Delta|$ we have

$$\begin{aligned} \operatorname{dist}\,(u(t, q), \gamma^+(p)) &= \inf_{t_1 \geq 0} |u(t, q) - u(t_1, p)| \\ &\leq |u(t, q) - u(t - \Delta, p)| . \end{aligned} \tag{2.47}$$

From (2.47) and (2.46) we obtain $\lim_{t\to+\infty} \operatorname{dist}\,(u(t, q), \gamma^+(p)) = 0$. □

Suppose that Eq. (2.45) has a T-period solution u. Consider the variational equation along this solution

$$\dot{v} = Df(u(t))v . \tag{2.48}$$

Let $\Phi(\cdot)$ be the fundamental matrix of (2.48) with $\Phi(0) = I$. Recall that $\Phi(T)$ is the *monodromy matrix* of (2.48) and the eigenvalues of $Y(T)$ are the *multipliers* of the period solution $u(\cdot)$. Denote them by $\rho_1, \rho_2, \ldots, \rho_n$ assuming that $|\rho_1| \geq |\rho_2| \geq \cdots \geq |\rho_n|$.

A basic result for multipliers is the following

Proposition 2.20 *At least one of the multipliers of the periodic solution u is equal to one.*

Proof The function $\xi(t) := f(u(t))$ is a solution of the variational equation (2.48). It follows that $\xi(t) = \Phi(t)\xi(0)$. Furthermore we have $\xi(0) = f(u(0)) = \xi(T)$. But this implies that $\xi(0) = \Phi(T)\xi(0)$. □

For completeness let us state the following Andronov-Vitt theorem [1].

Theorem 2.3 *Suppose that (2.45) has a T-periodic solution u with multipliers $\rho_1, \rho_2, \ldots, \rho_n$. If $\rho_1 = 1$ and $|\rho_j| < 1$, $j = 2, 3, \ldots, n$, then u is asymptotically orbitally stable and has an asymptotic phase.*

Denote by $\lambda_1(u) \geq \lambda_2(u) \geq \cdots \geq \lambda_n(u)$ the eigenvalues of the symmetrized Jacobi matrix

$$\frac{1}{2}\left[Df(u) + Df(u)^*\right].$$

Corollary 2.11 *(Poincaré criterion) Suppose $n = 2$ and u is a T-periodic solution of (2.45). If*

$$\int_0^T \left[\lambda_1(u(t)) + \lambda_2(u(t))\right] dt < 0\,, \tag{2.49}$$

then u is asymptotically orbitally stable and has an asymptotic phase.

Proof By the Liouville equation (2.36), we have

$$\rho_1\rho_2 = \det Y(T) = \exp\left(\int_0^T [\lambda_1(u(t)) + \lambda_2(u(t))]\, dt\right) < 1\,. \tag{2.50}$$

According to Proposition 2.20 one of the multipliers of u is equal to one. Hence, from (2.50) it follows that the second multiplier has an absolute value less than one. □

Now we state and prove a higher-dimensional analogy of the Poincaré criterion [5, 46].

Theorem 2.4 *Suppose that u is a T-period solution of (2.45). Suppose also that there exists a logarithmic norm Λ such that*

$$\int_0^T \Lambda(Df(u(t))^{[2]})\, dt < 0\,. \tag{2.51}$$

Then the solution u is asymptotically orbitally stable and has an asymptotic phase.

Proof Let $Y(\cdot)$ be the fundamental matrix of the variational system (2.48), let $\alpha_1(t) \geq \cdots \geq \alpha_n(t)$ be the singular values of $Y(t)$, and let $|\rho_1(t)| \geq \cdots \geq |\rho_n(t)|$ be the absolute values of the eigenvalues of $Y(t)$. Using Proposition 2.6, and Proposition 2.18, and also condition (2.51), we obtain

$$|\rho_1(mT)|\; |\rho_2(mT)| \leq \alpha_1(mT)\alpha_2(mT) < 1$$

for sufficiently large integer $m > 0$. Since $Y(mT) = Y(T)^m$, we have $|\rho_i(mT)| = |\rho_i(T)|^m$. Therefore, $|\rho_1(T)|\,|\rho_2(T)| < 1$ and, consequently, $|\rho_i(T)| < 1,\ i = 2, \ldots, n$. Now the result follows from Theorem 2.3. □

If the logarithmic norm is defined in terms of the Euclidean norm, then we obtain the following result [34, 46].

Corollary 2.12 *Let u be a T-periodic solution to the system (2.45), and let the following inequality hold:*

$$\int_0^T [\lambda_1(u(t)) + \lambda_2(u(t))]\, dt < 0 .$$

(Here $\lambda_1(u) \geq \lambda_2(u) \geq \cdots \geq \lambda_n(u)$ denote the eigenvalues of $\frac{1}{2}[Df(u) + Df(u)^*]$). Then the solution u is asymptotically orbitally stable and has an asymptotic phase.

We have noted above that in the case $n = 2$ inequality (2.49) is equivalent to the Andronov-Vitt condition. In higher-dimensional case this is not true as is shown in the next example.

Example 2.5 Let us consider Lanford's system which was derived in [23] for the simulation of turbulence in a fluid:

$$\begin{aligned} \dot{x} &= (\nu - 1)x - y + xy , \\ \dot{y} &= x + (\nu - 1)y + yz , \\ \dot{z} &= \nu z - (x^2 + y^2 + z^2) . \end{aligned} \tag{2.52}$$

Here ν denotes a parameter. One directly checks that this equation has for $\nu \in (\frac{1}{2}, 1)$ the periodic solution $u(t;\nu) = (R(\nu)\cos t,\ R(\nu)\sin t, \nu - 1)$, where $R(\nu) = \sqrt{-2\nu^2 + 3\nu - 1}$. It was shown in [23] that the Andronov-Vitt condition holds for all $\nu \in (\frac{1}{2}, \frac{2}{3})$, and consequently for each value of ν from this interval the solution $u(\cdot;\nu)$ is asymptotically orbitally stable and has an asymptotic phase.

Let us try to use Corollary 2.11: The symmetrized Jacobi matrix of the right-hand side of Lanford's system (2.52) with respect to the solution $u(\cdot;\nu)$ has the form

$$\begin{pmatrix} 2\nu - 2 & 0 & -\frac{1}{2}R\cos t \\ 0 & 0 & -\frac{1}{2}R\sin t \\ -\frac{1}{2}R\cos t & -\frac{1}{2}R\sin t & 3\nu - 2 \end{pmatrix} .$$

Its eigenvalues are 0 and $\frac{1}{2}(3\nu - 2) \pm \frac{1}{2}\sqrt{(3\nu - 2)^2 + R^2}$. Therefore, we have

$$\lambda_1 + \lambda_2 = 3\nu - 2 + \sqrt{7\nu^2 - 9\nu + 3} .$$

As it is easy to see, the inequality $\lambda_1 + \lambda_2 < 0$ is not true for any value $\nu \in (\frac{1}{2}, 1)$.

2.5 The Yakubovich-Kalman Frequency Theorem

2.5.1 The Frequency Theorem for ODE's

In the present subsection we shall discuss frequency-domain theorems for ODE's and some other related systems. Let us note that frequency-domain theorems can be considered as generalizations of the well-known Lyapunov theorems for the solvability of a matrix equation. For a proof see [55].

Theorem 2.5 (Lyapunov) *Suppose $A \in M_n(\mathbb{K})$ is Hurwitzian. Then for any matrix $G = G^* \in M_n(\mathbb{K})$ there exists a unique solution $P = P^* \in M_n(\mathbb{K})$ of the matrix equation $A^*P + PA = G$.*

If G is negative definite then P is positive definite.

Theorem 2.6 (Lyapunov) *Suppose $A \in M_n(\mathbb{C})$ is a given square matrix with eigenvalues $\lambda_1, \ldots, \lambda_n$ satisfying the inequalities $\lambda_k + \lambda_j \neq 0$ for all $k, j = 1, \ldots, n$. Then for any $G \in M_n(\mathbb{C})$ the equation*

$$A^*P + PA = G$$

has a unique solution $P \in M_n(\mathbb{C})$.

From Theorem 2.6 immediately follows:

Corollary 2.13 *Suppose A is as in Theorem 2.6. Then the inequality*

$$2\,\mathrm{Re}\,(Au, Pu)_n < 0, \quad \forall\, u \in \mathbb{C}^n$$

has a solution $P \in M_n(\mathbb{C})$.

Let us now consider pairs of matrices (A, B), where A is a (complex or real) matrix of order $n \times n$ and B is a (complex or real) matrix of order $n \times m$. Consider also an arbitrary Hermitian form $\mathcal{F}(u, \xi)$ of vectors $u \in \mathbb{C}^n$ and $\xi \in \mathbb{C}^m$, i.e.

$$\mathcal{F}(u, \xi) = (F_1u, u)_n + 2\mathrm{Re}\,(F_2\xi, u)_n + (F_3\xi, \xi)_m\,. \tag{2.53}$$

Here F_1, F_3 are Hermitian matrices of order $n \times n$, $m \times m$ respectively, F_2 is a matrix of order $n \times m$. The scalar products (norms) in $\mathbb{K}^n$ and $\mathbb{K}^m$ are denoted by $(\cdot,\cdot)_n(|\cdot|_n)$ and $(\cdot,\cdot)_m(|\cdot|_m)$, respectively. In applications the matrices A and B, and also the coefficients of the form $\mathcal{F}(u, \xi)$ are usually real. This case is called in the sequel the *real* one. The form $\mathcal{F}(u, \xi)$ is given in this case for real vectors $u \in \mathbb{R}^n$, $\xi \in \mathbb{R}^m$. Then the form $\mathcal{F}(u, \xi)$ defined by (2.53) is the extension of the real form to a Hermitian one.

We now introduce the following definitions [27, 55].

Definition 2.8 The pair (A, B) of matrices $A \in M_n(\mathbb{K})$ and $B \in M_{n,m}(\mathbb{K})$ is called *controllable*, if the rank of matrix $(B, AB, A^2B, \ldots, A^{n-1}B)$ is equal to n.

Definition 2.9 A pair (A, B) of matrices $A \in M_n(\mathbb{K})$ and $B \in M_{n,m}(\mathbb{K})$ is called *stabilizable* if there exists an $n \times m$ matrix E such that the matrix $A + BE^*$ is Hurwitzian, i.e. all its eigenvalues are located to the left of the imaginary axis.

The two following theorems are called *frequency-domain theorems of Yakubovich-Kalman* [27, 48, 54].

Theorem 2.7 *Let the pair (A, B) of matrices $A \in M_n(\mathbb{K})$ and $B \in M_{n,m}(\mathbb{K})$ be controllable. For the existence of a Hermitian $n \times n$ matrix P, real in the real case, such that for all $u \in \mathbb{C}^n$, $\xi \in \mathbb{C}^m$ the inequality*

$$2\,\mathrm{Re}\,(Au + B\xi, Pu)_n - \mathcal{F}(u, \xi) \le 0,$$

is satisfied it is necessary and sufficient that

$$\mathcal{F}[(i\omega I - A)^{-1}B\xi, \xi] \ge 0$$

for all $\xi \in \mathbb{C}^m$ and $\omega \in (-\infty, \infty)$ with $\det(i\omega I - A) \neq 0$.

Theorem 2.8 *Let the pair (A, B) of matrices $A \in M_n(\mathbb{K})$ and $B \in M_{n,m}(\mathbb{K})$ be stabilizable. For the existence of a Hermitian $n \times n$ matrix P, real in the real case, and such that for all $u \in \mathbb{C}^n$ and $\xi \in \mathbb{C}^m$ with $|u|_n + |\xi|_m \neq 0$ the inequality*

$$2\,\mathrm{Re}\,(Au + B\xi, Pu)_n - \mathcal{F}(u, \xi) < 0$$

is satisfied, it is necessary and sufficient that there exists an $\varepsilon > 0$ such that

$$\mathcal{F}(u, \xi) \ge \varepsilon(|u|_n^2 + |\xi|_m^2)$$

for all $u \in \mathbb{C}^n$, $\xi \in \mathbb{C}^m$ and for all $\omega \in (-\infty, \infty)$ with $\det(i\omega I - A) \neq 0$, *which are connected by the equality*

$$Au + B\xi = i\omega u.$$

In the case that the matrix A does not have pure imaginary eigenvalues, for the existence of a Hermitian matrix P real in the real case, it is necessary and sufficient that the inequality

$$\mathcal{F}[(i\omega I - A)^{-1}B\xi, \xi] > 0$$

holds for all $\xi \in \mathbb{C}^m$, $\xi \neq 0$, and all $\omega \in (-\infty, \infty)$ and

$$\lim_{\omega \to \pm\infty} \mathcal{F}(i\omega - A)^{-1}B\xi, \xi) > 0\,, \quad \forall \xi \in \mathbb{C}^m\,.$$

Theorems 2.7 and 2.8 give conditions for the existence of a Hermitian matrix P. But sometimes we need information about the spectrum of this matrix. For this purpose the two following lemmata seem to be useful.

Lemma 2.6 *Let an $n \times n$ matrix A and a Hermitian $n \times n$ matrix P satisfy the matrix inequality*

$$A^* P + PA < 0.$$

Then for the positive definiteness of the matrix P it is necessary and sufficient that the matrix A is Hurwitzian.

In order to formulate the second lemma let us introduce the notion of observability.

Definition 2.10 The pair (A, C) with $A \in M_n(\mathbb{K})$ and $C \in M_{n,m}(\mathbb{K})$ is called *observable*, if the rank of the matrix $(C, A^*C, (A^*)^2 C, \ldots, (A^*)^{n-1} C)$ is equal to n.

Lemma 2.7 *Let the pair (A, C) of matrices $A \in M_n(\mathbb{K})$ and $C \in M_{n,m}(\mathbb{K})$ be observable and the Hermitian $n \times n$ matrix P satisfy the inequality*

$$2\,\mathrm{Re}\,(Au, Pu)_n \leq -|C^* u|_m^2\,, \quad \forall u \in \mathbb{C}^n.$$

Then the given matrix A does not have eigenvalues on the imaginary axis, $\det P \neq 0$, *and the number of negative eigenvalues of the matrix P is equal to the number of eigenvalues of the matrix A located to the right of the imaginary axis. (Eigenvalues are counted with respect to their multiplicity.)*

The proof of Theorems 2.6 and 2.7 and also of Lemmata 2.6 and 2.7 can be found in [55].

2.5.2 *The Frequency Theorem for Discrete-Time Systems*

The first result in this subsection concerns the solvability of a matrix equation which is useful for the investigation of discrete-time systems. The proof of the following theorem can be reduced to Theorem 2.5.

Theorem 2.9 (Lyapunov) *Suppose $A \in M_n(\mathbb{K})$ is a matrix having all eigenvalues strongly inside the unit circle. Then for any matrix $G = G^* \in M_n(\mathbb{K})$ there exists a unique solution $P = P^* \in M_n(\mathbb{K})$ of the matrix equation*

$$A^* P A - P = G\,.$$

If G is negative definite then P is positive definite.

Let us state in this subsection a variant of the frequency-domain theorems which is useful for the investigation of discrete-time systems and which is also called the *Kalman–Szegö theorem* [4, 55].

Theorem 2.10 *Let the pair (A, B) of matrices $A \in M_n(\mathbb{K})$ and $B \in M_{n,m}(\mathbb{K})$ be controllable and let $\mathcal{F}$ be a Hermitian form. For the existence of a Hermitian $n \times n$ matrix P, satisfying the inequality*

$$(P(Au + B\xi), Au + B\xi)_n - (Pu, u)_n - \mathcal{F}(u, \xi) \leq 0 \tag{2.54}$$

for all $u \in \mathbb{C}^n$ and $\xi \in \mathbb{C}^m$, it is necessary and sufficient that the inequality

$$\mathcal{F}[(\lambda I - A)^{-1} B\xi, \xi] \geq 0$$

is satisfied for all $\xi \in \mathbb{C}^m$ and $\lambda \in \mathbb{C}$ such that $\det(A - \lambda I) \neq 0$.

Proof At first, we prove the sufficiency. Let us choose a number $z \in \mathbb{C}$, $|z| = 1$, such that $\det(A + zI) \neq 0$ and put

$$A_0 := (A - zI)(A + zI)^{-1}, \quad B_0 := \frac{1}{\sqrt{2}}(I - A_0)B, \tag{2.55}$$

$$u := z^{-1}(I - A_0)\left(\frac{\upsilon}{\sqrt{2}} - \frac{1}{2}B\xi\right). \tag{2.56}$$

Inequality (2.54) in this notation takes the form

$$2\,\mathrm{Re}\,(A_0\upsilon + B_0\xi, P\upsilon)_n - \mathcal{F}(u, \xi) \leq 0\ , \quad \forall\, \upsilon \in \mathbb{C}^n, \forall\, \xi \in \mathbb{C}^m\ .$$

It is easy to check that the pair (A_0, B_0) is controllable. According to Theorem 2.6, the existence of the desirable matrix P is equivalent to the inequality $\mathcal{F} \geq 0$ for all

$$\upsilon = (i\omega I - A_0)^{-1} B_0 \xi\ , \quad \forall\, \xi \in \mathbb{C}^m\ , \quad \forall\, \omega \in \mathbb{R} : \det(\omega I - A) \neq 0\ .$$

It can easily be seen that the last equality is equivalent to the relation

$$u = (\lambda I - A)^{-1} B\xi$$

for $\lambda = z\frac{1+i\omega}{1-i\omega}$, $\omega \in \mathbb{R}$. Thus, the sufficiency is proved.

The necessity follows immediately from the identity

$$\mathcal{F}(B_\lambda \xi, \xi) \equiv 0 \quad \text{for } |\lambda| = 1,$$

where $\mathcal{F}(u, \xi) = 2\,\mathrm{Re}\,(A_0 u + B_0 \xi, Pu)_n$ and $B_\lambda = (\lambda I - A)^{-1} B$. □

Using the following result one gets more information about the matrix P, determined by Theorem 2.10.

Lemma 2.8 *Let $P = P^*$, A and C be matrices of order $n \times n$, $n \times n$ and $n \times m$, respectively, and let the pair (A, C) be observable. Suppose that the inequality*

$$(PAu, Au)_n - (Pu, u)_n \leq -|C^* u|_m^2$$

is satisfied for all $u \in \mathbb{R}^n$. *Then* A *has no eigenvalues on the unit circle,* $\det P \neq 0$ *and the number of negative (resp. positive) eigenvalues of* P *is equal to the number of eigenvalues of* A*, which lie outside (resp. inside) the unit circle.*

Theorem 2.11 *Let the pair* (A, B) *of matrices* $A \in M_n(\mathbb{K})$ *and* $B \in M_{n,m}(\mathbb{K})$ *be stabilizable and let* $\mathcal{F}$ *be a Hermitian form. For the existence of a Hermitian* $n \times n$ *matrix* P*, satisfying the inequality*

$$(P(Au + B\xi), Au + B\xi)_n - (Pu, u)_n - \mathcal{F}(u, \xi) < 0 \quad , \\ \forall\, u \in \mathbb{C}^n,\ \forall\, \xi \in \mathbb{C}^m : |u|_n + |\xi|_m \neq 0\,, \tag{2.57}$$

it is necessary and sufficient that the inequality

$$\mathcal{F}((\lambda I - A)^{-1} B\xi, \xi) > 0\,,\quad \forall \xi \in \mathbb{C}^m,\ |\xi| \neq 0\,,\ \forall\, \lambda \in \mathbb{C} : \det(A - \lambda I) \neq 0 \tag{2.58}$$

is satisfied.

Proof The proof is similar to that of Theorem 2.10. □

2.6 Frequency-Domain Estimation of Singular Values

2.6.1 Linear Differential Equations

In this subsection we derive by frequency-domain techniques lower and upper estimates for the singular values $\alpha_1(t) \geq \alpha_2(t) \geq \cdots \geq \alpha_n(t)$ of the Cauchy matrix $\Phi(\cdot)$ to the linear differential equation

$$\dot{v} = A(t)v\,, \tag{2.59}$$

where $A(\cdot)$ is a continuous real $n \times n$ matrix valued function on $\mathbb{R}_+$. The main references for this subsection are [31, 33, 35].

Let us start with two lemmas. Consider a constant symmetric $n \times n$ matrix M_1 having at least $k \in \{1, 2, \ldots, n\}$ negative eigenvalues, the quadratic form $V_1(v) = (v, M_1 v)_n$ ($v \in \mathbb{R}^n$), and the set $\mathcal{G}_1 = \{v \in \mathbb{R}^n | V_1(v) < 0\}$.

Lemma 2.9 *If for a scalar continuous function* $\Theta_1(\cdot)$ *on* $\mathbb{R}_+$ *and any point* $z \in \mathcal{G}_1$ *the inequality*

$$(A(t)z, M_1 z)_n + \Theta_1(t)(z, M_1 z)_n \leq 0\,,\quad \forall t \geq 0\,, \tag{2.60}$$

is satisfied, then there exists a number $c_1 > 0$ *such that*

$$\alpha_k(t) \geq c_1 \exp\left(-\int_0^t \Theta_1(\tau)d\tau\right), \quad \forall t \geq 0. \tag{2.61}$$

Proof Let $\upsilon_0 \in \mathcal{G}_1$ be arbitrary and $\upsilon(\cdot)$ be the solution of (2.59) with initial condition $\upsilon(0) = \upsilon_0$. By (2.60) we have

$$\frac{d}{dt}(V_1(\upsilon(t)) \exp\left(2\int_0^t \Theta_1(\tau)d\tau)\right) \leq 0, \quad \forall t \geq 0,$$

and, consequently,

$$V_1(\upsilon(t)) \leq V_1(\upsilon_0) \exp\left(-2\int_0^t \Theta_1(\tau)d\tau\right), \quad \forall t \geq 0.$$

From this and from the assumptions on the eigenvalues of the matrix M_1 we conclude that there exists a k-dimensional linear subspace $\mathbb{L}_k$ of $\mathbb{R}^n$ and numbers c_1 such that for any $\upsilon_0 \in \mathbb{L}_k$ and any $t \geq 0$ the estimate

$$|\Phi(t)\upsilon_0| = |\upsilon(t)| \geq c_1|\upsilon_0| \exp\left(-\int_0^t \Theta_1(\tau)d\tau\right)$$

holds. To finish the proof it remains to refer to Corollary 2.3 of the Fischer-Courant theorem. □

Consider now a real constant symmetric $n \times n$ matrix M_2 having at least $k \in \{1, 2, \ldots, n\}$ positive eigenvalues, the quadratic form $V_2(\upsilon) = (M_2\upsilon, \upsilon)_n$ $(\upsilon \in \mathbb{R}^n)$, and the set $\mathcal{G}_2 = \{\upsilon \in \mathbb{R}^n | V_2(\upsilon) > 0\}$.

Lemma 2.10 *Let for some scalar continuous function $\Theta_2(\cdot)$ on $\mathbb{R}_+$ and for any point $w \in \mathcal{G}_2$ the inequality*

$$(M_2A(t)w, w)_n + \Theta_2(t)(M_2w, w)_n \leq 0, \quad \forall t \geq 0 \tag{2.62}$$

hold. Then there exists a number $c_2 > 0$ such that

$$\alpha_{n-k+1}(t) \leq c_2 \exp\left(-\int_0^t \Theta_2(\tau)d\tau\right), \quad \forall t \geq 0. \tag{2.63}$$

Proof Passing from system (2.59) to the adjoint system

$$\dot{w} = A(t)^*w, \tag{2.64}$$

and taking into account that the singular values of the Cauchy matrix of system (2.64) are the numbers $\alpha_n(t)^{-1} \geq \alpha_{n-1}(t)^{-1} \geq \cdots \geq \alpha_1(t)^{-1}$, on the basis of Lemma 2.9 we obtain the estimate

$$\alpha_{n-k+1}(t)^{-1} \geq c_2^{-1} \exp\left(\int_0^t \Theta_2(\tau)d\tau\right), \quad \forall t \geq 0,$$

where $c_2 > 0$ is a constant. From the last inequality the assertion (2.63) follows. □

Consider now system (2.59) with the matrix $A(t) = A + B\psi(t)$, i.e. the system

$$\dot{v} = [A + B\psi(t)]\, v\,. \tag{2.65}$$

Here A and B are constant real matrices of order $n \times n$ and $n \times l$, respectively, $\psi(\cdot)$ is a continuous on $\mathbb{R}_+$ real $l \times n$ matrix-valued function. Consider for $k = 1, 2, \ldots, n$ the Hermitian forms $\mathcal{F}_k(u, \xi)$ $(u \in \mathbb{C}^n, \xi \in \mathbb{C}^l)$. Suppose that for these forms there exist constant real $n \times l$ matrices C_k and a number $\varepsilon > 0$ such that the inequalities

$$\mathcal{F}_k(u, C_k^* u) \leq -\varepsilon |C_k^* u|_l^2\,, \quad \forall u \in \mathbb{R}^n, \quad k = 1, 2, \ldots, n\,, \tag{2.66}$$

hold. Suppose also that the inequalities

$$\mathcal{F}_k(u, \psi(t)u) \leq 0\,, \quad \forall u \in \mathbb{R}^n, \forall t \geq 0, \quad k = 1, 2, \ldots, n\,, \tag{2.67}$$

are satisfied.

Theorem 2.12 *Let the pair (A, B) be controllable, the pairs (A, C_k) be observable and let for some numbers $\beta_1 \leq \beta_2 \leq \cdots \leq \beta_n$ the following conditions be satisfied:*

(1) The matrix $A + BC_k^* + \beta_k I$ has at least k eigenvalues with positive real part;
(2) For any $k = 1, 2, \ldots, n$ and all $\omega \in \mathbb{R}$ the inequality

$$\mathcal{F}_k\big([(i\omega - \beta_k)I - A]^{-1} B\,\zeta, \zeta\big) \geq 0\,, \quad \forall \zeta \in \mathbb{C}^l \tag{2.68}$$

is true.

Then there exist numbers $c_k > 0$ such that the singular values $\alpha_k(t)$ of the Cauchy matrix of system (2.65) satisfy

$$\alpha_k(t) \geq c_k e^{-\beta_k t}\,, \quad \forall t \geq 0\,. \tag{2.69}$$

Proof From condition (2) of the theorem according to the Yakubovich-Kalman frequency theorem (Theorem 2.7) it follows that for any $k = 1, 2, \ldots, n$ there exists a constant symmetric $n \times n$ matrix P_k such

$$2\big((A + \beta_k I)u + B\xi, P_k u\big)_n - \mathcal{F}_k(u, \xi) \leq 0\,, \quad \forall u \in \mathbb{R}^n\,, \forall \xi \in \mathbb{R}^l\,. \tag{2.70}$$

Putting in (2.70) $\xi = C_k u$, by (2.66) we obtain the inequality

$$2\big((A + \beta_k I + BC_k^*)u, P_k u\big)_n \leq -\varepsilon |C_k^* u|_l^2\,, \quad \forall u \in \mathbb{R}^n\,. \tag{2.71}$$

From condition (1) of the present theorem and Lemma 2.7 it follows that the matrix P_k has at least k negative eigenvalues. From relations (2.67) and (2.70) it follows that for the function $V_k(v) := (P_k v, v)_n$ $(v \in \mathbb{R}^n)$ and for any solution $v(\cdot)$ of system (2.65) we have

$$\dot{V}_k(v(t)) + \beta_k V_k(v(t)) = 2\big((A + \beta_k I)v(t) + \; B\psi(t)v(t),\, P_k v(t)\big)_n \qquad (2.72)$$
$$- \mathcal{F}_k(v(t), \psi(t)v(t)) + \mathcal{F}_k(v(t), \psi(t)v(t)) \le 0\,, \quad \forall t \ge 0\,.$$

From (2.72) by Lemma 2.10 the statement of Theorem 2.12 follows. □

Using Lemma 2.10, by an analogous proof we obtain the following theorem.

Theorem 2.13 *Let the pair (A, B) be controllable, the pairs $(A, \mathcal{F}_k)$ be observable, and let for some numbers $\beta_1 \ge \beta_2 \ge \cdots \ge \beta_n$ the following conditions hold:*

(1) The matrix $A + BC_k^* + \beta_k I$ has at least k eigenvalues with negative real part;
(2) For any $k = 1, 2, \ldots, n$ and for all $\omega \in \mathbb{R}$ the inequalities

$$\mathcal{F}\big([(i\omega - \beta_k)I - A]^{-1}\, B\zeta, \zeta\big) \ge 0\,, \quad \forall \zeta \in \mathbb{C}^l$$

are satisfied.

Then there exist numbers $c_k > 0$, $k = 1, 2, \ldots, n$, such that the singular values of the Cauchy matrix of system (2.65) satisfy the inequalities

$$\alpha_{n-k+1}(t) \le c_k e^{-\beta_k t}\,, \quad \forall t \ge 0\,.$$

Consider now system (2.65) in the particular case $l = 1$, i.e.

$$\dot{v} = [A + b\,\psi(t)c^*]v\,, \qquad (2.73)$$

where b and c are constant n vectors and $\psi(\cdot)$ is again a continuous scalar-valued function. Let us suppose that ψ satisfies with some constant $\kappa > 0$ the inequality

$$0 \le \psi(t) \le \kappa\,, \quad \forall t \in \mathbb{R}_+\,, \qquad (2.74)$$

and consider the transfer function W given for all $z \in \mathbb{C}$, $\det(zI - A) \ne 0$, by

$$W(z) = c^*(zI - A)^{-1}b\,. \qquad (2.75)$$

Theorem 2.14 *Let in (2.73) the pair (A, b) be controllable, the pair (A, c) be observable, and let for some numbers $\beta_1 \le \beta_2 \le \cdots \le \beta_n$ the following conditions hold:*

(1) The matrix $A + \beta_k I$ has at least k eigenvalues with positive real part;
(2) For any $k = 1, 2, \ldots, n$ and all $\omega \in \mathbb{R}_+$ we have

$$\kappa^{-1} - \text{Re}\, W(i\omega - \beta_k) \geq 0 \, .$$

Then there exist constants $c_k > 0$, $k = 1, 2, \ldots, n$, such that for the singular values $\alpha_k(t)$ of the Cauchy matrix to system (2.69) the inequalities

$$\alpha_k(t) \geq c_k e^{-\beta_k t} \, , \quad \forall \, t \geq 0, \; k = 1, 2, \ldots, n \tag{2.76}$$

are satisfied.

Proof For the proof it is sufficient to use Theorem 2.12 with the quadratic forms

$$\mathcal{F}_k(u, \xi) = \xi^*(\kappa^{-1} - C_k^* u) + (\kappa^{-1}\xi - C_k^* u)\xi \, ,$$

where for any $k = 1, 2, \ldots, n$ we can put $C_k = \delta c$ with $\delta > 0$ sufficiently small. □

2.6.2 Linear Difference Equations

Consider now the linear difference equation

$$v_{t+1} = A(t) v_t \, , \; t = 0, 1, \ldots \, , \tag{2.77}$$

where $A(t)$ is for any $t = 0, 1, \ldots$ a real $n \times n$ matrix. We define by

$$\Phi(t) = \prod_{s=0}^{t} A(s) \, , \; t = 0, 1, \ldots \tag{2.78}$$

the Cauchy matrix of Eq. (2.77). Let $\alpha_1(t) \geq \alpha_2(t) \geq \cdots \geq \alpha_n(t)$ be its singular values.

Our aim is to find frequency-domain conditions for lower and upper estimates of these numbers. The results of this subsection have been obtained in [32, 33]. Suppose that M_1 is a constant real symmetric $n \times n$ matrix having at least $k \in \{1, 2, \ldots, n\}$ negative eigenvalues. Let us define the quadratic form $V_1(v) = (M_1 v, v)_n$ ($v \in \mathbb{R}^n$) and the set $\mathcal{G}_1 = \{v \in \mathbb{R}^n | V_1(v) < 0\}$.

Lemma 2.11 *Let for a certain positive number Θ_1 and an arbitrary point $w \in \mathcal{G}_1$ the inequality*

$$\frac{1}{\Theta_1^2} (M_1 A(t) w, A(t) w)_n \leq (M w, w)_n, \; t = 0, 1 \ldots \, ,$$

hold. Then there exists a number $c_1 > 0$ such that

$$\alpha_k(t) \geq c_1 \Theta_1^t \, , \; t = 0, 1, \ldots \, .$$

Proof The proof of this lemma is analogous to the proof of Lemma 2.9. □

Let us consider a constant real symmetric $n \times n$ matrix M_2 having at least $k \in \{1, 2, \dots, n\}$ positive eigenvalues.

Lemma 2.12 *Suppose that* $\det A(t) \neq 0$ *for all* $t = 0, 1, \dots$ *, and there is a positive number* Θ_2 *such that for all* $w \in \mathbb{R}^n$ *the inequality*

$$\frac{1}{\Theta_2^2}(M_2 A(t)w, A(t)w)_n \leq (M_2 w, w)_n, \ t = 0, 1 \dots ,$$

is satisfied. Then there exists a number $c_2 > 0$ *such that*

$$\alpha_{n-k+1}(t) \leq c_2 \Theta_2^t , \ t = 0, 1, \dots .$$

Proof The proof of this lemma is analogous to the proof of Lemma 2.10. □

Assume now that we have the Eq. (2.77) with

$$A(t) = A + B\psi(t)C^* , \ t = 0, 1, \dots . \tag{2.79}$$

Here A, B and C are constant real matrices of order $n \times n$, $n \times l$, and $n \times r$, respectively, $\psi(t)$ is for any $t = 0, 1, \dots$ an $l \times r$ matrix.

Let us introduce for $z \in \mathbb{C}$, $\det(pI - A) \neq 0$, the function

$$W(z) = C^*(zI - A)^{-1}B , \tag{2.80}$$

and the Hermitian form $\mathcal{F}_k(w, \xi)$ for $(w, \xi) \in \mathbb{C}^r \times \mathbb{C}^d$. Suppose that

$$\mathcal{F}_k(w, 0) \leq 0 , \quad \forall w \in \mathbb{R}^r , \tag{2.81}$$

and

$$\mathcal{F}_k(w, \psi(t)w) \leq 0 , \quad \forall w \in \mathbb{R}^r , \ t = 0, 1, \dots . \tag{2.82}$$

Theorem 2.15 *Suppose that there exist positive numbers* $0 < \rho_1 \leq \rho_2 \leq \dots \leq \rho_n$ *such that the following conditions are satisfied:*

(1) The matrix $\rho_k^{-1}A$ *has at least k eigenvalues outside the unit circle with center at the origin;*
(2) For all $z \in \mathbb{C}$ *with* $|z| = 1$ *the inequalities*

$$\mathcal{F}_k(W(\rho_k z)\zeta, \zeta) > 0 , \quad \forall\, \zeta \in \mathbb{C}^l \backslash \{0\}$$

hold. Then there exist positive numbers $c_k > 0$ *such that*

$$\alpha_k(t) \geq c_k \rho_k^t , \quad t = 0, 1, \dots .$$

Proof From condition (2) of the theorem it follows by the Kalman-Szegö theorem (Theorem 2.10) that there exists a constant symmetric matrix P_k such that

$$\frac{1}{\rho_k^2}(P_k(Au + B\xi), Au + B\xi)_n - (P_k u, u)_n - \mathcal{F}_k(C^* u, \xi) < 0 \tag{2.83}$$

for all $u \in \mathbb{R}^n$ and $\xi \in \mathbb{R}^l$ with $|u|_n + |\xi|_l \neq 0$. Putting in (2.83) $\xi = 0$ and taking into account (2.81) we obtain the inequality

$$\left(P_k\left(\frac{1}{\rho_k}Au\right), \frac{1}{\rho_k}Au\right)_n - (P_k u, u)_n < 0\,, \quad \forall u \in \mathbb{R}^n \setminus \{0\}\,. \tag{2.84}$$

From this and from condition (1) of the theorem it follows on the basis of Lemma 2.8 that the matrix P_k has at least k negative eigenvalues. From relations (2.82) and (2.83) we deduce that the inequalities needed in Lemma 2.11 are true. The reference to this lemma concludes the proof. □

Let us suppose now that instead of conditions (2.81) and (2.82) we have the following ones:

$$\mathcal{F}_k(0, \xi) \le 0\,, \quad \forall \xi \in \mathbb{R}^l, \tag{2.85}$$

$$\mathcal{F}_k(\psi(t)^* \xi, \xi) \le 0\,, \quad \forall \xi \in \mathbb{R}^l\,, \quad t = 0, 1, \ldots\,. \tag{2.86}$$

Analogously to the proof of the previous theorem the next one can be proved.

Theorem 2.16 *Suppose that* $\det A(t) \neq 0$*, for* $t = 0, 1, \ldots,$ *and there are numbers* $0 < \rho_1 \le \rho_2 \le \cdots \le \rho_n$ *such that the following conditions hold:*

(1) The matrix $\rho_k^{-1} A$ *has at least k eigenvalues inside the complex unit circle with center at the origin;*
(2) For all $z \in \mathbb{C}$ *with* $|z| = 1$ *the inequality*

$$\mathcal{F}_k(w, W(\rho_k z)^* w) > 0\,, \quad \forall\, w \in \mathbb{C}^r \setminus \{0\}\,,$$

is satisfied. Then there are numbers $c_k > 0$ *such the singular values* $\alpha_{n-k+1}(t)$ *of the Cauchy matrix of Eq. (2.77) satisfy the inequalities*

$$\alpha_{n-k+1}(t) \le c_k \rho_k^t\,, \quad t = 0, 1, \ldots\,.$$

Let us consider now the particular case $l = r = 1$ and suppose that for certain numbers $\kappa_1 \le \kappa_2$ such that $\kappa_1 \kappa_2 < 0$ the inequality

$$\kappa_1 \le \psi(t) \le \kappa_2\,, \quad t = 0, 1, \ldots\,,$$

is true.

Theorem 2.17 *Let for certain positive numbers $\rho_1 \geq \rho_2 \geq \cdots \geq \rho_n$ the following conditions hold:*

(1) The matrix $\rho_k^{-1} A$ has at least k eigenvalues outside the complex unit circle with center at the origin ;
(2) For all $z \in \mathbb{C}$ with $|z| = 1$ the inequality

$$\mathrm{Re}[(1 - \kappa_1 \overline{W(\rho_k z)})\,(1 - \kappa_2 W(\rho_k z))] > 0$$

is true. Then for any singular value $\alpha_k(\cdot)$ of the Cauchy matrix of equation (2.77) there exists a number $c_k > 0$ such that

$$\alpha_k(t) \geq c_k \rho_k^t\,, \quad t = 0, 1, \ldots\,.$$

Proof In order to prove this theorem it is sufficient to apply Theorem 2.10 with the Hermitian form

$$\mathcal{F}_k(w, \zeta) = (\zeta - \kappa_1 w)(\overline{\zeta} - \kappa_2 \overline{w}) + (\overline{\zeta} - \kappa_1 \overline{w})(\zeta - \kappa_2 w)\,.$$

□

2.7 Convergence in Systems with Several Equilibrium States

2.7.1 General Results

Suppose that $(\{\varphi^t\}_{t\in\mathbb{T}}, \mathcal{M}, \rho)$ is a dynamical system with more than one equilibrium state. This implies that none of them can be globally asymptotically stable. Nevertheless, it may be interesting to know whether every semi-orbit approaches one of the equilibrium points. For such systems an oscillatory behaviour is excluded. It is well-known that if the dynamical system is generated by a gradient field of a function $F(x)$ which tends to infinity as $|x| \to +\infty$ and has finitely many critical points then every solution of $\dot{x} = -\mathrm{grad} F(x)$, tends to one of these critical points. This leads to the following definition.

Definition 2.11 Suppose $(\{\varphi^t\}_{t\in\mathbb{T}}, \mathcal{M}, \rho)$ is a dynamical system with the set of equilibria $\mathcal{C}$. The positive semi-orbit $\gamma^+(p)$ is said to *converge to an equilibrium q* if $\varphi^t(p) \to q$ as $t \to +\infty$ and is said to *converge to the set* $\mathcal{C}$ if $\mathrm{dist}(\varphi^t(p), \mathcal{C}) \to 0$ as $t \to +\infty$.

Remark 2.4 The dynamical system is called quasi-gradient flow-like if each positive semi-orbit converges to the set $\mathcal{C}$, but is called gradient flow-like if each positive semi-orbit converges to an equilibrium [22, 36].

Let us start with a general Lyapunov-type result for the gradient-flow-behaviour [49]. For time-continuous systems in $\mathbb{R}^n$ this result has been obtained in [55].

Theorem 2.18 *Suppose that* $(\{\varphi^t\}_{t\in\mathbb{T}}, \mathcal{M}, \rho)$ *is a dynamical system generated in the continuous-time case by (1.5, Chap. 1) and in the discrete-time case by (1.6, Chap. 1). Suppose also that the set of equilibria* $\mathcal{C}$ *of the dynamical system consists of isolated points only and there exists a continuous function* $V : \mathcal{M} \to \mathbb{R}$ *such that the following conditions are satisfied:*

(a) V *is proper w.r.t.* $\mathcal{M}$*, i.e. for any compact set* $\mathcal{K} \subset \mathbb{R}$ *the set* $V^{-1}(\mathcal{K}) \subset \mathcal{M}$ *is compact;*
(b) $V(\varphi^t(p))$ *is non-increasing in* $\mathbb{T}_+$ *along any motion* $\varphi^{(\cdot)}(p)$*;*
(c) If $\varphi^{(\cdot)}(p)$ *is a motion for which there exists a* $\tau \in \mathbb{T}_+, \tau > 0$*, such that* $V(\varphi^\tau(p)) = V(p)$ *then* $\varphi^{(\cdot)}(p)$ *is an equilibrium.*

Then each positive semi-orbit of $(\{\varphi^t\}_{t\in\mathbb{T}}, \mathcal{M}, \rho)$ *converges to the set of equilibria. If the time is continuous then each positive semi-orbit converges to an equilibrium.*

Proof Suppose φ is an arbitrary motion. By assumption b) we have $V(\varphi^t(p)) \le V(p),\ \forall\, t \in \mathbb{T}_+$. From this and assumption a) it follows that the positive semi-orbit $\gamma^+(p)$ is bounded and the ω-limit set $\omega(p) \neq \emptyset$. Since the function $t \mapsto V(\varphi^t(p))$ is bounded and non-decreasing there exists the limit

$$\lim_{t\to+\infty} V(\varphi^t(p)) = c\ . \tag{2.87}$$

Suppose $q \in \omega(p)$ is arbitrary. It follows from Proposition 1.4, Chap. 1, that $\varphi^t(q) \in \omega(p),\ \forall\, t \in \mathbb{T}_+$. From (2.87) we conclude that $V(\varphi^t(q)) = c,\ \forall\, t \in \mathbb{T}_+$. From assumption c) we see that $\varphi^t(q) \equiv q$. This means that $\omega(p) \subset \mathcal{C}$. Suppose that for a motion φ we have $\varphi^t(p) \nrightarrow \mathcal{C}$ for $t \to +\infty$. This means that there exist an $\alpha > 0$ and a sequence $t_k \to +\infty$ as $k \to +\infty$ such that $\operatorname{dist}(\varphi^{t_k}(p), \mathcal{C}) > \alpha,\ k = 1, 2, \ldots$. From the last property it follows that φ has at least one ω-limit point outside $\mathcal{C}$. But this is a contradiction. If the time is continuous, the ω-limit set $\omega(p)$ is connected. Since $\omega(p) \subset \mathcal{C}$, this limit set is a single point. □

Corollary 2.14 *Under the assumptions of Theorem 2.18 the set* $\mathcal{C}$ *of equilibria is a global minimal attractor for the system.*

We derive now sufficient conditions for the convergence behaviour of the system

$$\dot{\upsilon} = w\ , \quad \dot{w} = -g(\upsilon, w) + z^* C f(\upsilon) - \phi(\upsilon)\ , \quad \dot{z} = Az + Bf(\upsilon)w\ , \tag{2.88}$$

where A is a constant Hurwitzian $n \times n$ matrix, i.e. a matrix whose eigenvalues have a negative real part. B and C are constant $n \times m$ matrices, $f : \mathbb{R} \to \mathbb{R}^m$ is a C^1-function, ϕ and g are scalar-valued C^1-functions on $\mathbb{R}$ resp. $\mathbb{R} \times \mathbb{R}$. Let us assume that with some $\kappa_0 > 0$ we have $g(\upsilon, w) \ge \kappa_0 w^2,\ \forall\, (\upsilon, w) \in \mathbb{R}^2$.

We suppose that system (2.87) has only isolated equilibrium states and any solution exists on $\mathbb{R}_+$.

Note that some models of an induction motor [42] can be written in the form (2.88). In case when f and ϕ are periodic system (2.88) describes also mathematical models of synchronous machines [36].

Let us show that the Lorenz system (1.12), Chap. 1 for $r > 1$ can be also written in the form (2.88). Recall that for $0 < r < 1$ the Lorenz system has the globally asymptotical equilibrium $(0, 0, 0)$. For $r > 1$ we can use the change of variables (with $\varepsilon = (r-1)^{-1/2}$)

$$(x, y, z) \mapsto \left(\frac{\sqrt{2\sigma}}{\varepsilon} \upsilon, \quad \frac{\sqrt{2}}{\varepsilon^2}(w + \varepsilon \upsilon), \quad \frac{1}{\varepsilon^2}(z + \frac{2\sigma}{b}\upsilon) \right)$$

and the new time $t \mapsto \frac{\varepsilon}{\sqrt{\sigma}} t$. We get a system (2.88) with

$$n = m := 1, \quad A := -\varepsilon \frac{b}{\sqrt{\sigma}}, \qquad B := 2\beta A^{-1}, \quad C := -1,$$
$$g(\upsilon, w) := \kappa_0 w, \quad f(\upsilon) := \upsilon, \quad \phi(\upsilon) := (1 - \beta A^{-1})\upsilon^3 - \upsilon,$$
$$\kappa_0 := \varepsilon \frac{\sigma + 1}{\sqrt{\sigma}}, \quad \text{and} \quad \beta := \varepsilon \frac{2\sigma - b}{\sqrt{\sigma}} .$$

Consider now the general system (2.87) and introduce for all $s \in \mathbb{C}$ with $\det(sI - A) \neq 0$ the matrix function

$$W(s) := C^*(sI - A)^{-1} B .$$

The next theorem is due to [30].

Theorem 2.19 *Suppose that there is a positive number κ such that the inequality*

$$\kappa_0 \kappa^{-1} I - \operatorname{Re} W(i\omega) > 0, \qquad \forall \omega \in [-\infty, +\infty] \tag{2.89}$$

is true. Then each bounded positive semi-orbit of a solution $u = (\upsilon, w, z)$ of (2.88) which satisfies the condition

$$\limsup_{t \to +\infty} |f(\upsilon(t))|^2 < \kappa$$

converges to an equilibrium.

Proof Let us show that with the quadratic form $\mathcal{F}(z, \xi) = \kappa_0 \kappa^{-1} |\xi|^2 - (C\xi, z)$ there exists an $n \times n$ matrix $P = P^*$ such that the inequality

$$2(Az + B\xi, Pz) - \mathcal{F}(z, \xi) < 0 , \tag{2.90}$$

holds for all $z \in \mathbb{R}^n$ and $\xi \in \mathbb{R}^m$ with $|z| + |\xi| \neq 0$. According to the Yakubovich-Kalman theorem (Theorem 2.7) for this it is sufficient that the pair (A, B) is stabilizable and the inequality

$$\mathcal{F}\big[(i\omega - A)^{-1}B\xi, \xi\big] > 0, \quad \forall \xi \in \mathbb{C}^m\,, \quad \xi \neq 0\,, \quad \forall \omega \in [-\infty, +\infty] \tag{2.91}$$

is true.

The pair (A, B) is stabilizable since A is a Hurwitzian matrix.

Condition (2.91) is equivalent to the inequality

$$\xi^*\big[\kappa_0\kappa^{-1}I - \operatorname{Re} W(i\omega)\big]\xi > 0\,, \quad \forall \xi \in \mathbb{C}^m\,, \quad \xi \neq 0, \quad \forall \omega \in [-\infty, +\infty]\,,$$

which is true by hypothesis (2.89).

Putting now $\xi = 0$ in (2.90) and taking into account that A is Hurwitzian, by Lemma 2.6, we get $P > 0$.

Inequality (2.90) implies the existence of a $\delta > 0$ such that

$$2(Az + B\xi, Pz) - \kappa_0\kappa^{-1}|\xi|^2 + (C\xi, z) \leq -\delta|z|^2\,, \ \forall z \in \mathbb{R}^n, \quad \forall \xi \in \mathbb{R}^m. \tag{2.92}$$

Introduce the function

$$V(\upsilon, w, z) = (Pz, z) + \frac{1}{2}w^2 + \int_0^\upsilon \phi(\sigma)\, d\sigma\,.$$

It follows from relation (2.92) that any solution $u(t) = \big(\upsilon(t), w(t), z(t)\big)$ of system (2.88) satisfies for $t \geq 0$ the estimate

$$\begin{aligned} \dot{V}(u(t)) &\leq -\delta|z(t)|^2 + \kappa_0\kappa^{-1}|f(\upsilon(t))w(t)|^2 - (Cf(\upsilon(t))w(t), z(t)) \\ &- \ g(\upsilon(t), w(t))w(t) + (Cf(\upsilon(t))w(t), z(t)) - \phi(\upsilon(t))w(t) + \phi(\upsilon(t))w(t) \\ &\leq \ -\delta|z(t)|^2 + \kappa_0\kappa^{-1}|f(\upsilon(t))w(t)|^2 - \kappa_0 w(t)^2 \\ &= -\delta|z(t)|^2 - \kappa_0 w(t)^2(1 - \kappa^{-1}|f(\upsilon(t))|^2)\,. \end{aligned} \tag{2.93}$$

This means that in the case when the positive semi-orbit of a solution $u = u(\cdot, p) = (\upsilon(\cdot, p), w(\cdot, p), z(\cdot, p))$ of (2.88) is bounded and satisfies

$$\limsup_{t\to+\infty} |f(\upsilon(t, p))|^2 < \kappa\,, \tag{2.94}$$

the function $V\big(u(t, p)\big)$ does not increase on a certain interval $(t_0, +\infty)$. From this and the boundedness of $V\big(u(t, p)\big)$ it follows that there exists a finite limit

$$\lim_{t\to+\infty} V\big(u(t, p)\big) = c\,.$$

Since the positive semi-orbit of $u(\cdot, p)$ is bounded the set ω of its ω-limit points is non-empty. Let $q \in \omega$ be arbitrary. Then the solution $u(\cdot, q)$ satisfies $u(t, q) \in \omega$ for all $t \geq 0$. Consequently, $V\big(u(t, q)\big) = c$ for all $t \geq 0$. Condition (2.94) also implies that $|f(\upsilon(t, q))|^2 < \kappa$ for $t \geq 0$. From this and (2.93) it follows that $z(t) \equiv 0$ and $w(t, q) \equiv 0$. From (2.88) and $w(t, q) \equiv 0$ we conclude that $\theta(t) \equiv$const. Thus, we see that ω consists of equilibrium states of system (2.88). But this is a contradiction, since we supposed that system (2.88) has only isolated equilibrium states. □

2.7.2 *Convergence in the Lorenz System*

Let us apply Theorem 2.19 to the Lorenz system. In the notation of system (2.88) we have $W(s) = 2\beta(A - s)^{-1}A^{-1}$ and, consequently,

$$W(i\omega) = \frac{2\beta}{A}\,\frac{A - i\omega}{A^2 + i\omega^2}\,, \qquad \omega \in \mathbb{R}.$$

Since $\operatorname{Re} W(i\omega) < 2\beta/A^2$, inequality (2.89) holds if

$$\beta < \kappa_0 A^2 (2\kappa)^{-1}\,.$$

The last inequality can be written as

$$2\sigma - b < \frac{(\sigma + b)b^2}{2\sigma(r - 1)}\,\kappa^{-1}\,. \tag{2.95}$$

Now, from (2.95), the dissipativity of the Lorenz system, and Theorem 2.19 we obtain the following result [30].

Theorem 2.20 *Suppose that a solution $u = \big(x, y, z\big)$ of the Lorenz system (1.12), Chap. 1 satisfies the condition*

$$(2\sigma - b) \limsup_{t \to +\infty} x(t)^2 < b^2(\sigma + 1)\,. \tag{2.96}$$

Then the positive semi-orbit of this solution converges to some equilibrium state.

The next result shows that the upper limit in Theorem 2.20 can be effectively estimated.

Lemma 2.13 *Suppose that l is defined by (1.20), Chap. 1.*
Then each solution $u = (x, y, z)$ of system (1.12), Chap. 1 satisfies the estimate

$$\limsup_{t \to +\infty} x(t)^2 \leq l^2 r^2\,.$$

Proof From (1.18, Chap. 1) we have the estimate

$$\limsup_{t\to+\infty}\big[y(t)^2+(z(t)-r)^2\big]\le l^2r^2\,.$$

Now we consider the two sets

$$\mathcal{G}_1 :=\big\{x,y,z\,|\,y^2+(z-r)^2\le l^2r^2,\quad x>lr\big\}$$
$$\text{and}\quad \mathcal{G}_2 :=\big\{x,y,z\,|\,y^2+(z-r)^2\le l^2r^2,\quad x<-lr\big\}\,.$$

Let us show that if at a certain time t we have $(x(t),y(t),z(t))\in\mathcal{G}_1$ then $\dot x(t)<0$. If we assume the opposite from the first equation of system (1.12), Chap. 1 it follows that $y(t)\ge x(t)>lr$. Thus, $y(t)^2+(z(t)-r)^2>l^2r^2$. But this inequality contradicts the assumption that the considered solution belongs to $\mathcal{G}_1$ at the time t.

Analogously one proves that if at a time t the solution belongs to $\mathcal{G}_2$ then $\dot x(t)>0$. □

From Theorem 2.20 and Lemma 2.13 we directly obtain that under the condition

$$2\sigma-b<\frac{b^2(\sigma+1)}{l^2r^2}$$

any positive semi-orbit of the Lorenz system converges to an equilibrium.

As a special case from this follows the well-known *Yudovich condition* [47] for the convergence of positive semi-orbits of the Lorenz system to an equilibrium point.

Corollary 2.15 *Suppose that* $2\sigma-b<0$ *then any positive semi-orbit of the Lorenz system converges to an equilibrium.*

Remark 2.5 For $\sigma=10$ and $b=8/3$ Theorem 2.20 guarantees the convergence of positive semi-orbits of the Lorenz system for $r<2$. This result will be essentially improved later on.

Consider now an example that demonstrates the convergence behaviour for a concrete physical system.

Example 2.6 The equation of motion of a director n in the dynamics of nematic liquid crystals can be written as the [21]

$$\frac{d}{dt}n=[\Omega,n]\,,\quad I\frac{d}{dt}\left[n,\frac{d}{dt}n\right]=-\gamma\,\Omega+\chi\,(H,n)\,[n,H]\,. \tag{2.97}$$

Here Ω and H are 3-dimensional vectors, I,γ and χ are positive constants, $(\cdot,\cdot)$ and $[\cdot,\cdot]$ denote the scalar and vector products, respectively. In [21] the Eqs. (2.97) were investigated numerically in the scalar variables $\xi=(H,n),\quad \eta=(H,[\Omega,n])$ and $\zeta=|\Omega|^2$. In these variables we can write the system (2.97) as

$$\begin{aligned} \dot{\xi} &= \eta\,, \\ \dot{\eta} &= -\gamma I^{-1}\eta - \zeta\xi + \chi I^{-1}\xi(|H|^2 - \xi^2)\,, \\ \dot{\zeta} &= -2\gamma I^{-1}\zeta + 2\chi I^{-1}\xi\eta\,. \end{aligned} \tag{2.98}$$

Making in (2.98) the change of variables [3]

$$(\xi, \eta, \zeta) \mapsto \left(\frac{\gamma}{4\sqrt{I\chi}}\, x\,,\ \ \frac{\gamma^2}{8\,I\sqrt{I\chi}}(y - x)\,,\ \ \frac{\gamma^2}{4\,I^2}(z - \frac{x^2}{4}) \right)$$

and $t \mapsto \frac{2I}{\gamma} t$, we obtain the new system

$$\dot{x} = -x + y, \quad \dot{y} = (1 + 4\,I\gamma^{-2}\chi|H|^2)x - y - xz, \quad \dot{z} = -4\,z + xy. \tag{2.99}$$

Thus, Eq. (2.98) is transformed into the Lorenz system (1.12), Chap. 1 with $\sigma = 1$, $b = 4$, and $r = 1 + 4I\gamma^{-1}\chi|H|^2$.

Obviously, the Yudovich condition $2\sigma - b < 0$ is satisfied for (2.99). This means that each positive semi-orbit of system (2.99) tends to one of the equilibria $(0, 0, 0)$ or $(\pm|H|, 0, 0)$. Chaotic behaviour as was found in numerical simulations in [21] for Eq. (2.99) is not possible.

Remark 2.6 The numerical study of the Lorenz system (1.12), Chap. 1 has shown a very complicated behaviour of orbits for certain values of parameters. In this remark we give some of the numerical results [40, 45] relating to bifurcations in system (1.12), Chap. 1.

The simplest analysis of system (1.12, Chap. 1) shows that for $0 < r \le 1$ the system has the unique equilibrium state $u_1 = (0, 0, 0)$. For $r > 1$ however two states $u_{2,3} = \big(\pm\sqrt{(r-1)b}, \pm\sqrt{(r-1)b}, r - 1\big)$ arise.

If we linearize system (1.12), Chap. 1 near the equilibrium u_1, the Jacobian matrix at u_1 will have eigenvalues, which can be found from the equation

$$(\lambda + b)\big[\lambda^2 + (\sigma + 1)\lambda + \sigma(1 - r)\big] = 0.$$

For $r < 1$ all three eigenvalues are negative, for $r > 1$ one value is negative and the two others are positive. Consequently, for $r > 1$ the point u_1 is a saddle equilibrium.

The Jacobian matrices of system (1.12), Chap. 1 being linearized near the equilibrium states $u_{2,3}$ have eigenvalues which are defined by the equation

$$\lambda^3 + (\sigma + b + 1)\lambda^2 + (r + \sigma)b\lambda + 2\sigma b(r - 1) = 0.$$

For $r > 1$ it has one negative root and two complex conjugate roots. The complex conjugate values become purely imaginary if the product of coefficients of λ^2 and λ is equal to the constant term, i.e., for $r = \frac{\sigma(\sigma+b+3)}{\sigma-b-1}$.

Lorenz [40] carried out his numerical experiments for $\sigma = 10$, $b = 8/3$, and changing r. Assuming that σ and b have the values mentioned, we consider the following sequence of bifurcations which are basic in the further account.

r < 1.
The equilibrium state u_1 is a global attractor. In other words, any orbit of system (1.12), Chap. 1 tends to the origin as $t \to +\infty$.

1 < r <≈ 13.926.
Two new equilibria u_2 and u_3 arise. Together with the equilibrium u_1 they form a global attractor, i.e., any orbit of system (1.12), Chap. 1 tends to one of these equilibrium states as $t \to +\infty$.

r ≈ 13.926.
Two homoclinic orbits arise, which issue out of the point u_1 by a one-dimensional unstable manifold and enter the point u_1 by a two-dimensional stable manifold. One of these orbits enters the half-space $x > 0$, the other enters the half-space $x < 0$. Each of these orbits is also called a *separatrix loop of the saddle*.

13.926 ≈< r < 24.74 $= \frac{\sigma(\sigma+b+3)}{\sigma-b-1}$.
Two homoclinic orbits turn into heteroclinic orbits. One of them, which issues out of the point u_1 into the subspace $x > 0$, enters u_2. The other, which issues out of the point u_1 into the subspace $x < 0$, enters u_3.

r > 24.74.
The so-called *Lorenz attractor* arises. This is a global attractor. In a simplified form, we may locally represent it as the product of a piece of smooth two-dimensional surface and a Cantor set. Due to its very complicated and unusual structure it is called a "strange" attractor.

Lorenz [40] has studied this attractor in detail by numerical methods in the case $r = 28$. At the present time the parameters values $\sigma = 10, b = 8/3$, and $r = 28$ have become canonical values. They are repeatedly used in publications connected with system (1.12), Chap. 1. We shall further refer to them as to *Lorenz's values*.

References

1. Andronov, A.A., Vitt, A.A., Khaikin, S.E.: Theory of Oscillations. Pergamon Press, Oxford (1966)
2. Binet, J.P.M.: Mémoire sur un systéme de fomules analytiques, et leŭr application à des considérations géometriques. J. École Polytech. **9** Cahier 16, 280–302 (1812)
3. Boichenko, V.A., Leonov, G.A.: Lorenz equations in dynamics of nematic liquid crystals. Dep. in VINITI 25.08.86, 6076–V86, Leningrad (1986). (Russian)
4. Boichenko, V.A., Leonov, G.A.: On orbital Lyapunov exponents of autonomous systems. Vestn. Leningrad Gos. Univ. Ser. 1, **3**, 7–10 (1988) (Russian, English transl. Vestn. Leningrad Univ. Math., **21**(3), 1–6, 1988)

5. Boichenko, V.A., Leonov, G.A.: Lyapunov functions, Lozinskii norms, and the Hausdorff measure in the qualitative theory of differential equations. Am. Math. Soc. Transl. **2**(193), 1–26 (1999)
6. Cauchy, A.L.: Mémoire sur les fouctions qui ne peuvent obteniv que deux valeurs égales et de signes contraires par suite des transpositions op'erées entre les variables quelles referment. J. École Polytech. **10** Cahier 17, 29–112 (1812)
7. Cesari, L.: Asymptotic Behavior and Stability Problems in Ordinary Differential Equations. Springer, Berlin (1959)
8. Chen, Z.-M.: A note on Kaplan-Yorke-type estimates on the fractal dimension of chaotic attractors. Chaos, Solitons Fractals **3**(5), 575–582 (1993)
9. Coppel, W.A.: Stability and Asymptotic Behavior of Differential Equations. D.C.Heath, Boston, Mass (1965)
10. Dahlquist, G.: Stability and error bounds in the numerical integration of ordinary differential equations. Trans. Roy. Inst. Tech. (Sweden) **130** (1959)
11. Demidovich, B.P.: Lectures on Mathematical Stability Theory. Nauka, Moscow (1967). (Russian)
12. Desoer, C.A., Vidyasagar, M.: Feedback Systems: Input-Output Properties. Academic Press, New York (1975)
13. Douady, A., Oesterlé, J.: Dimension de Hausdorff des attracteurs. C. R. Acad. Sci. Paris Ser. A **290**, 1135–1138 (1980)
14. Fan, K.: On a theorem of Weyl concerning eigenvalues of linear transformations I. Proc. Natl. Acad. Sci. (U.S.A.) **35**, 652–655 (1949)
15. Federer, H.: Geometric Measure Theory. Springer, New York (1969)
16. Fiedler, M.: Additive compound matrices and an inequality for eigenvalues of symmetric stochastic matrices. Czechoslovak Math. J. **24**, 392–402 (1974)
17. Forni, F., Sepulchre, R.: A differential Lyapunov framework for contraction analysis. IEEE Trans. Autom. Control **59**(3), 614–628 (2014)
18. Gantmacher, F.R.: The Theory of Matrices. Chelsea Publishing Company, New York (1959)
19. Ghidaglia, J.M., Temam, R.: Attractors for damped nonlinear hyperbolic equations. J. Math. Pures et Appl. **66**, 273–319 (1987)
20. Giesl, P.: Converse theorems on contraction metrics for an equilibrium. J. Math. Anal. Appl. **424**(2), 1380–1403 (2015)
21. Golo, V.L., Kats, E.I., Leman, A.A.: Chaos and long-lived modes in the dynamics of nematic liquid crystals. JETF **86**, 147–156 (1984). (Russian)
22. Hale, J.K., Rangel, G.: Lower semicontinuity of attractors of gradient systems and applications. Annali di Mat. Pura Appl. (IV)(CLIV), 281–326 (1989)
23. Hassard, B., Zhang, J.: Existence of a homoclinic orbit of the Lorenz system by precise shooting. SIAM J. Math. Anal. **25**, 179–196 (1994)
24. Horn, A.: On the singular values of a product of completely continuous operators. Proc. Natl. Acad. Sci. (U.S.A) **36**, 374–375 (1950)
25. Horn, R.A., Johnson, C.R.: Topics in Matrix Analysis. Cambridge University Press, Cambridge (1991)
26. Il'yashenko, Yu.S.: On the dimension of attractors of k-contracting systems in an infinite-dimensional space. Moskov Gos. Univ. Ser. 1, Mat. Mekh. **3**(3), 52–59 (1983). (Russian, English transl. Mosc. Univ. Math. Bull. **38**(3), 61–69, 1983)
27. Kalman, R.E.: Lyapunov functions for the problem of Lur'e in automatic control. Proc. Natl. Acad. Sci. U.S.A **49**(2), 201–205 (1963)
28. Ledrappier, F.: Some relations between dimension and Lyapunov exponents. Commun. Math. Phys. **81**, 229–238 (1981)
29. Lembcke, J., Reitmann, V.: Compound matrices and Hausdorff dimension estimates. DFG Priority Research Program "Dynamics: Analysis, efficient simulation, and ergodic theory", Preprint 07 (1995)
30. Leonov, G.A.: Global stability of the Lorenz system. Prikl. Mat. Mekh. **47**(5), 869–871 (1983). (Russian)

31. Leonov, G.A.: On the lower estimates of the Lyapunov exponents and the upper estimates of the Hausdorff dimension of attractors. Vestn. S. Peterburg Gos. Univ. Ser. 1, **29** (4), 18 – 24 (1996). (Russian)
32. Leonov, G.A.: Estimates of the Lyapunov exponents for discrete systems. Vestn. S. Peterburg Gos. Univ. Ser. 1, **30**(3), 49–56 (1997). (Russian)
33. Leonov, G.A.: Lyapunov Exponents and Problems of Linearization. From Stability to Chaos. St. Petersburg University Press, St. Petersburg (1997)
34. Leonov, G.A.: On a higher-dimensional analog of the Poincaré criterion of orbital stability. Diff. Urav. **24**, 1637–1639 (1988). (Russian, English transl. J. Diff. Equ. **24**, 1988)
35. Leonov, G.A., Noack, A., Reitmann, V.: Asymptotic orbital stability conditions for flows by estimates of singular values of the linearization. Nonlinear Anal. Theory Methods Appl. **44**, 1057–1085 (2001)
36. Leonov, G.A., Reitmann, V., Smirnova, V.B.: Non-local methods for pendulum-like feedback systems. Teubner-Texte zur Mathematik, Bd. 132, B. G. Teubner Stuttgart-Leipzig (1992)
37. Li, M.Y., Muldowney, J.S.: On Bendixson's criterion. J. Diff. Equ. **106**(1), 27–39 (1993)
38. Lidskii, V.B.: On the characteristic numbers of the sum and product of symmetric matrices. Dokl. Akad. Nauk, SSSR, **75**, 769–772 (1950). (Russian)
39. London, D.: On derivations arising in differential equations. Linear Multilinear Algebra **4**, 179–189 (1976)
40. Lorenz, E.N.: Deterministic nonperiodic flow. J. Atmos. Sci. **20**, 130–141 (1963)
41. Lozinskii, S.M.: Error estimation in the numerical integration of ordinary differential equations. I., Izv. Vuzov. Matematika **5**, 52–90 (1958). (Russian)
42. L'vovich, A.Yu., Rodynkov, F.F.: The Equations of Electrical Machines. St. Petersburg State University Press, St. Petersburg (1997). (Russian)
43. Lyapunov, A.M.: The general problem of the stability of motion. Kharkov (1892) (Russian, Engl. transl. Intern. J. Control (Centenary Issue), **55**, 531–572, 1992)
44. Marcus, M., Minc, H.: A Survey of Matrix Theory and Matrix Inequalities. Allyn and Bacon, Boston (1964)
45. Matsumoto, T., et al.: Bifurcations: Sights, Sounds, and Mathematics. Springer, Tokyo (1993)
46. Muldowney, J.S.: Compound matrices and ordinary differential equations. Rocky Mt. J. Math. **20**, 857–871 (1990)
47. Petrovskaya, N.V., Yudovich, V.I.: Homoclinic loops in the Salzman-Lorenz equation. Dep. at VINITI 28.06.79, No. 2, 380–79 (1979). (Russian)
48. Popov, V.M.: Absolute stability of nonlinear systems of automatic control. Avtomat. i Telemekh. **22**, 961–979 (1961). (Russian)
49. Reitmann, V., Kantz, H.: Generic analytical embedding methods for nonstationary systems based on control theory. In: Proceedings of International Conference on "Physics and Control", St. Petersburg (2005)
50. Reitmann, V., Zyryanov, D.: The global attractor of a multivalued dynamical system generated by a two-phase heating problem. In: Abstracts, 12th AIMS International Conference on Dynamical Systems, Differential Equations and Applications, Taipei, Taiwan, vol. 414 (2018)
51. Smith, R.A.: Some applications of Hausdorff dimension inequalities for ordinary differential equations. Proc. R. Soc. Edinb. **104A**, 235–259 (1986)
52. Temam, R.: Infinite-Dimensional Dynamical Systems in Mechanics and Physics. Springer, New York (1988)
53. Weyl, H.: Inequalities between the two kinds of eigenvalues of a linear transformation. Proc. Natl. Acad. Sci. (U.S.A) **35**, 408–411 (1949)
54. Yakubovich, V.A.: The solution of some matrix inequalities which appear in the automatic control theory. Dokl. Akad. Nauk, SSSR **143**(6), 1304– 1307 (1962). (Russian, English transl. Soviet Math. Dokl., **3**, 1962)
55. Yakubovich, V.A., Leonov, G.A., Gelig, AKh: Stability of Stationary Sets in Control Systems with Discontinuous Nonlinearities. World Scientific, Singapore (2004)
56. Zaks, M.A., Lyubimov, D.V., Chernatynsky, V.I.: On the influence of vibration upon the regimes of overcritical convection. Izv. Akad. Nauk SSSR, Fiz. Atmos. i Okeana, **19**, 312–314 (1983). (Russian)

Chapter 3
Introduction to Dimension Theory

Abstract In Chap. 2 the dimension of a vector space was defined as the maximal number of linearly independent vectors existing in it. The simplest example of an n-dimensional space, whose dimension is understood in this sense, is the space $\mathbb{R}^n$. The dimension theory, which was developed in the early 20th century, has extended this conception to more general classes of spaces and sets. In the following we give a short introduction into important notions of dimension for sets in general topological or metric spaces. We restrict ourselves to those dimensions and their properties which are especially useful in the investigation of ODE's.

3.1 Topological Dimension

We start our discussion of some elements of dimension theory with the definition of topological dimension. The notion of topological dimension in general topological spaces can be considered from different points of view. In the present section the topological dimension will be characterized in two ways:

The first one is connected with an inductive approach, the idea of which goes back to Poincaré [41] and the exact definition is given by Brouwer [7, 8], Urysohn [50] and Menger [34, 35]. On the basis of this inductive definition the main properties of topological dimension are obtained.

The second approach is due to Lebesgue [31]. The covering dimension is defined by the minimal order of arbitrary finess for the given topological space. Since for separable metric spaces both definitions coincide we call its common value briefly topological dimension.

All definitions and results stated in this section are standard in topological dimension theory. Most of them one can find in the monographs [2, 12, 24].

N. Kuznetsov and V. Reitmann, *Attractor Dimension Estimates for Dynamical Systems: Theory and Computation*, Emergence, Complexity and Computation 38,
https://doi.org/10.1007/978-3-030-50987-3_3

3.1.1 The Inductive Topological Dimension

Let $\mathcal{M}$ be a topological space. Denote by $\partial\mathcal{Z}$ the boundary of an arbitrary subset $\mathcal{Z} \subset \mathcal{M}$. By definition we have $\partial\mathcal{Z} = \overline{\mathcal{Z}} \cap (\overline{\mathcal{M} \setminus \mathcal{Z}})$, where $\overline{\mathcal{Z}}$ denotes the topological closure of a set $\mathcal{Z} \subset \mathcal{M}$. If $\mathcal{Z}$ is open, then it is obvious that $\partial\mathcal{Z} = \overline{\mathcal{Z}} \setminus \mathcal{Z}$.

In the following definition [7, 34, 35] $n \geq 0$ is an integer number.

Definition 3.1 The value $\operatorname{ind}\mathcal{M}$ is called *small inductive dimension* of the space $\mathcal{M}$, if the following conditions are satisfied:

(1) $\operatorname{ind}\mathcal{M} = -1$ if and only if $\mathcal{M} = \emptyset$;
(2) $\operatorname{ind}\mathcal{M} \leq n$ if for any point $p \in \mathcal{M}$ and any neighborhood $\mathcal{U}$ of p there exists an open set $\mathcal{V}$ with $p \in \mathcal{V} \subset \mathcal{U},\ mboxsuchthat$ $\operatorname{ind}\partial\mathcal{V} \leq n - 1$;
(3) $\operatorname{ind}\mathcal{M} = n$ if $\operatorname{ind}\mathcal{M} \leq n$ and not $\operatorname{ind}\mathcal{M} \leq n - 1$;
(4) $\operatorname{ind}\mathcal{M} = \infty$ if $\operatorname{ind}\mathcal{M} \leq n$ is not true for any n.

The topology of $\mathcal{M}$ induces in each subset $\mathcal{Z} \subset \mathcal{M}$ the *relative topology*: a set $\mathcal{U}' \subset \mathcal{Z}$ is by definition *open in* $\mathcal{Z}$ if there can be found a set $\mathcal{U}$ open in $\mathcal{M}$ and such that $\mathcal{U}' = \mathcal{U} \cap \mathcal{Z}$. It follows that for each subset $\mathcal{Z}$ of $\mathcal{M}$ considered as topological space the value $\operatorname{ind}\mathcal{Z}$ is also defined.

Example 3.1 It can be easily seen that $\mathbb{R}$ and any interval have the small inductive dimension 1.

Example 3.2 The space $\mathbb{R}^n$ has the small inductive dimension n. By induction on n it is easy to verify that the inequality $\operatorname{ind}\mathbb{R}^n \leq n$ holds. The proof of the opposite inequality is non-trivial and is not given here (see [12, 24]).

Let us consider now the main properties of the small inductive dimension. The standard references for this representation are [2, 12, 24].

The following statement immediately follows from Definition 3.1.

Proposition 3.1 *Let $\mathcal{M}_1$ and $\mathcal{M}_2$ be homeomorphic topological spaces. Then* $\operatorname{ind}\mathcal{M}_1 = \operatorname{ind}\mathcal{M}_2$.

Proposition 3.2 *Let $\mathcal{M}$ be a topological space and* $\operatorname{ind}\mathcal{M} = n < \infty$. *Then for any integer m, $-1 \leq m \leq n$, the space $\mathcal{M}$ contains some subset of small inductive dimension m.*

Proof Since $\operatorname{ind}\mathcal{M} \geq n - 1$, we can find a point $p_0 \in \mathcal{M}$ and a neighborhood $\mathcal{U}_0$ of this point which possesses the following property: if $\mathcal{V}$ is an arbitrary open set with $p_0 \in \mathcal{V} \subset \mathcal{U}_0$, then

$$\operatorname{ind}\partial\mathcal{V} \geq n - 1. \tag{3.1}$$

On the other hand, since $\operatorname{ind}\mathcal{M} \leq n$, there can be found an open set $\mathcal{V}_0$ such that $p_0 \in \mathcal{V}_0 \subset \mathcal{U}_0$ and $\operatorname{ind}\partial\mathcal{V}_0 \leq n - 1$. Since inequality (3.1) is satisfied for any neighborhood of the point p_0 which is in $\mathcal{U}_0$, it is also true for $\mathcal{V}_0$, i.e., $\operatorname{ind}\partial\mathcal{V}_0 \geq n - 1$.

Consequently, the boundary of $\mathcal{V}_0$ is a subset of the space $\mathcal{M}$ and has the small inductive dimension $n-1$.

Repeating this process if necessary we shall obtain the desired subset of the required dimension. □

Proposition 3.3 *Suppose that* $\mathcal{Z} \subset \mathcal{M}$ *and* $\operatorname{ind} \mathcal{M} \leq n$. *Then* $\operatorname{ind} \mathcal{Z} \leq n$.

Proof For $n=-1$ the statement is obvious. Suppose that it is true for $n-1, n \geq 0$ integer. Let $\operatorname{ind} \mathcal{M} \leq n$ and $p \in \mathcal{Z}$. Also let $\mathcal{U}'$ be a neighborhood of p in $\mathcal{Z}$. It means that there exists a neighborhood $\mathcal{U}$ of the point p in $\mathcal{M}$ such that $\mathcal{U}' = \mathcal{U} \cap \mathcal{Z}$. Since $\operatorname{ind} \mathcal{M} \leq n$, there can be found a set $\mathcal{V}$, which is open in $\mathcal{M}$ and such that

$$p \in \mathcal{V} \subset \mathcal{U} \quad \text{and} \quad \operatorname{ind} \partial \mathcal{V} \leq n-1.$$

Let $\mathcal{V}' = \mathcal{V} \cap \mathcal{Z}$. Then $\mathcal{V}'$ is open in $\mathcal{Z}$, $p \in \mathcal{V}' \subset \mathcal{U}'$. Suppose that $\mathcal{B}$ is the boundary of $\mathcal{V}$ in $\mathcal{M}$ and $\mathcal{B}'$ be the boundary of $\mathcal{V}'$ in $\mathcal{Z}$ ($\mathcal{B} = \overline{\mathcal{V}} \setminus \mathcal{V}$ and $\mathcal{B}' = (\overline{\mathcal{V}}' \setminus \mathcal{V}') \cap \mathcal{Z}$). Then $\mathcal{B}' \subset \mathcal{B} \cap \mathcal{Z}$. According to the induction hypothesis $\operatorname{ind} \mathcal{B}' \leq n-1$. □

For the next step, it is useful to consider in some detail a very important special case of a space having a zero small inductive dimension.

By Definition 3.1 the equality $\operatorname{ind} \mathcal{M} = 0$ means that any point $p \in \mathcal{M}$ has arbitrary small neighborhoods with empty boundary, i.e. for any neighborhood $\mathcal{U}$ of the point p there can be found an open set $\mathcal{V}$ such that

$$p \in \mathcal{V} \subset \mathcal{U} \quad \text{and} \quad \partial \mathcal{V} = \emptyset.$$

Proposition 3.4 *Let* $\mathcal{M}$ *be a metric space with metric* ρ. *If the space* $\mathcal{M}$ *is at most countable, then* $\operatorname{ind} \mathcal{M} = 0$.

Proof Let p be an arbitrary point of $\mathcal{M}$, $\mathcal{U}$ be an arbitrary neighborhood of this point. Choose a number $\varepsilon > 0$ such that $\mathcal{B}_\varepsilon(p) \subset \mathcal{U}$. Let $p_1, p_2, \ldots$ be the points of $\mathcal{M}$. Obviously, there can be found a positive number $\varepsilon' < \varepsilon$ such that $\varepsilon' \neq \rho(p_j, p)$ for all j. But in this case a ball $\mathcal{B}_{\varepsilon'}(p)$ is contained in $\mathcal{U}$ and its boundary is empty. □

Example 3.3 The set $\mathbb{Q}$ of rational numbers in $\mathbb{R}$ has the small inductive dimension zero.

Example 3.4 The set $\mathbb{I}$ of irrational numbers has the inductive dimension zero. Really, let $\mathcal{U}$ be an arbitrary neighborhood of an irrational point p. Let us choose the irrational numbers p_1 and p_2 such that $p_1 < p < p_2$ and a set $\mathcal{V}$ consisting of the irrational numbers which are from $\mathcal{U}$ between p_1 and p_2. In the space $\mathbb{I}$ of the irrational numbers the set $\mathcal{V}$ is open and has an empty boundary, since any irrational point, which is a limit point of $\mathcal{V}$, lies between p_1 and p_2 and, consequently, belongs to $\mathcal{V}$.

Example 3.5 The set $\mathbb{I}^2$ of all points in the plane with irrational coordinates has the small inductive dimension zero. More generally, the set $\mathbb{I}^n$ of all points of $\mathbb{R}^n$, with irrational coordinates, is zero-dimensional.

Really, any point of the plane is in arbitrary small rectangles bounded by straight lines, which are perpendicular to the axes and intersect them at the points with rational coordinates. The boundaries of these rectangles do not intersect $\mathbb{I}^2$. The same reasoning would also be valid in the general case $\mathbb{I}^n$.

The union of sets with small inductive dimension zero may be non zero-dimensional. This shows the decomposition of the straight line into sets of rational and irrational numbers

$$\mathbb{R} = \mathbb{Q} \cup \mathbb{I}.$$

On the other hand (Examples 3.1, 3.3 and 3.4) we have

$$\operatorname{ind} \mathbb{R} = 1, \ \operatorname{ind} \mathbb{Q} = 0 \quad \text{and} \quad \operatorname{ind} \mathbb{I} = 0.$$

This is the reason why we need an additional assumption in the following formula on dimension of the union of zero-dimensional sets. We give this formula without proof (see [24]).

Proposition 3.5 *Let $\mathcal{Z}_j \subset \mathcal{M}$, $j = 1, 2, \ldots,$ be some subsets of the topological space $\mathcal{M}$ and $\mathcal{M} = \cup_j \mathcal{Z}_j$. If the sets $\mathcal{Z}_j$ are closed and* $\operatorname{ind} \mathcal{Z}_j = 0$, *then* $\operatorname{ind} \mathcal{M} = 0$.

Example 3.6 Suppose that $0 < m \leq n$ are natural numbers. Denote by $\mathcal{Q}^n_m$ the set of points from $\mathbb{R}^n$ having exactly m rational coordinates. Then $\operatorname{ind} \mathcal{Q}^n_m = 0$. Indeed for any m indices $j_1, j_2, \ldots, j_m$, chosen from numbers $1, 2, \ldots, n$, and m rational numbers $r_1, r_2, \ldots, r_m$, the system of equations

$$u_{j_1} = r_1, \quad u_{j_2} = r_2, \ldots, \quad u_{j_m} = r_m$$

defines a $(n - m)$-dimensional linear subspace. Denote by $\mathcal{Z}_j$ the subset of this subspace consisting of points, all other coordinates of which are irrational. Each $\mathcal{Z}_j$ is isometric with $\mathbb{I}^{n-m}$ and, consequently, has the small inductive dimension 0 (Example 3.5). It is clear that $\mathcal{Z}_j$ is closed in $\mathcal{Q}^n_m$ and that the union of sets $\mathcal{Z}_j$ coincides with $\mathcal{Q}^n_m$. It follows that $\operatorname{ind} \mathcal{Q}^n_m = 0$ by virtue of Proposition 3.5.

Let us return to arbitrary topological spaces and consider the small inductive dimension of the union of sets in the general case. But at first we shall prove a useful proposition.

When investigating the inductive dimension of subsets in a space $\mathcal{M}$, it will sometimes be convenient to define their dimension by neighborhoods with respect to the whole subspace $\mathcal{M}$. In the next proposition we consider subspaces of completely normal spaces. Recall that a topological space $\mathcal{M}$ is called *completely normal* if for any subsets $\mathcal{A}$ and $\mathcal{B}$ with $\overline{\mathcal{A}} \cap \mathcal{B} = \emptyset$, $\mathcal{A} \cap \overline{\mathcal{B}} = \emptyset$ there can be found a closed set $\mathcal{C}$ and disjunct sets $\mathcal{U}$ and $\mathcal{V}$ such that $\mathcal{M} \setminus \mathcal{C} = \mathcal{U} \cup \mathcal{V}$, $\mathcal{A} \subset \mathcal{U}$, $\mathcal{B} \subset \mathcal{V}$. Any metric space $\mathcal{M}$ is completely normal (see, for example, [43]).

Proposition 3.6 *Suppose that $\mathcal{M}$ is a completely normal topological space and $\mathcal{Z} \subset \mathcal{M}$ is a subset. Then* $\operatorname{ind} \mathcal{Z} \leq n$ *if and only if any point from $\mathcal{Z}$ has arbitrary* *small neighborhoods in $\mathcal{M}$ such that the intersection of their boundaries with $\mathcal{Z}$ has dimension $\leq n-1$.*

Proof Suppose that $\mathcal{Z}$ satisfies the conditions of the proposition. Let us show that in this case $\operatorname{ind}\mathcal{Z} \leq n$. Let p be an arbitrary point of $\mathcal{Z}$, $\mathcal{U}'$ be a neighborhood of this point in $\mathcal{Z}$. Then there exists a neighborhood $\mathcal{U}$ of the point p in $\mathcal{M}$ such that $\mathcal{U}' = \mathcal{U} \cap \mathcal{Z}$. Consequently, in $\mathcal{M}$ there can be found an open set $\mathcal{V}$ such that

$$p \in \mathcal{V} \subset \mathcal{U}, \quad \operatorname{ind}(\partial\mathcal{V} \cap \mathcal{Z}) \leq n-1.$$

Let $\mathcal{V}' := \mathcal{V} \cap \mathcal{Z}$. Then $\mathcal{V}'$ is an open set in $\mathcal{Z}$, $p \in \mathcal{V}' \subset \mathcal{U}'$. Denote by $\mathcal{B}$ and $\mathcal{B}'$ the boundaries of $\mathcal{V}$ in $\mathcal{M}$ and the boundaries of a set $\mathcal{V}'$ in $\mathcal{Z}$, respectively. It is clear that $\mathcal{B}' \subset \mathcal{B} \cap \mathcal{Z}$. This means that $\operatorname{ind}\mathcal{B}' \leq n-1$ and it follows that $\operatorname{ind}\mathcal{Z} \leq n$.

Suppose now that $\operatorname{ind}\mathcal{Z} \leq n$ and let us show that $\mathcal{Z}$ satisfies the conditions of the proposition. Let $p \in \mathcal{Z}$, $\mathcal{U}$ be a neighborhood of this point in $\mathcal{M}$. Then $\mathcal{U}' = \mathcal{U} \cap \mathcal{Z}$ is a neighborhood of p in $\mathcal{Z}$. Therefore a neighborhood $\mathcal{V}'$ of the point p in $\mathcal{Z}$ can be found such that

$$p \in \mathcal{V}' \subset \mathcal{U}', \quad \operatorname{ind}\mathcal{B}' \leq n-1,$$

where $\mathcal{B}'$ is a boundary of $\mathcal{V}'$ in $\mathcal{Z}$. None of the disjunct sets $\mathcal{V}'$ and $\mathcal{Z} \setminus \overline{\mathcal{V}}'$ contains limit points of the other set. Since $\mathcal{M}$ is completely normal, there can be found an open set $\mathcal{W}$ such that

$$\mathcal{V}' \subset \mathcal{W}, \quad \mathcal{W} \cap (\mathcal{Z} \setminus \overline{\mathcal{V}}') = \emptyset.$$

Replacing, if it is necessary, $\mathcal{W}$ by the intersection $\mathcal{W} \cap \mathcal{U}$, one can suppose that $\mathcal{W} \subset \mathcal{U}$. The set $\overline{\mathcal{W}} \setminus \mathcal{W} = \partial\mathcal{W}$ does not contain any points from $\mathcal{V}'$ and from $\mathcal{Z} \setminus \overline{\mathcal{V}}'$. It follows that the intersection of $\mathcal{Z}$ with $\partial\mathcal{V}$ is contained in $\mathcal{B}'$ and, consequently (Proposition 3.3), has the small inductive dimension $\leq n-1$. Thus, the conditions of the proposition are satisfied. □

Let us now prove a proposition concerning the small inductive dimension of the union of two sets. As it was noted above (see Example 3.5), the small inductive dimension of $\mathcal{Z}_1 \cup \mathcal{Z}_2$ is in general not determined by the small inductive dimensions of the sets $\mathcal{Z}_1$ and $\mathcal{Z}_2$. Nevertheless the following proposition holds.

Proposition 3.7 *For any two subsets $\mathcal{Z}_1$, $\mathcal{Z}_2$ of the topological space $\mathcal{M}$ the inequality*

$$\operatorname{ind}(\mathcal{Z}_1 \cup \mathcal{Z}_2) \leq 1 + \operatorname{ind}\mathcal{Z}_1 + \operatorname{ind}\mathcal{Z}_2$$

is satisfied.

Proof Let us make use of a double induction on the dimensions of subsets $\mathcal{Z}_1$ and $\mathcal{Z}_2$. The proposition is obvious in the case of $\operatorname{ind}\mathcal{Z}_1 = \operatorname{ind}\mathcal{Z}_2 = -1$.

Let $\operatorname{ind}\mathcal{Z}_1 = m$, $\operatorname{ind}\mathcal{Z}_2 = n$. Suppose that the inequality holds if one of the two conditions

$$\begin{array}{llll} & \operatorname{ind} \mathcal{Z}_1 \leq m & \text{and} & \operatorname{ind} \mathcal{Z}_2 \leq n-1, \\ \text{or} & \operatorname{ind} \mathcal{Z}_1 \leq m-1 & \text{and} & \operatorname{ind} \mathcal{Z}_2 \leq n \end{array}$$

is satisfied. Let $p \in \mathcal{Z}_1 \cup \mathcal{Z}_2$. Suppose that $p \in \mathcal{Z}_1$. Let $\mathcal{U}$ be a neighborhood of p in $\mathcal{M}$. By Proposition 3.6 there can be found an open set $\mathcal{V}$ such that

$$p \in \mathcal{V} \subset \mathcal{U}, \quad \operatorname{ind}(\mathcal{W} \cap \mathcal{Z}_1) \leq m-1,$$

where $\mathcal{W}$ is the boundary of $\mathcal{V}$. Since $\mathcal{W} \cap \mathcal{Z}_2$ is a subset of $\mathcal{Z}_2$, it follows that $\operatorname{ind}(\mathcal{W} \cap \mathcal{Z}_2) \leq n$. By the induction hypothesis we have

$$\operatorname{ind}\left[\mathcal{W} \cap (\mathcal{Z}_1 \cup \mathcal{Z}_2)\right] \leq m+n.$$

By Proposition 3.6 it can be concluded that

$$\operatorname{ind}(\mathcal{Z}_1 \cup \mathcal{Z}_2) \leq m+n+1.$$

□

The following result is important for deducing a small inductive dimension estimate of a minimal set of a differential equation.

Proposition 3.8 *Let* $\mathcal{Z} \subset \mathbb{R}^n$. *Then* $\operatorname{ind} \mathcal{Z} = n$ *if and only if* $\mathcal{Z}$ *contains an inner point.*

In order to prove this proposition we need a lemma, whose proof will be omitted (see [24]).

Lemma 3.1 *For any two countable and dense subsets* $\mathcal{A}$ *and* $\mathcal{B}$ *in* $\mathbb{R}^n$ *there exists a homeomorphism of the space* $\mathbb{R}^n$, *which maps* $\mathcal{A}$ *on* $\mathcal{B}$.

Proof of Proposition 3.8 Suppose that p is an inner point of $\mathcal{Z}$. Let us choose $\varepsilon > 0$ such that $\mathcal{B}_\varepsilon(p) \subset \mathcal{Z}$. But $\mathcal{B}_\varepsilon(p)$ is obviously homeomorphic to $\mathbb{R}^n$ and, consequently, $\operatorname{ind} \mathcal{B}_\varepsilon(p) = n$. Therefore $\operatorname{ind} \mathcal{Z} = n$.

Let $\operatorname{ind} \mathcal{Z} = n$. Now let us show that $\mathcal{Z}$ has an inner point. Suppose the opposite. Then the supplement of the set $\mathcal{Z}$, which we denote by $\mathcal{G}$, is dense in $\mathbb{R}^n$. Since $\mathbb{R}^n$ can be represented as the union of rational balls (i.e. balls whose radii and coordinates of their centers are rational), it follows that having considered non-empty intersections of such balls with the set $\mathcal{G}$ and having chosen one point from $\mathcal{G}$ in each ball, we obtain a countable dense set in $\mathcal{G}$. Denote it by $\mathcal{G}_0$. From the density of $\mathcal{G}$ in $\mathbb{R}^n$ it follows that $\mathcal{G}_0$ is also dense in $\mathbb{R}^n$. By Lemma 3.1 the sets $\mathbb{R}^n \setminus \mathcal{G}_0$ and $\mathbb{R}^n \setminus \mathcal{Q}^n$ are homeomorphic. Here $\mathcal{Q}^n$ is a subset of $\mathbb{R}^n$, consisting of all points having rational coordinates. Having used the notation of Example 3.6, we represent the complement to the set $\mathcal{Q}^n$ as a union of zero-dimensional sets

$$\mathbb{R}^n \setminus \mathcal{Q}^n = \mathcal{Q}_0^n \cup \mathcal{Q}_1^n \cup \cdots \cup \mathcal{Q}_{n-1}^n.$$

Since each $\mathcal{Q}_j^n$ is zero-dimensional, by Proposition 3.7 we obtain $\operatorname{ind}(\mathbb{R}^n \setminus \mathcal{Q}^n) \leq n-1$. Further, we have

$$\operatorname{ind}\mathcal{Z} = \operatorname{ind}(\mathbb{R}^n \setminus \mathcal{G}) \leq \operatorname{ind}(\mathbb{R}^n \setminus \mathcal{G}_0) = \operatorname{ind}(\mathbb{R}^n \setminus \mathcal{Q}^n) \leq n-1.$$

Thus, we have arrived at a contradiction to the assumption that $\operatorname{ind}\mathcal{Z} = n$. This contradiction concludes the proof. □

Let us formulate two more propositions (see the proof in [2]).

Proposition 3.9 *For a topological space* $\mathcal{M} \neq \emptyset$ *we have* $\operatorname{ind}\mathcal{M} \leq n$ *if and only if* $\mathcal{M}$ *can be represented as the union of* $(n+1)$ *subspaces, i.e.*

$$\mathcal{M} = \cup_{j=0}^{n}\mathcal{M}_j$$

and $\operatorname{ind}\mathcal{M}_j = 0,\ j = 0, 1, \ldots, n.$

Example 3.7 In the notation of Example 3.6 we have the decomposition

$$\mathbb{R}^n = \cup_{j=0}^{n}\mathcal{Q}_j^n$$

with $\operatorname{ind}\mathcal{Q}_j^n = 0,\ j = 0, 1, \ldots, n.$

Proposition 3.10 *Suppose that* $(\mathcal{M}, \rho)$ *and* $(\mathcal{M}', \rho')$ *are two separable metric spaces one of which is compact. Then* $\operatorname{ind}(\mathcal{M} \times \mathcal{M}') \leq \operatorname{ind}\mathcal{M} + \operatorname{ind}\mathcal{M}'$.

3.1.2 The Covering Dimension

Let us consider now the definition of a topological dimension which is based on the use of covers of the given space. For the class of separable metric spaces this definition gives the same value as the inductive dimension. The proof of this fact can be found in the monograph [2]. The contents of Sect. 3.1.2 follows the standard representation [12, 24].

Let $\mathcal{M}$ be a topological space. Let $\mathfrak{U}$ be some set of subsets of $\mathcal{M}$. We say that $\mathfrak{U}$ is a *cover* of $\mathcal{M}$, if for any point $p \in \mathcal{M}$ there exists a subset $\mathcal{U} \in \mathfrak{U}$ such that $p \in \mathcal{U}$. A cover $\mathfrak{U}$ is said to be *open* (*closed*) if all its sets are open (closed).

Definition 3.2 A cover $\mathfrak{V}$ is called a *refinement* of $\mathfrak{U}$ (we say also that $\mathfrak{V}$ refines $\mathfrak{U}$) if for any $\mathcal{V} \in \mathfrak{V}$ there exists $\mathcal{U} \in \mathfrak{U}$ such that $\mathcal{V} \subset \mathcal{U}$. The property, that $\mathfrak{V}$ is a refinement of $\mathfrak{U}$, is written as $\mathfrak{U} \prec \mathfrak{V}$.

Definition 3.3 Let $n \geq 0$ be an integer number. We shall say that a cover $\mathfrak{U}$ has the *multiplicity* $\leq n$ if any $n+1$ sets from $\mathfrak{U}$ have empty intersection. If a cover $\mathfrak{U}$ has the multiplicity $\leq n$ but has not the multiplicity $\leq n-1$, then we say that $\mathfrak{U}$ has the multiplicity n.

Definition 3.4 For each topological space $\mathcal{M}$ the *covering dimension* $\operatorname{Cov} \mathcal{M}$ is a value which is characterized in the following way:

(1) $\operatorname{Cov} \mathcal{M} \leq n$ if any finite open cover of the space $\mathcal{M}$ has a finite open refinement with a multiplicity $\leq n+1$;
(2) $\operatorname{Cov} \mathcal{M} = n$ if the inequality $\operatorname{Cov} \mathcal{M} \leq n$ is true and $\operatorname{Cov} \mathcal{M} \leq n-1$ is not true;
(3) $\operatorname{Cov} \mathcal{M} = \infty$ if $\operatorname{Cov} \mathcal{M} \leq n$ is not true for any n.

It follows immediately from Definition 3.4 that if $\mathcal{M} = \emptyset$, then $\operatorname{Cov} \mathcal{M} = -1$ (since an order of a cover consisting of an empty set is zero).

Let us suppose that $(\mathcal{M}, \rho)$ is a compact metric space, i.e., consider a situation which is given in many applications.

We now prove a useful lemma which will be used not only in this section but also in section devoted to the topological entropy.

Lemma 3.2 *For an arbitrary open cover $\mathfrak{U}$ of the compact metric space $\mathcal{M}$ there exists a number $\varepsilon > 0$ such that any subset $\mathcal{U}$ of the space $\mathcal{M}$ having diameter less than ε is contained in some elements of the cover $\mathfrak{U}$.*

Proof If such number ε does not exist, then for any natural number n a subset $\mathcal{U}_n \subset \mathcal{M}$ with the diameter $< 1/n$ can be found, which is not contained in the elements of the cover $\mathfrak{U}$. When a point p_n is chosen arbitrarily in $\mathcal{U}_n$, consider the set $\{p_n\}$. Since $\mathcal{M}$ is compact, we see that this set has at least one limit point p_0. Let be $p_0 \in \mathcal{U}_0 \in \mathfrak{U}$ and let δ be a distance from p_0 to $\mathcal{M} \setminus \mathcal{U}_0$. If $n > 2/\delta$ and $\rho(p_0, p_n) < \delta/2$, then

$$\rho(p_0, p) \leq \rho(p_0, p_n) + \rho(p_n, p) \leq \frac{\delta}{2} + \frac{1}{n} < \delta$$

for any point $p \in \mathcal{U}_n$. It means, contrary to the supposition, that $\mathcal{U}_n \subset \mathcal{U}_0$. □

The number ε, which can be found in Lemma 3.2 is called the *Lebesgue number* of the cover $\mathfrak{U}$.

A cover $\mathfrak{U}$ of a metric space $\mathcal{M}$ is called an *ε-cover* if the diameter of each set from $\mathfrak{U}$ is less than ε.

Proposition 3.11 *The inequality* $\operatorname{Cov} \mathcal{M} \leq n$ *is true if and only if for any $\varepsilon > 0$ there exists a finite open ε-cover of the space $\mathcal{M}$ having the multiplicity $\leq n+1$.*

Proof According to Lemma 3.2, if ε is the Lebesgue number of the cover $\mathfrak{U}$, then any ε-cover $\mathfrak{V}$ is inscribed in $\mathfrak{U}$. □

Proposition 3.12 *The inequality* $\operatorname{Cov} \mathcal{M} \leq n$ *is true if and only if for any $\varepsilon > 0$ there exists a finite closed ε-cover of the space $\mathcal{M}$ with multiplicity $\leq n+1$.*

Proposition 3.12 immediately follows from Proposition 3.11 and the following lemma.

Lemma 3.3 *Let $\varepsilon > 0$ be an arbitrary positive number. For the existence of a finite open ε-cover of the space $\mathcal{M}$ with multiplicity $\leq n$ the existence of the finite closed ε-cover of the space $\mathcal{M}$ with multiplicity $\leq n$ is necessary and sufficient.*

Proof Let $\mathfrak{U} = \{\mathcal{U}_1, \mathcal{U}_2, \ldots, \mathcal{U}_m\}$ be an open ε-cover of the space $\mathcal{M}$ with multiplicity $\leq n$. Let us construct a finite closed ε-cover with multiplicity $\leq n$. For this purpose we consider the supplement to the open set $\mathcal{U}_2 \cup \mathcal{U}_3 \cup \cdots \cup \mathcal{U}_m$. Denote this closed set by $\mathcal{C}_1$. It is obvious that it can be covered by the set $\mathcal{U}_1$ and there can be found an open set $\mathcal{V}_1$ such that $\mathcal{C}_1 \subset \overline{\mathcal{V}}_1 \subset \mathcal{U}_1$. Assume that the open sets $\mathcal{V}_1, \mathcal{V}_2, \ldots, \mathcal{V}_{k-1}$ be already chosen. Let us construct a set $\mathcal{V}_k$. Denote by $\mathcal{C}_k$ the closed set, which is the supplement to the union

$$\mathcal{V}_1 \cup \cdots \cup \mathcal{V}_{k-1} \cup \mathcal{U}_{k+1} \cup \cdots \cup \mathcal{U}_m.$$

It is clear that $\mathcal{C}_k$ can be covered by the set $\mathcal{U}_k$ and it is possible to find an open set $\mathcal{V}_k$ such that $\mathcal{C}_k \subset \overline{\mathcal{V}}_k \subset \mathcal{U}_k$. Obviously, the cover $\{\overline{\mathcal{V}}_1, \overline{\mathcal{V}}_2, \ldots, \overline{\mathcal{V}}_m\}$ is a desirable closed ε-cover of multiplicity $\leq n$.

Conversely, assume that there exists a closed ε-cover $\mathfrak{U} = \{\mathcal{U}_1, \mathcal{U}_2, \ldots, \mathcal{U}_m\}$ of the space $\mathcal{M}$ with multiplicity $\leq n$. Let us show the existence of the corresponding open cover. For an arbitrary $\delta > 0$ denote by $\mathcal{U}_i(\delta)$ the open δ-neighborhoods of the sets $\mathcal{U}_i$, $i = 1, 2, \ldots, m$. We now choose an arbitrary subsystem

$$\mathfrak{S}_\delta = \{\overline{\mathcal{U}}_{i_1}(\delta), \overline{\mathcal{U}}_{i_2}(\delta), \ldots, \overline{\mathcal{U}}_{i_{n+1}(\delta)}\}, \quad 1 \leq i_1 < i_2 < \cdots < i_{n+1} \leq m.$$

Let us show that for sufficiently small δ the intersection of sets, which are a part of it, is empty. Indeed, suppose the opposite. Then for any natural number j and for $\delta_j = 1/j$ a point a_j can be found such that it belongs to all sets of the system $\mathfrak{S}_{\delta_j}$. By the compactness of the space $\mathcal{M}$ there exists a point a such that it is a limit point of the sequence $\{a_j\}$, and by virtue of the fact that the cover $\mathfrak{U}$ is closed, this limit point must belong to the intersection of the sets $\{\mathcal{U}_{i_1}, \mathcal{U}_{i_2}, \ldots, \mathcal{U}_{i_{n+1}}\}$. But this contradicts to the property that the multiplicity of $\mathfrak{U}$ is $\leq n$. The number of subsystems of the type $\mathfrak{S}_\delta$ is finite. Therefore, taking for δ_0 the least of numbers δ we obtain a cover $\{\overline{\mathcal{U}}_1(\delta_0), \overline{\mathcal{U}}_2(\delta_0), \ldots, \overline{\mathcal{U}}_m(\delta_0)\}$ of the space $\mathcal{M}$ with multiplicity $\leq n$. Since the diameter of each closed set $\mathcal{U}_i$ is less than ε, we can choose $\delta_0 > 0$ so small that the diameter of each set $\overline{\mathcal{U}}_i(\delta_0)$ will be, moreover, less then ε. It is clear that the cover $\{\mathcal{U}_1(\delta_0), \mathcal{U}_2(\delta_0), \ldots, \mathcal{U}_m(\delta_0)\}$ is the desired open ε-cover of multiplicity $\leq n$. □

Propositions 3.11 and 3.12 permit us to reformulate Definition 3.4 for compact metric spaces:

Definition 3.5 Suppose that $(\mathcal{M}, \rho)$ is a compact metric space. Then the *covering dimension* $\operatorname{Cov}\mathcal{M}$ is a number which is characterized by the following properties:

(1) $\operatorname{Cov}\mathcal{M} \leq n$ if for any $\varepsilon > 0$ there exists an open (closed) ε-cover of the space $\mathcal{M}$ with multiplicity $\leq n + 1$;

(2) $\operatorname{Cov}\mathcal{M} = n$ if $\operatorname{Cov}\mathcal{M} \le n$ and for some $\varepsilon > 0$ the space $\mathcal{M}$ does not have an open (closed) ε-cover of multiplicity $\le n$;
(3) $\operatorname{Cov}\mathcal{M} = \infty$ if $\operatorname{Cov}\mathcal{M} \le n$ is not true for any n.

Example 3.8 Consider the *Cantor set* $\mathcal{Z} := \bigcap_{j=0}^{\infty} \mathcal{Z}_j$, where

$$\begin{aligned}
\mathcal{Z}_0 &:= [0,\ 1],\\
\mathcal{Z}_1 &:= [0,\ 1]\setminus(1/3,\ 2/3) = [0,\ 1/3] \cup [2/3,\ 1],\\
\mathcal{Z}_2 &:= [0,\ 1/9] \cup [2/9,\ 3/9] \cup [6/9,\ 7/9] \cup [8/9,\ 1],\\
&\ \vdots
\end{aligned}$$

Thus we have $\mathcal{Z}_0 \supset \mathcal{Z}_1 \supset \cdots$ and each $\mathcal{Z}_n$, $n = 1, 2, \ldots$, is the union of 2^n intervals of length $1/3^n$ (Fig. 3.1). We shall show that $\operatorname{Cov}\mathcal{Z} = 0$.

Let $\varepsilon > 0$ be an arbitrary number. Choose the integer number $j > 0$ such that $3^{-j} < \varepsilon$. Then the closed intervals forming the set $\mathcal{Z}_j$ give a closed ε-covering of multiplicity ≤ 1. Therefore $\operatorname{Cov}\mathcal{Z} \le 0$. The inverse inequality is obvious since $\mathcal{Z} \ne \emptyset$.

We finish this section with a proposition which is proved, for example, in [2].

Proposition 3.13 *Suppose that* $(\mathcal{M}, \rho)$ *is a separable metric space. Then* $\operatorname{ind}\mathcal{M} = \operatorname{Cov}\mathcal{M}$.

This leads to the following:

Definition 3.6 Suppose $(\mathcal{M}, \rho)$ is a separable metric space. The *topological dimension* $\dim_T \mathcal{M}$ of $\mathcal{M}$ is given by $\dim_T \mathcal{M} := \operatorname{ind}\mathcal{M} = \operatorname{Cov}\mathcal{M}$.

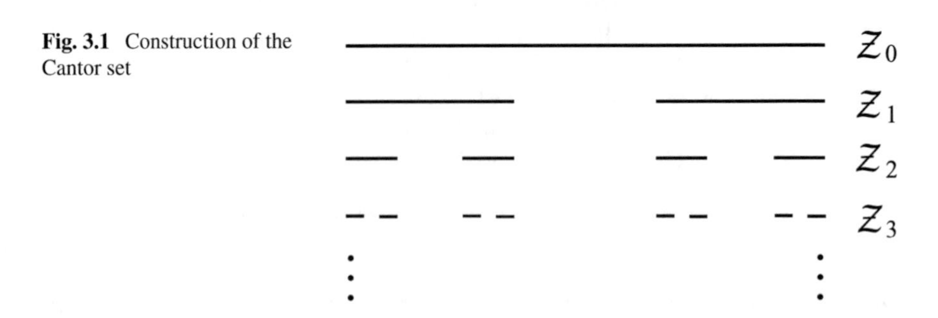

Fig. 3.1 Construction of the Cantor set

3.2 Hausdorff and Fractal Dimensions

3.2.1 The Hausdorff Measure and the Hausdorff Dimension

This subsection presents the main properties of the Hausdorff measure and the Hausdorff dimension introduced in [4, 20]. The treatment of this subsection may be found in the books [13, 14, 40, 44].

Suppose that $(\mathcal{M}, \rho)$ is a metric space, $\mathcal{Z}$ is an arbitrary subset of $\mathcal{M}$ and $d \geq 0$, $\varepsilon > 0$ are numbers. Let us cover $\mathcal{Z}$ by at most a set of countable many balls $\mathcal{B}_{r_j}$ of radii $r_j \leq \varepsilon$ and define the Hausdorff *premeasure* at level ε and of order d

$$\mu_H(\mathcal{Z}, d, \varepsilon) := \inf \left\{ \sum_{j \geq 1} r_j^d \,\middle|\, r_j \leq \varepsilon, \mathcal{Z} \subset \bigcup_{j \geq 1} \mathcal{B}_{r_j} \right\},$$

where the infimum is taken over all such countable ε-covers of $\mathcal{Z}$ and the convention that $\inf \emptyset = +\infty$. It is obvious that for fixed $\mathcal{Z}$ and d the function $\mu_H(\mathcal{Z}, d, \varepsilon)$ does not decrease with decreasing ε. Thus there exists the limit (which may be infinite)

$$\mu_H(\mathcal{Z}, d) := \lim_{\varepsilon \to 0+0} \mu_H(\mathcal{Z}, d, \varepsilon) = \sup_{\varepsilon > 0} \mu_H(\mathcal{Z}, d, \varepsilon).$$

In the following proposition we show that $\{\mu_H(\cdot, d)\}$ is a family of metric outer measures on $\mathcal{M}$.

Proposition 3.14 *For any fixed $d \geq 0$ the function $\mu_H(\cdot, d)$ is a metric outer measure on $\mathcal{M}$, i.e., it possesses the following properties:*

(1) $\mu_H(\emptyset, d) = 0$;
(2) $\mu_H(\mathcal{Z}_1, d) \leq \mu_H(\mathcal{Z}_2, d)$ *for all sets* $\mathcal{Z}_1 \subset \mathcal{Z}_2 \subset \mathcal{M}$;
(3) $\mu_H(\bigcup_{j \geq 1} \mathcal{Z}_j, d) \leq \sum_{j \geq 1} \mu_H(\mathcal{Z}_j, d)$ *for all sets* $\mathcal{Z}_j \subset \mathcal{M},\ j = 1, 2, \ldots$.

Proof Let us note that for any $d \geq 0$ and $\varepsilon > 0$ the function $\mu_H(\cdot, d, \varepsilon)$ is a metric outer measure on $\mathcal{M}$: We have $\mu_H(\emptyset, d, \varepsilon) = 0$ and for arbitrary sets $\mathcal{Z}_1 \subset \mathcal{Z}_2 \subset \mathcal{M}$ it follows that $\mu_H(\mathcal{Z}_1, d, \varepsilon) \leq \mu_H(\mathcal{Z}_2, d, \varepsilon)$ since any ε-cover of $\mathcal{Z}_2$ is also an ε-cover of $\mathcal{Z}_1$. In order to prove the third property of a metric outer measure we consider two cases.

Case 1: There exists an $i_0 \geq 1$ with $\mu_H(\mathcal{Z}_{i_0}, d, \varepsilon) = +\infty$. It follows that $\mu_H(\bigcup_{i \geq 1} \mathcal{Z}_i, d, \varepsilon) \geq \mu_H(\mathcal{Z}_{i_0}, d, \varepsilon) = +\infty$.

Case 2: For all $i \geq 1$ we have $\mu_H(\mathcal{Z}_i, d, \varepsilon) < +\infty$. Suppose that $\delta > 0$ is an arbitrary number. For any $i \geq 1$ we choose a countable ε-cover of $\mathcal{Z}_i$ consisting of balls with radij $r_{i,j} \leq \varepsilon$ for $j = 1, 2, \ldots$, such that $\mathcal{Z}_i \subset \bigcup_{j \geq 1} \mathcal{B}_{r_{i,j}}$ and $\sum_{j \geq 1} r_{i,j}^d \leq \mu_H(\mathcal{Z}_i, d, \varepsilon) + 2^{-i}\delta$. It follows that $\bigcup_{i \geq 1} \bigcup_{j \geq 1} \mathcal{B}_{r_{i,j}}$ is a countable ε-cover of $\bigcup_{i \geq 1} \mathcal{Z}_i$ and $\mu_H(\bigcup_{i \geq 1} \mathcal{Z}_i, d, \varepsilon) \leq$

$\sum\limits_{i,j\geq 1} r_{i,j}^d \leq \sum\limits_{i\geq 1}(\sum\limits_{j\geq 1} r_{i,j}^d) \leq \sum\limits_{i\geq 1}\mu_{_H}(\mathcal{Z}_i, d, \varepsilon) + \sum\limits_{i\geq 1} 2^{-i}\delta \leq \sum\limits_{i\geq 1}\mu_{_H}(\mathcal{Z}_i, d, \varepsilon) + \delta$. To finish the proof of the proposition let us note that if $\mathfrak{M} = \{\mathfrak{m}\}$ is a family of metric outer measures on $\mathcal{M}$ than $\mathcal{Z} \subset \mathcal{M} \mapsto \mu(\mathcal{Z}) := \sup\limits_{\mathfrak{m}\in\mathfrak{M}} \mathfrak{m}(\mathcal{Z})$ defines also a metric outer measure on $\mathcal{M}$. In the present situation this family of metric outer measures is given by $\mathfrak{M} := \{\mu_{_H}(\cdot, d, \varepsilon)\}$. □

Definition 3.7 The function $\mu_{_H}(\cdot, d)$ is called *Hausdorff-d-measure* on $\mathcal{M}$.

Proposition 3.15 *(1) For any fixed $\mathcal{Z} \subset \mathcal{M}$ the function $\mu_{_H}(\mathcal{Z}, \cdot)$ has exactly one critical value*
$d_{cr} = d_{cr}(\mathcal{Z}) \in [0, \infty]$ such that

$$\begin{aligned} \mu_{_H}(\mathcal{Z}, d) = \infty \quad &\textit{for any} \ \ 0 \leq d < d_{cr} \\ \textit{and} \ \ \mu_{_H}(\mathcal{Z}, d) = 0 \quad &\textit{for any} \ \ d > d_{cr}\,. \end{aligned} \tag{3.2}$$

(2) If $\mathcal{M} = \mathbb{R}^n$, then $d_{cr}(\mathbb{R}^n) \leq n$.

Proof (1) We distinguish two cases.

Case 1: There exists a value $d \geq 0$ with $\mu_{_H}(\mathcal{Z}, d) < +\infty$. Take arbitrary $\delta > 0$ and $\varepsilon > 0$ and consider $\mu_{_H}(\mathcal{Z}, d+\delta, \varepsilon) = \inf\{\sum\limits_{i\geq 1} r_i^{d+\delta}, r_i \leq \varepsilon, \bigcup \mathcal{B}_{r_i} \supset \mathcal{Z}\}$

$\leq \varepsilon^\delta \inf\left\{\sum\limits_{i\geq 1} r_i^d, r_i \leq \varepsilon, \bigcup \mathcal{B}_{r_i} \supset \mathcal{Z}\right\} = \varepsilon^\delta \mu_{_H}(\mathcal{Z}, d, \varepsilon)$. It follows that

$$\mu_{_H}(\mathcal{Z}, d') = 0 \quad \text{for all} \quad d' > d. \tag{3.3}$$

Thus, we can define the critical value satisfying (3.2) as

$$d_{cr}(\mathcal{Z}) := \inf\{d \mid \mu_{_H}(\mathcal{Z}, d) < +\infty\}.$$

Case 2: For any $d \geq 0$ we have $\mu_{_H}(\mathcal{Z}, d) = +\infty$. Here we define $d_{cr}(\mathcal{Z}) := +\infty$.

(2) Let $\mathcal{M} = \mathbb{R}^n$. Let us show that in this case $d_{cr} \leq n$. Denote by $\mathcal{C}$ an arbitrary cube in $\mathbb{R}^n$ with edges of length 1. Take a natural number $k > 0$ and divide $\mathcal{C}$ into k^n cubes with edges of length $1/k$. Each of these cubes is contained in a ball of radius $\frac{1}{2}k^{-1}\sqrt{n}$. Thus, if $\varepsilon \geq \frac{1}{2}k^{-1}\sqrt{n}$, then

$$\mu_{_H}(\mathcal{C}, n, \varepsilon) \leq k^n\left(\frac{1}{2}k^{-1}\sqrt{n}\right)^n = 2^{-n}n^{n/2}.$$

It follows that $\mu_{_H}(\mathcal{C}, n) < \infty$ and by (3.3) we have $\mu_{_H}(\mathcal{C}, d) = 0$ for all $d > n$. Since $\mathbb{R}^n$ can be represented as a countable union of such cubes, we obtain $\mu_{_H}(\mathbb{R}^n, d) = 0$ and, consequently, $\mu_{_H}(\mathcal{Z}, d) = 0$ (here the fact that $\mu_{_H}(\cdot, d)$ is an

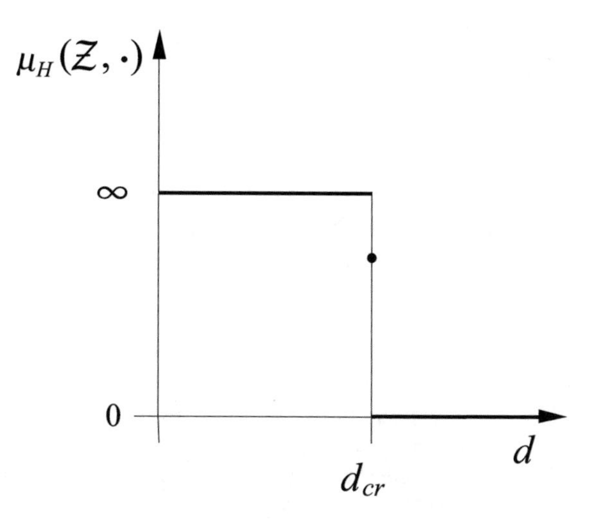

Fig. 3.2 The critical value of the Hausdorff measure

outer measure has been used). Then $d > d_{cr}$ and, since d is an arbitrary number greater than n, we get $d_{cr} \leq n$. □

Definition 3.8 For any set $\mathcal{Z} \subset \mathcal{M}$ the value

$$\dim_H \mathcal{Z} = d_{cr}(\mathcal{Z}) = \inf\{d \geq 0 \mid \mu_H(\mathcal{Z}, d) = 0\} = \sup\{d \geq 0 \mid \mu_H(\mathcal{Z}, d) = \infty\}$$

is called the *Hausdorff dimension* of the set $\mathcal{Z}$.

Remark 3.1 In the above definition of the Hausdorff dimension one can choose coverings of the set $\mathcal{Z}$ by at most countable many sets $\mathcal{U}_j$ of diameter $\leq \varepsilon$ to obtain the same value of the Hausdorff dimension.

By using Proposition 3.15, we can draw the graph of the function $\mu_H(\mathcal{Z}, \cdot)$ on $\mathbb{R}_+$ (for the case $0 < d_{cr} < +\infty$ and $\mu_H(\mathcal{Z}, d_{cr}) < +\infty$ see Fig. 3.2)

In practical dimension estimations the following properties of the Hausdorff dimension of a set $\mathcal{Z} \subset \mathcal{M}$ (metric space) are useful:

(P1) $\mu_H(\mathcal{Z}, d) > 0 \Rightarrow \dim_H \mathcal{Z} \geq d$;
(P2) $\dim_H \mathcal{Z} > d \Rightarrow \mu_H(\mathcal{Z}, d) = +\infty$;
(P3) $\mu_H(\mathcal{Z}, d) < +\infty \Rightarrow \dim_H \mathcal{Z} \leq d$;
(P4) $\dim_H \mathcal{Z} < d \Rightarrow \mu_H(\mathcal{Z}, d) = 0$.

Let us consider two examples in which the Hausdorff dimension can be immediately calculated by Definition 3.8.

Example 3.9 Consider again the Cantor set $\mathcal{Z}$ from Example 3.8

Let us show that $\mu_H(\mathcal{Z}, d) = 1$, where $d = \log 2/\log 3 = 0.6309\ldots$ and, consequently, $\dim_H \mathcal{Z} = \log 2/\log 3$.

Since $\mathcal{Z}$ can be covered by 2^j closed intervals of length 3^{-j}, which form the set $\mathcal{Z}_j$, taking into account that $3^d = 2$, we have

$$\mu_H(\mathcal{Z}, d, 3^{-1}) \leq 2^j (3^{-j})^d = 2^j 2^{-j} = 1.$$

Thus, we get $\mu_H(\mathcal{Z}, d) \leq 1$ as $j \to +\infty$.

We now prove the inverse inequality. For this purpose we need to show that if $\mathfrak{J}$ is an arbitrary set of intervals covering $\mathcal{Z}$, then

$$1 \leq \sum_{\mathcal{J} \in \mathfrak{J}} |\mathcal{J}|^d. \tag{3.4}$$

Here $|\cdot|$ is the length of the interval $\mathcal{J}$. Because of the compactness of $\mathcal{Z}$, it is sufficient to prove that inequality (3.4) is satisfied under the assumption that $\mathfrak{J}$ consists of the finite number of intervals. Let $\mathcal{J}$ be some interval from $\mathfrak{J}$. Without loss of generality it can be supposed that $\mathcal{J}$ contains intervals, a pair J, J', which take part in the construction of $\mathcal{Z}$. Further we shall suppose that in $\mathcal{J}$ the intervals J and J' are the intervals of maximal length. Let $J'' = \mathcal{J} \setminus (J \cup J')$. Obviously, $|J''| \geq |J|\,|J'|$. Therefore,

$$|\mathcal{J}|^d = (|J| + |J'| + |J''|)^d \geq \left(\frac{3}{2}(|J| + |J'|)\right)^d = 2\left(\frac{1}{2}|J|^d + \frac{1}{2}|J'|^d\right) \geq |J|^d + |J'|^d.$$

Here we make use of the facts that the function $t \mapsto t^d$ is concave and $3^d = 2$. Thus, the replacement of $\mathcal{J}$ in (3.4) by two closed intervals J and J' does not increase the sum in (3.4). Continuing in the same way, i.e., replacing $\mathcal{J}$ by intervals J and J', after a finite number of steps we obtain a covering of $\mathcal{Z}$ by intervals of the same length, for example, 3^{-j}. This covering must include all intervals which compose $\mathcal{Z}_j$ and therefore we have

$$\sum_{\mathcal{J} \in \mathfrak{J}} |\mathcal{J}|^d \geq 2^j (3^{-j})^d = 1.$$

Remark 3.2 It remains to note that the Cantor set is the simplest example of a set having a non-integer Hausdorff dimension, the value of which can be computed precisely. Sets of a more complicated structure, such as the strange attractor of a dynamical system, also have non-integer Hausdorff dimension, but in this case often there can be found only estimates of its dimension.

For 'regular' sets the Hausdorff dimension has an integer value which is intuitively clear. This can be shown by the following example.

Example 3.10 Let in $\mathbb{R}^3$ a smooth surface $\mathcal{S}$ be given by

$$\mathcal{S} = \{(x, y, z) \mid z = f(x, y), \quad (x, y) \in \mathcal{C}\}$$

where $f(\cdot,\cdot)$ is a continuously differentiable function on the unit square $\mathcal{C}$ of the plane. We shall show that $\dim_H \mathcal{S} = 2$.

Let $\{\mathcal{B}_{r_j}\}$ be an ε-cover of the surface $\mathcal{S}$. The projection of balls $\mathcal{B}_{r_j}$ on the (x, y)-plane form an ε-cover of the square $\mathcal{C}$ consisting of discs. A ball $\mathcal{B}_{r_j}$ of radius r_j is projected on a disc having the area πr_j^2. Thus, we have $\sum \pi r_j^2 \geq 1$, i.e., $\sum r_j^2 \geq \pi^{-1}$. Consequently, $\mu_H(\mathcal{S}, 2) > 0$ and $\dim_H \mathcal{S} \geq 2$.

Let us now show that $\dim_H \mathcal{S} \leq 2$. Since f is continuously differentiable, there can be found a number $M > 0$ such that

$$|f(u') - f(u'')| \leq M|u' - u''|, \qquad \forall u', u'' \in \mathcal{C}. \tag{3.5}$$

We choose an integer number $k > 0$ and divide in an obvious way the square $\mathcal{C}$ into k^2 equal squares with sides $1/k$. By (3.5), each part of the surface $\mathcal{S}$ being placed above by such small square, may be embedded into a cube with the edge length M/k, which is included in a ball of radius $\frac{\sqrt{3}}{2}M/k$. Thus, the surface $\mathcal{S}$ is covered by cubes and, if $\frac{\sqrt{3}}{2}M/k \leq \varepsilon$, then

$$\mu_H(\mathcal{S}, 2, \varepsilon) \leq k^2 \left(\frac{\sqrt{3}}{2}M/k\right)^2 = \frac{9}{4}M^2.$$

It follows that $\mu_H(\mathcal{S}, 2) < \infty$ and, consequently, $\dim_H \mathcal{S} \leq 2$.

The following proposition gives some basic properties of the Hausdorff dimension.

Proposition 3.16 *Suppose $(\mathcal{M}, \rho)$ is a metric space. Then the following statements are true:*

(1) $\dim_H \emptyset = 0$;
(2) $\dim_H \mathcal{Z} \geq 0$ *for any* $\mathcal{Z} \subset \mathcal{M}$; *if* $\mathcal{M} = \mathbb{R}^n$ *and* $\mathcal{Z} \subset \mathbb{R}^n$ *then* $\dim_H \mathcal{Z} \leq n$;
(3) $\dim_H \mathcal{Z}_1 \leq \dim_H \mathcal{Z}_2$ *for any sets* $\mathcal{Z}_1 \subset \mathcal{Z}_2 \subset \mathcal{M}$;
(4) $\dim_H(\bigcup_{j\geq 1} \mathcal{Z}_j) = \sup_{j\geq 1} \dim_H \mathcal{Z}_j$ *for any sets* $\mathcal{Z}_j \subset \mathcal{M}$, $j = 1, 2, \ldots$;
(5) *If the set* $\mathcal{Z} \subset \mathcal{M}$ *is at most countable, then* $\dim_H \mathcal{Z} = 0$.

Proof The validity of (1)–(3) follows directly from the definition of the Hausdorff dimension. Assertion (5) results from (4).

It remains to show the validity of (4). Let $d' = \sup_j \dim \mathcal{Z}_j$. Suppose that $d' < \infty$ and we take an arbitrary $d > d'$. Then $\dim \mathcal{Z}_j < d$ and, consequently, $\mu_H(\mathcal{Z}_j, d) = 0$ for $j \geq 1$. Using the fact that $\mu_H(\cdot, d)$ is an outer measure, we obtain

$$\mu_H\Big(\bigcup_{j\geq 1} \mathcal{Z}_j, d\Big) \leq \sum_{j\geq 1} \mu_H(\mathcal{Z}_j, d) = 0.$$

Then $\dim_H(\bigcup_{j\geq 1} \mathcal{Z}_j) \leq d$. Therefore, by virtue of arbitrariness of $d > d'$, we have $\dim_H(\bigcup_{j\geq 1} \mathcal{Z}_j) \leq d'$. The validity of the inverse inequality follows directly from (3), since $\mathcal{Z}_j \subset \bigcup_{i\geq 1} \mathcal{Z}_i$ for any $j \geq 1$. Thus, property (4) is proved under the assumption $d' < \infty$. The case $d' = \infty$ is trivial. □

In studying ordinary differential equations the change of variables is often used. The following proposition shows that the Hausdorff dimension of a compact set in the phase space of such a system is invariant with respect to a smooth transformation of the phase variables.

Proposition 3.17 *Suppose that $\mathcal{K} \subset \mathbb{R}^n$ is a compact set, $\mathcal{U}$ with $\mathcal{K} \subset \mathcal{U}$ is an open set and $\Phi \colon \mathcal{U} \subset \mathbb{R}^n \to \Phi(\mathcal{U})$ is a diffeomorphism. Then*

$$\dim_H \mathcal{K} = \dim_H \Phi(\mathcal{K}) .$$

Proof Let $d \geq 0$, $\delta > \varepsilon > 0$ be fixed numbers. Denote by $\mathcal{B}_r(u)$ the ball with radius r and center u in $\mathbb{R}^n$ and let $\mathcal{K}_\delta$ be a closed δ-neighborhood of $\mathcal{K}$ such that $\mathcal{K}_\delta \subset \mathcal{U}$.

Suppose that $\{\mathcal{B}_{r_j}(u_j)\}$ is an arbitrary ε-cover of $\mathcal{K}$. Then, since Φ is a diffeomorphism, there can be found a constant $M_1 > 0$ such that

$$|\Phi(u'') - \Phi(u')| \leq M_1 |u'' - u'| \quad \text{for all} \quad u', u'' \in \mathcal{K}_\delta .$$

In particular, for $u' \in \mathcal{B}_{r_j}(u_j)$ we have

$$|\Phi(u') - \Phi(u_j)| \leq M_1 |u' - u_j| \leq M_1 r_j .$$

Consequently, the set $\Phi(\mathcal{B}_{r_j}(u_j))$ is contained in the ball $\mathcal{B}_{M_1 r_j}(\Phi(u_j))$ and we have $\mathcal{K} \subset \bigcup_{j\geq 1} \mathcal{B}_{r_j}(u_j)$. It follows that $\Phi(\mathcal{K}) \subset \bigcup_{j\geq 1} \mathcal{B}_{M_1 r_j}(\Phi(u_j))$.

Thus, for an arbitrary ε-cover of $\mathcal{K}$ there can be found an $(M_1\varepsilon)$-cover of the set $\Phi(\mathcal{K})$. Since

$$\sum_j r_j^d = M_1^{-d} \sum_j (M_1 r_j)^d ,$$

we have

$$\mu_H(\mathcal{K}, d, \varepsilon) \geq M_1^{-d} \mu_H(\Phi(\mathcal{K}), d, M_1 \varepsilon).$$

Therefore

$$\mu_H(\mathcal{K}, d) \geq M_1^{-d} \mu_H(\Phi(\mathcal{K}), d).$$

Using the same line of reasoning as above, but taking the set $\Phi(\mathcal{K})$ vice $\mathcal{K}$ and mapping Φ^{-1} vice Φ, we obtain

$$\mu_H(\Phi(\mathcal{K}), d) \geq M_2^{-d} \mu_H(\mathcal{K}, d),$$

where $M_2 > 0$ is a corresponding constant for Φ^{-1}.

Finally, we get

$$M_1^{-d}\mu_H(\Phi(\mathcal{K}), d) \le \mu_H(\mathcal{K}, d) \le M_2^d \mu_H(\Phi(\mathcal{K}), d).$$

Therefore $\dim_H \mathcal{K} = \dim_H \Phi(\mathcal{K})$. □

Remark 3.3 Proposition 3.17 follows from a more general result. Suppose that $(\mathcal{M}, \rho)$ and $(\mathcal{M}', \rho')$ are metric spaces, $\Phi : \mathcal{M} \to \mathcal{M}'$ is a Lipschitz continuous map, i.e., there exists a constant $L > 0$ such that $\rho'(\Phi(u), \Phi(v)) \le L\rho(u, v)$ for any $u, v \in \mathcal{M}$. Then for any set $\mathcal{Z} \subset \mathcal{M}$ we have $\dim_H \Phi(\mathcal{Z}) \le \dim_H \mathcal{Z}$.

Further, everywhere in this section, we shall suppose that $\mathcal{M} = \mathbb{R}^n$.

Earlier we introduced the Hausdorff d-measure, based on covers by balls. The Hausdorff measure so defined is often called *spherical*. However, instead of spherical balls in the covers it is also possible to use balls of another type, as for example cubes.

Let $\mathcal{Z}$ be an arbitrary subset of $\mathbb{R}^n$ and $d \ge 0$, $\varepsilon > 0$ numbers. Cover the set $\mathcal{Z}$ by n-dimensional cubes having edges with length $l_j \le \varepsilon$ and denote

$$\widehat{\mu}_H(\mathcal{Z}, d, \varepsilon) = \inf \sum_j l_j^d,$$

where the infimum is taken over all such ε-covers of $\mathcal{Z}$. Let us define now the function $\widehat{\mu}_H(\mathcal{Z}, d)$ (*cubical Hausdorff d-measure*) in the same way as the spherical Hausdorff d-measure was introduced above.

Proposition 3.18 *For any $\mathcal{Z} \subset \mathbb{R}^n$ and $d \ge 0$ the equality*

$$\mu_H(\mathcal{Z}, d) = \widehat{\mu}_H(\mathcal{Z}, d)$$

holds.

Proof Any cube $\mathcal{C}$ with edges of length l can be included in a ball of radius $\sqrt{n}l/2$. On the other hand any ball of radius $l/2$ is contained in an included in this cube $\mathcal{C}$. Therefore the following inequalities

$$\mu_H(\mathcal{Z}, d, \sqrt{n}l/2) \le \widehat{\mu}_H(\mathcal{Z}, d, l) \le \mu_H(\mathcal{Z}, d, l/2)$$

are true. Going to the limit as $l \to 0$, we get the desirable equality. □

Taking into account Proposition 3.18, we shall use further the notation $\mu_H(\cdot, \cdot)$ both for the spherical and as cubical Hausdorff d-measure.

Denote by $\mu_L(\mathcal{Z}, n)$ the *n-dimensional Lebesgue measure* of a measurable set $\mathcal{Z} \subset \mathbb{R}^n$. From the following proposition it follows that any open set in $\mathbb{R}^n$ has the Hausdorff dimension n.

Proposition 3.19 *If $\mu_L(\mathcal{Z}, n) > 0$ then $\dim_H \mathcal{Z} = n$.*

Proof Let $\varepsilon > 0$ be fixed and $\{\mathcal{B}_{r_j}\}$ be a countable cover of $\mathcal{Z}$ by balls with radius $r_j \leq \varepsilon$ such that

$$\sum_j r_j^n < \mu_H(\mathcal{Z}, n) + \varepsilon.$$

Since $\mu_L(\mathcal{B}_{r_j}, n) = c_n r_j^n$, where c_n is a constant depending only on n (the volume of the n-dimensional unit ball), we have

$$\mu_L(\mathcal{Z}, n) \leq \sum_j \mu_L(\mathcal{B}_{r_j}, n) = c_n \sum_j r_j^n < c_n \mu_H(\mathcal{Z}, n) + c_n \varepsilon.$$

Therefore $\mu_H(\mathcal{Z}, n) \geq c_n^{-1} \mu_L(\mathcal{Z}, n)$ and, consequently, $\mu_H(\mathcal{Z}, n) > 0$. From this it follows that $\dim_H \mathcal{Z} \geq n$. The reverse inequality results from Proposition 3.15. □

Remark 3.4 In Example 3.10, starting directly from the definition of the Hausdorff dimension, we found that the dimension of any piece of a smooth surface is equal to 2. Now it is clear that this fact is a simple corollary of Propositions 3.17 and 3.19.

Corollary 3.1 *(a)* $\dim_H \mathbb{R}^n = n$;
(b) *For any open set* $\mathcal{Z} \subset \mathbb{R}^n$ *we have* $\dim_H \mathcal{Z} = n$;
(c) *If* $(\mathcal{M}, g)$ *is a smooth n-dimensional Riemannian manifold then* $\dim_H \mathcal{M} = n$.

Proposition 3.20 *Suppose that for a set* $\mathcal{Z}$ *in the separable metric space* $(\mathcal{M}, \rho)$ *we have* $\dim_T \mathcal{Z} = n$ $(0 \leq n < \infty)$. *Then* $\mu_H(\mathcal{Z}, n) > 0$ *and, consequently,*

$$\dim_T \mathcal{Z} \leq \dim_H \mathcal{Z}. \tag{3.6}$$

Remark 3.5 The Cantor set $\mathcal{Z}$ gives an example which shows that inequality (3.6) may be strictly. A *fractal* is a space whose Hausdorff dimension is greater than its topological dimension. For any non-empty open set $\mathcal{Z} \subset \mathbb{R}^n$ inequality (3.6) goes over in an equality. This property follows from Proposition 3.8 and from Proposition 3.19. The following theorem from [37] establishes an important link of the topological dimension with the Hausdorff dimension. Let $(\mathcal{M}, \rho)$ be a separable metric space. Then $\dim_T \mathcal{M} = \inf \dim_H \mathcal{M}'$, where the infimum is taken over all separable metric spaces homeomorphic to $\mathcal{M}$.

To prove Proposition 3.20 we need the following lemma. Denote by $\mathcal{S}_r(p)$ the boundary of a ball $\mathcal{B}_r(p)$. Put $\mathcal{S}'_r(p) = \mathcal{S}_r(p) \cap \mathcal{Z}$.

Lemma 3.4 *Suppose that* $\mu_H(\mathcal{Z}, l+1) = 0$ *for some integer* $l \geq 0$. *Then if p is an arbitrary point of* $\mathcal{Z}$, *then for almost all r we have* $\mu_H\big(\mathcal{S}'_r(p), l\big) = 0$.
(Almost all means for all, excluding some set of Lebesgue measure zero).

Proof Let us fix a point $p \in \mathcal{Z}$. For any natural m a collection of balls $\{\mathcal{B}_j^{(m)}\}_{j \in \mathfrak{I}(m)}$ with some index set $\mathfrak{I}(m)$ can be found, which forms a cover of $\mathcal{Z}$ such that

$$\sum_{j\in\mathfrak{I}(m)} \left(r_j^{(m)}\right)^{l+1} < \frac{1}{m}. \tag{3.7}$$

Here $r_j^{(m)}$ is the radius of ball $\mathcal{B}_j^{(m)}$.

Put

$$h_j^{(m)}(r) = \begin{cases} (r_j^{(m)})^l, & \text{if } \mathcal{B}_j^{(m)} \cap \mathcal{S}_r(p) \neq \emptyset, \\ 0, & \text{if } \mathcal{B}_j^{(m)} \cap \mathcal{S}_r(p) = \emptyset, \end{cases}$$

and

$$h^{(m)}(r) = \sum_{j\in\mathfrak{I}(m)} h_j^{(m)}(r).$$

From the definition of $h_j^{(m)}(r)$ it follows that

$$\int_0^\infty h_j^{(m)}(r)\,dr \le \int_{\delta_j^{(m)}}^{\delta_j^{(m)}+2r_j^{(m)}} h_j^{(m)}(r)\,dr = 2\left(r_j^{(m)}\right)^{l+1},$$

where $\delta_j^{(m)}$ is the distance from p to the ball $\mathcal{B}_j^{(m)}$. Therefore (3.7)

$$\int_0^\infty h^{(m)}(r)\,dr \le 2 \sum_{j\in\mathfrak{I}(m)} \left(r_j^{(m)}\right)^{l+1}.$$

By virtue of it follows that the sequence of functions $h^{(m)}(r)$ converges in mean to zero. Therefore [49] there exists a subsequence $h^{(m_k)}(r)$ converging to zero for Lebesgue-almost all r.

If $\mathcal{B}_j^{(m)} \cap \mathcal{S}_r(p) \neq \emptyset$, then

$$\left(r_j^{(m)}\right)^l = h_j^{(m)}(r).$$

Hence for almost all r

$$\sum_{j\in\mathfrak{I}'(m_k)} \left(r_j^{(m_k)}\right)^l \to 0 \quad as \quad k \to +\infty. \tag{3.8}$$

Here $\mathfrak{I}'(m_k)$ denotes a subset of $\mathfrak{I}(m_k)$ consisting of all that indices j, for which $\mathcal{B}_j^{(m)} \cap \mathcal{S}_r(p) \neq \emptyset$.

Since

$$\mathcal{S}'_r(p) \subset \bigcup_{j\in\mathfrak{I}'(m_k)} \mathcal{B}_j^{(m_k)},$$

relation (3.8) means that $\mu_H\left(\mathcal{S}'_r(p), l\right) = 0$ for almost all r. □

Proof of Proposition 3.20 Suppose that the assertion of the proposition is not true, i.e., $\mu_H(\mathcal{Z}, n) = 0$. If we prove that this implies the inequality $\dim_T \mathcal{Z} \leq n - 1$, then we obtain a contradiction to the assumptions of Proposition 3.20, which proves the assertion.

Thus, it is sufficient to show that $\mu_H(\mathcal{Z}, n) = 0$ implies $\dim_T \mathcal{Z} \leq n - 1$.

We shall prove it by induction. Let $n = 0$. The condition $\mu_H(\mathcal{Z}, 0) = 0$ means that $\mathcal{Z} = \emptyset$. Indeed, if $\mathcal{Z}$ contains at least one point, then $\mu_H(\mathcal{Z}, 0) \geq 1$. Since $\mathcal{Z} = \emptyset$, we see that $\dim_T \mathcal{Z} = -1$.

Supposing that the statement is proved for n, we show the trueness for $n + 1$. Suppose that $\mu_H(\mathcal{Z}, n + 1) = 0$. We have to prove that $\dim_T \mathcal{Z} \leq n$. On the base of Proposition 3.6 it is sufficient to state that any point $p \in \mathcal{Z}$ possesses in $\mathcal{M}$ arbitrarily small neighborhoods such that the intersection of their boundaries with $\mathcal{Z}$ has the topological dimension $\leq n - 1$. But this directly follows from the lemma proved above and the induction hypothesis. □

3.2.2 *Fractal Dimension and Lower Box Dimension*

In this subsection we take a closer look at the fractal dimension and lower box dimension of a set [12, 13, 29, 42].

Let $(\mathcal{M}, \rho)$ be a metric space and $\mathcal{Z} \subset \mathcal{M}$ be an arbitrary totally bounded set. Recall that a set $\mathcal{Z} \subset \mathcal{M}$ is said to be *totally bounded* if for each $\varepsilon > 0$ the set $\mathcal{Z}$ can be written as finite union of subsets of $\mathcal{M}$ with a diameter smaller than ε. According to Hausdorff's theorem [53] a set $\mathcal{Z}$ of a complete metric space $\mathcal{M}$ is totally bounded if and only if $\mathcal{Z}$ is relatively compact. Denote by $N_\varepsilon(\mathcal{Z})$ the minimal number of balls of radius $\varepsilon > 0$ which are necessary to cover $\mathcal{Z}$.

Definition 3.9 The *fractal dimension* or *upper box dimension* of the set $\mathcal{Z}$ is given by

$$\dim_F \mathcal{Z} \equiv \overline{\dim}_B \mathcal{Z} := \limsup_{\varepsilon \to 0} \frac{\log N_\varepsilon(\mathcal{Z})}{\log 1/\varepsilon}.$$

The *lower box dimension* of the set $\mathcal{Z}$ is defined by

$$\underline{\dim}_B \mathcal{Z} := \liminf_{\varepsilon \to 0} \frac{\log N_\varepsilon(\mathcal{Z})}{\log 1/\varepsilon}.$$

If $\underline{\dim}_B \mathcal{Z} = \overline{\dim}_B \mathcal{Z} := \dim_B \mathcal{Z}$ the value $\dim_B \mathcal{Z}$ is called *box dimension* of $\mathcal{Z}$.

Let us give now an equivalent description of the fractal dimension of the set $\mathcal{Z}$. Let $d \geq 0$, $\varepsilon > 0$ be numbers and put $\mu_B(\mathcal{Z}, d, \varepsilon) := \varepsilon^d N_\varepsilon(\mathcal{Z})$ (capacitive d-measure at level ε), $\mu_F(\mathcal{Z}, d) \equiv \overline{\mu}_B(\mathcal{Z}, d) := \limsup_{\varepsilon \to 0} \mu_B(\mathcal{Z}, d, \varepsilon)$ (*upper capacitive d-measure*) and $\underline{\mu}_B(\mathcal{Z}, d) := \liminf_{\varepsilon \to 0} \mu_B(\mathcal{Z}, d, \varepsilon)$ (*lower capacitive d-measure*). It is easy to see that for the functions $\mu_F(\mathcal{Z}, \cdot)$ and $\underline{\mu}_B(\mathcal{Z}, \cdot)$ there exists, as in the Hausdorff measure case, a critical value d_{cr}.

Proposition 3.21 *For any totally bounded set $\mathcal{Z} \subset \mathcal{M}$ we have*

$$\overline{\dim}_B \mathcal{Z} = \inf\{d \geq 0 | \, \mu_F(\mathcal{Z}, d) = 0\} \text{ and } \underline{\dim}_B \mathcal{Z} = \inf\{d \geq 0 \,|\, \underline{\mu}_B(\mathcal{Z}, d) = 0\}.$$

Proof Let us prove the first statement and put

$$d_1 := \inf\{d | \, \mu_F(\mathcal{Z}, d) = 0\}, \quad d_2 := \limsup_{\varepsilon \to 0} \frac{\log N_\varepsilon(\mathcal{Z})}{\log 1/\varepsilon}.$$

Firstly, we show that $d_1 \leq d_2$. If $d_1 = 0$, then this statement is obvious. Then we shall suppose that $d_1 > 0$. For a given $\gamma \in (0, \, d_1)$, choose a sequence of numbers $\varepsilon_m > 0$ such that $\varepsilon \to 0+0$ and

$$\infty = \mu_F(\mathcal{Z}, d_1 - \gamma) = \lim_{m \to +\infty} \mu_F(\mathcal{Z}, d_1 - \gamma, \varepsilon_m).$$

Then for all sufficiently large m we have

$$\varepsilon_m^{d_1 - \gamma} N_{\varepsilon_m}(\mathcal{Z}) \geq 1. \tag{3.9}$$

Passing if necessary to a subsequence, one can suppose that there exists the following limit

$$\lim_{m \to +\infty} \frac{\log N_{\varepsilon_m}(\mathcal{Z})}{\log 1/\varepsilon_m} \leq d_2. \tag{3.10}$$

By (3.9) we have

$$d_1 \leq \gamma + \frac{\log N_{\varepsilon_m}(\mathcal{Z})}{\log 1/\varepsilon_m}.$$

Passing to the limit as $m \to +\infty$ and taking into account (3.10), we obtain $d_1 \leq d_2 + \gamma$. By virtue of arbitrariness of γ it follows that $d_1 \leq d_2$.

Now we show that $d_1 \geq d_2$. Let $\gamma > 0$ be an arbitrary number. Let us choose a sequence of numbers $\varepsilon'_m > 0$ such that $\varepsilon'_m \to 0+0$ and

$$d_2 = -\lim_{m \to +\infty} \frac{\log N_{\varepsilon'_m}(\mathcal{Z})}{\log \varepsilon'_m}. \tag{3.11}$$

Passing if necessary to a subsequence, one can suppose that

$$0 = \mu_F(\mathcal{Z}, d_1 + \gamma) = \lim_{m \to +\infty} \mu_F(\mathcal{Z}, d_1 + \gamma, \varepsilon'_m).$$

Then for all sufficiently great m we obtain

$$\varepsilon'^{\, d_1 + \gamma}_m N_{\varepsilon'_m}(\mathcal{Z}) \leq 1.$$

Then $d_1 \geq -\gamma - \log N_{\varepsilon'_m}(\mathcal{Z})/\log \varepsilon'_m$. Passing to the limit as $m \to +\infty$ and taking into account (3.11), we get $d_1 \geq d_2 - \gamma$. Since γ is arbitrary, then $d_1 \geq d_2$. □

Remark 3.6 Since the inequality $\mu_H(\mathcal{Z}, d, \varepsilon) \leq \mu_B(\mathcal{Z}, d, \varepsilon)$ is obvious, we have

$$\begin{aligned}\mu_H(\mathcal{Z}, d) &= \lim_{\varepsilon\to 0} \mu_H(\mathcal{Z}, d, \varepsilon) = \lim_{\varepsilon\to 0} \inf \left\{ \sum r_i^d \,\middle|\, r_i \leq \varepsilon, \bigcup B_{r_i} \supset \mathcal{Z} \right\} \\ &\leq \liminf_{\varepsilon\to 0} N_\varepsilon(\mathcal{Z})\varepsilon^d = \underline{\mu}_B(\mathcal{Z}, d) \leq \limsup_{\varepsilon\to 0} N_\varepsilon(\mathcal{Z})\varepsilon^d = \mu_F(\mathcal{Z}, d)\,.\end{aligned}$$

Therefore

$$\dim_H \mathcal{Z} \leq \underline{\dim}_B \mathcal{Z} \leq \dim_F \mathcal{Z}. \tag{3.12}$$

Example 3.11 Returning now to Examples 3.8 and 3.10 of the previous subsection, we can easily find that the fractal dimension of the Cantor set is $\log 2/\log 3$ and the fractal dimension of a smooth surface in $\mathbb{R}^3$ is 2. Let us demonstrate the first fact. Suppose $\{\varepsilon_n\}_{n=1}^\infty$ is a sequence of numbers with

$$\frac{1}{2}\,\frac{1}{3^n} < \varepsilon_n \leq \frac{1}{2}\,\frac{1}{3^{n-1}}\,, \quad n = 1, 2, \ldots.$$

It follows from the construction of the Cantor set $\mathcal{Z}$ in Example 3.8 that $N_{\varepsilon_n}(\mathcal{Z}) \leq 2^n$, $n = 1, 2, \ldots$, and thus $\dim_F \mathcal{Z} \leq \limsup_{n\to\infty} \frac{\log 2^n}{\log(2\cdot 3^{n-1})} = \frac{\log 2}{\log 3}$. Assume now that $\{\widetilde{\varepsilon}_n\}_{n=1}^\infty$ is a sequence with $\frac{1}{2}\,\frac{1}{3^{n+1}} \leq \widetilde{\varepsilon}_n < \frac{1}{2}\frac{1}{3^n}$, $n = 1, 2, \ldots$. It is easy to see that each ball $\mathcal{B}_{\widetilde{\varepsilon}_n}$ intersects at most one of the intervals $\mathcal{Z}_n$ from Example 3.8. It follows that

$$N_{\widetilde{\varepsilon}_n}(\mathcal{Z}) \geq 2^n\,, \; n = 1, 2, \ldots,$$

and thus $\underline{\dim}_B \mathcal{Z} \geq \liminf_{n\to\infty} \frac{\log 2^n}{\log(2\cdot 3^{n+1})} = \frac{\log 2}{\log 3}$. Putting together the two estimates we get

$$\frac{\log 2}{\log 3} \leq \underline{\dim}_B \mathcal{Z} \leq \overline{\dim}_B \mathcal{Z} \leq \frac{\log 2}{\log 3}\,.$$

This gives, together with the result from Example 3.8, for the Cantor set the dimensions

$$\dim_H \mathcal{Z} = \dim_B \mathcal{Z} = \frac{\log 2}{\log 3}\,.$$

Let us consider an example which demonstrates that the Hausdorff dimension and the fractal dimension may not coincide.

Example 3.12 Let $\mathcal{M} = [0, 1]$ with the usual metric from $\mathbb{R}$ and $\mathcal{Z} = \{0, 1, \frac{1}{2}, \frac{1}{3}, \ldots\}$. We show that $\dim_H \mathcal{Z} = 0$ and $\underline{\dim}_B \mathcal{Z} = \dim_F \mathcal{Z} = 1/2$. Indeed, since $\mathcal{Z}$ is a countable set, we have by Proposition 3.16 $\dim_H \mathcal{Z} = 0$.

Put $p_0 := 0$ and $p_j := 1/j$ for $j > 0$. Let us show at first that $\underline{\dim}_B \mathcal{Z} \geq 1/2$. Assuming an arbitrary integer number $m > 0$ to be fixed we put $\varepsilon_m := \frac{1}{2m(m+1)}$. Since $\frac{1}{j} - \frac{1}{j+1} = \frac{1}{j(j+1)}$, it follows that none of two points p_j, where $1 \leq j \leq m$, can be covered by one interval of length $2\varepsilon_m$. Consequently, $N_{\varepsilon_m}(\mathcal{Z}) \geq m$. Thus,

$$\frac{\log N_{\varepsilon_m}(\mathcal{Z})}{\log 1/\varepsilon_m} \geq \frac{\log m}{\log[2m(m+1)]} = \frac{\log m}{\log 2 + 2\log m + \log(1+\frac{1}{m})} \stackrel{m\to+\infty}{\longrightarrow} \frac{1}{2}.$$

It follows that $\underline{\dim}_B \mathcal{Z} \geq 1/2$.

Let us show now that $\dim_F \mathcal{Z} \leq 1/2$. Suppose that $\varepsilon \in (0, 1)$ is an arbitrary number. Let us choose an integer $m > 0$ such that $\frac{1}{2m(m+1)} < \varepsilon \leq \frac{1}{m^2}$.

The points p_0, p_j with $j \geq m(m+1)$ belong to the interval $[0, \frac{1}{m(m+1)}]$, which can be covered with one interval of length 2ε. The points p_j with j satisfying $m < j < m(m+1)$ belong to the interval $[\frac{1}{m(m+1)}, \frac{1}{m}]$, which, obviously, can be covered by $m+1$ intervals of length 2ε. Consequently $N_\varepsilon(\mathcal{Z}) \leq 1 + m + 1 + m = 2(m+1)$. Therefore we have

$$\frac{\log N_\varepsilon(\mathcal{Z})}{\log 1/\varepsilon} \leq \frac{\log[2(m+1)]}{\log m^2} = \frac{\log 2 + \log m + \log(1+\frac{1}{m})}{2\log m} \stackrel{m\to+\infty}{\longrightarrow} \frac{1}{2}.$$

Together with the inequality from above we have $1/2 \leq \underline{\dim}_B \mathcal{Z} \leq \dim_F \mathcal{Z} \leq 1/2$.

Example 3.13 Let $\mathbb{H}$ be a separable Hilbert space with scalar product $\langle \cdot, \cdot \rangle$ and an associated norm $\|\cdot\|$. Let $\{e_j\}_{j=1}^{\infty}$ be an orthonormal basis in $\mathbb{H}$. Consider the set

$$\mathcal{Z} := \{0\} \cup \left\{ \frac{1}{\log j} e_j \mid j = 2, 3, \ldots \right\}.$$

Since the compact set $\mathcal{Z}$ is countable, we have $\dim_H \mathcal{Z} = 0$ (Proposition 3.16). Let us prove that $\dim_F \mathcal{Z} = \infty$.

Write for shortness $p_j := \frac{1}{\log j} e_j$, $j > 1$. Further, fix an arbitrary integer m, and put

$$\varepsilon_m := \frac{1}{\sqrt{2}\log m}.$$

For the distance between arbitrary points p_i and p_j we have

$$|p_i - p_j| = \sqrt{\left(\frac{1}{\log i}\right)^2 + \left(\frac{1}{\log j}\right)^2} > \frac{\sqrt{2}}{\log j} \geq 2\varepsilon_m,$$

assuming that $2 \leq i < j \leq m$. Therefore, $\|p_j - p_{j+1}\| > 2\varepsilon_m$, $j = 2, 3, \ldots, m-1$. This means that in any cover of $\mathcal{Z}$ with sets of radius ε_m each of the points $p_2, p_3, \ldots, p_m$ must be covered by an extra ball. It follows, that $N_{\varepsilon_m}(\mathcal{Z}) \geq m - 1$ and therefore we have

$$\frac{\log N_{\varepsilon_m}(\mathcal{Z})}{\log 1/\varepsilon_m} \geq \frac{\log(m-1)}{\log(\sqrt{2}\log m)} \overset{m\to+\infty}{\longrightarrow} \infty.$$

Hence $\dim_F \mathcal{Z} = \infty$.

The main properties of the lower and upper box dimension can be found in the next proposition [39, 40].

Proposition 3.22 *Suppose that $(\mathcal{M}, \rho)$ is a metric space. Then it holds:*

(1) $\underline{\dim}_B \emptyset = \dim_F \emptyset = 0$*;*
(2) $\underline{\dim}_B \mathcal{Z}_1 \leq \underline{\dim}_B \mathcal{Z}_2$, $\dim_F \mathcal{Z}_1 \leq \dim_F \mathcal{Z}_2$*, if* $\mathcal{Z}_1 \subset \mathcal{Z}_2 \subset \mathcal{M}$ *are totally bounded sets;*
(3) $\dim_F \left(\bigcup_{j\geq 1} \mathcal{Z}_j\right) \geq \sup_{j\geq 1}\{\dim_F \mathcal{Z}_j\}$, $\underline{\dim}_B(\bigcup_{j\geq 1} \mathcal{Z}_j) \geq \sup_{j\geq 1} \underline{\dim}_B \mathcal{Z}_j$*, where* $\mathcal{Z}_j \subset \mathcal{M}$, $j = 1, 2, \ldots$ *are totally bounded sets;*
(4) $\dim_F \left(\bigcup_{j=1}^k \mathcal{Z}_j\right) = \max_{1\leq j\leq k}\{\dim_F \mathcal{Z}_j\}$ *for arbitrary totally bounded sets* $\mathcal{Z}_j \subset \mathcal{M}$, $j = 1, 2, \ldots, k$*;*
(In general $\underline{\dim}_B(\bigcup_{j=1}^k \mathcal{Z}_j) = \max_{1\leq j\leq k}\{\underline{\dim}_B \mathcal{Z}_j\}$ *is* *not* *true.)*
(5) If the set $\mathcal{Z} \subset \mathcal{M}$ *is finite, then* $\underline{\dim}_B \mathcal{Z} = \dim_F \mathcal{Z} = 0$*;*
(6) If $(\mathcal{M}', \rho')$ *is a second metric space and* $\Phi : \mathcal{M} \to \mathcal{M}'$ *is a bijection such that* Φ *and* Φ^{-1} *are Lipschitz then for any totally bounded set* $\mathcal{Z} \subset \mathcal{M}$ *we have* $\underline{\dim}_B \mathcal{Z} = \underline{\dim}_B \Phi(\mathcal{Z})$ *and* $\dim_F \mathcal{Z} = \dim_F \Phi(\mathcal{Z})$*;*
(7) If $\mathcal{Z} \subset \mathcal{M}$ *is a totally bounded set,* $\overline{\mathcal{Z}}$ *is the closure in* $\mathcal{M}$*, then* $\underline{\dim}_B \mathcal{Z} = \underline{\dim}_B \overline{\mathcal{Z}}$ *and* $\dim_F \mathcal{Z} = \dim_F \overline{\mathcal{Z}}$*, i.e. the lower and upper box dimensions do not distinguish between a set and its closure;*
(8) If $(\mathcal{M}, g)$ *is an n-dimensional compact Riemannian manifold, then* $\dim_B \mathcal{M} = n$*.*

Proof Let us demonstrate (2). For any $\varepsilon > 0$ we have $N_\varepsilon(\mathcal{Z}_1) \leq N_\varepsilon(\mathcal{Z}_2)$. It follows that for all ε which are sufficiently small

$$\frac{\log N_\varepsilon(\mathcal{Z}_1)}{\log 1/\varepsilon} \leq \frac{\log N_\varepsilon(\mathcal{Z}_2)}{\log 1/\varepsilon}$$

and

$$\limsup_{\varepsilon\to 0} \frac{\log N_\varepsilon(\mathcal{Z}_1)}{\log 1/\varepsilon} \leq \limsup_{\varepsilon\to 0} \frac{\log N_\varepsilon(\mathcal{Z}_2)}{\log 1/\varepsilon}.$$

Let us show the first assertion of (4) and consider for simplicity the case $k = 2$. Since $\mathcal{Z}_1 \subset \mathcal{Z}_1 \cup \mathcal{Z}_2$ and $\mathcal{Z}_2 \subset \mathcal{Z}_1 \cup \mathcal{Z}_2$ we conclude by (3) that $\max_{i=1,2} \dim_F(\mathcal{Z}_i) \leq \dim_F(\mathcal{Z}_1 \cup \mathcal{Z}_2)$. In order to prove the opposite inequality we use for $\mathcal{Z} := \mathcal{Z}_1 \cup \mathcal{Z}_2$ and arbitrary $\varepsilon > 0$ the inequality

$$N_\varepsilon(\mathcal{Z}) \leq N_\varepsilon(\mathcal{Z}_1) + N_\varepsilon(\mathcal{Z}_2).$$

We know that there exists a sequence $\{\varepsilon_n\}$, $\varepsilon_n \to 0$ such that $\dim_F \mathcal{Z} = \lim_{n\to\infty} \frac{\log N_{\varepsilon_n}(\mathcal{Z})}{\log 1/\varepsilon_n}$.

If necessary we choose a subsequence of $\{\varepsilon_n\}$ (with the same notation) in order to have also the two limits

$$\beta_i := \lim_{n\to\infty} \frac{\log N_{\varepsilon_n}(\mathcal{Z}_i)}{\log 1/\varepsilon_n} \qquad (i = 1, 2)\,.$$

Clearly, $\beta_i \le \dim_F \mathcal{Z}_2\,, \quad i = 1, 2\,.$ Define $a_n := \frac{\log N_{\varepsilon_n}(\mathcal{Z}_1)}{\log 1/\varepsilon_n} - \frac{\log N_{\varepsilon_n}(\mathcal{Z}_2)}{\log 1/\varepsilon_n}\,,$ $n = 1, 2, \ldots$. It follows that for all these n

$$a_n \log 1/\varepsilon_n = \log N_{\varepsilon_n}(\mathcal{Z}_1) - \log N_{\varepsilon_n}(\mathcal{Z}_2)$$

and, consequently,

$$N_{\varepsilon_n}(\mathcal{Z}_2) = N_{\varepsilon_n}(\mathcal{Z}_1)\varepsilon_n^{a_n}\,,\; N_{\varepsilon_n}(\mathcal{Z}) \le N_{\varepsilon_n}(\mathcal{Z}_1)\,[1 + \varepsilon_n^{a_n}] \qquad \text{and}$$
$$\log N_{\varepsilon_n}(\mathcal{Z}) \le N_{\varepsilon_n}(\mathcal{Z}_1) + \log(1 + \varepsilon_n^{a_n})\,.$$

W.l.o.g. we consider $\beta_1 > \beta_2$. It follows that there exists an $a > 0$ and $n_0 > 0$, $n_0 \in \mathbb{N}$, such that $a_n \ge a$, $n = n_0, n_0 + 1, \ldots$. Thus we can choose a constant $c_1 > 0$ such that

$$\log N_{\varepsilon_n}(\mathcal{Z}) \le \log N_{\varepsilon_n}(\mathcal{Z}_1) + c_1\varepsilon_n^{a_n}\,,\; n = n_0, n_0 + 1, \ldots.$$

It follows that

$$\frac{\log N_{\varepsilon_n}(\mathcal{Z})}{\log 1/\varepsilon_n} \le \frac{\log N_{\varepsilon_n}(\mathcal{Z}_1)}{\log 1/\varepsilon_n} + \frac{c_1\varepsilon_n^{a_n}}{\log 1/\varepsilon_n} \text{and} \dim_F \mathcal{Z} \le \beta_1 \le \max\{\dim_F \mathcal{Z}_1, \dim_F \mathcal{Z}_2\}\,.$$

□

Example 3.14 (a) Assume $\mathcal{Z} := \mathbb{Q} \cap [0, 1]$ denotes the set of rational numbers in $[0, 1]$. It follows from Proposition 3.22, that $\dim_F \mathcal{Z} = \dim_F \overline{\mathcal{Z}} = \dim_F([0, 1]) = 1\,.$

Since the rational numbers are countable we can write $\mathcal{Z} = \{q_i\}_{i=1}^{\infty}$.
Using Proposition (3.16), we conclude that $\dim_H \mathcal{Z} = 0$. This implies that $\dim_H \mathcal{Z} \ne \dim_H \overline{\mathcal{Z}} = \dim_H([0, 1]) = 1$.

Since $\sup_{i\ge1} \dim_F\{q_i\} = \sup_{i\ge1} 0 = 0$, we see that assertion (3) in Proposition 3.22 is really an inequality.

Remark 3.7 Suppose that $\mathcal{M}$ is an arbitrary topological space. A map $\Phi : \mathcal{M} \to \mathbb{R}^N$ is called a *topological* or *C^0-embedding* if the restriction $\Phi : \mathcal{M} \to \Phi(\mathcal{M})$ is a homeomorphism.

It is a classical result (Menger [34], Nöbeling [37], and Hurewicz [24]) that a space with $\dim_T \mathcal{M} = n < \infty$ can be topologically embedded in the space $\mathbb{R}^{2n+1}$ and the set of such homeomorphisms forms a dense set G_δ (i.e., a countable intersection of open sets.) Because of inequality (3.12), any set $\mathcal{Z}$ in a metric space with finite Hausdorff dimension $\dim_H \mathcal{Z}$ can thus topologically be embedded in $\mathbb{R}^{[2\dim_H \mathcal{Z}+1]}$. (Here, $[k]$ denotes for a real number k the smallest integer greater than or equal to k.) Important embedding results and their relation to dimensions are presented in the book by Whitney [52]. One can construct subsets of a Hilbert space with finite Hausdorff dimension that cannot be embedded by an orthogonal projection into $\mathbb{R}^N$ for any N [47]. In 1981 R. Mané [32] showed that "most" projections of a set $\mathcal{M}$ in a Banach space with a finite fractal dimension $\dim_F \mathcal{M}$ onto subspaces of dimension $> \dim_F \mathcal{M} + 1$ are injective. "Most" means here that the set of injective projections contains a countable intersection of dense sets in the space of all projections endowed with the norm topology. However this does not allowed any statement on the fractal dimension of the projected set. Ben-Artzi et al. [3] showed in the finite dimensional case the existence of such projections whose inverse is in edition Hölder continuum on the projected set.

This result has been generalized by Foias and Olsen [17], Robinson [46] to Hilbert spaces. In the works of Hunt and Kaloshin [22] and Hunt et al. [23] it was shown that the former results also can be expected generically in a probabilistic sense which is called prevalence. In the paper of Okon [38] such an embedding result for compacta with finite fractal dimension for Banach spaces is generalized to metric spaces. Embedding results for dynamical systems on infinite-dimensional manifolds are shown in Reitmann and Popov [45].

3.2.3 Self-similar Sets

Let $(\mathcal{M}, \rho)$ be a separable complete metric space. If $\Phi : \mathcal{M} \to \mathcal{M}$ is a map, then the Lipschitz constant of Φ is

$$\operatorname{Lip} \Phi := \sup_{p \neq q} \frac{\rho\,(\Phi(p), \Phi(q))}{\rho\,(p, q)}\,.$$

The map Φ is *Lipschitz* if $\operatorname{Lip} \Phi < \infty$ and Φ is a *contraction* if $\operatorname{Lip} \Phi < 1$.

Let $\{\Phi_1, \dots, \Phi_m\} (m \geq 2)$ be contractions on $\mathcal{M}$. A non-empty compact set $\mathcal{K} \subset \mathcal{M}$ is called *self-similar* with respect to $\{\Phi_1, \dots, \Phi_m\}$ if $\mathcal{K}$ satisfies the equation

$$\mathcal{K} = \Phi_1(\mathcal{K}) \cup \cdots \cup \Phi_m(\mathcal{K})\,. \tag{3.13}$$

One can show that (3.13) has for Lipschitz maps a unique non-empty compact solution $\mathcal{K} = \mathcal{K}(\Phi_1, \dots, \Phi_m)$. The Hausdorff dimension and the topological dimension of $\mathcal{K}(\Phi_1, \dots, \Phi_m)$ can be estimated by the following theorem [19].

Theorem 3.1 *Let $\delta = \delta(\Phi_1, \dots, \Phi_m)$ be the unique root of the equation $\sum_{j=1}^{m} (\operatorname{Lip} \Phi_j)^{\delta} = 1$. Then we have*

$$\dim_H \mathcal{K}(\Phi_1, \dots, \Phi_m) \leq \delta\,.$$

Proof We follow the proof in [19]. Let W_n^m be the set of all finite words of length $n \in \mathbb{N}$ of m symbols $\{1, \ldots, m\}$. For any $w = (w_1, \ldots, w_m) \in W_m^n$ we consider the set

$$\mathcal{K}_w := \Phi_{w_1} \circ \Phi_{w_2} \circ \cdots \circ \Phi_{w_n}(\mathcal{K}(\Phi_1, \ldots, \Phi_m)).$$

It follows from (3.13) that

$$\mathcal{K}(\Phi_1, \ldots, \Phi_m) = \bigcup_{w \in W_n^m} \mathcal{K}_w . \tag{3.14}$$

Moreover, denoting by diam $\mathcal{Z}$ the diameter of a set $\mathcal{Z}$, we have

$$\operatorname{diam} \mathcal{K}_w \le \lambda^n \operatorname{diam} \mathcal{K}(\Phi_1, \ldots, \Phi_m) =: \varepsilon_n ,$$

where $\lambda := \max_{j=1,\ldots,m} \operatorname{Lip} \Phi_j < 1$. Since $\{\mathcal{K}_w\}_{w \in W_n^m}$ is a finite ε_n-covering of $\mathcal{K}(\Phi_1, \ldots, \Phi_m)$ by (3.14), it follows for the corresponding outer Hausdorff measure at level ε_n and of order δ that

$$\mu_H(\mathcal{K}(\Phi_1, \ldots, \Phi_m), \delta, \varepsilon_n) \le \sum_{w \in W_n^m} (\operatorname{diam} \mathcal{K}_w)^\delta$$

$$\le (\operatorname{diam} \mathcal{K}(\Phi_1, \ldots, \Phi_m))^\delta \left(\sum_{j=1}^m (\operatorname{Lip} \Phi_j)^\delta \right)^n = (\operatorname{diam} \mathcal{K}(\Phi_1, \ldots, \Phi_m))^\delta < \infty .$$

Letting $n \to \infty$ we have $\mu_H(\mathcal{K}(\Phi_1, \ldots, \Phi_m), \delta) < \infty$ which implies that $\dim_H \mathcal{K}(\Phi_1, \ldots, \Phi_m) \le \delta$. □

A map $\Phi : \mathcal{M} \to \mathcal{M}$ is a *similitude* if there is a fixed $0 < r < 1$ such that

$$\rho(\Phi(p), \Phi(q)) = r\rho\,(p, q), \ \forall p, q \in \mathcal{M}.$$

Let the similitude Φ have a fixed point a, let $\operatorname{Lip} \Phi = r$, and let T be the orthonormal transformation given by $T(p) := \frac{1}{r}[\Phi(p + a) - a]$. Then $\Phi(p) = rT(p - a) + a$. According to [25] we write $\Phi = (a, r, T)$ for this representation. A set of maps $\{\Phi_1, \ldots, \Phi_m\}$ on $\mathcal{M}$ satisfies the *open set condition* if there exists a non-empty open set $\mathcal{U}$ such that

(i) $\bigcup_{j=1}^m \Phi_j(\mathcal{U}) \subset \mathcal{U}$, and
(ii) $\Phi_i\,(\mathcal{U}) \cap \Phi_j(\mathcal{U}) = \emptyset \quad \text{if } i \neq j$.

Suppose that there is a closed set $\mathcal{C} \subset \mathcal{M}$ with the interior int $\mathcal{C} \neq \emptyset$ such that

(a) $\Phi_i(\mathcal{C}) \subset \mathcal{C} \quad \text{if} \quad i = 1, \ldots, m$, and
(b) $\operatorname{int} \Phi_i(\mathcal{C}) \cap \operatorname{int} \Phi_j(\mathcal{C}) = \emptyset \quad \text{if} \quad i \neq j$.

Then the open set condition is satisfied [25]. The following theorem was proved by Moran [36] and Hutchinson [25].

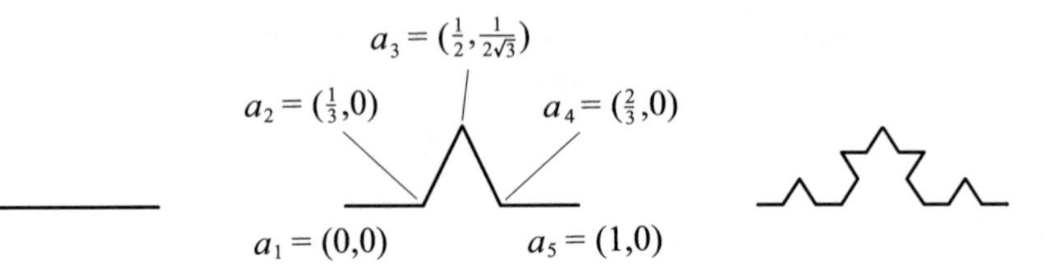

Fig. 3.3 The Koch curve

Theorem 3.2 *If* $\{\Phi_1, \ldots, \Phi_m\}$ *is a set of similitudes of* $\mathbb{R}^n$ *and if this set satisfies the open set condition, then*

$$\dim_H \mathcal{K}(\Phi_1, \ldots, \Phi_m) = \dim_B \mathcal{K}(\Phi_1, \ldots, \Phi_m) = \delta,$$

where δ *is the unique solution of* $\sum_{j=1}^{m} (\mathrm{Lip}\, \Phi_j)^\delta = 1$.

Proof The estimate $\dim_H \mathcal{K}(\Phi_1, \ldots, \Phi_m) \le \delta$ follows from Theorem 3.1. For the lower estimate $\dim_H \mathcal{K}(\Phi_1, \ldots, \Phi_m) \ge \delta$ see [36] or [25]. From (3.12) it follows that $\delta \le \underline{\dim}_B \mathcal{K}(\Phi_1, \ldots, \Phi_m) \le \overline{\dim}_B \mathcal{K}(\Phi_1, \ldots, \Phi_m)$. Using the same technique as in the proof of Theorem 3.1 one shows that $\overline{\dim}_B \mathcal{K}(\Phi_1, \ldots, \Phi_m) \le \delta$. □

Example 3.15 (a) Let $\{\Phi_1, \Phi_2\}$ be similitudes $\Phi_i : \mathbb{R} \to \mathbb{R}$ given by $\Phi_1 = (0, \frac{1}{3}, I)$ and $\Phi_2 = (1, \frac{1}{3}, I)$ (I is the identity map). Then $\mathcal{K}(\Phi_1, \Phi_2) =: \mathcal{Z} = \Phi_1(\mathcal{Z}) \cup \Phi_2(\mathcal{Z})$, where $\mathcal{Z}$ is the Cantor set (see Examples 3.8 and 3.9).

(b) Let a_1, a_2, a_3, a_4, a_5 be points as shown in Fig. 3.3. Let $\{\Phi_1, \Phi_2, \Phi_3, \Phi_4\}$ be similitudes of $\mathbb{R}^2$, where $\Phi_i : \mathbb{R}^2 \to \mathbb{R}^2, i = 1, \ldots, 4$, is the unique similitude mapping $\overrightarrow{a_1 a_5}$ to $\overrightarrow{a_i a_{i+1}}$ and having positive determinant. The set $\mathcal{K} = \mathcal{K}\{\Phi_1, \Phi_2, \Phi_3, \Phi_4\}$ is called *Koch curve*. Figure 3.3 shows the approximation of $\mathcal{K}$.

(c) The open set condition holds for (a) with $\mathcal{C} = [0, 1]$ and for (b) with $\mathcal{C}$ the triangle (a_1, a_5, a_3). It follows from Theorem 3.2 that for the set $\mathcal{Z} = \mathcal{K}(\Phi_1, \Phi_2)$ from (a) we have $\dim_H \mathcal{Z} = \dim_B \mathcal{K}(\Phi_1, \Phi_2) = \log 2/ \log 3$, where $\delta = \log 2/ \log 3$ is the unique solution of $(1/3)^\delta + (1/3)^\delta = 1$.

It follows from the same theorem that for the set $\mathcal{Z} = \mathcal{K}(\Phi_1, \Phi_2, \Phi_3, \Phi_4)$ from (b) we have $\dim_H \mathcal{Z} = \dim_B \mathcal{Z} = \log 4/ \log 3$, where $\delta = \log 4/ \log 3$ is the unique solution of

$$(1/3)^\delta + (1/3)^\delta + (1/3)^\delta + (1/3)^\delta = 1\,.$$

3.2.4 Dimension of Cartesian Products

The subject of this subsection is the study of dimension properties of Cartesian products. We follow the representation in [14, 21, 51]. Let $(\mathcal{M}, \rho)$ and $(\mathcal{M}', \rho')$ be metric spaces. We consider in the Cartesian product $\mathcal{M} \times \mathcal{M}'$ the metric $\widetilde{\rho}$ given for arbitrary $(p, p'), (q, q') \in \mathcal{M} \times \mathcal{M}'$ by $\widetilde{\rho}((p, p'), (q, q')) := \sqrt{\rho(p, q)^2 + \rho'(p', q')^2}$.

It follows that for arbitrary $r > 0, r' > 0$ and points $p_0 \in \mathcal{M}$, $p_0' \in \mathcal{M}'$

$$\begin{aligned} &\mathcal{B}_r(p_0) \times \mathcal{B}'_{r'}(p_0') \subset \widetilde{\mathcal{B}}_{\sqrt{r^2+r'^2}}(p_0, p_0') \\ &= \left\{(p, p') \mid \widetilde{\rho}\,((p, p'), (p_0, p_0')) \le \sqrt{r^2 + r'^2}\right\}. \end{aligned}$$

Example 3.16 (a) Suppose that $\mathcal{M}$ and $\mathcal{M}'$ are n-resp. m-dimensional smooth Riemannian manifolds. Then we have $\dim_H \mathcal{M} = n$, $\dim_H \mathcal{M}' = m$, and $\dim_H(\mathcal{M} \times \mathcal{M}') = m + n = \dim_H \mathcal{M} + \dim_H \mathcal{M}'$.

(b) It is shown in [14] that there exist sets $\mathcal{Z}, \mathcal{Z}' \subset \mathbb{R}$ with $\dim_H \mathcal{Z} = \dim_H \mathcal{Z}' = 0$ but $\dim_H \mathcal{Z} \times \mathcal{Z}' \ge 1$.

Proposition 3.23 *Suppose that $(\mathcal{M}, \rho)$ and $(\mathcal{M}', \rho')$ are metric spaces, $\mathcal{Z} \subset \mathcal{M}$ and $\mathcal{Z}' \subset \mathcal{M}'$ are compact sets. Then*

$$\dim_F(\mathcal{Z} \times \mathcal{Z}') \le \dim_F(\mathcal{Z}) + \dim_F(\mathcal{Z}').$$

Proof We show the mean idea of the proof. Suppose that for $\varepsilon > 0$ the expressions $N_\varepsilon(\mathcal{Z})$ and $N_\varepsilon(\mathcal{Z}')$ denote the minimal number of balls of radius ε, necessary for the covering of $\mathcal{Z}$ and $\mathcal{Z}'$, respectively. It follows that $\mathcal{Z} \times \mathcal{Z}'$ is already covered by $N_\varepsilon(\mathcal{Z})N_\varepsilon(\mathcal{Z}')$ balls of radius $\sqrt{2}\,\varepsilon$ and

$$\begin{aligned} &\dim_F(\mathcal{Z} \times \mathcal{Z}') \le \limsup_{\varepsilon \to 0} \frac{\log[N_\varepsilon(\mathcal{Z})N_\varepsilon(\mathcal{Z}')]}{-\log(\sqrt{2}\varepsilon)} \\ &= \limsup_{\varepsilon \to 0} \left[\frac{\log N_\varepsilon(\mathcal{Z})}{-\log(\sqrt{2}\varepsilon)} + \frac{\log N_\varepsilon(\mathcal{Z}')}{-\log(\sqrt{2}\varepsilon)}\right] \\ &\le \limsup_{\varepsilon \to 0} \frac{\log N_\varepsilon(\mathcal{Z})}{-\log(\sqrt{2}\varepsilon)} + \limsup_{\varepsilon \to 0} \frac{\log N_\varepsilon(\mathcal{Z}')}{-\log(\sqrt{2}\varepsilon)} \\ &= \dim_F \mathcal{Z} + \dim_F \mathcal{Z}'. \end{aligned}$$

□

Proposition 3.24 *Suppose that $(\mathcal{M}, \rho)$ and $(\mathcal{M}', \rho')$ are metric spaces, $\mathcal{Z} \subset \mathcal{M}$ and $\mathcal{Z}' \subset \mathcal{M}'$ are compact sets. Then*

$$\dim_H(\mathcal{Z} \times \mathcal{Z}') \le \dim_H \mathcal{Z} + \dim_F \mathcal{Z}'.$$

Proof Suppose that $s > \dim_H \mathcal{Z}$ and $t > \dim_F \mathcal{Z}'$ are arbitrary numbers. Let us choose $\varepsilon_0 > 0$ so small that

$$\varepsilon^t N_\varepsilon(\mathcal{Z}') \le 1\,, \quad \forall \varepsilon \in (0, \varepsilon_0)\,. \tag{3.15}$$

Suppose that $\{B_{r_i}\}$ is for $\varepsilon \in (0, \varepsilon_0)$ an ε-cover of $\mathcal{Z}$ such that $\sum_i r_i^2 < 1$. For each i let $\{\mathcal{B}'_{r_i}(p_{ij})\}$ be a cover of $\mathcal{Z}'$ with the minimal number $N_{r_i}(\mathcal{Z}')$ of balls with radius r_i.

It follows that in $\mathcal{M} \times \mathcal{M}'$ each set $\mathcal{B}_{r_i} \times \mathcal{Z}'$ is covered by $N_{r_i}(\mathcal{Z}')$ balls $\widetilde{\mathcal{B}}_{\sqrt{2}r_i} \supset \mathcal{B}_{r_i} \times \mathcal{B}'_{r_i}(p_{ij})$. But this implies that $\mathcal{Z} \times \mathcal{Z}' \subset \bigcup_i \bigcup_j \widetilde{\mathcal{B}}_{\sqrt{2}r_i}$ and, using (3.15),

$$\mu_H(\mathcal{Z} \times \mathcal{Z}', s+t, \sqrt{2}\,\varepsilon) \le \sum_i \sum_j (\sqrt{2}\,r_i)^{s+t}$$

$$\le \sum_i N_{r_i}(\mathcal{Z}')(\sqrt{2})^{s+t} r_i^{s+t} \le (\sqrt{2})^{s+t} \sum_i r_i^s < +\infty.$$

Thus we have a constant $c > 0$ such that $\mu_H(\mathcal{Z} \times \mathcal{Z}', s+t, \sqrt{2}\,\varepsilon) \le c$ and, consequently, $\mu_H(\mathcal{Z} \times \mathcal{Z}', s+t) = \sup_{\varepsilon>0} \mu_H(\mathcal{Z} \times \mathcal{Z}', s+t, \sqrt{2}\,\varepsilon) \le c$. It follows that $\dim_H(\mathcal{Z} \times \mathcal{Z}') \le s+t$. Since $s > \dim_H \mathcal{Z}$ and $t > \dim_F \mathcal{Z}'$ are arbitrary, we get $\dim_H(\mathcal{Z} \times \mathcal{Z}') \le \dim_H \mathcal{Z} + \dim_F \mathcal{Z}'$. □

The next proposition is shown in [51] and we state it without proof.

Proposition 3.25 *Suppose that $(\mathcal{M}, \rho)$, $(\mathcal{M}', \rho')$ are metric spaces, $\mathcal{Z} \subset \mathcal{M}$ and $\mathcal{Z}' \subset \mathcal{M}'$ are arbitrary subsets. Then we have*

$$\dim_H(\mathcal{Z} \times \mathcal{Z}') \ge \dim_H \mathcal{Z} + \dim_H \mathcal{Z}'.$$

Corollary 3.2 *Suppose that $(\mathcal{M}, \rho)$, $(\mathcal{M}', \rho')$ are metric spaces, $\mathcal{Z} \subset \mathcal{M}$ and $\mathcal{Z}' \subset \mathcal{M}'$ are totally bounded sets, and $\dim_H \mathcal{Z}' = \dim_F \mathcal{Z}'$. Then*

$$\dim_H(\mathcal{Z} \times \mathcal{Z}') = \dim_H \mathcal{Z} + \dim_H \mathcal{Z}'.$$

Proof From Proposition 3.25 it follows immediately that

$$\dim_H \mathcal{Z} + \dim_H \mathcal{Z}' \le \dim_H(\mathcal{Z} \times \mathcal{Z}')$$

and from Proposition 3.24 and the assumption of the corollary we conclude that

$$\dim_H(\mathcal{Z} \times \mathcal{Z}') \le \dim_H \mathcal{Z} + \dim_F \mathcal{Z}' = \dim_H \mathcal{Z} + \dim_H \mathcal{Z}'.$$

□

Example 3.17 (a) Suppose $(\mathcal{M}, \rho)$ is a metric space, $\mathcal{Z} \subset \mathcal{M}$ is totally bounded and $\mathcal{Z}' := [a, b] \subset \mathbb{R}$. Then we have $\dim_H \mathcal{Z}' = \dim_F \mathcal{Z}' = 1$ and, by Corollary 3.2, $\dim_H(\mathcal{Z} \times [a, b]) = \dim_H \mathcal{Z} + 1$.

(b) Suppose $\mathcal{Z}$ is as in a) and $\mathcal{Z}'$ is the Cantor set. We have shown in Examples 3.8 and 3.11 that $\dim_H \mathcal{Z}' = \dim_F \mathcal{Z}' = \frac{\log 2}{\log 3}$. It follows from Corollary 3.2 and also that $\dim_H(\mathcal{Z} \times \mathcal{Z}') = \dim_H \mathcal{Z} + \dim_H \mathcal{Z}' = \dim_H \mathcal{Z} + \frac{\log 2}{\log 3}$.

3.3 Topological Entropy

In 1965 Adler et al. [1] defined the topological entropy as an analogon of the metric entropy introduced previously by Kolmogorov and Sinai [28, 48]. In contrast to the last one, the topological entropy is defined without using any invariant measure. An equivalent definition of topological entropy was introduced by Bowen [5, 6] and Dinaburg [10] using spanning and separated sets. The application of the concept of topological entropy to one-dimensional maps is one of the best-investigated cases. Here, for example, it is shown that the positiveness of entropy of a map is equivalent to its random behaviour. The applications of topological entropy to the study of high-dimensional discrete and continuous-time dynamical systems is less well understood. However also in this case there were found some interesting connections with dimension-like characteristics of attractors.

3.3.1 *The Bowen-Dinaburg Definition*

In the present subsection we shall give a definition of topological entropy and describe its essential properties. The representation is based on the papers [5, 6, 33]. The analysis is done in a compact metric space $\mathcal{M}$ with a metric ρ. We also suppose that a continuous map $\varphi : \mathcal{M} \to \mathcal{M}$ is given. Let us recall that by $\mathcal{B}_r(p)$ there is denoted a ball of radius r with its centre p in the space $\mathcal{M}$

$$\mathcal{B}_r(p) = \big\{q \in \mathcal{M} |\ \rho(p,q) < r\big\}.$$

For any integer number $m > 0$ we define in $\mathcal{M}$ the *Bowen ball* with centre in p and radius r by the equality

$$\mathcal{B}_r(p,m) := \big\{q \in \mathcal{M} |\ \max_{0 \le j \le m-1} \rho\big(\varphi^j(p), \varphi^j(q)\big) < r\big\}.$$

Since φ is continuous, it follows that for any $\varepsilon > 0$ and any integer $m > 0$ there can be found a number $\delta > 0$ such that $\mathcal{B}_\delta(p) \subset \mathcal{B}_\varepsilon(p,m)$. Therefore the Bowen ball is an open set. Consequently, in virtue of compactness of $\mathcal{M}$, the number of the Bowen balls which are necessary to cover $\mathcal{M}$ is finite.

Let $N_\varepsilon(\mathcal{M}, m)$ be a minimal number of such balls. It is clear that $N_{\varepsilon'}(\mathcal{M}, m) \le N_\varepsilon(\mathcal{M}, m)$ for $\varepsilon' \ge \varepsilon$. Whence it follows that the following definition is correct.

Definition 3.10 *The topological entropy* of the continuous map $\varphi : \mathcal{M} \to \mathcal{M}$ of the compact metric space $(\mathcal{M}, \rho)$ is given by

$$h_{\text{top}}(\varphi, \mathcal{M}) \equiv h_{\text{top}}(\varphi) = \lim_{\varepsilon \to 0+} \limsup_{m \to +\infty} \frac{1}{m} \log N_\varepsilon(\mathcal{M}, m).$$

The set $\mathcal{P} \subset \mathcal{M}$ is said to be *(m, ε)-spanning set* for $\mathcal{M}$ with respect to φ, if for any $p \in \mathcal{M}$ there exists an $q \in \mathcal{P}$ such that $p \in \mathcal{B}_\varepsilon(q, m)$. The number $N_\varepsilon(\mathcal{M}, m)$ introduced above may be considered as the smallest cardinality of an (m, ε)-spanning set for $\mathcal{M}$ with respect to φ.

A set $\mathcal{R} \subset \mathcal{M}$ is said to be *(m, ε)-separated* with respect to φ if for any $p, q \in \mathcal{R}$ with $p \neq q$ the inequality

$$\max_{0 \le j \le m-1} \rho\big(\varphi^j(p), \varphi^j(q)\big) > \varepsilon$$

is satisfied. Let $S_\varepsilon(\mathcal{M}, m)$ denote the largest cardinality of an (m, ε)-separated set $\mathcal{R} \subset \mathcal{M}$ with respect to φ. Actually, it follows from the next lemma that both $S_\varepsilon(\mathcal{M}, m)$ as $N_\varepsilon(\mathcal{M}, m)$ are finite numbers.

Lemma 3.5 *Suppose that $\varepsilon > 0$ and $m \in \mathbb{N}$ are arbitrary. Then the following inequalities hold:*

(1) $N_\varepsilon(\mathcal{M}, m) \le S_\varepsilon(\mathcal{M}, m) \le N_{\varepsilon/2}(\mathcal{M}, m)$*;*
(2) $S_{\varepsilon'}(\mathcal{M}, m) \le S_\varepsilon(\mathcal{M}, m)$ *for any* $\varepsilon' \ge \varepsilon$.

Proof Prove that $S_\varepsilon(\mathcal{M}, m) \le N_{\varepsilon/2}(\mathcal{M}, m)$. Let $\mathcal{R}$ be an (ε, m)-separated set, and $\mathcal{P}$ be an $(\varepsilon/2, m)$-spanning set for $\mathcal{M}$ with respect to φ. We define a map $\psi : \mathcal{R} \to \mathcal{P}$ in the following way. For $p \in \mathcal{R}$ we choose some point $\psi(p) \in \mathcal{P}$ such that $p \in \mathcal{B}_{\varepsilon/2}\big(\psi(p), m\big)$. If $\psi(p) = \psi(q)$ for $p, q \in \mathcal{R}$, then

$$\max_{0 \le j \le m-1} \rho\big(\varphi^j(p), \varphi^j(q)\big) \le \frac{\varepsilon}{2} + \frac{\varepsilon}{2} = \varepsilon.$$

Therefore $p = q$. Consequently, ψ is a bijective map from $\mathcal{R}$ on $\psi(\mathcal{R}) \subset \mathcal{P}$. It follows that the cardinality of $\mathcal{R}$ is not greater than the cardinality of $\mathcal{P}$.

We now prove that $N_\varepsilon(\mathcal{M}, m) \le S_\varepsilon(\mathcal{M}, m)$. Let $\mathcal{R}$ be an (m, ε)-separated set of a maximal cardinality $S_\varepsilon(\mathcal{M}, m)$. Let us show that $\mathcal{R}$ is an (m, ε)-spanning set. Assuming the opposite, that there exists an $p \in \mathcal{M}$ such that

$$p \notin \mathcal{B}_\varepsilon(q, m), \quad \text{for all} \quad q \in \mathcal{R}.$$

Then $\mathcal{R} \cup \{p\}$ is also an (m, ε)-separated set which contradicts the choice of $\mathcal{R}$. Thus, part (1) of the lemma is proved. Assertion (2) is obvious. □

The following proposition is an immediate corollary of the lemma proved above.

Proposition 3.26 *Suppose $(\mathcal{M}, \rho)$ is a compact metric space and $\varphi : \mathcal{M} \to \mathcal{M}$ is continuous. Then*

$$h_{\text{top}}(\varphi, \mathcal{M}) = \lim_{\varepsilon \to 0+} \limsup_{m \to +\infty} \frac{1}{m} \log S_\varepsilon(\mathcal{M}, m).$$

Example 3.18 Suppose $\varphi : \mathcal{M} \to \mathcal{M}$ is an isometry, i.e. for all $p, q \in \mathcal{M}$ one has $\rho\big(\varphi(\mathcal{M}), \varphi(q)\big) = \rho(p, q)$. Then $h_{\text{top}}(\varphi, \mathcal{M}) = 0$.

Indeed, for an isometry we have $\mathcal{B}_\varepsilon(p, m) = \mathcal{B}_\varepsilon(p)$ for any $p \in \mathcal{M}$, $\varepsilon > 0$ and $m \in \mathbb{N}$. In other words, each Bowen ball coincides with the usual metric ball. Therefore $N_\varepsilon(\mathcal{M}, m)$ does not depend on m and the result follows immediately from definition.

Example 3.19 Consider Example 1.5, Chap. 1, i.e. the dynamical system $\big(\{\vartheta^m\}_{m\in\mathbb{Z}_+}, \Omega_2^+, \rho\big)$. Let us show that $h_{\text{top}}(\vartheta, \mathcal{M}) = \log 2$.

Denote by $\mathcal{P}^k$ the set of finite sequences $(\omega_1, \omega_2, \ldots, \omega_k)$, consisting of 0 and 1. Obviously, the number of elements in $\mathcal{P}^k$ is 2^k. Let l be an arbitrary natural number. From the inequality $\rho(\omega, \omega') > 2^{-l}$ it follows that there can be found a $j \in \{1, 2, \ldots, l\}$ such that $\omega_j \neq \omega'_j$ or, more generally, the inequality $\rho\big(\vartheta^m(\omega), \vartheta^m(\omega')\big) > 2^{-l}$ implies the existence of $j \in \{1, 2, \ldots, m+l\}$ such that $\omega_j \neq \omega'_j$. Consequently, the number of elements in any $(m, 2^{-l})$-separated set $\mathcal{R}$ of the space $\mathcal{M}$ does not exceed the number of elements in the set $\mathcal{P}^{m+l}$. Therefore $S_{2^{-l}}(\mathcal{M}, m) \le 2^{m+l}$.

Let us enumerate all elements of the set $\mathcal{P}^m$ and to each its j-th element ($j = 1, 2, \ldots, 2^m$) assign an element from the space $\mathcal{M}$ in accordance with the rule

$$(\omega_1, \ldots, \omega_m) \to (\omega_1, \ldots, \omega_m, 0, \ldots, 0, 1, 0, \ldots),$$

i.e., on the $(m + j)$-th place is 1, and the other elements of a 'tail' are zero. It is clear that the subset of $\mathcal{M}$, constructed above, is an $(\frac{1}{2}, m)$-separated one. It follows that $S_{\frac{1}{2}}(\mathcal{M}, m) \ge 2^m$.

Let $\varepsilon \in (0, \frac{1}{2})$ be an arbitrary number. Let us choose a natural l such that $2^{-l} < \varepsilon$. Using the monotonicity of the value $S_\varepsilon(\mathcal{M}, m)$ with respect to ε, we obtain

$$S_{\frac{1}{2}}(\mathcal{M}, m) \le S_\varepsilon(\mathcal{M}, m) \le S_{2^{-l}}(\mathcal{M}, m).$$

Consequently,

$$2^m \le S_\varepsilon(\mathcal{M}, m) \le 2^{m+l}.$$

It follows that

$$\log 2 \le \frac{1}{m} \log S_\varepsilon(\mathcal{M}, m) \le \left(1 + \frac{l}{m}\right) \log 2 \quad \text{and}$$

$$\limsup_{m\to+\infty} \frac{1}{m} \log S_\varepsilon(\mathcal{M}, m) = \log 2.$$

3.3.2 The Characterization by Open Covers

In this subsection we give a characterization of topological entropy suggested in [1]. Suppose that for an open cover $\mathfrak{U}$ of the compact metric space $(\mathcal{M}, \rho)$ the expression $N(\mathfrak{U})$ is the minimal number of elements of $\mathfrak{U}$ which is necessary to cover $\mathcal{M}$. For any continuous map $\varphi : \mathcal{M} \to \mathcal{M}$ and any open covers $\mathfrak{U}$ and $\mathfrak{V}$ of $\mathcal{M}$ introduce the notion

$$\mathfrak{U} \vee \mathfrak{V} := \{\mathcal{U} \cap \mathcal{V} \mid \mathcal{U} \in \mathfrak{U},\ \mathcal{V} \in \mathfrak{V}\},$$
$$\varphi^{-1}\mathfrak{U} := \{\varphi^{-1}(\mathcal{U}) \mid \mathcal{U} \in \mathfrak{U}\} \quad \text{and} \quad H(\mathfrak{U}) := \log N(\mathfrak{U}).$$

Note that the product $\mathfrak{U} \vee \mathfrak{V}$ of two covers and the preimage $\varphi^{-1}(\mathfrak{U})$ are open covers of $\mathcal{M}$. Define the *entropy of the map* φ *with respect to the cover* $\mathfrak{U}$ by

$$h(\varphi, \mathfrak{U}) := \lim_{m \to +\infty} \frac{1}{m} \log N(\mathfrak{U} \vee \varphi^{-1}\mathfrak{U} \vee \cdots \vee \varphi^{-(m-1)}\mathfrak{U}). \tag{3.16}$$

In order to verify the correctness of the given definition, it is necessary to prove the existence of the limit in the right-hand side of (3.16). This follows from the following lemmas.

Lemma 3.6 *Suppose that* $\varphi : \mathcal{M} \to \mathcal{M}$ *is continuous and* $\mathfrak{U}$ *and* $\mathfrak{V}$ *are open covers of* $\mathcal{M}$. *Then the following relations are true:*

(1) $\varphi^{-1}(\mathfrak{U} \vee \mathfrak{V}) = \varphi^{-1}\mathfrak{U} \vee \varphi^{-1}\mathfrak{V}$;
(2) $H(\mathfrak{U} \vee \mathfrak{V}) \le H(\mathfrak{U}) + H(\mathfrak{V})$;
(3) $H(\varphi^{-1}\mathfrak{U}) \le H(\mathfrak{U})$;
(4) $H(\varphi^{-1}\mathfrak{U}) = H(\mathfrak{U})$, *if* φ *is surjective.*

Proof The validity of this lemma follows directly from the definitions of $\mathfrak{U} \vee \mathfrak{V}$, $\varphi^{-1}\mathfrak{U}$ and $H(\mathfrak{U})$. □

Lemma 3.7 *Let* $\{a_m\}_{m \ge 1}$ *be a sequence of real numbers such that* $a_{m+i} \le a_m + a_i$ *for any* $m, i \in \mathbb{N}$. *Then the limit*

$$\lim_{m \to +\infty} \frac{a_m}{m}$$

exists and is equal to $\inf\limits_{m \in \mathbb{N}} \frac{a_m}{m}$.

Proof Let us consider a natural i to be fixed and represent m in the form $m = ki + j$, where k, j are non-negative integer numbers and $0 \le j < i$. We have

$$\frac{a_m}{m} = \frac{a_{ki+j}}{ki + j} \le \frac{a_{ki}}{ki} + \frac{a_j}{ki} \le \frac{ka_i}{ki} + \frac{a_j}{ki}.$$

If $m \to +\infty$, then also $k \to +\infty$. Therefore

$$\limsup_{m\to+\infty} \frac{a_m}{m} \le \frac{a_i}{i}.$$

Consequently,

$$\limsup_{m\to+\infty} \frac{a_m}{m} \le \inf_i \frac{a_i}{i}.$$

On the other hand

$$\inf_i \frac{a_i}{i} \le \liminf_{m\to+\infty} \frac{a_m}{m}.$$

□

Lemma 3.8 *The limit in the right-hand side of (3.16) exists.*

Proof By Lemma 3.6 for any natural numbers m and i we have

$$\begin{aligned} &H(\mathfrak{U} \vee \varphi^{-1}\mathfrak{U} \vee \cdots \vee \varphi^{-(m+i-1)}\mathfrak{U}) \\ &\quad = H\big(\mathfrak{U} \vee \cdots \vee \varphi^{-(m-1)}\mathfrak{U} \vee \varphi^{-m}(\mathfrak{U} \vee \cdots \vee \varphi^{-(i-1)}\mathfrak{U})\big) \\ &\quad \le H(\mathfrak{U} \vee \cdots \vee \varphi^{-(m-1)}\mathfrak{U}) + H\big(\varphi^{-m}(\mathfrak{U} \vee \cdots \vee \varphi^{-(i-1)}\mathfrak{U})\big) \\ &\quad \le H(\mathfrak{U} \vee \cdots \vee \varphi^{-(m-1)}\mathfrak{U}) + H(\mathfrak{U} \vee \cdots \vee \varphi^{-(i-1)}\mathfrak{U}). \end{aligned}$$

Therefore the validity of Lemma 3.8 follows from Lemma 3.7, if we put for any $m \in \mathbb{N}$

$$a_m = H(\mathfrak{U} \vee \varphi^{-1}\mathfrak{U} \vee \cdots \vee \varphi^{-(m-1)}\mathfrak{U}).$$

□

Thus, the correctness of the definition of the entropy of φ with respect to $\mathfrak{U}$, i.e. $h(\varphi, \mathfrak{U})$ is proven.

Proposition 3.27 *Suppose $(\mathcal{M}, \rho)$ is a compact metric space and $\varphi : \mathcal{M} \to \mathcal{M}$ is continuous. Then $h_{\text{top}}(\varphi, \mathcal{M}) = \sup h(\varphi, \mathfrak{U})$, where the supremum is taken over all finite open covers $\mathfrak{U}$ of $\mathcal{M}$.*

Proof Denote for brevity $h^* := \sup_{\mathfrak{U}} h(\varphi, \mathfrak{U})$, where the supremum is taken over all finite open covers $\mathfrak{U}$ of $\mathcal{M}$.

We show at first that $h^* \le h_{\text{top}}(\varphi, \mathcal{M})$. Let $\mathfrak{U} := \{\mathcal{U}_1, \ldots, \mathcal{U}_r\}$ be some open cover of $\mathcal{M}$ and δ be the Lebesgue number of $\mathfrak{U}$ (see Sect. 3.1.2). Let also $\mathcal{P}$ be an $(m, \delta/2)$-spanning set of minimal cardinality for $\mathcal{M}$. For $q' \in \mathcal{P}$, choose $\mathcal{U}_{j_0}(q'), \ldots, \mathcal{U}_{j_{m-1}}(q')$ in $\mathfrak{U}$ such that

$$\overline{\mathcal{B}_{\delta/2}\big(\varphi^k(q')\big)} \subset \mathcal{U}_{j_k}(q'), \quad k = 0, 1, \ldots, m-1.$$

It follows that

$$\mathcal{C}(q') := \mathcal{U}_{j_0}(q') \cap \varphi^{-1}(\mathcal{U}_{j_1}(q')) \cap \cdots \cap \varphi^{-(m-1)}(\mathcal{U}_{j_{m-1}}(q'))$$

is an element of the cover $\mathfrak{U} \vee \varphi^{-1}\mathfrak{U} \vee \cdots \vee \varphi^{-(m-1)}\mathfrak{U}$.

We have

$$\mathcal{M} = \bigcup_{q' \in \mathcal{P}} \mathcal{C}(q'),$$

since for every $p \in \mathcal{M}$, there can be found a $q' \in \mathcal{P}$ such that $p \in \mathcal{B}_{\delta/2}(q', m)$ and, consequently, $p \in \varphi^{-k}(\mathcal{B}_{\delta/2}(\varphi^k(q'))) \subset \varphi^{-k}(\mathcal{U}_{j_k}(q'))$, $0 \le k \le m-1$. Therefore $p \in \mathcal{C}(q')$.

Thus,

$$N(\mathfrak{U} \vee \varphi^{-1}\mathfrak{U} \vee \cdots \vee \varphi^{-(m-1)}\mathfrak{U}) \le \operatorname{card}\mathcal{P} = N_{\delta/2}(\mathcal{M}, m) \tag{3.17}$$

and

$$h(\varphi, \mathfrak{U}) \le \limsup_{m \to +\infty} \frac{1}{m} \log N_{\delta/2}(\mathcal{M}, m).$$

If we let δ tend to 0, then we obtain $h^*(\varphi) \le h^*_{\text{top}}(\varphi, \mathcal{M})$.

In order to prove the opposite equality, suppose that a $\delta > 0$ is given. Let us choose for $\mathcal{M}$ an open cover $\mathfrak{U} = \{\mathcal{U}_1, \ldots, \mathcal{U}_r\}$ such that diam $(\mathcal{U}_j) < \delta$ for all $j = 1, 2, \ldots, r$. Let $\mathcal{R}$ be an (m, δ)-separated subset of $\mathcal{M}$ with maximal cardinality. It should be noted that two elements $p, q \in \mathcal{R}$, $p \neq q$ can not belong to one and the same element of $\mathfrak{U} \vee \varphi^{-1}\mathfrak{U} \vee \cdots \vee \varphi^{-(m-1)}\mathfrak{U}$, since if

$$p, q \in \bigcap_{k=0}^{m-1} \varphi^{-k}(\mathcal{U}_{j_k}),$$

then

$$\max_{0 \le k \le m-1} \rho(\varphi^k(p), \varphi^k(q)) < \delta$$

and therefore $p = q$. Thus,

$$N(\mathfrak{U} \vee \varphi^{-1}\mathfrak{U} \vee \cdots \vee \varphi^{-(m-1)}\mathfrak{U}) \ge \operatorname{card}\mathcal{R} = S_\delta(\mathcal{M}, m).$$

Thus,

$$h^*(\varphi) \ge h(\varphi, \mathfrak{U}) \ge \limsup_{m \to +\infty} \frac{1}{m} \log S_\delta(\mathcal{M}, m).$$

If we let δ tend to 0, then we obtain $h^*(\varphi) \ge h_{\text{top}}(\varphi, \mathcal{M})$. □

3.3.3 Some Properties of the Topological Entropy

In this subsection we shall consider some important properties of the topological entropy of a map or of a flow. Most of these properties can be found in [1, 27, 30].

Proposition 3.28 *Suppose if $\varphi : \mathcal{M} \to \mathcal{M}$ is a homeomorphism of the compact metric space $(\mathcal{M}, \rho)$. Then*

$$h_{\text{top}}(\varphi, \mathcal{M}) = h_{\text{top}}(\varphi^{-1}, \mathcal{M}) .$$

Proof By the properties (4) and (1) of Lemma 3.6 we have for any finite open cover $\mathfrak{U}$ of $\mathcal{M}$ and any natural m

$$H(\mathfrak{U} \vee \varphi^{-1}\mathfrak{U} \vee \cdots \vee \varphi^{-(m-1)}\mathfrak{U}) = H\big(\varphi^{m-1}(\mathfrak{U} \vee \varphi^{-1}\mathfrak{U} \vee \cdots \vee \varphi^{-(m-1)}\mathfrak{U})\big)$$
$$= H(\mathfrak{U} \vee \varphi\mathfrak{U} \vee \cdots \vee \varphi^{m-1}\mathfrak{U}) = H\big(\mathfrak{U} \vee (\varphi^{-1})^{-1}\mathfrak{U} \vee \cdots \vee (\varphi^{-1})^{-(m-1)}\mathfrak{U}\big).$$

Therefore $h(\varphi, \mathfrak{U}) = h(\varphi^{-1}, \mathfrak{U})$ and the assertion follows from Proposition 3.27.
□

Proposition 3.29 *Suppose that $(\mathcal{M}_1, \rho_1)$ and $(\mathcal{M}_2, \rho_2)$ are compact metric spaces, $\varphi_j : \mathcal{M}_j \to \mathcal{M}_j$ $(j = 1, 2)$ are continuous maps and $\chi : \mathcal{M}_1 \to \mathcal{M}_2$ is a homeomorphism such that $\chi \circ \varphi_1 = \varphi_2 \circ \chi$, i.e. χ is a conjugacy between φ_1 and φ_2. Then $h_{\text{top}}(\varphi_1, \mathcal{M}_1) = h_{\text{top}}(\varphi_2, \mathcal{M}_2)$. In particular, the topological entropy does not depend on the choice of the metric.*

Proof By relations (4) and (1) of Lemma 3.6 we have for an arbitrary finite open cover $\mathfrak{U}$ of $\mathcal{M}$ and arbitrary natural m

$$H(\mathfrak{U} \vee \varphi_2^{-1}\mathfrak{U} \vee \cdots \vee \varphi_2^{-(m-1)}\mathfrak{U}) = H\big(\chi^{-1}(\mathfrak{U} \vee \varphi_2^{-1}\mathfrak{U} \vee \cdots \vee \varphi_2^{-(m-1)}\mathfrak{U})\big)$$
$$= H(\chi^{-1}\mathfrak{U} \vee \varphi_1^{-1}\chi^{-1}\mathfrak{U} \vee \cdots \vee \varphi_1^{-(m-1)}\chi^{-1}\mathfrak{U}).$$

Therefore, $h(\varphi_2, \mathfrak{U}) = h(\varphi_1, \chi^{-1}\mathfrak{U})$ and the assertion follows from Proposition 3.27. □

Proposition 3.30 *Suppose that $\varphi : \mathcal{M} \to \mathcal{M}$ is a continuous map of the compact metric space $(\mathcal{M}, \rho)$ and $\mathcal{Z} \subset \mathcal{M}$ is a closed and φ-invariant subset. Then $h_{\text{top}}(\varphi, \mathcal{Z}) \leq h_{\text{top}}(\varphi, \mathcal{M})$.*

Proof Let $\widetilde{\mathfrak{U}}$ be some finite open cover of $\mathcal{Z}$. For each $\widetilde{\mathcal{U}} \in \widetilde{\mathfrak{U}}$ there can be found an open subset $\mathcal{U}$ such that $\widetilde{\mathcal{U}} = \mathcal{U} \cap \mathcal{Z}$. A family of such sets $\mathcal{U}$ together with the open set $\mathcal{M} \setminus \mathcal{Z}$ forms a finite open cover $\mathfrak{U}$ of the set $\mathcal{M}$.

Suppose that $\widetilde{\varphi} = \varphi_{|\mathcal{Z}}$ and let m be an arbitrary natural number. Then

$$H(\widetilde{\mathfrak{U}} \vee \widetilde{\varphi}^{-1}\widetilde{\mathfrak{U}} \vee \cdots \vee \widetilde{\varphi}^{-(m-1)}\widetilde{\mathfrak{U}}) \leq H(\mathfrak{U} \vee \varphi^{-1}\mathfrak{U} \vee \cdots \vee \varphi^{-(m-1)}\mathfrak{U}).$$

Therefore $h(\widetilde{\varphi}, \widetilde{\mathfrak{U}}) \leq h(\varphi, \mathfrak{U})$ and the assertion follows from Proposition 3.27. □

Proposition 3.31 *Suppose $(\mathcal{M}, \rho)$ is a compact metric space and $\varphi : \mathcal{M} \to \mathcal{M}$ is continuous. Then for any natural k*

$$h_{\text{top}}(\varphi^k, \mathcal{M}) = k\, h_{\text{top}}(\varphi, \mathcal{M}). \tag{3.18}$$

Proof If k is an arbitrary natural and $\mathfrak{U}$ is an arbitrary finite open cover of $\mathcal{M}$, then

$$h_{\text{top}}(\varphi^k, \mathcal{M}) \geq h(\varphi^k, \mathfrak{U} \vee \varphi^{-1}\mathfrak{U} \vee \cdots \vee \varphi^{-(k-1)}\mathfrak{U})$$
$$= \lim_{m\to+\infty} k\, H(\mathfrak{U} \vee \varphi^{-1}\mathfrak{U} \vee \cdots \vee \varphi^{-k+1}\mathfrak{U} \vee \varphi^{-k}\mathfrak{U} \vee \cdots \vee \varphi^{-(mk-1)}\mathfrak{U})/mk = k\, h(\varphi, \mathfrak{U}).$$

Consequently, $h_{\text{top}}(\varphi^k, \mathcal{M}) \geq k\, h_{\text{top}}(\varphi, \mathcal{M})$.
On the other hand, since there is the refinement

$$\mathfrak{U} \vee (\varphi^k)^{-1}\mathfrak{U} \vee \cdots \vee (\varphi^k)^{-(m-1)}\mathfrak{U} \prec \mathfrak{U} \vee \varphi^{-1}\mathfrak{U} \vee \cdots \vee \varphi^{-(mk-1)}\mathfrak{U},$$

by (3.16) we have

$$h(\varphi, \mathfrak{U}) = \lim_{m\to+\infty} H(\mathfrak{U} \vee \varphi^{-1}\mathfrak{U} \vee \cdots \vee \varphi^{-(mk-1)}\mathfrak{U})/mk \geq h(\varphi^k, \mathfrak{U})/k.$$

And the assertion again follows the Proposition 3.27. □

Let us consider on the metric space $(\mathcal{M}, \rho)$ a family of continuous maps $\{\varphi^t\}_{t\in\mathbb{R}}$, having the following properties:

(1) φ^0 is the identity map on $\mathcal{M}$;
(2) $\varphi^{t+s}(p) = \varphi^t\big(\varphi^s(p)\big)$ for any $t, s \in \mathbb{R}$ and $p \in \mathcal{M}$;
(3) The map $(t, p) \mapsto \varphi^t(p)$ is continuous on $\mathbb{R} \times \mathcal{M}$.

The family $\{\varphi^t\}_{t\in\mathbb{R}}$, satisfying (1)–(3) is called C^0-*flow* in $\mathcal{M}$.
For such a C^0-flow the following generalization of formula (3.18) is valid [6].

Proposition 3.32 *Suppose $\{\varphi^t\}_{t\in\mathbb{R}}$ is a C^0-flow on the compact metric space $(\mathcal{M}, \rho)$. Then for any $t \in \mathbb{R}$*

$$h_{\text{top}}(\varphi^t, \mathcal{M}) = |t|\, h_{\text{top}}(\varphi^1, \mathcal{M}).$$

Proof Let us show that for any $t, s > 0$

$$h_{\text{top}}(\varphi^t, \mathcal{M}) \leq \frac{t}{s}\, h_{\text{top}}(\varphi^s, \mathcal{M}). \tag{3.19}$$

From the continuity of the flow it follows that

$$\forall \varepsilon > 0 \;\; \exists \delta > 0 \;\; \forall p, q \in \mathcal{M}, \quad \rho(p, q) \leq \delta \, : \rho(\varphi^r(p), \varphi^r(q)) \leq \varepsilon, \; 0 \leq r \leq s.$$

Suppose $\mathcal{P} \subset \mathcal{M}$ is an (n, δ)-spanning set for $\mathcal{M}$ w.r.t. φ^s, i.e. assume that

$$\forall p \in \mathcal{M} \quad \exists q \in \mathcal{P} \quad \rho(p, q) \leq \delta, \ldots, \rho\Big(\varphi^{s(n-1)}(p), \varphi^{s(n-1)}(q)\Big) \leq \delta.$$

It follows that for all $0 \leq r \leq s$

$$\rho(\varphi^r(p), \varphi^r(q)) \leq \varepsilon, \ldots, \rho\Big(\varphi^{s(n-1)+r}(p), \varphi^{s(n-1)+r}(q)\Big) \leq \varepsilon\,.$$

But this implies that $\mathcal{P}$ is an (m, ε)-spanning set for $\mathcal{M}$ with respect to φ^t if $mt \leq s(n-1) + s = ns$, and

$$N_\varepsilon(\mathcal{M}, m, \varphi^t) \leq N_\delta\Big(\mathcal{M}, \Big[\frac{mt}{s}\Big] + 1, \varphi^s\Big).$$

(Here $[\cdot]$ denotes the integer part.) As a consequence we have

$$\limsup_{m \to \infty} \frac{1}{m} \log N_\varepsilon(\mathcal{M}, m, \varphi^t) \leq \frac{t}{s} h_{\text{top}}(\varphi^s, \mathcal{M}),$$

and, consequently, the inequality (3.19).

Now we can write with $t > 0$ and $s = 1$

$$h_{\text{top}}(\varphi^t, \mathcal{M}) \leq t\, h_{\text{top}}(\varphi^1, \mathcal{M}) \leq t\, \frac{1}{t} h_{\text{top}}(\varphi^t, \mathcal{M}) = h_{\text{top}}(\varphi^t, \mathcal{M}).$$

For negative t consider the map $\psi := \varphi^{-1}$ and write for $s = -t$

$$h_{\text{top}}(\psi^s, \mathcal{M}) = s\, h_{\text{top}}(\psi^1, \mathcal{M}) = (-t) h_{\text{top}}(\varphi^{-1}, \mathcal{M}) = (-t) h_{\text{top}}(\varphi, \mathcal{M})$$

(Propositions 3.28 and 3.31).

Since $\varphi^0 = \mathrm{id}_{\mathcal{M}}$ we conclude from Example 3.18 that $h_{\text{top}}(\varphi^0, \mathcal{M}) = 0$. □

Consider again a metric space $(\mathcal{M}, \rho)$, a continuous map $\varphi : \mathcal{M} \to \mathcal{M}$ and the associated dynamical system $\{\varphi^k\}_{k \in \mathbb{Z}_+}$. A point $p \in \mathcal{M}$ is said to be *non-wandering* with respect to the map φ (or the dynamical system $\{\varphi^k\}_{k \in \mathbb{Z}}$) if for any open neighborhood $\mathcal{U}$ of the point p and arbitrary $k > 0$ it can be found a natural number $m > k$, $m \in \mathbb{N}$, such that $\varphi^m(\mathcal{U}) \cap \mathcal{U} \neq \emptyset$.

We have the following result [5, 6].

Proposition 3.33 *Suppose that $\mathcal{NW}(\varphi)$ is the set of all non-wandering points of φ. Then $h_{\text{top}}(\varphi, \mathcal{M}) = h_{\text{top}}\big(\varphi, \mathcal{NW}(\varphi)\big)$.*

In the previous sections, when introducing different dimensions, we found relations between their values and finally got the chain

$$\dim_T \mathcal{Z} \leq \dim_H \mathcal{Z} \leq \dim_F \mathcal{Z}. \tag{3.20}$$

It is natural to find a connection between the topological entropy of a map φ on $\mathcal{M}$ and the dimension of the space $\mathcal{M}$, in which φ acts. Such results are found for all three types of dimensions in (3.20) (see, for example, [11, 15, 18, 26] and references in these works). But in these cases on the continuous map φ the different additional constraints are imposed. The following simple example gives a representation about results of this kind.

Example 3.20 Suppose that $\varphi : \mathcal{M} \to \mathcal{M}$ is a Lipschitz continuous map, i.e., there exists a constant $\lambda > 0$ such that

$$\rho\big(\varphi(p), \varphi(q)\big) \leq \lambda\, \rho(p, q), \quad \text{for all } p, q \in \mathcal{M}.$$

Let us show that under the condition that $\dim_F \mathcal{M} < +\infty$

$$h_{\text{top}}(\varphi, \mathcal{M}) \leq \max\{0, \log \lambda\} \dim_F \mathcal{M}.$$

Choose an arbitrary $d > \dim_F \mathcal{M}$. Then

$$\frac{\log N_\delta(\mathcal{M})}{\log \delta^{-1}} < d$$

for all sufficiently small $\delta > 0$. Let us also choose an arbitrary $\gamma > \lambda$ and $\varepsilon > 0$. Consider at first the case $\lambda \geq 1$.

The condition $\rho(p, q) < \gamma^{-m}\varepsilon$ results in the following relation

$$\rho\big(\varphi^j(p), \varphi^j(q)\big) < \lambda^j \gamma^{-m}\varepsilon \leq \gamma^{-(m-j)}\varepsilon < \varepsilon$$

for $j = 0, 1, \ldots, m-1$. Therefore $\mathcal{B}_\delta(p) \subset \mathcal{B}_\varepsilon(p, m)$ with $\delta = \gamma^{-m}\varepsilon$. Thus, $N_\varepsilon(\mathcal{M}, m) \leq N_\delta(\mathcal{M})$. Then for all sufficiently large m we have

$$\frac{\log N_\varepsilon(\mathcal{M}, m)}{m} \leq \frac{\log N_\delta(\mathcal{M})}{\log \delta^{-1}} \frac{\log \delta^{-1}}{m} < d \left(\log \gamma - \frac{\log \varepsilon}{m}\right).$$

Consequently $h_{\text{top}}(\varphi, \mathcal{M}) \leq d \log \gamma$ and $h_{\text{top}}(\varphi, \mathcal{M}) \leq \log \lambda \dim_F \mathcal{M}$.

If φ is Lipschitz with $\lambda < 1$ it also satisfies a Lipschitz condition with $\lambda = 1$. From the considered case above it follows that $h_{\text{top}}(\varphi, \mathcal{M}) = 0$.

Let us return again to Examples 3.18 and 3.19. If φ is an isometry, then φ is a Lipschitz map with $\lambda = 1$. Therefore from Example 3.20 we immediately obtain $h_{\text{top}}(\varphi, \mathcal{M}) = 0$. If φ is a shift operator on the space $\mathcal{M}$ from Example 3.19, then φ is a Lipschitz map with $\lambda = 2$. Since $h_{\text{top}}(\varphi, \mathcal{M}) = \log 2$, it follows from Example 3.20 that $\dim_F \mathcal{M} \geq 1$.

The next proposition which is taken from [27] shows that the degree of a map can give useful information on a lower bound for the topological entropy.

Proposition 3.34 *Suppose that $(\mathcal{M}, g)$ is an n-dimensional compact orientable Riemannian C^k-manifold and $\varphi : \mathcal{M} \to \mathcal{M}$ is a C^r-map ($k \geq r \geq 1$).*

Then $h_{\text{top}}(\varphi, \mathcal{M}) \geq \log |\deg \varphi|$.

Proof Suppose μ is a volume form on $\mathcal{M}$, $\beta \in (0, 1)$ is a number and define for $p \in \mathcal{M}$

$$J\varphi(p) := \det D\varphi(p)\,,\ L := \sup_{p \in \mathcal{M}} |J\varphi(p)|, \quad \varepsilon := L^{-\frac{\beta}{\beta-1}},$$

$\mathcal{V} := \{ p \in \mathcal{M} \,|\, |J\varphi(p)| \geq \varepsilon\}$. Choose an open cover $\mathfrak{U}$ with Lebesgue number $\delta > 0$ of $\mathcal{V}$ such that φ is injective on the elements of $\mathfrak{U}$. It follows that if $p, q \in \mathcal{V}$, $\rho(p, q) \leq \delta$ then $\varphi(p) \neq \varphi(q)$. Fix now $n \in \mathbb{N}$ and define the set

$$\mathcal{U} := \{p \in \mathcal{M} \,|\, \text{card}(\mathcal{V} \cap \{p, \dots, \varphi^{n-1}(p)\} \leq \beta n\}.$$

This implies that if $p \in \mathcal{U}$ then

$$|J\varphi^n(p)| = \prod_{j=0}^{n-1} |J\varphi(\varphi^j(p))| < L^{\beta n(1-\beta)n}\varepsilon = (\varepsilon^{1-\beta}L^{\beta})^n = 1.$$

It follows that $\text{vol}_n(\varphi^n(\mathcal{U})) = \int_{\varphi^n(\mathcal{U})} \mu = \int_{\mathcal{U}} J\varphi^n \mu < \text{vol}_{\text{n}}(\mathcal{U})$.

Sard's theorem (Theorem A.3, Appendix A) guarantees us the existence of a regular value $p \in \mathcal{M} \setminus \varphi^n(\mathcal{U})$ of φ^n. Take now an (n, δ)-separated set in $\varphi^{-n}(p)$. Since p is a regular point for φ, the point p has w.r.t. φ at least $N := \deg \varphi$ preimages (see Sect. B.3, Appendix B). If all N points belong to $\mathcal{V}$ we put $\mathcal{R}_1 := \varphi^{-n}(p) \cap \mathcal{V}$. In the other case let the set $\mathcal{R}_1$ contain exactly one preimage outside $\mathcal{V}$. It follows that $\mathcal{R}_1 \subset \varphi^{-n}(p)$ consists of regular points of φ.

The same procedure will be repeated for each point $q \in \mathcal{R}_1$ in order to get the inclusions $\mathcal{R}_2 \subset \varphi^{-2}(p), \dots, \mathcal{R}_n \subset \varphi^{-n}(p)$.

Let us show that the set $\mathcal{R}_n$ is (n, δ)-separated for φ. Suppose $q_1, q_2 \in \mathcal{R}_n$ are arbitrary points and $\rho(\varphi^k(q_1), \varphi^k(q_2)) \leq \delta$ for $k = 0, 1, \dots, n-1$. Then we have $\varphi^{n-1}(q_1) = \varphi^{n-1}(q_2)$. This follows from the fact that by construction $\varphi^{n-1}(q_1) \in \mathcal{V}$, $\varphi^{n-1}(q_2) \in \mathcal{V}$, δ is the Lebesgue-number, and $p = \varphi(\varphi^{n-1}(q_1)) \neq \varphi(\varphi^{n-1}(q_2)) = p$ gives a contradiction.

In the same way one shows that $\varphi^{n-2}(q_1) = \varphi^{n-2}(q_2), \dots, q_1 = q_2$. Since $\mathcal{R}_n \subset \varphi^{-n}(p) \subset \varphi^{-n}(\mathcal{M} \setminus \varphi^n(A)) \subset \mathcal{M} \setminus \mathcal{U}$ it follows that $\mathcal{R}_n \cap \mathcal{U} = \emptyset$.

But this implies that for each $q \in \mathcal{R}_n$ there exist at least βn time-values $k \in \{0, \dots, n-1\}$ for which $\varphi^k(q) \in \mathcal{V}$. In this case $\text{card}\mathcal{R}_n \geq N^m \geq N^{\beta n}$ and $S_\delta(\mathcal{M}, n, \varphi) \geq N^{\beta n}$. By definition we have

$$h_{\text{top}}(\varphi, \mathcal{M}) = \lim_{\delta \to 0} \limsup_{n \to \infty} \frac{1}{n} S_\delta(\mathcal{M}, n, \varphi) \geq \beta \log N, \forall \beta \in (0, 1).$$

□

3.4 Dimension-Like Characteristics

In this section the basic ideas of dimension-like characteristics introduced by Pesin [39, 40] are described. This approach is based on the notion of Carathéodory measure [9, 16]. It will be shown that Hausdorff and fractal dimensions, are special types of such dimension-like characteristics.

It is demonstrated in [39] that the topological entropy and the topological pressure can also be considered as dimension-like characteristics.

In Chap. 5 dimension-like characteristics are used for the estimation of Hausdorff dimension of invariant sets. All constructions and statements, represented in this section, can be found, together with complete proofs, in [39, 40].

3.4.1 Carathéodory Measure, Dimension and Capacity

Let $\mathcal{M}$ be an arbitrary set, $\mathfrak{F}$ be some set of subsets of $\mathcal{M}$, $\mathbb{P} := [d^*, +\infty)$ for finite d^* or $\mathbb{P} := \mathbb{R}$ be a parameter set, and let the following three functions $\xi : \mathfrak{F} \times \mathbb{P} \to \mathbb{R}_+$, $\eta : \mathfrak{F} \times \mathbb{R} \to \mathbb{R}_+$ and $\psi : \mathfrak{F} \to \mathbb{R}_+$ be given. We shall suppose that the following conditions are satisfied:

(A1) $\emptyset \in \mathfrak{F},\ \xi(\emptyset, d) = 0,\ \psi(\emptyset) = 0$;

(A2) $\xi(\mathcal{U}, d + d') = \eta(\mathcal{U}, d')\xi(\mathcal{U}, d), \quad \forall \mathcal{U} \in \mathfrak{F}, \forall d, d' \in \mathbb{P}$;

(A3) For any $\varepsilon > 0$ there exists $\delta > 0$ such that for any $\mathcal{U} \in \mathfrak{F}$ with $\psi(\mathcal{U}) \le \delta$, it holds $\eta(\mathcal{U}, d) \le \varepsilon$ if $d > 0$ and $\eta(\mathcal{U}, d) \ge \varepsilon^{-1}$ if $d < 0$;

(A4) For any $\varepsilon > 0$ there exists $\mathcal{U} \in \mathfrak{F}$ for which $\psi(\mathcal{U}) = \varepsilon$.

We say that $\mathcal{Z} \subset \mathcal{M}$ is an *admissible* set if for any $\varepsilon > 0$ there can be found some finite or countable set $\mathfrak{U}$ of subsets from $\mathfrak{F}$, forming a cover of $\mathcal{Z}$. Moreover we have $\psi(\mathcal{U}) = \varepsilon$ for any $\mathcal{U} \in \mathfrak{U}$.

We shall also suppose that one more condition is satisfied:

(A5) Any subset $\mathcal{Z} \subset \mathcal{M}$ is admissible.

In analogy to [39] we call such a collection $(\mathfrak{F}, \mathbb{P}, \xi, \eta, \psi)$ which satisfies **(A1)**–**(A5)** a *Carathéodory (dimension) structure on* $\mathcal{M}$.

Let $\mathcal{Z}$ be an arbitrary subset of $\mathcal{M}$ and $d \ge d^*$, $\varepsilon > 0$ numbers. Define the *Carathéodory d-measure at level* ε *of* $\mathcal{Z}$ *with respect to* $(\mathfrak{F}, \mathbb{P}, \xi, \eta, \psi)$ by

$$\mu_c(\mathcal{Z}, d, \varepsilon) := \inf \sum_{\mathcal{U} \in \mathfrak{U}} \xi(\mathcal{U}, d),$$

where the infimum is taken over all finite or countable sets $\mathfrak{U}$ of subsets from $\mathfrak{F}$ covering $\mathcal{Z}$ and such that $\psi(\mathcal{U}) \le \varepsilon$ for all $\mathcal{U} \in \mathfrak{U}$. It is obvious that $\mu_c(\mathcal{Z}, d, \varepsilon)$ for fixed $\mathcal{Z}$ and d does not decrease with decreasing ε. It follows that there exists a limit

$$\mu_C(\mathcal{Z}, d) := \lim_{\varepsilon \to 0} \mu_C(\mathcal{Z}, d, \varepsilon).$$

A simple check shows the validity of the following assertion.

Proposition 3.35 *Assuming $d \in \mathbb{P}$ to be fixed, the function $\mu_C(\cdot, d)$ is an* outer measure on $\mathcal{M}$, *i.e. it possesses the following properties:*

(1) $\mu_C(\emptyset, d) = 0$;
(2) $\mu_C(\mathcal{Z}_1, d) \le \mu_C(\mathcal{Z}_2, d)$ *if* $\mathcal{Z}_1 \subset \mathcal{Z}_2 \subset \mathcal{M}$;
(3) $\mu_C(\bigcup_{j\ge 1} \mathcal{Z}_j, d) \le \sum_{j\ge 1} \mu_C(\mathcal{Z}_j, d)$ *for all* $\mathcal{Z}_j \subset \mathcal{M}$ $(j = 1, 2, \dots)$.

Definition 3.11 The function $\mu_C(\cdot, d)$ is called *Carathéodory d-measure* with respect to $(\mathfrak{F}, \mathbb{P}, \xi, \eta, \psi)$.

From assumptions **(A1)–(A5)** it can be easily seen that the following proposition is true.

Proposition 3.36 *If $\mathcal{Z} \subset \mathcal{M}$ is kept fixed then for the function $\mu_C(\mathcal{Z}, \cdot)$ there exists $d_{cr}(\mathcal{Z}) \in \overline{\mathbb{P}}$ such that for $d \in \mathbb{P}$*

$$\mu_C(\mathcal{Z}, d) = \infty \quad for \quad d < d_{cr}(\mathcal{Z}), \quad \mu_C(\mathcal{Z}, d) = 0 \quad for \quad d > d_{cr}(\mathcal{Z}) .$$

Definition 3.12 The value $\dim_C \mathcal{Z} := d_{cr}(\mathcal{Z})$ is called *Carathéodory dimension* of the set $\mathcal{Z}$ with respect to the structure $(\mathfrak{F}, \mathbb{P}, \xi, \eta, \psi)$.

Example 3.21 For a standard Carathéodory structure let $\mathcal{M}$ be a separable metric space, $\mathfrak{F}$ the family consisting of open balls $\mathcal{B}(u, r)$ in $\mathcal{M}$ with center u and radius r and the empty set, $\mathbb{P} = \mathbb{R}_+$, $\xi(\mathcal{B}(u, r), d) = r^d$, $\eta(\mathcal{B}(u, r), s) = r^s$, $\psi(\mathcal{B}(u, r)) = r$, $\xi(\emptyset, d) = \psi(\emptyset) = 0$, and $\eta(\emptyset, s) = 1$ for each $u \in \mathcal{M}$, $r > 0$ and each $d \ge 0$, $s \in \mathbb{R}$. It is easy to see that such a system $(\mathfrak{F}, \mathbb{P}, \xi, \eta, \psi)$ defines a Carathéodory structure on $\mathcal{M}$. We denote by $\mu_H(\cdot, d, r)$, $\mu_H(\cdot, d)$ and $\dim_H$ the resulting Carathéodory measures and Carathéodory dimension which are in fact the Hausdorff d-measure at level r, the Hausdorff d-measure and the Hausdorff dimension, respectively, introduced in Sect. 3.2.1.

Let $\mathcal{Z}$ be an arbitrary subset of $\mathcal{M}$ and $d \in \mathbb{P}$, $\varepsilon > 0$ numbers. Put

$$\mathfrak{m}_C(\mathcal{Z}, d, \varepsilon) := \inf \sum_{\mathcal{U} \in \mathfrak{U}} \xi(\mathcal{U}, d),$$

where the infimum is taken over all finite or countable systems $\mathfrak{U}$ of subsets from $\mathfrak{F}$, forming a cover of $\mathcal{Z}$ and such that $\psi(\mathcal{U}) = \varepsilon$ for all $\mathcal{U} \in \mathfrak{U}$.

Let us put

$$\overline{\mathfrak{m}}_C(\mathcal{Z}, d) := \limsup_{\varepsilon \to 0} m_C(\mathcal{Z}, d, \varepsilon).$$

and $\underline{\mathfrak{m}}_C(\mathcal{Z}, d) := \liminf_{\varepsilon \to 0} m_C(\mathcal{Z}, d, \varepsilon)$.

Definition 3.13 The function $\overline{\mathfrak{m}}_C(\cdot, d)(\underline{\mathfrak{m}}_C(\cdot, d))$ is called upper (lower) capacitive *Carathéodory d-measure* with respect to the structure $(\mathfrak{F}, \mathbb{P}, \xi, \eta, \psi)$.

The functions $\overline{\mathfrak{m}}_C(\cdot, d)$ and $\underline{\mathfrak{m}}_C(\cdot, d)$ for fixed d satisfy properties similar to these which are described in Proposition 3.35. It follows that the values

$$\overline{\text{cap}}_C \mathcal{Z} := \inf \{d \in \mathbb{P} \mid \overline{\mathfrak{m}}_C(\mathcal{Z}, d) = 0\} \quad \text{and}$$
$$\underline{\text{cap}}_C \mathcal{Z} := \inf \{d \in \mathbb{P} \mid \underline{\mathfrak{m}}_C(\mathcal{Z}, d) = 0\}$$

are defined.

Definition 3.14 The value $\overline{\text{cap}}_C \mathcal{Z}(\underline{\text{cap}}_C \mathcal{Z})$ is called upper (lower) *Carathéodory capacity of the set $\mathcal{Z}$* with respect to the structure $(\mathfrak{F}, \mathbb{P}, \xi, \eta, \psi)$.

Example 3.22 Let $\mathcal{M}$ be a compact metric space, $\mathfrak{F}$ be the family of all open balls in $\mathcal{M}$, and $\mathbb{P} := [0, \infty)$. For an arbitrary $\mathcal{B}_\varepsilon \in \mathfrak{F}$ with $\varepsilon > 0$ and arbitrary $d \in \mathbb{P}$ put

$$\xi(\mathcal{B}_\varepsilon, d) \equiv \eta(\mathcal{B}_\varepsilon, d) := \varepsilon^d \qquad \text{and}$$
$$\psi(\mathcal{B}_\varepsilon) \equiv \Phi(\mathcal{B}_\varepsilon) := \varepsilon\,.$$

It is clear that the structure $(\mathfrak{F}, \mathbb{P}, \xi, \psi, \phi)$ so introduced satisfies the conditions **(A1)–(A5)** and the upper Carathéodory capacity $\overline{\text{cap}}\, \mathcal{Z}$ of a set $\mathcal{Z} \subset \mathcal{M}$ obtained in this case coincides with the fractal dimension $\dim_F \mathcal{Z}$, introduced in Sect. 3.2.2.

Table 3.1 shows the dimension-like characteristics and their symbols.

Table 3.1 Symbols of dimension-like characteristics

Symbol	Dimension-like characteristics	Sections
$\text{ind}\, \mathcal{M}$	Small inductive dimension	3.1.1
$\text{Cov}\, \mathcal{M}$	Covering dimension	3.1.2
$\dim_T \mathcal{M}$	Topological dimension	3.1.2
$\dim_H \mathcal{Z}$	Hausdorff dimension	3.2.1
$\dim_F \mathcal{Z}$	Fractal dimension	3.2.2
$\underline{\dim}_B \mathcal{Z}$	Lower box dimension	3.2.2
$\overline{\dim}_B \mathcal{Z}$	Upper box dimension	3.2.2
$\dim_B \mathcal{Z}$	Box dimension	3.2.2
$h_{\text{top}}(\varphi, \mathcal{M})$	Topological entropy	3.3.1
$\dim_C \mathcal{Z}$	Carathéodory dimension	3.4.1
$\overline{\text{cap}}_C \mathcal{Z}$	Upper Carathéodory capacity	3.4.1
$\underline{\text{cap}}_C \mathcal{Z}$	Lower Carathéodory capacity	3.4.1

3.4.2 Properties of the Carathéodory Dimension and Carathéodory Capacity

The following result on the main properties of the Carathéodory dimension is a direct corollary of Proposition 3.35. The proof goes parallel to the proofs of similar results for the Hausdorff dimension in Sect. 3.2.1.

Proposition 3.37 *Suppose that $(\mathfrak{F}, \mathbb{P}, \xi, \eta, \psi)$ is a Carathéodory structure on $\mathcal{M}$ which satisfies* **(A1)–(A5)**. *Then the following properties are true:*

(1) $\dim_C \emptyset = d^*$;
(2) $\dim_C \mathcal{Z}_1 \le \dim_C \mathcal{Z}_2$ *for any* $\mathcal{Z}_1 \subset \mathcal{Z}_2 \subset \mathcal{M}$;
(3) $\dim_C(\bigcup_{j\ge 1} \mathcal{Z}_j) = \sup_{j\ge 1}\{\dim_C \mathcal{Z}_j\}$ *for any* $\mathcal{Z}_j \subset \mathcal{M},\ j = 1, 2, \dots$.

Proposition 3.38 *Suppose that $(\mathfrak{F}, \mathbb{P}, \xi, \eta, \psi)$ and $(\mathfrak{F}', \mathbb{P}', \xi', \eta', \psi')$ are two Carathéodory structures satisfying* **(A1)–(A5)** *on $\mathcal{M}$ and $\mathcal{M}'$, respectively with $\mathbb{P} = \mathbb{P}'$.*

Suppose also that there exists a bijective map $\chi : \mathcal{M} \to \mathcal{M}'$ and a constant $c > 0$ such that for any $\mathcal{U} \in \mathfrak{F}$, $\mathcal{U}' \in \mathfrak{F}'$, $d \in \mathbb{P}$

(1) $\chi^{-1}(\mathcal{U}') \in \mathfrak{F}$, $\chi(\mathcal{U}) \in \mathfrak{F}'$;
(2) $c^{-1}\xi'(\mathcal{U}', d) \le \xi\big(\chi^{-1}(\mathcal{U}'), d\big) \le c\xi'(\mathcal{U}', d)$;
(3) $c^{-1}\eta'(\mathcal{U}', d) \le \eta\big(\chi^{-1}(\mathcal{U}'), d\big) \le c\eta'(\mathcal{U}', d)$;
(4) $c^{-1}\psi'(\mathcal{U}') \le \psi\big(\chi^{-1}(\mathcal{U}')\big) \le c\eta'(\mathcal{U}')$.

Then for any $\mathcal{Z}' \subset \mathcal{M}'$ we have

$$\begin{aligned} \dim_{C,\mathfrak{F}',\xi',\eta',\psi'} \mathcal{Z}' &= \dim_{C,\mathfrak{F},\xi,\eta,\psi} \chi^{-1}(\mathcal{Z}'), \\ \overline{cap}_{C,\mathfrak{F}',\xi',\eta',\psi'} \mathcal{Z}' &= \overline{cap}_{C,\mathfrak{F},\xi,\eta,\psi} \chi^{-1}(\mathcal{Z}'), \\ \underline{cap}_{C,\mathfrak{F}',\xi',\eta',\psi'} \mathcal{Z}' &= \underline{cap}_{C,\mathfrak{F},\xi,\eta,\psi} \chi^{-1}(\mathcal{Z}'), \end{aligned}$$

where $\dim_{C,\mathfrak{F},\xi,\eta,\psi}$, $\overline{\mathrm{cap}}_{C,\mathfrak{F},\xi,\eta,\psi}$, *and* $\underline{\mathrm{cap}}_{C,\mathfrak{F},\xi,\eta,\psi}$ ($\dim_{C,\mathfrak{F}',\xi',\eta',\psi'}$, $\overline{\mathrm{cap}}_{C,\mathfrak{F}',\xi',\eta',\psi'}$, *and* $\underline{\mathrm{cap}}_{C,\mathfrak{F}',\xi',\eta',\psi'}$) *denote the Carathéodory dimension, upper and lower Carathéodory capacity of a set w.r.t. the structures $(\mathfrak{F}, \mathbb{P}, \xi, \eta, \psi)$ and $(\mathfrak{F}', \mathbb{P}', \xi', \eta', \psi')$ respectively.*

Proof Let us show the result for the Carathéodory dimension.
Suppose that $\mathcal{Z}' \subset \mathcal{M}'$ is a set and that $\mathfrak{G}' := \{\mathcal{U}'\}$ is an arbitrary ε-cover of $\mathcal{Z}'$. It follows that $\mathfrak{G} := \{\chi^{-1}(\mathcal{U}')\,|\,\mathcal{U}' \in \mathfrak{G}'\}$ is a $c\,\varepsilon$-covering of $\chi^{-1}(\mathcal{Z}')$ with $\psi(\chi^{-1}(\mathcal{U}')) \le c\,\psi'(\mathcal{U}) \le c\,\varepsilon$, $\xi(\chi^{-1}(\mathcal{U}'), d) \le c\,\xi'(\mathcal{U}', d)$. Furthermore we have

$$\sum_{\substack{\chi^{-1}(\mathcal{U}')\\ \psi'(\mathcal{U}')\le\varepsilon}} \xi(\chi^{-1}(\mathcal{U}'), d) \le c \sum_{\substack{\chi^{-1}(\mathcal{U}')\\ \psi'(\mathcal{U}')\le\varepsilon}} \xi'(\mathcal{U}', d)\,.$$

It follows that

$$\mu_c(\chi^{-1}(\mathcal{Z}'), d, c\,\varepsilon) = \inf_{\substack{\mathfrak{G} \\ \psi(\mathcal{U}) \le c\,\varepsilon}} \sum_{\mathcal{U}\in\mathfrak{G}} \xi(\mathcal{U}, d) \le \inf_{\substack{\tilde{\mathfrak{G}}:\chi^{-1}(\mathcal{U}') \\ \psi(\mathcal{U}')\le\varepsilon}} \sum \xi(\chi^{-1}(\mathcal{U}'), d)$$

$$\le c \inf_{\substack{\mathfrak{G}' \\ \psi(\mathcal{U}')\le\varepsilon}} \sum_{\mathcal{U}'\in\mathfrak{G}'} \xi'(\mathcal{U}', d) = c\,\mu_c(\mathcal{Z}', d, \varepsilon).$$

This implies that for any $d \in \mathbb{P}$ we have $\mu_c(\chi^{-1}(\mathcal{Z}'), d) \le c\mu_c(\mathcal{Z}', d)$ and, consequently, $\dim_{C,\mathfrak{F},\xi,\eta,\psi}(\chi^{-1}(\mathcal{Z}')) \le \dim_{C,\mathfrak{F}',\xi',\eta',\psi'}(\mathcal{Z}')$. In the same way one shows the opposite inequality. □

The simplest properties of capacity are given by the next proposition.

Proposition 3.39 *Suppose that* $(\mathfrak{F}, \mathbb{P}, \xi, \eta, \psi)$ *is a Carathéodory structure on* $\mathcal{M}$ *satisfying* **(A1)–(A5)**. *Then the following properties are true:*

(1) $\overline{cap}_c\,\emptyset = \underline{cap}_c\,\emptyset = d^*$;
(2) $\dim_c \mathcal{Z} \le \underline{cap}_c\,\mathcal{Z} \le \overline{cap}_c\,\mathcal{Z}$ *for any* $\mathcal{Z} \subset \mathcal{M}$;
(3) $\overline{cap}_c\,\mathcal{Z}_1 \le \overline{cap}_c\,\mathcal{Z}_2$, $\underline{cap}_c\,\mathcal{Z}_1 \le \underline{cap}_c\,\mathcal{Z}_2$ *for any* $\mathcal{Z}_1 \subset \mathcal{Z}_2 \subset \mathcal{M}$;
(4) $\overline{cap}_c(\bigcup_{j\ge1}\mathcal{Z}_j) \ge \sup_{j\ge1}\{\overline{cap}_c\,\mathcal{Z}_j\}$, $\underline{cap}_c(\bigcup_{j\ge1}\mathcal{Z}_j) \ge \sup_{j\ge1}\{\underline{cap}_c\,\mathcal{Z}_j\}$ *for any* $\mathcal{Z}_j \subset \mathcal{M}$, $j = 1, 2, \ldots$.

We shall consider further some conditions, under which the inequality in statement (4) of Proposition 3.39 turns out to be an equality.

Suppose also that in addition to conditions **(A1)–(A5)** the following condition is satisfied:

(A6) There exist two functions $\varkappa$, $\Phi : \mathfrak{F} \to \mathbb{R}_+$ such that

$$\xi(\mathcal{U}, d) = \varkappa(\mathcal{U})\Phi(\mathcal{U})^d, \quad \eta(\mathcal{U}, d) = \Phi(\mathcal{U})^d, \ \forall d \in \mathbb{P}, \forall \mathcal{U} \in \mathfrak{F}, \tag{3.21}$$

and the equality

$$\Phi(\mathcal{U}_1) = \Phi(\mathcal{U}_2), \quad \text{for } \mathcal{U}_1, \mathcal{U}_2 \in \mathfrak{F} \quad \text{is satisfied if} \quad \psi(\mathcal{U}_1) = \psi(\mathcal{U}_2)\,. \tag{3.22}$$

Using relation (3.22) we can define for $\varepsilon > 0$ the function

$$\phi(\varepsilon) := \Phi(\mathcal{U}), \quad \text{where} \quad \mathcal{U} \in \mathfrak{F} \text{ is an arbitrary set with} \quad \psi(\mathcal{U}) = \varepsilon.$$

Let us put for arbitrary $\varepsilon > 0$ and $\mathcal{Z} \subset \mathcal{M}$

$$\Upsilon(\mathcal{Z}, \varepsilon) := \inf \sum_{\mathcal{U}\in\mathfrak{U}} \varkappa(\mathcal{U}), \tag{3.23}$$

where the infimum is taken over all finite or countable sets $\mathfrak{U}$ of subsets from the family $\mathfrak{F}$ forming a covering of $\mathcal{Z}$ such that $\psi(\mathcal{U}) = \varepsilon$, $\forall \mathcal{U} \in \mathfrak{U}$.

For the function $\phi(\varepsilon)$ the following lemma is true.

Lemma 3.9 *Under the above conditions we have*

$$\lim_{\varepsilon\to 0}\phi(\varepsilon)=0\,.$$

Proof Suppose the opposite. Then it can be found a number $\gamma>0$ and a sequence $\varepsilon_m\to 0$ such that

$$\phi(\varepsilon_m)\geq\gamma \quad \text{for all} \quad m\geq 1.$$

By condition **(A3)** it can be found $\delta>0$ such that for any $\mathcal{U}\in\mathfrak{F}$, for which $\psi(\mathcal{U})\leq\delta$, the inequality $\Phi(\mathcal{U})\leq\gamma/2$ is satisfied. Choose m to be so large that $\varepsilon_m\leq\delta$ and take $\mathcal{U}\in\mathfrak{F}$ such that $\psi(\mathcal{U})=\varepsilon_m$. Then $\phi(\varepsilon_m)\leq\gamma/2$. Thus, we obtain a contradiction. □

The statement to be proved below generalizes the results of the previous section on two equivalent definitions of fractal dimension.

Proposition 3.40 *Suppose* $(\mathfrak{F},\mathbb{P},\xi,\eta,\psi,\varkappa,\phi)$ *is a Carathéodory structure for* $\mathcal{M}$ *satisfying* **(A1)–(A6)**. *Then for any* $\mathcal{Z}\subset\mathcal{M}$

$$\overline{\operatorname{cap}}_C\,\mathcal{Z}=\limsup_{\varepsilon\to 0}\frac{\log\Upsilon(\mathcal{Z},\varepsilon)}{\log\bigl(1/\phi(\varepsilon)\bigr)} \quad \textit{and} \quad \underline{\operatorname{cap}}_C\,\mathcal{Z}=\liminf_{\varepsilon\to 0}\frac{\log\Upsilon(\mathcal{Z},\varepsilon)}{\log\bigl(1/\phi(\varepsilon)\bigr)}\,.$$

Proof Put

$$d_1:=\operatorname{cap}_C\mathcal{Z},\quad d_2:=\limsup_{\varepsilon\to 0}\frac{\log\Upsilon(\mathcal{Z},\varepsilon)}{\log\bigl(1/\phi(\varepsilon)\bigr)}.$$

For a given $\gamma>0$, choose a sequence $\varepsilon_m\to 0$ such that

$$\infty=\mathfrak{m}_C(\mathcal{Z},d_1-\gamma)=\lim_{m\to+\infty}\mathfrak{m}_C(\mathcal{Z},d_1-\gamma,\varepsilon_m).$$

Then $\mathfrak{m}_C(\mathcal{Z},d_1-\gamma,\varepsilon_m)\geq 1$ for all sufficiently large m. Therefore by (3.21)–(3.23) for all sufficiently large m we obtain

$$\phi(\varepsilon_m)^{d_1-\gamma}\,\Upsilon(\mathcal{Z},\varepsilon_m)\geq 1. \tag{3.24}$$

Passing, if necessary, to a subsequence, we can suppose that there exists a limit

$$\lim_{m\to+\infty}\frac{\log\Upsilon(\mathcal{Z},\varepsilon_m)}{\log\bigl(1/\phi(\varepsilon_m)\bigr)}\leq d_2. \tag{3.25}$$

According to (3.24)

$$d_1\leq\gamma+\frac{\log\Upsilon(\mathcal{Z},\varepsilon_m)}{\log\bigl(1/\phi(\varepsilon_m)\bigr)}.$$

Passing to the limit as $m \to +\infty$ and taking into account (3.25), we obtain

$$d_1 \le d_2 + \gamma. \tag{3.26}$$

We now choose a sequence $\varepsilon'_m \to 0$ such

$$d_2 = -\lim_{m\to+\infty} \frac{\log \Upsilon(\mathcal{Z}, \varepsilon'_m)}{\log \phi(\varepsilon'_m)}. \tag{3.27}$$

Passing, if necessary, to a subsequence, we can suppose that

$$0 = \mathfrak{m}_c(\mathcal{Z}, d_1 + \gamma) = \lim_{m\to+\infty} \mathfrak{m}_c(\mathcal{Z}, d_1 + \gamma, \varepsilon'_m).$$

Then $\mathfrak{m}_c(\mathcal{Z}, d_1 + \gamma, \varepsilon'_m) \le 1$ for all sufficiently large m. Therefore by virtue of (3.27) we have for all sufficiently large m

$$\phi(\varepsilon'_m)^{d_1+\gamma}\Upsilon(\mathcal{Z}, \varepsilon'_m) \le 1.$$

It follows that $d_1 \ge -\gamma - \log \Upsilon(\mathcal{Z}, \varepsilon'_m)/\log \phi(\varepsilon'_m)$. Passing to the limit as $m \to +\infty$ and taking into account (3.27), we get

$$d_1 \ge d_2 - \gamma. \tag{3.28}$$

Since γ is an arbitrary number, it follows from (3.26) and (3.28) that $d_1 = d_2$. The second statement is proved similarly. □

Now we can refine statement (4) of Proposition 3.39.

Proposition 3.41 *Suppose* $(\mathfrak{F}, \mathbb{P}, \xi, \eta, \psi, \varkappa, \phi)$ *is a Carathéodory structure for* $\mathcal{M}$ *satisfying* **(A1)–(A6)**. *Let* $\mathcal{Z}_j \subset \mathcal{M}$, $j = 1, 2, \ldots, k$ *be arbitrary sets. Then*

$$\overline{\text{cap}}_c\Big(\bigcup_{j=1}^{k} \mathcal{Z}_j\Big) = \max_{1\le j\le k}\{\overline{\text{cap}}_c \mathcal{Z}_j\}.$$

$\Big($*In general it is not true that* $\underline{\text{cap}}_c\big(\bigcup_{j=1}^{k} \mathcal{Z}_j\big) = \max_{1\le j\le k}\{\underline{\text{cap}}_c \mathcal{Z}_j\}\Big)$.

Proof Let us show the first assertion. Obviously, it is sufficient to consider the case $k = 2$. Suppose that $\mathcal{Z} = \mathcal{Z}_1 \cup \mathcal{Z}_2$. Then from (3.23) for all $\varepsilon > 0$ we have

$$\Upsilon(\mathcal{Z}, \varepsilon) \le \Upsilon(\mathcal{Z}_1, \varepsilon) + \Upsilon(\mathcal{Z}_2, \varepsilon). \tag{3.29}$$

By Proposition 3.39 there can be found the sequence $\varepsilon_m \to 0$ such that

$$\operatorname{cap}_{\mathrm{c}} \mathcal{Z} = - \lim_{\mathrm{m} \to +\infty} \frac{\log \Upsilon(\mathcal{Z}, \varepsilon_{\mathrm{m}})}{\log \phi(\varepsilon_{\mathrm{m}})}.$$

Passing, if it is necessary, to a subsequence, we shall suppose that there exist the limits

$$d_j = - \lim_{m \to +\infty} \frac{\log \Upsilon(\mathcal{Z}_j, \varepsilon_m)}{\log \phi(\varepsilon_m)} \leq \operatorname{cap}_{\mathrm{c}} \mathcal{Z}_j, \quad j = 1, 2.$$

Put

$$a_m = \frac{\log \Upsilon(\mathcal{Z}_1, \varepsilon_m)}{\log\big(1/\phi(\varepsilon_m)\big)} - \frac{\log \Upsilon(\mathcal{Z}_2, \varepsilon_m)}{\log\big(1/\phi(\varepsilon_m)\big)}.$$

Whence it follows that

$$\frac{\Upsilon(\mathcal{Z}_1, \varepsilon_m)}{\Upsilon(\mathcal{Z}_2, \varepsilon_m)} = \phi(\varepsilon_m)^{-a_m}.$$

Therefore by (3.29) we get

$$\log \Upsilon(\mathcal{Z}, \varepsilon_m) \leq \log \Upsilon(\mathcal{Z}_1, \varepsilon_m) + \log\big(1 + \phi(\varepsilon_m)^{a_m}\big). \tag{3.30}$$

Let us consider three cases.

Case 1: $d_1 > d_2$. There exists $a > 0$ such that $a_m \geq a$ for all sufficiently large m.

Since $\log(1+t) \leq t$ for $t \geq t_0$, from (3.30) it follows that for such m we have

$$\log \Upsilon(\mathcal{Z}, \varepsilon_m) \leq \log \Upsilon(\mathcal{Z}_1, \varepsilon_m) + \phi(\varepsilon_m)^{a_m},$$

Therefore

$$\frac{\log \Upsilon(\mathcal{Z}, \varepsilon_m)}{\log(1/\phi(\varepsilon_m))} \leq \frac{\log \Upsilon(\mathcal{Z}_1, \varepsilon_m)}{\log(1/\phi(\varepsilon_m))} - a_m \frac{\phi(\varepsilon_m)^{a_m}}{\log(\phi(\varepsilon_m)^{a_m})}.$$

Passing to the limit as $m \to +\infty$, taking into account Lemma 3.9,

$$\operatorname{cap}_c \mathcal{Z} \leq d_1 \leq \max\{\operatorname{cap}_c \mathcal{Z}_1,\ \operatorname{cap}_c \mathcal{Z}_2\}. \tag{3.31}$$

Case 2: $d_1 = d_2$. Suppose that it can be found a number q such that $\phi(\varepsilon_m)^{a_m} \leq q \leq 1$ for all sufficiently large m. Then since $\log(1+t) \leq c \log 1/t$ for all $t \in (0, q)$ with a certain constant $c > 0$, we have

$$\log\big(1 + \phi(\varepsilon_m)^{a_m}\big) \leq c a_m \log\big(1/\phi(\varepsilon_m)\big),$$

Therefore, using inequality (3.30) for $\log(1/\phi(\varepsilon_m))$ and passing to the limit as $m \to +\infty$, we arrive at (3.31).

If such number q does not exist, then, passing to a subsequence, if it is necessary, we may consider that $\phi(\varepsilon_m)^{a_m} \to 1$. By inequality (3.30) for $\log(1/\phi(\varepsilon_m))$ and passing to the limit for $m \to +\infty$, we again obtain at (3.31).

Case 3: $d_1 < d_2$. This case is analogous to case (1), but the values $\Upsilon(\mathcal{Z}_1, \varepsilon)$ and $\Upsilon(\mathcal{Z}_2, \varepsilon)$ change over.

Thus, in all cases inequality (3.31) holds. Then the validity of the proposition being proved follows from statement (4) of Proposition 3.39. □

References

1. Adler, R.A., Konheim, A., McAndrew, M.: Topological entropy. Trans. Am. Math. Soc. **114**, 309–319 (1965)
2. Alexandrov, P.S., Pasynkov, B.A.: Introduction to Dimension Theory. Nauka, Moscow (1973). (Russian)
3. Ben-Artzi, A., Eden, A., Foias, C., Nicolaenko, B.: Hölder continuity for the inverse of the Mañé projection. J. Math. Anal. Appl. **178**, 22–29 (1993)
4. Besicovitch, A.S.: Sets of fractional dimensions. Part I. Math. Ann. **101**, 161–193 (1929)
5. Bowen, R.: Topological entropy and Axiom A. Global Analysis. In: Proceedings of Symposia in Pure Mathematics, vol. 14, pp. 23–41 (1968). (Am. Math. Soc.)
6. Bowen, R.: Entropy for group endomorphisms and homogenous spaces. Trans. Am. Math. Soc. **153**(171), 401–414 (1971)
7. Brouwer, L.E.J.: Beweis der Invarianz der Dimensionszahl. Math. Ann. **70**, 161–165 (1911)
8. Brouwer, L.E.J.: Über den natürlichen Dimensionsbegriff. J. f. reine u. angew. Math. **142**, 146–152 (1913)
9. Carathéodory, C.: Über das lineare Mass von Punktmengen-eine Verallgemeinerung des Längenbegriffs. Göttinger Nachrichten, pp. 406–426 (1914)
10. Dinaburg, E.I.: The relation between topological entropy and metric entropy. Dokl. Akad. Nauk SSSR **190**, 19–22 (1970). (Russian)
11. Eden, A., Foias, C., Temam, R.: Local and global Lyapunov exponents. J. Dynam. Diff. Equ. **3**, 133–177 (1991). (Preprint No. 8804, The Institute for Applied Mathematics and Scientific Computing, Indiana University, 1988)
12. Edgar, G.A.: Measure. Topology and Fractal Geometry. Springer, Berlin (1990)
13. Falconer, K.J.: The geometry of fractal sets. In: Cambridge Tracts in Mathematics, vol. 85. Cambridge University Press (1985)
14. Falconer, K.J.: Fractal Geometry: Mathematical Foundations and Applications. Wiley, Chichester (1990)
15. Fathi, A.: Expansiveness, hyperbolicity, and Hausdorff dimension. Commun. Math. Phys. **126**, 249–262 (1989)
16. Federer, H.: Geometric Measure Theory. Springer, New York (1969)
17. Foias, C., Olsen, E.J.: Finite fractal dimension and Hölder-Lipschitz parameterization. Ind. Univ. Math. J. **45**, 603–616 (1996)
18. Gu, X.: An upper bound for the Hausdorff dimension of a hyperbolic set. Nonlinearity **4**(3), 927–934 (1991)
19. Hata, M.: Topological aspects of self-similar sets and singular functions. In: Bélairc, J., Dubuc, S. (eds.) Fractal Geometry and Analysis. Canada, Kluwer (1991)
20. Hausdorff, F.: Dimension und äußeres Maß. Math. Ann. **79**, 157–179 (1919)
21. Howroyd, J.D.: On dimension and on the existence of sets of finite positive Hausdorff measure. Proc. Lond. Math. Soc. **70**, 581–604 (1995)
22. Hunt, B.R., Kaloshin, VYu.: Regularity of embeddings of infinite-dimensional fractal sets into finite-dimensional spaces. Nonlinearity **12**, 1263–1275 (1999)

23. Hunt, B.R., Sauer, T., James, J.A.: Prevalence: a translation-invariant "almost every" on infinite-dimensional spaces. Bull. Am. Math. Soc. **27**(2), 217–238 (1992)
24. Hurewicz, W., Wallman, H.: Dimension Theory. Princeton University Press, Princeton (1948)
25. Hutchinson, J.E.: Fractals and self-similarity. Ind. Univ. Math. J. **30**, 713–747 (1981)
26. Ito, S.: An estimate from above for the entropy and the topological entropy of a C^1-diffeomorphism. Proc. Jpn. Acad. **46**, 226–230 (1970)
27. Katok, A., Hasselblatt, B.: Introduction to the Modern Theory of Dynamical Systems. (Encyclopedia of Mathematics and its Applications), vol. 54. Cambridge University Press, Cambridge (1995)
28. Kolmogorov, A.N.: A new metric invariant of transient dynamical systems and automorphisms of Lebesgue spaces. Dokl. Akad. Nauk, SSSR **119**, 861–864 (1958). (Russian)
29. Kolmogorov, A.N., Tihomirov, V.M.: ε-entropy and ε-capacity of sets in function spaces. Uspekhi Mat. Nauk **14**(2), 3–86 (1960). (Russian, Trans. Am. Math. Soc. Transl. Ser. 2 **17**, 277–364, 1960)
30. Kornfeld, I.P., Sinai, Ya.G., Fomin, S.V.: Ergodic Theory. Nauka, Moscow (1980). (Russian)
31. Lebesgue, H.: Sur la non applicabilité de deux domaines appartemant à deux espaces de n et n+p dimensions. Math. Ann. **70**, 166–168 (1911)
32. Mané, R.: On the dimension of the compact invariant sets of certain non-linear maps. Lecture Notes in Mathematics, vol. 898, pp. 230–241. Springer, Berlin (1981)
33. Mané, R.: Ergodic Theory and Differentiable Dynamics. Springer, Berlin (1987)
34. Menger, K.: Über umfassendste n-dimensionale Mengen. Proc. Akad. Wetensch. Amst. **29**, 1125–1128 (1926)
35. Menger, K.: Dimensionstheorie. B. G. Teubner, Leipzig (1928)
36. Moran, P.: Additive functions of intervals and Hausdorff measure. Math. Proc. Camb. Phil. Soc. **42**, 15–23 (1946)
37. Nöbeling, G.: Über eine n-dimensionale Universalmenge in R_{2n+1},. Math. Ann. **104**, 71–80 (1930)
38. Okon, T.: Dimension estimate preserving embeddings for compacta in metric spaces. Archiv der Mathematik **78**, 36–42 (2002)
39. Pesin, Ya.B.: Dimension type characteristics for invariant sets of dynamical systems. Uspekhi Mat. Nauk **43**(4), 95–128 (1988). (Russian, English Transl. Russian Math. Surveys **43**(4), 111–151, 1988)
40. Pesin, Ya.B.: Dimension Theory in Dynamical Systems: Contemporary Views and Applications. Chicago Lectures in Mathematics, The University of Chicago Press, Chicago and London (1997)
41. Poincaré, H.: Pourquoi l'espace a trois dimensions. Revue de Métaphysique et de Morale **20**, 484 (1912)
42. Pontryagin, L.S., Shnirelman, L.G.: On a metric property of dimension. Appendix to the Russian Translation of Hurewitz, W. and H. Wallman, Dimension Theory. Izdat. Inostr. Lit., Moscow (1948)
43. Postnikov, M.M.: Smooth Manifolds. Nauka, Moscow (1987). (Russian)
44. Reitmann, V.: Regular and Chaotic Dynamics. Teubner Verlagsgesellschaft, Stuttgart-Leipzig, B. G (1996). (German)
45. Reitmann, V., Popov, S.: Embedding of compact invariant sets of dynamical systems on infinite-dimensional manifolds into finite-dimensional spaces. In: Abstracts, 9th AIMS International Conference on Dynamical Systems, Differential Equations and Applications, Orlando, USA, p. 247 (2012)
46. Robinson, J.C.: Dimensions, Embeddings, and Attractors, p. 186. Cambridge University Press, Cambridge (2010)
47. Sauer, T., Yorke, J.A., Casdagli, M.: Embedology. J. Stat. Phys. **65**, 579–616 (1991)
48. Sinai, Ya.G.: On the concept of entropy of a dynamical system. Dokl. Akad. Nauk, SSSR **124**, 768–771 (1959). (Russian)
49. Titchmarsh, E.C.: The Theory of Functions. Oxford (1932)

50. Urysohn, P.S.: Mémoire sur les multiplicités cantoriennes. Fund. Math. **7/8**, 30–139, 225–359 (1925)
51. Wegmann, H.: Die Hausdorff-Dimension von kartesischen Produktmengen in metrischen Räumen. J. Reine und Angew. Math. **234**, 163–171 (1969)
52. Whitney, H.: Differentiable manifolds. Ann. Math., II. Ser. **37**, 645–680 (1936)
53. Zeidler, E.: Nonlinear Functional Analysis and its Applications. Springer, New York (1986)

Part II
Dimension Estimates for Almost Periodic Flows and Dynamical Systems in Euclidean Spaces

Chapter 4
Dimensional Aspects of Almost Periodic Dynamics

Abstract The first part (Sects. 4.2, 4.3, 4.5 and 4.6) of the present chapter contains several approaches to the investigation of the Fourier spectrum of almost periodic solutions to various differential equations. The core element here is the Cartwright theorem [6] that links the topological dimension of the orbit closure of an almost periodic flow and the algebraic dimension of its frequency module (Theorem 4.8). The next step is an extension of this theorem to non-autonomous differential equations (Theorem 4.11) originally presented in [7]. Applications of Cartwright's theorems are given for almost periodic ODEs based on the approach due to R. A. Smith (Theorem 4.12) and for DDEs based on results of Mallet-Paret from [16] (Theorem 4.14). In Sect. 4.7 we develop a method for studying fractal dimensions of forced almost periodic oscillations using some kind of recurrence properties. This approach differs from the one due to Douady and Oesterlé and highly relies on almost periodicity. Some fundamental ideas firstly appeared in the works of Naito (see [17, 18]) and then were developed in [1, 2]. In Sect. 4.8 we study forced almost periodic oscillations in Chua's circuit and compare the analytical upper estimates of the fractal dimension of their trajectory closures with numerical simulations given by the standard box-counting algorithm.

4.1 Introduction

Almost periodic differential equations naturally appear in many fields of science including physics, chemistry, biology and ecology. The simplest models describe periodically or almost periodically forced oscillations in mechanics, behaviour of chemical reactions under the influence of periodic or almost periodic perturbations or population dynamics with time-dependent seasonal effects in ecology. A nice list of references on quasi-periodicity phenomena discovered in applied problems is given in [9].

N. Kuznetsov and V. Reitmann, *Attractor Dimension Estimates for Dynamical Systems: Theory and Computation*, Emergence, Complexity and Computation 38,
https://doi.org/10.1007/978-3-030-50987-3_4

A possible way to study such systems is based on the method of small parameter [14, 20]. On the other hand, nonlocal results can be derived with the use of topological (see, for example, [3, 6, 7, 16, 26]) and operator (see [1, 2, 5, 14, 15, 17–19]) methods.

In this chapter we are mainly interested in dimension-like properties of almost periodic solutions to various differential equations. Namely, those are the properties established by Cartwright's theorems and estimation of the fractal dimension of almost trajectories closures.[1] Unlike the former that has purely topological nature, the latter problem (i.e. the study of fractal dimensions) is closely related to a method proving the existence of almost periodic solutions. We present here an application based on a method of Krasnosel'skii (Theorem 4.17 and [5, 14]). One more approach which we do not discuss here is based on the method of strongly monotone operators that often leads to the existence of a globally exponentially stable almost periodic solution [15, 19, 29]. It can be applied to study variational inequalities [19] and provides (under suitable conditions) all the required information about the solution: its regularity and estimation of the Diophantine dimension (see Theorem 4.16, and the discussion in [2]).

4.2 Topological Dimension of Compact Groups

In this section we give an introduction to topological groups theory (see [21]). We omit most of the proofs and concentrate on the role of the Lebesgue covering dimension in the proof of Theorem 4.2, which will be used later to prove the Cartwright theorem (Theorem 4.8). Note that many results in the case of our interest, i.e. for compact and discrete groups, can be shown in a much easier manner.

Recall that the rank of an abelian group $\mathbb{G}$ is the maximal number of linearly independent elements in $\mathbb{G}$. It is denoted by rank $\mathbb{G}$.

Remark 4.1 If the group is torsion-free then the rank is the minimum dimension of $\mathbb{Q}$-vector space in which the group $\mathbb{G}$ can be embedded. To see this suppose $x_1, \ldots, x_k$ is a maximal linearly independent system in $\mathbb{G}$. For each $x \in \mathbb{G}$ the family $x, x_1, \ldots, x_k$ is linearly dependent and we have $ax = a_1x_1 + \cdots + a_kx_k$ with integer coefficients $a, a_1, \ldots, a_k$ and $a \neq 0$. The map $x \mapsto (a_1/a, \ldots, a_k/a)$ from $\mathbb{G}$ to $\mathbb{Q}^k$ is well-defined and realizes the required embedding.

Suppose that $\mathbb{G}$ is a topological group, i.e. there is a topology and a group structure on $\mathbb{G}$ such that the operations of multiplication ($(x, y) \mapsto xy$) and the inverse ($x \mapsto x^{-1}$) are continuous maps. In the sequel we will deal only with abelian Hausdorff topological groups. Simple examples such groups are given by the flat torus $\mathbb{T}^m = \mathbb{R}^m/\mathbb{Z}^m$ and $\mathbb{R}^m$. The *character group* (or *dual group*) $\widehat{\mathbb{G}}$ is defined by the set of all continuous homomorphisms from $\mathbb{G}$ to the circle group $\mathbb{T}^* = \{z \in \mathbb{C} : |z| = 1\}$.

---

[1] That can be regarded as estimation of the fractal dimension of minimal sets consisting of almost periodic orbits of skew-product flows which is an extension of an almost periodic minimal flow.

There is a natural abelian group structure on $\widehat{\mathbb{G}}$ given by the pointwise product. Since $\widehat{\mathbb{G}}$ is a subset in the space of all complex-valued continuous maps on $\mathbb{G}$ we endow $\widehat{\mathbb{G}}$ with the compact-open topology. If $\mathbb{G}$ is locally compact then it can be shown that $\widehat{\mathbb{G}}$ is also locally compact. For $x \in \mathbb{G}$ consider the map $\alpha_x : \widehat{\mathbb{G}} \to \mathbb{T}^*$ defined as $\alpha_x(\chi) := \chi(x)$ for $\chi \in \widehat{\mathbb{G}}$. Clearly, α_x is a character of $\widehat{\mathbb{G}}$, i.e. $\alpha_x \in \widehat{\widehat{\mathbb{G}}}$. The following fundamental fact is known as the Pontryagin duality.

Theorem 4.1 *For every locally compact abelian group $\mathbb{G}$ the homomorphism $x \mapsto \alpha_x$ is a topological isomorphism*[2] *between $\mathbb{G}$ and $\widehat{\widehat{\mathbb{G}}}$.*

The dual group $\widehat{\mathbb{G}}$ is discrete provided that $\mathbb{G}$ is compact and $\widehat{\mathbb{G}}$ is compact provided that $\mathbb{G}$ is discrete. This fact along with the Pontryagin duality allows us to express topological properties of a compact group in purely algebraic terms of its discrete character group.

For a closed subgroup $\mathbb{H} \subset \mathbb{G}$ we define the *annihilator* of $\mathbb{H}$ as the set $\operatorname{Ann}(\mathbb{H}) := \{\chi \in \widehat{\mathbb{G}} \mid \chi(x) = 1\ \forall x \in \mathbb{H}\}$. Clearly, $\operatorname{Ann}(\mathbb{H})$ is a subgroup of $\widehat{\mathbb{G}}$. In order to find the characters group of $\mathbb{H}$ or $\mathbb{G}/\mathbb{H}$ one has the following facts:

(1) $\widehat{\mathbb{H}} \cong \widehat{\mathbb{G}}/\operatorname{Ann}(\mathbb{H})$;
(2) $\widehat{\mathbb{G}/\mathbb{H}} \cong \operatorname{Ann}(\mathbb{H})$.

For our purposes we also need the following lemma.

Lemma 4.1 *Suppose $\mathbb{G}$ is discrete. Then for every neighborhood $\mathcal{U}$ of zero in $\widehat{\mathbb{G}}$ with positive Lebesgue measure there exists a finitely generated group $\mathbb{H} \subset \mathbb{G}$ such that* $\operatorname{Ann}(\mathbb{H}) \subset \mathcal{U}$.

Suppose we have a continuous map $f : \mathcal{M} \to \mathcal{N}$ between two topological spaces. We say that f *refines* the covering $\mathfrak{U}$ of $\mathcal{M}$ if for every $q \in \mathcal{N}$ the preimage $f^{-1}(q)$ entirely lies in some element of $\mathfrak{U}$.

Lemma 4.2 *Let $\mathcal{M}$ be a compact Hausdorff space. Let* $\dim_T \mathcal{M} = n < \infty$. *Then there exists a finite open cover $\mathfrak{U}_0$ of $\mathcal{M}$ with the property that for arbitrary Hausdorff space $\mathcal{N}$ and arbitrary continuous mapping $f : \mathcal{M} \to \mathcal{N}$, if f refines the covering $\mathfrak{U}_0$ then it is necessary* $\dim_T \mathcal{N} \geq n$.

Proof Let $\mathfrak{U}_0$ be an open cover of $\mathcal{M}$ such that any refinement of it has order $\geq n+1$. Suppose f refines $\mathfrak{U}_0$. For any $w \in \mathcal{N}$ we have $w = \bigcap \overline{\mathcal{V}}_w$, where the intersection is taken over all open neighborhoods $\mathcal{V}_w$ of w. Let $\mathcal{U} \in \mathfrak{U}_0$ be such that $f^{-1}(w) \subset \mathcal{U}$. Since $f^{-1}(w) = f^{-1}\left(\bigcap \overline{\mathcal{V}_w}\right) = \bigcap f^{-1}\left(\overline{\mathcal{V}}_w\right) \subset \mathcal{U}$ we have a system of closed subsets whose intersection lies in an open set. Due to compactness of $\mathcal{M}$ for some $\mathcal{V}_w$ we have $f^{-1}(\overline{\mathcal{V}}_w) \subset \mathcal{U}$. In particular, for every $w \in \mathcal{N}$ there is an open neighborhood $\mathcal{V}_w$ such that $f^{-1}(\mathcal{V}_w)$ entirely lies in some element of $\mathfrak{A}_0$. Let $\mathfrak{V}$ be a finite open cover of $\mathcal{N}$ by such neighborhoods $\mathcal{V}_w$, $w \in \mathcal{N}$. Suppose that

[2] That is, the mentioned mapping is an isomorphism of groups and a homeomorphism of topological spaces.

$\dim_T \mathcal{N} < n$. Then there exists a refinement, say $\mathfrak{V}_0$, of $\mathfrak{V}$ with order $\leq n$. Since $f^{-1}(\mathfrak{V})$ is a refinement of $\mathfrak{U}_0$ the same holds for $f^{-1}(\mathfrak{V}_0)$. The order of $\mathfrak{V}_0$ is not lesser than the order of $f^{-1}(\mathfrak{V}_0)$ that is $\geq n+1$ by the particular choose of $\mathfrak{U}_0$. This is a contradiction. □

Theorem 4.2 *Let $\mathbb{G}$ be compact. We have*

$$\dim_T \mathbb{G} = \operatorname{rank} \widehat{\mathbb{G}}. \tag{4.1}$$

Proof For convenience, we put $n := \dim_T \mathbb{G}$ and $r := \operatorname{rank} \widehat{\mathbb{G}}$. By Theorem 4.1 we may consider $\mathbb{G}$ as the character group of $\widehat{\mathbb{G}}$.

(1) Lets show that $n \leq r$. Let $\mathfrak{U}$ be a finite open cover of $\mathbb{G}$ (below we will emphasize an additional property for $\mathfrak{U}$). For every $x \in \mathbb{G}$ there exists an open neighborhood of zero $\mathcal{V}_x$ such that $x + 2\mathcal{V}_x$ entirely lies in some element from $\mathfrak{U}$. Let $\mathcal{V}_{x_1}, \ldots, \mathcal{V}_{x_m}$ be a finite cover of $\mathbb{G}$ by such sets and put $\mathcal{V} := \mathcal{V}_{x_1} \cap \ldots \cap \mathcal{V}_{x_m}$. Since $\widehat{\mathbb{G}}$ is discrete, by Lemma 4.1 there exists a finitely generated subgroup $\mathbb{X} \subset \widehat{\mathbb{G}}$ with $\operatorname{Ann}(\mathbb{X}) \subset \mathcal{V}$. By the fundamental theorem of finitely generated abelian groups $\mathbb{X} \cong \mathbb{Z}^k \oplus \mathbb{F}$, where $k \leq r$ and $\mathbb{F}$ is a finite abelian group. Clearly, $\widehat{\mathbb{X}} \cong (\mathbb{T}^*)^k \oplus \mathbb{F}$ and since $\widehat{\mathbb{X}} \cong \mathbb{G}/\operatorname{Ann}(\mathbb{X})$ we have $\dim_T(\mathbb{G}/\operatorname{Ann}(\mathbb{X})) = k \leq r$. Now let $\mathfrak{U} = \mathfrak{U}_0$ be an open cover given by Lemma 4.2. Since the natural projection $\pi : \mathbb{G} \to \mathbb{G}/\operatorname{Ann}(\mathbb{X})$ refines the covering $\mathcal{U}$ we have $n = \dim_T \mathbb{G} \leq \dim_T(\mathbb{G}/\operatorname{Ann}(\mathbb{X}))$ and, consequently, $n \leq r$.
(2) Consider an arbitrary $k \leq r$ (or $k < r$ if $r = \infty$) and suppose that $\mathcal{S}$ is a maximal system of linearly independent elements in $\widehat{\mathbb{G}}$. Then $\mathcal{S}$ contains at least k elements, say $\chi_1, \ldots, \chi_k$. Put $\mathcal{S}' = \mathcal{S} \backslash \{\chi_1, \ldots, \chi_k\}$. Let $\mathbb{X}$ be the subgroup of all elements $\chi \in \widehat{\mathbb{G}}$ such that the family $\chi \cup \mathcal{S}'$ is linearly dependent. Clearly, the factor $\widehat{\mathbb{G}}/\mathbb{X}$ is torsion-free and $[\chi_1], \ldots, [\chi_k]$ is a maximal linearly independent system in $\widehat{\mathbb{G}}$. Consider the annihilator $\operatorname{Ann}(\mathbb{X})$ as a subgroup of $\mathbb{G}$. The group $\operatorname{Ann}(\mathbb{X})$ is the character group of $\widehat{\mathbb{G}}/\mathbb{X}$. We will show that $\dim_T \operatorname{Ann}(\mathbb{X}) \geq k$ that implies $\dim_T \mathbb{G} \geq k$. Consider the cube $\mathcal{Q}^k = \{x \in \mathbb{R}^k \mid |x_j| \leq 1/3,\ j = 1, \ldots, k\}$. For every $x \in \mathbb{Q}^k$ we define a character ξ_x of $\widehat{\mathbb{G}}/\operatorname{Ann}(\mathbb{X})$ as follows. For every $[\chi] \in \widehat{\mathbb{G}}/\operatorname{Ann}(\mathbb{X})$ the family $[\chi], [\chi_1], \ldots, [\chi_k]$ is linearly dependent so we have $a[\chi] = a_1[\chi_1] + \cdots + a_k[\chi_k]$ for some integers $a, a_1, \ldots, a_k$ with $a \neq 0$. Consider

$$\xi_x([\chi]) := e^{i2\pi\left(\frac{a_1}{a}x_1 + \cdots + \frac{a_k}{a}x_k\right)}. \tag{4.2}$$

Clearly, ξ_x is a character (see Remark 4.1). Moreover, the map $x \mapsto \xi_x$ defines a homeomorphism between $\mathcal{Q}^k$ and a subset of $\operatorname{Ann}(\mathbb{X})$. Therefore, $\dim_T \operatorname{Ann}(\mathbb{X}) \geq k$ that finishes the proof. □

4.3 Frequency Module and Cartwright's Theorem on Almost Periodic Flows

Many presented facts of the theory of almost periodic functions may be found in [8, 15, 19]. The proof of Cartwright's theorem (see [6]) based on the Pontryagin duality was borrowed from [30].

Frequency Spectrum First of all, we will introduce the concept of the Fourier spectrum for general almost periodic functions.

Let $\mathbb{E}$ be a Banach space (over $\mathbb{R}$ or $\mathbb{C}$) with the norm $\|\cdot\|$ and let $u\colon \mathbb{R} \to \mathbb{E}$ be a continuous function. For a given $\varepsilon > 0$ denote by $\mathcal{T}_\varepsilon(u)$ the set of $\tau \in \mathbb{R}$ such that $\|u(\cdot+\tau) - u(\cdot)\|_\infty \leq \varepsilon$, where $\|\cdot\|_\infty$ stands for the uniform norm. Such a number τ is called an ε-almost period of $u(\cdot)$. Remind that a subset $\mathcal{A} \subset \mathbb{R}$ is *relatively dense* if there is a number $L > 0$ such that the intersection $\mathcal{A} \cap [a, a+L]$ is not empty for all $a \in \mathbb{R}$. The function $u(\cdot)$ is called $\mathbb{E}$-*almost periodic* (or simply, almost periodic) if the set $\mathcal{T}_\varepsilon(u)$ is relatively dense for every $\varepsilon > 0$. From the definition it follows that almost periodic functions are uniformly continuous and compact. It is clear that the set of almost periodic functions is a closed subset of $C_b(\mathbb{R}; \mathbb{E})$.

The *mean value* of $u(\cdot)$ is the limit

$$\mathrm{M}\{u(\cdot)\} := \lim_{T\to+\infty} \frac{1}{2T} \int_{-T}^{T} u(t)dt. \tag{4.3}$$

Its existence for almost periodic functions is known as the Bohr theorem. For $\nu \in \mathbb{R}$ consider the *Fourier transform* of $u(\cdot)$:

$$U(\nu) := \mathrm{M}\left\{u(\cdot)e^{-i\nu\cdot}\right\} = \lim_{T\to+\infty} \frac{1}{2T} \int_{-T}^{T} u(t)e^{-i\nu t}dt. \tag{4.4}$$

It is known that $U(\nu) \neq 0$ for an at most countable set of ν's, say $\{\nu_1, \nu_2, \ldots\}$. Call this set the *spectrum* of u and denote this set by $\mathrm{Sp}(u)$ and put $U_k := U(\nu_k)$. Then there is a formal *Fourier series* of $u(\cdot)$:

$$u(t) \sim \sum_{k=1}^{\infty} U_k e^{i\nu_k t}. \tag{4.5}$$

The smallest additive subgroup of reals containing the set $\mathrm{Sp}(u)$ is called $\mathbb{Z}$-*module* and denoted by $\mathrm{mod}_{\mathbb{Z}}(u)$. The linear subspace of $\mathbb{R}$ over $\mathbb{Q}$ spanned by $\mathrm{Sp}(u)$ is called $\mathbb{Q}$-*module* and denoted by $\mathrm{mod}_{\mathbb{Q}}(u)$. The real numbers $\omega_1, \ldots, \omega_m$ are called *rational base* for $u(\cdot)$ if for every $\nu_k \in \mathrm{Sp}(u)$, $k = 1, 2, \ldots$, there is a unique representation

$$\nu_k = \sum_{j=1}^{m} r_j^{(k)} \omega_j, \tag{4.6}$$

with $r_j^{(k)} \in \mathbb{Q}$. If (4.6) holds with $r_j^k \in \mathbb{Z}$ then $\omega_1, \ldots, \omega_m$ are an *integral base* or *frequencies*. In the latter case the function $u(\cdot)$ is called *quasi-periodic*. The uniqueness of (4.6) is equivalent to linear independence of $\omega_1, \ldots, \omega_m$ over $\mathbb{Q}$ (*rational independence*).

The following theorem is due to Bochner and gives a characterization of almost periodic functions.

Theorem 4.3 *A bounded continuous function $u(\cdot)$ is almost periodic if and only if the set of its translates $\{u(\cdot + s)\}_{s \in \mathbb{R}}$ is relatively compact in the topology of uniform convergence.*

From Theorem 4.3 it is clear that a sum of two almost periodic functions is also almost periodic.

Almost Periodic Flows Suppose we have a continuous flow $\{\varphi^t\}_{t \in \mathbb{R}}$ on a subset $\mathcal{M}$ of a Banach space $\mathbb{E}$ such that $\varphi^t(\mathcal{M}) \subset \mathcal{M}$ for every $t \in \mathbb{R}$. For $u_0 \in \mathcal{M}$ the motion $t \to \varphi^t(u_0)$ is called *almost periodic* if the function $u(t) := \varphi^t(u_0)$ is $\mathbb{E}$-almost periodic. Let $\mathcal{M}_u$ be the closure of $\gamma(u_0) := \{\varphi^t(u_0) \mid t \in \mathbb{R}\} = u(\mathbb{R})$ in $\mathbb{E}$.

Theorem 4.4 *The flow $\{\varphi^t\}_{t \in \mathbb{R}}$ can be uniquely extended from $\gamma(u_0)$ to $\mathcal{M}_u$ in such a way that*

(1) The set $\mathcal{M}_u$ is minimal and every motion $t \to \varphi^t(v)$, $v \in \mathcal{M}_u$, is almost periodic with the same frequencies.
(2) The family $\{\varphi^t\}_{t \in \mathbb{R}}$ is equicontinuous on $\mathcal{M}_u$.

Proof (1) Since $\{\varphi^t\}$ is continuous on $\gamma(u_0)$ for every $\varepsilon > 0$ and $T > 0$ there exists $\delta > 0$ such that for $v_1, v_2 \in \gamma(u_0)$ we have

$$\|\varphi^t(v_1) - \varphi^t(v_2)\| \le \varepsilon, \ 0 \le t \le T,$$

provided that $\|v_1 - v_2\| < \delta$. Let $L > 0$ be a number such that $\mathcal{T}_\varepsilon(u) \cap [a, a + L]$ is non-empty for every $a \in \mathbb{R}$ and take $T > L$. Then for every t there is an ε-almost period τ such that $t = \tau + r$, $0 \le r \le L$. Hence, for $t \in \mathbb{R}$ we have

$$\begin{aligned} \|\varphi^t(v_1) - \varphi^t(v_2)\| &\le \|\varphi^t(v_1) - \varphi^r(v_1)\| + \|\varphi^r(v_1) - \varphi^r(v_2)\| \\ &+ \|\varphi^r(v_2) - \varphi^t(v_2)\| \le 3\varepsilon. \end{aligned} \tag{4.7}$$

So the flow $\{\varphi^t\}$ is equicontinuous on $\gamma(u_0)$.

(2) Now suppose that $u_m = u(t_m) = \varphi^{t_m}(u_0) \to v \in \mathcal{M}_u$ for some sequence t_m, $m = 1, 2 \ldots$. We have from (1) that for $\varepsilon > 0$ there is $M = M(\varepsilon)$ such that

$$\|\varphi^t(u_m) - \varphi^t(u_{m+p})\| < \varepsilon, \ p = 1, 2, \ldots \tag{4.8}$$

provided that $m > M$. In other words, the definition $\varphi^t(v) := \lim_{m \to +\infty} \varphi^t(u_m)$ is correct and the limit exists uniformly in $t \in \mathbb{R}$. The motion $t \mapsto \varphi^t(v) =: v(t)$ is almost periodic. Indeed, for $\tau \in \mathcal{T}_\varepsilon(u)$ we have

$$\|\upsilon(t+\tau)-\upsilon(t)\| \le \|\upsilon(t+\tau)-\varphi^{t+\tau}(u_m)\| + \|\varphi^{t+\tau}(u_m)-\varphi^{t}(u_m)\| \\ + \|\varphi^{t}(u_m)-\upsilon(t)\| \le 3\varepsilon. \tag{4.9}$$

Now the minimality of $\mathcal{M}_u$ is obvious. The equicontinuity of $\{\varphi^t\}$ on $\mathcal{M}_u$ follows from (1) and (2). Since we have

$$u(t+t_m) \sim \sum_{k=1}^{\infty} U_k e^{i\nu_k t_m} e^{i\nu_k t}, \tag{4.10}$$

it follows that $\mathrm{Sp}(\upsilon) = \mathrm{Sp}(u)$. Thus, the theorem is proved. □

Remark 4.2 Further, for an almost periodic function $u(\cdot)$ we will study some dimensional-like properties of the set $\mathcal{M}_u := Cl(u(\mathbb{R}))$. In the case $u(\cdot)$ is an almost periodic motion of a flow we say that $\mathcal{M}_u$ is the closure of an almost periodic *orbit*, otherwise (i.e. when the curve given by $u(\cdot)$ has self-intersections) we refer to $\mathcal{M}_u$ as the closure of an almost periodic *trajectory*.

Now we define a group structure on $\mathcal{M}_u$. For $\upsilon_1, \upsilon_2 \in \mathcal{M}_u$ such that $\varphi^{t_n}(u_0) \to \upsilon_1$ and $\varphi^{s_n}(u_0) \to \upsilon_2$ we put

$$\upsilon_1 \overset{u_0}{+} \upsilon_2 := \lim_{n\to\infty} \varphi^{t_n+s_n}(u_0). \tag{4.11}$$

Theorem 4.5 *The above definition of* $\overset{u_0}{+}$ *is correct and* $(\mathcal{M}_u, \overset{u_0}{+})$ *is a compact connected abelian group.*

Proof The family $\{\varphi^t\}_{t\in\mathbb{R}}$ is equicontinuous on $\gamma(u_0)$ and, therefore, the uniformly continuous on a dense subset $\gamma(u_0) \times \gamma(u_0)$ function $\overset{u_0}{+}$ can be uniquely extended to a continuous function on $\mathcal{M}_u \times \mathcal{M}_u$ in the way we did in (4.11). Now it is clear from (4.11) that the operation $\overset{u_0}{+}$ is associative, commutative and u_0 is the zero element. If $\varphi^{t_n}(u_0) \to \upsilon$ then, by Theorem 4.3, t_n can be chosen such that the sequence $\varphi^{-t_n}(u_0)$ is also convergent and its limit $\overset{u_0}{-}\upsilon$ is the inverse of υ, i.e. $\upsilon \overset{u_0}{+} (\overset{u_0}{-}\upsilon) = u_0$. Topological properties of $\mathcal{M}_u$ are obvious. □

The Cartwright Theorem Now our purpose is to show that for every almost periodic function there is a corresponding almost periodic flow. Consider the *hull* $\mathcal{H}(u)$ of an $\mathbb{E}$-almost periodic function $u(\cdot)$ defined by the set $Cl\{u(\cdot+s) \mid s \in \mathbb{R}\}$, where the closure is taken in the uniform topology of the space $C_b(\mathbb{R};\mathbb{E})$. By Theorem 4.3, the set $\mathcal{H}(u)$ is compact. The family of shift operators $\vartheta^s : C_b(\mathbb{R};\mathbb{E}) \to C_b(\mathbb{R};\mathbb{E})$, where $s \in \mathbb{R}$ and $\vartheta^s(\upsilon) := \upsilon(\cdot+s)$ for $\upsilon \in C_b(\mathbb{R};\mathbb{E})$, restricted to $\mathcal{H}(u)$ defines an almost periodic flow $(\{\vartheta^s\}_{s\in\mathbb{R}}, \mathcal{H}(u))$ for which the set $\mathcal{H}(u)$ is minimal, i.e. $\mathcal{H}(u) = \mathcal{H}(\upsilon)$ for every $\upsilon \in \mathcal{H}(u)$. In particular, the function $\mathbb{U}(t) := \vartheta^t(u)$ is $C_b(\mathbb{R};\mathbb{E})$-almost periodic. The following proposition can be shown by straightforward calculations, which we omit here.

Proposition 4.1 *For $\mathbb{U}(\cdot)$ we have*

$$\mathbb{U}(t) \sim \sum_{k=1}^{\infty} \mathbb{U}_k e^{i\nu_k t}, \tag{4.12}$$

where $\mathbb{U}_k(s) = e^{i\nu_k s} U_k$ with U_k and ν_k from (4.5). *In particular, $Sp(\mathbb{U}) = Sp(u)$.*

As it was shown before there is a group structure on $\mathcal{H}(u)$ given by the operation $\overset{u}{+}$. The following theorem shows that if $u(\cdot)$ is given by an almost periodic motion then the group structures on $\mathcal{M}_u$ and $\mathcal{H}(u)$ are topologically isomorphic.

Theorem 4.6 *Suppose the motion $t \to \varphi^t(u_0)$ is almost periodic and $u(t) := \varphi^t(u_0)$; then the groups $(\mathcal{M}_u, \overset{u_0}{+})$ and $(\mathcal{H}(u), \overset{u}{+})$ are topologically isomorphic.*

Proof We define a map $i : \mathcal{H}(u) \to \mathcal{M}_u$ as follows. Put $i(\vartheta^t(u)) := \varphi^t(u_0)$ for every $t \in \mathbb{R}$ and then extend it by continuity. Note that if $\vartheta^{t_n}(u) \to v$ in $C_b(\mathbb{R}; \mathbb{E})$ then $\varphi^{t_n}(u_0) \to v(0)$ in $\mathbb{E}$. Thus, $i(v) = v(0)$ for any $v \in \mathcal{H}(u)$ and $i(\vartheta^t(v)) = \varphi^t(v(0))$. In particular, i is continuous. In order to show the injectivity of i note that $i(v) = u_0$ implies $v(0) = u_0$ and, consequently, $v \equiv u$. If $\varphi^{t_n}(u_0) \to v_0 \in \mathcal{M}_u$ then the Bochner theorem guarantees there is a subsequence $\{t_n'\} \subset \{t_n\}$ such that $\vartheta^{t_n'}(u)$ converges uniformly to some $v \in \mathcal{H}(u)$. It is obvious that $i(v) = v_0$ and this shows the surjectivity of i. Finally, the map i is a continuous bijective map between compact metric spaces and, therefore, is a homeomorphism. □

In order to apply results from the previous section we need to describe the character group of $\mathcal{H}(u)$ to calculate its rank.

Theorem 4.7 *For any $\mathbb{E}$-almost periodic function $u(\cdot)$ we have*[3]

$$\widehat{\mathcal{H}(u)} \cong \mathrm{mod}_{\mathbb{Z}}(u). \tag{4.13}$$

Proof Suppose $u(\cdot)$ has the Fourier series as in (4.5). Then for any $v \in \mathcal{H}(u)$ we have that $v(t) \sim \sum_{k=1}^{\infty} U_k e^{i\theta_k(v)} e^{i\nu_k t}$, where $\theta_k(v)$ is defined modulo 2π. Define $\chi_k(v) := e^{i\theta_k(v)}$. From (4.11) it is clear that $\chi(v_1 \overset{u}{+} v_2) = \chi(v_1)\chi(v_2)$. The homomorphism $\chi_k(v)$ is continuous and, therefore, is a character of $\mathcal{H}(u)$. For any finite set of integers $a_1, \ldots, a_m$ the homomorphism $\chi_1^{a_1} \cdot \ldots \cdot \chi_m^{a_m}$ is also a character. It turns out that there is no other characters (for a proof see [30]). Any character $\chi(\cdot)$ satisfies $\chi(\vartheta^t(u)) = e^{i\theta t}$ with $\theta = \sum_{k=1}^{m} a_k \nu_k$ for some integers $a_1, \ldots, a_m$. The map $\chi \mapsto \theta$ defines a group isomorphism. □

Now the Cartwright theorem can be formulated as follows.

---

[3]Here by $\widehat{\mathcal{H}(u)} \cong \mathrm{mod}_{\mathbb{Z}}(u)$ we mean a group isomorphism. In general, it is not a homeomorphism since the character group $\widehat{\mathcal{H}(u)}$ is discrete and $\mathrm{mod}_{\mathbb{Z}}(u)$ can be a dense subgroup of $\mathbb{R}$.

Theorem 4.8 *For any $\mathbb{E}$-almost periodic function $u(\cdot)$ we have*

$$\dim_T \mathcal{H}(u) = \dim \operatorname{mod}_{\mathbb{Q}}(u). \tag{4.14}$$

In particular, if $u(t) = \varphi^t(u_0)$, where $t \mapsto \varphi^t(u_0)$ is an almost periodic motion, then $\dim_T \mathcal{M}_u = \dim \operatorname{mod}_{\mathbb{Q}}(u)$.

Proof From Theorem 4.2 we have

$$\dim_T \mathcal{H}(u) = \operatorname{rank} \widehat{\mathcal{H}(u)}. \tag{4.15}$$

By Theorem 4.7 and since $\operatorname{mod}_{\mathbb{Q}}(u)$ is the least $\mathbb{Q}$-vector space containing $\operatorname{mod}_{\mathbb{Z}}(u)$ it follows (see Remark 4.1) that $\operatorname{rank} \widehat{\mathcal{H}(u)} = \dim \operatorname{mod}_{\mathbb{Q}}(u)$. The second part of the theorem directly follows from Theorem 4.6. □

4.4 Minimal Sets in Euclidean Spaces

In this subsection we shall prove Hilmy's theorem (see [11]) on the estimation of the topological dimension for a minimal set $\mathcal{M}_{\min}$ of a dynamical system $(\{\varphi^t\}_{t\in\mathbb{R}}, \mathcal{M}, \rho)$ on the open subset $\mathcal{M} \subset \mathbb{R}^n$ with the Euclidean distance ρ and with a group as time set, i.e. $\mathbb{T} \in \{\mathbb{R}, \mathbb{Z}\}$.

Theorem 4.9 *Suppose $\mathcal{M}_{\min} \subset \mathcal{M}$ is a compact minimal set of $(\{\varphi^t\}_{t\in\mathbb{T}}, \mathcal{M}, \rho)$. Then $\dim_T \mathcal{M}_{\min} \le n-1$.*

For the proof of Theorem 4.9 we need two lemmas.

Lemma 4.3 *Let $\mathcal{S} \subset \mathcal{M}$ be an invariant set of $(\{\varphi^t\}_{t\in\mathbb{T}}, \mathcal{M}, \rho)$. If the boundary $\partial\mathcal{S}$ is non-empty then $\partial\mathcal{S}$ is also invariant.*

Proof Suppose that $\partial\mathcal{S} \neq \emptyset$ and $u_0 \in \partial\mathcal{S}$. Consider an arbitrary $t \in \mathbb{T}$ and an arbitrary $\varepsilon > 0$. By the continuity of the dynamical system w.r.t. the map $u_0 \mapsto \varphi^t(u_0)$ it is possible to find a number $\delta > 0$ such that $\mathcal{B}_\delta(u_0) \subset \mathcal{M}$ and

$$\varphi^t(\mathcal{B}_\delta(u_0)) \subset \mathcal{B}_\varepsilon(\varphi^t(u_0)),$$

where $\mathcal{B}_\delta(u_0)$ (resp. $\mathcal{B}_\varepsilon(\varphi^t(u_0))$) denotes the ball of radius δ (resp. ε) and centrum at u_0 (resp. $\varphi^t(u_0)$). Since $\mathcal{B}_\delta(u_0)$ contains points of $\mathcal{S}$ as well as points of $\mathcal{M}\backslash\mathcal{S}$, it follows that $\varphi^t(\mathcal{B}_\delta(u_0))$ and, consequently, $\mathcal{B}_\varepsilon(\varphi^t(u_0))$, posses the same property. But ε was an arbitrary positive number. Therefore, $\varphi^t(u_0) \in \partial\mathcal{S}$. □

Lemma 4.4 *If $\mathcal{M}_{\min}$ is a minimal set of $(\{\varphi^t\}_{t\in\mathbb{T}}, \mathcal{M}, \rho)$ then all points of $\mathcal{M}_{\min}$ are either boundary points or all are inner points.*

Proof Suppose to the contrary that $\mathcal{M}_{\min}$ contains boundary points as well as inner points. Since $\mathcal{M}_{\min}$ is closed, we have $\partial\mathcal{M}_{\min} \subset \mathcal{M}_{\min}$. From Lemma 4.3 it follows that $\partial\mathcal{M}_{\min}$ is an invariant set. This shows that $\partial\mathcal{M}_{\min}$ is a proper subset of $\mathcal{M}_{\min}$ which is also invariant. But this contradicts the fact that $\mathcal{M}_{\min}$ is minimal. □

Proof of Theorem 4.9 Suppose that $\dim_T \mathcal{M}_{\min} = n$. Then by Proposition 3.8, Chap. 3, the set $\mathcal{M}_{\min}$ must contain inner points. But according to Lemma 4.4 this is impossible, since from the compactness of $\mathcal{M}_{\min}$ it follows that $\partial\mathcal{M}_{\min} \neq \emptyset$. □

Suppose that $(\{\varphi^t\}_{t\in\mathbb{R}}, \mathcal{M}_u)$ is an almost periodic flow defined on the closure $\mathcal{M}_u \subset \mathbb{R}^n$ of an almost periodic motion $t \mapsto \varphi^t(u_0) = u(t)$. Recall that its frequencies is the Fourier exponents $Sp(u)$ of $u(\cdot)$. Since the set $\mathcal{M}_u$ is minimal, Theorems 4.8 and 4.9 guarantee that $\dim_T \mathcal{M}_u = \dim \operatorname{mod}_{\mathbb{Q}}(u) \leq n-1$, i.e. the frequencies of the flow have a rational base with no more that $n-1$ terms. We omit a proof of the following theorem that is a mix of results of Cartwright and Kodaira and Abe (see [6]).

Theorem 4.10 *If* $\dim_T \mathcal{M}_u = n-1$ *then* $(\mathcal{M}_u, \overset{u_0}{+})$ *is isomorphic to the* $(n-1)$-*dimensional torus group and the frequencies of the flow have an integral base with* $n-1$ *terms.*

The second part of Theorem 4.10 says that almost periodic flows in $\mathbb{R}^n$ having the highest possible dimension (i.e. $n-1$) are quasi-periodic.

4.5 Almost Periodic Solutions of Almost Periodic ODEs

Structure of the Frequency Spectrum In what follows we will deal with the following ODE in $\mathbb{R}^n$

$$\dot{u} = f(t, u), \tag{4.16}$$

where $f(t, u)$ is continuous. We also assume that

(A1) $f(t, u)$ is almost periodic in t uniformly in u from compact subsets of $\mathbb{R}^n$.

(A2) The solutions to (4.16) are unique.

Remark 4.3 Condition **(A1)** means that f is almost periodic as a function $t \mapsto f(t, \cdot) \in C(\mathcal{K}; \mathbb{R}^n)$ for every compact $\mathcal{K} \subset \mathbb{R}^n$. We say that the corresponding ε-almost periods are the *ε-almost periods of $f(\cdot, u)$ uniformly in $u \in \mathcal{K}$*. Since $C(\mathcal{K}; \mathbb{R}^n)$ is a Banach space, we have all the introduced Fourier theory for such functions, namely, one can write

$$f(t, u) \sim \sum_{k=1}^{\infty} F_k(u) e^{i\nu_k^* t}, \tag{4.17}$$

where $F_k(\cdot), k = 1, 2, \ldots$, are continuous functions that is not identically zero on $\mathbb{R}^n$. Thus, we can consider the $\mathbb{Z}$-module $\mathrm{mod}_{\mathbb{Z}}(f)$ of f generated by all the exponents from (4.17).

The following lemma is due to E. Kamke (see Theorem 3.2 in [10]).

Lemma 4.5 *Let $\mathcal{G} \subset \mathbb{R}^n$ be an open subset and consider a sequence $g_k \in C(\mathbb{R} \times G; \mathbb{R}^n)$, $k = 1, 2, \ldots$. Let $u_k(\cdot)$ be a maximal solution to $\dot{u} = g_k(t, u)$. Suppose that g_k converges to g in the compact-open topology and $u_k(0)$ converges to some $u_0 \in \mathcal{G}$; then*

(A) there is a subsequence of $u_k(\cdot)$ converging to a solution $u(\cdot)$ of $\dot{u} = g(t, u)$ with $u(0) = u_0$. The convergence is uniform on the compact intervals on which u exists.
(B) If the solution to $\dot{u} = g(t, u)$ with $u(0) = u_0$ is unique then the entire sequence $u_k(\cdot)$ converges to $u(\cdot)$.

Condition **(A2)** is essential for the following lemma which will be used in the future.

Lemma 4.6 *Suppose* **(A1)–(A2)** *hold and u is an almost periodic solution to* (4.16). *Let the sequence t_m, $m = 1, 2, \ldots$, be such that $(\vartheta^{t_m}(f))(t, v) = f(t + t_m, v) \to f(t, v)$ uniformly in $(t, v) \in \mathbb{R} \times \mathcal{M}(u)$, where $\mathcal{M}(u) = Cl(u(\mathbb{R}))$. If $u(t_m)$ converges to some $\tilde{v}_0$ then $u(\cdot + t_m)$ converges to the solution $\tilde{v}$ of* (4.16) *with $\tilde{v}(0) = \tilde{v}_0$ which is almost periodic.*

Proof By Lemma 4.5 we get the convergence of $u(\cdot + t_m)$ to $\tilde{v}(\cdot)$ on compact subsets in the interval of its existence. Since $u(\cdot)$ is bounded we deduce that $\tilde{v}$ is defined on the whole real axis and, consequently, the convergence of $u(\cdot + t_m)$ is uniform and $\tilde{v}$ is almost periodic. □

It may happen that an almost periodic solution, say u, to (4.16) may have the Fourier exponents (frequencies) that do not belong to the $\mathbb{Q}$-module of f. We call these frequencies *additional*. For example, if f is independent of t then any frequency of u is additional. In the latter case we know from Theorems 4.8 and 4.9 that $\dim \mathrm{mod}_{\mathbb{Q}}(u) \le n - 1$. It turns out that a similar fact holds within the non-autonomous situation. Namely, the dimension of the subspace generated by additional frequencies is always bounded from above by $n - 1$. This is also a result of M. L. Cartwright (see [7]) and can be formulated as follows.

Theorem 4.11 *Suppose* **(A1)–(A2)** *hold and u is an almost periodic solution to* (4.16). *Denote the set of additional exponents by $Sp^C(u)$. If $Sp^C(u)$ is not empty then there exists an almost periodic flow $\{\varphi_0^t\}$ defined on the set $\mathcal{A}_0 \subset \mathbb{R}^n$ consisting of initial conditions v_0 such that the solution $v(t) = v(t, 0, v_0)$ to* (4.16) *is almost periodic. Moreover, the $\mathbb{Q}$-module of this flow is the subspace generated by $Sp^C(u)$. In particular,*

$$\dim \frac{\mathrm{mod}_{\mathbb{Q}}(u, f)}{\mathrm{mod}_{\mathbb{Q}}(f)} = \dim \mathrm{span}(Sp^C(u)) \le n - 1. \tag{4.18}$$

Remark 4.4 The first assertion of Theorem 4.11 (i.e. the existence of an almost periodic flow) can be proved for cocycles in a Banach space for which the driving system is an almost periodic minimal flow. The ideas of the proof is almost identical to the ones presented below. Then some reduction principle as in Theorem 4.12 can be used to show a similar inequality as in (4.18).

In order to prove Theorem 4.11 we need to establish some auxiliary facts. Firstly, we choose a maximal linearly independent set $\{\overline{\nu}_l^*\} \subset Sp(f)$ by a standard procedure, i.e. we put $\overline{\nu}_1^* := \nu_1^*$ and then $\overline{\nu}_2^* := \nu_{k_0}^*$ where k_0 is the smallest number k such that ν_k is linearly independent with ν_1 and by induction we define $\overline{\nu}_l^*$ as $\nu_{k_0}^*$ where k_0 is the smallest number k such that ν_k is linearly independent with $\overline{\nu}_1^*, \ldots, \overline{\nu}_{l-1}^*$. By similar procedure we choose a subset $\{\overline{\nu}_j\} \subset Sp(u)$ which complements the set $\{\overline{\nu}_l^*\}$ to a basis of $\mathrm{mod}_{\mathbb{Q}}(u, f)$. Then any exponent of u can be uniquely represented as

$$\nu_k := \sum_{j=1}^{J(k)} r_{j,k}\overline{\nu}_j + \sum_{l=1}^{L(k)} r_{l,k}^*\overline{\nu}_l^*, \tag{4.19}$$

where $r_{j,k}$ and $r_{l,k}^*$ are some rational numbers. Let P_ε be a sequence of trigonometric polynomials approximating u with an error of $\leq \varepsilon$. The polynomial P_ε can be written as

$$P_\varepsilon(t) = \sum_{k=1}^{N(\varepsilon)} P_k^{(\varepsilon)} e^{i\sum_{j=1}^{J(k)} r_{j,k}\overline{\nu}_j t} \cdot e^{i\sum_{l=1}^{L(k)} r_{l,k}^*\overline{\nu}_j^* t}. \tag{4.20}$$

Along with (4.20) we consider the family of functions $\Phi_\varepsilon(t, \theta_1, \ldots, \theta_{M(\varepsilon)})$ defined as

$$\Phi_\varepsilon\left(t, \theta_1, \ldots, \theta_{M(\varepsilon)}\right) := \sum_{k=1}^{N(\varepsilon)} P_k^{(\varepsilon)} e^{i\sum_{j=1}^{J(k)} r_{j,k}\overline{\nu}_j\theta_j} \cdot e^{i\sum_{l=1}^{L(k)} r_{l,k}^*\overline{\nu}_l^* t}, \tag{4.21}$$

where $M(\varepsilon) := \max_{1\leq k\leq N(\varepsilon)} J(k)$. The following proposition holds.

Proposition 4.2 *The limit*[4]

$$\Phi(t, \theta_1, \theta_2, \ldots) := \lim_{\varepsilon\to 0} \Phi_\varepsilon\left(t, \theta_1, \ldots, \theta_{M(\varepsilon)}\right) \tag{4.22}$$

exists uniformly in $t, \theta_1, \theta_2, \ldots \in \mathbb{R}$. *Moreover,*

(1) $u(t) = \Phi(t, t, t, \ldots)$;
(2) For every $\theta_1, \theta_2, \ldots$ *and* $v_0 = \Phi(0, \theta_1, \theta_2, \ldots)$ *the solution* $v(t) = v(t, 0, v_0)$ *to* (4.16) *is almost periodic and* $v(t) = \Phi(t, \theta_1 + t, \theta_2 + t, \ldots)$.

[4]Note that at the current moment we know nothing about the number of variables of Φ so we do not exclude the case when this number may be infinite.

Proof To get the uniformity in (4.22) choose $0 < \varepsilon_1, \varepsilon_2 \le \varepsilon$ and consider the corresponding functions Φ_{ε_1} and Φ_{ε_2}. Fix $t \in \mathbb{R}$ and $\theta_1, \ldots, \theta_{M^+} \in \mathbb{R}$, where $M^+ = \max\{M(\varepsilon_1), M(\varepsilon_2)\}$. We will show that $\tilde{t} \in \mathbb{R}$ can be chosen such that the differences

$$\left|\Phi_{\varepsilon_1}\left(t, \theta_1, \ldots, \theta_{M(\varepsilon_1)}\right) - P_{\varepsilon_1}(\tilde{t})\right| \text{ and } \left|\Phi_{\varepsilon_1}\left(t, \theta_1, \ldots, \theta_{M(\varepsilon_2)}\right) - P_{\varepsilon_2}(\tilde{t})\right| \quad (4.23)$$

will be arbitrarily small. It is clear that from this we immediately get that

$$\left|\Phi_{\varepsilon_1}\left(t, \theta_1, \ldots, \theta_{M(\varepsilon_1)}\right) - \Phi_{\varepsilon_2}\left(t, \theta_1, \ldots, \theta_{M(\varepsilon_2)}\right)\right| \le \varepsilon_1 + \varepsilon_2 \le 2\varepsilon \quad (4.24)$$

for all $t, \theta_1, \ldots, \theta_{M^+} \in \mathbb{R}$ and, consequently, the limit in (4.22) is uniform. We get what we need if the sequence t_m, $m = 1, 2, \ldots$, is chosen to satisfy the following conditions

(I) $\sum_{j=1}^{J(k)} r_{j,k}\overline{\nu}_j\theta_j - \sum_{j=1}^{J(k)} r_{j,k}\overline{\nu}_j t_m \to 0 \pmod{2\pi}$ as $m \to \infty$;

(II) $\sum_{l=1}^{L(k)} r^*_{l,k}\overline{\nu}^*_l t - \sum_{l=1}^{L(k)} r^*_{l,k}\overline{\nu}^*_l t_m \to 0 \pmod{2\pi}$ as $m \to \infty$

for all $k = 1, \ldots, N^+$, where $N^+ = \max\{N(\varepsilon_1), N(\varepsilon_2)\}$. Finally, **(I)** and **(II)** will be satisfied if as $m \to \infty$

$$\begin{aligned} \overline{\nu}_j t_m - \overline{\nu}_j\theta_j &\to 0 \quad (\mathrm{mod}\ 2\pi Q),\ j = 1, \ldots, J^+ := \max_{1\le k\le N^+}\{J(k)\}, \\ \overline{\nu}^*_l t_m - \overline{\nu}^*_l t &\to 0 \quad (\mathrm{mod}\ 2\pi Q),\ l = 1, \ldots, L^+ := \max_{1\le k\le N^+}\{L(k)\} \end{aligned} \quad (4.25)$$

for a proper choose of Q.[5] Since $\overline{\nu}_1, \ldots, \overline{\nu}_{J^+}, \overline{\nu}^*_1, \ldots, \overline{\nu}^*_{L^+}$ are linearly independent the winding on the $(J^+ + L^+)$-dimensional torus $\mathbb{R}^{J^+ + L^+}/2\pi Q\mathbb{Z}$ in the corresponding direction is dense and, consequently, the required sequence $\{t_m\}$ exists.

Item (1) of the theorem is obvious.

To get (2) put $\tilde{\upsilon}(t) := \Phi(t, \theta_1 + t, \theta_2 + t, \ldots)$. It is clear that $\tilde{\upsilon}$ is almost periodic as the uniform limit of the functions $\Phi_\varepsilon(t, \theta_1 + t, \ldots, \theta_{M(\varepsilon)} + t)$ which are trigonometric polynomials for fixed $\theta_1, \theta_2, \ldots$. As above we can find a sequence $\{t_m\}$ such that $P_\varepsilon(t + t_m) \to \Phi_\varepsilon(t, \theta_1 + t, \ldots, \theta_{M(\varepsilon)} + t)$ uniformly in t. From this we conclude

$$\begin{aligned} |u(t + t_m) - \tilde{\upsilon}(t)| \le{}& |u(t + t_m) - P_\varepsilon(t + t_m)| \\ &+ \left|P_\varepsilon(t + t_m) - \Phi_\varepsilon\left(t, \theta_1 + t, \ldots, \theta_{M(\varepsilon)} + t\right)\right| \\ &+ \left|\Phi_\varepsilon\left(t, \theta_1 + t, \ldots, \theta_{M(\varepsilon)} + t\right) - \tilde{\upsilon}(t)\right| \end{aligned} \quad (4.26)$$

[5] For example, if Q is the least common multiple of the denominators of all $r_{j,k}$ and $r^*_{l,k}$ for $1 \le j \le J^+, 1 \le l \le L^+, 1 \le k \le N^+$.

that shows $u(\cdot + t_m) \to \tilde{\upsilon}(\cdot)$ uniformly. Moreover, from the arithmetic nature of almost periods, the sequence $\{t_m\}$ can be chosen such that $(\vartheta^{t_m}(f))(t, \upsilon) = f(t + t_m, \upsilon) \to f(t, \upsilon)$ uniformly in $(t, \upsilon) \in \mathbb{R} \times \mathcal{M}_u$. Therefore, we are in the situation of Lemma 4.6 and $\tilde{\upsilon} \equiv \upsilon(t)$. □

Now consider the set $\mathcal{A}_0 := \{\Phi(0, t, t, \ldots) \mid t \in \mathbb{R}\}$. For $p = \Phi(0, s, s, \ldots) \in \mathcal{A}_0$ we put $\varphi_0^t(p) := \Phi(0, t + s, t + s, \ldots)$. The following proposition shows the correctness of such a definition and, as a consequence, $\{\varphi_0^t\}$ is an almost periodic flow on $\mathcal{A}_0$.

Proposition 4.3 *Suppose for some $t_1, t_2 \in \mathbb{R}$ we have the identity $\Phi(0, t_1, t_1, \ldots) = \Phi(0, t_2, t_2, \ldots)$. Then either $t_1 = t_2$ or* span $Sp^C(u) =$ span$\{\overline{\nu}_1\}$ *and $t_1 - t_2 = 0$* (mod $2\pi Q/\overline{\nu}_1$) *for some integer Q. Moreover, in the latter case the flow $\{\varphi_0^t\}$ is $2\pi Q/\overline{\nu}_1$-periodic.*

Proof Since we have, by Proposition 4.2, $\upsilon_1(t) := \Phi(t, t_1 + t, t_1 + t, \ldots)$ and $\upsilon_2(t) := \Phi(t, t_2 + t, t_2 + t, \ldots)$ are almost periodic solutions to (4.16) from $\upsilon_1(0) = \upsilon_2(0)$ and **(A2)** we get that $\upsilon_1 \equiv \upsilon_2$. Considering the Fourier expansions of υ_1 and υ_2 we get

$$U_k e^{\sum_{j=1}^{J(k)} r_{j,k} \overline{\nu}_j t_1} = U_k e^{\sum_{j=1}^{J(k)} r_{j,k} \overline{\nu}_j t_2}, \tag{4.27}$$

where U_k is the Fourier coefficient of u corresponding to ν_k from (4.19). In virtue the choice of $\{\overline{\nu}_j\}$ for every j there exists k such that (4.27) takes the form

$$\overline{\nu}_j(t_1 - t_2) = 0 \pmod{2\pi}. \tag{4.28}$$

If there are more than one of such j's then from (4.28) it follows that $t_1 = t_2$. Otherwise $\overline{\nu}_1$ forms a basis for the set of additional exponents and

$$r_{1,k}\overline{\nu}_1(t_1 - t_2) = 0 \pmod{2\pi}, \ k = 1, 2, \ldots. \tag{4.29}$$

It is easy to see that (4.29) is satisfied if $t_1 - t_2 = 2\pi Q/\overline{\nu}_1$ for a proper integer Q. For such a choice of t_1 and t_2 in virtue of the uniqueness of Fourier expansions we have $\Phi(t, t_1 + t, t_1 + t, \ldots) = \Phi(t, t_2 + t, t_2 + t, \ldots)$. So the flow $\{\varphi_0^t\}$ is $2\pi Q/\overline{\nu}_1$-periodic. □

Now we can finish the proof.

Proof of Theorem 4.11 We have defined the flow $\{\varphi_0^t\}$ on the set $\mathcal{A}_0$ consisting of initial conditions of almost periodic solutions to (4.16). For $p = \Phi(0, 0, 0, \ldots) \in \mathcal{A}_0$ the function $\upsilon(t) = \varphi_0^t(p) = \Phi(0, t, t, \ldots)$ is almost periodic and it has the Fourier expansion

$$\upsilon(t) \sim \sum_{k=1}^{\infty} U_k e^{i \sum_{j=1}^{J(k)} r_{j,k} \overline{\nu}_j t}. \tag{4.30}$$

By Theorem 4.4 the flow $\{\varphi_0^t\}$ can be extended to a minimal almost periodic flow on the compact set $\mathcal{M}_\upsilon$ and from Theorems 4.8 and 4.9 we get dim span$(\Lambda^C(u)) = \dim_T \mathcal{M}_\upsilon \leq n-1$. □

Sharper Estimates under the Squeezing Property. Results in this paragraph are based on ideas of Smith ([26], see also [3]).

Let $\mathcal{S} \subset \mathbb{R}^n$ be some subset and consider for system (4.16) the following conditions

(A3) There exist constants $\kappa > 0$, $\varepsilon > 0$ and a constant real symmetric matrix P such that for all $t \in \mathbb{R}$ and all $u_1, u_2 \in \mathcal{S}$

$$(P[f(t,u_1) - f(t,u_2) + \kappa(u_1 - u_2)], u_1 - u_2) \leq -\varepsilon|u_1 - u_2|^2; \quad (4.31)$$

(A4) P has j negative eigenvalues and $n-j$ positive eigenvalues;

If $V(u) := (Pu, u)$ and $u(\cdot)$, $\upsilon(\cdot)$ are solutions of (4.16), (**A3**) gives

$$\begin{aligned}
&\frac{d}{dt}\left[e^{2\kappa t}(V(u(t)) - V(\upsilon(t)))\right] \\
&= 2e^{2\kappa t}\left(P[f(t,u(t)) - f(t,\upsilon(t)) + \kappa(u(t) - \upsilon(t))], u(t) - \upsilon(t)\right) \\
&\leq -2\varepsilon|u(t) - \upsilon(t)|^2 e^{2\kappa t}, \ \text{ for all } t \text{ such that } u(t), \upsilon(t) \in \mathcal{S}.
\end{aligned} \quad (4.32)$$

Remark 4.5 Inequality (4.32) is often called *squeezing property*. In inertial manifold theory the eigenvalue properties of P in **(A4)** are connected with the *gap condition* [23].

If $\kappa \geq 0$ is the constant in **(A3)** then a solution $u(\cdot)$ of (4.16) is said to be *amenable* in $(-\infty; \tau]$, for some $\tau \in \mathbb{R}$, if $u(t) \in \mathcal{S}$ for $-\infty < t \leq \tau$ and $\int_{-\infty}^{\tau} e^{2\kappa t}|u(t)|^2 dt < +\infty$. For each τ let $\mathfrak{A}_\tau$ denote the subset of $\mathcal{S}$ consisting of points $u(\tau)$ taken over all solution $u(\cdot)$ of (4.16) which are amenable in $(-\infty; \tau]$. Then $\mathfrak{A}_\tau$ is called an *amenable set* of (4.16) in $\mathcal{S}$.

Lemma 4.7 *Let u and υ be two amenable in* $(-\infty; \tau]$ *solutions of* (4.16)*; then* $V(u(t) - \upsilon(t)) \leq 0$ *for every* $t \in (-\infty, \tau]$.

Proof Integrating inequality (4.32) on $[r, t]$, $r \leq t \leq \tau$, we get

$$e^{2\kappa t} V(u(t) - \upsilon(t)) \leq e^{2\kappa r} V(u(r) - \upsilon(r)) - 2\varepsilon \int_r^t e^{2\kappa s}|u(s) - \upsilon(s)|^2 ds. \quad (4.33)$$

Since $\int_{-\infty}^{t} e^{2\kappa s}|u(s) - \upsilon(s)|^2 ds < \infty$ there exists a sequence $s_k \to -\infty$ such that $e^{2\kappa s_k}|u(s_k) - \upsilon(s_k)|^2 \to 0$ as $k \to \infty$. Putting $r = s_k$ in (4.33) and taking it to the limit as $k \to \infty$ we get that $V(u(t) - \upsilon(t)) \leq 0$. □

If P satisfies **(A3)** there exists an invertible real $n \times n$ matrix Q such that $Q^* P Q = \operatorname{diag}(-I_j, I_{n-j})$, where I_r denotes the unit $r \times r$-matrix. The quadratic form $V(u) = (Pu, u)$ is therefore reduced to the canonical form $V(u) = |\eta|^2 - |\zeta|^2$ by the linear substitution $x = Q(\zeta, \eta)^T$ in which $\zeta \in \mathbb{R}^j$ and $\eta \in \mathbb{R}^{n-j}$. Let $\Pi : \mathbb{R}^n \to \mathbb{R}^j$ be the linear map defined by $\Pi u := \zeta$ for all $u \in \mathbb{R}^n$. Since $|Q^{-1}u| = |\zeta|^2 + |\eta|^2$ we have

$$V(u) + 2|\Pi u|^2 = |Q^{-1}u|^2 \geq |\Pi u|^2, \ \forall u \in \mathbb{R}^n. \tag{4.34}$$

Consider two arbitrary amenable in $(-\infty; \tau]$ solutions u and v of (4.16). Under the assumption that **(A3)** and **(A4)** are satisfied from Lemma 4.7 it follows that $V(u(t) - v(t)) \leq 0$, $t \leq \tau$, and (4.34) implies that

$$2|\Pi(u(\tau) - v(\tau))|^2 \geq |Q^{-1}(u(\tau) - v(\tau))|^2 \geq |\Pi(u(\tau) - v(\tau))|^2. \tag{4.35}$$

Thus, for any $p_1, p_2 \in \mathfrak{A}_\tau$ we have

$$2|\Pi p_1 - \Pi p_2|^2 \geq |Q^{-1}(p_1 - p_2)|^2 \geq |\Pi p_1 - \Pi p_2|^2. \tag{4.36}$$

This shows, that the restricted mapping $\Pi : \mathfrak{A}_\tau \to \Pi\mathfrak{A}_\tau$ gives a homeomorphism of $\mathfrak{A}_\tau$ onto the subset $\Pi\mathfrak{A}_\tau$ of $\mathbb{R}^j$.

Theorem 4.12 *Suppose that* (4.16) *satisfies **(A1)–(A4)**. If u is an almost periodic solution to* (4.16) *such that $\mathcal{M}_u \subset \mathcal{S}$ then*

$$\dim \frac{\operatorname{mod}_{\mathbb{Q}}(u, f)}{\operatorname{mod}_{\mathbb{Q}}(f)} = \dim \operatorname{span}(Sp^C(u)) \leq j - 1. \tag{4.37}$$

Proof Since $\kappa > 0$ any bounded on the entire line solution which lies in $\mathcal{S}$ is amenable and, in particular, so is u. From Theorem 4.11 we get the flow φ_0^t on the set $\mathcal{A}_0 \subset \mathcal{M}_u \subset \mathcal{S}$ which consists of initial conditions corresponding to almost periodic solutions of (4.16). In particular, $\mathfrak{A}_0 \supset \mathcal{A}_0$. Since by (4.36) the map $\Pi \colon \mathfrak{A}_0 \to \Pi\mathfrak{A}_0 \subset \mathbb{R}^j$ is a homeomorphism it takes the flow $\{\varphi_0^t\}$ to the flow $\{\psi_0^t\}$ on the set $\mathcal{B}_0 := \Pi\mathcal{A}_0$ defined by $\psi_0^t(w) := (\Pi \circ \varphi_0^t \circ \Pi^{-1})(w)$ for all $w \in \mathcal{B}_0$ and $t \in \mathbb{R}$. Let $z = \Pi w$. Note that $\zeta(t) := \psi_0^t(z)$ is almost periodic and $\operatorname{mod}_{\mathbb{Z}}(\zeta) = \operatorname{mod}_{\mathbb{Z}}(v)$, where $v(t) = \varphi_0^t(w)$. In particular the dimensions of the $\mathbb{Q}$-modules of the flows coincide. From this and Theorems 4.8 and 4.9 we establish that $\dim \operatorname{mod}_{\mathbb{Q}}(v) = \dim \operatorname{mod}_{\mathbb{Q}}(\zeta) \leq j - 1$ that is (4.37). □

Frequency-Domain Conditions. Let us consider a generalized feedback control system

$$\begin{cases} \dot{u} = Au + B\xi + g(t), \ v = Cu, \\ \xi = \phi(v), \end{cases} \tag{4.38}$$

in which $\phi \colon \mathbb{R} \times \mathbb{R}^s \to \mathbb{R}^r$ is a locally Lipschitz continuous function; A, B, C are constant real matrices of order $n \times n$, $n \times r$, $s \times n$, respectively and $g(t)$ is an $\mathbb{R}^n$-

almost periodic function. It is assumed that there exists a quadratic form $\mathcal{F}: \mathbb{R}^s \times \mathbb{R}^r \to \mathbb{R}$ such that

$$\mathcal{F}(\upsilon, 0) \geq 0, \forall \upsilon \in \mathbb{R}^s, \tag{4.39}$$

$$\mathcal{F}(\upsilon_1 - \upsilon_2, \phi(t, \upsilon_1) - \phi(t, \upsilon_2)) \geq 0, \forall t \in \mathbb{R}, \forall \upsilon_1, \upsilon_2 \in \mathbb{R}^s. \tag{4.40}$$

Denote by $\mathcal{F}_{\mathbb{C}}$ the Hermitian extension of $\mathcal{F}$ onto $\mathbb{C}^s \times \mathbb{C}^r$.

Theorem 4.13 *Suppose that there exists parameters $\kappa > 0$ and $\delta > 0$ such that the following conditions hold:*

(1) The pair $(A + \kappa I, B)$ is stabilizable;
(2) The matrix $A + \kappa I$ has $j \geq 1$ eigenvalues with positive real part and $n - j$ eigenvalues with negative real part;
(3) $\mathcal{F}_{\mathbb{C}}(C\upsilon, \xi) \leq -\delta(|\upsilon|^2 + |\xi|^2)$, $\forall \upsilon \in \mathbb{C}^n$, $\forall \xi \in \mathbb{C}^r$, $\forall \omega \in \mathbb{R}$ such that $i\omega\upsilon = (A + \kappa I)\upsilon + B\xi$.

*Then **(A1)–(A4)** holds for* (4.38) *with $\mathcal{S} = \mathbb{R}^n$. In particular, any almost periodic solution $u(\cdot)$ of* (4.38) *satisfy* (4.37).

Proof Since $(A + \kappa I, B)$ is stabilizable and the frequency-domain condition (3) is satisfied, we conclude from the Yakubovich-Kalman frequency theorem (Theorem 2.7, Chap. 2) that there exists a real $n \times n$ matrix $P = P^*$ and $\delta > 0$ such that for all $u \in \mathbb{R}^n$, $\xi \in \mathbb{R}^r$

$$2\,(u, P[(A + \kappa I)u + B\xi]) + \mathcal{F}(Cu, \xi) \leq -\delta\left[|u|^2 + |\xi|^2\right]. \tag{4.41}$$

If we put in (4.41) $\xi = 0$ we get, using (4.39), the inequality

$$2\,(u, P(A + \kappa I)u) \leq -\delta|u|^2, \ \forall u \in \mathbb{R}^n. \tag{4.42}$$

It follows from Lemma 2.8, Chap. 2 and the assumption (2) of the theorem that the matrix P has exactly j negative and $(n - j)$ positive eigenvalues. Consider (4.41) with $u = u_1 - u_2$ and $\xi = \phi(t, Cu_1) - \phi(t, Cu_2)$, where $u_1, u_2 \in \mathbb{R}^n$ are arbitrary. It follows that

$$\begin{aligned} 2\,(u_1 - u_2, P[A(u_1 - u_2) + B(\phi(t, u_1) - \phi(t, u_2)) + \kappa(u_1 - u_2)]) \\ +\mathcal{F}(Cu_1 - Cu_2, \phi(t, Cu_1) - \phi(t, Cu_2)) \leq -\delta|u_1 - u_2|^2. \end{aligned} \tag{4.43}$$

From the assumption (3) we have

$$\mathcal{F}(Cu_1 - Cu_2, \phi(t, Cu_1) - \phi(t, Cu_2)) \geq 0. \tag{4.44}$$

This and (4.43) imply that for $f(t, u) := Au + B\phi(t, Cu) + g(t)$ the hypothesis **(A3)** is satisfied. Thus Theorem 4.12 is applicable. □

4.6 Frequency Spectrum of Almost Periodic Solutions for DDEs

The main results of this section are due to Mallet-Paret [16].

In what follows we study the Fourier spectrum of almost periodic solutions to the following delay differential equation

$$\dot{u}(t) = f(u(t), u(t-h_1), \dots, u(t-h_N)), \tag{4.45}$$

where $u \in \mathbb{R}^n$ and $f: \mathbb{R}^{n(N+1)} \to \mathbb{R}^n$ is continuously differentiable function with bounded on $\mathbb{R}^{n(N+1)}$ derivative; $h_j \in (0, 1]$ are constants. We choose as our phase space the Hilbert space

$$\mathbb{H} = L^2([-1, 0], \mathbb{R}^n) \times \mathbb{R}^n.$$

We write $(\psi(\cdot), \psi(0))$ for the elements of $\mathbb{H}$. Then the norm is given by

$$\|(\psi(\cdot), \psi(0))\|^2 = \int_{-1}^{0} |\psi^2(s)| ds + |\psi(0)|^2.$$

Given $(\psi, \psi(0)) \in \mathbb{H}$ we consider (4.45) along with the initial value problem

$$\begin{aligned} u_0(s) &= u(s) = \psi(s), \ -1 \le s < 0, \\ u(0) &= \psi(0). \end{aligned} \tag{4.46}$$

Standard techniques show that the initial value problem (4.45), (4.46) has a unique solution $u(\cdot)$ defined on $[0, +\infty)$. So we obtain a semiflow $\{\varphi^t\}_{t \ge 0}$ on $\mathbb{H}$ such that

$$\varphi^t(\psi(\cdot), \psi(0)) := (u(\cdot + t)\big|_{[-1,0]}, u(t)), \ t \ge 0. \tag{4.47}$$

Put $\varphi := \varphi^1 : \mathbb{H} \to \mathbb{H}$. We omit the proof that φ is compact (i.e. it takes bounded sets into the sets of compact closure) and has a continuous Fréchet derivative. For $b > 0$ consider the set

$$\begin{aligned} \Gamma_b := &\{(\psi(\cdot), \psi(0)) \in \mathbb{H} \mid \text{ the problem (4.45), (4.46) has a solution } u(t), \\ &t \le 0, \text{ such that } |u(t)| \le b \text{ for } t \le 0\}. \end{aligned}$$

Suppose $u(\cdot)$ is an almost periodic solution to (4.45). It is clear that each $v(\cdot) \in \mathcal{H}(u)$ is also an almost periodic solution to (4.45). Put $b := \sup_{t \in \mathbb{R}} |u(t)|$. The map $v \mapsto (v\big|_{[-1,0]}, v(0))$ maps $\mathcal{H}(u)$ into Γ_b continuously. If $v_1\big|_{[-1,0]} = v_2\big|_{[-1,0]}$ then, since the solution to (4.45), (4.46) is unique, $v_1(t) = v_2(t)$ for $t \ge 0$. It follows that the Fouries series of $v_1(\cdot)$ and $v_2(\cdot)$ coincide and, therefore, $v_1 \equiv v_2$. Hence, $\mathcal{H}(u)$ is homeomorphic to a subset of Γ_b and, in particular, $\dim_T \mathcal{H}(u) \le \dim_T \Gamma_b$. Thus, by Theorem 4.8, the algebraic dimension of the frequency module $\operatorname{mod}_{\mathbb{Q}}(u)$ is majorized by the topological dimension of Γ_b. We will show the following

Theorem 4.14 *The set Γ_b has finite topological dimension for each $b > 0$. In particular, every almost periodic solution to* (4.45) *has a finite rational frequency base.*

In order to prove Theorem 4.14 we need to establish some properties of compact operators and their negatively invariant sets. In the further text $\mathbb{H}$ denotes a separable Hilbert space.

The set Γ_b defined above is compact and negatively invariant (i.e. $\varphi(\Gamma_b) \supset \Gamma_b$). This is due to the following

Proposition 4.4 *Suppose $\varphi \colon \mathcal{U} \to \mathbb{H}$ is continuous and we have $\mathcal{V} \subset \mathcal{U} \subset \mathbb{H}$ with $\mathcal{U}$ open and $\mathcal{V}$ closed. Let $\varphi(\mathcal{V})$ have a compact closure. Then the set*

$$\Gamma := \{x_0 \in \mathcal{V} \mid \exists \{x_k\}_{k=1}^{\infty} \subset \mathcal{V} : \varphi(x_k) = x_{k-1}\}$$

is compact and negatively invariant.

Proof It is clear that $\varphi(\Gamma) \supset \Gamma$. To prove Γ is compact assume that there is a sequence $\{x_0^{(m)}\}_{m=1}^{\infty} \subset \Gamma$. For every m there is a sequence $\{x_k^{(m)}\}_{k=1}^{\infty} \subset \mathcal{V}$ with $\varphi(x_k^{(m)}) = x_{k-1}^{(m)}$, $k \geq 1$. Consider the following diagram

$$\begin{array}{ccccccccc}
x_0^{(1)} & \overset{\varphi}{\leftarrow} & x_1^{(1)} & \overset{\varphi}{\leftarrow} & \ldots & \overset{\varphi}{\leftarrow} & x_k^{(1)} & \ldots \\
x_0^{(2)} & \overset{\varphi}{\leftarrow} & x_1^{(2)} & \overset{\varphi}{\leftarrow} & \ldots & \overset{\varphi}{\leftarrow} & x_k^{(2)} & \ldots \\
\vdots & & \vdots & & & & \vdots & \\
x_0^{(m)} & \overset{\varphi}{\leftarrow} & x_1^{(m)} & \overset{\varphi}{\leftarrow} & \ldots & \overset{\varphi}{\leftarrow} & x_k^{(m)} & \ldots
\end{array}$$

Since each point lies in $\varphi(\mathcal{V})$ by using diagonal procedure we can assume that the sequence in k-th vertical section converges as $m \to \infty$ to some $x_k^* \in \mathcal{V}$. From the continuity of φ it follows that $\varphi(x_k^*) = x_{k-1}^*$ for $k \geq 1$ and, therefore, $x_0^* \in \Gamma$ and Γ is compact. □

Proposition 4.5 *Suppose $\mathcal{U} \subset \mathbb{H}$ is an open set and $\varphi \colon \mathcal{U} \to \mathbb{H}$ has a continuous Fréchet derivative. If the closure cl $(\varphi(\mathcal{U}))$ is compact then for every $x \in \mathcal{U}$ the differential $D\varphi(x)$ is a compact operator.*

Proof Suppose that for some $x \in \mathcal{U}$ the differential $D\varphi(x)$ is non-compact. Then there is a weakly convergent sequence y_m such that $D\varphi(x)y_m$ does not converge strongly. We may assume that $y_m \rightharpoonup 0$, $\|y_m\| = 1$ and $\|D\varphi(x)y_m\| \geq \varepsilon$ for some $\varepsilon > 0$. Let $\delta > 0$ be such that for every $\widetilde{x} \in \mathcal{U}$ with $\|x - \widetilde{x}\| \leq \delta$ we have $\|D\varphi(x) - D\varphi(\widetilde{x})\| \leq \frac{\varepsilon}{3}$. Consider

$$\begin{aligned} z_m := \varphi(x + \delta y_m) &= \varphi(x) + \int_0^{\delta} D\varphi(x + \alpha y_m) y_m d\alpha \\ &= \varphi(x) + \delta D\varphi(x) y_m + r_m, \end{aligned} \tag{4.48}$$

where $||r_m|| \leq \frac{\delta\varepsilon}{3}$. Since $z_m \in \varphi(\mathcal{U})$ we may assume that z_m converges strongly to some $z \in \mathbb{H}$. From

$$\|z_m - \varphi(x)\| \geq \|\delta D\varphi(x) y_m\| - \|r_m\| \geq \frac{2\delta\varepsilon}{3}$$

we have $\|z - \varphi(x)\| \geq \frac{2\delta\varepsilon}{3}$. On the other hand from (4.48) it is clear that $z_m - \varphi(x)$ converges weakly to the same limit as r_m and, therefore, $\|z - \varphi(x)\| \leq \limsup_{m\to\infty} \|r_m\| \leq \frac{\delta\varepsilon}{3}$. This is a contradiction. □

Lemma 4.8 *Suppose* $S\colon \mathbb{H} \to \mathbb{H}$ *is a linear compact operator. Then for each* $\delta > 0$ *there is a subspace* $\mathbb{E} \subset \mathbb{H}$ *of finite codimension such that* $\|S\big|_{\mathbb{E}}\| \leq \delta$.

Proof Suppose the opposite, i.e. there is a $\delta > 0$ such that for every subspace $\mathbb{E} \subset \mathbb{H}$ of finite codimension we have $\|S\big|_{\mathbb{E}}\| \geq 2\delta$. Let $\{e_k\}_{k=1}^{\infty}$ be an orthonormal basis for $\mathbb{H}$. Consider the subspaces $\mathbb{E}_m := \text{span}\{e_{m+1}, e_{m+2}, \ldots\}$. By assumption there is $x_m \in \mathbb{E}_m$, $\|x_m\| = 1$ such that $\|S(x_m)\| \geq \delta$. Note that x_m weakly converges to zero and since T is compact $S(x_m)$ strongly converges to zero. But the norm of $S(x_m)$ is bounded from below and this is a contradiction. □

Proposition 4.6 *Suppose* $\mathcal{U} \subset \mathbb{H}$ *is an open set and* $\varphi\colon \mathcal{U} \to \mathbb{H}$ *has a continuous Fréchet derivative. Let* $\Gamma \subset \mathcal{U}$ *be a compact subset and* $D\varphi(x)$ *is compact for each* $x \in \Gamma$. *Then there is a subspace* $\mathbb{E}$ *of finite codimension such that*

$$\sup_{x\in\Gamma} \|D\varphi(x)\big|_{\mathbb{E}}\| < 1. \tag{4.49}$$

Proof By Lemma 4.8 for every $x \in \Gamma$ there is a subspace $\mathbb{E}(x) \subset \mathbb{H}$ of finite codimension with $\|D\varphi(x)\big|_{\mathbb{E}(x)}\| \leq \frac{1}{3}$. Let $\delta(x) > 0$ be such that $\|D\varphi(\widetilde{x})\big|_{\mathbb{E}(x)}\| \leq \frac{2}{3}$ provided by $|\widetilde{x} - x| < \delta$. Suppose that Γ is covered by open balls of radii $\delta(x_j)$, $j = 1, 2, \ldots, J$. Take $\mathbb{E} := \bigcap_{j=1}^{J} \mathbb{E}(x_j)$. Clearly, $\mathbb{E}$ has finite codimension and for every $y \in \Gamma$ there is x_j with $\|y - x_j\| < \delta(x_j)$ and, therefore,

$$\|D\varphi(y)\big|_{\mathbb{E}}\| \leq \|D\varphi(y)\big|_{\mathbb{E}(x_j)}\| \leq \frac{2}{3}.$$

□

In what follows $d(\mathcal{A})$ denotes the diameter of $\mathcal{A} \subset \mathbb{H}$. For given $\delta > 0$ and $d \geq 0$ we consider the corresponding Hausdorff pre-measure $\mu_H(\cdot, d, \delta)$ giving by the covers of arbitrary sets with diameter $\leq \delta$. The final ingredient is

Theorem 4.15 *Suppose* $\Gamma \subset \mathcal{U} \subset \mathbb{H}$, *where* Γ *is compact and* $\mathcal{U}$ *is open. Let* $\varphi\colon \mathcal{U} \to \mathbb{H}$ *have a continuous Fréchet derivative and* $\varphi(\Gamma) \supset \Gamma$. *Suppose further there is a linear subspace* $\mathbb{E} \subset \mathbb{H}$ *with finite codimension and such that*

$$\sup_{x\in\Gamma} \|D\varphi(x)\big|_{\mathbb{E}}\| < 1. \tag{4.50}$$

Then the topological dimension of Γ is finite.

Before giving a proof, we need the following lemma.

Lemma 4.9 *Suppose $\mathcal{Z} \subset \mathbb{R}^n$ has diameter $\eta < \infty$ and let $p > 0$ be an integer. Then there exists a partition of $\mathcal{Z}$ into not more than p^n sets, with each set of diameter at most $2\eta n^{1/2} p^{-1}$.*

Proof Enclose $\mathcal{Z}$ in a ball of diameter 2η, and the ball in a cube of edge 2η. Partition the cube into p^n subcubes $\mathcal{K}_i$, each of edge $2\eta p^{-1}$, and, therefore, having diameter $2\eta n^{1/2} p^{-1}$. Clearly, the sets $\mathcal{K}_i \cap \mathcal{Z}$ form the required partition. □

Proof of Theorem 4.15 We will show that there exist constants $\alpha, \beta \in (0, 1)$, $\delta_0 > 0$ and $N > 0$ such that for given sets $\{\mathcal{S}_i\}$ with

$$\Gamma = \bigcup_{i=1}^{\infty} \mathcal{S}_i \text{ and } \operatorname{diam}(\mathcal{S}_i) \le \delta < \delta_0 \tag{4.51}$$

there are sets $\mathcal{Q}_{ij} \subset \mathbb{H}$, $i \ge 1$, $1 \le j \le J_i$, such that

$$\begin{aligned} \Gamma \subset \bigcup_{i=1}^{\infty} \bigcup_{j=1}^{J_i} \mathcal{Q}_{ij}, \ \operatorname{diam}(\mathcal{Q}_{ij}) \le \alpha\delta \\ \sum_{ij} \operatorname{diam}(\mathcal{Q}_{ij})^N \le \beta \sum_i \operatorname{diam}(\mathcal{S}_i)^N. \end{aligned} \tag{4.52}$$

By appropriate choosing of $\mathcal{S}_i$ we can make the sum $\sum_i \operatorname{diam}(\mathcal{S}_i)^N$ arbitrarily close to $\mu_H(\Gamma, N, \delta)$. So we obtain

$$\mu_H(\Gamma, N, \alpha\delta) \le \beta\mu_H(\Gamma, N, \delta), \text{ for } \delta \le \delta_0. \tag{4.53}$$

Iterating (4.53) we get

$$\mu_H(\Gamma, N, \alpha^m\delta) \le \beta^m \mu_H(\Gamma, N, \delta), \text{ for } m = 1, 2, \ldots, \tag{4.54}$$

which implies that $\mu_H(\gamma, N) = 0$ since for compact sets any pre-measure is finite. Therefore by Proposition 3.20, Chap. 3, $\dim_T \Gamma \le \dim_H \Gamma < \infty$. Now we are going to show the existence of α, β, δ_0 and N, and construct $\mathcal{Q}_{ij}$'s.

Since Γ is compact there are constants $K > 0$, $a \in (0, 1)$ and an open neighborhood (without less of generality $\mathcal{U}$) of Γ such that

$$\|D\varphi(x)\| \le K, \ \|D\varphi(x)\big|_{\mathbb{E}}\| \le a, \text{ for all } x \in \mathcal{U}. \tag{4.55}$$

Let $\delta_0 > 0$ be the distance between Γ and $\mathbb{H}\backslash\mathcal{U}$. Suppose $\mathcal{S}_i$ and δ are as in (4.51). Let $\mathbb{F}$ be the orthogonal complement of $\mathbb{E}$ and set $n := \dim \mathbb{F}$. Since $\mathbb{H}$ is a Hilbert space, the projections $\pi_{\mathbb{E}}$ and $\pi_{\mathbb{F}}$ on $\mathbb{E}$ and $\mathbb{F}$ respectively have norm one. Hence

$$\begin{aligned}\xi_i &:= \operatorname{diam}(\pi_{\mathbb{E}}(\mathcal{S}_i)) \leq \operatorname{diam}(\mathcal{S}_i) \leq \delta, \\ \eta_i &:= \operatorname{diam}(\pi_{\mathbb{F}}(\mathcal{S}_i)) \leq \operatorname{diam}(\mathcal{S}_i) \leq \delta.\end{aligned} \tag{4.56}$$

Suppose $p_i > 0$ is an integer (to be determined later) and by Lemma 4.9 there are subsets $\mathcal{Z}_{ij} \subset \mathcal{S}_i$ with $1 \leq j \leq J_i \leq p_i^n$ such that

$$\pi_{\mathbb{F}}(\mathcal{S}_i) = \bigcup_{j=1}^{J_i} \mathcal{Z}_{ij} \text{ and } \operatorname{diam}(\mathcal{Z}_{ij}) \leq 2\eta_i n^{1/2} p_i^{-1} \text{ for all } j.$$

Let $\mathcal{P}_{ij} := \pi_{\mathbb{F}}^{-1}(\mathcal{Z}_{ij}) \cap \mathcal{S}_i$ and $\mathcal{Q}_{ij} := \varphi(\mathcal{P}_{ij})$. We have

$$\begin{aligned}\bigcup_j \mathcal{Q}_{ij} &= \varphi\left(\bigcup_j \mathcal{P}_{ij}\right) = \varphi(\mathcal{S}_i) \text{ and} \\ \bigcup_{i,j} \mathcal{Q}_{ij} &= \varphi\left(\bigcup_i \mathcal{S}_i\right) = \varphi(\Gamma) \supset \Gamma.\end{aligned}$$

Denote the convex hull of $\mathcal{S}_i$ by $\operatorname{conv}(\mathcal{S}_i)$. Clearly,

$$\operatorname{diam}(\mathcal{S}_i) = \operatorname{diam}(\operatorname{conv}(\mathcal{S}_i)) \leq \delta < \delta_0,$$

so that $\mathcal{S}_i \subset \operatorname{conv}(\mathcal{S}_i) \subset \mathcal{U}$. Now we will estimate $\operatorname{diam}(\mathcal{Q}_{ij})$ to show (4.52). For $u \in \mathbb{H}$ we write $u = (v, w)$, where $v \in \mathbb{E}$, $w \in \mathbb{F}$. By definition, for $u_1, u_2 \in \mathcal{P}_{ij}$ the points $T(u_1)$ and $T(u_2)$ are arbitrary points in $\mathcal{Q}_{ij}$. We have immediately

$$\begin{aligned}\|v_1 - v_2\| &\leq \operatorname{diam}(\pi_{\mathbb{E}}(\mathcal{S}_i)) = \xi_i, \\ \|w_1 - w_2\| &\leq \operatorname{diam}(\pi_{\mathbb{F}}(\mathcal{P}_{ij})) = \operatorname{diam}(\mathcal{Z}_{ij}) \leq 2\eta_i n^{1/2} p_i^{-1}.\end{aligned}$$

Since $\operatorname{conv}(\mathcal{P}_{ij}) \subset \operatorname{conv}(\mathcal{S}_i) \subset \mathcal{U}$, we have

$$\begin{aligned}\|\varphi(u_2) - \varphi(u_1)\| &= \int_0^1 D\varphi(u_1 + t(u_2 - u_1))(u_2 - u_1)dt \\ &= \int_0^1 D\varphi(u_1 + t(u_2 - u_1))(v_2 - v_1, 0)dt \\ &+ \int_0^1 D\varphi(u_1 + t(u_2 - u_1))(0, w_2 - w_1)dt.\end{aligned}$$

From (4.55) we conclude

$$\begin{aligned}\|\varphi(u_2) - \varphi(u_1)\| &\leq a\|v_2 - v_1\| + K\|w_2 - w_1\| \\ &\leq a\xi_i + 2K\eta_i n^{1/2} p_i^{-1},\end{aligned}$$

and hence using (4.56) we get

$$\operatorname{diam}(\mathcal{Q}_{ij}) \le a\xi_i + 2n^{1/2}\eta_i p_i^{-1} \le (a + 2Kn^{1/2}p_i^{-1}) \operatorname{diam}(\mathcal{S}_i). \quad (4.57)$$

Put $\alpha := (1+a)/2 < 1$ and let $p_i = p$ be a large integer such that $a + 2Kn^{1/2}p_i^{-1} \le \alpha$. We have

$$\operatorname{diam}(\mathcal{Q}_{ij}) \le \alpha \operatorname{diam}(\mathcal{S}_i) \le \alpha\delta \text{ and } \sum_j \operatorname{diam}(\mathcal{Q}_{ij})^N \le p^n \alpha^N \operatorname{diam}(\mathcal{S}_i)^N, \quad (4.58)$$

and for N being sufficiently large

$$\sum_{i,j} \operatorname{diam}(\mathcal{Q}_{ij})^N \le p^n\alpha^N \sum_i \operatorname{diam}(\mathcal{S}_i)^N \le \frac{1}{2}\sum_i \operatorname{diam}(\mathcal{S}_i)^N.$$

Thus, we can put $\beta := \frac{1}{2}$ and finish the proof. □

Summing up the above things we get

Proof of Theorem 4.15 Consider the sets

$$\mathcal{V}_b := \{(\psi(\cdot), \psi(0)) \in \mathbb{H} \mid |\psi(t)| \le b \text{ a.e. and } |\psi(0)| \le b\},$$
$$\mathcal{U}_b := \{(\psi(\cdot), \psi(0)) \in \mathbb{H} \mid \|\psi, \psi(0)\|^2 < 3b^2\}.$$

Clearly, $\mathcal{U}_b$ is open and $\mathcal{V}$ is closed. Thus, the trio $\Gamma_b \subset \mathcal{V} \subset \mathcal{U}$ along with the operator φ satisfies Propositions 4.4, 4.5, 4.6 and, therefore, Theorem 4.15 is applicable. □

4.7 Fractal Dimensions of Almost Periodic Trajectories and The Liouville Phenomenon

Investigating fractal dimensions of the closures of quasi-periodic trajectories we face the following obstacle. Since a quasi-periodic function $u(\cdot)$ is the restriction of a periodic function of several, say m, variables to a dense winding on the torus, i.e. $u(t) = \Phi_u(\omega t)$, $\omega \in \mathbb{R}^m$, the dimensional properties of its closure $\mathcal{M}_u = \Phi_u(\mathbb{T}^m)$ depends on how Φ_u affects the diameters of sets, i.e. on the Hölder-like properties. But it seems impossible to get some regularity results for Φ_u in the case of nonlinear equations since the original differential equation determines the behaviour of Φ_u only in the direction of the winding ωt on $\mathbb{T}^m$ and do not restrict it in transversal directions. It appears that this problem can be avoided if we have additional information about $u(\cdot)$, for example, if $u(\cdot)$ is a forced quasi-periodic oscillation.

In this section we derive several results concerning recurrence properties of abstract almost periodic trajectories and its link with the fractal dimension. The results are then applied to study the fractal dimension of forced almost periodic oscillations in a class of control systems. The main results is based on [1, 2, 4].

The Fundamental Theorem on Diophantine Dimension. First of all, note that the definition of an almost periodic function given in Sect. 4.3 can be extended to the functions with values in some metric space $\mathcal{X}$. Suppose $u(\cdot)$ is $\mathcal{X}$-almost periodic. The set of ε-almost periods $\mathcal{T}_\varepsilon(u)$ is relatively dense, i.e. for some $L(\varepsilon) > 0$ the intersection $[a, a + L(\varepsilon)] \cap \mathcal{T}_\varepsilon(u)$ is not empty for all $a \in \mathbb{R}$. Let $l_u(\varepsilon)$ be the minimum of such numbers $L(\varepsilon)$. The limits

$$\begin{aligned} \mathfrak{Di}(u) &:= \limsup_{\varepsilon \to 0+} \frac{\ln l_u(\varepsilon)}{\ln 1/\varepsilon} \\ \mathfrak{di}(u) &:= \liminf_{\varepsilon \to 0+} \frac{\ln l_u(\varepsilon)}{\ln 1/\varepsilon} \end{aligned} \tag{4.59}$$

are called the *Diophantine dimension* and *lower Diophantine dimension* respectively.

Suppose that $\chi : \mathcal{X} \to \mathcal{Y}$ is a map between metric spaces $\mathcal{X}$ and $\mathcal{Y}$ satisfying the *Hölder condition* on $\mathcal{M}_u$ with some exponent $a \in (0, 1]$, i.e. there is a constant $C > 0$ such that for all $x, y \in \mathcal{M}_u$ we have $\rho_{\mathcal{Y}}(\chi(x), \chi(y)) \leq C\rho_{\mathcal{X}}(x, y)^\alpha$. Note that $\chi \circ u$ is $\mathcal{Y}$-almost periodic. The following proposition can be proved by straightforward calculations.

Proposition 4.7 *In the above conditions we have*

$$\mathfrak{Di}(\chi \circ u) \leq \frac{\mathfrak{Di}(u)}{\alpha}. \tag{4.60}$$

We know that an almost periodic function is uniformly continuous. Let $\delta(\varepsilon)$ be such that $\rho_{\mathcal{X}}(u(t_1), u(t_2)) \leq \varepsilon$ provided by $\|t_1 - t_2\|_\infty \leq \delta(\varepsilon)$, where $\|\cdot\|_\infty$ is the sup-norm in $\mathbb{R}^n$. Let $\delta^*(\varepsilon)$ be the supremum of such numbers $\delta(\varepsilon)$. Consider the value

$$\overline{\Delta}(u) := \limsup_{\varepsilon \to 0+} \frac{\ln \delta^*(\varepsilon)}{\ln \varepsilon}. \tag{4.61}$$

It is clear that if $u(\cdot)$ satisfies the Hölder condition with an exponent $\alpha \in (0, 1]$ then $\overline{\Delta}(u) \leq \frac{1}{\alpha}$.

Theorem 4.16

$$\begin{aligned} \overline{\dim}_B \mathcal{H}(u) &\leq \mathfrak{Di}(u) + \overline{\Delta}(u), \\ \underline{\dim}_B \mathcal{H}(u) &\leq \mathfrak{di}(u) + \overline{\Delta}(u). \end{aligned} \tag{4.62}$$

Proof

$$\rho_\infty(u(\cdot + t) - u(\cdot + \overline{t})) \leq \varepsilon. \tag{4.63}$$

Indeed, there is an ε-almost period $\tau \in [-t, -t + l_u(\varepsilon)]$ for $u(\cdot)$. Then $\overline{t} := t + \tau$ is what we wanted. Now for arbitrary $\upsilon \in \mathcal{H}(u)$ there exists $t \in \mathbb{R}^n$ such that $\rho_\infty(\upsilon(\cdot), u(\cdot + t)) \leq \varepsilon$ and, consequently,

$$\rho_\infty\left(v(\cdot), u(\cdot+\bar{t})\right) \le \rho_\infty\left(v(\cdot), u(\cdot+t)\right) + \rho_\infty\left(u(\cdot+t), u(\cdot+\bar{t})\right) \le 2\varepsilon. \quad (4.64)$$

For convenience' sake for $\mathcal{Q} \subset \mathbb{R}$ let $\mathcal{Q}_u := \{u(\cdot+t) \mid t \in \mathcal{Q}\} \subset \mathcal{H}(u)$. It follows from (4.64) that it is sufficient to cover the set $[0, l_u(\varepsilon)]_u$ by open balls. Let $\mathcal{B}_\varepsilon(u(\cdot+t))$ be the open ball centered at $u(\cdot+t)$ with radius ε. It is clear that for $t \in \mathbb{R}$

$$\mathcal{B}_\varepsilon(u(\cdot+t)) \supset \left[t - \frac{\delta^*(\varepsilon)}{2}, t + \frac{\delta^*(\varepsilon)}{2}\right]_u. \quad (4.65)$$

Thus, the set $[0, l_u(\varepsilon)]_u$ can be covered by $\frac{l_u(\varepsilon)}{\delta^*(\varepsilon)} + 1$ open balls of radius ε and, consequently, the set $\mathcal{H}(u)$ can be covered by the same number of balls of radius 3ε. Therefore, $N_{3\varepsilon}(\mathcal{H}(u)) \le \frac{l_u(\varepsilon)}{\delta^*(\varepsilon)} + 1 \le 2 \cdot \frac{l_u(\varepsilon)}{\delta^*(\varepsilon)}$ and

$$\frac{\ln N_{3\varepsilon}(\mathcal{H}(u))}{\ln 1/\varepsilon} \le \frac{\ln\left(2 \cdot \frac{l_u(\varepsilon)}{\delta^*(\varepsilon)}\right)}{\ln 1/\varepsilon}. \quad (4.66)$$

For every $\delta > 0$ there exists $\varepsilon_0 > 0$ such that $\frac{\ln \delta^*(\varepsilon)}{\ln \varepsilon} \le \overline{\Delta}(u) + \delta$ for $\varepsilon \in (0, \varepsilon_0)$. From (4.66) we have

$$\frac{\ln N_{3\varepsilon}(\mathcal{H}(u))}{\ln 3 + \ln 1/(3\varepsilon)} \le \frac{\ln 2}{\ln 1/\varepsilon} + \frac{\ln l_u(\varepsilon)}{\ln 1/\varepsilon} + \overline{\Delta}(u) + \delta. \quad (4.67)$$

Taking it to the lower/upper limit in (4.67) and using an arbitrary choice of δ we finish the proof. □

Remark 4.6 Consider the map $i\colon \mathcal{H}(u) \to \mathcal{M}_u$, $i(v) := v(0)$, from Theorem 4.6. In the case of an arbitrary almost periodic function $u(\cdot)$ the map i is not a homeomorphism (and the set $\mathcal{M}_u$ does not have a natural group structure). Anyway, i is still surjective and Lipschitz. As a consequence of this and Theorem 4.16 we have $\overline{\dim}_B \mathcal{M}_u = \overline{\dim}_B i(\mathcal{H}(u)) \le \overline{\dim}_B \mathcal{H}(u) \le \mathfrak{Di}(u) + \overline{\Delta}(u)$ and a similar estimate for $\underline{\dim}_B \mathcal{M}_u$. In what follows we will use this fact to obtain some estimates of dimensions for $\mathcal{M}_u$ in the case when $u(\cdot)$ is a forced almost periodic oscillation.

Almost Periodic Regimes in Control Systems Consider the following control system

$$\begin{aligned} \dot{u} &= Au + b\phi(v) + f(t), \\ v &= (c, u). \end{aligned} \quad (4.68)$$

where A is a $n \times n$-matrix; b and c are n-vectors; $f(\cdot)$ is a $\mathbb{R}^n$-almost periodic function and $\phi(\cdot)$ is a C^2 scalar function satisfying with $\kappa_0 \le +\infty$ the inequality

$$0 \le \phi(v)v \le \kappa_0 v^2, \quad \forall v \in \mathbb{R}. \quad (4.69)$$

In [5] I. M. Burkin and V. A. Yakubovich, using a method of M. A. Krasnoselskii (see Theorem 12.2 in [14]), have obtained frequency domain conditions (see below) for the existence of exactly two almost periodic solutions to (4.68) one of which is exponentially stable and one is unstable. We continue this investigation with the following theorem. Note that the frequency domain conditions for the existence of Bohr almost periodic solutions in general evolution equations are obtained in [12, 22].

Theorem 4.17 *Under assumptions* ***(C1)–(C7)*** *below for the exponentially stable almost periodic solution* $u(\cdot)$ *to* (4.68) *we have*

$$\mathrm{mod}_{\mathbb{Z}}(u) = \mathrm{mod}_{\mathbb{Z}}(f). \tag{4.70}$$

Moreover,

$$\mathfrak{Di}(u) \le \mathfrak{Di}(f). \tag{4.71}$$

The estimate in (4.71) allows us to give an upper bound for the fractal dimension of $\mathcal{M}_u$ (see Theorem 4.21 below).

Denote by $W(z) = (c, (A - zI)^{-1}b)$ the transfer function of system (4.68). In what follows we assume that the pair (A, b) is completely controllable. Consider the following conditions

(C1) The matrix A is Hurwitz.

(C2) The matrix A has a leading eigenvalue, i.e. a simple real eigenvalue λ_0 such that $\lambda_0 > \mathrm{Re}\,\lambda_i$, where λ_i are the other eigenvalues.

(C3) There exists a number $\varepsilon > 0$ satisfying the conditions $\mathrm{Re}\,\lambda_i < -\varepsilon < \lambda_0$ and such that

$$\mathrm{Re}\,W(\varepsilon + i\omega) > 0 \text{ for } \omega \ge 0, \quad \lim_{\omega \to +\infty} \omega^2\,\mathrm{Re}\,W(\varepsilon + i\omega) > 0.$$

(C4) We have the inequality $\lim_{z \to \infty} [-zW(z)] = (c, b) \ge 0$.

(C5) The function $\phi(\cdot)$ is monotonically increasing and convex, i.e.

$$\phi'(\upsilon) > 0, \phi''(\upsilon) > 0 \text{ for any } \upsilon \in \mathbb{R},$$

with

$$\phi'(0) < -W(0)^{-1}.$$

(C6) The following limits are valid:

$$\lim_{\upsilon \to +\infty} \frac{\phi(\upsilon)}{\upsilon} = \kappa_1, \text{ where } \begin{cases} -W(0)^{-1} < \kappa_1 < \kappa_0 \ (\kappa_0 < +\infty), \\ -W(0)^{-1} < \kappa_1 \le \kappa_0 \ (\kappa_0 = +\infty); \end{cases}$$

$$\lim_{\upsilon \to -\infty} \frac{\phi(\upsilon)}{\upsilon} = \kappa_2 < -W(0)^{-1}.$$

From **(C1)–(C4)** it follows that $W(0) < 0$ (Lemma 1.1. in [5]) so the formulation of **(C5)–(C6)** is correct. Put $\alpha_0 := -W(0) = -(A^{-1}b, c) > 0$. Since

$-A^{-1} = \int_0^\infty e^{As} ds$ we have $\alpha_0 = \int_0^\infty (e^{As}b, c) ds > 0$. Let $\psi(\upsilon) := \upsilon - \alpha_0\phi(\upsilon)$. Under assumptions **(C5)–(C6)** there is a unique maximum of $\psi(\cdot)$ at υ^+. Put $\beta^+ := \psi(\upsilon^+)$. Consider the function

$$g(t) := \int_{-\infty}^{t} \left(e^{A(t-s)} f(s), c\right) ds. \tag{4.72}$$

The last assumption is

(C7) $\sup_{t\in\mathbb{R}} g(t) < \beta^+$.

Remark 4.7 The method of Krasnoselskii is based on the fact that there is a cone $\mathcal{K} \subset \mathbb{R}^n$ such that the family of operators e^{As}, $s > 0$, is strictly monotone w.r.t. $\mathcal{K}$ (i.e. they map the points of $\mathcal{K}$ into its interior). Define by $\mathcal{K}^+$ the part of $\mathcal{K}$ lying in the half-plane $\{(c, x) \geq 0\}$. From **(C1)–(C4)** it follows that $b \in \mathcal{K}^+$. Conditions **(C1)–(C4)** guarantee the existence of a strictly invariant cone $\mathcal{K}$ such that the interior of $\mathcal{K}$ has empty intersection with the plane $\{(c, x) = 0\}$ (Lemma 2.2 in [5]). Therefore, since $e^{As}b \in \operatorname{Int}\mathcal{K}$ we have $(e^{As}b, c) \geq 0$.

Let $f(\cdot)$ be an $\mathbb{E}$-almost periodic function ($\mathbb{E}$ is a Banach space). A sequence $\{t_k\} \subset \mathbb{R}$ is called *f-returning* if the sequence $\{f(\cdot + t_k)\}$ converges uniformly to $f(\cdot)$. The following lemma can be found in [8].

Lemma 4.10 *For two almost periodic functions $u(\cdot)$ and $f(\cdot)$ the following conditions are equivalent:*

(1) $\operatorname{mod}_{\mathbb{Z}}(u) \subset \operatorname{mod}_{\mathbb{Z}}(f)$.
(2) For every $\varepsilon > 0$ there is a $\delta > 0$ such that $\mathcal{T}_\varepsilon(f) \subset \mathcal{T}_\varepsilon(u)$.
(3) For every f-returning sequence there is an u-returning subsequence.

Proof of Theorem 4.17 From the proof of Theorem 12.2 in [14] it follows that the exponentially stable almost periodic solution $u(\cdot)$ is given by the formula

$$u(t) = \int_{-\infty}^{t} e^{A(t-s)} \left(b\phi(\upsilon^*(s)) + f(s)\right) ds, \tag{4.73}$$

where the scalar almost periodic function $\upsilon^*(\cdot)$ satisfy the equation

$$\upsilon^*(t) = \int_{-\infty}^{t} \left(e^{A(t-s)}b, c\right) \phi(\upsilon^*(s)) ds + \int_{-\infty}^{t} \left(e^{A(t-s)} f(s), c\right) ds, \tag{4.74}$$

and $\upsilon^*(\cdot)$ is the unique almost periodic solution of (4.74) such that $\sup_{t\in\mathbb{R}} \upsilon^*(t) < \upsilon^+$.

Let us show the inclusion $\mathrm{mod}_{\mathbb{Z}}(u) \subset \mathrm{mod}_{\mathbb{Z}}(f)$. Let $\{t_k\}$, $k = 1, 2, \ldots$, be f-returning, i.e. $f(\cdot + t_k)$ converges to $f(\cdot)$ uniformly. Let a subsequence $\{t_k'\} \subset \{t_k\}$ be such that the sequence $\{\upsilon^*(\cdot + t_k')\}$ converges to some almost periodic function $\hat{\upsilon}(\cdot)$. It is easy to see that

$$\upsilon^*(t + t_k') = \int_{-\infty}^{t} \big(e^{A(t-s)}b, c\big)\, \phi(\upsilon^*(s + t_k'))ds + \int_{-\infty}^{t} \big(e^{A(t-s)} f(s + t_k'), c\big)\, ds. \tag{4.75}$$

Since $\phi(\cdot)$ is continuously differentiable and $\upsilon^*(\cdot)$ is bounded, the sequence $\{\upsilon^*(\cdot + t_k')\}$ converges to a solution of (4.74) and $\sup_{t\in\mathbb{R}} \hat{\upsilon}(t) = \sup_{t\in\mathbb{R}} \upsilon^*(t) < \upsilon^+$. Thus, by the uniqueness, $\hat{\upsilon} = \upsilon^*$ and the sequence $\{t_k'\}$ is υ^*-returning. Therefore, $\mathrm{mod}_{\mathbb{Z}}(\upsilon^*) \subset \mathrm{mod}_{\mathbb{Z}}(f)$. The right-hand side of (4.73) for $t \leftrightarrow t + t_k'$, that is $u(t + t_k')$, converges uniformly (by the same argument) to $u(\cdot)$ and, consequently, $\{t_k'\}$ is u-returning. Now by Lemma 4.10 we get the inclusion $\mathrm{mod}_{\mathbb{Z}}(u) \subset \mathrm{mod}_{\mathbb{Z}}(f)$.

To prove the inverse inclusion $\mathrm{mod}_{\mathbb{Z}}(u) \supset \mathrm{mod}_{\mathbb{Z}}(f)$ note that the function $\dot{u}(\cdot)$ is $\mathbb{R}^n$-almost periodic with the same Fourier exponents as for $u(\cdot)$ (due to the fact that Fourier series of $\dot{u}(\cdot)$ is the formal differentiation of the Fourier series for $u(\cdot)$). In particular, $\mathrm{mod}_{\mathbb{Z}}(u) = \mathrm{mod}_{\mathbb{Z}}(\dot{u})$. Now express $f(\cdot)$ from (4.68) and the required inclusion after what we have said is obvious.

Now we will show (4.71). For convenience, we write υ instead of υ^*. Suppose $\varepsilon > 0$ and $\tau \in \mathcal{T}_\varepsilon(f)$. Put $\upsilon_{\max} := \sup_{t\in\mathbb{R}} \upsilon(t) < \upsilon^+$, $M_\upsilon(\tau; t) := |\upsilon(t + \tau) - \upsilon(t)|$ and $M_\upsilon(\tau) := \sup_{t\in\mathbb{R}} M_\upsilon(\tau; t)$. Then for some t_0 we have $|M_\upsilon(\tau; t_0) - M_\upsilon(\tau)| \le \frac{\varepsilon}{2}$. From (4.74) we have

$$\begin{aligned} \upsilon(t_0 + \tau) - \upsilon(t_0) - \int_{-\infty}^{t} \big(e^{A(t-s)}b, c\big) \big[\phi(\upsilon^*(s + \tau)) - \phi(\upsilon^*(s))\big]\, ds \\ = \int_{-\infty}^{t} \big(e^{A(t-s)}[f(s + \tau) - f(s)], c\big)\, ds. \end{aligned} \tag{4.76}$$

Put

$$I(\tau) := \left| \int_{-\infty}^{t_0} \big(e^{A(t_0-s)}b, c\big) \big[\phi(\upsilon^*(s + \tau)) - \phi(\upsilon^*(s))\big]\, ds \right|. \tag{4.77}$$

From (4.76) we have

$$|M_\upsilon(\tau) - I(\tau)| \le C\varepsilon + \varepsilon. \tag{4.78}$$

Note that (see Remark 4.7)

$$I(\tau) \le -W(0)\phi'(\upsilon_{\max}) \cdot M_\upsilon(\tau), \tag{4.79}$$

where $\alpha_0 = -W(0) = \int_0^\infty (e^{As}b, c)ds > 0$. From **(C5)–(C6)** we have $\alpha_0\phi'(\upsilon_{\max}) < \alpha_0\phi'(\upsilon^+) = 1$. From this and (4.78)–(4.79) we deduce there is a constant $\widetilde{C} > 0$ such that $M_\upsilon(\tau) \le \widetilde{C}\varepsilon$, i.e. τ-is an $\widetilde{C}\varepsilon$ almost period for $\upsilon(\cdot)$. Now from (4.73) it is evident that τ is an $\hat{C}\varepsilon$-almost period for $u(\cdot)$ with some constant $\hat{C} > 0$. In particular, $l_u(\hat{C}\varepsilon) \le l_f(\varepsilon)$ and, hence, (4.71) holds. □

Estimates of the Diophantine Dimension for Quasi-Periodic Trajectories. For $\theta \in \mathbb{R}^m$ we denote by $|\theta|_m$ the distance from θ to $\mathbb{Z}^m$. Clearly, $|\cdot|_m$ defines a metric on m-dimensional flat torus $\mathbb{T}^m = \mathbb{R}^m/\mathbb{Z}^m$. An equivalent definition of a $\mathbb{E}$-quasi-periodic function $u(\cdot)$ is that there is a continuous function $\Phi_u \colon \mathbb{T}^m \to \mathbb{E}$ and a vector of rationally independent numbers $\omega = (\omega_1, \ldots, \omega_m)$ such that $u(t) = \Phi_u(\omega t)$. In this case the numbers $\omega_1, \ldots, \omega_m$ are called 1-*frequencies* and the function Φ_u is called the *parametrization* of $u(\cdot)$ given by 1-frequencies vector ω. Since the motion $t \mapsto \omega t$ densely fills the torus $\mathbb{T}^m$ the function $\Phi_u(\cdot)$ is uniquely determined by ω. Note that 1-frequencies have to be multiplied by 2π to become the frequencies defined in Sect. 4.3. We will also call these 1-frequencies (or the vector ω) an integral base for $u(\cdot)$.

The integral base $\omega = (\omega_1, \ldots, \omega_m)$ is a *base of true size m* if there is no integral bases for $u(\cdot)$ with less than m frequencies. Suppose that $u(t) = \Phi_u(\omega t)$. Then Φ induces a map $\hat{\Phi} \colon \mathbb{T}^m \to \mathcal{H}(u)$ given by $\hat{\Phi}(\theta) := \Phi(\omega \cdot + \theta) \in \mathcal{H}(u)$. The base $\omega_1, \ldots, \omega_m$ is called *maximal* if the map $\hat{\Phi}$ is a homeomorphism. It is clear that any maximal base of m frequencies is a base of true size m. The following fundamental theorem can be found in [24].

Theorem 4.18 *For every $\mathbb{E}$-quasi-periodic function there exists a maximal base.*

Corollary 4.1 *The hull $\mathcal{H}(u)$ of a quasi-periodic function having a base of true size m is homeomorphic to $\mathbb{T}^m$.*

Since by (3.12), Chap. 3, $\underline{\dim}_B \mathcal{H}(u) \ge \dim_T \mathcal{H}(u)$ we can combine Corollary 4.1 and Theorem 4.16 to get the following proposition.

Proposition 4.8 *Suppose that $u(\cdot)$ is a $\mathbb{E}$-quasi-periodic function having a base of true size m and satisfying the Hölder condition with an exponent $\alpha \in (0, 1]$; then*

$$\mathfrak{Di}(u) \ge m - \frac{1}{\alpha}. \tag{4.80}$$

Now we are going to investigate the link between the Diophantine dimension of $u(t) = \Phi_u(\omega t)$ and rational approximations of the m-tuple ω.

We say that an m-tuple $\omega = (\omega_1, \ldots, \omega_m)$ of real numbers satisfy the *Diophantine condition* of order $\beta \ge 0$ if for some $C > 0$ and all natural q the inequality

$$|\omega q|_m \ge C \left(\frac{1}{q}\right)^{\frac{1+\beta}{m}} \tag{4.81}$$

holds. In the next paragraph we will study metric properties of the Diophantine-like numbers. Now we need to formulate a result from the geometry of numbers.

A *lattice* $\mathbb{L}$ in $\mathbb{R}^n$ is an additive subgroup generated by n linearly independent vectors. For example, $\mathbb{L} = \mathbb{Z}^n$ is generated by standard basis $e_1, \ldots, e_n$. Suppose that $\mathcal{K}$ is a closed convex centrally symmetric body in $\mathbb{R}^n$ and 0 is an interior point of $\mathcal{K}$. For $k = 1, \ldots, n$ let $\mathfrak{s}_k$ be the minimal number $\mathfrak{s}$ such that $\mathfrak{s} \cdot \mathcal{K}$ contains k linearly independent vectors of $\mathbb{L}$. The numbers $\mathfrak{s}_1, \ldots, \mathfrak{s}_n$ are called *successive minima* of $\mathcal{K}$ w.r.t. $\mathbb{L}$. Clearly, $\mathfrak{s}_1 \leq \mathfrak{s}_2 \leq \ldots \leq \mathfrak{s}_n$. We have the following fundamental theorem of Minkowski (see [25]).

Theorem 4.19 *In the above constructions for the successive minima of $\mathcal{K}$ w.r.t. $\mathbb{L}$ we have*

$$\frac{2^n}{n!} \operatorname{vol}(\mathbb{R}^n/\mathbb{L}) \leq \mathfrak{s}_1 \cdot \ldots \cdot \mathfrak{s}_n \cdot \operatorname{vol}(\mathcal{K}) \leq 2^n \operatorname{vol}(\mathbb{R}^m/\mathbb{L}). \tag{4.82}$$

A subset $\mathcal{D} \subset \mathbb{R}^n$ is a *fundamental domain* of $\mathbb{L}$ if the natural projection $\pi : \mathbb{R}^n \to \mathbb{R}^n/\mathbb{L}$, $x \mapsto [x]$, is bijective on $\mathcal{D}$. From Theorem 4.19 we deduce the following lemma.

Lemma 4.11 *Let $\omega_1, \ldots, \omega_m$ satisfy the Diophantine condition of order $\beta \geq 0$ with $\beta(m-1) < 1$. Then there is $K > 0$ such that the system of inequalities*

$$|\omega_1 \tau|_1 \leq \varepsilon, \ \ldots, \ |\omega_m \tau|_1 \leq \varepsilon \tag{4.83}$$

has an integer solution τ in each interval of length $L(\varepsilon)$, where $L(\varepsilon) = K \left(\frac{1}{\varepsilon}\right)^d$ with $d = \frac{(1+\beta)m}{1-\beta(m-1)}$.

Proof For convenience put $\gamma := \frac{1+\beta}{m}$. For $T \geq 1$ consider the parallelepiped

$$\Pi_T := \left\{(x, y_1, \ldots, y_m) \in \mathbb{R}^{m+1} \ : \ \max_{1 \leq j \leq m} |\omega_j x - y_j| \leq \frac{C'}{T^\gamma}, |x| \leq T\right\}, \tag{4.84}$$

for any fixed $C_1 < C$, where C is from (4.81). Consider the successive minima $\mathfrak{s}_1, \ldots, \mathfrak{s}_{m+1}$ of Π_T w.r.t. $\mathbb{L} = \mathbb{Z}^{m+1}$. It follows that there is no non-zero integer points in Π_T and, consequently, $\mathfrak{s}_1 \geq 1$. Since the volume of Π_T is proportional to $T^{m\gamma - 1}$ for some constant $C_2 > 0$ we have

$$\mathfrak{s}_{m+1} \leq \mathfrak{s}_1 \cdot \ldots \cdot \mathfrak{s}_{m+1} \leq C_2 T^{m\gamma-1}. \tag{4.85}$$

Therefore, the parallelepiped $(n+1)C_2 T^{m\gamma-1} \cdot \Pi_T$ contains a fundamental domain of the lattice $\mathbb{Z}^{m+1}$ and, consequently, any translate of it contains an integer point. So, it follows that for some constants $C_3 > 0$ and $C_4 > 0$ the system

$$|\omega x - \theta|_m \leq C_3 T^{(m-1)\gamma - 1} \tag{4.86}$$

has an integer solution x with $A \le x \le A + C_4 T^{m\gamma}$ for arbitrary number A and $\theta = (\theta_1, \dots, \theta_m) \in \mathbb{R}^m$. Note that $(m-1)\gamma - 1 = \frac{1}{m}(\beta(m-1) - 1) < 0$. For all sufficiently small $\varepsilon > 0$ choose T such that $\varepsilon = C_3 T^{\gamma(m-1)-1}$ and put $\theta = 0$ in (4.86). Since $T^{m\gamma} = C_5 \left(\frac{1}{\varepsilon}\right)^d$, where $d = \frac{m\gamma}{1-\gamma(m-1)} = \frac{(1+\beta)m}{1-\beta(m-1)}$ and $C_5 > 0$ is a proper constant, we have an integer solution τ to (4.83) in each interval of length $L(\varepsilon) = K\left(\frac{1}{\varepsilon}\right)^d$ for an appropriate $K > 0$ and, thus, the proof is finished. □

From Lemma 4.11 we have

Theorem 4.20 *Let $u(t) = \Phi_u(\omega_0 t, \omega_1 t, \dots, \omega_m t)$ be an $\mathbb{E}$-quasi-periodic function with the 1-frequencies $\omega_0, \omega_1, \dots, \omega_m$, $m \ge 1$. Suppose that Φ_u satisfies the Hölder condition with an exponent $\alpha \in (0, 1]$ and the m-tuple $\omega' = (\frac{\omega_1}{\omega_0}, \dots, \frac{\omega_m}{\omega_0})$ satisfies the Diophantine condition of order $\beta \ge 0$ with $\beta(m-1) < 1$. Then we have*

$$\mathfrak{Di}(u) \le \frac{1}{\alpha} \cdot \frac{(1+\beta)m}{1-\beta(m-1)}. \tag{4.87}$$

Proof Put $\chi := \Phi_u$ and $\upsilon(t) := (\omega t)$. Now we can use Proposition 4.7 to get that $\mathfrak{Di}(u) = \mathfrak{Di}(\chi \circ \upsilon) \le \frac{1}{\alpha}\mathfrak{Di}(\upsilon)$. Thus the problem is reduced to the estimation of the Diophantine dimension for the linear flow on torus.

<u>Case 1</u>: $\omega_0 = 1$. Every ε-almost period of $\upsilon(\cdot)$ is a solution to the system

$$|\tau|_1 \le \varepsilon,\ |\omega_1 \tau|_1 \le \varepsilon,\ \dots,\ |\omega_m \tau|_1 < \varepsilon. \tag{4.88}$$

We omit the first condition by looking for τ being integer. Then by Lemma 4.11 there is an integer τ in every interval of length $L(\varepsilon) = \left(\frac{1}{\varepsilon}\right)^d$ with $d = \frac{(1+\beta)m}{1-\beta(m-1)}$, satisfying 4.88. From this we immediately get (4.87).
<u>Case 2</u>: $\omega = (\omega_0, \dots, \omega_m)$ does not have 1 as a frequency. Then for $\varepsilon > 0$ any ε-almost period τ of $\upsilon(\cdot)$ is a solution to

$$|\omega_0 \tau|_1 \le \varepsilon,\ \dots,\ |\omega_m \tau|_1 < \varepsilon. \tag{4.89}$$

Put $\zeta := \omega_0 \tau$. Then system (4.89) becomes

$$|\zeta|_1 \le \varepsilon,\ |\omega_1' \zeta|_1 \le \varepsilon,\ \dots,\ |\omega_m' \zeta|_1 \le \varepsilon, \tag{4.90}$$

where $\omega_j' = \frac{\omega_j}{\omega_0}$, $j = 1, \dots, m$. Denote $\upsilon'(t) := (1, \omega_1' t, \dots, \omega_m' t)$. It is clear that $\mathfrak{Di}(\upsilon) = \mathfrak{Di}(\upsilon')$ as their almost periods are proportional. Thus, the theorem is proved. □

Metric Properties of Diophantine Numbers. Denote the set of m-tuples satisfying the Diophantine condition of a given order $\beta \ge 0$ by $\mathcal{D}_m(\beta)$. Let $\mathcal{D}_m(\cap) := \bigcap_{\beta>0} \mathcal{D}_m(\beta) \cup \mathcal{D}_m(0)$. By a theorem of Khinchine (see [25]), the set $\mathcal{D}_m(\beta)$ has full measure (= its complement has Lebesgue measure zero) for every $\beta > 0$ and, therefore, since $\mathcal{D}_n(\beta_1) \subset \mathcal{D}_m(\beta_2)$ for $\beta_1 < \beta_2$, $\mathcal{D}_m(\cap)$ is a set of full measure. Let

$\mathring{\mathcal{D}}_m$ be the set of m-tuples in $\mathcal{D}_m(\cap)$ that are linearly independent. Note that the set $\mathring{\mathcal{D}}_m$ is also a set of full measure.

For $1 \leq j \leq m$ let $\mathring{\mathcal{D}}_m^j$ be the set of linearly independent m-tuples $\omega = (\omega_1, \ldots, \omega_m)$ such that $(\omega_1, \ldots, \hat{\omega}_j, \ldots, \omega_m) \in \omega_j \mathring{\mathcal{D}}_{m-1}$.

Lemma 4.12 *The set $\mathring{\mathcal{D}}_m^j$ has full measure.*

Proof Let $\mathcal{C}_m := \mathbb{R}^m \setminus \mathring{\mathcal{D}}_m^j$. For $\xi \in \mathbb{R}$ consider the sections of $\mathcal{C}_m$ along the $(m-1)$-dimensional plane $\{\omega_j = \xi\}$. By Cavalieri's principle,

$$\mu_L^{(m)}(\mathcal{C}_m) = \int\limits_{-\infty}^{+\infty} \mu_L^{(m-1)}\left(\{\omega_j = \xi\} \cap \mathcal{C}_m\right) d\xi, \tag{4.91}$$

where $\mu_L^{(m)}$ stands for the m-dimensional Lebesgue measure. Clearly, for $\xi \neq 0$ we have $\mu_L^{(m-1)}\left(\{\omega_j = \xi\} \cap \mathcal{C}_m\right) = \mu_L^{(m-1)}\left(\mathbb{R}^{m-1} \setminus \xi \mathring{\mathcal{D}}_{m-1}\right) = 0$ since the set $\xi \mathring{\mathcal{D}}_{m-1}$ has full measure. From this and (4.91) it follows that $\mu_L^{(m)}(\mathcal{C}_m) = 0$. □

Liouville Phenomenon in Estimates of Fractal Dimensions. Suppose that $f(\cdot)$ in (4.68) is an $\mathbb{R}^n$-quasi-periodic function with m frequencies $\omega = (\omega_1, \ldots, \omega_m)$, i.e. $f(t) = \Phi_f(\omega t)$. By Eq. (4.70) from Theorem 4.17 the exponentially stable almost periodic solution is quasi-periodic with the same frequencies. While $u(\cdot)$ is C^1 we can not say the same about the parametrization Φ_u (see also the introduction to this section). Hence, the results similar to Theorem 4.20 can not be directly applicable. In this case the fact that Theorem 4.16 requires only the smoothness of $u(\cdot)$ is essential for our investigations. Suppose that $f(\cdot)$ in (4.68) is an $\mathbb{R}^n$-quasi-periodic function with m frequencies $\omega = (\omega_1, \ldots, \omega_m)$, i.e. $f(t) = \Phi_f(\omega t)$. By equation (4.70) from Theorem 4.17 the exponentially stable almost periodic solution is quasi-periodic with the same frequencies. While $u(\cdot)$ is C^1 we can not say the same about the parametrization Φ_u (see also the introduction to this section). Hence, the results similar to Theorem 4.20 can not be directly applicable. In this case the fact that Theorem 4.16 requires only the smoothness of $u(\cdot)$ is essential for our investigations.

Theorem 4.21 *Under assumptions* **(C1)–(C9)** *for* (4.68) *suppose that* $f(t) = \Phi_f(\omega t)$, *where* $\omega = (\omega_0, \omega_1, \ldots, \omega_m)$, $m \geq 1$, *and* Φ_f *satisfies the Hölder condition with an exponent* $\alpha \in (0, 1]$. *Suppose also that* $(\frac{\omega_1}{\omega_0}, \ldots, \frac{\omega_m}{\omega_0})$ *satisfy the Diophantine condition of order* $\beta \geq 0$ *and* $\beta(m-1) < 1$. *Then the exponentially stable almost periodic solution* $u(\cdot)$ *to* (4.68) *is quasi-periodic and we have*

$$\overline{\dim}_B \mathcal{M}_u \leq \frac{1}{\alpha} \cdot \frac{(1+\beta)m}{1-\beta(m-1)} + 1. \tag{4.92}$$

Proof Inequality (4.71) from Theorem 4.17 gives us $\mathfrak{Di}(u) \leq \mathfrak{Di}(f)$. Since $\dot{u}$ is almost periodic and, in particular, bounded we have $\overline{\Delta}(u) \leq 1$. In virtue of Theorem 4.20 we get $\mathfrak{Di}(f) \leq \frac{1}{\alpha} \cdot \frac{(1+\beta)m}{1-\beta(m-1)}$. Thus from 4.16 we obtain (4.92). □

The dependence on the quality of rational approximations β in (4.92) is because of inability to control any regularity of Φ_u in nonlinear systems. If the frequency-vector ω is in some sense well-approximable it seems that an appearance of the *Liuoville phenomenon* is possible. Its resulting effect is in that we can not control the fractal dimension of forced quasi-periodic oscillations with well-approximable frequencies. Note that in virtue of results in the previous paragraph for almost all ω we may put $\beta = 0$ in (4.92).

4.8 Fractal Dimensions of Forced Almost Periodic Regimes in Chua's Circuit

In this section we show the existence of forced almost periodic oscillations in Chua's circuit using the approach of Krasnosel'skii (partially described in Sect. 4.7) and compare the analytical upper estimates of the fractal dimension of the trajectory closures with numerical simulations given by the standard box-counting algorithm. The main results are borrowed from [4].

Existence of Almost Periodic Regimes Consider the perturbed Chua's circuit [27]

$$\begin{cases} \dot{x} = \eta_1(y - x + h(x)) + f_1(t), \\ \dot{y} = x - y + z + f_2(t), \\ \dot{z} = -(\eta_2 y + \eta_3 z) + f_3(t), \end{cases} \tag{4.93}$$

where $h(x) = \kappa_1 x + \frac{1}{2}(\kappa_0 - \kappa_1)(|x+1| - |x-1|)$ and $\eta_1, \eta_2, \eta_3, \kappa_0, \kappa_1$ are parameters. For certain values of the parameters system (4.93) may demonstrate a regular behaviour as well as the chaotic one; there is the possibility of presence of hidden chaotic attractors and limit cycles (see [27] and the links therein). In the almost periodically perturbed system the appearance of the so-called strange non-chaotic attractors is possible [28]. Here we use the previously discussed method of Krasnosel'ski to obtain conditions for existence of an exponentially stable almost periodic solution to (4.93). Next, we study the fractal dimensions of its closure and compare the analytical upper estimates with the estimates provided by numerical experiments. All the main results is borrowed from [4].

We write system (4.93) as a control system (4.68), where

$$A = \begin{bmatrix} -\eta_1 & \eta_1 & 0 \\ 1 & -1 & 1 \\ 0 & -\eta_2 & -\eta_3 \end{bmatrix}, \quad b = \begin{bmatrix} \eta_1 \\ 0 \\ 0 \end{bmatrix}, \quad c = \begin{bmatrix} 1 \\ 0 \\ 0 \end{bmatrix},$$

and $\phi(\upsilon) = \kappa_1 \upsilon + \frac{1}{2}(\kappa_0 - \kappa_1)(|\upsilon + 1| - |\upsilon - 1|)$.

The perturbation $f(t) = (f_1(t), f_2(t), f_3(t))$ is supposed to be almost periodic.

For two elements u, v of a closed subspace $\mathbb{E} \subset C_b(\mathbb{R}; \mathbb{R})$ we write $u \prec v$, if $v - u \geq 0$. A *cone segment* $\langle u, v \rangle$ in $\mathbb{E}$ is the set of all $w \in \mathbb{E}$ such that $u \prec w \prec v$.

Theorem 4.22 *Let (C1)–(C4) (defined in the previous section) be satisfied and in addition suppose that*

(CH1) $0 < \kappa_0 < -W(0)^{-1} < \kappa_1$

(CH2) *For* $v_0 \in (0, 1]$ *consider* $M := W(0)\phi(v_0) + v_0 = (1 + W(0)\kappa_0)v_0 > 0$ *and for* $g(t)$ *from* (4.72) *we have*

$$- M < \sup_{t \in \mathbb{R}} g(t) < M. \tag{4.94}$$

Then system (4.93) *has an exponentially stable almost periodic solution* $u^*(\cdot)$ *which lies in* $\{-v_0 < c^* u < v_0\}$ *and satisfies*

$$\operatorname{mod}_{\mathbb{Z}}(u^*) = \operatorname{mod}_{\mathbb{Z}}(f) \text{ and } \mathfrak{Di}(u^*) \leq \mathfrak{Di}(f). \tag{4.95}$$

Proof Consider the integral operator

$$[\Pi v](t) := \int_{-\infty}^{t} \left(e^{A(t-s)} b, c\right) \phi(v(s)) ds + \int_{-\infty}^{t} \left(e^{A(t-s)} f(s), c\right) ds \tag{4.96}$$

in the space $\mathcal{AP}(\mathbb{R}; \mathbb{R})$ of scalar almost periodic functions with the uniform norm. From **(C1)**–**(C4)** we have $(e^{As} b, c) \geq 0$ for all $s \geq 0$ (see Remark 4.7). Since the inequality in (4.94) is strict, we may assume that $v_0 \in (0, 1)$. Consider two constant function $v_1(t) \equiv -v_0$ and $v_2(t) \equiv v_0$. We have

$$\begin{aligned} &[\Pi v_1](t) + v_0 = -W(0)\phi(-v_0) + v_0 + g(t) = M + g(t) > 0, \\ &v_0 - [\Pi v_2](t) = v_0 + W(0)\phi(v_0) - g(t) = M - g(t) > 0. \end{aligned} \tag{4.97}$$

From the monotonicity of ϕ and (4.97) it follows that the operator Π is monotone on the cone segment $\langle -v_0, v_0 \rangle$ and leaves it invariant. Now suppose $v_1, v_2 \in \mathcal{AP}(\mathbb{R}; \mathbb{R})$ and $-v_0 \prec v_1 \prec v_2 \prec v_0$. We have

$$0 \prec [\Pi v_1] - [\Pi v_2](t) = m_0 \int_{-\infty}^{t} \left(e^{A(t-s)} b, c\right) (v_1(s) - v_2(s)) ds \prec S(v_1 - v_2), \tag{4.98}$$

where the linear operator S is defined on $\mathcal{AP}(\mathbb{R}; \mathbb{R})$ as

$$[Sv](t) = \kappa_0 \int_{-\infty}^{t} \left(e^{A(t-s)} b, c\right) v(s) ds. \tag{4.99}$$

It is obvious that $\|S\| = -W(0)\kappa_0 < 1$. In virtue of Theorem 10.2 from [14] the operator Π has a unique fixed point v^* on $\langle -v_0, v_0 \rangle$. It is clear that the formula

$$u^*(t) = \int_{-\infty}^{t} e^{A(t-s)} b\phi(\upsilon^*(s))ds + \int_{-\infty}^{t} e^{A(t-s)} f(s)ds \tag{4.100}$$

defines an almost periodic solution to (4.93) and $(u^*(t), c) = \upsilon^*(t) \in (-\upsilon_0, \upsilon_0)$ is satisfied for all $t \in \mathbb{R}$. The exponential stability of u^* follows from the fact that it is a solution of a linear system with the Hurwitz matrix. A proof of (4.95) can be carried out analogously to the proof of Theorem 4.17. □

Now, for simplicity, we put $f(t) = \Phi_f(\omega_1 t, \omega_2 t)$, where $\Phi_f : \mathbb{T}^2 \to \mathbb{R}^3$ satisfies the Hölder conditions with an exponent $\alpha \in (0, 1]$, and $\omega = \frac{\omega_1}{\omega_2}$ is an irrational number. Suppose that the k-th convergent of ω, say q_k, $k = 1, 2, \dots,$ for some $\beta \geq 0$ satisfy $q_{k+1} = O(q_k^{1+\beta})$ for all k. From basic properties of continued fractions (see, for example, [13]) it follows the latter is equivalent to the existence of a constant $C > 0$ such that for all integer p and positive integer q we have

$$|\omega q - p| \geq \frac{C}{q^{1+\beta}}, \tag{4.101}$$

i.e. ω satisfies the Diophantine conditions of order $\beta \geq 0$. Analogously to Theorem 4.21 we have

Theorem 4.23 *Suppose the assumptions of Theorem 4.22 and the above conditions on f and ω hold. Then for the exponentially stable almost periodic solution u^* we have*

$$\overline{\dim}_B \mathcal{M}_{u^*} \leq \frac{1+\beta}{\alpha} + 1. \tag{4.102}$$

It is well-known that the numbers ω which satisfies the Diophantine condition of order 0 (= *badly approximable* numbers) have bounded terms in its continued fraction expansion and vice versa. In particular, these are all the quadratic irrationals (see [13]), i.e. $\sqrt{2}$, $\sqrt{3}$ or the golden mean $\varphi_0 := \frac{1+\sqrt{5}}{2}$.

Numerical Experiments Figure 4.1 shows a numerically constructed domain in the space of parameters $\eta_2 \in [0, 5]$, $\eta_3 \in [0, 5]$, and $\eta_1 = 1.4$ for which system (4.93) satisfies **(C1)**–**(C4)**. In the sequel we give a more formal investigation of the conditions of Theorem 4.22 for a certain parameters.

We consider system (4.93) with parameters $\eta_1 = 1.4$, $\eta_2 = 2.2$, and $\eta_3 = 4.8$. For these parameters the characteristic polynomial of A has 3 real roots: $\lambda_0 \approx -0.258$, $\lambda_1 \approx -3.125$, and $\lambda_2 \approx -3.817$. We put $\varepsilon := 1.5$. By straightforward calculations we get

$$W(z) = -\eta_1 \cdot \frac{z^2 + (\eta_3 + 1)z + \eta_2 + \eta_3}{z^3 + (1 + \eta_1 + \eta_3)z^2 + (\eta_2 + \eta_3 + \eta_1\eta_3)z + \eta_1\eta_2} \tag{4.103}$$

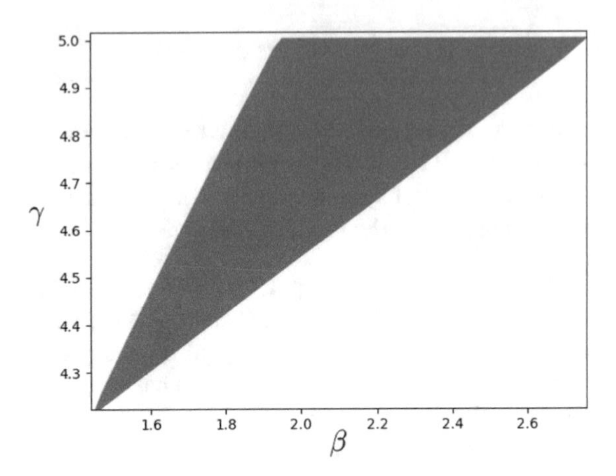

Fig. 4.1 The domain (black) of η_2 (horizontally) and η_3 (vertically) for which system (4.93) with $\eta_1 = 1.4$ satisfies **(C1)**–**(C4)**. Taken from [4]

and, as a corollary,

$$\operatorname{Re} W(i\omega - \varepsilon) = 1.4 \cdot \frac{(2.7\omega^2 + 4.675) \cdot (-\omega^2 + 0.55) + 2.8\omega \cdot (\omega^3 + 1.13\omega)}{(2.7\omega^2 + 4.675)^2 + (\omega^3 + 1.13\omega)^2}. \tag{4.104}$$

The simplest analysis of the numerator in (4.104) shows that $W(i\omega - \varepsilon) > 0$ for all ω. Next, it is clear that $\operatorname{Re} W(i\omega - \varepsilon) \sim \frac{0.14}{\omega^2}$ as $\omega \to \infty$. Thus conditions **(C1)**–**(C4)** for the linear part is satisfied. From (4.103) it follows that

$$W(0) = -1 - \frac{\eta_3}{\eta_2} = -\frac{35}{11}.$$

Therefore, to satisfy condition **(CH1)** we may take $\kappa_0 < \frac{11}{35} < \kappa_1$. We choose $\kappa_0 := \frac{1}{5}$ and $\kappa_1 := 1$. Then for $\upsilon_0 = 1$ we have

$$M = (1 + W(0)\kappa_0)\upsilon_0 = 1 - \frac{7}{11} = \frac{4}{11}.$$

Therefore, if for the chosen parameters the perturbation $f(t) = (f_1(t), f_2(t), f_3(t))$ in (4.93) satisfy

$$-\frac{4}{11} < \sup_{t \in \mathbb{R}} \int_{-\infty}^{t} \left(e^{A(t-s)} f(s), c\right) ds < \frac{4}{11}, \tag{4.105}$$

Theorem 4.22 gives us the exponentially stable almost periodic solution lying in $\{-1 < (c, u) < 1\}$. Since the matrix A is diagonalisable for the operator norm $\|\cdot\|$ associated with the Euclidean norm $|\cdot|$ in $\mathbb{R}^3$ we have $\|e^{At}\| = \|e^{Jt}\| = e^{\lambda_0 t}$, where J is the Jordan form of A. Put $\sup_{s \in \mathbb{R}} |f(s)| =: \kappa$. We have

$$\left| \int_{-\infty}^{t} \left(e^{A(t-s)} f(s), c \right) ds \right| \leq \kappa \int_{0}^{+\infty} e^{\lambda_0 s} ds = \frac{\kappa}{|\lambda_0|}.$$

In particular, the inequality in (4.105) is satisfied if $\kappa < \frac{1}{11}$.

Consider the Weierstrass function

$$w(t) = \sum_{k=1}^{\infty} a^k \cos(b^k t), \tag{4.106}$$

where $b \in \mathbb{Z}$ and $a \in \mathbb{R}$ are parameters. We take $a = b^{-\alpha}$, where $\alpha \in (0, 1]$. It is known (see [31] p. 47) that the function in (4.106) satisfies the Hölder condition with the exponent $-\frac{\ln a}{\ln b} = \alpha$. We will use this function for numerical simulations.

Let the function $\Phi_f \colon \mathbb{T}^2 \to \mathbb{R}^3$ satisfy the Hölder condition with an exponent $\alpha \in (0, 1]$ and consider $f(t) := \Phi_f(\omega_1 t, \omega_2 t)$, where $\frac{\omega_1}{\omega_2} = \varphi_0 = \frac{1+\sqrt{5}}{2}$. Suppose that $\kappa = \sup_{s \in \mathbb{R}} |f(s)| < \frac{1}{11}$. Then for the fractal dimension of $\mathcal{M}_{u^*}$, where u^* is the exponentially stable almost periodic solution of (4.93) given by Theorems 4.22, 4.23 gives the estimate

$$\overline{\dim}_B \mathcal{M}_{u^*} \leq \frac{1}{\alpha} + 1. \tag{4.107}$$

In what follows we will compare this upper estimate with an estimates given by numerical simulations.

In numerical experiments we use the standard box-counting algorithm. The coordinates x, y, z in (4.93) are stretched in 25 times to prevent possible problems while counting the boxes of large diameter. The set $\mathcal{M}_{u^*}$ is approximated by a part of the trajectory $u(\cdot)$ with initial value $u(0) = 0$ (which is attracted to u^*) considered on $[0, T]$. We calculate the values of the solution in 10^8 points, which are uniformly distributed on $[0, T]$, with the use of Runge-Kutta method of 4–5th order.[6] We calculate ε-boxes required to cover $u([0, T])$ (denote its number $N(\varepsilon, T)$) for $\varepsilon = \varepsilon_k = 2^{-k/2}$, where $k = 10, \ldots, 14$. Next, by observations $N(\varepsilon, T)$ we have to estimate $N(\varepsilon)$. Note that the estimate of the Diophantine dimension in (4.95) allows to estimate the time T for which the set $\mathcal{M}_{u^*}$ will lie in the δ-neighborhood of the set $u^*([0, T])$. For such T we may calculate the number of ε-boxes for $\varepsilon \ll \delta$. In our case we have $T \leq C \left(\frac{1}{\delta}\right)^{1/\alpha}$. Thus, for $\alpha = \frac{3}{2}$ and $\delta = 2^{-9}$ the estimate for T has the order of 10^4. From this it follows that in simulations for large enough T the numbers $N(\varepsilon, T)$ stop to change significantly and therefore they can be considered as an estimate of $N(\varepsilon)$.

For the observations $(-\ln \varepsilon, \ln N(\varepsilon))$, where $N(\varepsilon)$ is the estimated number of ε-boxes, the dependence $\ln N(\varepsilon)$ on $-\ln \varepsilon$ is approximated in two ways. The first one is the least squares method to find the parameters of the linear model $y = dx + \upsilon$. The coefficient d is then considered as a estimate of the fractal dimension of $\mathcal{M}_{u^*}$. For the second approximation we use the nonlinear model $y = dx + \beta e^{-x} + \upsilon$ and

[6] We use the implementation of the method within the procedure solve_ivp of package scipy.integrate of programming language Python 3.7.1. Parameter max_step of the procedure is chosen to be 2^{-9}.

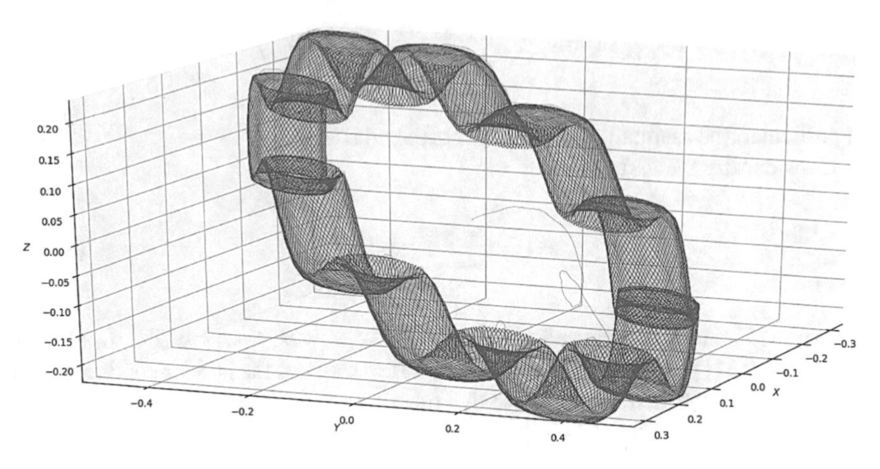

Fig. 4.2 The set $\mathcal{M}_{u^*}$ from Example 4.1. Taken from [4]

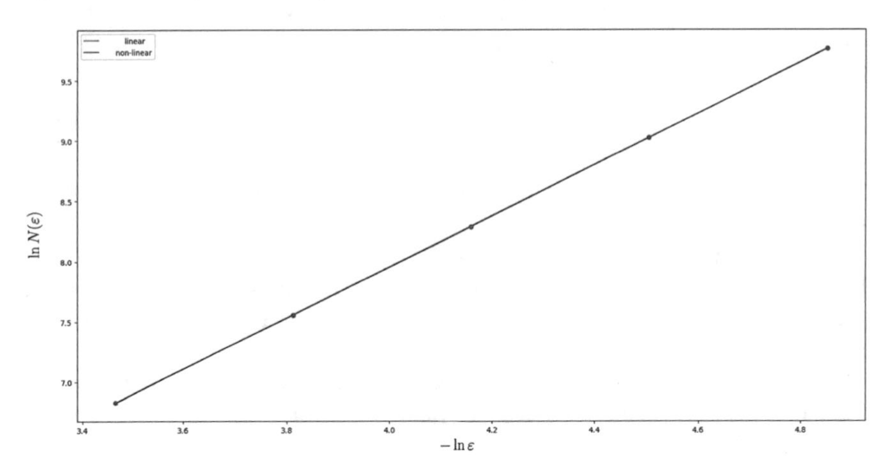

Fig. 4.3 Plots of the linear and nonlinear regressions expressing the dependence of $\ln N(\varepsilon)$ (vertically) on $-\ln\varepsilon$ (horizontally) considered in Example 4.1. The estimates of $\overline{\dim}_B\mathcal{M}_{u^*}$ is equal to 2.11 and 2.16 for the linear and nonlinear models respectively. Taken from [4]

again the method of least squares to estimate its parameters, where the coefficient d is considered as an estimate of the fractal dimension. Below we consider these two numerical estimates of $\overline{\dim}_B\mathcal{M}_{u^*}$ with the analytical upper estimate provided by Theorem 4.23.

Example 4.1 Consider $\Phi_f(\theta_1,\theta_2) := (0, C_2\cdot w(2\pi\theta_1) + C_1\cdot\cos(2\pi\theta_2), C_2\cdot w(2\pi\theta_2))$ with $C_2 = \frac{1}{7}$, $C_1 = \frac{1}{25}$ and put $f(t) := \Phi_f(\omega_1 t, \omega_2 t)$, where $\omega_1 = \frac{1}{2\pi}$ and $\omega_2 = \frac{\varphi_0}{2\pi}$ (we recall that $\varphi_0 = \frac{1+\sqrt{5}}{2}$). The Weierstrass function (4.106) is considered for $b = 10$ and $\alpha = \frac{2}{3}$. It can be shown that $\kappa = \sup_{s\in\mathbb{R}}|f(s)| < \frac{1}{11}$ and therefore all the conditions of Theorem 4.22 are satisfied. The set $\mathcal{M}_{u^*}$ is shown in Fig. 4.2.

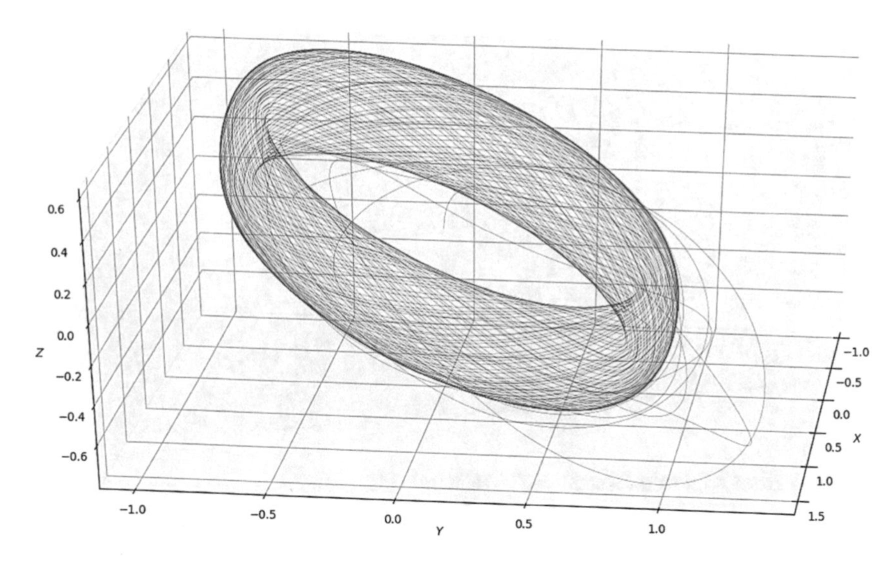

Fig. 4.4 The set $\mathcal{M}_{u^*}$ from Example 4.2. Taken from [4]

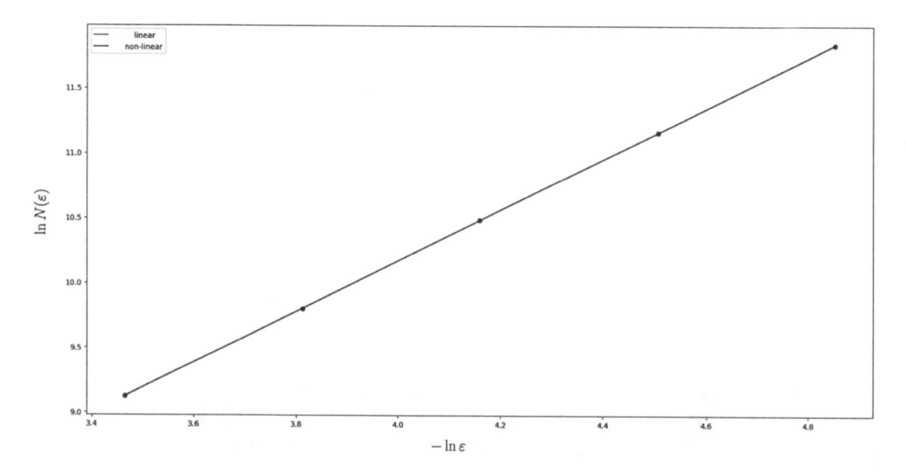

Fig. 4.5 Plots of the linear and nonlinear regressions expressing the dependence of $\ln N(\varepsilon)$ (vertically) on $-\ln\varepsilon$ (horizontally) considered in Example 4.2. The estimates of $\overline{\dim}_B \mathcal{M}_{u^*}$ is equal to 1.96 and 1.96 for the linear and nonlinear models respectively. Taken from [4]

The values $N(\varepsilon, T)$ were close enough for T from 1000 to 20,000. Both models provided an estimate of $\overline{\dim}_B \mathcal{M}_{u^*}$ approximately equal to 2.15 ± 0.05 (Fig. 4.3) that is largely differs from the analytical upper estimate of 2.5 given by Theorem 4.23.

Example 4.2 Consider $\Phi_f(\theta_1, \theta_2) := (0, C_1 \cdot \sin(2\pi t), C_2 \cdot (\cos(2\pi t) + \sin(2\pi t)))$ with $C_1 = C_2 = \frac{1}{20}$ and put $f(t) := \Phi_f(\omega_1 t, \omega_2 t)$, where $\omega_1 = \frac{1}{2\pi}$ and $\omega_2 = \frac{\varphi_0}{2\pi}$. It can be shown that $\kappa = \sup_{s \in \mathbb{R}} |f(s)| < \frac{1}{11}$ and therefore all the conditions of

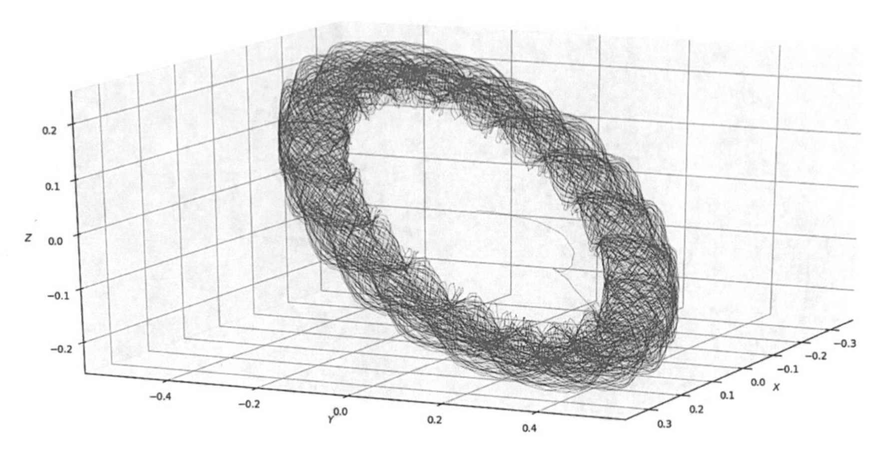

Fig. 4.6 The set $\mathcal{M}_{u^*}$ from Example 4.3. Taken from [4]

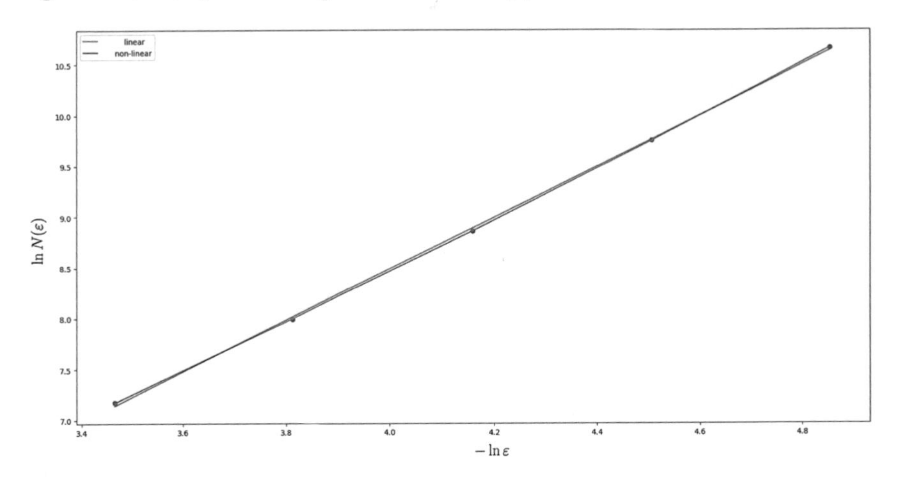

Fig. 4.7 Plots of the linear and nonlinear regressions expressing the dependence of $\ln N(\varepsilon)$ (vertically) on $-\ln \varepsilon$ (horizontally) considered in Example 4.3. The estimates of $\overline{\dim}_B \mathcal{M}_{u^*}$ is equal to 2.51 and 2.76 for the linear and nonlinear models respectively. Taken from [4]

Theorem 4.22 are satisfied. The set $\mathcal{M}_{u^*}$ is shown in Fig. 4.4. As an estimate for $N(\varepsilon)$ we chose $N(\varepsilon, T)$ for $T = 20{,}000$. Both models provided an estimate of $\overline{\dim}_B \mathcal{M}_{u^*}$ approximately equal to 2 (Fig. 4.5) that coincide with the analytical upper estimate given by Theorem 4.23.

Example 4.3 Consider $\Phi_f(\theta_1, \theta_2) := (0, C_2 \cdot w(2\pi\theta_1) + C_1 \cdot \cos(2\pi\theta_2), C_2 \cdot w(2\pi\theta_2) \cdot \sin(2\pi\theta_1))$ with $C_2 = \frac{1}{7}$, $C_1 = \frac{1}{25}$ and put $f(t) := \Phi_f(\omega_1 t, \omega_2 t)$, where $\omega_1 = \frac{1}{2\pi}$ and $\omega_2 = \frac{\varphi_0}{2\pi}$. Consider the Weierstrass function (4.106) with the parameters $b = 10$ and $\alpha = \frac{2}{3}$. It can be shown that $\kappa = \sup_{s \in \mathbb{R}} |f(s)| < \frac{1}{11}$ and therefore all the conditions of Theorem 4.22 are satisfied. The set $\mathcal{M}_{u^*}$ is shown in Fig. 4.6. The values $N(\varepsilon, T)$ were close enough for T from 1000 to 20,000. The linear regression gave the estimate of $\overline{\dim}_B \mathcal{M}_{u^*}$ approximately equal to 2.51 (see Fig. 4.7) that almost coincide with the analytical upper estimate given by Theorem 4.23. The nonlinear model gave the estimate of 2.76 that largely exceeds the analytical upper estimate of 2.5 and with the other estimated parameters ($\beta = 14.72$ and $\upsilon = -2.86$) suggests its unusability in this situation.

References

1. Anikushin, M.M.: Dimension theory approach to the complexity of almost periodic trajectories. Int. J. Evol. Equ. **10**(3–4), 215–232 (2017)
2. Anikushin, M.M.: On the Liouville phenomenon in estimates of fractal dimensions of forced quasi-periodic oscillations. Vestn. St. Petersburg Univ. Math. **52**(3) (2019)
3. Anikushin, M.M.: On the Smith reduction theorem for almost periodic ODEs satisfying the squeezing property. Russ. J. Nonlinear Dyn. **15**(1), 97–108 (2019)
4. Anikushin, M.M., Reitmann, V., Romanov, A.O.: Analytical and numerical estimates of fractal dimensions of forced quasi-periodic oscillations in control systems. Electron. J. Diff. Equ. Contr. Process. **85**(2) (2019) (Russian)
5. Burkin, I.M., Yakubovich, V.A.: Frequency conditions of existence of two almost periodic solutions in a nonlinear control system. Sibirsk. Mat. Zh. **16**(5), 916–924 (1975) (Russian); English transl. Siberian Math. J. **16**(5), 699–705 (1975)
6. Cartwright, M.L.: Almost periodic flows and solutions of differential equations. Proc. Lond. Math. Soc. **17**, 355–380 (1967)
7. Cartwright, M.L.: Almost periodic differential equations and almost periodic flows. J. Diff. Equ. **5**(1), 167–181 (1969)
8. Fink, A.M.: Almost Periodic Differential Equations. Springer (2006)
9. Glazier, J.A., Libchaber, A.: Quasi-periodicity and dynamical systems: an experimentalist's view. IEEE Trans. Circuits Syst. **35**(7), 790–809 (1988)
10. Hartman, P.: Ordinary Differential Equations. SIAM, Philadelphia (2002)
11. Hilmy, G.F.: On a property of minimal sets. Dokl. Akad. Nauk SSSR **14**, 261–262 (1937). (Russian)
12. Kalinin, Y.N., Reitmann, V.: Almost periodic solutions in control systems with monotone nonlinearities. Electron. J. Diff. Equ. Contr. Process. **4**, 40–68 (2012)
13. Khinchin, A.I.: Continued Fractions. P. Noordhoff (1963)
14. Krasnoselśkii, M.A., Burd, V.S., Kolesov, Y.S: Nonlinear Almost Periodic Oscillations. Wiley, New York (1973)
15. Levitan, B.M., Zhikov, V.V.: Almost Periodic Functions and Differential Equations. Cambridge University Press, Cambridge (1982)
16. Mallet-Paret, J.: Negatively invariant sets of compact maps and an extension of a theorem of Cartwright. J. Diff. Equ. **22**(2), 331–348 (1976)
17. Naito, K.: On the almost periodicity of solutions of a reaction diffusion system. J. Diff. Equ. **44**(1), 9–20 (1982)
18. Naito, K.: Dimension estimate of almost periodic attractors by simultaneous Diophantine approximation. J. Diff. Equ. **141**(1), 179–200 (1997)
19. Pankov, A.A.: Bounded and Almost Periodic Solutions of Nonlinear Operator Differential Equations, vol. 55. Springer Science & Business Media (2012)
20. Pliss, V.A.: Integral Sets of Periodic Systems of Differential Equations. Nauka, Moscow (1977). (Russian)
21. Pontryagin, L.S.: Topological Groups, Moscow (1973) (Russian)

22. Reitmann, V.: Frequency domain conditions for the existence of Bohr almost periodic solutions in evolution equations. IFAC Proceedings Volumes, St. Petersburg, **40**(14), 240–244 (2007)
23. Robinson, J.C.: Inertial manifolds and the cone condition. Dyn. Syst. Appl. **2**, 311–330 (1993)
24. Samoilenko, A.M.: Elements of the Mathematical Theory of Multi-Frequency Oscillations. Springer Science & Business Media (2012)
25. Siegel, C.L.: Lectures on the Geometry of Numbers. Springer Science & Business Media (2013)
26. Smith, R.A.: Massera's convergence theorem for periodic nonlinear differential equations. J. Math. Anal. Appl. **120**(2), 679–708 (1986)
27. Stankevich, N.V., Kuznetsov, N.V., Leonov, G.A., Chua, L.O.: Scenario of the birth of hidden attractors in the Chua circuit. Int. J. Bifurc. Chaos **27**(12), 1–18 (2017)
28. Suresh, K., Prasad, A., Thamilmaran, K.: Birth of strange nonchaotic attractors through formation and merging of bubbles in a quasiperiodically forced Chua's oscillator. Phys. Lett. A **377**(8), 612–621 (2013)
29. Yakubovich, V.A.: Method of matrix inequalities in theory of nonlinear control systems stability. I. Forced oscillations absolute stability. Avtom. Telemekh. **25**(7), 1017–1029 (1964)
30. Zinchenko, I.L.: The group of characters on the closure of the almost periodic trajectory of an autonomous system of differential equations. Diff. Urav. **24**(6), 1043–1045 (1988) (Russian)
31. Zygmund, A.: Trigonometric Series. Cambridge University Press, Cambridge (2002)

Chapter 5
Dimension and Entropy Estimates for Dynamical Systems

Abstract In the present chapter various approaches to estimate the fractal dimension and the Hausdorff dimension, which involve Lyapunov functions, are developed. One of the main results of this chapter is a theorem called by us the *limit theorem for the Hausdorff measure* of a compact set under differentiable maps. One of the sections of Chap. 5 is devoted to applications of this theorem to the theory of ordinary differential equations. The use of Lyapunov functions in the estimates of fractal dimension and of topological entropy is also considered. The representation is illustrated by examples of concrete systems.

5.1 Upper Estimates for the Hausdorff Dimension of Negatively Invariant Sets

5.1.1 The Limit Theorem for Hausdorff Measures

In this subsection we shall derive an upper estimate for the Hausdorff dimension of negatively invariant sets which is a generalization of the well-known Douady-Oesterlé theorem [18]. For other generalizations see [1, 35]. We start with a statement for differentiable maps, apply this result in Sect. 5.1.2 to the description of fixed points and invariant curves of maps and deduce in Sect. 5.1.3 a first upper estimate for the Hausdorff dimension of the invariant sets of the Hénon map.

Let $\mathcal{U}$ be an open set in $\mathbb{R}^n$, $\mathcal{K} \subset \mathcal{U}$ be a compact set, and

$$\varphi : \mathcal{U} \to \mathbb{R}^n \tag{5.1}$$

be a C^1-map. It follows that for any point $u \in \mathcal{U}$ and each $h \in \mathbb{R}^n$ with $u + h \in \mathcal{U}$ the Taylor expansion

$$\varphi(u + h) - \varphi(u) = D\varphi(u)h + o(h) \tag{5.2}$$

N. Kuznetsov and V. Reitmann, *Attractor Dimension Estimates for Dynamical Systems: Theory and Computation*, Emergence, Complexity and Computation 38,
https://doi.org/10.1007/978-3-030-50987-3_5

holds, where $D\varphi(u)$ denotes the differential of φ at u. For any $d \in [0, n]$ we denote, according to (2.4), Chap. 2, by $\omega_d(D\varphi(u))$ the singular value function of order d for $D\varphi(u)$ and according to Sect. 3.2, Chap. 3, by $\mu_H(\mathcal{Z}, d)$ the outer Hausdorff d-measure of a compact set $\mathcal{Z}$.

Theorem 5.1 *Suppose that for (5.1) there exist a sequence $\{\mathcal{U}_m\}_{m=0}^{\infty}$ of open subsets of $\mathcal{U}$, a compact set $\widetilde{\mathcal{K}} \subset \overline{\mathcal{U}}$, a sequence of compact sets $\{\mathcal{K}_m\}_{m=0}^{\infty}$ and a number $d \in (0, n]$ such that the following conditions are satisfied:*

(1) For any $m = 1, 2, \ldots$ the m-th iterate φ^m is defined on $\mathcal{U}_m$;
(2) For any $m = 1, 2, \ldots$ we have $\mathcal{K}_m \subset \mathcal{U}_m$ and $\varphi^j(\mathcal{K}_m) \subset \widetilde{\mathcal{K}}$ for $j = 0, 1, \ldots, m-1$;
(3) There exists a continuous positive scalar function $\varkappa : \widetilde{\mathcal{K}} \to \mathbb{R}_+$ and a number $d \in [0, n]$ such that

$$\sup_{u \in \widetilde{\mathcal{K}}} \left[\frac{\varkappa(\varphi(u))}{\varkappa(u)} \, \omega_d \left(D\varphi(u) \right) \right] < 1. \tag{5.3}$$

Then it holds:
(a) If $\mu_H(\mathcal{K}_m, d) \le \text{const}$ for $m = 1, 2, \ldots$, then $\lim_{m\to+\infty} \mu_H\left(\varphi^m(\mathcal{K}_m), d\right) = 0$;
(b) If $\mathcal{K}_m \subset \mathcal{K} \subset \varphi^m(\mathcal{K}_m)$ for $m = 1, 2, \ldots$, then $\dim_H \mathcal{K} < d$ and $\mu_H\left(\varphi^m(\mathcal{K}_m), d\right) = 0$ for all sufficiently large m.

Further, Theorem 5.1 [29–31] is called by us the *limit theorem for the Hausdorff measure*. It's proof will be given below. In the following subsection three important corollaries of this theorem will be obtained.

Remark 5.1 The function $\varkappa$ in (5.3) can be considered as a *regulating function* for the contraction property of $\omega_d(D\varphi(\cdot))$ in $\widetilde{\mathcal{K}}$. If we define $V(u) := \log \varkappa(u)$, $u \in \widetilde{\mathcal{K}}$, the inequality (5.3) is equivalent to

$$\sup_{u \in \widetilde{\mathcal{K}}} [\log \omega_d(D\varphi(u)) + V(\varphi(u)) - V(u)] < 0\,. \tag{5.4}$$

The inequality (5.4) contains the first difference of V with respect to the dynamical system $\{\varphi^k\}_{k\in\mathbb{N}_0}$. Thus $V = \log \varkappa$ can be considered as Lyapunov function for this system.

In order to prove Theorem 5.1 we need the following lemma. Recall that for a compact set $\mathcal{Z} \subset \mathbb{R}^n$, a number $d \in (0, n]$ and a number $\delta > 0$ the expression $\mu_H(\mathcal{Z}, d, \varepsilon)$ denotes the Hausdorff outer measure of $\mathcal{Z}$ at level ε and of order d. For an ellipsoid $\mathcal{E} \subset \mathbb{R}^n$ and a number $d \in (0, n]$ the symbol $\omega_d(\mathcal{E})$ denotes the d-dimensional ellipsoid measure (cf. (2.5), Chap. 2).

Lemma 5.1 *Let $d \in (0, n]$ and $\varepsilon > 0$ be numbers and $\mathcal{E} \subset \mathbb{R}^n$ be an ellipsoid such that $\left(\omega_d(\mathcal{E})\right)^{1/d} \le \varepsilon$. Then the inequality*

$$\mu_H(\mathcal{E}, d, \lambda\varepsilon) \le M\omega_d(\mathcal{E})$$

holds, where $\lambda := \sqrt{d_0 + 1}$, $M := 2^{d_0}(d_0 + 1)^{d/2}$ *and* $d = d_0 + s$ *with* $d_0 \in \{0, 1, \dots, n-1\}$, $s \in (0, 1]$.

Proof Let $\mathcal{E}$ be given by

$$\mathcal{E} := \left\{ (u_1, \dots, u_n) \in \mathbb{R}^n \,\middle|\, \sum_{j=1}^{n} (u_j/a_j)^2 \le 1 \right\},$$

where $a_1 \ge a_2 \ge \dots \ge a_n > 0$ are the lengths of the semi-axes of $\mathcal{E}$. Introduce the notions $\mathcal{E}_0 := \mathcal{E} \cap \mathbb{R}^{d_0}$, $\mathbb{R}^{d_0} := \{(u_1, \dots, u_{d_0})\}$, and $\varsigma := a_{d_0+1}$. We have $\varsigma \le \left(\omega_d(\mathcal{E})\right)^{1/d} \le \varepsilon$. In this case we can inscribe $\mathcal{E}_0$ into the parallelopiped $\mathcal{P} := [-a_1, a_1] \times [-a_2, a_2] \times \dots \times [-a_{d_0}, a_{d_0}]$ and cover $\mathcal{P}$ by N cubes with edge of length 2ς, where

$$N := \prod_{j=1}^{d_0} \left([a_j/\varsigma] + 1\right).$$

Since $a_j/\varsigma \ge 1$ for $j \le d_0$, we have

$$N \le 2^{d_0} \prod_{j=1}^{d_0} (a_j/\varsigma) = (2/\varsigma)^{d_0} \prod_{j=1}^{d_0} a_j.$$

The ellipsoid $\mathcal{E}$ is contained in $\mathcal{E}_0 \times \widetilde{\mathcal{B}}_\varsigma(0)$, where $\widetilde{\mathcal{B}}_\varsigma(0) := \mathcal{B}_\varsigma(0) \cap \mathbb{R}^{n-d_0}$. Consequently, $\mathcal{E}$ can be covered by sets of the form $\mathcal{C} \times \widetilde{\mathcal{B}}_\varsigma(0)$, where $\mathcal{C}$ is a cube with edges of length 2ς. If we introduce a rectangular coordinate system with the origin in the center of such set and choose the first d_0 of coordinate axes to be parallel to edges of the cube, then the coordinates of points of this set which are the most distant from the center satisfy the following relations

$$|v_1| = \dots = |v_{d_0}| = \varsigma, \qquad v_{d_0+1}^2 + \dots + v_n^2 = \varsigma^2.$$

It means that such a set is contained in a ball of radius $\varsigma\sqrt{d_0 + 1}$.

Consequently

$$\mu_H(\mathcal{E}, d, \sqrt{d_0 + 1}\varepsilon) \le N(d_0 + 1)^{d/2}\varsigma^d \le 2^{d_0}(d_0 + 1)^{d/2} a_1 \dots a_{d_0} \varsigma^s = M\omega_d(\mathcal{E}).$$

□

Proof of Theorem 5.1 *From the hypothesis of Theorem 5.1 it follows that there exists a positive number* $\kappa_1 < 1$ *such that*

$$\sup_{u\in\widetilde{\mathcal{K}}}\left[\frac{\varkappa(\varphi(u))}{\varkappa(u)}\omega_d\big(D\varphi(u)\big)\right]\le\kappa_1. \tag{5.5}$$

For an arbitrary integer $m\ge 1$ we introduce the notation

$$\kappa(m):=\kappa_1^m\sup_{u\in\mathcal{K}_m}\frac{\varkappa(u)}{\varkappa\big(\varphi^m(u)\big)}. \tag{5.6}$$

Let $l>0$ be an arbitrary number. Obviously, there can be found a number $m_0>0$ such that for all $m>m_0$ the inequality

$$\kappa(m)<l \tag{5.7}$$

is true. Let us fix $m>m_0$ and denote by $D\varphi^m(u)$ the differential of the map φ^m at the point $u\in\mathcal{K}_m$. The chain rule gives

$$D\varphi^m(u)=D\varphi(\varphi^{m-1}(u))\cdot D\varphi(\varphi^{m-2}(u))\cdots D\varphi(u)\,. \tag{5.8}$$

Applying Horn's inequality (Proposition 2.4, Chap. 2), it follows from (5.8) for $u\in\mathcal{K}_m$ that

$$\omega_d\big(D\varphi^m(u)\big)\le\prod_{j=1}^m\omega_d\big(D\varphi(\varphi^{m-j}(u))\big).$$

Using this and taking into account (5.5) and (5.6), we get

$$\omega_d\big(D\varphi^m(u)\big)\le\prod_{j=1}^m\kappa_1\frac{\varkappa\big(\varphi^{m-j}(u)\big)}{\varkappa\big(\varphi^{m-j+1}(u)\big)}=\kappa_1^m\frac{\varkappa(u)}{\varkappa\big(\varphi^m(u)\big)}\le\kappa(m).$$

Thus,

$$\sup_{u\in\mathcal{K}_m}\omega_d\big(D\varphi^m(u)\big)\le\kappa(m). \tag{5.9}$$

Let us use Lemma 2.1, Chap. 2, with $\kappa=\kappa(m)$ and a number δ such that

$$\sup_{u\in\mathcal{K}_m}|D\varphi^m(u)|\le\delta,\qquad\kappa(m)\le\delta^d,$$

and choose $\eta>0$ such that $(1+c\eta)^d\kappa(m)<l$ (this is possible by (5.7)). Let $\varepsilon>0$ be so small that an ε-neighborhood of the compact set $\mathcal{K}_m$ is contained in $\mathcal{D}_m$ and the inequality

$$|\varphi^m(v)-\varphi^m(u)-D\varphi^m(u)(v-u)|\le\eta|v-u| \tag{5.10}$$

is true for all $v\in\mathcal{B}_r(u)$ with $r\le\varepsilon$. The following inclusion

$$\varphi^m\big(\mathcal{B}_r(u)\big) \subset \varphi^m(u) + \mathcal{E} + \mathcal{B}_{\eta r}(0),$$

where $\mathcal{E} := D\varphi^m(u)\mathcal{B}_r(0)$, follows from (5.10). By virtue of Proposition 2.2, Chap. 2, the set $\mathcal{E}$ is an ellipsoid in $\mathbb{R}^n$ whose semi-axes have the lengths $r\alpha_j\big(D\varphi^m(u)\big)$.

Taking into account (5.9), we have

$$\omega_d(\mathcal{E}) = r^d\omega_d\Big(\frac{1}{r}\mathcal{E}\Big) = r^d\omega_d\big(D\varphi^m(u)\big) \leq \kappa(m)r^d.$$

According to Lemma 2.1, Chap. 2, the set $\mathcal{E} + \mathcal{B}_{\eta r}(0)$ is included into an ellipsoid $\mathcal{E}'$ for which

$$\omega_d(\mathcal{E}') \leq (1 + \varkappa\eta)^d\kappa(m)r^d < lr^d.$$

Thus, if $\{\mathcal{B}_{r_j}(u_j)\}$ is a covering of $\mathcal{K}_m$ by balls of radii $r_j \leq \varepsilon$, then we can construct a covering of $\varphi^m(\mathcal{K}_m)$ by ellipsoids with $\big(\omega_d(\mathcal{E}_j')\big)^{1/d} \leq l^{1/d}r_j$ and

$$\sum_j \omega_d(\mathcal{E}_j') \leq l\sum_j r_j^d. \tag{5.11}$$

For an arbitrary compact set $\mathcal{K}' \subset \mathbb{R}^n$ we put

$$\widetilde{\mu}(\mathcal{K}', d, \varepsilon) := \inf\sum_j \omega_d(\mathcal{E}_j),$$

where the infimum is taken over all finite coverings of $\mathcal{K}'$ by ellipsoids $\mathcal{E}_j$, for which $\big(\omega_d(\mathcal{E}_j)\big)^{1/d} \leq \varepsilon$. From (5.6) it follows that

$$\widetilde{\mu}\big(\varphi^m(\mathcal{K}_m), d, l^{1/d}\varepsilon\big) \leq l\mu_H(\mathcal{K}_m, d, \varepsilon). \tag{5.12}$$

From Lemma 5.1 for an arbitrary compact set $\mathcal{K}' \subset \mathbb{R}^n$ we obtain the inequality

$$\mu_H(\mathcal{K}', d, \lambda\varepsilon) \leq c\,\widetilde{\mu}(\mathcal{K}', d, \varepsilon). \tag{5.13}$$

Indeed, for a finite covering of the compact set $\mathcal{K}'$ by ellipsoids $\{\mathcal{E}_j\}$ with $\big(\omega_d(\mathcal{E}_j)\big)^{1/d} \leq \varepsilon$ we have

$$\mu_H(\mathcal{K}', d, \lambda\varepsilon) \leq \mu_H\Big(\bigcup_j \mathcal{E}_j, d, \lambda\varepsilon\Big) \leq \sum_j \mu_H(\mathcal{E}_j, d, \lambda\varepsilon) \leq M\sum_j \omega_d(\mathcal{E}_j).$$

From this it follows that (5.13) is true.

Using (5.13) with $\mathcal{K}' := \varphi^m(\mathcal{K}_m)$ and then (5.12), we obtain

$$\mu_H\big(\varphi^m(\mathcal{K}_m), d, \lambda l^{1/d}\varepsilon\big) \leq M\widetilde{\mu}\big(\varphi^m(\mathcal{K}_m), d, l^{1/d}\varepsilon\big) \leq Ml\mu_H(\mathcal{K}_m, d, \varepsilon). \tag{5.14}$$

Suppose that $\mu_{H}(\mathcal{K}_m, d) \le \mu_0 < \infty$ *for* $m = 1, 2, \ldots$ *. Passing to the limit in (5.14) as* $\varepsilon \to 0$*, we obtain*

$$\mu_{H}\big(\varphi^m(\mathcal{K}_m), d\big) \le Ml\mu_{H}(\mathcal{K}_m, d) \le Ml\mu_0. \tag{5.15}$$

From (5.7) it follows that for sufficiently large m the number l and, consequently, the right-hand side of (5.15) will be as small as desired. So, assertion (a) is proved.

To prove assertion (b), let us assume that an arbitrary number l satisfies the conditions $\lambda l^{1/d} < 1$ *and* $Ml < 1$*. In this case* $\mu_{H}(\mathcal{K}_m, d, \varepsilon) \le \mu_{H}(\mathcal{K}_m, d, \lambda l^{1/d}\varepsilon)$*, and from (5.14) we obtain the inequality*

$$\mu_{H}\big(\varphi^m(\mathcal{K}_m), d, \lambda l^{1/d}\varepsilon\big) \le Ml\mu_{H}(\mathcal{K}_m, d, \lambda l^{1/d}\varepsilon).$$

Taking into account the inclusion $\mathcal{K}_m \subset \mathcal{K} \subset \varphi^m(\mathcal{K}_m)$ *for* $m = 1, 2, \ldots,$ *from the last inequality we can see that*

$$\mu_{H}(\mathcal{K}, d, \lambda l^{1/d}\varepsilon) \le Ml\mu_{H}(\mathcal{K}, d, \lambda l^{1/d}\varepsilon).$$

Therefore $\mu_{H}(\mathcal{K}, d, \lambda l^{1/d}\varepsilon) = 0$ *and, consequently,* $\mu_{H}\big(\varphi^m(\mathcal{K}_m), d, \lambda l^{1/d}\varepsilon\big) = 0$*. Passing to the limit as* $\varepsilon \to 0$ *in these two equalities, we get* $\mu_{H}\big(\varphi^m(\mathcal{K}_m), d\big) = 0$ *and* $\mu_{H}(\mathcal{K}, d) = 0$*. From the last equality it follows that* $\dim_{H} \mathcal{K} \le d$*. Note that inequality (5.3) is also satisfied for* $d - \varepsilon$ *with* $\varepsilon > 0$ *sufficiently small. This implies that* $\dim_{H} \mathcal{K} \le d - \varepsilon$*, i.e.* $\dim_{H} \mathcal{K} < d$*.* □

5.1.2 Corollaries of the Limit Theorem for Hausdorff Measures

In this subsection we prove three corollaries from Theorem 5.1 originating from [9].

Corollary 5.1 *Suppose that for (5.1) on a compact set* $\mathcal{K} \subset \mathcal{U}$ *there exists a continuous positive function* $\varkappa : \mathcal{K} \to \mathbb{R}_+$ *and a number* $d \in [0, n]$ *such that*

$$\sup_{u \in \mathcal{K}} \left[\frac{\varkappa\big(\varphi(u)\big)}{\varkappa(u)} \omega_d\big(D\varphi(u)\big) \right] < 1.$$

Then it holds:

(a) If $\mu_{H}(\mathcal{K}, d) < \infty$*, then* $\lim_{m \to +\infty} \mu_{H}\big(\varphi^m(\mathcal{K}), d\big) = 0$*;*
(b) If $\mathcal{K} \subset \varphi(\mathcal{K})$*, then* $\dim_{H} \mathcal{K} < d$*.*

Proof For the proof it is sufficient to put in Theorem 5.1 $\widetilde{\mathcal{K}} := \mathcal{K}$ and to define recurrently the sequences of sets $\{\mathcal{U}_m\}$ and $\{\mathcal{K}_m\}$ by

$$\mathcal{U}_1 := \mathcal{U},\ \mathcal{U}_{m+1} := \varphi^{-1}(\mathcal{U}_m),\ \mathcal{K}_1 := \mathcal{K},\ \mathcal{K}_{m+1} := \mathcal{K} \cap \varphi^{-1}(\mathcal{K}_m),\ m = 1, 2, \ldots .$$

Here $\varphi^{-1}(\cdot)$ denotes the preimage of a given set under the map φ. □

In the following corollaries we denote by $\alpha_1(u) \geq \alpha_2(u) \geq \cdots \geq \alpha_n(u)$ the singular values of $D\varphi(u)$.

Corollary 5.2 *Suppose for (5.1) that $\mathcal{U}$ is an open simply connected domain in $\mathbb{R}^n$ and there exist a compact set $\widetilde{\mathcal{K}}$ and a continuous positive on $\widetilde{\mathcal{K}}$ function $\varkappa : \widetilde{\mathcal{K}} \to \mathbb{R}_+$ such that the relations*

$$\varphi(\mathcal{U}) \subset \widetilde{\mathcal{K}} \subset \mathcal{U}, \tag{5.16}$$

$$\alpha_1(u)\alpha_2(u) < \frac{\varkappa(u)}{\varkappa\big(\varphi(u)\big)}, \quad \forall u \in \widetilde{\mathcal{K}}. \tag{5.17}$$

are true.

Then in $\mathcal{U}$ there can not exist any smooth and closed curve Γ which satisfies $\varphi(\Gamma) = \Gamma$.

In order to prove Corollary 5.2 we need the following lemma which is connected with the existence problem of minimal surfaces spanning a given curve, the so-called *Plateau problem* [16, 23, 42].

Lemma 5.2 *Given a closed smooth curve Γ in $\mathbb{R}^n (n \geq 3)$. Then there exists a rectifiable surface $\mathcal{S}$ of finite Lebesgue area which spans it.*

Proof (Sketch) Suppose that $\gamma : [0, 1] \to \mathbb{R}^n$ is a smooth parameterization of Γ such that $\gamma(0) = \gamma(1)$ and $\gamma([0, 1]) = \Gamma$. Choose a point $z \notin \Gamma$ and connect any point of Γ with z by the segment of a straight line. In the result we get a piecewise smooth surface $\mathcal{S}$ with finite two-dimensional Lebesgue measure since the lengths of segments connecting Γ with z are bounded from above by a constant and the length of Γ is finite. Note that $\mathcal{S}$ can have self-intersections. It is clear that the piecewise smoothness of $\mathcal{S}$ is preserved. □

Proof of Corollary 5.2 Suppose the opposite, i.e. suppose that there exists such a curve Γ. Let us span on Γ some piecewise smooth surface $\mathcal{S} \equiv \mathcal{K} \subset \mathcal{U}$ of finite area with $\mu_H(\mathcal{S}, 2) < \infty$. The existence of such a surface is guaranteed by Lemma 5.2. Since $\varphi^m(\Gamma) = \Gamma$ for $m = 1, 2, \ldots$ we have

$$\inf_{m \geq 0} \mu_{_H}\big(\varphi^m(\mathcal{K}), 2\big) > 0. \tag{5.18}$$

On the other hand by Theorem 5.1 with $\mathcal{U}_m := \mathcal{U}$ and $\mathcal{K}_m := \varphi(\mathcal{K})$ we obtain

$$\lim_{m \to +\infty} \mu_{_H}\big(\varphi^m(\mathcal{K}), 2\big) = 0,$$

which contradicts the inequality (5.18) □

The following corollary is a generalization of a result in [40].

Corollary 5.3 *Suppose that $\mathcal{D}$ is an open simply connected domain and $\varphi : \mathcal{D} \to \mathcal{D}$ is an analytic map. Suppose also that there exist a compact set $\widetilde{\mathcal{K}}$ and a continuous on $\widetilde{\mathcal{K}}$ positive function $\varkappa : \widetilde{\mathcal{K}} \to \mathbb{R}_+$ such that*

$$\varphi(\mathcal{D}) \subset \widetilde{\mathcal{K}} \subset \mathcal{D}, \tag{5.19}$$

$$\alpha_1(u)\alpha_2(u) < \frac{\varkappa(u)}{\varkappa(\varphi(u))}, \quad \forall u \in \widetilde{\mathcal{K}}. \tag{5.20}$$

Then the map φ has in $\mathcal{D}$ at most a finite number of fixed points.

Proof First of all notice that by Corollary 5.2 in $\mathcal{D}$ there is no smooth closed curve Γ satisfying the relation $\varphi(\Gamma) = \Gamma$.

Let us suppose now that the conclusion is not valid, i.e. that $\mathcal{D}$ contains an infinite sequence $\{u_m\}$ of different fixed points of φ. By (5.19) we have for all $u_m \in \widetilde{\mathcal{K}}$. Since $\widetilde{\mathcal{K}}$ is a compact set, there exists a limit point υ of the sequence $\{u_m\}$ and $\upsilon - \varphi(\upsilon) = \lim_{m\to\infty}\big[u_m - \varphi(u_m)\big] = 0$ (we remain the old indices). Since $u - \varphi(u)$ is an analytic function we see that in the neighborhood of υ the representation

$$u - \varphi(u) = (I - D\varphi(\upsilon))\,(u - \upsilon) + h(u), \tag{5.21}$$

is possible, where $h(u)$ is analytic in $\mathcal{D}$ and $\|h(u)\| = O\big(\|u - \upsilon\|^2\big)$ as $u \to \upsilon$.

Denote by $\lambda_1 = \lambda_1(\upsilon), \dots, \lambda_n = \lambda_n(\upsilon)$ the eigenvalues of $D\varphi(\upsilon)$ numbered so that $|\lambda_1| \ge |\lambda_2| \ge \cdots \ge |\lambda_n|$. Using Weyl's inequality (Proposition 2.6, Chap. 2)

$$|\lambda_1 \dots \lambda_k| \le \alpha_1(\upsilon) \dots \alpha_k(\upsilon), \quad k = 1, \dots, n$$

and inequality (5.20) we obtain

$$|\lambda_1\lambda_2| < \frac{\varkappa(\upsilon)}{\varkappa(\varphi(\upsilon))} = 1.$$

If the number $|\lambda_1|$ is less than 1, then the fixed point υ is asymptotically stable for the iterations of φ and, consequently, it is an isolated point. Since υ is not an isolated one, we have $|\lambda_1| \ge 1 > |\lambda_2|$. Thus, it follows that λ_1 is a real number. By Jordan's theorem there exists a regular real $n \times n$ matrix M such that

$$M^{-1} D\varphi(\upsilon) M = \operatorname{diag}(\lambda_1, C), \tag{5.22}$$

where C is a real $(n-1) \times (n-1)$ matrix with eigenvalues all have modulus smaller than 1. In order to find all fixed points of the map φ in the neighborhood of υ let us perform in (5.21) the change $u = \upsilon + M\,(x, y)^T$ with $x \in \mathbb{R}$ and $y \in \mathbb{R}^{n-1}$. Using (5.22), we replace the fixpoint equation $u - \varphi(u) = 0$ by the pair of equations

$$x - \lambda_1 x + h_1(x, y) = 0, \quad (I - C)y + h_2(x, y) = 0, \tag{5.23}$$

where h_1, h_2 are analytic functions such that $|h_1|, |h_2| = O(|x|^2 + |y|^2)$ as $(x, y) \to (0, 0)$. Since $\det(I - C) \neq 0$ from the implicit function theorem it follows that the second equation in (5.23) has a solution $y(x)$, which is analytic in some interval $x \in (-\delta, \delta)$, and $y(0) = 0$. Therefore each solution of system (5.23) in a neighborhood of $(0, 0)$ can be represented in the form $\big(x, y(x)\big)$, where in a neighborhood of 0 the number x is a root of the equation

$$x - \lambda_1 x + h_1\big(x, y(x)\big) = 0. \tag{5.24}$$

Since υ is a non-isolated fixed point and the right-hand side of (5.24) is analytic for $x \in (-\delta, \delta)$ it is identically equal to zero in all the interval. Thus, the solution of system (5.23) in a neighborhood of $(0, 0)$ is an analytic arc $\big(x, y(x)\big)$ with $x \in (-\delta, \delta)$.

Thus, it follows that any non-isolated fixed point υ of the map φ lies on an analytic curve passing through υ and entirely consists of fixed points of φ. Moreover all fixed points in the neighborhood of υ belong to this curve. Let Γ be one of these analytic curves. Suppose that Γ is not the part of a bigger curve of such kind. It is obvious that all limit points Γ belong to it. Consequently Γ is a compact set. It follows by Milnor's theorem (Theorem A.2, Appendix A) that a smooth one-dimensional simply connected manifold is diffeomorphic either to an interval of the real axis $\mathbb{R}$ or to the circle $\mathcal{S}^1$. Consequently Γ is a closed curve. But any point Γ is a fixed point of φ. It means that $\varphi(\Gamma) = \Gamma$. But this contradicts to the fact which was stated at the beginning of the proof. □

The following results (Lemma 5.3 and Theorem 5.2) were obtained by Smith [40].

Lemma 5.3 *For each simple closed contour $\gamma \subset \mathbb{R}^n$ there exist a plane $\mathbb{E} \subset \mathbb{R}^n$ and a closed circular disc $\mathbb{D} \subset \mathbb{E}$ such that $\mathbb{D} \cap \pi\gamma$ is a simple arc joining distinct points p and q on the boundary $\partial\mathbb{D}$ of $\mathbb{D}$, where $\pi : \mathbb{R}^n \to \mathbb{E}$ denotes the orthogonal projection on $\mathbb{E}$.*

Proof Let $\upsilon = u_2 - u_1$, where u_1, u_2 are distinct points on γ. If $c \in \mathbb{R}$ then $\mathbb{L}_c = \{u \in \mathbb{R}^n | (\upsilon, u) = c\}$ is a hyperplane in $\mathbb{R}^n$ which is orthogonal to υ with respect to the usual inner product in $\mathbb{R}^n$. If $c_1 = (\upsilon, u_1)$ and $c_2 = (\upsilon, u_2)$, then $u_1 \in \mathbb{L}_{c_1}$, $u_2 \in \mathbb{L}_{c_2}$ and $c_2 - c_1 = |\upsilon|^2 > 0$. Since γ is connected, the set $\gamma \cap \mathbb{L}_c$ is non-empty when $c_1 \le c \le c_2$. If $\gamma \cap \mathbb{L}_c$ is an infinite set, then $\mathbb{L}_c$ is tangential to one of the regular arcs constituting γ. The values of c for which this property is satisfied form a set of Lebesgue measure zero by Sard's theorem (Theorem A.3, Appendix A). We can therefore choose $\overline{c}$ with $c_1 < \overline{c} < c_2$ such that $\gamma \cap \mathbb{L}_{\overline{c}}$ is a finite set. At each point p of this finite set we can assume that γ has a tangent vector $\tau(p)$ with $(\upsilon, \tau(p)) \neq 0$.

Choose a point p_0 in the finite set $\gamma \cap \mathbb{L}_{\overline{c}}$ and choose an $(n-1)$-flat $\mathbb{F} \subset \mathbb{L}_\tau$ such that $p_0 \in \mathbb{F}$ and $p \notin \mathbb{F}$ for all other points p in $\gamma \cap \mathbb{L}_{\overline{c}}$. If $\mathbb{E}$ is a plane in $\mathbb{R}^n$ orthogonal to $\mathbb{F}$ and $\pi : \mathbb{R}^n \to \mathbb{E}$ is the orthogonal projection, then $\pi p_0 \in \pi\gamma$ and $\pi p \neq \pi p_0$ for all points $p \neq p_0$ on γ. Since $(\upsilon, \tau(p_0)) \neq 0$, the plane curve $\pi\gamma$ has a unique tangent line $\mathbb{L}_0$ at the point πp_0. We can choose a small open arc $\gamma_0 \subset \gamma$ such that $\pi\gamma_0$ is approximately equal to a line segment of $\mathbb{L}_0$ which has πp_0 as its centre. Since $\gamma' = \gamma \setminus \gamma_0$ is closed and $\pi p_0 \in \pi\gamma'$ we can choose a small circular disc $\mathbb{D} \subset \mathbb{E}$ with centre at πp_0 such that $\mathbb{D} \cap \pi\gamma'$ is empty. Then $\mathbb{D} \cap \pi\gamma = \mathbb{D} \cap \pi\gamma_0$ and this is a simple are joining distinct points p, q on $\partial\mathbb{D}$. □

Remark 5.2 In Lemma 5.3 sufficient smoothness of the simple closed contour, i.e. of a parameterized 1-boundary in $\mathbb{R}^n$, is assumed. This allows the use of Sard's theorem to show the existence of an orthogonal projection onto a plane. This plays a role similar to that of the rigid rotation R used in the proof of Proposition B.3, Appendix B.

We now come to a central result which connects for a given map an upper Hausdorff dimension bound for invariant sets with the non-existence of invariant closed curves. The direct proof given below uses Lemma 5.3 which is based on Sard's theorem and properties of the winding number of a vector field in the plane. We shall note that this result also follows from the more general result of Muldowney and Li (Theorem 10.12, Chap. 10).

Theorem 5.2 *Suppose that $\mathcal{U} \subset \mathbb{R}^n$ is simply-connected and $\varphi : \mathcal{U} \to \mathcal{U}$ is a continuous map which satisfies $\varphi(\mathcal{U}) \subset \mathcal{K}_0 \subset \mathcal{U}$, where $\mathcal{K}_0$ is a compact set. If the maximal invariant set $\mathcal{K}_1$ of φ in $\mathcal{U}$ has Hausdorff dimension $\dim_H \mathcal{K}_1 < 2$, then $\varphi(\gamma) \neq \gamma$ for every simple closed contour $\gamma \subset \mathcal{U}$, supposed that φ is one-to-one on γ.*

Proof Assume that the theorem is false, i.e. suppose that $\varphi(\gamma) = \gamma$ for some simple closed contour $\gamma \subset \mathcal{U}$. With this γ, Lemma 5.3 associates an orthogonal projection $\pi : \mathbb{R}^n \to \mathbb{E}$ and a circular disc $\mathbb{D} \subset \mathbb{E}$. For each point q in $\mathbb{E} \setminus \pi\gamma$ let $w(q)$ the winding number about q (see Sect. B.3, Appendix B) of the projected vector field along the closed plane curve $\pi\gamma$. Then $w(q)$ is an integer which remains constant as q varies over any connected component of $\mathbb{E} \setminus \pi\gamma$. In particular $w(q)$ is constant in each of the components $\mathbb{D}_1, \mathbb{D}_2$ of $\mathbb{D} \setminus \pi\gamma$. Furthermore $w(q_2) = w(q_1) \pm 1$ when $q_1 \in \mathbb{D}_1$ and $q_2 \in \mathbb{D}_2$. Hence $w(q) \neq 0$ in at least one of the components $\mathbb{D}_1, \mathbb{D}_2$.

Since $\dim_H \mathcal{K}_1 < 2$, we have $\dim_H \pi\mathcal{K}_1 < 2$ and therefore neither $\mathbb{D}_1$, nor $\mathbb{D}_2$ lies wholly within $\pi\mathcal{K}_1$. Hence, a point q_0 can be chosen in $\mathbb{D}_1 \cup \mathbb{D}_2$ such that $q_0 \notin \pi\mathcal{K}_1$ and $w(q_0) \neq 0$. Then $q_0 \notin \pi\varphi^j(\mathcal{K}_0)$ for some integer $j \geqq 0$ because $\pi\mathcal{K}_1 = \bigcap_{j\geq 0} \pi\varphi^j\mathcal{K}_0$.

Since γ can be continuously contracted to a point within the simply-connected set $\mathcal{U}$, the plane curve $\pi\varphi^{k+1}\gamma$ can be contracted to a point within the set $\pi\varphi^{k+1}(\mathcal{U})$. Since $\gamma = \varphi^{k+1}(\gamma)$ and $\varphi(\mathcal{U}) \subset \mathcal{K}_0$ it follows that $\pi\gamma$ can be contracted to a point within $\pi\varphi^k(\mathcal{K}_0)$. Since $q_0 \notin \pi\varphi^k(\mathcal{K}_0)$ the winding number about q_0 of the contracting curve remains constant throughout this continuous deformation. Its initial value is $w(q_0)$ and its final value is zero because $\pi\gamma$ is contracted to a point distinct from q_0. However, q_0 was chosen so that $w(q_0) \neq 0$. □

5.1.3 Application of the Limit Theorem to the Hénon Map

Consider the Hénon map ([4, 21, 27, 35, 37]) $\varphi : \mathbb{R}^2 \to \mathbb{R}^2$ given by

$$(x, y) \in \mathbb{R}^2 \longmapsto (a + by - x^2, x), \tag{5.25}$$

where $a > 0$ and $0 < b < 1$ are parameters.

Suppose that $\mathcal{K} \subset \mathbb{R}^2$ is a compact set, which is invariant with respect to $\{\varphi^m\}$. In order to get Hausdorff dimension estimates for $\mathcal{K}$, we want to use Corollary 5.1. This corollary says that $\dim_H \mathcal{K} < 1 + s$ for some $s \in [0, 1]$, if

$$\alpha_1(u)\,\alpha_2^s(u)\frac{\varkappa(\varphi(u))}{\varkappa(u)} < 1, \qquad \forall\, u \in \mathcal{K}, \tag{5.26}$$

where $\alpha_1(u) \geq \alpha_2(u)$ are the singular values of the differential of φ at $u = (x, y)$ and $\varkappa(u)$ is a continuous positive scalar function on $\mathcal{K}$. We demonstrate two variants: the first employs the function $\varkappa(u) \equiv 1$, the second is based on a function $\varkappa(u) \not\equiv 1$.

It is easy to see that the Jacobian matrix of φ at an arbitrary point $u = (x, y)$ is given by

$$J(x, y) = \begin{pmatrix} -2x & b \\ 1 & 0 \end{pmatrix} \tag{5.27}$$

and admits the maximal singular value

$$\alpha_1(u) \equiv \alpha_1(x) = \frac{1}{2}\left[\sqrt{4x^2 + (1+b)^2} + \sqrt{4x^2 + (1-b)^2}\right].$$

Since $\alpha_1(x)\,\alpha_2(x) \equiv |\det J(x, y)| = b$ we can write for the second singular value

$$\alpha_2(x) = \frac{b}{\alpha_1(x)}. \tag{5.28}$$

Suppose that $\delta > 0$ is a number such that

$$\mathcal{K} \subset \{(x, y) \big| |x| \leq \delta\}. \tag{5.29}$$

The following results (Theorems 5.3 and 5.4) have been obtained in [12].

Theorem 5.3 *Suppose that $\mathcal{K}$ is a compact and invariant with respect to $\{\varphi^m\}$ set and the inclusion (5.29) is satisfied. Then*

$$\dim_H \mathcal{K} < 1 + \frac{1}{1 - \log b / \log \alpha_1(\delta)}.$$

Proof Under consideration of (5.28) and (5.29) the condition (5.26) is satisfied if

$$b^s \alpha_1^{1-s}(x) < 1, \quad \forall x \in [-\delta, \delta].$$

Since the function $\alpha_1(x)$ takes on $[-\delta, \delta]$ the maximum for $|x| = \delta$, the last inequality is satisfied if

$$b^s \alpha_1^{1-s}(\delta) < 1. \tag{5.30}$$

Using the fact that $\alpha_1(x) \geq 0$ for all x, this is equivalent to the inequality

$$s > \frac{\log \alpha_1(\delta)}{\log \alpha_1(\delta) - \log b}.$$

□

Remark 5.3 The Hénon map was investigated numerically in [27]. It follows from the results of this paper that for $a = 1.4$, $b = 0.3$ the parameter δ in (5.29) can be taken as $\delta = 1.8$. Theorem 5.3 gives with such a δ the estimate $\dim_H \mathcal{K} \leq 1.523$.

Our next Hausdorff dimension bound for a compact invariant set of the Hénon map improves the result of Theorem 5.3 and does not need any information from numerical experiments.

Let us introduce the values

$$\tau := \frac{2}{1 - b + \sqrt{(1-b)^2 + 4a}}, \quad \kappa := \frac{-4\tau \log(a\tau^2)}{(1-b)[4 + (1+b)^2\tau^2]}.$$

Theorem 5.4 *Suppose that $\mathcal{K}$ is an invariant compact set for the Hénon map and the inequality*

$$\kappa \leq \min\left\{\frac{4}{(1+b)^2}, \frac{\tau}{1 - b + \tau b}\right\} \tag{5.31}$$

is satisfied. Then

$$\dim_H \mathcal{K} \leq 1 + \frac{1}{1 - 2\log b / M},$$

where $M := 2\log 2 - \log \kappa - 1 + \kappa\left(\frac{1-b}{\tau} + \frac{(1+b)^2}{4} + a\right)$.

Proof Since $a\tau^2 < 1$ we have $\kappa > 0$. Let $s \in (0, 1)$ be arbitrary and define for $u = (x, y) \in \mathbb{R}^2$ the scalar function

$$\varkappa(u) := e^{(1-s)\frac{\kappa}{2}(x+by)}.$$

A direct calculation shows that

$$\frac{\varkappa(\varphi(u))}{\varkappa(u)} = e^{(1-s)\frac{\kappa}{2}[-x^2-(1-b)x+a]}$$

and the condition (5.26) with this function has the form

$$b^s 2^{-(1-s)} \left[\sqrt{4x^2+(1+b)^2}+\sqrt{4x^2+(1-b)^2}\right]^{1-s} e^{(1-s)\frac{\kappa}{2}[-x^2-(1-b)x+a]} < 1$$

or

$$\begin{aligned} &s\log b-(1-s)\log 2+(1-s)\\ &\times\left\{\log\left[\sqrt{4x^2+(1+b)^2}+\sqrt{4x^2+(1-b)^2}\right]+\frac{\kappa}{2}[-x^2-(1-b)x+a]\right\} < 0\,. \end{aligned} \tag{5.32}$$

The last inequality is satisfied if

$$s\log b+(1-s)\frac{1}{2}\,h\,(x)<0,\quad \forall x\in \operatorname{Proj}\mathcal{K},$$

where Proj denotes the projection on the x-axis and

$$h(x):=\log\,[4x^2+(1+b)^2]-\kappa[x^2+(1-b)x-a].$$

It follows from Corollary 5.1 that the theorem is proved if $h(x)\le M$, $\forall x\in\operatorname{Proj}\mathcal{K}$. We want to show that in fact this inequality is satisfied for all $x\in\mathbb{R}$. Let us consider the following three cases.

Case 1: Suppose $x\in[0,+\infty)$. It follows that $4x^2<4\left(x+\frac{1-b}{2}\right)^2$ and $h(x)<h_1(x)$, where

$$h_1(x):=\log\left[4\left(x+\frac{1-b}{2}\right)^2+(1+b)^2\right]-\kappa\left[\left(x+\frac{1-b}{2}\right)^2-\frac{(1-b)^2}{4}-a\right].$$

Since

$$h_1'(x)=\frac{8\left(x+\frac{1-b}{2}\right)}{4\left(x+\frac{1-b}{2}\right)^2+(1+b)^2}-2\kappa\left(x+\frac{1-b}{2}\right)$$

the points with $h_1'(x)=0$ satisfy the equation

$$\left(x+\frac{1-b}{2}\right)\left\{4-\kappa\left[4\left(x+\frac{1-b}{2}\right)^2+(1+b)^2\right]\right\}=0.$$

Under the condition (5.30) is $\frac{1}{\kappa}-(1+b)^2/4\ge 0$. It follows that the function $h_1(x)$ has the global maximum on $\mathbb{R}$ for x satisfying

$$\left(x+\frac{1-b}{2}\right)^2=\frac{1}{\kappa}-\frac{(1+b)^2}{4}.$$

Consequently, $h_1(x) \le 2\log 2 - \log\kappa - 1 + \kappa\left(\frac{1+b^2}{2} + a\right)$. From $\frac{1}{\tau} > (1-b)/4$ it follows that

$$\frac{1+b^2}{2} + a \le \frac{1-b}{\tau} + \frac{(1+b)^2}{4} + a.$$

This shows that $h_1(x) \le M, \forall x \in \mathbb{R}$, so that $h(x) \le M,\ \forall x \in [0, +\infty)$.

Case 2: Suppose $x \in [-1/\tau, 0)$. It follows that

$$4x^2 + (1+b)^2 = 4\left(x + \frac{1-b}{2}\right)^2 - 4(1-b)x - (1-b)^2 + (1+b)^2$$
$$\le 4\left(x + \frac{1-b}{2}\right)^2 + 4\left(\frac{1-b}{\tau} + b\right).$$

This demonstrates that $h(x) \le h_2(x)$, where

$$h_2(x) := \log\left[4\left(x + \frac{1-b}{2}\right)^2 + 4B\right] - \kappa\left[\left(x + \frac{1-b}{2}\right)^2 - \frac{(1-b)^2}{4} - a\right]$$

and $B := (1-b)/\tau + b$.

Because of

$$h_2'(x) = \frac{8\left(x + \frac{1-b}{2}\right)}{4\left(x + \frac{1-b}{2}\right)^2 + 4B} - 2\kappa\left(x + \frac{1-b}{2}\right),$$

the points with $h_2'(x) = 0$ satisfy

$$\left(x + \frac{1-b}{2}\right)\left\{1 - \kappa\left[\left(x + \frac{1-b}{2}\right)^2 + B\right]\right\} = 0.$$

Because of (5.31), the inequality $\frac{1}{\kappa} - B \ge 0$ holds. From this it follows that $h_2(x)$ is maximal on $\mathbb{R}$ for such $x \in \mathbb{R}$ satisfying

$$\left(x + \frac{1-b}{2}\right)^2 - \frac{1}{\kappa} = B.$$

This implies

$$h_2(x) \le 2\log 2 \log\kappa - 1 + \kappa(B + \frac{(1-b)^2}{4} + a)$$

for all x and $h_2(x) \le M,\ \forall x \in \mathbb{R}$. Using this we get $h(x) \le M,\ \forall x \in [-\frac{1}{\tau}, 0)$.

Case 3: Suppose $x \in (-\infty, -\frac{1}{\tau})$. It follows that $x > -\tau x^2$ and

$$x^2 + (1-b)x > [1 - \tau(1-b)]x^2.$$

With this we get for all $x \in \mathbb{R}$ $h(x) \le h_3(x)$, where

$$h_3(x) := \log[4x^2 + (1+b)^2] - \kappa(Cx^2 - a)$$

and $C := 1 - \tau(1-b) = a\tau^2$. Since

$$h_3'(x) = \frac{8x}{4x^2 + (1+b)^2} - 2\kappa C x,$$

all points with $h_3'(x) = 0$ satisfy

$$x\{4 - \kappa C\left[4x^2 + (1+b)^2\right]\} = 0.$$

Employing $C = a\tau^2 < 1$, we conclude from (5.31) that $4/(\kappa C) - (1+b)^2 \ge 0$. This guaranties that $h_3(x)$ is maximal on $\mathbb{R}$ for x satisfying

$$x^2 = \frac{1}{\kappa C} - \frac{(1+b)^2}{4}.$$

This means that

$$h_3(x) \le 2\log 2 - \log\kappa - 1 - \log C + \kappa\left(C\frac{(1+b)^2}{4} + a\right).$$

Using the representation of κ through τ we see, that the right-hand side of the last inequality is equal to M. It follows that $h_3(x) \le M, \forall x \in \mathbb{R}$, and, consequently, $h(x) \le M, \forall x \in (-\infty, -1/\tau)$. □

Consider in (5.25) again the parameters $a = 1.4, b = 0.3$. A direct calculation gives $\tau = 0.63$ and $\kappa = 0.450$. Since (5.30) is satisfied, we conclude on the base of Theorem 5.4 that $\dim_H \mathcal{K} < 1.510$. Suppose that $(x_\pm, y_\pm)$ are the fixed points of φ. The first coordinates are given by the roots of

$$x^2 + (1-b)x - a = 0.$$

Thus we have

$$x_\pm = \frac{1}{2}\left[-(1-b) \pm \sqrt{(1-b)^2 + 4a}\right].$$

If the compact and invariant set $\mathcal{K}$ is a global attractor, all fixed points are contained in $\mathcal{K}$. However for these points u we have

$$\frac{\varkappa(\varphi(u))}{\varkappa(u)} = 1.$$

As a consequence we see that, independently of the choice of $\varkappa(u)$ in (5.26), there is no better estimate than

$$\dim_H \mathcal{K} \le 1 + \frac{1}{1 - \log b / \log \alpha_1(x_-)}. \tag{5.33}$$

For the values $a = 1.4$, $b = 0.3$ we get from (5.33) $\dim_H \mathcal{K} \le 1.500\ldots$.

Note that the Hausdorff dimension bounds from Theorems 5.3 and 5.4 are also bounds for the fractal dimension. This fact follows from the equality (5.28) which is in fact Chen's condition (see Sect. 5.4).

5.2 The Application of the Limit Theorem to ODE's

5.2.1 *An Auxiliary Result*

Consider the system

$$\frac{d\varphi}{dt} = f(t, \varphi) \tag{5.34}$$

in $\mathcal{J} \times \mathcal{U}$. Here $\mathcal{J} \subset \mathbb{R}$ is an open interval containing $\mathbb{R}_+$, $\mathcal{U} \subset \mathbb{R}^n$ is an open set and $f : \mathcal{J} \times \mathcal{U} \to \mathbb{R}^n$ is a continuously differentiable function. Let

$$\lambda_1(t, u) \ge \lambda_2(t, u) \ge \cdots \ge \lambda_n(t, u)$$

be the eigenvalues of the symmetrized Jacobian matrix of the right-hand side of (5.34), i.e. of the matrix $[D_2 f(t, u) + D_2 f(t, u)^*]$ at a point $(t, u) \in \mathcal{J} \times \mathcal{U}$.

For a continuously differentiable scalar function $V : \mathcal{U} \to \mathbb{R}$ we shall use the derivative of V w.r.t. (5.34)

$$\dot{V}(t, u) = (f(t, u), \operatorname{grad} V(u)) .$$

For a number $d = d_0 + s$ with $d_0 \in [0, n-1]$ integer and $s \in [0, 1]$ and a logarithmic Λ we introduce the *partial d-trace w.r.t.* Λ of $D_2 f : \mathcal{J} \times \mathcal{U} \to \mathbb{R}$ by

$$\operatorname{tr}_{d,\Lambda} D_2 f(t, u) := s\Lambda\big(D_2 f(t, u)^{[d_0+1]}\big) + (1 - s)\Lambda\big(D_2 f(t, u)^{[d_0]}\big).$$

Suppose that there exist a number $\tau > 0$ and an open set $\mathcal{U}_0$ ($\overline{\mathcal{U}}_0 \subset \mathcal{U}$) satisfying the following condition: if $u \in \overline{\mathcal{U}}_0$, then the solution $\varphi(\cdot, u)$ of (5.34) with $\varphi(0, u) = u$ satisfies $\varphi(t, u) \in \mathcal{U}$ for $t \in [0, \tau]$. Denote by $\varphi^\tau : \overline{\mathcal{U}}_0 \to \mathcal{U}$ the time-τ-map of system (5.34) which is defined by $\varphi^\tau(u) := \varphi(\tau, u)$.

The following lemma [11] will be used to introduce logarithmic norms and Lyapunov functions into dimension estimates.

Lemma 5.4 *Let V be a continuously differentiable scalar function in the domain $\mathcal{U}$, $d = d_0 + s$, with integer $d_0 \in [0,\ n-1]$ and $s \in (0,\ 1]$, and Λ be some logarithmic norm. Then there exists a continuous positive on $\mathcal{U}$ function $\varkappa$ which satisfies the inequality*

$$\frac{\varkappa\big(\varphi^\tau(u)\big)}{\varkappa(u)}\omega_d\big(D\varphi^\tau(u)\big) \\ \leq \int_0^\tau \big[\mathrm{tr}_{d,\Lambda} D_2 f(t, \varphi(t,u)) + \dot{V}(t, \varphi(t,u))\big]\,dt, \quad \forall u \in \mathcal{U}. \tag{5.35}$$

Proof The derivative $D\varphi^t$ of the map φ^t is the Cauchy matrix solution of the system

$$\frac{dv}{dt} = D_2 f(t, \varphi(t,u))v\,.$$

Therefore by Proposition 2.18, Chap. 2, we have with a continuous positive function $\beta_1(\cdot; k)$

$$\omega_k\big(D\varphi^\tau(u)\big) \leq \beta_1\big(\varphi(\tau,u); k\big) \exp \int_0^\tau \Lambda\big(D_2 f(t, \varphi(t,u))^{[k]}\big)\,dt,$$

$k = 1, 2, \ldots, n$. (The fact that $\beta_1(\cdot; k)$ is continuous as function of the phase variable follows from the construction of β_1 in the proof of Proposition 2.18, Chap. 2). Let us take $\varkappa(u) := \kappa(u)^{-1} \exp V(u)$, where

$$\kappa(u) := \beta_1^{1-s}(u; d_0)\,\beta_1^s(u; d_0+1).$$

To finish the proof, take into account the relations

$$\frac{\varkappa\big(\varphi^\tau(u)\big)}{\varkappa(u)} = \frac{\kappa(u)}{\kappa\big(\varphi^\tau(u)\big)} \exp\big\{V(\varphi^\tau(u)) - V(u)\big\} \\ = \frac{1}{\kappa\big(\varphi(\tau,u)\big)} \exp\Big\{\int_0^\tau \dot{V}\big(t, \varphi(t,u)\big)\,dt\Big\} \tag{5.36}$$

and recall that

$$\omega_d\big(D\varphi^\tau(u)\big) = \omega_{d_0}^{1-s}\big(D\varphi^\tau(u)\big)\omega_{d_0+1}^s\big(D\varphi^\tau(u)\big).$$

□

5.2.2 Estimates of the Hausdorff Measure and of Hausdorff Dimension

Let $\mathcal{K} \subset \mathcal{U}_0$ be a compact set satisfying for the time-τ-map of (5.34) the relation $\mathcal{K} \subset \varphi^\tau(\mathcal{K})$ for some $\tau > 0$. Recall that $\varphi^\tau(\cdot) \equiv \varphi(\tau, \cdot)$ is the solution of (5.34) with $\varphi(0, u) = u$.

By using Lemma 5.4 and Corollary 5.1 from the limit theorem we directly obtain the following theorem [11].

Theorem 5.5 *Suppose that there exist an integer number $d_0 \in [0, n-1]$, a real $s \in (0, 1]$, a logarithmic norm Λ, and a continuously differentiable on $\mathcal{K}$ function V satisfying with $d = d_0 + s$ the inequality*

$$\int_0^\tau \big[\mathrm{tr}_{d,\Lambda} D_2 f(t, \varphi(t, u)) + \dot{V}(t, \varphi(t, u))\big]\, dt < 0, \quad \forall u \in \mathcal{K}.$$

Then $\dim_H \mathcal{K} < d_0 + s$.

Choosing the logarithmic norm defined by the Euclidean vector norm, we obtain

Corollary 5.4 *Suppose that there exist an integer number $d_0 \in [0, n-1]$, a real number $s \in (0, 1]$, and a continuously differentiable on $\mathcal{K}$ function V such*

$$\int_0^\tau \big[\lambda_1(t, \varphi(t, u)) + \cdots + \lambda_{d_0}(t, \varphi(t, u)) + s\lambda_{d_0+1}(t, \varphi(t, u)) \\ + \dot{V}(t, \varphi(t, u))\big] dt < 0, \quad \forall u \in \mathcal{K}.$$

Then $\dim_H \mathcal{K} < d_0 + s$.

The following corollary (see [40]) plays an essential role in the estimation of Hausdorff dimension.

Corollary 5.5 *Suppose that there exist a constant real symmetric positive definite $n \times n$ matrix Q and a continuous function $\Theta : \mathbb{R}_+ \times \mathcal{U} \to \mathbb{R}$ such that*

$$J(t, u)^* Q + Q J(t, u) + 2\Theta(t, u) Q \geq 0, \ \forall (t, u) \in \mathbb{R}_+ \times \mathcal{U}. \tag{5.37}$$

Here $J(t, u) := D_2 f(t, u)$ denotes the Jacobian matrix of f at $(t, u) \in \mathbb{R}_+ \times \mathcal{U}$.

Suppose also that there exists a number $T > 0$, period of f with respect to t, and a number $d \in (0, n]$ such that

$$\int_0^T \big[(n - d)\Theta(\tau, \varphi(\tau, u)) + \mathrm{tr} J(\tau, \varphi(\tau, u)\big]\, d\tau < 0, \ \forall u \in \mathcal{K} \subset \mathcal{U}, \tag{5.38}$$

where $\mathcal{K}$ is an invariant compact set, i.e. $\varphi^T(\mathcal{K}) = \mathcal{K}$.

Then $\dim_H \mathcal{K} < d$.

Proof Consider the map $(\varphi^T)^m := \varphi(mT, \cdot)$. Since $(\varphi^T)^m \mathcal{K} \subset \mathcal{U}$ there exists an open set $\mathcal{U}_m, \mathcal{K} \subset \mathcal{U}_m \subset \mathcal{U}$ with $\varphi(t, u) \in \mathcal{U}$ for all (t, u) in $[0, mT] \times \mathcal{U}_m$. Then $(\varphi^T)^m(u) = \varphi(mT, u) = \varphi^{mT}(u)$, $\forall u \in \mathcal{U}_m$. If we put $A(t) := J(t, \varphi(t, u))$, $\Theta(t) := \Theta(t, \varphi(t, u))$ then (5.37) implies that inequality (2.43) of Corollary 2.10, Chap. 2, holds for $0 \le t \le mT$. Corollary 2.10, Chap. 2, then gives

$$\alpha_1(mT, u)\alpha_2(mT, u) \cdots \alpha_k(mT, u) \tag{5.39}$$
$$\le \lambda_1(Q)^{k/2}\lambda_1(Q^{-1})^{k/2} \exp \int_0^{mT} \big[(n-k)\Theta(\tau, \varphi(\tau, u)) + \operatorname{tr} J(\tau, \varphi(\tau, u))\big] d\tau \,.$$

Write $d = d_0 + s$ with $d_0 \in \{0, \dots, n-1\}$, $s \in (0, 1]$. Then it follows from (5.39) that

$$\alpha_1(mT, u) \cdots \alpha_{d_0}(mT, u)\alpha_{d_0+1}(mT, u)^s$$
$$\le \lambda_1(Q)^{d/2}\lambda_1(Q^{-1})^{d/2} \exp \int_0^{mT} \big[(n-d)\Theta + \operatorname{tr} J\big] d\tau \,. \tag{5.40}$$

Choose

$$\beta := \max_{p \in \mathcal{K}} \int_0^T \big[(n-d)\Theta(\tau, \varphi(\tau, u)) + \operatorname{tr} J(\tau, \varphi(\tau, u)\big] d\tau < 0 \,,$$

which is possible because of (5.38). Then

$$\sup_{u \in \mathcal{K}} \big[\alpha_1(mT, u) \dots \alpha_{d_0}(mT, u)\alpha_{d_0+1}(mT, u)^s\big]$$
$$\le \lambda_1(Q)^{d/2}\lambda_2(Q^{-1})^{d/2} e^{-m\beta} < 1 \,,$$

if m is sufficiently large.

The statement of the corollary follows now from Theorem 5.1. □

Example 5.1 Consider the forced Duffing equation

$$\ddot{x} + 2\delta\dot{x} + x^3 - x = \varepsilon \cos \omega t \,, \tag{5.41}$$

where $\delta, \varepsilon, \omega$ are positive parameters. Introducing the new variable $y := \dot{x} + \delta x$, we get $\dot{y} = \ddot{x} + \delta\dot{x} = (1 + \delta^2)x - x^3 - \delta y + \varepsilon \cos \omega t$. Thus (5.41) is equivalent to the planar system

$$\dot{x} = y - \delta x, \quad \dot{y} = (1 + \delta^2)x - x^3 - \delta y + \varepsilon \cos \omega t \,. \tag{5.42}$$

It was shown in [26] by numerical calculations that for certain parameter values $\delta, \varepsilon, \omega$ there exists a strange attractor in the system generated by (5.42).

Since (5.42) is point-dissipative, there exists a compact invariant set. In the next theorem [40] we give an upper estimate of the Hausdorff dimension of such a set.

Theorem 5.6 *Suppose $\mathcal{K}$ is a compact invariant set of (5.42). Then*

$$\dim_H \mathcal{K} \le 2\,(1 - \delta/\Theta) \tag{5.43}$$

with

$$\left.\begin{aligned} \Theta &:= \delta + 3\big(1+\delta^2+2^{-1/2}c\big)/4a^{1/2}\,, \\ a &:= 1/2(1+\delta^2+2^{-1/2}\,3c\big)\,, \\ c &:= \delta^{-1}\varepsilon + (\delta^{-2}\varepsilon^2 + 2(1+\delta^2)^2)^{1/2}\,. \end{aligned}\right\} \tag{5.44}$$

Proof Let us consider in $\mathbb{R}^2$ the real valued function V given by $V(x, y) := 2y^2 + (1+\delta^2 - x^2)^2$. If $(x(\cdot), y(\cdot))$ is any solution of (5.42), then

$$\begin{aligned} \frac{d}{dt} V(x(t), y(t)) &= 4\,y\varepsilon\cos\omega t - 2\,\delta x^4 - 2\,\delta\big[V(x,y)\big] \\ &= 4\,\varepsilon y\cos\omega t - 4\,\delta\big[y^2 + x^4 - (1+\delta^2)x^2\big] \\ &= 4\,\varepsilon y\cos\omega t - 2\,\delta x^4 - 2\,\delta\big[V(x,y) - (1+\delta^2)^2\big] \\ &\le 2\,\varepsilon[2\,V(x,y)]^{1/2} - 2\,\delta\big[V(x,y) - (1+\delta^2)^2\big] \\ &= \frac{1}{\delta}\varepsilon^2 + 2\,\delta(1+\delta^2)^2 - \delta\left[\frac{1}{\delta}\varepsilon - (2\,V(x,y))^{1/2}\right]^2. \end{aligned}$$

This inequality shows that $\dot{V}(x, y) < 0$ if

$$V(x,y) \ge \frac{1}{2}\left[\,1\delta\varepsilon + (\delta^{-2}\varepsilon^2 + 2(1+\delta^2)^2)^{1/2}\right]^2 = \,12\,c^2\,.$$

Introduce the set $\mathcal{D} := \{(x, y) \in \mathbb{R}^2 \mid V(x, y) \le \frac{1}{2}\,c^2\}$. Then $\mathcal{D}$ is absorbing and positively invariant for (5.42). Hence $\mathcal{D}$ contains every compact invariant set $\mathcal{K}$ of (5.42). If $V(x, y) < \;12\,c^2$ then $(1+\delta^2-x^2)^2 < \;12c^2$ and $0 \le x^2 < 1+\delta^2+2^{-1/2}c$. It follows that

$$(1+\delta^2+2^{-1/2}c - 2x^2)^2 \le (1+\delta^2+2^{-1/2}c)^2\,, \quad \forall (x,y) \in \mathcal{D} \tag{5.45}$$

Suppose Θ is a constant and $Q := \begin{pmatrix} a & 0 \\ 0 & 1 \end{pmatrix}$ is a 2×2 matrix with $a \in \mathbb{R}$.

The Jacobi matrix J of the right-hand side of (5.42) at an arbitrary point $(t, x, y) \in \mathbb{R}_+ \times \mathbb{R}^2$ is given by $J(t, x, y) = \begin{pmatrix} -\delta & 1 \\ 1+\delta^2-3x^2 & -\,\delta \end{pmatrix}$.

It follows that

$$J^*Q + QJ + 2\,\Theta\,Q = \begin{pmatrix} 2\,a(\Theta-\delta) & a+1+\delta^2-3x^2 \\ a+1+\delta^2-3x^2 & 2\,(\Theta-\delta) \end{pmatrix}. \tag{5.46}$$

The Routh-Hurwitz criterion shows that if

$$\Theta > \delta \quad \text{and} \quad 4\,a^2(\Theta-\delta)^2 > (a+1+\delta^2-3x^2)^2\,, \tag{5.47}$$

then the matrix (5.46) is positive-definite.

These conditions are equivalent to (5.37) provided that we choose Θ and a as in (5.44) .

Since $\operatorname{tr} J \equiv -2\,\delta$, the condition (5.38) reduces to $[2-d]\Theta - 2\delta < 0$.

Corollary 5.5 then shows that every compact invariant set $\mathcal{K} \subset \mathcal{D}$ has $\dim_{_H} \mathcal{K} < d$. Hence

$$\dim_{_H} \mathcal{K} \le 2(1-\Theta^{-1}\delta)\,.$$

□

The following simple proposition [11] resulting also from Theorem 5.5 shows the connection between the divergence of the vector field, the zero Lebesgue measure of the considered set, and the Hausdorff dimension of this set.

Theorem 5.7 *Suppose that there exists a positive continuously differentiable on* $\mathcal{K}$ *function* $\varkappa(\cdot)$ *such that*

$$\operatorname{div}(\varkappa f) \equiv \frac{\partial(\varkappa f_1)}{\partial u_1} + \frac{\partial(\varkappa f_2)}{\partial u_2} + \cdots + \frac{\partial(\varkappa f_n)}{\partial u_n} < 0$$

for all $(t,u) \in [0,\ \tau] \times \mathcal{K}$. *Then* $\mathcal{K}$ *has zero Lebesgue measure.*

Proof Since

$$\frac{\partial(\varkappa f_1)}{\partial u_1} + \cdots + \frac{\partial(\varkappa f_n)}{\partial u_n} = \varkappa \operatorname{tr} D_2 f + \dot{\varkappa},$$

the condition of Theorem 5.5 is satisfied with $d_0 = n-1$, $s = 1$ and the auxiliary function $V(u) := \log \varkappa(u)$. Consequently, $\dim_{_H} \mathcal{K} < n$, and by Proposition 3.19, Chap. 3, $\mu_{_L}(\mathcal{K}, n) = 0$.

□

Now we suppose that $\mathcal{U}_0$ is a bounded set, and the relations $\mathcal{K} \subset \overline{\mathcal{U}}_0 \subset \mathcal{U}, \varphi(t,u) \in \mathcal{U}\ \forall u \in \overline{\mathcal{U}}_0$ and $t \in [0,\ \infty)$, $(\varphi^\tau)^m(u) = \varphi(m\tau, u) \in \overline{\mathcal{U}}_0$ are true for any $u \in \mathcal{K}$ and $m \ge 0$.

The proof of the following theorem [11] is analogous to the proof of Theorem 5.5. It is also based on the first corollary from the limit theorem.

Theorem 5.8 *Suppose the existence of an integer number* $d_0 \in [0,\ n-1]$*, a real number* $s \in (0,\ 1]$*, a logarithmic norm* Λ*, and a continuously differentiable on* $\mathcal{U}_0$ *function* $V(\cdot)$ *such that*

$$\int_0^\tau \left[\operatorname{tr}_{d,\Lambda} D_2 f(t,\varphi(t,u)) + \dot{V}(t,\varphi(t,u))\right] dt < 0, \quad \forall u \in \overline{\mathcal{U}}_0.$$

Then, if for $d = d_0 + s$ the Hausdorff d-measure satisfies $\mu_H(\mathcal{K}, d) < \infty$, then

$$\lim_{m\to+\infty} \mu_H\big[(\varphi^\tau)^m(\mathcal{K}), d\big] = 0.$$

Choosing the logarithmic norm defined by the Euclidean vector norm, we obtain

Corollary 5.6 *Suppose that there exist an integer number $d_0 \in [0, n-1]$, a real number $s \in (0, 1]$, and a continuously differentiable on $\mathcal{U}_0$ function V such that*

$$\int_0^\tau \big[\lambda_1(t, \varphi(t, u)) + \cdots + \lambda_{d_0}(t, \varphi(t, u)) + s\lambda_{d_0+1}(t, \varphi(t, u)) \\ + \dot V(t, \varphi(t, u))\big]dt < 0, \quad \forall u \in \overline{\mathcal{U}}_0.$$

Then, if for $d = d_0 + s$ the Hausdorff d-measure satisfies $\mu_H(\mathcal{K}, d) < \infty$, then

$$\lim_{m\to+\infty} \mu_H\big[(\varphi^\tau)^m(\mathcal{K}), d\big] = 0.$$

5.2.3 The Generalized Bendixson Criterion

If we put in Corollary 5.2 from the limit theorem $\varphi := \varphi^\tau$ (the time-τ-map of system (5.34)), $\mathcal{U} = \mathcal{U}_0$, $\widetilde{\mathcal{K}} := \varphi^\tau(\overline{\mathcal{U}}_0)$, and use Lemma 5.4, then we obtain [11]:

Theorem 5.9 *Suppose, there exist a logarithmic norm Λ and a continuously differentiable on $\mathcal{U}$ function V such that the inequality*

$$\int_0^\tau \big[\Lambda(D_2 f(t, \varphi(t, u)) + \dot V(t, \varphi(t, u))\big]\,dt < 0, \quad \forall u \in \overline{\mathcal{U}}_0.$$

holds. Then in $\mathcal{U}_0$ there is no smooth closed curve Γ which satisfies $\varphi^\tau(\Gamma) = \Gamma$.

Choosing again the logarithmic norm defined by the Euclidean norm, we get:

Corollary 5.7 *Suppose there exists a continuously differentiable function V such that*

$$\int_0^\tau \big[\lambda_1(t, \varphi(t, u)) + \lambda_2(t, \varphi(t, u)) + \dot V(t, \varphi(t, u))\big]dt < 0, \quad \forall u \in \overline{\mathcal{U}}_0.$$

Then in $\mathcal{U}_0$ there is no smooth closed curve Γ which satisfies $\varphi^\tau(\Gamma) = \Gamma$.

5.2.4 On the Finiteness of the Number of Periodic Solutions

The proof of the following theorem [11] is based on Corollary 5.3 of the limit theorem.

Theorem 5.10 *Suppose that system (5.34) with $\mathcal{J} := \mathbb{R}$ and $\mathcal{U} := \mathbb{R}^n$ is dissipative, analytic, and T-periodic in t. If a logarithmic norm Λ and a continuously differentiable function $V(\cdot)$ can be found such that the inequality*

$$\Lambda\left(D_2 f(t,u)^{[2]}\right) + \dot{V}(t,u) < 0, \quad \forall (t,u) \in \mathbb{R} \times \mathbb{R}^n, \tag{5.48}$$

is satisfied, then system (5.34) has only a finite number of T-periodic solutions.

Proof Put $\mathcal{D} := \{u \mid |u| < \varsigma + 1\}$. By virtue of dissipativity there exist a number $\varsigma > 0$ and an integer number $k > 0$ such that for any $u \in \mathcal{D}$ the solution $\varphi(t,u)$ of system (5.34) exists for all $t \geq 0$ and $|\varphi(kT,u)| \leq \varsigma$.

Let us define the map $\varphi : \mathcal{D} \to \mathcal{D}$ by $\varphi(u) = \varphi(kT,u)$. Then for $\widetilde{\mathcal{K}} = \{u \mid |u| \leq \varsigma\}$ the inclusions

$$\varphi(\mathcal{D}) \subset \widetilde{\mathcal{K}} \subset \mathcal{D}$$

take place. Obviously, there exists a bounded open set $\mathcal{U} \subset \mathbb{R}^n$ such that $\varphi(t,u) \in \mathcal{U}$ for all (t,u) with $t \in [0,kT]$ and $u \in \overline{\mathcal{D}}$. Condition (5.48) implies

$$\sup\left\{\Lambda\left(D_2 f(t,u)^{[2]}\right) + \dot{V}(t,u)\right\} < 0$$

on the compact set $[0,kT] \times \overline{\mathcal{U}}$. Thus, by Lemma 5.4 we can find a continuous positive on $\widetilde{\mathcal{K}}$ function $\varkappa$, such that

$$\alpha_1(u)\alpha_2(u) < \frac{\varkappa(u)}{\varkappa(\varphi(u))}, \quad \forall u \in \widetilde{\mathcal{K}}.$$

Since the vector function f is analytic, we see that the map φ is also analytic. Thus, by Corollary 5.3 of the limit theorem φ has only a finite number of fixed points in $\mathcal{D}$. It means that system (5.34) has a finite number of kT-periodic solutions. But each T-periodic solution is kT-periodic too. □

Taking the logarithmic matrix norm defined by the Euclidean vector norm, we obtain from Theorem 5.10 the following result.

Corollary 5.8 *Suppose, system (5.34) with $\mathcal{U} := \mathbb{R}^n$ is dissipative, analytic, and T-periodic with respect to t. If there is a continuously differentiable function V such that the inequality*

$$\lambda_1(t,u) + \lambda_2(t,u) + \dot{V}(t,u) < 0, \quad \forall (t,u) \in \mathbb{R} \times \mathbb{R}^n,$$

is true, then system (5.34) has only a finite number of T-periodic solutions.

5.2.5 *Convergence Theorems*

Now let system (5.34) be autonomous, i.e.

$$\frac{d\varphi}{dt} = f(\varphi), \tag{5.49}$$

where $f : \mathcal{U} \to \mathbb{R}^n$ is C^1 and $\mathcal{U} \subset \mathbb{R}^n$ is open. Suppose that there exists a bounded simply connected open set $\mathcal{U}_0$ in $\mathbb{R}^n$ such that $\overline{\mathcal{U}}_0 \subset \mathcal{U}$ and any positive semi-orbit of (5.49) which meets the boundary of $\mathcal{U}_0$ crosses it strictly inwards $\mathcal{U}_0$. We also suppose that system (5.49) has only a finite number of equilibrium states in the domain $\mathcal{U}_0$. A positive semi-orbit $\gamma_+(u)$ of a solution $\varphi(\cdot, u)$ of (5.49) with $u \in \mathcal{U}_0$ is said to converge to an equilibrium state υ if $\varphi(t, u) \to \upsilon$ as $t \to +\infty$. Then the following theorem [31] is true.

Theorem 5.11 *Suppose that there exist a logarithmic norm Λ and a continuously differentiable on $\mathcal{U}_0$ function $V(\cdot)$ such that*

$$\Lambda\big(Df(u)^{[2]}\big) + \dot{V}(u) < 0, \quad \forall u \in \overline{\mathcal{U}}_0. \tag{5.50}$$

Then any positive semi-orbit of system (5.49) in $\mathcal{U}_0$ converges to one of the equilibrium states, i.e. the system is gradient-like.

Proof Let $\varphi(\cdot, u)$ be a solution of (5.49) whose positive semi-orbit for $t \geq 0$ is located in $\mathcal{U}_0$. From the boundedness of $\mathcal{U}_0$ it follows that the ω-limit set $\omega(u)$ of $\varphi(\cdot, u)$ is not empty.

Let us show that any ω-limit point $\upsilon' \in \omega(u)$ is an equilibrium state.

Suppose the opposite. Then by Pugh's closing lemma (Theorem A.4, Appendix A) for any $\varepsilon > 0$ a C^1-vector function $g(\cdot)$ can be found such that on $\mathcal{U}_0$ we have $\|f - g\|_{C^1} < \varepsilon$, and system (5.49) with the vector field g instead of f has a closed trajectory Γ passing trough the point υ'. Take ε so small that

$$\Lambda\big(Dg(u)^{[2]}\big) + \dot{V}(u) < 0, \quad \forall u \in \overline{\mathcal{U}}_0. \tag{5.51}$$

and, besides, any trajectory of g meeting the boundary of $\mathcal{U}_0$ crosses it strictly inwards $\mathcal{U}_0$. But in this case by Theorem 5.9 in $\mathcal{U}_0$ there is no closed trajectories of system (5.49). The contradiction obtained shows that the set $\omega(u)$ consists of equilibrium states.

Since we suppose that system (5.49) has only isolated equilibrium states, we see that from the connectness of the ω-limit set it follows that $\omega(u)$ consists of a unique equilibrium state. □

Choosing the logarithmic matrix norm defined by Euclidean vector norm, we obtain

Corollary 5.9 *Suppose that there exists a continuously differentiable on $\mathcal{U}_0$ function V such that*

$$\lambda_1(u) + \lambda_2(u) + \dot{V}(u) < 0, \quad \forall u \in \overline{\mathcal{U}}_0. \tag{5.52}$$

Then each positive semi-orbit of system (5.49) in $\mathcal{U}_0$ converges to one of the equilibrium states.

5.3 Convergence in Third-Order Nonlinear Systems Arising from Physical Models

5.3.1 The Generalized Lorenz System

In this subsection we shall prove a theorem on the convergence behaviour of the generalized Lorenz system (see system (1.59), Chap. 1)

$$\dot{x} = -\sigma x + \sigma y - axy, \quad \dot{y} = rx - y - xz, \quad \dot{z} = -bz + xy\,, \tag{5.53}$$

where σ, b, r are positive parameters and a is an arbitrary real parameter.

For any $\varsigma > 0$ we define the functions

$$P(\varsigma) :=(\sigma + b)\Big(\varsigma - \frac{a}{\varsigma}\Big)^2 - (\sigma + 1)\Big(\varsigma + \frac{a}{\varsigma}\Big)^2 \quad \text{and} \tag{5.54}$$

$$Q(\varsigma) := \frac{1}{\varsigma}(\sigma + 1)(\sigma - ar)\Big(\varsigma + \frac{a}{\varsigma}\Big). \tag{5.55}$$

The following theorem [8] on convergence in the generalized Lorenz system (5.53) is the main result of this subsection.

Theorem 5.12 *Each positive semi-orbit of system (5.53) converges to an equilibrium if for some $\varsigma > 0$ one of the following conditions holds:*

(1) $lrP(\varsigma) \le |Q(\varsigma)|$ *and* $4(\sigma + b)(b + 1) - \Big[lr\Big|\varsigma + \frac{a}{\varsigma}\Big| + \frac{1}{\varsigma}\Big|\sigma - ar\Big|\Big]^2 > 0;$

(2) $lrP(\varsigma) > |Q(\varsigma)|$ *and* $4(\sigma + 1)(b + 1)P(\varsigma)$

$$-\Big(\varsigma - \frac{a}{\varsigma}\Big)^2 l^2 r^2 p(\varsigma) - \frac{1}{\varsigma^2}(\sigma + 1)\Big(\varsigma - \frac{a}{\varsigma}\Big)^2 (\sigma - ar)^2 > 0.$$

Here the number l is defined by (1.20), Chap. 1.

For the proof of Theorem 5.12 we need some notions and an auxiliary proposition. Put for any $z \in \mathbb{R}$

$$h(z) := c_1 z^2 + 2c_2 z,$$

where c_1, c_2 are arbitrary real numbers.

The following simple lemma from [5] (whose proof we omit) defines the minimum of $h(z)$ on the segment $[-\Delta_1, \Delta_2]$, where $\Delta_1 = (l-1)r$ and $\Delta_2 = (l+1)r$.

Lemma 5.5 *Let*

$$m := \min_{[-\Delta_1, \Delta_2]} h(z).$$

Then

$$m = \begin{cases} h(\Delta_2)\,, & \text{if } \ a)\ \ c_1 \le 0 \text{ and } c_2 + c_1 r \le 0 \ \ or \\ & \qquad b)\ \ c_1 > 0 \text{ and } c_2 + c_1 \Delta_2 \le 0\,, \\ h(-\Delta_1)\,, & \text{if } \ c)\ \ c_1 \le 0 \text{ and } c_2 + c_1 r \ge 0 \ \ or \\ & \qquad d)\ \ c_1 > 0 \text{ and } c_2 - c_1 \Delta_1 \ge 0\,, \\ h(-c_2/c_1)\,, & \text{if } \ e)\ \ c_1 > 0 \text{ and } -c_1\Delta_2 \le c_2 \le c_1 \Delta_1\,. \end{cases}$$

Proof of Theorem 5.12 *Let us make in (5.53) the change of variables $(x, y, z) \mapsto (\varsigma x, y, z)$. Then we obtain the system*

$$\dot{x} = -\sigma x + \frac{\sigma}{\varsigma} y - \frac{a}{\varsigma} yz, \quad \dot{y} = \varsigma x(r - z) - y, \quad \dot{z} = -bz + \varsigma xy. \tag{5.56}$$

Denote by $J(x, y, z)$ the Jacobian matrix of the right-hand side of (5.56) at (x, y, z) and let $\lambda_1(y, z) \ge \lambda_2(y, z) \ge \lambda_3(y, z)$ be the eigenvalues of the symmetrized matrix $\frac{1}{2}(J(x, y, z)^* + J(x, y, z))$.

The proof of the theorem is based on Corollary 5.9. Therefore we must check the inequality $\lambda_1(y, z) + \lambda_2(y, z) < 0$ in the dissipativity region of system (5.53).

We have $\lambda_1(y, z) + \lambda_2(y, z) = \operatorname{tr} J(x, y, z) - \lambda_3(y, z)$. In order to verify the above condition, it is sufficient to prove that the matrix

$$\frac{1}{2}(J(x, y, z)^* + J(x, y, z)) + \lambda I$$
$$= \begin{pmatrix} \lambda - \sigma & \frac{1}{2}[\frac{\sigma - az}{\varsigma} + \varsigma(r - z)] & \frac{1}{2}(\varsigma - \frac{a}{\varsigma})y \\ \frac{1}{2}[\frac{\sigma - az}{\varsigma} + \varsigma(r - z)] & \lambda - 1 & 0 \\ \frac{1}{2}(\varsigma - \frac{a}{\varsigma})y & 0 & \lambda - b \end{pmatrix}$$

is positive-definite for $\lambda = -\operatorname{tr} J(x, y, z) = \sigma + b + 1$ on the set

$$\overline{\mathcal{D}}_1 := \{(y, z) \mid y^2 + (z - r)^2 \le l^2 r^2\}.$$

Using the Sylvester criterion for this purpose, it is sufficient to state the inequality

$$\psi(y,z) := \det\left(\frac{1}{2}(J(x,y,z)^* + J(x,y,z)) + \lambda I\right) > 0 \qquad (5.57)$$

for $(y,z) \in \overline{\mathcal{D}}_1$.

We have

$$\begin{aligned}\psi(y,z) &= c - \frac{1}{4}(\sigma+b)\Big(\varsigma - \frac{a}{\varsigma}\Big)^2 y^2 - \frac{1}{4}(\sigma+1)\Big[\frac{\sigma - az}{\varsigma} + \varsigma(r-z)\Big]^2 \\ &\geq c - \frac{1}{4}(\sigma+b)\Big(\varsigma - \frac{a}{\varsigma}\Big)^2 l^2r^2 + \frac{1}{4}(\sigma+b)\Big(\varsigma - \frac{a}{\varsigma}\Big)^2 (z-r)^2 \qquad (5.58) \\ &- \frac{1}{4}(\sigma+1)\Big[\frac{\sigma - az}{\varsigma} + \varsigma(r-z)\Big]^2 = \frac{1}{4}[h(z) + c_3],\end{aligned}$$

where $c := (\sigma+b)(\sigma+1)(b+1)$ *and the coefficients of the polynomial* $h(z)$ *and the constant* c_3 *are defined by*

$$\begin{aligned}c_1 &:= (\sigma+b)\Big(\varsigma - \frac{a}{\varsigma}\Big)^2 - (\sigma+1)\Big(\varsigma + \frac{a}{\varsigma}\Big)^2, \\ c_2 &:= (\sigma+1)\Big(\varsigma + \frac{a}{\varsigma}\Big)\Big(\frac{\sigma}{\varsigma} + \varsigma r\Big) - (\sigma+b)\Big(\varsigma - \frac{a}{\varsigma}\Big)^2 r, \\ c_3 &:= 4c - (\sigma+b)\Big(\varsigma - \frac{a}{\varsigma}\Big)^2 (l^2-1)r^2 - (\sigma+1)\Big(\frac{\sigma}{\varsigma} + \varsigma r\Big)^2.\end{aligned}$$

For the estimation of $h(z)$ *from below we shall use Lemma 5.5. Note previously the following obvious relations:*

$$c_2 + c_1 r = Q(\varsigma), \qquad (5.59)$$

$$c_2 + c_1\Delta_2 = Q(\varsigma) + lrc_1, \qquad (5.60)$$

$$c_2 - c_1\Delta_1 = Q(\varsigma) - lrc_1. \qquad (5.61)$$

Besides, we get

$$h(\Delta_2) + c_3 = 4c - (\sigma+1)\Big[lr\Big(\varsigma + \frac{a}{\varsigma}\Big) - \frac{1}{\varsigma}(\sigma - ar)\Big]^2, \qquad (5.62)$$

$$h(-\Delta_1) + c_3 = 4c - (\sigma+1)\Big[lr\Big(\varsigma + \frac{a}{\varsigma}\Big) + \frac{1}{\varsigma}(\sigma - ar)\Big]^2. \qquad (5.63)$$

Therefore if $Q(\varsigma) \leq 0$, *then by (5.62)*

$$h(\Delta_2) + c_3 = 4c - (\sigma+1)\Big[lr\Big|\varsigma + \frac{a}{\varsigma}\Big| + \frac{1}{\varsigma}\Big|\sigma - ar\Big|\Big]^2. \qquad (5.64)$$

And if $Q(\varsigma) \geq 0$, *then by (5.63)*

$$h(-\Delta_1) + c_3 = 4c - (\sigma + 1)\Big[lr\Big|\varsigma + \frac{a}{\varsigma}\Big| + \frac{1}{\varsigma}\Big|\sigma - ar\Big|\Big]^2. \tag{5.65}$$

Finally,

$$\begin{aligned} h(-c_2/c_1) + c_3 = \frac{1}{c_1}\Big[&4cc_1 - (\sigma + b)\Big(\varsigma - \frac{a}{\varsigma}\Big)^2 l^2 r^2 c_1 \\ &- \frac{1}{\varsigma^2}(\sigma + b)(\sigma + 1)\Big(\varsigma - \frac{a}{\varsigma}\Big)^2 (\sigma - ar)^2\Big]. \end{aligned} \tag{5.66}$$

Let us suppose that condition (1) of the theorem is satisfied, i.e. $lrc_1 \leq |Q(\varsigma)|$.

Assume that $c_1 \leq 0$. *If* $Q(\varsigma) \leq 0$, *then the validity of the theorem follows from (5.59), (5.64) and Lemma 5.5 (case (a)). If* $Q(\varsigma) \geq 0$, *then the validity of the theorem follows from (5.59), (5.65) and Lemma 5.5 (case (b)).*

Assume that $c_1 > 0$. *If* $Q(\varsigma) \leq -lrc_1$, *then the theorem is true according to (5.60), (5.64) and Lemma 5.5 (case (c)). If* $Q(\varsigma) \geq lrc_1$, *then the theorem holds by (5.61), (5.65) and Lemma 5.5 (case (d)). Thus, under condition (1) the theorem is proved.*

Now we suppose that condition (2) of the theorem is satisfied, i.e. $lrc_1 > |Q(\varsigma)|$. *In this case the theorem follows from Lemma 5.5 (case (e)) and relations (5.62) and (5.66).* □

From Theorem 5.12 we obtain the following simple condition [5] for convergence of the Lorenz system (1.12), Chap. 1.

Corollary 5.10 *Let* $a = 0$. *Each positive semi-orbit of system (5.53) converges to an equilibrium if*

$$r < (b + 1)(b/\sigma + 1) \quad \textit{and} \quad b \leq 2$$

or

$$r < \frac{2\sqrt{b-1}}{\sigma b}(\sigma + b)(\sigma + 1)(b + 1)\min\left\{\frac{1}{\sigma + 1}, \frac{1}{b - 1}\right\} \quad \textit{and} \quad b \geq 2.$$

Proof We have $P(\varsigma) := (b-1)\varsigma^2$, $Q(\varsigma) := \sigma(\sigma + 1)$. Condition (1) of Theorem 5.12 is reduced to the inequalities

$$lr(b-1)\varsigma^2 \leq \sigma(\sigma + 1) \quad \text{and} \quad \varsigma lr + \sigma/\varsigma < 2\sqrt{(\sigma + b)(b + 1)}.$$

Take $\varsigma := \sigma/\sqrt{(\sigma + b)(b + 1)}$. Then these inequalities take the form

$$lr(b-1)\sigma \leq (\sigma + b)(\sigma + 1)(b + 1) \quad \text{and} \quad r < \frac{(\sigma + b)(b + 1)}{l\sigma}.$$

Thus, it follows that each positive semi-orbit converges to an equilibrium if we have

$$r < (b+1)(b/\sigma+1) \quad \text{and} \quad b \leq 1$$

or

$$r < \frac{(\sigma+b)(\sigma+1)(b+1)}{l\sigma} \min\left\{\frac{1}{\sigma+1}, \frac{1}{b-1}\right\} \quad \text{and} \quad b > 1.$$

Recalling now the definition of l, we obtain the conditions formulated in Corollary 5.10. □

Remark 5.4 For $\sigma = 10$, $b = 8/3$ Corollary 5.10 guarantees the convergence in the Lorenz system (1.12), Chap. 1, for $r < 4.4$.

The following second corollary [5] of Theorem 5.12 can be useful for the study of some concrete systems reducible to the generalized Lorenz system.

Corollary 5.11 *Suppose that* $\sigma = ar$. *Then each positive semi-orbit of system (5.53) converges to an equilibrium if*

$$r < (b+1)(b/\sigma+1)/l^2. \tag{5.67}$$

Proof We have $Q(\varsigma) \equiv 0$. Condition (1) of Theorem 5.12 is reduced to the inequalities

$$P(\varsigma) \leq 0 \quad \text{and} \quad 4(\sigma+b)(b+1) - \left(\varsigma + \frac{a}{\varsigma}\right)^2 l^2 r^2 > 0. \tag{5.68}$$

Choose $\varsigma := \sqrt{a}$. Then $P(\varsigma) = -4(\sigma+1)a < 0$, and the second of the inequalities (5.68) can be written as

$$r^2 < \frac{(\sigma+b)(b+1)}{l^2 a}.$$

Substituting here $a := \sigma/r$, we obtain

$$r < \frac{(\sigma+b)(b+1)}{l^2\sigma}.$$

□

In the proof of Theorem 5.12 for the estimation of the dissipativity region $\mathcal{D}$ of the generalized Lorenz system the inclusion

$$\mathcal{D} \subset \{x| -\infty < x < \infty\} \times \overline{\mathcal{D}}_1$$

was used. A further improvement of conditions for the convergence of system (5.53) is possible if one uses a better inclusion. In some cases it is possible to use Lemma 1.6, Chap. 1. To show this let us formulate the following result for the Lorenz system (1.12), Chap. 1.

Theorem 5.13 *Let $a = 0$ and $b > 1$. If*

$$r < \frac{4(\sigma + b)(b + 1)}{\sigma\left[2 + \sqrt{\frac{(\sigma+b)b^2 - 4(b-1)^2}{(\sigma+1)(b-1)}}\,\right]},$$

then each positive semi-orbit of system (5.53) converges to an equilibrium.

For $\sigma = 10,\ b = 8/3$ Theorem 5.13 guarantees the convergence in the Lorenz system (1.12), Chap. 1, for $r < 4.5$.

Further, for these values of σ and b by using Lyapunov functions, we shall obtain a stronger result for the Lorenz system.

5.3.2 Euler's Equations Describing the Rotation of a Rigid Body in a Resisting Medium

Let us investigate now some concrete physical systems which can be transformed to the generalized Lorenz system. Consider in a resisting medium the equation for a rotating rigid body attached to the center of mass with no torque and a constant angular moment directed along one of the main axes.

Euler's equations describing such a motion are given by [17]

$$\begin{aligned} A_1\dot{\omega}_1 &= (A_2 - A_3)\omega_2\omega_3 - s_1\omega_1 + m, \\ A_2\dot{\omega}_2 &= (A_3 - A_1)\omega_1\omega_3 - s_2\omega_2, \\ A_3\dot{\omega}_3 &= (A_1 - A_2)\omega_1\omega_2 - s_3\omega_3, \end{aligned} \tag{5.69}$$

where A_i are the moments of inertia of the body, ω_i are the components of the angular velocity vector, m is the constant moment of external forces, s_i are the coefficients of resistance ($i = 1, 2, 3$).

In order to have several equilibrium states in (5.69) we shall suppose that the inequality

$$(A_1 - A_2)(A_3 - A_1) > 0, \tag{5.70}$$

is satisfied. (Under the opposite inequality system (5.69) has only one equilibrium state for any value of m.) The change of variables [24]

$$\begin{aligned} &(\omega_1, \omega_2, \omega_3) \mapsto \left(\frac{m}{s_1} - s_2 s_3 (A_3 - A_1)^{-1} T^{-1} z,\quad s_2 s_3 S^{-1} T^{-1} y,\quad s_2 S^{-1} x\right), \\ &t \mapsto \frac{1}{s_2} A_2 t, \end{aligned} \tag{5.71}$$

where

$$S := \left[A_1^{-1}A_2|(A_3 - A_1)(A_2 - A_3)|\right]^{\frac{1}{2}}, \qquad T := \frac{m}{s_1}(A_1 - A_2),$$

transfers system (5.69) into the generalized Lorenz system (5.53) with parameters

$$\sigma := \frac{s_3}{s_2}A_2A_3^{-1}, \quad b := \frac{s_1}{s_2}A_1^{-1}A_2, \quad r := \frac{m}{s_1s_2s_3}(A_3 - A_1)T, \tag{5.72}$$

$$a := s_3^2A_2A_3^{-1}(A_1 - A_2)(A_3 - A_1)^{-1}T^{-2}. \tag{5.73}$$

It is easy to see that the relation $\sigma = ar$ holds. Thus, it is possible to apply to system (5.69) Corollary 5.11 from the theorem on convergence in the generalized Lorenz system. An immediate result of the application of this corollary and Theorem 1.5, Chap. 1, on global asymptotic stability of the equilibrium (0, 0, 0) in the generalized Lorenz system is the following proposition [7].

Theorem 5.14 *Suppose that* $(A_1 - A_2)(A_3 - A_1) > 0$.
Then it holds: The unique equilibrium of system (5.69) is asymptotically globally stable if

$$m^2(A_1 - A_2)(A_3 - A_1) < s_1^2s_2s_3.$$

(b) Each positive semi-orbit of system (5.69) converges to an equilibrium if

$$m^2(A_1 - A_2)(A_3 - A_1) < s_1^2A_1^{-2}(s_1A_2 + s_2A_1)(s_1A_3 + s_3A_1),$$
$$s_1A_2 \le 2s_2A_1$$

or

$$m^2(A_1 - A_2)(A_3 - A_1) < 4s_2A_1^{-1}A_2^{-2}(s_1^2A_2^2 - s_2^2A_1^2)(s_1A_3 + s_3A_1),$$
$$s_1A_2 \ge 2s_2A_1.$$

5.3.3 A Nonlinear System Arising from Fluid Convection in a Rotating Ellipsoid

In the book [25] the convection of a fluid within a rotating ellipsoid is considered. The axis of rotation coincides with one of the main axes of the ellipsoid and the angle between this axis and the gravity vector is different from zero. Convective motion is generated by an outer horizontal-irregular heating.

The system of differential equations that appears in this model has the form

$$
\begin{aligned}
\dot{x} &= \sigma(y - x) - \frac{\delta\sigma^2}{(\delta R + 1)^2} yz, \\
\dot{y} &= \frac{R}{\sigma}(\delta R + 1)x - y - xz, \\
\dot{z} &= -z + xy,
\end{aligned}
\tag{5.74}
$$

where σ, R, δ are positive parameters. We call it *Glukhovsky-Dolzhansky-system*.

System (5.74) coincides with the generalized Lorenz system (5.53) if we put

$$
b := 1, \qquad a := \frac{\delta\sigma^2}{(\delta R + 1)^2}, \qquad r := \frac{R}{\sigma}(\delta R + 1).
$$

For $\sigma = 4$, $\delta = 0.04$, $R = 250$ it is found by numerical simulation that system (5.74) has a strange attractor.

The following convergence theorem holds [5].

Theorem 5.15 *Each positive semi-orbit of system (5.74) converges to an equilibrium if*

$$
R < \frac{1}{2\delta}\left(\sqrt{8\delta(\sigma + 1) + 1} - 1\right). \tag{5.75}
$$

Proof The proof is based on Condition (1) of Theorem 5.12 on convergence of the generalized Lorenz system. In the notation of this theorem we have $l := 1$, $P(\varsigma) := -4(\sigma + 1)a$. Condition (1) is reduced to the inequality

$$
8(\sigma + 1) - \left[\frac{R}{\sigma}(\delta R + 1)\varsigma + \frac{\sigma}{\varsigma}\right]^2 > 0,
$$

which holds for

$$
R < \frac{1}{2\delta}\left[\sqrt{4\frac{\delta\sigma}{\varsigma}\sqrt{8(\sigma + 1)} - 4\frac{\delta\sigma^2}{\varsigma^2} + 1} - 1\right].
$$

Choosing

$$
\varsigma = \frac{\sigma}{\sqrt{2(\sigma + 1)}},
$$

we arrive at condition (5.75). □

From Theorem 1.5, Chap. 1, on global asymptotic stability of the unique equilibrium (0, 0, 0) of the generalized Lorenz system it follows that for $\sigma = 4$ and $\delta = 0.04$ the unique equilibrium of system (5.74) is globally asymptotically stable if $R < 3.5$.

From Theorem 5.15 it follows that each positive semi-orbit of system (5.74) is converging to an equilibrium if $R < 7.6$.

5.3.4 A System Describing the Interaction of Three Waves in Plasma

The systems of the two physical examples considered above were reduced to the generalized Lorenz system with a positive parameter a. Now we shall investigate a system that, by a change of variables, will be reduced to the generalized Lorenz system with $a < 0$.

In the book [34] (see also [36]), studying waves in plasma, the following system of equations was deduced and analysed:

$$\dot{x} = hy - \kappa_1 x - yz, \quad \dot{y} = hx - \kappa_2 y + xz, \quad \dot{z} = -z + xy. \tag{5.76}$$

This system (we call it *Rabinovich's system*) describes the interaction of three reasonably coupled waves, two of them being parameterically excited. Here, the parameter h is proportional to the amplitude of pumping, parameters κ_1, κ_2 are normed damping coefficients.

The change of variables

$$(x, y, z) \mapsto (\kappa_1\kappa_2 h^{-1} y\,,\ \kappa_1 x\,,\ \kappa_1\kappa_2 h^{-1} z)\,, \quad t \mapsto \kappa_1^{-1} t$$

transforms system (5.76) into the generalized Lorenz system (5.53) with parameters $\sigma := \kappa_1^{-1}\kappa_2$, $b := \kappa_1^{-1}$, $a := -\kappa_2^2 h^{-2}$, $r := \kappa_1^{-1}\kappa_2^{-1}h^2$.

System (5.76) was studied by numerical methods for fixed $\kappa_1 = 1$, $\kappa_2 = 4$, and various parameter h. For $h = 4.92$ the existence of a strange attractor was stated. We shall suppose further that $\kappa_1 = 1$ and drop, for brevity, the index of parameter κ_2.

Let ς be a positive number. Denote by ς_0 a positive number such that

$$\varsigma_0^2 := \frac{8\kappa^2}{9(\kappa + 1)}.$$

Now take into consideration the polynomial

$$P(\lambda) := \lambda^3 + 3\frac{\kappa^2}{\varsigma^2}\lambda^2 + \left[3\frac{\kappa^4}{\varsigma^4} - 8(\kappa + 1)\frac{\kappa^2}{\varsigma^2}\right]\lambda + \frac{\kappa^6}{\varsigma^6}.$$

Denote by $\lambda(\varsigma)$ the maximal real root of the equation $P(\lambda) = 0$ and put

$$\lambda_0 := \sup_{\varsigma \geq \varsigma_0} \lambda(\varsigma).$$

Theorem 5.16 ([6]) *Each positive semi-orbit of system (5.76) converges to an equilibrium if $h^2 < \lambda_0$.*

We would like to remark that the result of Theorem 5.16 will be strengthened further (see Theorem 5.27) due to the application of a Lyapunov function. But the proof of this theorem, in contrast to the two previous examples, uses not only condition (1) but also condition (2) of the Theorem 5.12 on convergence of the generalized Lorenz system.

Proof of Theorem 5.16 *In the notation of Theorem 5.12 on convergence we have*

$$l = 1, \qquad P(\varsigma) := 4(\kappa+1)\frac{\kappa^2}{h^2}, \qquad Q(\varsigma) := \frac{2\kappa}{\varsigma}(\kappa+1)(\varsigma - \frac{\kappa^2}{\varsigma h^2}).$$

Define

$$\tau_1(\varsigma) := \frac{\kappa}{\varsigma}[3\frac{\kappa}{\varsigma} - \sqrt{8(\kappa+1)}\,], \qquad \tau_2(\varsigma) := \frac{\kappa^2}{3\varsigma^2}.$$

Condition (1) is reduced to the inequalities

$$2\varsigma \le \left|\varsigma - \frac{\kappa^2}{\varsigma h^2}\right|, \qquad 8(\kappa+1) - \left[\frac{h^2}{\kappa}\left|\varsigma - \frac{\kappa^2}{\varsigma h^2}\right| + \frac{2\kappa}{\varsigma}\right]^2 > 0,$$

which may be written in the form

$$\tau_1(\varsigma) < h^2 \le \tau_2(\varsigma).$$

Note that for $\varsigma > \varsigma_0$ we have the interval $\big(\tau_1(\varsigma), \tau_2(\varsigma)\big] \neq \emptyset$. Consequently, for $h^2 \in \big(\tau_1(\varsigma), \tau_2(\varsigma)\big]$, $\varsigma > \varsigma_0$ each positive semi-orbit of system (5.76) converges to an equilibrium.

Condition (2) is reduced to the inequalities

$$2\varsigma > \left|\varsigma - \frac{\kappa^2}{\varsigma h^2}\right|, \qquad P(h^2) < 0. \tag{5.77}$$

The first of them is equivalent to $h^2 > \tau_2(\varsigma)$. Further, since

$$P(\kappa^2/3\varsigma^2) = \frac{8\kappa^4}{3\varsigma^4}\left[\frac{8\kappa^2}{9\varsigma^2} - (\kappa+1)\right],$$

we have $P(\kappa^2/3\varsigma^2) < 0$ for $\varsigma > \varsigma_0$. Taking into account the relations $P(-\infty) = -\infty$, $P(+\infty) = +\infty$ and $P(0) = \kappa^6/\varsigma^6$, we can conclude that the inequalities (5.77) for $\varsigma > \varsigma_0$ are equivalent to

$$\tau_2(\varsigma) < h^2 < \lambda(\varsigma),$$

where for $\varsigma > \varsigma_0$ we have an interval $\big(\tau_2(\varsigma), \lambda(\varsigma)\big) \neq \emptyset$. Consequently, for $h^2 \in \big(\tau_2(\varsigma), \lambda(\varsigma)\big)$ and $\varsigma > \varsigma_0$ system (5.76) is convergent.

Let us show that (5.76) is convergent for $h^2 < \lambda(\varsigma)$ and arbitrary $\varsigma > \varsigma_0$. This will prove the theorem. Let us remark that for $h^2 < \kappa$ the unique equilibrium of the system *is globally asymptotically stable, since in this case condition (2) of Theorem 1.5, Chap. 1, on global asymptotic stability of the equilibrium* $(0, 0, 0)$ *of the generalized Lorenz system is satisfied. Consequently, if* $\tau_1(\varsigma) < \kappa$*, then the proof is complete.*

Suppose that $\tau_1(\varsigma) \geq \kappa$ *and choose* $\varsigma_1 > \varsigma$ *so large that* $\tau_1(\varsigma_1) < \kappa$*. It is clear that (5.76) is convergent if* $h^2 \in \mathcal{Z}$*, where*

$$\mathcal{Z} := \bigcup_{\beta \in [\varsigma, \varsigma_1]} \big(\tau_1(\beta), \tau_2(\beta)\big].$$

But

$$\big(0, \lambda(\varsigma)\big) \subset (0, \kappa) \cup \mathcal{Z} \cup \big(\tau_2(\varsigma), \lambda(\varsigma)\big).$$

□

For $\kappa_1 = 1$ and $\kappa_2 = 4$ Condition (2) of Theorem 1.5, Chap. 1, on global asymptotic stability of the equilibrium $(0, 0, 0)$ of the generalized Lorenz system is reduced for system (5.76) to the inequality $h < 2$, which guarantees the global asymptotic stability of the equilibrium $(0, 0, 0)$ of system (5.76). From Theorem 5.12 the convergence of the system follows for $h < 2.4$.

5.4 Estimates of Fractal Dimension

5.4.1 Maps with a Constant Jacobian

In this subsection we derive for a class of differentiable maps in $\mathbb{R}^n$ an upper estimate of the fractal dimension of an invariant set. Our representation in Sects. 5.4.1 and 5.4.2 is based on the results of [13, 14].

Let us recall that, in various kinds of applications, for chaotic attractors, their fractal dimension is of higher significance than their Hausdorff dimension. One example are embedding strategies for dynamical systems with a high-dimensional phase space, which answers the question how many degrees of freedom for a model system are sufficient to represent the essential dynamics faithfully. If for such a system, we have given an attractor of fractal dimension d, Sauer, Yorke and Casdagli [39] show that in "almost all cases" it can be mapped injectively via a linear transformation into $\mathbb{R}^n$ provided $n > 2d$. A counterexample by I. Kan in the appendix of [39] points out that the fractal dimension may not be replaced by the Hausdorff dimension. (See also Remark 3.7, Chap. 3.) Another example are noisy systems where the volume of the attractor scales with the magnitude of the noise, with a scaling factor depending on the fractal dimension of the noiseless attractor [33].

Let

$$\varphi : \mathbb{R}^n \to \mathbb{R}^n \tag{5.78}$$

be a continuously differentiable map. Suppose that the compact set $\mathcal{K} \subset \mathbb{R}^n$ is an invariant set for φ, i.e. $\varphi(\mathcal{K}) = \mathcal{K}$.

Denote by $\alpha_1(u) \geq \cdots \geq \alpha_n(u)$ the singular values of the differential $D\varphi(u)$ of the map φ at u.

Theorem 5.17 *Suppose that*

$$\alpha_1(u) \cdots \alpha_n(u) \equiv \text{const} \neq 0, \quad \forall u \in \mathcal{K}, \tag{5.79}$$

and there exist a real $s \in [0,\ 1]$ and a continuous positive on $\mathcal{K}$ function $\varkappa(\cdot)$ such that

$$\alpha_1(u) \cdots \alpha_{n-1}(u)\alpha_n^s(u) < \frac{\varkappa(u)}{\varkappa\big(\varphi(u)\big)}, \quad \forall u \in \mathcal{K}. \tag{5.80}$$

Then $\dim_F \mathcal{K} \leq n - 1 + s$.

Remark 5.5 Conditions analogous to (5.79) are considered in [14] for invertible maps as the Hénon system. In contrast to the results in this section the fractal dimension estimates in [14] are given in terms of Lyapunov exponents and without use of a "regulating function" $\varkappa$. In the following we call (5.79) *Chen's condition* [14] for maps. It will be shown in Sect. 8.2, Chap. 8 that the fractal dimension estimate (5.80) is true under more general assumptions.

Before we come to the proof of Theorem 5.17 let us formulate and prove several auxiliary results.

Let us introduce the notation

$$\varsigma := \min_{u \in \mathcal{K}} \alpha_n(u), \quad \omega_n(u) := \alpha_1(u) \cdots \alpha_n(u),\ u \in \mathbb{R}^n. \tag{5.81}$$

Since by (5.79) the quantity $\omega_n(u)$ is constant on $\mathcal{K}$, we put for $u \in \mathcal{K}$

$$\overline{\omega}_n := \omega_n(u). \tag{5.82}$$

Lemma 5.6 *Consider the compact invariant set $\mathcal{K}$ of the map (5.78) and the numbers ς and $\overline{\omega}_n$ defined in (5.81) and (5.82), respectively.*

Suppose that $\varsigma < (\sqrt{n})^{-1}$ and $d > 0$ is a number such that

$$\overline{\omega}_n \varsigma^{d-n} \leq 4^{-n} n^{-d/2}. \tag{5.83}$$

Then $\dim_F \mathcal{K} \leq d$.

Proof Take an $\eta \in (0,\ \varsigma)$. Let $r_0 > 0$ be so small that

$$\big|\varphi(\upsilon) - \varphi(u) - D\varphi(u)(\upsilon - u)\big| \le \eta|\upsilon - u|, \quad \forall u,\ \upsilon \in \mathcal{K},\ |u - \upsilon| < r_0. \tag{5.84}$$

Let us fix an arbitrary $r \in (0,\ r_0)$. By the compactness and invariance of $\mathcal{K}$ points $u_j \in \mathcal{K}$, $j = 1, \dots, N_{\mathcal{K}}(r)$, can be found such that

$$\mathcal{K} = \bigcup_j \mathcal{B}_r(u_j) \cap \mathcal{K} = \bigcup_j \varphi\big(\mathcal{B}_r(u_j) \cap \mathcal{K}\big). \tag{5.85}$$

Denote $\mathcal{E}_j := D\varphi(u_j)\mathcal{B}_1(0)$. Let $\mathcal{E}'_j$ be an ellipsoid corresponding to the ellipsoid $\mathcal{E}'_j$ according to Lemma 2.2, Chap. 2. Using (5.84), we get

$$\varphi\big(\mathcal{B}_r(u_j) \cap \mathcal{K}\big) \subset \varphi(u_j) + r\big(\mathcal{E}_j + \mathcal{B}_\eta(0)\big) \subset \varphi(u_j) + r\mathcal{E}'_j. \tag{5.86}$$

Now put $\sigma := \sqrt{n}\,\varsigma$. From (5.85), (5.86) we have

$$N_{\sigma r}(\mathcal{K}) \le N_r(\mathcal{K}) \max_j N_{\sigma r}(r\mathcal{E}'_j). \tag{5.87}$$

Since $\varsigma \le \alpha_n(u_j) = \alpha_n(\mathcal{E}_j) < \alpha_n(\mathcal{E}'_j)$, by Lemma 2.3, Chap. 2, we have

$$\begin{aligned} N_{\sigma r}(r\mathcal{E}'_j) &\le 2^n \omega_n(r\mathcal{E}'_j)(r\varsigma)^{-n} = 2^n \omega_n(\mathcal{E}'_j)\varsigma^{-n} \le 2^n(1 + \eta/\varsigma)^n \omega_n(\mathcal{E}_j)\varsigma^{-n} \\ &\le 4^n \omega_n(u_j)\varsigma^{-n} = 4^n \overline{\omega}_n \varsigma^{-n}. \end{aligned} \tag{5.88}$$

From the previous inequality, inequalities (5.87) and (5.83) we obtain

$$N_{\sigma r}(\mathcal{K}) \le N_r(\mathcal{K}) 4^n \overline{\omega}_n \varsigma^{-n+d} n^{d/2} \sigma^{-d} \le \sigma^{-d} N_r(\mathcal{K}). \tag{5.89}$$

Since $\sigma < 1$, we can find for $\varepsilon \in (0,\ r_0)$ an integer number $l \ge 0$ such that $\sigma^{l+1} r_0 \le \varepsilon < \sigma^l r_0$. Consequently, applying l times inequality (5.88), we have

$$N_\varepsilon(\mathcal{K}) \le \sigma^{-ld} N_{\sigma^{-l}\varepsilon}(\mathcal{K}) \le \left(\frac{r_0}{\varepsilon}\right)^d N_{\sigma^{-l}\varepsilon}(\mathcal{K}) \le \left(\frac{r_0}{\varepsilon}\right)^d N_{\sigma r_0}(\mathcal{K}).$$

It follows that

$$\dim_F \mathcal{K} = \limsup_{\varepsilon \to 0} \frac{\log N_\varepsilon(\mathcal{K})}{\log(1/\varepsilon)} \le d.$$

□

Lemma 5.7 *Let $c > 0$ be an arbitrary number, and let $\mathcal{K}$ be a compact and invariant set for (5.78) for which the relations (5.79) and (5.80) are true. Then it holds:*

(a) There exists an integer $m > 0$ such that

$$\omega_d\big(D\varphi^m(u)\big) \le c, \quad \forall x \in \mathcal{K};$$

(b) For an integer $m > 0$ we have

$$\alpha_1\big(D\varphi^m(u)\big) \cdots \alpha_n\big(D\varphi^m(u)\big) = \text{const} \ne 0, \quad \forall u \in \mathcal{K}.$$

Proof Let us prove (a). By (5.80) we can find a number $0 < \kappa < 1$ such that

$$\max_{x\in\mathcal{K}} \left[\frac{\varkappa(\varphi(u))}{\varkappa(u)} \omega_d\big(D\varphi(u)\big) \right] \le \kappa.$$

Using the chain rule

$$D\varphi^m(u) = D\varphi\big(\varphi^{m-1}(u)\big) \cdots D\varphi\big(\varphi(u)\big) D\varphi(u), \tag{5.90}$$

we obtain

$$\omega_d\big(D\varphi^m(u)\big) \le \prod_{j=1}^{m} \omega_d\big(D\varphi\big(\varphi^{m-j}(u)\big)\big) \le \prod_{j=1}^{m} \kappa \frac{\varkappa\big(\varphi^{m-j}(u)\big)}{\varkappa\big(\varphi^{m-j+1}(u)\big)} = \kappa^m \frac{\varkappa(u)}{\varkappa\big(\varphi^m(u)\big)} < c$$

for all sufficiently large m.

Let us show (b). Since $|\det D\varphi(u)| = \overline{\omega}_n$, $\forall u \in \mathcal{K}$, by (5.90) we have $|\det D\varphi^m(u)| = {\overline{\omega}_n}^m$, $\forall u \in \mathcal{K}$. □

Proof of Theorem 5.17 *Put $c := 4^{-n} n^{-n/2}$. Without loss of generality we can suppose that the following inequalities hold:*

$$\alpha_1(u) \cdots \alpha_{n-1}(u) \alpha_n^s(u) \le c, \quad \forall u \in \mathcal{K}, \tag{5.91}$$

$$\varsigma < (\sqrt{n})^{-1}. \tag{5.92}$$

In the opposite case, taking into account Lemma 5.7, we can use for the dimension estimation of $\mathcal{K}$ the map φ^m instead of the map φ.

According to Lemma 5.6 for the proof of the theorem it is sufficient to check the inequality

$$\overline{\omega}_n \varsigma^{s-1} \le c.$$

Let $u_0 \in \mathcal{K}$ be such that $\alpha_n(u_0) = \varsigma$. Then, by (5.89) we obtain

$$\overline{\omega}_n \varsigma^{s-1} = \alpha_1(u_0) \cdots \alpha_n(u_0) \alpha_n^{s-1}(u_0) \le c.$$

□

5.4.2 *Autonomous Differential Equations Which are Conservative on the Invariant Set*

Let us consider now some applications of Theorem 5.17 to the differential equation

$$\frac{d\varphi}{dt} = f(\varphi). \tag{5.93}$$

Here $\mathcal{U} \subset \mathbb{R}^n$ is an open set, and $f : \mathcal{U} \to \mathbb{R}^n$ is a continuously differentiable function. Let

$$\lambda_1(u) \geq \lambda_2(u) \geq \cdots \geq \lambda_n(u)$$

be the eigenvalues of the symmetrized Jacobian matrix $\frac{1}{2}[Df(u) + Df(u)^*]$ at a point $u \in \mathcal{U}$. For a continuously differentiable function V we shall use again the notation $\dot{V}(u) = (f(u), \operatorname{grad} V(u))$. Denote by $\varphi(\cdot, u)$ the unique maximal solution of (5.93) starting in $u \in \mathcal{U}$ at $t = 0$. Suppose that we can find a number $\tau > 0$ and an open set $\mathcal{U}_0$ ($\overline{\mathcal{U}}_0 \subset \mathcal{U}$) such that $u \in \overline{\mathcal{U}}_0$ implies that $\varphi(t, u) \in \mathcal{U}$ for $t \in [0, \tau]$. Denote by $\varphi^\tau : \overline{\mathcal{U}}_0 \to \mathcal{U}$ the solution operator of system (5.93) defined by the equality $\varphi^\tau(u) = \varphi(\tau, u)$.

Theorem 5.18 *Let $\mathcal{K} \subset \mathcal{U}_0$ be a compact set satisfying $\mathcal{K} = \varphi^\tau(\mathcal{K})$ for the solution operator $\varphi^\tau(\cdot)$ of (5.93).*

Suppose that

$$\operatorname{tr} Df(u) = \text{const}, \quad \forall u \in \mathcal{K}, \tag{5.94}$$

and there exists a real $s \in [0, 1]$, a logarithmic norm Λ, and continuously differentiable on $\mathcal{K}$ function $V(\cdot)$ such that

$$\operatorname{tr}_{n-1+s,\Lambda}\big(Df(u)\big) + \dot{V}(u) < 0, \quad \forall u \in \mathcal{K}.$$

Then we have the estimate $\dim_F \mathcal{K} \leq n - 1 + s$.

Proof The proof follows with $d_0 = n - 1$ immediately from Theorem 5.17 and Lemma 5.4.

□

Remark 5.6 In analogy with (5.79) we call (5.94) *Chen's condition* [14] for differential equations.

Taking the logarithmic matrix norm defined by Euclidean vector norm we get

Corollary 5.12 *Suppose that $\mathcal{K} \subset \mathcal{U}_0$ is a compact set satisfying $\varphi^\tau(\mathcal{K}) = \mathcal{K}$,*

$$\operatorname{tr} Df(u) = \text{const}, \quad \forall u \in \mathcal{K}$$

and there exist a real $s \in [0, 1]$ *and a continuously differentiable on* $\mathcal{K}$ *function* $V(\cdot)$ *such that*

$$\lambda_1(u) + \cdots + \lambda_{n-1}(u) + s\lambda_n(u) + \dot{V}(u) < 0, \quad \forall u \in \mathcal{K}.$$

Then the estimate $\dim_F \mathcal{K} \le n - 1 + s$ *is true.*

5.5 Fractal Dimension Estimates for Invariant Sets and Attractors of Concrete Systems

In this section we will estimate with the help of Lyapunov functions the fractal dimension of invariant sets or attractors of the following concrete systems: Rössler's system, the Lorenz system, equations of the third order and a system describing the interaction of waves in plasma.

All these systems can be considered in $\mathbb{R}^3$ and satisfy Chen's condition. Therefore, in order to find upper fractal dimension estimates, we can use Corollary 5.12, and the main goal is to verify the inequality

$$\lambda_1(x, y, z) + \lambda_2(x, y, z) + s\lambda_3(x, y, z) + \dot{V}(x, y, z) < 0, \quad \forall (x, y, z) \in \mathcal{K}, \tag{5.95}$$

where $\lambda_1(x, y, z) \ge \lambda_2(x, y, z) \ge \lambda_3(x, y, z)$ are the eigenvalues of the symmetrized Jacobian matrix of the right side of the system at the point (x, y, z), s is some number from the interval $[0, 1]$, and $\dot{V}(x, y, z)$ is the derivative of a certain continuously differentiable function V with respect to the system at a point $(x, y, z) \in \mathcal{K}$.

5.5.1 *The Rössler System*

Let us consider the Rössler system [38]

$$\frac{dx}{dt} = -y - z, \quad \frac{dy}{dt} = x, \quad \frac{dz}{dt} = -bz + a(y - y^2), \tag{5.96}$$

where a and b are positive parameters. Computer simulation shows that for certain positive values of these parameters system (5.96) has a compact invariant set $\mathcal{K}$ with non-integer fractal dimension.

The following theorem is due to [31].

Theorem 5.19 *Let* $\mathcal{K} \subset \mathbb{R}^3$ *be a compact and invariant set of system (5.96). Then*

$$\dim_F \mathcal{K} \le 3 - \frac{2b}{b + \sqrt{(a + 2b)^2 + b^2 + 1}}. \tag{5.97}$$

Proof It is easy to see that the eigenvalues of the symmetrized Jacobian matrix for the right-hand side of system (5.96) are

$$0 \quad \text{and} \quad \frac{1}{2}\left\{-b \pm \sqrt{b^2+1+a^2(1-2y)^2}\right\}.$$

Consequently, condition (5.95) with respect to a Lyapunov function V and a parameter $s \in [0, 1)$ can be written as

$$-(1+s)b + (1-s)\sqrt{b^2+1+a^2(1-2y)^2} + 2\dot{V} < 0, \; \forall t \in \mathbb{R}. \tag{5.98}$$

Let us choose the function V as

$$V(x, z) := \frac{1}{2}(1-s)\kappa(z-bx),$$

where κ is a varying parameter. A direct calculation shows that

$$\dot{V} = \frac{1}{2}(1-s)\kappa\left[(a+b)y - ay^2\right]$$

and inequality (5.98) is equivalent to

$$-(1+s)b + (1-s)h(y;\kappa) < 0, \quad y \in \mathbb{R}, \tag{5.99}$$

where

$$h(y;\kappa) = \sqrt{b^2+1+a^2(1-2y)^2} + \kappa\left[(a+b)y - ay^2\right].$$

Let us put

$$m := \inf_{\kappa \in \mathbb{R}} \max_{y \in \mathbb{R}} h(y;\kappa).$$

From (5.99) by virtue of Corollary 5.12 we obtain

$$\dim_F \mathcal{K} \le 2 + \frac{m-b}{m+b} = 3 - \frac{2b}{m+b}. \tag{5.100}$$

We can write h as

$$h(y;\kappa) = -\left(\theta\sqrt{b^2+1+a^2(1-2y)^2} - \frac{1}{2\theta}\right)^2 + \theta^2\left[b^2+1+a^2(1-2y)^2\right]$$
$$+\frac{1}{4\theta^2} + \kappa\left[(a+b)y - ay^2\right],$$

where $\theta \neq 0$ is a new varying parameter. Using this representation, we can estimate h for arbitrary $y \in \mathbb{R}$ by

$$\begin{aligned}h(y;\kappa) &\le \theta^2\big[b^2+1+a^2(1-2y)^2\big]+\frac{1}{4\theta^2}+\kappa\big[(a+b)y-ay^2\big]\\ &\le \theta^2(a^2+b^2+1)+\frac{1}{4\theta^2}-(\kappa a-4\theta^2a^2)y^2-\big[4\theta^2a^2-\kappa(a+b)\big]y\\ &= -(\kappa a-4\theta^2a^2)\left[y+\frac{4\theta^2a^2-\kappa(a+b)}{2(\kappa a-4\theta^2a^2)}\right]^2+\frac{\big[4\theta^2a^2-\kappa(a+b)\big]^2}{4(\kappa a-4\theta^2a^2)}\\ &+\theta^2(a^2+b^2+1)+\frac{1}{4\theta^2}.\end{aligned}$$

Let us take κ and θ so that $\kappa a-4\theta^2a^2>0$. Then we get from the above

$$h(y;\kappa)\le\frac{\big[4\theta^2a^2-\kappa(a+b)\big]^2}{4(\kappa a-4\theta^2a^2)}+\theta^2(a^2+b^2+1)+\frac{1}{4\theta^2}.$$

Now if we choose

$$\kappa:=4\theta^2a\frac{a+2b}{a+b},\qquad \theta^2:=\frac{1}{2\sqrt{(a+2b)^2+b^2+1}},$$

we receive the inequality

$$h(y;\kappa)\le\sqrt{(a+2b)^2+b^2+1},\quad y\in\mathbb{R}.$$

The previous estimate shows that (5.100) implies (5.97). □

Example 5.2 For $a=0.386$ and $b=0.2$ numerically was obtained [38] a chaotic invariant set $\mathcal{K}$ in (5.96). From estimate (5.97) we obtain for this set $\dim_F K\le 2.731$.

5.5.2 *Lorenz Equation*

Consider again the Lorenz system

$$\dot{x}=-\sigma x+\sigma y,\qquad \dot{y}=rx-y-xz,\qquad \dot{z}=-bz+xy. \tag{5.101}$$

Recall that σ, b, r are positive parameters. Firstly we shall get a fractal dimension estimate for an arbitrary attractor of system (5.101) without use of a Lyapunov function V, i.e. in inequality (5.95) $\dot{V}$ will be identically equal to zero. Consider the following cubic equation

$$\zeta^3+a_1\zeta^2+a_2\zeta+a_3=0 \tag{5.102}$$

with coefficients

$$
\begin{aligned}
a_1 &:= -(\sigma + b + 1),\\
a_2 &:= \sigma + b(\sigma + 1) - \frac{1}{4}\varsigma^2 l^2 r^2 - \frac{1}{4}\sigma^2/\varsigma^2 - \frac{1}{2}\sigma r,\\
a_3 &:= -\sigma b + \frac{1}{4}\varsigma^2 l^2 r^2 + \frac{1}{4}\sigma^2 b/\varsigma^2 + \frac{1}{2}\sigma b r + \frac{1}{4}\varsigma^2 r^2 (b-1),
\end{aligned}
$$

where $\varsigma \neq 0$ is a varying parameter and the number l is defined by

$$
l = \begin{cases} 1, & \text{if } \ b \le 2, \\ 0.5b/\sqrt{b-1}, & \text{if } \ b \ge 2. \end{cases} \tag{5.103}
$$

Denote by ζ_0 the maximal real root of equation (5.102) and put

$$
k := \inf_{\varsigma} \max \{\sigma + b + 1,\ b + \varsigma^2 r(b-1)/\sigma,\ \zeta_0\}.
$$

The next theorem and Corollary 5.15 are obtained in [7].

Theorem 5.20 *Let $\mathcal{A}$ be an attractor of system (5.101). Suppose that $b > 1$. Then*

$$
\dim_F \mathcal{A} \le 3 - (\sigma + b + 1)/k.
$$

Proof By the change of variables $(x, y, z) \mapsto (\varsigma x, y, z)$ we can transform system (5.101) into the form

$$
\dot{x} = -\sigma(x - \varsigma^{-1} y), \qquad \dot{y} = (r - z)\varsigma x - y, \qquad \dot{z} = -bz + \varsigma x y. \tag{5.104}
$$

Recall that system (5.101) is dissipative with a dissipativity domain $\mathcal{U}_0$ satisfying

$$
\mathcal{U}_0 \subset \{x \mid -\infty < x < \infty\} \times \left(\overline{\mathcal{D}}_1 \cap \{z \mid z \ge 0\}\right),
$$

where

$$
\mathcal{D}_1 := \{y, z \mid y^2 + (z - r)^2 < l^2 r^2\}.
$$

Let us take an arbitrary number w satisfying the inequality

$$
w > \max \{\sigma + b + 1,\ b + \varsigma^2 r(b-1)/\sigma,\ \zeta_0\}. \tag{5.105}
$$

Denote by $J(y, z)$ the Jacobian matrix of the right-hand side of system (5.104). It is sufficient to prove that the matrix

$$
\frac{1}{2}\Big(J(y,z)^* + J(y,z)\Big) + \zeta I = \begin{pmatrix} -\sigma + \zeta & \frac{1}{2}[\varsigma(r - z) + \sigma/\varsigma] & \frac{1}{2}\varsigma y \\ \frac{1}{2}[\varsigma(r - z) + \sigma/\varsigma] & -1 + \zeta & 0 \\ \frac{1}{2}\varsigma y & 0 & -b + \zeta \end{pmatrix}
$$

is positive definite in $\mathcal{U}_0$, since in this case Corollary 5.4 with $V(x, y, z) \equiv \text{const}$ gives the estimate $\dim_F \mathcal{K} < d$. Here d is an arbitrary non-negative number satisfying the inequality $(3 - d)\zeta - (\sigma + b + 1) < 0$. The previous inequality is equivalent to $d > 3 - (\sigma + b + 1)/\zeta$. Whence, by (5.105), the proof of the theorem is completed.

In order to prove the positive definiteness, it is sufficient, by virtue of the Sylvester criterion and the choice of w, to verify that

$$\psi(y, z) := \det\left[\frac{1}{2}(J(y, z)^* + J(y, z)) + \zeta I\right] > 0,$$

when $(y, z) \in \overline{\mathcal{D}}_1$. But for $(y, z) \in \overline{\mathcal{D}}_1$ we have

$$\psi(y, z) \geq (\zeta - \sigma)(\zeta - 1)(\zeta - b) - \frac{1}{4}\varsigma^2 l^2 r^2 + \frac{1}{4}\varsigma^2(b - 1)\left[z - r + \frac{\sigma(\zeta - b)}{\varsigma^2(b - 1)}\right]^2 - \frac{\sigma^2(\zeta - b)(\zeta - 1)}{4\varsigma^2(b - 1)}. \tag{5.106}$$

Since $\zeta > b + \varsigma^2 r(b - 1)/\sigma$ and $z \geq 0$, we see that the expression in square brackets attains its minimum for $z = 0$. Taking this fact into account and also the form of coefficients of equation (5.102), we obtain

$$\psi(y, z) \geq \zeta^3 + a_1\zeta^2 + a_2\zeta + a_3.$$

Consequently, by virtue of the choice of w, $\psi(y, z) > 0$ in $\overline{\mathcal{D}}_1$. □

Example 5.3 For values of parameters $\sigma = 10, b = 8/3, r = 28$ Theorem 5.20 with $\varsigma = 0.6$ gives for an attractor $\mathcal{A}$ of (5.101) the estimate $\dim_F \mathcal{A} < 2.405\ldots$.

Corollary 5.13 *Let $\mathcal{A}$ be an attractor of system (5.101). Suppose, $1 < b \leq 2$. Then*

$$\dim_F \mathcal{A} \leq 3 - \frac{2(\sigma + b + 1)}{\sigma + 1 + \sqrt{(\sigma - 1)^2 + 4\sigma r}}. \tag{5.107}$$

Proof Without loss of generality we can assume that the inequality

$$r \geq (b + 1)(b/\sigma + 1) \tag{5.108}$$

is satisfied, since in the case of validity of the opposite inequality system (5.101) is gradient-like by Corollary 5.10.

According to hypothesis $b \leq 2$, we have $l = 1$ and equation (5.102) can be written in the form

$$(\zeta - b)\left[\zeta^2 - (\sigma + 1)\zeta + \sigma - (\varsigma r + \sigma/\varsigma)^2\right] = 0. \tag{5.109}$$

Let us show that for $\varsigma := \sqrt{\sigma/r}$ the number

$$\zeta_0 := \frac{1}{2}\left(\sigma + 1 + \sqrt{(\sigma - 1)^2 + 4\sigma r}\right)$$

is the maximum root of equation (5.102). This follows from the inequality

$$\zeta_0 \geq \sigma + b + 1 \tag{5.110}$$

which is equivalent to

$$\sqrt{(\sigma - 1)^2 + 4\sigma r} \geq 2b + \sigma + 1\,.$$

The previous inequality is equivalent to hypothesis (5.108).

From (5.102) it follows also that for $\varsigma = \sqrt{\sigma/r}$ we have

$$\zeta_0 > b + \varsigma^2 r(b - 1)/\sigma = 2b - 1,$$

since $\sigma + b + 1 > 2b - 1$ for $b \leq 2$. Thus, by virtue of Theorem 5.20

$$\dim_F \mathcal{K} \leq 3 - (\sigma + b + 1)/\zeta_0.$$

□

The region of parameters for which the simpler estimate, given by Corollary 5.13 is true, does not include Lorenz's values of parameters ($\sigma = 10, b = 8/3, r = 28$).

The following theorem [31, 32] which is proved with the use of a Lyapunov function $V(x, y, z) \not\equiv \text{const}$ shows the validity of the estimate (5.107) in another domain of parameters that includes in particular the Lorenz's values of parameters.

Theorem 5.21 *Let $\mathcal{A}$ be an attractor of system (5.101) and assume that $b > 1$ and the inequalities*

$$\sigma + 1 - 2b \geq 0, \quad \sigma^2 r(4 - b) + 2\sigma(b - 1)(2\sigma - 3b) - b(b - 1)^2 \geq 0 \tag{5.111}$$

hold. Then the estimate (5.107) is true.

Proof By the linear change of variables

$$(x, y, z) \mapsto (\varsigma x, \frac{\varsigma(b - 1)}{\sigma}x + y, z)\,,$$

where $\varsigma > 0$ is a varying parameter, system (5.101) takes the form

$$\begin{aligned}
\dot{x} &= (b - \sigma - 1)x + \frac{\sigma}{\varsigma}y, \\
\dot{y} &= \varsigma\left[r + \frac{(b - 1)(\sigma - b)}{\sigma}\right]x - by - \varsigma xz, \\
\dot{z} &= -bz + \frac{\varsigma^2(b - 1)}{\sigma}x^2 + \varsigma xy.
\end{aligned} \tag{5.112}$$

For the symmetrized Jacobian matrix of the right-hand side of system (5.112)

$$\begin{pmatrix} b-\sigma-1 & \frac{\varsigma}{2}\left[r-z+\frac{(b-1)(\sigma-b)}{\sigma}+\frac{\sigma}{\varsigma^2}\right] & \frac{\varsigma^2(b-1)}{\sigma}x+\frac{1}{2}\varsigma y \\ \frac{\varsigma}{2}\left[r-z+\frac{(b-1)(\sigma-b)}{\sigma}+\frac{\sigma}{\varsigma^2}\right] & -b & 0 \\ \frac{\varsigma^2(b-1)}{\sigma}x+\frac{1}{2}\varsigma y & 0 & -b \end{pmatrix}$$

the eigenvalues at a point (x, y, z) are

$$\lambda_{1,3}(x,y,z) = \frac{1}{2}\left[-(\sigma+1) \pm \sqrt{P(x,y,z)}\right], \qquad \lambda_2 = -b,$$

where

$$P(x,y,z) =$$
$$\left[\varsigma\Big(r-z+\frac{(b-1)(\sigma-b)}{\sigma}\Big)+\frac{\sigma}{\varsigma}\right]^2 + \left[2\frac{\varsigma^2(b-1)}{\sigma}x+\varsigma y\right]^2 + (\sigma+1-2b)^2.$$

Note that the first of the inequalities (5.111) guarantees the relations $\lambda_1(x,y,z) \ge \lambda_2 \ge \lambda_3(x,y,z)$. Condition (5.95) now takes the form

$$-(\sigma+1+2b) - (\sigma+1)s + (1-s)\sqrt{P}(x,y,z) + 2\dot{V}(x,y,z) < 0. \quad (5.113)$$

In order to exclude the square root, we consider the inequality

$$\sqrt{P} \le \tau^2 P + \frac{1}{4\tau^2}, \quad (5.114)$$

where $\tau \ne 0$ is a new varying parameter.

Let us take a Lyapunov function V in the form

$$V(x,y,z) = \frac{1}{2}(1-s)\tau^2 V_1(x,y,z), \quad (5.115)$$

where

$$V_1(x,y,z) := \tau_1 x^2 + \tau_2\left[y^2+z^2+2\frac{\varsigma(b-1)}{\sigma}xy\right] + \tau_3 z$$

and τ_1, τ_2, τ_3 are also varying parameters.

From (5.114) and (5.115) it follows that (5.113) is valid if the inequality

$$-(\sigma+1+2b) - (\sigma+1)s + \frac{1-s}{4\tau^2} + (1-s)\tau^2(P+\dot{V}) < 0, \; (x,y,z) \in \mathcal{K} \quad (5.116)$$

is satisfied. We have

$$P(x, y, z) + \dot{V}(x, y, z) = A_1 x^2 + A_2 y^2 + A_3 z^2 + A_4 xy + A_5 z + P_0,$$

where

$$\begin{aligned}
A_1 &:= 2\tau_1(b - \sigma - 1) + 2\tau_2 \frac{\varsigma^2(b-1)}{\sigma}\Big(r + \frac{(b-1)(\sigma - b)}{\sigma}\Big) \\
&\quad + \tau_3 \frac{\varsigma^2(b-1)}{\sigma} + 4\frac{\varsigma^4}{\varsigma^2}(b-1)^2, \\
A_2 &:= -2\tau_2 + \varsigma^2, \qquad A_3 := -2\tau_2 b + \varsigma^3, \\
A_4 &:= 2\tau_1 \frac{\sigma}{\varsigma} + 2\tau_2 \frac{\varsigma}{\sigma}(\sigma r - b^2 + 1) + \tau_3 \varsigma + 4\frac{\varsigma^3(b-1)}{\sigma}, \\
A_5 &:= -\tau_3 b - 2\varsigma\Big[\varsigma\Big(r + \frac{(b-1)(\sigma - b)}{\sigma}\Big) + \frac{\sigma}{\varsigma}\Big], \qquad P_0 := P(0,0,0).
\end{aligned}$$

Let us show now that we can to choose the parameters τ_1, τ_2, and τ_3 such that

$$P(x, y, z) + \dot{V}(x, y, z) \le P_0\,, \quad (x, y, z) \in \mathbb{R}^3\,. \tag{5.117}$$

Put

$$\tau_3 = -\frac{2\varsigma}{b}\left[\varsigma\left(r + \frac{(b-1)(\sigma - b)}{\sigma}\right) + \frac{\sigma}{\varsigma}\right]. \tag{5.118}$$

Then $A_5 = 0$.

In order to annul the coefficients A_1 and A_4, we choose the parameters τ_1 and τ_2 such that the following system of equations hold:

$$\begin{aligned}
&(b - \sigma - 1)\tau_1 + \frac{\varsigma^2(b-1)}{\sigma}\left(r + \frac{(b-1)(\sigma - b)}{\sigma}\right)\tau_2 \\
&\qquad\qquad = -2\frac{\varsigma^4}{\sigma^2}(b-1)^2 - \frac{1}{2}\tau_3 \frac{\varsigma^2}{\sigma}(b-1), \\
&\frac{\sigma}{\varsigma}\tau_1 + \frac{\varsigma}{\sigma}(\sigma r - b^2 + 1)\tau_2 = -2\frac{\varsigma^3}{\sigma}(b-1) - \frac{1}{2}\tau_3 \varsigma.
\end{aligned}$$

The determinant of this system is

$$\Delta := \frac{\varsigma}{\sigma}\big[2b(\sigma + 1) - (\sigma^2 r + 2\sigma + b^2 + 1)\big].$$

Without loss of generality we can assume that $r \ge 1$, since in the opposite case the equilibrium $(0, 0, 0)$ of system (5.101) is globally asymptotically stable. Therefore

$$\Delta \le \frac{\varsigma}{\sigma}\big[2b(\sigma + 1) - (\sigma^2 + 2\sigma + b^2 + 1)\big] = \frac{\varsigma}{\sigma}\big[-\sigma(\sigma + 1 - 2b) - \sigma - (b-1)^2\big].$$

Taking into account the first of inequalities (5.111), we conclude that $\Delta < 0$ and, consequently, system (5.119) is uniquely solvable with respect to τ_1 and τ_2.

Since by the hypothesis of the theorem $b > 1$, we have

$$\varsigma^2 < 2\tau_2 , \tag{5.119}$$

$A_2 \leq 0$ and $A_3 \leq 0$. From (5.119) it follows that

$$\tau_2 = \frac{\left[4\varsigma^2(b-1) + \sigma\tau_3\right]\sigma}{2\left[2b(\sigma+1) - (\sigma^2 r + 2\sigma + b^2 + 1)\right]}. \tag{5.120}$$

Substituting into (5.119) the values of τ_2 and τ_3 defined by equalities (5.120) and (5.118), respectively, we obtain the condition

$$\sigma^2 r(2-b) + 2\sigma(b-1)(\sigma-2b) - b(b-1) - 2\frac{\sigma^3}{\varsigma^2} \geq 0. \tag{5.121}$$

Thus, we have seen that if inequality (5.121) holds, then we can choose the parameters τ_1, τ_2, τ_3 such that inequality (5.117) is satisfied.

Supposing that (5.121) is valid we change inequality (5.116) to

$$-(\sigma+1+2b) - (\sigma+1)s + \frac{1-s}{4\tau^2} + (1-s)\tau^2 P_0 < 0.$$

Defining $\tau^2 := \frac{1}{2}(P_0)^{-1/2}$ we rewrite the last inequality in the form

$$-(\sigma+1+2b) - (\sigma+1)s + (1-s)\sqrt{P_0} < 0.$$

For

$$\varsigma := \sigma/\sqrt{\sigma r + (b-1)(\sigma-b)} \tag{5.122}$$

we have

$$-(\sigma+1+2b) - (\sigma+1)s + (1-s)\sqrt{(\sigma-1)^2 + 4\sigma r} < 0,$$

which is true if

$$s > 1 - \frac{2(\sigma+b+1)}{\sigma+1+\sqrt{(\sigma-1)^2+4\sigma r}}.$$

Whence it follows that estimate (5.107) holds, since condition (5.121) for ς defined by equality (5.122) coincides with the second of inequalities (5.111). □

Example 5.4 Let $\mathcal{A}$ be an attractor of the Lorenz system (5.101) for $\sigma = 10$, $b = 8/3$, $r = 28$. For these parameters the inequalities (5.111) are satisfied and the estimate (5.107) gives $\dim_F \mathcal{A} < 2.401\ldots$.

Using Corollary 5.12 and the proof of the previous theorem, we can obtain the following result.

Theorem 5.22 *Let inequality (5.111) be satisfied. If*

$$r < (b+1)(b/\sigma + 1), \tag{5.123}$$

then each positive semi-orbit of system (5.101) converges to an equilibrium.

Example 5.5 For $\sigma = 10,\ b = 8/3$ condition (5.123) guarantees the convergence of all positive semi-orbits of the Lorenz system for $r < 4.64$.

We recall that for $b \le 2$ condition (5.123) was proved in Sect. 3.4, Chap. 3.

5.5.3 Equations of the Third Order

In this subsection we will derive an upper fractal dimension estimate for the compact sets, which are invariant with respect to the flow, generated by the equation

$$\dddot{x} + a\ddot{x} + b\dot{x} + f(x) = 0\,, \tag{5.124}$$

where $a,\ b$ are positive numbers and $f : \mathbb{R} \to \mathbb{R}$ is a continuously differentiable function. Assume for this that for any $p = (p_1, p_2, p_3) \in \mathbb{R}^3$ the solution $x(\cdot, p)$ of (5.124) satisfying $x(0, p) = p_1, \dot{x}(0, p) = p_2, \ddot{x}(0, p) = p_3$, exists on $\mathbb{R}$. Define by

$$\varphi^t(p) := (x(t, p),\ \dot{x}(t, p),\ \ddot{x}(t, p)), \quad t \in \mathbb{R}, p \in \mathbb{R}^3\,,$$

the flow in $\mathbb{R}^3$ generated by (5.124).

The deduction of the estimate is based on Corollary 5.12 and on the construction of a Lyapunov function of the form "quadratic + linear forms". The introduction of functions of this kind makes it possible to obtain dimension estimates without localization of attractors in the three-dimensional phase space of the associated flow. All results of Sect. 5.5.3 are due to [6].

Consider the auxiliary function

$$h(x; \kappa_1, \kappa_2) := \sqrt{a^2 + b + \left(f'(x)/b - a\right)^2} + (\kappa_1 x + \kappa_2) f(x),$$

where $\kappa_1 \ge 0$ and κ_2 are arbitrary real numbers.

Theorem 5.23 *Suppose that $\mathcal{K}$ is a compact set which is invariant with respect to the flow generated by Eq. (5.124). Then the estimate*

$$\dim_F \mathcal{K} \le 3 - 2a/(a+k) \tag{5.125}$$

with $k := \inf_{\kappa_1 \ge 0, \kappa_2 \in \mathbb{R}} \sup_{x \in \mathbb{R}} h(x; \kappa_1, \kappa_2)$ *is true.*

Proof Let us make in (5.124) a change of the time variable by $t \mapsto b^{-1/2} t$ and rewrite the equation in the form

$$\dddot{x} + a_1 \ddot{x} + \dot{x} + f_1(x) = 0, \tag{5.126}$$

where $a_1 := ab^{-1/2}$, and $f_1(x) := b^{-3/2} f(x)$. If we put $y = \dot{x}$, $z = -x - \ddot{x}$ the Eq. (5.126) will be transformed to the system

$$\dot{x} = y, \qquad \dot{y} = -x - z, \qquad \dot{z} = -a_1 x - a_1 z + f_1(x). \tag{5.127}$$

(Note that this transformation, different to the standard one, does not change the fractal dimension of the associated invariant sets.) The symmetrized Jacobian matrix of the right-hand side of this system in an arbitrary point (x, y, z) has the eigenvalues

$$\lambda_{1,3}(x) = \frac{1}{2}\Big[-a_1 \pm \sqrt{a_1^2 + 1 + (f_1'(x) - a_1)^2} \Big], \quad \lambda_2 = 0.$$

Now we introduce the function

$$V(x, y, z) := \frac{1}{2}(1-s)\Big[\tau_1 (xz - a_1 xy + \frac{1}{2}x^2 + \frac{1}{2}y^2) + \tau_2 (z - a_1 y) \Big],$$

where s is some number from the interval $(0, 1)$, $\tau_1 \ge 0$ and τ_2 are arbitrary real numbers. A direct computation shows that

$$\dot{V} = \frac{1}{2}(1-s)\big[\tau_1 x f_1(x) - \tau_1 a_1 y^2 + \tau_2 f_1(x) \big],$$

where $\dot{V}$ is the derivative of V with respect to system (5.127). Condition (5.95) results now from of the inequality

$$-(1+s)a_1 + (1-s)\big[\sqrt{a_1^2 + 1 + (f_1' - a_1)^2} + (\tau_1 x + \tau_2) f \big] < 0, \quad \forall x \in \mathbb{R},$$

which is true if

$$-(1+s)a + (1-s) \sup_{x \in \mathbb{R}} h(x; \kappa_1, \kappa_2) < 0,$$

where $\kappa_1 = \tau_1 / b$ and $\kappa_2 = \tau_2 / b$. It follows that estimate (5.125) is valid. □

Let us illustrate the use of Theorem 5.17 by some examples.

(a) Equations with a quadratic nonlinearity: Consider Eq. (5.124) with $f(x) := cx - dx^2$, i.e.

$$\dddot{x} + a\ddot{x} + b\dot{x} + cx - dx^2 = 0\,, \tag{5.128}$$

where c, d are real non-zero parameters.

Theorem 5.24 *Let $\mathcal{K}$ be a set which is compact and invariant with respect to the flow generated by equation (5.128). Then we have*

$$\dim_F \mathcal{K} \le 3 - 2a/\big[a + \sqrt{a^2 + b + (ab + |c|)^2/b^2}\,\big]. \tag{5.129}$$

Proof Take the function h from Theorem 5.23 with $\kappa_1 = 0$

$$h(x; 0, \kappa) = \sqrt{a^2 + b + (f'(x)/b - a)^2} + \kappa f(x)$$

(we omit for brevity the index of the parameter κ_2) and represent h in the form

$$\begin{aligned} h(x; 0, \kappa) = &-\left[\tau\sqrt{a^2 + b + (f'(x)/b - a)^2} - \frac{1}{2\tau}\right]^2 \\ &+ \tau^2\big[a^2 + b + (f'(x)/b - a)^2\big] + \frac{1}{4\tau^2} + \kappa f(x), \end{aligned}$$

where $\tau \neq 0$ is a new varying parameter. Without loss of generality we can further suppose that $d = 1$, since in the opposite case we can make in (5.128) the transformation $x \mapsto x/d$. We obtain for arbitrary x (which is omitted in h) the inequality

$$\begin{aligned} h &\le \tau^2\big[a^2 + b + (c/b - a - 2/bx)^2\big] + \frac{1}{4\tau^2} + \kappa c x - \kappa x^2 \\ &= -\,(\kappa - 4\tau^2/b^2)\left[x + \frac{4\tau^2(c/b - a)/b - \kappa c}{2(\kappa - 4\tau^2/b^2)}\right]^2 + \frac{\big[4\tau^2(c/b - a)/b - \kappa c\big]^2}{4(\kappa - 4\tau^2/b^2)} \\ &+ \tau^2\big[a^2 + b + (c/b - a)^2\big] + \frac{1}{4\tau^2}. \end{aligned}$$

If we choose κ and τ such that $\kappa > 4\tau^2/b^2$, then

$$h \le \frac{\big[4\tau^2(c/b - a)/b - \kappa c\big]^2}{4(\kappa - 4\tau^2/b^2)} + \tau^2\big[a^2 + b + (c/b - a)^2\big] + \frac{1}{4\tau^2}.$$

Finally we put

$$\kappa := \frac{4\tau^2(ab + |c|)}{b^2|c|}, \qquad \tau^2 := \frac{1}{2}\big[a^2 + b + (ab + |c|)^2/b^2\big]^{-1/2}.$$

Since we have in this case the inequality

$$h \leq \sqrt{a^2 + b + (ab + |c|)^2/b^2}\,,$$

the estimate (5.129) follows by Theorem 5.23. □

Example 5.6 For the values $a = 0.1$, $b = 1$, $c = d = -0.58$ which are considered in numerical simulations, the estimate (5.129) gives $\dim_F \mathcal{K} < 2.848$.

It is obvious that the system

$$\dot{x} = y - ax, \qquad \dot{y} = z - bx, \qquad \dot{z} = -cx + dx^2, \tag{5.130}$$

can be transformed to an Eq. (5.128). This system is used in [20] for the simulation of the propagation of excitations in nerve fibers. The study of (5.130) by numerical simulation in [20] shows the existence of a strange attractor in a certain part of the parameter space. In particular, a strange attractor of system (5.130) was found for $a = 1$, $b = 2$, $c = 3.5$, $d = 2$. In this case we obtain from (5.129) the estimate $\dim_F \mathcal{K} < 2.530$.

The Rössler system (5.96), i.e.

$$\dot{x} = -y - z, \qquad \dot{y} = x, \qquad \dot{z} = -\beta z + \alpha(y - y^2), \tag{5.131}$$

where α and β are positive parameters can also be reduced to an equation of the form (5.128). Differentiating the second equation of this system twice and using now the other two equations, we obtain the third order equation

$$\dddot{y} + \beta\ddot{y} + \dot{y} + (\alpha + \beta)y - \alpha y^2 = 0\,. \tag{5.132}$$

From Theorem 5.24 it follows now that for any compact set $\mathcal{K}$ which is invariant with respect to the flow generated by equation (5.132)

$$\dim_F \mathcal{K} \leq 3 - 2\beta/\big[\beta + \sqrt{\beta^2 + 1 + (\alpha + 2\beta)^2}\,\big].$$

(b) Equations with a cubic nonlinearity: Consider Eq. (5.128) with $f(x) = cx - dx^3$, where c, d are positive constants, i.e.

$$\dddot{x} + a\ddot{x} + b\dot{x} + cx - dx^3 = 0. \tag{5.133}$$

Theorem 5.25 *Let $\mathcal{K}$ be a compact set which is invariant with respect to the flow generated by equation (5.133). Then the estimate*

$$\dim_F \mathcal{K} \leq 3 - 2a/\big[a + \sqrt{a^2 + b + (ab + 2c)^2/b^2}\,\big]$$

is true.

Proof The proof is analogous to the proof of the previous theorem if we take the function h with $\kappa_2 = 0$. □

Example 5.7 For the values of parameters $a = 1$, $b = 3.5$, $c = 9.6$, $d = 1$ in the paper [1, 3] it was stated by numerical integration, that a strange attractor for the flow, generated by equation (5.133), exists. If follows from Theorem 5.25 that its fractal dimension is not greater than 2.745.

Remark 5.7 Let us remark that in contrast to the case of a quadratic nonlinearity we obtained in Theorem 5.25 the dimension estimate only for $d > 0$. This constraint is due to our method. There are examples [3, 15], which show that for $d < 0$ the flow, generated by equation (5.131) has a strange attractor. The condition $c > 0$ is not really a constraint, since for $c < ab$ and $d > 0$ equation (5.133) has no compact invariant sets different from the equilibrium states. This follows from the next result.

Lemma 5.8 *Suppose that the flow $\{\varphi^t\}_{t\in\mathbb{R}}$, generated by equation (5.124), has only isolated equilibrium states. Then each bounded positive semi-orbit of a motion $t \mapsto \varphi^t(p) = (x(t,p), \dot{x}(t,p), \ddot{x}(t,p))$ which satisfies one of the conditions*

$$(1)\ \limsup_{t\to+\infty} f'\big(x(t,p)\big) < ab \quad \textit{or} \quad (2)\ \liminf_{t\to+\infty} f'\big(x(t,p)\big) > ab,$$

tends to an equilibrium state.

Proof Analogously to the proof of Theorem 5.23 let us consider equation (5.126) instead of equation (5.124). Assuming $y = \dot{x}$, $z = a_1\dot{x} + \ddot{x}$, $f_1(x) = b^{-3/2} f(x)$, we can write (5.126) in the form of the following system

$$\dot{x} = y, \qquad \dot{y} = -a_1 y + z, \qquad \dot{z} = -y - f_1(x)\,. \tag{5.134}$$

Suppose that condition (1) is satisfied and consider the function

$$V(x,y,z) := \frac{1}{2}(y^2 + z^2) + y f_1(x) + a_1 \int_0^x f_1(\zeta)\, d\zeta.$$

For the derivative of V with respect to (5.134) we have

$$\dot{V} = -\big(a_1 - f_1'(x)\big)y^2 = -b^{-3/2}\big(ab - f'(x)\big)y^2. \tag{5.135}$$

Therefore, if we denote by $u(t,u_0) = \big(x(t,u_0), y(t,u_0), z(t,u_0)\big)$ that solution of system (5.134) which corresponds to the bounded solution $\varphi^t(p)$, then the function $V\big(u(t,u_0)\big)$ does not increase in t on some interval (τ,∞). From this and from the boundedness of $V\big(u(t,u_0)\big)$ it follows that there exists the finite limit $\lim_{t\to+\infty} V\big(u(t,u_0)\big) = l$. From the boundedness of the orbit $u(t,u_0)$ on $(0,\infty)$ it follows that the set $\omega(u_0)$ of its ω-limit points is not empty. Let $q \in \omega(u_0)$ be arbitrary and let $u(t,q) = (x(t,q), y(t,q), z(t,q))$ be the solution of (5.134) passing through q. By Proposition 1.4, Chap. 1, $u(t,q) \in \omega(u_0)$ $\forall t \in \mathbb{R}$. Therefore $V\big(u(t,q)\big) \equiv l$

$\forall t \in \mathbb{R}$. From condition (1) of the lemma it follows also that $f'\big(x(t,q)\big) < ab\,\forall t \in \mathbb{R}$. But then, using (5.135), we obtain the identity $y(t,q) \equiv 0$. Taking this into account, we get from the second equation (5.134) $z(t,q) \equiv 0$ and from the first equation $x(t,q) \equiv \text{const}$. Consequently, all points of $\omega(u_0)$ are equilibrium states. Since the equilibrium states are supposed to be isolated, Lemma 5.8 is proved under the assumption that condition (1) is satisfied.

In the case that condition (2) is satisfied the proof of the lemma is analogous if we use instead of V the function $-V$. □

5.5.4 Equations Describing the Interaction Between Waves in Plasma

We have shown above that the introduction of a Lyapunov function V in the estimates of dimension makes it possible to obtain a result without localization of attractors in the phase space. But if a system is dissipative and for the estimation of its domain of dissipation some Lyapunov function V is used, then the same function V or, more generally, a function $\psi(V)$, where ψ is some continuously differentiable function, may be used as V. The example considered in the present subsection illustrates this situation.

Let us return to Rabinovich's system (5.76) describing the interaction of waves in plasma

$$\dot{x} = hy - x - yz, \qquad \dot{y} = hx - \omega y + xz, \qquad \dot{z} = -z + xy, \tag{5.136}$$

where h, ω are positive numbers. In Sect. 5.3.4, considering (5.136) in the frames of the generalized Lorenz system, we obtained a sufficient condition for gradient flow-like behaviour of this system. Further, this result will be improved. Besides, the following estimate of attractor dimension [7] of system (5.136) will be obtained.

Theorem 5.26 *Let $\mathcal{A}$ be an attractor of system (5.136). Then*

$$\dim_F \mathcal{A} \le 3 - \frac{2(\omega+2)}{\omega + 1 + \sqrt{(\omega-1)^2 + k_1 h^2}} \tag{5.137}$$

with $k_1 = (13\sqrt{13} + 35)/18$.

Proof Let us make the change of variables in (5.136) by $(x, y, z) \mapsto (x, \varsigma y, z)$, where $\varsigma \neq 0$ is a varying parameter. In the new variable system (5.136) takes the form

$$\dot{x} = \varsigma hy - x - \varsigma yz, \qquad \dot{y} = \frac{h}{\varsigma}x - \omega y + \frac{1}{\varsigma}xz, \qquad \dot{z} = -z + \varsigma xy. \tag{5.138}$$

The eigenvalues of the symmetrized Jacobian matrix of the right-hand side of system (5.138) are

$$\lambda_{1,3}(x,z) = \frac{1}{2}\{-(\omega+1) \pm \sqrt{P(x,z)}\}, \qquad \lambda_2 = -1,$$

where

$$P(x,z) := (\omega-1)^2 + \left(\frac{1}{\varsigma}+\varsigma\right)^2 x^2 + \left[\left(\frac{1}{\varsigma}+\varsigma\right)h + \left(\frac{1}{\varsigma}-\varsigma\right)z\right]^2.$$

Condition (5.95) can be written in the form

$$-(\omega+3) - (\omega+1)s + (1-s)\sqrt{P} + 2\dot{V} < 0. \tag{5.139}$$

In order to exclude the square root let us consider the inequality

$$\sqrt{P} \le \tau^2 P + \frac{1}{4\tau^2}, \tag{5.140}$$

where $\tau \neq 0$ is a varying parameter.

From the results of Sect. 1.4, Chap. 1, it follows that: (1) system (5.138) is dissipative, i.e. in the phase space there is an ellipsoid $\mathcal{E}(\varsigma)$ such that any trajectory of the system arrives to it in a finite time and further remains there; 2) as a dissipativity domain one can choose also the more "narrow" set $\mathcal{G}_0$, given by

$$\mathcal{G}_0 := \mathcal{E}(\varsigma) \cap \{x, z |\, V_1(x,z) \le h^2\},$$

where $V_1(x,z) := x^2 + (z-h)^2$. We choose a Lyapunov function V in the form

$$V(x,z) := \frac{1}{2}(1-s)\tau^2 \kappa V_1(x,z),$$

where κ is a varying parameter. Then, taking into account (5.140), we can conclude that inequality (5.139) will be satisfied if

$$-(\omega+3) - (\omega+1)s + \frac{1-s}{4\tau^2} + (1-s)\tau^2(P + \kappa \dot{V}_1) < 0. \tag{5.141}$$

We have

$$P + \kappa \dot{V}_1 = \phi(x,z) + P_0,$$

where $\phi(x,z) := Ax^2 + Bz^2 + Chz$, $P_0 := P(0,0)$, and

$$A := \left(\frac{1}{\varsigma}+\varsigma\right)^2 - \kappa, \quad B := \left(\frac{1}{\varsigma}-\varsigma\right)^2 - \kappa, \quad C := 2\left(\frac{1}{\varsigma^2}-\varsigma^2\right) + \kappa.$$

Since $\mathcal{K} \subset \mathcal{G}_0$ and for the points of $\mathcal{G}_0$ the relation $x^2 + z^2 - 2hz \le 0$ is satisfied, we have for $A \ge 0$, i.e. if

$$\kappa \le \left(\frac{1}{\varsigma} + \varsigma\right)^2, \tag{5.142}$$

is true, the inequality $Ax^2 \le -Az^2 + 2Ahz$. Consequently, for points in $\mathcal{G}_0$

$$\phi \le -(A - B)z^2 + (2A + C)hz = -(A - B)\left(z - \frac{(2A + C)h}{2(A - B)}\right)^2 + \frac{(2A + C)^2 h^2}{4(A - B)}.$$

Since $A - B = 4$, in the case of validity of (5.142) we have $\phi \le (2A + C)^2 h^2/16$. Further, $2A + C = 4(1/\varsigma^2 + 1) - \kappa$. Choose now

$$\kappa := 4(1/\varsigma^2 + 1)\tau,$$

where τ is a new varying parameter. Then $\phi \le (1 - \tau)^2 (1/\varsigma^2 + 1)^2 h^2$.

Thus, (5.141) will be satisfied if

$$-(\omega + 3) - (\omega + 1)s + \frac{1 - s}{4\tau^2} + (1 - s)\tau^2 \left[(1 - \tau)^2 \left(\frac{1}{\varsigma}^2 + 1\right)^2 h^2 + P_0\right] < 0. \tag{5.143}$$

Let us take $\tau^2 := 1/(2\iota)$, where

$$\iota := \sqrt{(1 - \tau)^2 \left(\frac{1}{\varsigma^2} + 1\right)^2 h^2 + P_0}.$$

If (5.142) is true, then (5.143) takes the form

$$-(\omega + 3) - (\omega + 1)s + (1 - s)\iota < 0. \tag{5.144}$$

Whence

$$s > \frac{\iota - \omega - 3}{\iota + \omega + 1} = 1 - \frac{2(\omega + 2)}{\iota + \omega + 1}. \tag{5.145}$$

Inequality (5.142) is satisfied for $\varsigma^2 \ge 4\tau - 1$. Take

$$\tau := \frac{1}{4}(\sqrt{13} - 1), \qquad \varsigma^2 := \sqrt{13} - 2.$$

Then from (5.145) we obtain estimate (5.137). □

Recall [34] that for $\omega = 4$, $h = 4.92$ system (5.136) has a strange attractor $\mathcal{A}$. Theorem 5.26 gives the estimate $\dim_F \mathcal{A} < 2.246$.

Theorem 5.27 *Each positive semi-orbit of system (5.136) converges to an equilibrium if*

$$h^2 < k_2(\omega + 1), \tag{5.146}$$

where $k_2 := 4(13\sqrt{13} - 35)/27$.

Proof For the proof of this theorem [7] we can use Corollary 5.9. Using the proof of the previous theorem with $s = 0$, we get inequality (5.144), which is equivalent to condition (5.146) if the varying parameters is the same as in the proof of Theorem 5.26. □

For $\omega = 4$ Theorem 5.27 guarantees the convergence of all positive semi-orbits of system (5.136) for $h < 2.96$.

5.6 Estimates of the Topological Entropy

5.6.1 Ito's Generalized Entropy Estimate for Maps

In this subsection we derive upper estimates of the topological entropy of a continuous map acting in a compact metric space $(\mathcal{M}, \rho)$ in terms of asymptotic Lipschitz constants and the fractal dimension or the lower box dimension of $\mathcal{M}$. These estimates involve a known result of Ito [28]. The generalization consists of using in the bound asymptotic local Lipschitz constants and "regulating functions". Our representation in this and the next subsection follows [9, 10].

Let $(\mathcal{M}, \rho)$ be a compact metric space,

$$\varphi : \mathcal{M} \to \mathcal{M} \tag{5.147}$$

be a continuous map, $\varkappa : \mathcal{M} \times \mathcal{M} \to (0, +\infty)$ be a positive continuous function.

Suppose that ρ' is another metric on $\mathcal{M}$ that is equivalent to ρ, i.e. for certain $c_1 > 0$ and $c_2 > 0$ we have

$$c_1\rho(p, q) \le \rho'(p, q) \le c_2\rho(p, q), \quad \forall p, q \in \mathcal{M}. \tag{5.148}$$

Denote

$$k_j := \limsup_{\varepsilon \to 0} \sup_{\rho'(p,q)<\varepsilon} \left[\frac{\rho'\big(\varphi^j(p), \varphi^j(q)\big)}{\rho'(p, q)} \cdot \frac{\varkappa\big(\varphi^j(p), \varphi^j(q)\big)}{\varkappa(p, q)} \right] \tag{5.149}$$

for $j = 1, 2, \ldots$ and

$$k := \inf_{j \ge 1} k_j^{1/j}. \tag{5.150}$$

Theorem 5.28 *Suppose $k < \infty$ and $\dim_F \mathcal{M} < \infty$. Then we have for the map (5.147)*

$$h_{\text{top}}(\varphi, \mathcal{M}) \le \max\{0, \log k\} \dim_F \mathcal{M}. \tag{5.151}$$

Proof Let us take an arbitrary $\kappa > k$ and denote by j_0 an integer number such that $j_0 \ge 1$, $\kappa > k_{j_0}^{1/j_0} \ge k$. If $j_0 > 1$, then we also choose an arbitrary $\kappa_j > k_j^{1/j}$, $j = 1, \dots, j_0 - 1$. Let $\kappa_0 := 1$ and $\kappa_{j_0} := \kappa$.

By virtue of (5.149) we can find an $\varepsilon > 0$ such that

$$\rho'\big(\varphi^j(p), \varphi^j(q)\big) \le \kappa_j^j \frac{\varkappa(p,q)}{\varkappa\big(\varphi^j(p), \varphi^j(q)\big)} \rho'(p,q),$$
$$j = 0, \dots, j_0, \quad \forall p, q \in \mathcal{M},\ \rho'(p,q) < c_2\varepsilon. \tag{5.152}$$

Take an arbitrary number $\varsigma > \dim_F \mathcal{M}$. For all sufficiently small $\delta > 0$

$$\frac{\log N_\delta(\mathcal{M})}{\log \delta^{-1}} < \varsigma. \tag{5.153}$$

Denote

$$\varkappa_1 := \max_{p,q \in \mathcal{M}} \varkappa(p,q) \qquad \text{and} \qquad \varkappa_2 := \min_{p,q \in \mathcal{M}} \varkappa(p,q).$$

We consider now separately the cases $k \ge 1$ and $k < 1$.

<u>Case $k \ge 1$</u> Choose an integer $l > 0$ such that

$$\kappa_{j_0}^{-l} \kappa_s^s \varkappa_1 / \varkappa_2 < c_1/c_2, \quad s = 1, \dots, j_0.$$

Let us fix an arbitrary integer $m > 0$. We are going to show that

$$\rho'(p,q) < \kappa_{j_0}^{-(m+l)} c_2 \varepsilon$$

implies the inequality

$$\rho'\big(\varphi^j(p), \varphi^j(q)\big) \le \kappa_{j_0}^{q j_0} \kappa_s^s \frac{\varkappa(p,q)}{\varkappa\big(\varphi^j(p), \varphi^j(q)\big)} \rho'(p,q) \tag{5.154}$$

for any integer $j \in \{0, \dots, m-1\}$. Here $q = [j/j_0]$ and $s = j - q j_0$.

From (5.154) we have

$$\rho'\big(\varphi^j(p), \varphi^j(q)\big) \le \kappa_{j_0}^{q j_0} \kappa_s^s \frac{\varkappa_1}{\varkappa_2} \kappa_{j_0}^{-(m+l)} c_2 \varepsilon \le \kappa_{j_0}^{-(m-j)} c_1 \varepsilon. \tag{5.155}$$

In particular,

$$\rho'\big(\varphi^j(p), \varphi^j(q)\big) < c_2 \varepsilon \tag{5.156}$$

since $\kappa_{j_0} > k \geq 1$. Now we shall prove (5.154). Suppose that $j \leq j_0 \leq m-1$. Then (5.154) is true by (5.152).

Suppose now that $j = j_0 + s \leq m-1$ and $s \in \{1, \dots, j_0\}$. Then

$$\begin{aligned}\rho'\big(\varphi^j(p), \varphi^j(q)\big) &\leq \kappa_s^s \frac{\varkappa\big(\varphi^{j_0}(p), \varphi^{j_0}(q)\big)}{\varkappa\big(\varphi^j(p), \varphi^j(q)\big)} \rho'\big(\varphi^{j_0}(p), \varphi^{j_0}(q)\big) \text{according to (5.156)} \\ &\leq \kappa_s^s \frac{\varkappa\big(\varphi^{j_0}(p), \varphi^{j_0}(q)\big)}{\varkappa\big(\varphi^j(p), \varphi^j(q)\big)} \kappa_{j_0}^{j_0} \frac{\varkappa(p,q)}{\varkappa\big(\varphi^{j_0}(p), \varphi^{j_0}(q)\big)} \rho'(p,q) \\ &\leq \kappa_{j_0}^{j_0} \kappa_s^s \frac{\varkappa(p,q)}{\varkappa\big(\varphi^j(p), \varphi^j(q)\big)} \rho'(p,q).\end{aligned}$$

Suppose that $j = 2j_0 + s \leq m-1$ and $s \in \{1, \dots, j_0\}$. Then again we have

$$\rho'\big(\varphi^j(p), \varphi^j(q)\big) \leq \kappa_s^s \frac{\varkappa\big(\varphi^{2j_0}(p), \varphi^{2j_0}(q)\big)}{\varkappa\big(\varphi^j(p), \varphi^j(q)\big)} \rho'\big(\varphi^{2j_0}(p), \varphi^{2j_0}(q)\big)$$

according to (5.156)

$$\begin{aligned} &\leq \kappa_s^s \frac{\varkappa\big(\varphi^{2j_0}(p), \varphi^{2j_0}(q)\big)}{\varkappa\big(\varphi^j(p), \varphi^j(q)\big)} \kappa_{j_0}^{2j_0} \frac{\varkappa(p,q)}{\varkappa\big(\varphi^{2j_0}(p), \varphi^{2j_0}(q)\big)} \rho'(p,q) \\ &\leq \kappa_{j_0}^{2j_0} \kappa_s^s \frac{\varkappa(p,q)}{\varkappa\big(\varphi^j(p), \varphi^j(q)\big)} \rho'(p,q).\end{aligned}$$

Thus, if we continue this procedure (5.154) is proved.

From (5.155) and (5.148) it follows that if $\rho(p,q) < \kappa_{j_0}^{-(m+l)} \varepsilon$, then

$$\rho\big(\varphi^j(p), \varphi^j(q)\big) < \varepsilon, \quad j = 0, \dots, m-1.$$

Consequently, the inclusion $\mathcal{B}_\delta(p) \subset \mathcal{B}_\varepsilon(p, m)$ holds with $\delta := \kappa_{j_0}^{-(m+l)} \varepsilon$. But then $N_\varepsilon(\mathcal{M}, m) \leq N_\delta(\mathcal{M})$. Using (5.153), we obtain

$$\frac{\log N_\varepsilon(\mathcal{M}, m)}{m} \leq \frac{\log N_\delta(\mathcal{M})}{\log \delta^{-1}} \frac{\log \delta^{-1}}{m} < \varsigma \left[(1 + \frac{l}{m}) \log \kappa_{j_0} - \frac{\log \varepsilon}{m} \right].$$

Passing to the upper limit as $m \to +\infty$, we get

$$h_{\text{top}}(\varphi, \mathcal{M}) \leq \varsigma \log \kappa_{j_0}.$$

Since ς and $\kappa = \kappa_{j_0}$ are arbitrary numbers, the assertion of our theorem is proved for $k \geq 1$.

Case $k < 1$: In addition suppose now that $\kappa = \kappa_{j_0} < 1$. Let us choose an integer $l > 0$ such that

$$\kappa_{j_0}^l \varkappa_1/\varkappa_2 < c_1/c_2 \quad \text{and} \quad \kappa_{j_0}^l \kappa_s^s \varkappa_1/\varkappa_2 < c_1/c_2, \quad s = 1, \ldots, j_0.$$

In the same way we can show that if $\rho(p, q) < \kappa_{j_0}^l \varepsilon$, then

$$\rho\big(\varphi^j(p), \varphi^j(q)\big) < \varepsilon, \quad \forall j \geq 0.$$

Therefore $N_\varepsilon(\mathcal{M}, m) \leq N_\delta(\mathcal{M})$ with $\delta = \kappa_{j_0}^l \varepsilon$. Using again (5.153), we obtain

$$\frac{\log N_\varepsilon(\mathcal{M}, m)}{m} < \varsigma\Big(-\frac{l \log \kappa_{j_0}}{m} - \frac{\varepsilon}{m}\Big)$$

and, consequently, $h_{\text{top}}(\varphi) \leq 0$. □

Let us consider a C^0-semi-flow $(\{\varphi^t\}_{t\geq 0}, \mathcal{M}, \rho)$ (see Sect. 3.3.3, Chap. 3). We assume that any map $\varphi^t : \mathcal{M} \to \mathcal{M}$ is Lipschitz and the Lipschitz constants are bounded on an interval $[0, t_0]$ which implies that they are bounded on any interval $[0, t]$. Let

$$\nu(t) := \inf\{\overline{\nu} \in \mathbb{R}_+ \mid \rho(\varphi^t(p), \varphi^t(q)) \leq \overline{\nu}\rho(p, q), \ \forall p, q \in \mathcal{M}\}.$$

Since $\rho(\varphi^{t_1+t_2}(p), \varphi^{t_1+t_2}(q)) \leq \nu(t_1)\nu(t_2)\rho(p, q)$ we get $\nu(t_1 + t_2) \leq \nu(t_1)\nu(t_2)$ for any $t_1, t_2 \geq 0$, i.e. ν is a subexponential function. It follows [19, 41] that there exists the limit

$$\nu := \lim_{t\to\infty} \frac{1}{t} \log \nu(t)\,.$$

Now we can deduce from Theorem 5.28 with $\rho(p, q) \equiv 1$, $\rho'(p, q) \equiv \rho(p, q)$ and $\varphi = \varphi^1$ the following inequality obtained in [19].

Corollary 5.14 *The topological entropy of φ^1 on $\mathcal{M}$ satisfies*

$$h_{\text{top}}(\varphi^1, \mathcal{M}) \leq \max\{0, \nu\} \dim_F \mathcal{M}\,. \tag{5.157}$$

Proof It is clear that $k_j \leq \nu(j)$. Hence, $k \leq [\nu(j)]^{1/j}$ or, equivalently, $\log k \leq \frac{1}{j}\log \nu(j)$. Setting $j \to \infty$, we find that $\log k \leq \nu$. The validity of (5.157) follows now from (5.151). □

Using the notion of lower box dimension we now derive a result which sharpens the estimate (5.151).

Theorem 5.29 *Consider the map (5.147). Suppose that for k, defined by (5.150), we have $k < \infty$ and let $\underline{\dim}_B \mathcal{M} < \infty$. Then*

$$h_{\text{top}}(\varphi, \mathcal{M}) \leq \max\{0, \log k\} \underline{\dim}_B \mathcal{M}\,. \tag{5.158}$$

Proof The proof is similar to the one of the previous theorem. But instead of the Bowen-Dinaburg definition of topological entropy we will use the characterization of the topological entropy by open covers. Let $\mathfrak{U}$ be an open cover of $\mathcal{M}$. Then there exists an $\varepsilon > 0$ such that

$$N\left(\mathfrak{U} \vee \varphi^{-1}\mathfrak{U} \vee \cdots \vee \varphi^{-(m-1)}\mathfrak{U}\right) \le N_{\varepsilon}(\mathcal{M}, m) \tag{5.159}$$

(see the Proof of Proposition 3.27, Chap. 3, inequality 3.17).

Decreasing ε, if necessary, we can assume that ε from (5.159) is such that (5.152) is valid. Consider again separately the two cases $k \ge 1$ and $k < 1$.

Case $k \ge 1$: By the same way as above we obtain the inequality $N_{\varepsilon}(\mathcal{M}, m) \le N_{\delta}(\mathcal{M})$ with $\delta = \kappa_{j_0}^{-(m+l)}\varepsilon$. Hence

$$\frac{1}{m} \log N\left(\mathfrak{U} \vee \varphi^{-1}\mathfrak{U} \vee \cdots \vee \varphi^{-(m-1)}\mathfrak{U}\right) \le \frac{\log N_{\delta}(\mathcal{M})}{\log \delta^{-1}} \frac{\log \delta^{-1}}{m}. \tag{5.160}$$

Since for a decreasing function $\psi : \mathbb{R}_+ \to \mathbb{R}_+$, $\theta \in (0, 1)$ and $\varepsilon_0 > 0$ the equality

$$\liminf_{\varepsilon \to 0} \frac{\log \psi(\varepsilon)}{-\log \varepsilon} = \liminf_{m \to \infty} \frac{\log \psi(\theta^m \varepsilon_0)}{-\log(\theta^m \varepsilon_0)}$$

is satisfied ([22], Lemma 6.2) we get

$$\underline{\dim}_B \mathcal{M} = \liminf_{m \to \infty} \frac{\log N_{\delta_m}(\mathcal{M})}{\log \delta_m^{-1}}, \tag{5.161}$$

where $\delta_m := \theta^m \varepsilon_0$, $\theta := 1/\kappa_{j_0}$ and $\varepsilon_0 := \kappa_{j_0}^{-l}\varepsilon$. Let us choose an arbitrary $\varsigma > \underline{\dim}_B \mathcal{M}$. Due to (5.161) there exists a subsequence $m' \to \infty$ such that for all sufficiently large m'

$$\frac{\log N_{\delta_{m'}}(\mathcal{M})}{\log \delta_{m'}^{-1}} < \varsigma. \tag{5.162}$$

Setting in (5.160) $m = m' \to \infty$ and taking into account that for the left-hand side there exists the limit (see Sect. 3.3, Chap. 3), we obtain

$$h(\varphi, \mathfrak{U}) < \varsigma \log \kappa_{j_0}.$$

Since the cover $\mathfrak{U}$, numbers ς and $\kappa = \kappa_{j_0}$ where arbitrary we have proved (5.158) for $k \ge 1$.

Case $k < 1$: As in the proof of Theorem 5.28 we have $N_{\varepsilon}(\mathcal{M}, m) \le N_{\delta}(\mathcal{M})$ with $\delta := \kappa_{j_0}^{l}\varepsilon$.

Consequently, (5.160) is valid. Since δ now does not depend on m, we let $m \to +\infty$ in (5.160) and find that $h(\varphi, \mathfrak{U}) \le 0$. □

Remark 5.8 The estimate (5.158) involves for $\rho(p, q) \equiv 1$, $k = k_1$ and $\rho'(p, q) \equiv \rho(p, q)$ the related result in [22, 28]. A generalization of this result is given in [2].

5.6.2 Application to Differential Equations

We apply now Theorem 5.28 to the differential equation

$$\frac{d\varphi}{dt} = f(\varphi), \tag{5.163}$$

where $f : \mathbb{R}^n \to \mathbb{R}^n$ is a continuously differentiable function such that the flow $\{\varphi^t\}_{t\in\mathbb{R}}$ of (5.163) exists. Let

$$\lambda_1(p) \geq \lambda_2(p) \geq \cdots \geq \lambda_n(p)$$

be the eigenvalues of its symmetrized Jacobian matrix $\frac{1}{2}[Df(p) + Df(p)^*]$ at the point $p \in \mathbb{R}^n$. For a continuously differentiable function $V : \mathcal{U} \to \mathbb{R}$ ($\mathcal{U} \subset \mathbb{R}^n$ an open set) we shall use as above the notation $\dot{V}(p) := (f(p), \operatorname{grad} V(p))$.

Suppose that (5.163) has an invariant compact convex set $\mathcal{K} \subset \mathbb{R}^n$. Recall that the invariance means that the equality $\mathcal{K} = \varphi^t(\mathcal{K})$ holds for all $t \in \mathbb{R}$.

According to the definition of topological entropy of a dynamical system, the topological entropy of system (5.163) with respect to $\mathcal{K}$ is the entropy of the operator φ^1 acting on $\mathcal{K}$. Let us denote it by $h_{\text{top}}(\varphi^1, \mathcal{K})$.

To estimate the asymptotic Lipschitz constant k entering in the estimate given by Theorem 5.28, we use logarithmic norms. Besides, we replace the "regulating" function $\varkappa$, which is used in the definition of k_j, by a Lyapunov function.

Theorem 5.30 *Let Λ be some logarithmic norm in $M_n(\mathbb{R})$ and $V : \mathcal{U} \to \mathbb{R}$ be a function which is continuously differentiable on a neighborhood $\mathcal{U}$ of $\mathcal{K}$. Then the estimate*

$$h_{\text{top}}(\varphi^1, \mathcal{K}) \leq \max\{0, \widetilde{k}\} \dim_F \mathcal{K}$$

is true, where the constant $\widetilde{k}$ is defined by

$$\widetilde{k} := \max_{p\in\mathcal{K}} \left\{\Lambda(Df(p)) + \dot{V}(p)\right\}.$$

Proof By virtue of Theorem 5.28 it is sufficient to show that for any $\varsigma > 0$ the inequality $k_1 < \exp(\widetilde{k} + \varsigma)$ is true, where k_1 is defined by (5.149). For this purpose it is sufficient to check that for all $\varsigma > 0$ a number $\varepsilon > 0$ can be found such that with $\varphi \equiv \varphi^1$

$$\frac{\|\varphi^1(p_1) - \varphi^1(p_2)\|}{\|p_1 - p_2\|} \cdot \frac{\varkappa(\varphi^1(p_1), \varphi^1(p_2))}{\varkappa(p_1, p_2)} < \exp(\widetilde{k} + \varsigma),$$
$$\forall p_1, p_2 \in \mathcal{U},\ 0 < \|p_1 - p_2\| < \varepsilon.$$

Here $\|\cdot\|$ denotes the vector norm in $\mathbb{R}^n$, used in the definition of the logarithmic matrix norm Λ. Now let us fix $\varsigma > 0$ and choose a number $\varepsilon > 0$ to be so small that $\|\Lambda(Df(\varphi(\tau, p_1))) - \Lambda(Df(\varphi(\tau, p_2)))\| < \frac{1}{2}\varsigma |\dot{V}(\varphi(\tau, p_1)) - \dot{V}(\varphi(\tau, p_2))| < \varsigma$, $\forall\, p_1, p_2 \in \mathcal{K}$, $\|p_1 - p_2\| < \varepsilon$, and $\forall \tau \in [0, 1]$. We have

$$\begin{aligned}\varphi^1(p_1) - \varphi^1(p_2) &\equiv \varphi(1, p_1) - \varphi(1, p_2) = \int_0^1 \frac{d}{d\xi}\varphi\big(1, \xi p_1 + (1-\xi)p_2\big)\, d\xi \\ &= \int_0^1 D_2\varphi\big(1, \xi p_1 + (1-\xi)p_2\big)\, d\xi\, (p_1 - p_2).\end{aligned} \tag{5.164}$$

Therefore

$$\frac{\|\varphi^1(p_1) - \varphi^1(p_2)\|}{\|p_1 - p_2\|} < \|D_2\varphi(1, p')\|, \tag{5.165}$$

where $p' := \xi' p_1 + (1 - \xi')p_2$ and ξ' is some number from $[0, 1]$. Since the matrix D_2u is the Cauchy matrix solution of the system

$$\frac{d\upsilon}{dt} = Df(\varphi(t, p'))\upsilon,$$

we can apply the Lozinskii estimate (Theorem 2.2, Chap. 2) to obtain

$$\frac{\|\varphi^1(p_1) - \varphi^1(p_2)\|}{\|p_1 - p_2\|} \le \exp \int_0^1 \Lambda\big(Df(\varphi(\tau, p'))\big)\, d\tau. \tag{5.166}$$

Put

$$\varkappa(p_1, p_2) := \exp\Big\{\frac{1}{2}\big[V(p_1) + V(p_2)\big]\Big\}.$$

Then

$$\frac{\varkappa(\varphi^1(p_1), \varphi^1(p_2))}{\varkappa(p_1, p_2)} = \exp\left\{\frac{1}{2}\int_0^1 \big[\dot{V}(\varphi(\tau, p_1)) + \dot{V}(\varphi(\tau, p_2))\big]\, d\tau\right\}. \tag{5.167}$$

We have

$$\frac{1}{2}\big[\dot{V}(\varphi(\tau, p_1)) + \dot{V}(\varphi(\tau, p_2))\big] \le \dot{V}(\varphi(\tau, p_1)) + \frac{1}{2}\big|\dot{V}(\varphi(\tau, p_2)) - \dot{V}(\varphi(\tau, p_1))\big|,$$

$$\Lambda\big(Df(\varphi(\tau, p'))\big) \le \Lambda\big(Df(\varphi(\tau, p_1))\big) + \big\|Df(\varphi(\tau, p')) - Df(\varphi(\tau, p_1))\big\|$$

and $\|p' - p_1\| \le \|p_1 - p_2\|$. Therefore, if $p_1, p_2 \in \mathcal{K}$ and $0 < \|p_1 - p_2\| < \varepsilon$, then from (5.165)–(5.167) it follows that

$$\frac{\|\varphi^1(p_1)-\varphi^1(p_2)\|}{\|p_1-p_2\|}\cdot\frac{\varkappa\big(\varphi^1(p_1),\varphi^1(p_2)\big)}{\varkappa(p_1,p_2)}$$
$$< \exp\left\{\int_0^1\big[\Lambda\big(Df\big(\varphi(\tau,p_1)\big)\big)+\dot V\big(\varphi(\tau,p_1)\big)\big]\,d\tau+\varsigma\right\}\le\exp(\widetilde{k}+\varsigma).$$

□

Choosing the logarithmic norm defined by the Euclidean norm, we obtain from Theorem 5.30 the following:

Corollary 5.15 *Let V be a real valued continuously differentiable on a neighborhood of the compact flow-invariant set $\mathcal{K}$ of (5.163) function. Then the estimate*

$$h_{top}(\varphi^1,\mathcal{K})\le\max\{0,\widetilde{k}\}\dim_F\mathcal{K}$$

is true where the constant $\widetilde{k}$ is defined by

$$\widetilde{k}:=\max_{p\in\mathcal{K}}\big[\lambda_1(p)+\dot V(p)\big].$$

References

1. Almeida, J., Barreira, L.: Hausdorff dimension in convex bornological spaces. J. Math. Anal. Appl. **266**, 590–601 (2002)
2. Anikushin, M.M., Reitmann, V.: Development of the topological entropy conception for dynamical systems with multiple time. Electr. J. Diff. Equ. Contr. Process. **4** (2016). (Russian); English transl. J. Diff. Equ. **52**(13), 1655–1670 (2016)
3. Arnédo, A., Coullet, P.H., Spiegel, E.A.: Chaos in a finite macroscopic system. Phys. Lett. **92**A, 369–373 (1982)
4. Benedicks, M. Carleson, L.: The dynamics of the Hénon map. Ann. Math. II. Ser. **133**(1), 73–169 (1991)
5. Boichenko, V.A.: Frequency theorems on estimates of the dimension of attractors and global stability of nonlinear systems. Leningrad, Dep. at VINITI 04.04.90, No. 1832–B90 (1990) (Russian)
6. Boichenko, V.A.: The Lyapunov function in estimates of the Hausdorff dimension of attractors of an equation of third order. Diff. Urav. **30**, 913–915 (1994). (Russian)
7. Boichenko, V.A., Leonov, G.A.: On estimates of attractors dimension and global stability of generalized Lorenz equations. Vestn. Lening. Gos. Univ. Ser. 1, **2**, 7–13 (1990) (Russian); English transl. Vestn. Lening. Univ. Math. **23**(2), 6–12 (1990)
8. Boichenko, V.A., Leonov, G.A.: On Lyapunov functions in estimates of the Hausdorff dimension of attractors. Leningrad, Dep. at VINITI 28.10.91, No. 4 123–B 91 (1991) (Russian)
9. Boichenko, V.A., Leonov, G.A.: Lyapunov functions in the estimation of the topological entropy. Preprint, University of Technology Dresden, Dresden (1994)
10. Boichenko, V.A., Leonov, G.A.: Lyapunov's direct method in estimates of topological entropy. Zap. Nauchn. Sem. POMI **231**, 62–75 (1995) (Russian); English transl. J. Math. Sci. **91**(6), 3370–3379 (1998)
11. Boichenko, V.A., Leonov, G.A.: Lyapunov functions, Lozinskii norms, and the Hausdorff measure in the qualitative theory of differential equations. Am. Math. Soc. Transl. **2**(193), 1–26 (1999)

12. Boichenko, V.A., Leonov, G.A.: On dimension estimates for the attractors of the Hénon map. Vestn. S. Peterburg Gos. Univ. Ser. 1, Matematika, (3), 8–13 (2000) (Russian); English transl. Vestn. St. Petersburg Univ. Math. **33** (3), 5–9 (2000)
13. Boichenko, V.A., Leonov, G.A., Franz, A., Reitmann, V.: Hausdorff and fractal dimension estimates of invariant sets of non-injective maps. Zeitschrift für Analysis und ihre Anwendungen (ZAA) **17**(1), 207–223 (1998)
14. Chen, Z.-M.: A note on Kaplan-Yorke-type estimates on the fractal dimension of chaotic attractors. Chaos, Solitons & Fractals **3**(5), 575–582 (1993)
15. Coullet, P., Tresser, C., Arnédo, A.: Transition to stochasticity for a class of forced oscillators. Phys. Lett. 72 A, 268–270 (1979)
16. Courant, R.: Dirichlet's Principle, Conformed Mappings, and Minimal Surfaces. Springer, New York (1977)
17. Denisov, G.G.: On the rotation of a solid body in a resisting medium. Izv. Akad. Nauk SSSR Mekh. Tverd. Tela **4**, 37–43 (1989) (Russian)
18. Douady, A., Oesterlé, J.: Dimension de Hausdorff des attracteurs. C. R. Acad. Sci. Paris, Ser. A **290**, 1135–1138 (1980)
19. Eden, A., Foias, C., Temam, R.: Local and global Lyapunov exponents. J. Dyn. Diff. Equ. **3**, 133–177 (1991) [Preprint No. 8804, The Institute for Applied Mathematics and Scientific Computing, Indiana University, 1988]
20. Ermentrout, G.B.: Periodic doublings and possible chaos in neural models. SIAM J. Appl. Math. **44**, 80–95 (1984)
21. Falconer, K.J.: The Geometry of Fractal sets. Cambridge Tracts in Mathematics 85, Cambridge University Press (1985)
22. Fathi, A.: Some compact invariant subsets for hyperbolic linear automorphisms of Torii. Ergod. Theory Dyn. Syst. **8**, 191–202 (1988)
23. Giusti, E.: Minimal Surfaces and Functions of Bounded Variation. Birkhäuser, Basel (1984)
24. Glukhovsky, A.B.: Nonlinear systems in the form of superposition of gyrostats. Dokl. Akad. Nauk SSSR **266** (7) (1982) (Russian)
25. Glukhovsky, A.B., Dolzhanskii, F.V: Three-component geostrophic model of convection in rotating fluid. Izv. Akad. Nauk SSSR, Fiz. Atmos. i Okeana **16**, 451–462 (1980) (Russian)
26. Guckenheimer, J., Holmes, P.: Nonlinear Oscillations, Dynamical Systems, and Bifurcations of Vector Fields. Springer, New York (1983)
27. Hénon, M.A.: A two-dimensional mapping with a strange attractor Commun. Math. Phys. **50**, 69–77 (1976)
28. Ito, S.: An estimate from above for the entropy and the topological entropy of a C^1-diffeomorphism. Proc. Jpn. Acad. **46**, 226–230 (1970)
29. Leonov, G.A.: On a method for investigating global stability of nonlinear systems. Vestn. Leningr. Gos. Univ. Mat. Mekh. Astron. **4**, 11–14 (1991) (Russian); English transl. Vestn. Leningr. Univ. Math. **24**(4), 9–11 (1991)
30. Leonov, G.A.: On estimates of the Hausdorff dimension of attractors. Vestn. Leningr. Gos. Univ. Ser. 1 **15**, 41–44 (1991) (Russian); English transl. Vestn. Leningr. Univ. Math. **24**(3), 38–41 (1991)
31. Leonov, G.A., Boichenko, V.A.: Lyapunov's direct method in the estimation of the Hausdorff dimension of attractors. Acta Appl. Math. **26**, 1–60 (1992)
32. Leonov, G.A., Lyashko, S.A.: Surprising property of Lyapunov dimension for invariant sets of dynamical systems. DFG Priority Research Program "Dynamics: Analysis, Efficient Simulation, and Ergodic Theory". Preprint 04 (2000)
33. Ott, E., Yorke, E.D., Yorke, J.A.: A scaling law: how an attractor's volume depends on noise level. Physica D **16**(1), 62–78 (1985)
34. Pikovsky, A.S., Rabinovich, M.I., Trakhtengerts, VYu.: Appearance of stochasticity on decay confinement of parametric instability. JTEF **74**, 1366–1374 (1978). (Russian)
35. Pochatkin, M.A., Reitmann, V.: The Douady-Oesterlé theorem for dynamical systems in convex bornological spaces. Preprint, St. Petersburg State University, St. Petersburg (2017). (Russian)

36. Rabinovich, M.I.: Stochastic self-oscillations and turbulence. Uspekhi Fiz. Nauk **125**, 123–168 (1978). (Russian)
37. Reitmann, V.: Dynamical Systems, Attractors and their Dimension Estimates. St. Petersburg State University Press, St. Petersburg (2013). (Russian)
38. Rössler, O.E.: Different types of chaos in two simple differential equations. Z. Naturforsch. **31** A, 1664–1670 (1976)
39. Sauer, T., Yorke, J.A., Casdagli, M.: Embedology. J. Stat. Phys. **65**, 579–616 (1991)
40. Smith, R.A.: Some applications of Hausdorff dimension inequalities for ordinary differential equations. Proc. R. Soc. Edinburgh **104A**, 235–259 (1986)
41. Temam, R.: Infinite-Dimensional Dynamical Systems in Mechanics and Physics. Springer, New York, Berlin (1988)
42. Thi, D.T., Fomenko, A.T.: Minimal Surfaces, Stratified Multivarifolds, and the Plateau Problem, Nauka, Moscow (1987) (Russian)

Chapter 6
Lyapunov Dimension for Dynamical Systems in Euclidean Spaces

Abstract Nowadays there is a number of surveys and theoretical works devoted to Lyapunov exponents and Lyapunov dimension, however most of them are devoted to infinite dimensional systems or rely on special ergodic properties of a system. At the same time the provided illustrative examples are often finite dimensional systems and the rigorous proof of their ergodic properties can be a difficult task. Also the Lyapunov exponents and Lyapunov dimension have become so widespread and common that they are often used without references to the rigorous definitions or pioneering works. This chapter is devoted to the finite dimensional dynamical systems in Euclidean space and its aim is to explain, in a simple but rigorous way, the connection between the key works in the area: Kaplan and Yorke (the concept of Lyapunov dimension, 1979), Douady and Oesterlé (estimation of Hausdorff dimension via the Lyapunov dimension of maps, 1980), Constantin, Eden, Foias, and Temam (estimation of Hausdorff dimension via the Lyapunov exponents and Lyapunov dimension of dynamical systems, 1985–90), Leonov (estimation of the Lyapunov dimension via the direct Lyapunov method, 1991), and numerical methods for the computation of Lyapunov exponents and Lyapunov dimension. In this chapter a concise overview of the classical results is presented, various definitions of Lyapunov exponents and Lyapunov dimension are discussed. An effective analytical method for the estimation of the Lyapunov dimension is presented, its application to self-excited and hidden attractors of well-known dynamical systems is demonstrated, and analytical formulas of exact Lyapunov dimension are obtained.

6.1 Singular Value Function and Invariant Sets of Maps of Dynamical Systems

Consider the autonomous differential equation

$$\dot{u} = f(u), \tag{6.1}$$

N. Kuznetsov and V. Reitmann, *Attractor Dimension Estimates for Dynamical Systems: Theory and Computation*, Emergence, Complexity and Computation 38,
https://doi.org/10.1007/978-3-030-50987-3_6

where $f : \mathcal{U} \subset \mathbb{R}^n \to \mathbb{R}^n$ is a continuously differentiable vector-function. Suppose that any solution $u(t, p)$ of (6.1) such that $u(0, p) = p \in \mathcal{U}$ exists for $t \in [0, \infty)$, is unique, and stays in $\mathcal{U}$. Then the evolutionary operator $\varphi^t(p) := u(t, p)$ is continuously differentiable and satisfies the semigroup property

$$\varphi^{t+s}(p) = \varphi^t(\varphi^s(p)), \ \varphi^0(p) = p \quad \forall\, t, s \geq 0, \ \forall p \in \mathcal{U}. \tag{6.2}$$

Thus $\{\varphi^t\}_{t\geq 0}$ is a smooth dynamical system in the phase space $(\mathcal{U}, |\cdot|)$: $\big(\{\varphi^t\}_{t\geq 0}, (\mathcal{U} \subset \mathbb{R}^n, |\cdot|)\big)$. Here $|u| = \sqrt{u_1^2 + \cdots + u_n^2}$ is the Euclidean norm of the vector $u = (u_1, \ldots, u_n) \in \mathbb{R}^n$. Similarly, we can consider a dynamical system generated by the difference equation

$$u_{t+1} = \varphi(u_t), \quad t = 0, 1, .., \tag{6.3}$$

where $\varphi : \mathcal{U} \subset \mathbb{R}^n \to \mathcal{U}$ is a continuously differentiable vector-function. Here $\varphi^t(u) = \underbrace{(\varphi \circ \varphi \circ \cdots \varphi)}_{t-\text{times}}(u)$, $\varphi^0(u) = u$, and the existence and uniqueness (in the forward-time direction) take place for all $t \geq 0$. Further $\{\varphi^t\}_{t\geq 0}$ denotes a smooth dynamical system with continuous or discrete time.

Consider the linearizations of systems (6.1) and (6.3) along the solution $\varphi^t(u)$

$$\dot{v} = J(\varphi^t(u))v, \quad J(u) = Df(u), \tag{6.4}$$

$$v_{t+1} = J(\varphi^t(u))v_t, \quad J(u) = D\varphi(u), \tag{6.5}$$

where $J(u)$ is the $n \times n$ Jacobian matrix, all elements of which are continuous functions of u. Consider the fundamental matrix

$$D\varphi^t(u) = \big(v^1(t), ..., v^n(t)\big), \quad D\varphi^0(u) = I, \tag{6.6}$$

which consists of linearly independent solutions $\{v^i(t)\}_{i=1}^n$ of the linearized system. An important cocycle property of the fundamental matrix (6.6) is

$$D\varphi^{t+s}(u) = D\varphi^t\big(\varphi^s(u)\big)D\varphi^s(u), \ \forall t, s \geq 0, \ \forall u \in \mathcal{U} \subset \mathbb{R}^n. \tag{6.7}$$

Consider the singular values of the matrix $D\varphi^t(u)$ sorted by descending for each $t \in [0, +\infty)$ and $u \in \mathcal{U} \subset \mathbb{R}^n$:

$$\alpha_i(t, u) := \alpha_i(D\varphi^t(u)), \ \alpha_1(t, u) \geq ... \geq \alpha_n(t, u) \geq 0, \quad \forall t \geq 0, u \in \mathcal{U} \subset \mathbb{R}^n. \tag{6.8}$$

Similar to Chap. 2, we introduce the singular value function of $D\varphi^t(u)$ of order d by $\omega_d(D\varphi^t(u))$.

For a fixed $t \geq 0$ one can consider the map defined by the evolutionary operator $\varphi^t : \mathcal{U} \subset \mathbb{R}^n \to \mathcal{U}$.

Further we need the following auxiliary statements.

Lemma 6.1 *From formula (2.4), Chap. 2 it follows that for any $u \in \mathcal{U}$ and $t \geq 0$ the function $d \mapsto \omega_d(D\varphi^t(u))$ is a left-continuous function.*

Lemma 6.2 *For any $d \in [0, n]$ and $t \geq 0$ the function $u \mapsto \omega_d(D\varphi^t(u))$ is continuous on $\mathcal{U}$ (see, e.g. [30]). Therefore for a compact set $\mathcal{K} \subset \mathcal{U}$ and $t \geq 0$ we have*

$$\sup_{u\in\mathcal{K}} \omega_d(D\varphi^t(u)) = \max_{u\in\mathcal{K}} \omega_d(D\varphi^t(u)). \tag{6.9}$$

Proof It follows from the continuity of the functions $u \mapsto \alpha_i(D\varphi(u))$ $i = 1, 2, ..., n$ on $\mathcal{U}$. □

Next, unless otherwise stated, the invariance of the set $\mathcal{K} \subset \mathcal{U} \subset \mathbb{R}^n$ is considered with respect to the dynamical system $\big(\{\varphi^t\}_{t\geq 0}, (\mathcal{U} \subset \mathbb{R}^n, |\cdot|)\big)$: $\varphi^t(\mathcal{K}) = \mathcal{K}, \ \forall t \geq 0$.

Lemma 6.3 *For a compact invariant set $\mathcal{K}$ and any $d \in [0, n]$, the function $t \mapsto \max_{u\in\mathcal{K}} \omega_d(D\varphi^t(u))$ is sub-exponential, i.e.,*

$$\max_{u\in\mathcal{K}} \omega_d(D\varphi^{t+s}(u)) \leq \max_{u\in\mathcal{K}} \omega_d(D\varphi^t(u)) \max_{u\in\mathcal{K}} \omega_d(D\varphi^s(u)), \quad \forall t, s \geq 0; \tag{6.10}$$

If $\max_{u\in\mathcal{K}} \omega_d(D\varphi^t(u)) > 0$ for $t \geq 0$, then $\log \max_{u\in\mathcal{K}} \omega_d(D\varphi^{t+s}(u))$ is subadditive, i.e.,

$$\log \max_{u\in\mathcal{K}} \omega_d(D\varphi^{t+s}(u)) \leq \log \max_{u\in\mathcal{K}} \omega_d(D\varphi^t(u)) + \log \max_{u\in\mathcal{K}} \omega_d(D\varphi^s(u)).$$

Proof By (6.7) and (6.2)

$$\max_{u\in\mathcal{K}} \omega_d(D\varphi^{t+s}(u)) = \max_{u\in\mathcal{K}} \big(\omega_d(D\varphi^t(\varphi^s(u))D\varphi^s(u))\big)$$
$$\leq \max_{u\in\mathcal{K}} \omega_d(D\varphi^t(\varphi^s(u))) \max_{u\in\mathcal{K}} \omega_d(D\varphi^s(u)) \leq \max_{u\in\mathcal{K}} \omega_d(D\varphi^t(u)) \max_{u\in\mathcal{K}} \omega_d(D\varphi^s(u)).$$

□

Corollary 6.1 *For an equilibrium point $u_{eq} \equiv \varphi^t(u_{eq})$ we have*

$$\omega_d(D\varphi^t(u_{eq})) = \big(\omega_d(D\varphi(u_{eq}))\big)^t, \quad \forall t \geq 0. \tag{6.11}$$

Corollary 6.2 *Remark that for a compact invariant set $\mathcal{K}$*

$$\inf_{t>0} \max_{u\in\mathcal{K}} \omega_d(D\varphi^t(u)) < 1 \Leftrightarrow \liminf_{t\to+\infty} \max_{u\in\mathcal{K}} \omega_d(D\varphi^t(u)) < 1.$$

In this case[1]

$$\inf_{t>0} \max_{u\in\mathcal{K}} \omega_d(D\varphi^t(u)) = \liminf_{t\to+\infty} \max_{u\in\mathcal{K}} \omega_d(D\varphi^t(u)) = 0. \tag{6.12}$$

[1]Considering additional properties of the dynamical system and the singular value function, one could get $\lim_{t\to+\infty}$ instead of $\liminf_{t\to+\infty}$, but we do not need it for our further consideration.

Proof Let $\inf_{t>0} \max_{u \in \mathcal{K}} \omega_d(D\varphi^t(u)) = M < 1$. There are $\delta > 0$ and $t_0 = t_0(\delta)$ such that $\max_{u \in \mathcal{K}} \omega_d(D\varphi^{t_0}(u)) \leq 1 - \delta$. Thus by (6.10) we have for $u \in \mathcal{K}$ and $n \geq 0$

$$0 \leq \omega_d(D\varphi^{nt_0}(u)) \leq \max_{u \in \mathcal{K}} \omega_d(D\varphi^{nt_0}(u)) \leq (\max_{u \in \mathcal{K}} \omega_d(D\varphi^{t_0}(u)))^n \leq (1-\delta)^n \to_{n\to+\infty} 0$$

and therefore $M = \liminf_{n\to+\infty} \omega_d(D\varphi^{nt_0}(u)) = \liminf_{t\to+\infty} \max_{u \in \mathcal{K}} \omega_d(D\varphi^t(u)) = 0$. The same is true if we consider $\liminf_{t\to+\infty} \max_{u \in \mathcal{K}} \omega_d(D\varphi^t(u)) < 1$ first. □

Corollary 6.3 *If for fixed $t > 0$ and $d \in [0, n]$ we have $\max_{u \in \mathcal{K}} \omega_d(D\varphi^t(u)) < 1$, then*

$$\liminf_{t\to+\infty} \max_{u \in \mathcal{K}} \omega_d(D\varphi^t(u)) = \liminf_{t\to+\infty} \omega_d(D\varphi^t(u)) = 0, \quad \forall u \in \mathcal{K}.$$

Lemma 6.4 ([15, 98]) *From the sub-exponential behavior of the singular value function (see* (6.10)*) on a compact invariant set $\mathcal{K}$ it follows that*

$$\inf_{t>0} \big(\max_{u \in \mathcal{K}} \omega_d(D\varphi^t(u)) \big)^{1/t} = \lim_{t\to+\infty} \big(\max_{u \in \mathcal{K}} \omega_d(D\varphi^t(u)) \big)^{1/t}, \quad \forall t > 0. \tag{6.13}$$

Proof The proof of this result follows from Fekete's lemma for subadditive functions [41].[2] □

Corollary 6.4 *If $\omega_d(D\varphi^t(u)) > 0$, then*

$$\inf_{t>0} \max_{u \in \mathcal{K}} \frac{1}{t} \log \big(\omega_d(D\varphi^t(u))\big) = \lim_{t\to+\infty} \max_{u \in \mathcal{K}} \frac{1}{t} \log \big(\omega_d(D\varphi^t(u))\big). \tag{6.14}$$

For a compact set $\mathcal{K}$, $t > 0$, $u \in \mathcal{K}$, and $d \in [0, n]$ we consider two scalar functions $g_d(t, u)$ and $f_d(t, u)$. Suppose that $g_0(t, u) = f_0(t, u) \equiv c$, therefore the following expressions

$$d_g^+(t, u) = \sup\{d \in [0, n] : g_d(t, u) \geq c\}, \quad d_f^+(t, u) = \sup\{d \in [0, n] : f_d(t, u) \geq c\}$$

are well defined. Also we consider

$$d_g^-(t, u) = \inf\{d \in [0, n] : g_d(t, u) < c\}, \quad d_f^-(t, u) = \inf\{d \in [0, n] : f_d(t, u) < c\}.$$

Here and further if the infimum on the empty set is considered, then we assume that the infimum is equal to n. Define

$$d_f^+(t) = \sup\{d \in [0, n] : \sup_{u \in \mathcal{K}} f_d(t, u) \geq c\}, \quad d_f^-(t) = \inf\{d \in [0, n] : \sup_{u \in \mathcal{K}} f_d(t, u) < c\}.$$

---

[2] If $f : \mathbb{R}^n \to \mathbb{R}$ is a measurable subadditive function, then for every $u \in \mathbb{R}^n$ there exists the limit $\lim_{t\to\infty} \frac{f(tu)}{t}$.

Lemma 6.5 *We have the following properties:*

(P1) *If for fixed $t > 0$ and $u \in \mathcal{K}$ the implication $\big(g_d(t,u) < c \Rightarrow f_d(t,u) < c\big)$ holds $\forall d \in [0,n]$, then*

$$\inf\{d \in [0,n] : f_d(t,u) < c\} \le \inf\{d \in [0,n] : g_d(t,u) < c\}; \tag{6.15}$$

(P2) *If for fixed $t > 0$ and $u \in \mathcal{K}$ the inequality $f_d(t,u) \le g_d(t,u)$ holds $\forall d \in [0,n]$, then*

$$\inf\{d \in [0,n] : f_d(t,u) < c\} \le \inf\{d \in [0,n] : g_d(t,u) < c\}; \tag{6.16}$$

$$\sup\{d \in [0,n] : f_d(t,u) \ge c\} \le \sup\{d \in [0,n] : g_d(t,u) \ge c\}; \tag{6.17}$$

(P3) *If*

$$\sup\{d \in [0,n] : f_d(t,u) \ge c\} = \inf\{d \in [0,n] : f_d(t,u) < c\}, \tag{6.18}$$

then

$$\sup\{d \in [0,n] : \sup_{u \in \mathcal{K}} f_d(t,u) \ge c\} = \sup_{u \in \mathcal{K}} \sup\{d \in [0,n] : f_d(t,u) \ge c\}; \tag{6.19}$$

(P4) *If for fixed $t > 0$ the equality*

$$\sup_{u \in \mathcal{K}} f_d(t,u) = \max_{u \in \mathcal{K}} f_d(t,u) \quad \forall d \in [0,n] \tag{6.20}$$

is valid and (6.18) *holds, then*

$$\inf\{d \in [0,n] : \sup_{u \in \mathcal{K}} f_d(t,u) < c\} = \sup_{u \in \mathcal{K}} \inf\{d \in [0,n] : f_d(t,u) < c\}; \tag{6.21}$$

(P5)

$$\inf\{d \in [0,n] : \inf_{t>0} f_d(t,u) < c\} = \inf_{t>0} \inf\{d \in [0,n] : f_d(t,u) < c\}. \tag{6.22}$$

Proof **(P1), (P2)**: Since in **(P1)** and **(P2)** the set of possible d, considered in the left-hand side of the expression, involves the set of possible d, considered in the right-hand side of the expression, we have the corresponding inequalities for the infimums of the sets. Similarly we get the relation for supremums.

(**P3**): Since $f_d(t,u) \le \sup_{u\in\mathcal{K}} f_d(t,u)$, by (6.17) in (**P2**) we have $\sup_{u\in\mathcal{K}} d_f^+(t,u) \le d_f^+(t)$.
Let $\sup_{u\in\mathcal{K}} d_f^+(t,u) < d_f^+(t) \Rightarrow$ by (6.18) $\exists d_0 \in \big(\sup_{u\in\mathcal{K}} d_f^+(t,u), d_f^+(t)\big)$: $f_{d'}(t,u) < c \ \forall d' \in [d_0, n] \ \forall u \in \mathcal{K} \Rightarrow d_f^+(t) \le d_0$. Thus we get the contradiction.

(**P4**): Since $\sup_{u\in\mathcal{K}} f_d(t,u) < c$ implies $f_d(t,u) < c$ for all $u \in \mathcal{K}$, by (**P1**) we have $d_f^-(t,u) \le d_f^-(t)$ for all $u \in \mathcal{K} \Rightarrow \sup_{u\in\mathcal{K}} d_f^-(t,u) \le d_f^-(t)$.

Let $\sup_{u\in\mathcal{K}} d_f^-(t,u) < d_f^-(t)$. Then $\exists d_0 \in \big(\sup_{u\in\mathcal{K}} d_f^-(t,u),\ d_f^-(t)\big)$. Since $d_0 < d_f^-(t)$, we have $\sup_{u\in\mathcal{K}} f_{d_0}(t,u) \ge c$. Therefore, from condition (6.20), $\exists u_0 : f_{d_0}(t,u_0) \ge c$. Finally, according to condition (6.18), we have $d_0 \le \sup\{d \in [0,n] : f_d(t,u_0) \ge c\} = \inf\{d \in [0,n] : f_d(t,u_0) < c\} = d_f^-(t,u_0)$ $\le \sup_{u\in\mathcal{K}} d_f^-(t,u)$. Thus we get a contradiction.

(**P5**): Since $\inf_{t>0} f_d(t,u) \le f_d(t,u)$, by (6.16) from (**P2**) we have $d_f^-(u) \le d_f^-(t,u)$ and, thus, $d_f^-(u) \le \inf_{t>0} d_f^-(t,u)$.
Let $d_f^-(u) < \inf_{t>0} d_f^-(t,u) \Rightarrow \exists d_0 \in \big[d_f^-(u), \inf_{t>0} d_f(t,u)\big) : \inf_{t>0} f_{d_0}(t,u) < c$ $\Rightarrow \exists t_0 : f_{d_0}(t_0,u) < c \Rightarrow d_0 \ge d_f^-(t_0,u) \ge \inf_{t>0} d_f^-(t,u)$. Thus we get a contradiction. □

Theorem 6.1 ([25, 26]) *Let $\omega_d(D\varphi^t(u)) > 0$. For a compact invariant set $\mathcal{K}$ and $d \in [0,n]$ there is a point $u^{cr} = u^{cr}(d) \in \mathcal{K}$ (it may be not unique) such that*

$$\frac{1}{t}\log\omega_d(D\varphi^t(u^{cr}(d))) \ge \limsup_{t\to+\infty}\sup_{u\in\mathcal{K}}\frac{1}{t}\log\omega_d(D\varphi^t(u)) \quad \forall t > 0. \tag{6.23}$$

Relation (6.23) is presented in [25, 26] and its proof is based on the theory of positive operators [12] (see also [32]).

Corollary 6.5 (see, e.g. [25])

$$\begin{aligned}\sup_{u\in\mathcal{K}}\limsup_{t\to+\infty}\frac{1}{t}\log\omega_d(D\varphi^t(u)) &= \lim_{t\to+\infty}\frac{1}{t}\log\omega_d(D\varphi^t(u^{cr}(d)))\\ &= \max_{u\in\mathcal{K}}\limsup_{t\to+\infty}\frac{1}{t}\log\omega_d(D\varphi^t(u))\\ &= \limsup_{t\to+\infty}\sup_{u\in\mathcal{K}}\frac{1}{t}\log\omega_d(D\varphi^t(u)).\end{aligned} \tag{6.24}$$

Proof It is easy to check that (see, e.g. [15])

$$\sup_{u\in\mathcal{K}}\limsup_{t\to+\infty}\frac{1}{t}\log\omega_d(D\varphi^t(u)) \le \limsup_{t\to+\infty}\sup_{u\in\mathcal{K}}\frac{1}{t}\log\omega_d(D\varphi^t(u)). \tag{6.25}$$

Thus, taking into account (6.23), we get (6.24). □

6.2 Lyapunov Dimension of Maps

The concept of the Lyapunov dimension had been suggested in the seminal paper by Kaplan and Yorke [39] and later it was developed in a number of papers (see, e.g. [15, 29]).

The following two definitions are inspirited by Douady-Oesterlé [22], see also Chap. 5.

Definition 6.1 The *(local) Lyapunov dimension*[3] of a continuously differentiable map $\varphi : \mathcal{U} \subset \mathbb{R}^n \to \mathbb{R}^n$ at the point $u \in \mathcal{U}$ is defined as

$$\dim_L(\varphi, u) := \sup\{d \in [0, n] : \omega_d(D\varphi(u)) \geq 1\}.$$

For any $u \in \mathcal{U}$ this value is well-defined since $\omega_0(D\varphi(u)) \equiv 1$.

By Lemma 6.1 we get

$$\dim_L(\varphi, u) = \max\{d \in [0, n] : \ \omega_d(D\varphi(u)) \geq 1\}. \tag{6.26}$$

Additionally, since the singular values in (1) are ordered by decreasing, we have

$$\dim_L(\varphi, u) = \max\{d \in [0, n] : \ \omega_d(D\varphi(u)) \geq 1\} = \inf\{d \in [0, n] : \ \omega_d(D\varphi(u)) < 1\} \tag{6.27}$$

if the infimum exists (i.e., there exists $d \in (0, n]$ such that $\omega_d(D\varphi(u)) < 1$). Here and further in the similar constructions if the infimum does not exist, we assume that the infimum and considered dimension are taken equal to n.

Definition 6.2 *The Lyapunov dimension of a* (continuously differentiable) *map* $\varphi : \mathcal{U} \subset \mathbb{R}^n \to \mathbb{R}^n$ of the compact set $\mathcal{K} \subset \mathcal{U} \subset \mathbb{R}^n$ is defined as

$$\dim_L(\varphi, \mathcal{K}) := \sup_{u \in \mathcal{K}} \dim_L(\varphi, u) = \sup_{u \in \mathcal{K}} \sup\{d \in [0, n] : \omega_d(D\varphi(u)) \geq 1\}.$$

Remark that by Lemma 6.5 (property (6.19)) and Lemma 6.2 we have

$$\begin{aligned} \dim_L(\varphi, \mathcal{K}) &= \sup_{u \in \mathcal{K}} \sup\{d \in [0, n] : \omega_d(D\varphi(u)) \geq 1\} \\ &= \sup\{d \in [0, n] : \max_{u \in \mathcal{K}} \omega_d(D\varphi(u)) \geq 1\}. \end{aligned} \tag{6.28}$$

Additionally, by (6.27) and Lemma 6.5 (property (6.21)), we have

[3]This is not a dimension in a rigorous sense (see, e.g. [35, 36, 43]).

$$\begin{aligned}\dim_L(\varphi, \mathcal{K}) &= \sup_{u\in\mathcal{K}} \inf\{d \in [0,n] : \ \omega_d(D\varphi(u)) < 1\} \\ &= \inf\{d \in [0,n] : \ \max_{u\in\mathcal{K}} \omega_d(D\varphi(u)) < 1\}\end{aligned} \tag{6.29}$$

if the infimum exists (i.e., there exists $d \in (0, n]$ such that $\max_{u\in\mathcal{K}} \omega_d(D\varphi(u)) < 1$).

Theorem 6.2 (Douady-Oesterlé, [22]; see also [8, 94, 98], and Chap. 5) *If the continuously differentiable map $\varphi : \mathcal{U} \subset \mathbb{R}^n \to \mathbb{R}^n$ has a negatively invariant or invariant compact set $\mathcal{K} \subset \mathcal{U}$, i.e.,*

$$\varphi(\mathcal{K}) \supset \mathcal{K},$$

then

$$\dim_H \mathcal{K} \le \dim_L(\varphi, \mathcal{K}).$$

Remark that under the assumptions of Theorem 6.2 if $\omega_d(D\varphi(u)) < 1$ for some $d \le 1$, then $\dim_H \mathcal{K} = 0$ (see, e.g. [98]). Thus, taking into account Lemma 6.2, we have

Lemma 6.6 (see, e.g. [30]) *The functions $u \mapsto \dim_L(\varphi, u)$ is continuous on $\mathcal{U}$ except at a point u, which satisfies $\alpha_1(D\varphi(u)) = 1$, where it is still upper semi-continuous.*

Corollary 6.6 *By the Weierstrass extreme value theorem for upper semi-continuous functions, there exists a critical point u_L (it may be not unique) such that*

$$\sup_{u\in\mathcal{K}} \dim_L(\varphi, u) = \max_{u\in\mathcal{K}} \dim_L(\varphi, u) = \dim_L(\varphi, u_L). \tag{6.30}$$

For an invariant compact set $\mathcal{K}$ of the dynamical system $\left(\{\varphi^t\}_{t\ge0}, (\mathcal{U} \subset \mathbb{R}^n, |\cdot|)\right)$ one may consider for a fixed t the evolutionary operator $\varphi^t(u)$, then

$$\varphi^t(\mathcal{K}) = \mathcal{K}$$

and the corresponding Lyapunov dimension (*finite time Lyapunov dimension*)

$$\dim_L(\varphi^t, \mathcal{K}) = \sup_{u\in\mathcal{K}} \dim_L(\varphi^t, u) = \inf\{d \in [0,n] : \max_{u\in\mathcal{K}} \omega_d(D\varphi^t(u)) < 1\}. \tag{6.31}$$

Example 6.1 If for a non-empty compact set $\mathcal{K} \subset \mathcal{U} \subset \mathbb{R}^n$ it is considered the identical map $\varphi = \text{id}$, then $D\varphi(u) = I$ and by the definition of the Lyapunov dimension we have $\dim_L(\text{id}, \mathcal{K}) = n$. Remark that for $t = 0$ we have $\varphi^0 = \text{id}$ and $\dim_L(\varphi^0, \mathcal{K}) = n$, thus we further consider $t > 0$.

Remark 6.1 For the numerical estimations of dimension, the following remark is important: for any $t > 0$ the equality (6.12) for a compact invariant set $\mathcal{K}$ implies the existence of a $s = s(t) > 0$ such that

$$\dim_L(\varphi^{t+s}, \mathcal{K}) \le \dim_L(\varphi^t, \mathcal{K}). \tag{6.32}$$

While in computations we can consider only a finite time $t \le T$ and the evolutionary operator $\varphi^T(u)$, from a theoretical point of view, it is interesting to study the limit behavior of the dynamical system $\{\varphi^t\}_{t\ge 0}$ as $t \to +\infty$. Next, unless otherwise stated, $\mathcal{K} \subset \mathcal{U} \subset \mathbb{R}^n$ denotes a compact invariant set with respect to the dynamical system $\big(\{\varphi^t\}_{t\ge 0}, (\mathcal{U} \subset \mathbb{R}^n, |\cdot|)\big)$: $\varphi^t(\mathcal{K}) = \mathcal{K},\ \forall t > 0$.

6.3 Lyapunov Dimensions of a Dynamical System

According to Definition 6.2 it is natural to give the following generalization for dynamical systems.

Definition 6.3 ([44]) The *Lyapunov dimension of the dynamical system* $\{\varphi^t\}_{t\ge 0}$ with respect to a compact invariant set $\mathcal{K}$ is defined as

$$\dim_L(\{\varphi^t\}_{t\ge 0}, \mathcal{K}) := \inf_{t>0} \dim_L(\varphi^t, \mathcal{K}) = \inf_{t>0} \sup\{d \in [0, n] : \max_{u\in\mathcal{K}} \omega_d(D\varphi^t(u)) \ge 1\}. \tag{6.33}$$

By Theorem 6.2 we have

$$\dim_H \mathcal{K} \le \dim_L(\{\varphi^t\}_{t\ge 0}, \mathcal{K}) \le \dim_L(\varphi^t, \mathcal{K}). \tag{6.34}$$

By (6.31) and Lemma 6.5 (property (6.22)) we have[4]

$$\dim_L(\{\varphi^t\}_{t\ge 0}, \mathcal{K}) = \inf_{t>0} \dim_L(\varphi^t, \mathcal{K}) = \inf\{d \in [0, n] : \inf_{t>0} \max_{u\in\mathcal{K}} \omega_d(D\varphi^t(u)) < 1\} \tag{6.35}$$

and, finally, by (6.12) we have

$$\dim_L(\{\varphi^t\}_{t\ge 0}, \mathcal{K}) = \inf_{t>0} \dim_L(\varphi^t, \mathcal{K}) = \inf\{d \in [0, n] : \liminf_{t\to+\infty} \max_{u\in\mathcal{K}} \omega_d(D\varphi^t(u)) = 0\}. \tag{6.36}$$

---

[4]While inf and sup give the same values for $\omega_d(D\varphi^t(u))$ in (6.28) and (6.29), for $\inf_{t>0} \max_{u\in\mathcal{K}} \omega_d(D\varphi^t(u))$ we need consider
$\sup\{d \in [0, n] : \forall \widetilde{d} \in [0, d] \quad \inf_{t>0} \max_{u\in\mathcal{K}} \omega_{\widetilde{d}}(D\varphi^t(u)) \ge 1\} = \inf\{d \in [0, n] : \inf_{t>0} \max_{u\in\mathcal{K}} \omega_d (D\varphi^t(u)) < 1\}$..

It is interesting to consider a critical point $u_L(T) \in \mathcal{K}$ such that the supremum of the local finite time Lyapunov dimension $\dim_L(\varphi^T, u)$ is achieved at this point[5]

Proposition 6.1 *Suppose that for a certain $t = T > 0$ the supremum of the local finite time Lyapunov dimensions $\dim_L(\varphi^T, u)$ is achieved at one of the equilibrium points:*

$$\dim_L(\varphi^T, u_{eq}^{cr}) = \sup_{u \in \mathcal{K}} \dim_L(\varphi^T, u), \quad \varphi^t(u_{eq}^{cr}) \equiv u_{eq}^{cr} \in \mathcal{K}. \tag{6.37}$$

Then

$$\dim_H \mathcal{K} \le \dim_L(\varphi^T, u_{eq}^{cr}) = \dim_L(\{\varphi^t\}_{t \ge 0}, \mathcal{K}) = \dim_L(\varphi^T, \mathcal{K}). \tag{6.38}$$

Proof From (6.11) we have

$$\omega_d(D\varphi^t(u_{eq}^{cr})) = \big(\omega_d(D\varphi(u_{eq}^{cr}))\big)^t, \quad \forall t \ge 0.$$

Therefore

$$\big(\omega_d(D\varphi^T(u_{eq}^{cr})) < 1\big) \Leftrightarrow \big(\omega_d(D\varphi(u_{eq}^{cr})) < 1\big) \Leftrightarrow \big(\omega_d(D\varphi^t(u_{eq}^{cr})) < 1, \ \forall t > 0\big)$$

and

$$\big(\liminf_{t \to +\infty} \max_{u \in \mathcal{K}} \omega_d(D\varphi^t(u)) < 1\big)$$
$$\Rightarrow \big(\liminf_{t \to +\infty} \omega_d(D\varphi^t(u_{eq}^{cr})) < 1\big) \Leftrightarrow \big(\omega_d(D\varphi^T(u_{eq}^{cr})) < 1\big).$$

By Lemma 6.5 (property (6.15)) we obtain

$$\inf\{d \in [0, n] : \omega_d(D\varphi^T(u_{eq}^{cr})) < 1\} = \dim_L(\varphi^T, u_{eq}^{cr}) \le \dim_L(\{\varphi^t\}_{t \ge 0}, \mathcal{K}).$$

Finally, by from (6.34) we get the assertion of the proposition. □

Further, to consider $\log \omega_d(D\varphi^t(u))$, we suppose that $\det J(u) \neq 0 \ \forall u \in \mathcal{U}$ and thus

$$\alpha_i(t, u) > 0, \quad i = 1, \ldots, n. \tag{6.39}$$

[5]If there exists $\lim_{t \to +\infty} \max_{u \in \mathcal{K}} \omega_d(D\varphi^t(u)) = 0$ then it is interesting to study the existence of a critical point $u_0 \in \mathcal{K}$ such that $\lim_{t \to +\infty} \max_{u \in \mathcal{K}} \omega_d(D\varphi^t(u)) = \limsup_{t \to +\infty} \omega_d(D\varphi^t(u_0))$, and to compare $\inf\{d \in [0, n] : \sup_{u \in \mathcal{K}} \limsup_{t \to +\infty} \omega_d(D\varphi^t(u)) < 1\}$ or $\sup_{u \in \mathcal{K}} \limsup_{t \to +\infty} \sup\{d \in [0, n] : \omega_d(D\varphi^t(u)) \ge 1\}$ with $\dim_H \mathcal{K}$. Remark, it is clear that $\lim_{t \to +\infty} \max_{u \in \mathcal{K}} \omega_d(D\varphi^t(u)) \ge \sup_{u \in \mathcal{K}} \limsup_{t \to +\infty} \omega_d(D\varphi^t(u))$. From (6.30) it follows the existence of a critical point $u_L(t)$ such that $\dim_L(\varphi^t, u_L(t)) = \max_{u \in \mathcal{K}} \dim_L(\varphi^t, u)$. Taking into account (6.32) we can consider a sequence $t_k \to +\infty$ such that $\dim_L(\varphi^{t_k}, u_L(t_k))$ is monotonically converging to $\inf_{t \ge 0} \max_{u \in \mathcal{K}} \dim_L(\varphi^t, u)$. Since $\mathcal{K}$ is a compact set, we can consider a subsequence $t_m = t_{k_m} \to +\infty$ such that there exists a limit critical point u_L^{cr}: $u_L(t_m) \to u_L^{cr} \in \mathcal{K}$ as $t_m \to +\infty$. Thus we have $\dim_L(\varphi^{t_m}, u_L(t_m)) \searrow \dim_L(\{\varphi^t\}_{t \ge 0}, \mathcal{K})$ and $u_L(t_m) \to u_L^{cr} \in \mathcal{K}$ as $m \to +\infty$.

The following definitions of Lyapunov dimension are inspirited by Constantin, Foias, Temam [15], and Eden [25].[6]

Definition 6.4 *The (global) Lyapunov dimension* of the dynamical system $\{\varphi^t\}_{t\geq 0}$ with respect to a compact invariant set $\mathcal{K}$ is defined as[7]

$$\dim_L^{\mathrm{E}}(\{\varphi^t\}_{t\geq 0}, \mathcal{K}) := \inf\{d \in [0, n] : \lim_{t\to+\infty} \max_{u\in\mathcal{K}} \frac{1}{t} \log \omega_d(D\varphi^t(u)) < 0\}. \quad (6.40)$$

The correctness of the definition follows from (6.14).

By (6.39) we have

$\big(\liminf_{t\to+\infty} \max_{u\in\mathcal{K}}(\omega_d(D\varphi^t(u))) < 1 \Leftrightarrow \liminf_{t\to+\infty} \max_{u\in\mathcal{K}} \log(\omega_d(D\varphi^t(u))) < 0\big)$ and $\left(\liminf_{t\to+\infty} \max_{u\in\mathcal{K}} \frac{1}{t} \log\left(\omega_d(D\varphi^t(u))\right) < 0 \Rightarrow \liminf_{t\to+\infty} \max_{u\in\mathcal{K}} \log(\omega_d(D\varphi^t(u))) < 0\right)$.

Thus, taking into account (6.36) and (6.14), by Lemma 6.5 (property (6.15)) we have

$$\begin{aligned} &\dim_L(\{\varphi^t\}_{t\geq 0}, \mathcal{K}) \\ &= \inf\{d \in [0, n] : \liminf_{t\to+\infty} \max_{u\in\mathcal{K}} \log \omega_d(D\varphi^t(u)) < 0\} \\ &\leq \inf\{d \in [0, n] : \lim_{t\to+\infty} \max_{u\in\mathcal{K}} \frac{1}{t} \log \omega_d(D\varphi^t(u)) < 0\} = \dim_L^{\mathrm{E}}(\{\varphi^t\}_{t\geq 0}, \mathcal{K}). \end{aligned} \quad (6.41)$$

Since for fixed $t > 0$ and $d \in [0, n]$

$$\begin{aligned} &\left(1 > \max_{u\in\mathcal{K}} \left(\omega_d(D\varphi^t(u))\right)\right) \\ \Rightarrow &\left(0 > \max_{u\in\mathcal{K}} \frac{1}{t} \log \omega_d(D\varphi^t(u)) \geq \lim_{t\to+\infty} \max_{u\in\mathcal{K}} \frac{1}{t} \log \omega_d(D\varphi^t(u))\right), \end{aligned}$$

by Lemma 6.5 (property (6.15)) and (6.34) we have

Proposition 6.2

$$\dim_H \mathcal{K} \leq \dim_L(\{\varphi^t\}_{t\geq 0}, \mathcal{K}) \leq \dim_L^{\mathrm{E}}(\{\varphi^t\}_{t\geq 0}, \mathcal{K}) \leq \dim_L(\varphi^t, \mathcal{K}) \quad \forall t > 0. \quad (6.42)$$

---

[6] In [15] Constantin, Foias, Temam stated that if $\sup_{u\in\mathcal{K}} \limsup_{t\to+\infty} \left(\omega_d(D\varphi^t(u))\right)^{1/t} < 1$ or $\limsup_{t\to+\infty} \sup_{u\in\mathcal{K}} \left(\omega_d(D\varphi^t(u))\right)^{1/t} < 1$, then $\dim_H \mathcal{K} \leq d$. In [25] Eden considered the value $\dim_L^{\mathrm{DO}}(\mathcal{K}) = \inf\{d > 0 : \sup_{u\in\mathcal{K}} \omega_d(D\varphi^t(u))$ converges to zero exponentially as $t \to \infty\}$ and called it *the Douady-Oesterlé dimension of* $\mathcal{K}$.

[7] Comparing the expressions in the definitions (6.33) and (6.40), remark that we can change $\frac{1}{t}$ in (6.40) to another scalar positive monotonically decreasing function $q(t)$ such that $\inf_{t>0} q(t) \max_{u\in\mathcal{K}} \omega_d(D\varphi^t(u)) = \lim_{t\to+\infty} q(t) \max_{u\in\mathcal{K}} \omega_d(D\varphi^t(u))$. The last relation is important from a computational point of view.

Corollary 6.7 *Taking* $\inf_{t>0}$ *in* (6.42), *we obtain*

$$\dim_L(\{\varphi^t\}_{t\geq 0}, \mathcal{K}) = \dim_L^{\mathrm{E}}(\{\varphi^t\}_{t\geq 0}, \mathcal{K}). \tag{6.43}$$

Definition 6.5 The local Lyapunov dimension of the dynamical system $\{\varphi^t\}_{t\geq 0}$ at the point u is defined as

$$\dim_L^{\mathrm{E}}(\{\varphi^t\}_{t\geq 0}, u) := \inf\{d \in [0, n] : \limsup_{t\to+\infty} \frac{1}{t} \log \omega_d(D\varphi^t(u)) < 0\}. \tag{6.44}$$

By (6.24) and Lemma 6.5 (property (6.21)) we have

$$\begin{aligned} &\sup_{u\in\mathcal{K}} \inf\{d \in [0, n] : \limsup_{t\to+\infty} \frac{1}{t} \log \omega_d(D\varphi^t(u)) < 0\} \\ &= \inf\{d \in [0, n] : \sup_{u\in\mathcal{K}} \limsup_{t\to+\infty} \frac{1}{t} \log \omega_d(D\varphi^t(u)) < 0\}. \end{aligned} \tag{6.45}$$

Therefore, by (6.25) and Lemma 6.5 (property (6.16)) we get

$$\begin{aligned} \sup_{u\in\mathcal{K}} \dim_L^{\mathrm{E}}(\{\varphi^t\}_{t\geq 0}, u) &= \sup_{u\in\mathcal{K}} \inf\{d \in [0, n] : \limsup_{t\to+\infty} \frac{1}{t} \log \omega_d(D\varphi^t(u)) < 0\} \\ &= \inf\{d \in [0, n] : \sup_{u\in\mathcal{K}} \limsup_{t\to+\infty} \frac{1}{t} \log \omega_d(D\varphi^t(u)) < 0\} \\ &\leq \inf\{d \in [0, n] : \limsup_{t\to+\infty} \sup_{u\in\mathcal{K}} \frac{1}{t} \log \omega_d(D\varphi^t(u)) < 0\} \\ &= \dim_L^{\mathrm{E}}(\{\varphi^t\}_{t\geq 0}, \mathcal{K}). \end{aligned} \tag{6.46}$$

Proposition 6.3 *If there is a critical equilibrium point* u_{eq}^{cr} *such that* (6.38) *is valid, then*

$$\dim_L(\varphi^T, u_{eq}^{cr}) = \dim_L^{\mathrm{E}}(\{\varphi^t\}_{t\geq 0}, u_{eq}^{cr}) = \sup_{u\in\mathcal{K}} \dim_L^{\mathrm{E}}(\{\varphi^t\}_{t\geq 0}, \mathcal{K})$$

and from (6.42) *it follows that*

$$\begin{aligned} \dim_H \mathcal{K} &\leq \dim_L(\{\varphi^t\}_{t\geq 0}, \mathcal{K}) = \dim_L^{\mathrm{E}}(\{\varphi^t\}_{t\geq 0}, \mathcal{K}) = \dim_L^{\mathrm{E}}(\{\varphi^t\}_{t\geq 0}, u_{eq}^{cr}) \\ &= \dim_L(\varphi^T, \mathcal{K}). \end{aligned} \tag{6.47}$$

In this case for the estimation of the Hausdorff dimension by (6.47) we need only the Douady-Oesterlé theorem (see Theorem 6.2). In the general case the existence of a critical point u_L^{E} (it may be not unique) such that

$$\dim_L^{\mathrm{E}}(\{\varphi^t\}_{t\geq 0}, u_L^{\mathrm{E}}) = \sup_{u\in\mathcal{K}} \dim_L^{\mathrm{E}}(\{\varphi^t\}_{t\geq 0}, u) = \dim_L^{\mathrm{E}}(\{\varphi^t\}_{t\geq 0}, \mathcal{K}) \tag{6.48}$$

follows from (6.24). The so-called *Eden conjecture* states that u_L^{E} corresponds to an equilibrium point or to a periodic orbit [23].

Finally, from (6.42) and (6.46) we have

Theorem 6.3

$$\begin{aligned}\dim_H \mathcal{K} &\le \dim_L(\{\varphi^t\}_{t\ge0}, \mathcal{K}) = \dim_L^{\mathrm{E}}(\{\varphi^t\}_{t\ge0}, \mathcal{K}) \\ &= \sup_{u\in\mathcal{K}} \dim_L^{\mathrm{E}}(\{\varphi^t\}_{t\ge0}, u) \le \dim_L(\varphi^t, \mathcal{K}) = \sup_{u\in\mathcal{K}} \dim_L(\varphi^t, u).\end{aligned} \tag{6.49}$$

6.3.1 Lyapunov Exponents: Various Definitions

Definition 6.6 The *Lyapunov exponent functions* of singular values (also called *finite-time Lyapunov exponents* [1]) of the dynamical system $\big(\{\varphi^t\}_{t\ge0},\ (\mathcal{U} \subset \mathbb{R}^n, |\cdot|)\big)$ at the point $u \in \mathcal{U}$ are denoted by

$$\begin{aligned}&\nu_i(t,u) = \nu_i(D\varphi^t(u)), \quad i = 1, 2, ..., n,\\ &\nu_1(t,u) \ge \cdots \ge \nu_n(t,u),\ \forall t > 0,\end{aligned}$$

and defined as

$$\nu_i(t,u) := \frac{1}{t}\log\alpha_i(t,u).$$

Definition 6.7 The *Lyapunov exponents (LEs) of singular values*[8] of the dynamical system $\{\varphi^t\}_{t\ge0}$ at the point u are defined (see, e.g. [15, 79]) as

$$\nu_i(u) := \limsup_{t\to+\infty} \nu_i(t,u) = \limsup_{t\to+\infty} \frac{1}{t}\log(\alpha_i(t,u)), \quad i = 1, 2, .., n.$$

Often $\nu_i(u)$ are called upper LEs and denoted as $\overline{\nu}_i(u)$, while $\underline{\nu}_i(u) := \liminf_{t\to+\infty} \nu_i(t,u)$ are called lower LEs. Remark that the Lyapunov exponents of singular values are the same for any fundamental matrices of the linearized systems (6.4) or (6.5).

Proposition 6.4 (see, e.g. [46]) *For the matrix $D\varphi^t(u)P$, where P is a non-singular $n \times n$ matrix (i.e.,* $\det P \ne 0$*), one has*

$$\lim_{t\to+\infty}\Big(\nu_i(D\varphi^t(u)) - \nu_i(D\varphi^t(u)P)\Big) = 0, \quad i = 1, 2, ..., n.$$

[8] We add "of singular value" to distinguish this definition from other definitions of Lyapunov exponents; if the differences in the definitions are not significant for the presentation, we use the term "Lyapunov exponents" or "LEs".

Definition 6.8 The Lyapunov exponent functions of the fundamental matrix columns $(v^1(t,u), ..., v^n(t,u)) = D\varphi^t(u)$

$$\nu^{\mathrm{L}}{}_i(t,u) = \nu^{\mathrm{L}}{}_i(D\varphi^t(u)),\ i = 1, 2, ..., n,\ u \in \mathcal{U}$$

are defined as

$$\nu^{\mathrm{L}}{}_i(t,u) := \frac{1}{t} \log |v^i(t,u)|.$$

The ordered *Lyapunov exponent functions of the fundamental matrix columns* at the point u (also called finite-time Lyapunov characteristic exponents) are given by the ordered set (for all $t > 0$) of $\nu^{\mathrm{L}}{}_i(t,u)$:

$$\nu^{\mathrm{L}^o}_1(t,u) \geq \cdots \geq \nu^{\mathrm{L}^o}_n(t,u),\ \forall t \geq 0.$$

Definition 6.9 The Lyapunov exponents of the fundamental matrix columns[9] are defined (see [75]) as

$$\nu^{\mathrm{L}}{}_i(u) := \limsup_{t\to+\infty} \nu^{\mathrm{L}^o}_i(t,u), \quad i = 1, 2, .., n.$$

Remark 6.2 The Lyapunov exponents of the fundamental matrix columns may be different for different fundamental matrices in contrast to the definition of Lyapunov exponents of singular values (see, e.g. Proposition 6.4). To get the set of all possible values of Lyapunov exponents of the fundamental matrix columns (the set with the minimal sum of values), one has to consider the so-called normal fundamental matrices (see [60, 75]).

Definition 6.10 The *relative Lyapunov exponents of singular value functions* of the dynamical system $\{\varphi^t\}_{t\geq 0}$ at the point u are defined (see, e.g. [79]) as

$$\begin{aligned}
\widetilde{\nu}_1(u) &:= \limsup_{t\to+\infty}(\nu_1(t,u)),\\
\widetilde{\nu}_{i+1}(u) &:= \limsup_{t\to+\infty}(\nu_1(t,u) + \cdots + \nu_{i+1}(t,u)) -\\
&\quad - \limsup_{t\to+\infty}(\nu_1(t,u) + \cdots + \nu_i(t,u)),\ i = 1, ..., n-1.
\end{aligned}$$

For $k = 1, 2, .., n$ we have

$$\widetilde{\nu}_1(u) + \cdots + \widetilde{\nu}_k(u) = \limsup_{t\to+\infty}(\nu_1(t,u) + \cdots + \nu_k(t,u)),$$

[9]Often they are called Lyapunov characteristic exponents (LCE) [60]. In [75] these values are defined with the opposite sign and called *characteristic exponents* at the point u.

$$\underline{\nu}_k(u) \le \widetilde{\nu}_k(u) = \limsup_{t\to+\infty} \sum_{i=1}^{k} \nu_i(t,u) - \limsup_{t\to+\infty} \sum_{i=1}^{k-1} \nu_i(t,u) \le \limsup_{t\to+\infty} \nu_k(t,u) = \nu_k(u).$$

From the Fischer-Courant theorem ([34], Theorem 2.1, Chap. 2) it follows (see, e.g. [3])[10] that

$$\nu_i(t,u) \le \nu^{\mathrm{L}}{}_i(t,u),\ \forall t \ge 0,\ \forall u \in \mathcal{U},\quad i = 1, 2, .., n. \tag{6.50}$$

Definition 6.11 ([14, 15]) The relative global (or uniform) Lyapunov exponents of singular value functions of the dynamical system $\{\varphi^t\}_{t\ge0}$ with respect to the compact invariant set $\mathcal{K} \subset \mathcal{U}$ are defined as

$$\begin{aligned}\widetilde{\nu}_1(\mathcal{K}) &:= \limsup_{t\to+\infty} \sup_{u\in\mathcal{K}} \nu_1(t,u),\\ \widetilde{\nu}_{i+1}(\mathcal{K}) &:= \limsup_{t\to+\infty} \sup_{u\in\mathcal{K}} \big(\nu_1(t,u) + \cdots + \nu_{i+1}(t,u)\big)\\ &\quad - \limsup_{t\to+\infty} \sup_{u\in\mathcal{K}} \big(\nu_1(t,u) + \cdots + \nu_i(t,u)\big),\ i = 1, ..., n-1.\end{aligned}$$

For $i = 1, 2, ..., n$ we have

$$\widetilde{\nu}_1(\mathcal{K}) + \cdots + \widetilde{\nu}_i(\mathcal{K}) = \limsup_{t\to+\infty} \sup_{u\in\mathcal{K}} \big(\nu_1(t,u) + \cdots + \nu_i(t,u)\big).$$

By (6.13) and (6.9) we get (see, e.g. [98])

$$\begin{aligned}\widetilde{\nu}_1(\mathcal{K}) &= \lim_{t\to+\infty} \max_{u\in\mathcal{K}} \nu_1(t,u),\\ \widetilde{\nu}_{i+1}(\mathcal{K}) &= \lim_{t\to+\infty} \max_{u\in\mathcal{K}} \big(\nu_1(t,u) + \cdots + \nu_{i+1}(t,u)\big)\\ &\quad - \lim_{t\to+\infty} \max_{u\in\mathcal{K}} \big(\nu_1(t,u) + \cdots + \nu_i(t,u)\big),\ i = 1, ..., n-1.\end{aligned}$$

From (6.25) (see, e.g. [23, 26]) for $u \in \mathcal{K}$ we obtain the following inequality

$$\widetilde{\nu}_1(u) + \cdots + \widetilde{\nu}_i(u) \le \widetilde{\nu}_1(\mathcal{K}) + \cdots + \widetilde{\nu}_i(\mathcal{K}),\ i = 1, 2, ..., n. \tag{6.51}$$

At the same time, according to (6.24), there exists $u^{cr}(m) \in \mathcal{K}$ (it may be not unique) such that the above expressions in (6.51) coincide [23, 24, 26]:

---

[10] For example [46], for the matrix $u(t) = \begin{pmatrix} 1 & g(t) - g^{-1}(t) \\ 0 & 1 \end{pmatrix}$ we have the following ordered values: $\nu^{\mathrm{L}}{}_1 = \max\big(\limsup_{t\to+\infty} \frac{1}{t}\log|g(t)|, \limsup_{t\to+\infty} \frac{1}{t}\log|g^{-1}(t)|\big)$, $\nu^{\mathrm{L}}{}_2 = 0$; $\nu_{1,2} = \max, \min\big(\limsup_{t\to+\infty} \frac{1}{t}\log|g(t)|, \limsup_{t\to+\infty} \frac{1}{t}\log|g^{-1}(t)|\big)$..

$$\begin{aligned}\widetilde{\nu}_1(\mathcal{K}) + \cdots + \widetilde{\nu}_m(\mathcal{K}) &= \widetilde{\nu}_1(u^{cr}(m)) + \cdots + \widetilde{\nu}_m(u^{cr}(m)) \\ &= \max_{u\in\mathcal{K}} \big(\widetilde{\nu}_1(u) + \cdots + \widetilde{\nu}_m(u)\big).\end{aligned} \tag{6.52}$$

Various characteristics of chaotic behavior are based on Lyapunov exponents (e.g., LEs are used in the Kaplan-Yorke formula of the Lyapunov dimension and the sum of positive LEs may be used [76, 82] as the characteristic of the Kolmogorov-Sinai entropy rate [40, 93]). The properties of Lyapunov exponents and their various generalizations are studied, e.g., in [2, 4, 10, 17, 37, 42, 46, 47, 58, 60, 75, 79, 82].

6.3.2 *Kaplan-Yorke Formula of the Lyapunov Dimension*

Consider the dynamical system $\big(\{\varphi^t\}_{t\geq 0}, (\mathcal{U} \subset \mathbb{R}^n, |\cdot|)\big)$.

6.3.2.1 Kaplan-Yorke Formula with Respect to the Finite Time Lyapunov Exponents

For $t > 0$ we have

$$\frac{1}{t}\log(\omega_d(D\varphi^t(u))) = \begin{cases} 0, & \text{if } d = 0, \\ \sum_{i=1}^{\lfloor d\rfloor} \nu_i(t,u), & \text{if } d = \lfloor d\rfloor \in \{1,\ldots,n\}, \\ \sum_{i=1}^{\lfloor d\rfloor} \nu_i(t,u) + (d - \lfloor d\rfloor)\,\nu_{\lfloor d\rfloor+1}(t,u), & \text{if } d \in (0,n). \end{cases} \tag{6.53}$$

If $\log(\omega_n(D\varphi^t(u))) \leq 0$ for fixed $t > 0$ and a given point $u \in \mathcal{K}$, then by (6.27) for $d(t,u) := \dim_L(\varphi^t, u)$ we have

$$\frac{1}{t}\log(\omega_{d(t,u)}(D\varphi^t(u))) = 0, \quad \frac{1}{t}\log(\omega_{d(t,u)+\delta}(D\varphi^t(u))) < 0 \quad \forall\delta \in (0, n - d(t,u)]. \tag{6.54}$$

Let for $t > 0$ be

$$\begin{aligned} d_0(t,u) &= d_0\big(\{\nu_i(t,u)\}_1^n\big) := \lfloor d(t,u)\rfloor \in \{0,..,n\}, \\ s(t,u) &= s\big(\{\nu_i(t,u)\}_1^n\big) := d(t,u) - d_0(t,u) \in [0,1). \end{aligned}$$

Then for $d_0(t,u) \leq n-1$ from (6.54) it follows that

$$\sum_{i=1}^{d_0(t,u)} \nu_i(t,u) \geq 0, \quad \sum_{i=1}^{d_0(t,u)+1} \nu_i(t,u) < 0. \tag{6.55}$$

We have

$$\frac{1}{t}\log(\omega_{d(t,u)}(D\varphi^t(u))) = \frac{1}{t}\log\left((\omega_{d_0(t,u)}(D\varphi^t(u)))^{(1-s(t,u))}(\omega_{d_0(t,u)+1}(D\varphi^t(u)))^{s(t,u)}\right)$$
$$= (1-s(t,u))\sum_{i=1}^{d_0(t,u)}\nu_i(t,u) + s(t,u)\sum_{i=1}^{d_0(t,u)+1}\nu_i(t,u) = 0.$$

Therefore we get

$$s(t,u) = \begin{cases} \dfrac{\nu_1(t,u)+\cdots+\nu_{d_0(t,u)}(t,u)}{|\,\nu_{d_0(t,u)+1}(t,u)|} < 1, & \text{if } d_0(t,u) \in \{1,\ldots,n-1\}, \\ 0, & \text{if } d_0(t,u) = 0 \text{ or } d_0(t,u) = n. \end{cases}$$

The expression

$$\dim_{_L}^{\mathrm{KY}}(\{\nu_i(t,u)\}_1^n) := d_0(t,u) + \frac{\nu_1(t,u)+\cdots+\nu_{d_0(t,u)}(t,u)}{|\,\nu_{d_0(t,u)+1}(t,u)|} \tag{6.56}$$

corresponds to the Kaplan-Yorke formula [39] with respect to the finite time Lyapunov exponents, i.e., the ordered set $\{\nu_i(t,u)\}_1^n$. The idea of the $\dim_{_L}^{\mathrm{KY}}$ construction may be used with other types of Lyapunov exponents (see below).

Further we assume that the relation $s(t,u) = 0$ for $d_0(t,u) = 0$ and $d_0(t,u) = n$ follows from the first expression for $s(t,u)$. Since $\frac{1}{t}\log(\omega_{d(t,u)}(D\varphi^t(u))) \le 0 \Leftrightarrow \omega_{d(t,u)}(D\varphi^t(u)) \le 1$ for $t > 0$, from (6.31) we have

Proposition 6.5

$$\begin{aligned} \dim_{_L}(\varphi^t,\mathcal{K}) &= \sup_{u\in\mathcal{K}}\dim_{_L}(\varphi^t,u) = \sup_{u\in\mathcal{K}}\dim_{_L}^{\mathrm{KY}}(\{\nu_i(t,u)\}_1^n) \\ &= \sup_{u\in\mathcal{K}}\left(d_0(t,u) + \frac{\nu_1(t,u)+\cdots+\nu_{d_0(t,u)}(t,u)}{|\,\nu_{d_0(t,u)+1}(t,u)|}\right). \end{aligned} \tag{6.57}$$

While in computing we can consider only finite time $t \le T$, from a theoretical point of view, it may be interesting to study the limit behavior of $\sup_{u\in\mathcal{K}}\dim_{_L}^{\mathrm{KY}}(\{\nu_i(t,u)\}_1^n)$ as $t \to +\infty$.

6.3.2.2 Kaplan-Yorke Formula with Respect to the Relative Global Lyapunov Exponents of Singular Value Functions

Let us define $d_0 = d_0\big(\{\widetilde{\nu}_i(\mathcal{K})\}_1^n\big) := \max\{m \in \{0,\ldots,n\} : \sum_{i=1}^{m}\widetilde{\nu}_i(\mathcal{K}) \ge 0\}$,

$$s = s\big(\{\widetilde{\nu}_i(\mathcal{K})\}_1^n\big) := \begin{cases} 0 \le \dfrac{\widetilde{\nu}_1(\mathcal{K})+\cdots+\widetilde{\nu}_{d_0}(\mathcal{K})}{|\,\widetilde{\nu}_{d_0+1}(\mathcal{K})|} < 1, & \text{if } d_0 \in \{1,\ldots,n-1\}, \\ 0, & \text{if } d_0 = 0 \text{ or } d_0 = n. \end{cases}$$

The expression $\dim_L^{\mathrm{KY}}(\{\widetilde{\nu}_i(\mathcal{K})\}_1^n) := d_0 + \frac{\widetilde{\nu}_1(\mathcal{K})+\cdots+\widetilde{\nu}_{d_0}(\mathcal{K})}{|\widetilde{\nu}_{d_0+1}(\mathcal{K})|}$ is the Kaplan-Yorke formula of Lyapunov dimension with respect to the relative global Lyapunov exponents of the singular value function. Then we have

$$\lim_{t\to+\infty}\max_{u\in\mathcal{K}}\frac{1}{t}\log\left(\omega_{d_0+s}(D\varphi^t(u))\right) = \lim_{t\to+\infty}\max_{u\in\mathcal{K}}\left(\sum_{i=1}^{d_0}\nu_i(t,u) + s\,\nu_{d_0+1}(t,u)\right)$$

$$= \lim_{t\to+\infty}\max_{u\in\mathcal{K}}\left((1-s)\sum_{i=1}^{d_0}\nu_i(t,u) + s\sum_{i=1}^{d_0+1}\nu_i(t,u)\right)$$

(since, in general, the maximums may be achieved at different points u)

$$\leq \lim_{t\to+\infty}\max_{u\in\mathcal{K}}(1-s)\sum_{i=1}^{d_0}\nu_i(t,u) + \lim_{t\to+\infty}\max_{u\in\mathcal{K}} s\sum_{i=1}^{d_0+1}\nu_i(t,u)$$

$$= (1-s)\lim_{t\to+\infty}\max_{u\in\mathcal{K}}\sum_{i=1}^{d_0}\nu_i(t,u) + s\lim_{t\to+\infty}\max_{u\in\mathcal{K}}\sum_{i=1}^{d_0+1}\nu_i(t,u) = 0.$$

Thus, for any $\overline{s}: \ s < \overline{s} < 1$, $\lim\limits_{t\to+\infty}\max\limits_{u\in\mathcal{K}}\frac{1}{t}\log(\omega_{d_0+\overline{s}}(D\varphi^t(u))) < 0$ and from Definition 6.4 we have

Proposition 6.6 (see, e.g. [15])

$$\dim_L^{\mathrm{E}}(\{\varphi^t\}_{t\geq 0}, \mathcal{K}) \leq \dim_L^{\mathrm{KY}}(\{\widetilde{\nu}_i(\mathcal{K})\}_1^n).$$

Under some conditions we can obtain the equality.

Corollary 6.8 *If the critical points $u^{cr}(d_0)$ and $u^{cr}(d_0+1)$ from* (6.24) *coincide, i.e., $u^{cr} = u^{cr}(d_0) = u^{cr}(d_0+1)$, then*

$$\begin{aligned}\lim_{t\to+\infty}\sum_{k=1}^{d_0}\nu_k(t,u^{cr}) &= \lim_{t\to+\infty}\max_{u\in\mathcal{K}}\sum_{k=1}^{d_0}\nu_k(t,u),\\ \lim_{t\to+\infty}\sum_{k=1}^{d_0+1}\nu_k(t,u^{cr}) &= \lim_{t\to+\infty}\max_{u\in\mathcal{K}}\sum_{k=1}^{d_0+1}\nu_k(t,u),\end{aligned} \tag{6.58}$$

and

$$\dim_L(\{\varphi^t\}_{t\geq 0}, \mathcal{K}) = \dim_L^{\mathrm{KY}}(\{\widetilde{\nu}_i(\mathcal{K})\}_1^n).$$

In [30] systems, having property (6.58), are called “typical systems”.

6.3.2.3 Kaplan-Yorke Formula with Respect To Relative Lyapunov Exponents of Singular Value Functions

Let

$$d_0(u) = d_0\big(\{\widetilde{\nu}_i(u)\}_1^n\big) := \max\{m \in \{0, \dots, n\} : \sum_{i=1}^{m} \widetilde{\nu}_i(u) \geq 0\}$$

$$s(u) = s\big(\{\widetilde{\nu}_i(u)\}_1^n\big) := \begin{cases} 0 \leq \dfrac{\widetilde{\nu}_1(u) + \cdots + \widetilde{\nu}_{d_0(u)}(u)}{|\,\widetilde{\nu}_{d_0(u)+1}(u)|} < 1, & \text{if } d_0(u) \in \{1, \dots, n-1\}, \\ 0, & \text{if } d_0(u) = 0 \text{ or } d_0(u) = n. \end{cases}$$

The expression $\dim_L^{\mathrm{KY}}(\{\widetilde{\nu}_i(u)\}_1^n) := d_0(u) + \frac{\widetilde{\nu}_1(u)+\cdots+\widetilde{\nu}_{d_0(u)}(u)}{|\,\widetilde{\nu}_{d_0(u)+1}(u)|}$ is the Kaplan-Yorke formula of the Lyapunov dimension with respect to the relative Lyapunov exponents of singular value functions. We have

$$\begin{aligned}
&\limsup_{t\to+\infty} \frac{1}{t} \log\big(\omega_{d_0(u)+s(u)}(D\varphi^t(u))\big) = \limsup_{t\to+\infty}\left(\sum_{i=1}^{d_0(u)} \nu_i(t,u) + s(u)\,\nu_{d_0(u)+1}(t,u)\right) \\
&= \limsup_{t\to+\infty}\left((1-s(u))\sum_{i=1}^{d_0(u)} \nu_i(t,u) + s(u)\sum_{i=1}^{d_0(u)+1} \nu_i(t,u)\right) \\
&\leq \limsup_{t\to+\infty}(1-s(u))\sum_{i=1}^{d_0(u)} \nu_i(t,u) + \limsup_{t\to+\infty} s(u)\sum_{i=1}^{d_0(u)+1} \nu_i(t,u) \\
&= (1-s(u))\limsup_{t\to+\infty}\sum_{i=1}^{d_0(u)} \nu_i(t,u) + s(u)\limsup_{t\to+\infty}\sum_{i=1}^{d_0(u)+1} \nu_i(t,u) = 0.
\end{aligned} \tag{6.59}$$

Thus, for any $d_0(u) < n$ and $\overline{s} : s(u) < \overline{s} < 1, \limsup_{t\to+\infty} \frac{1}{t}\log\big(\omega_{d_0(u)+\overline{s}}(D\varphi^t(u))\big) < 0$ and from Definition 6.5 we get

Proposition 6.7

$$\sup_{u\in\mathcal{K}} \dim_L(\{\varphi^t\}_{t\geq 0}, u) \leq \sup_{u\in\mathcal{K}} \dim_L^{\mathrm{KY}}(\{\widetilde{\nu}_i(u)\}_1^n) = \sup_{u\in\mathcal{K}}\left(d_0(u) + \frac{\widetilde{\nu}_1(u)+\cdots+\widetilde{\nu}_{d_0(u)}(u)}{|\,\widetilde{\nu}_{d_0(u)+1}(u)|}\right).$$

Proposition 6.8 (see, e.g. [23])

$$\sup_{u\in\mathcal{K}}(\dim_L^{\mathrm{KY}}(\{\widetilde{\nu}_i(u)\}_1^n)) \leq \dim_L^{\mathrm{KY}}(\{\widetilde{\nu}_i(\mathcal{K})\}_1^n). \tag{6.60}$$

Proof The assertion follows from the relation (see [23])

$$\sup_{u\in\mathcal{K}} d_0\big(\{\widetilde{\nu}_i(u)\}_1^n\big) = d_0\big(\{\widetilde{\nu}_i(\mathcal{K})\}_1^n\big) \tag{6.61}$$

and inequality (6.51). □

Remark that there are examples in which inequality (6.60) is strict (see, e.g. [23][11]).

6.3.2.4 Kaplan-Yorke Formula with Respect To the Lyapunov Exponents of Singular Values

Let (see, e.g. [15]) us introduce

$$d_0(u) = d_0\big(\{\nu_i(u)\}_1^n\big) := \max\{m \in \{0,\dots,n\} : \sum_{i=1}^{m} \nu_i(u) \geq 0\},$$

$$s(u) = s\big(\{\nu_i(u)\}_1^n\big) := \begin{cases} 0 \leq \dfrac{\nu_1(u) + \cdots + \nu_{d_0(u)}(u)}{|\nu_{d_0(u)+1}(u)|} < 1, & \text{if } d_0(u) \in \{1,\dots,n-1\}, \\ 0, & \text{if } d_0(u) = 0 \text{ or } d_0(u) = n. \end{cases}$$

The expression $\dim_L^{\text{KY}}(\{\nu_i(u)\}_1^n) := d_0(u) + \frac{\nu_1(u)+\cdots+\nu_{d_0(u)}(u)}{|\nu_{d_0(u)+1}(u)|}$ is the Kaplan-Yorke formula of the Lyapunov dimension with respect to the Lyapunov exponents of singular values.

Then it follows that

$$\begin{aligned} &\limsup_{t\to+\infty} \frac{1}{t} \log\big(\omega_{d_0(u)+s(u)}(D\varphi^t(u))\big) \\ &\leq (1-s(u)) \limsup_{t\to+\infty} \sum_{i=1}^{d_0(u)} \nu_i(t,u) + s(u) \limsup_{t\to+\infty} \sum_{i=1}^{d_0(u)+1} \nu_i(t,u) \\ &\leq (1-s(u)) \sum_{i=1}^{d_0(u)} \nu_i(u) + s(u) \sum_{i=1}^{d_0(u)+1} \nu_i(u) = 0. \end{aligned}$$

For $d_0(u) < n$ and any $\overline{s} : \ s(u) < \overline{s} < 1$, $\limsup\limits_{t\to+\infty} \frac{1}{t} \log\big(\omega_{d_0(u)+\overline{s}}(D\varphi^t(u))\big) < 0$ and from Definition 6.5 we have

Proposition 6.9 *Under the above assumptions it is true that*

$$\sup_{u\in\mathcal{K}} \dim_L(\{\varphi^t\}_{t\geq 0}, u) \leq \sup_{u\in\mathcal{K}} \dim_L^{\text{KY}}(\{\nu_i(u)\}_1^n) = \sup_{u\in\mathcal{K}} \left(d_0(u) + \frac{\nu_1(u) + \cdots + \nu_{d_0(u)}(u)}{|\nu_{d_0(u)+1}(u)|}\right).$$

---

[11]Let $\nu_1(t,u) = (e^u)^t$, $\nu_2(t,u) = (\frac{1}{2}(1-u))^t$ for all $u \in \mathcal{K} = [0,1]$. Thus $\nu_1(u) = \widetilde{\nu}_1(u) = u$, $\nu(u) = \widetilde{\nu}_2 = \log(1-u) - \log 2$; $\widetilde{\nu}_1(\mathcal{K}) = 1$, $\widetilde{\nu}_2 = -1 - \log 2$. Here $u^{cr}(1) = 1$: $\widetilde{\nu}_1(1) == \widetilde{\nu}_1(\mathcal{K}) = 1$; $u^{cr}(2) = 0$: $\widetilde{\nu}_1(0) + \widetilde{\nu}_2(0) = \widetilde{\nu}_1(\mathcal{K}) + \widetilde{\nu}_2(\mathcal{K}) = -\log 2$. Then $\sup_{u\in[0,1]} \dim_L^{\text{KY}}(\{\widetilde{\nu}_i(u)\}_1^2) = \frac{u}{\log 2 - \log(1-u)} < 1 + \frac{1}{1+\log 2} = \dim_L^{\text{KY}}(\{\widetilde{\nu}_i(\mathcal{K})\}_1^2)$.

6.3.2.5 Kaplan-Yorke Formula with Respect To the Lyapunov Exponents of Fundamental Matrix Columns

Let us define

$$d_0(u) = d_0\big(\{\nu^{\mathrm{L}}{}_i(u)\}_1^n\big) := \max\{m \in \{0,\dots,n\} : \sum_{k=1}^{m} \nu^{\mathrm{L}}{}_k(u) \geq 0\},$$

$$s(u) = s\big(\{\nu^{\mathrm{L}}{}_i(u)\}_1^n\big) := \begin{cases} 0 \leq \dfrac{\nu^{\mathrm{L}}{}_1(u) + \cdots + \nu^{\mathrm{L}}{}_{d_0(u)}(u)}{|\,\nu^{\mathrm{L}}{}_{d_0(u)+1}(u)|} < 1, & d_0(u) \in \{1,\dots,n-1\}, \\ 0, & d_0(u) = 0 \text{ or } d_0(u) = n. \end{cases}$$

The expression $\dim_{L}^{\mathrm{KY}}(\{\nu^{\mathrm{L}}{}_i(u)\}_1^n) := d_0(u) + \frac{\nu^{\mathrm{L}}{}_1(u)+\cdots+\nu^{\mathrm{L}}{}_{d_0(u)}(u)}{|\,\nu^{\mathrm{L}}{}_{d_0(u)+1}(u)|}$ is the Kaplan-Yorke formula of the Lyapunov dimension with respect to the Lyapunov exponents of the fundamental matrix columns.

Then, similar to (6.59), by (6.50) we obtain

$$\begin{aligned} &\limsup_{t\to+\infty} \frac{1}{t} \log\big(\omega_{d_0(u)+s(u)}(D\varphi^t(u))\big) \\ &\leq (1-s(u)) \sum_{i=1}^{d_0(u)} \nu^{\mathrm{L}}{}_i(u) + s(u) \sum_{i=1}^{d_0(u)+1} \nu^{\mathrm{L}}{}_i(u) = 0. \end{aligned}$$

Thus, for $d_0(u) < n$ and any $\overline{s} : \ s(u) < \overline{s} < 1$, $\limsup\limits_{t\to+\infty} \frac{1}{t} \log\big(\omega_{d_0(u)+\overline{s}}(D\varphi^t(u))\big) < 0$ and from Definition 6.5 we get

Proposition 6.10

$$\begin{aligned} &\sup_{u\in\mathcal{K}} \dim_{L}(\{\varphi^t\}_{t\geq 0}, u) \leq \sup_{u\in\mathcal{K}} \dim_{L}^{\mathrm{KY}}(\{\nu^{\mathrm{L}}{}_i(u)\}_1^n) \\ &= \sup_{u\in\mathcal{K}} \left(d_0(u) + \frac{\nu^{\mathrm{L}}{}_1(u) + \cdots + \nu^{\mathrm{L}}{}_{d_0(u)}(u)}{|\,\nu^{\mathrm{L}}{}_{d_0(u)+1}(u)|} \right). \end{aligned}$$

6.3.2.6 Computation by the Kaplan-Yorke Formulas

For a given invariant set $\mathcal{K}$ and a given point $p \in \mathcal{K}$ there are two essential questions related to the computation of Lyapunov exponents and the use of the Kaplan-Yorke formulas of local Lyapunov dimension $\sup_{u\in\mathcal{K}} \dim_{L}^{\mathrm{KY}}(\{\nu_i(u)\}_1^n)$ and $\sup_{u\in\mathcal{K}} \dim_{L}^{\mathrm{KY}}(\{\widetilde{\nu}_i(u)\}_1^n)$:

(a) $\limsup\limits_{t\to+\infty} \nu_i(t,p) \overset{?}{=} \lim\limits_{t\to+\infty} \nu_i(t,p)$ or $\limsup\limits_{t\to+\infty}(\sum_1^m \nu_i(t,u)) \overset{?}{=} \lim\limits_{t\to+\infty}(\sum_1^m \nu_i(t,u))$

(b) if the above limits do not exist, then
$\sup_{u\in\mathcal{K}} \dim_{L}^{\mathrm{KY}}(\{\nu_m(u)\}_1^n) \overset{?}{=} \sup_{u\in\mathcal{K}\setminus\{\varphi^t(p),t\geq 0\}} \dim_{L}^{\mathrm{KY}}(\{\nu_i(u)\}_1^n)$
or
$\sup_{u\in\mathcal{K}} \dim_{L}^{\mathrm{KY}}(\{\widetilde{\nu}_i(u)\}_1^n) \overset{?}{=} \sup_{u\in\mathcal{K}\setminus\{\varphi^t(p),t\geq 0\}} \dim_{L}^{\mathrm{KY}}(\{\widetilde{\nu}_i(u)\}_1^n)$.

In order to get rigorously a positive answer to these questions, from a theoretical point of view, one may use various ergodic properties of the dynamical system $\{\varphi^t\}_{t\geq 0}$ (see, Oseledec [79], Ledrappier [52], and some auxiliary results in [6, 19]). However, from a practical point of view, the rigorous use of the above results is a challenging task (see, e.g. the corresponding discussions in [5, 16, 81, 102] and the works on Perron effects of the largest Lyapunov exponent sign reversals [47, 60]). For an example of the effective rigorous use of ergodic theory for the estimation of the Hausdorff and Lyapunov dimensions see, e.g. [87].

Thus, in the general case, from a practical point of view, one cannot rely on the above relations (a) and (b) and shall use $\limsup_{t\to+\infty}$ in the definitions of local Lyapunov exponents and the corresponding formulas for the Lyapunov dimension (see, e.g. [98]).

However, if p is an equilibrium point, then the expression "$\limsup_{t\to+\infty}$" in Definitions 6.7, 6.9, and 6.10 can be replaced by "$\lim_{t\to+\infty}$" and we have

Lemma 6.7 *Let $\varphi^t(p)$ be a stationary point, i.e. $\varphi^t(p) \equiv p$. Then for $i = 1, 2, ..., n$ we have*

$$\lim_{t\to+\infty} \nu_i(t, p) = \nu_i(p) = \widetilde{\nu}_i(p) = \nu^{\mathrm{L}}{}_i(p).$$

Thus, for $d_0 = d_0\big(\{\widetilde{\nu}_i(\mathcal{K})\}_1^n\big)$, we get

Proposition 6.11 *If the critical points in* (6.48) *and* (6.58) *coincide with an equilibrium point u_{eq}^{cr}, i.e. $\varphi^t(u_{eq}^{cr}) \equiv u_{eq}^{cr} = u_L^{cr} \equiv u^{cr}(d_0) = u^{cr}(d_0+1)$, then*

$$\dim_{L}(\{\varphi^t\}_{t\geq 0}, u_{eq}^{cr}) = \dim_{L}(\{\varphi^t\}_{t\geq 0}, \mathcal{K}) = \dim_{L}^{\mathrm{KY}}(\{\widetilde{\nu}_i(\mathcal{K})\}_1^n)$$

and

$$\dim_{L}(\{\varphi^t\}_{t\geq 0}, u_{eq}^{cr}) = \sup_{u\in\mathcal{K}} \dim_{L}^{\mathrm{KY}}(\{\widetilde{\nu}_i(u)\}_1^n) = \sup_{u\in\mathcal{K}} \dim_{L}^{\mathrm{KY}}(\{\nu_i(u)\}_1^n)$$
$$= \sup_{u\in\mathcal{K}} \dim_{L}^{\mathrm{KY}}(\{\nu^{\mathrm{L}}{}_i(u)\}_1^n) = \lim_{t\to+\infty} \max_{u\in\mathcal{K}} \dim_{L}^{\mathrm{KY}}(\{\nu_i(t,u)\}_1^n).$$

If $u_L^{cr} = u^{cr}(d_0) = u^{cr}(d_0+1)$ belongs to a periodic orbit with period T, then the same reasoning can be applied for $(\varphi^T)^t$.

The last section of this chapter is devoted to the examples in which the maximum of the local Lyapunov dimension achieves at an equilibrium point.

Taking into account the existence of different definitions of Lyapunov dimension and related formulas and following [16], we recommend that *whatever you call your Lyapunov dimension, please state clearly how is it being computed.*

6.4 Analytical Estimates of the Lyapunov Dimension and its Invariance with Respect to Diffeomorphisms

Along with widely used numerical methods for estimating and computing the Lyapunov dimension (see, e.g. MATLAB realizations of the methods based on QR and SVD decompositions in [51, 66]) there is an effective analytical approach, proposed by G.A. Leonov in 1991 [54] (see also [8, 55–57, 59, 62, 66]). The Leonov method is based on the direct Lyapunov method with special Lyapunov-like functions. The advantage of this method is that it allows one to estimate the Lyapunov dimension of an invariant set without localization of the set in the phase space and in many cases to get effectively exact Lyapunov dimension formula [55, 58, 62–64, 67, 71, 72].

Following [44], next the invariance of Lyapunov dimension with respect to diffeomorphisms and its relation with the Leonov method are discussed. An analog of the Leonov method for discrete time dynamical systems is suggested.

While topological dimensions are invariant with respect to Lipschitz homeomorphisms, the Hausdorff dimension is invariant with respect to Lipschitz diffeomorphisms and non-integer Hausdorff dimension is not invariant with respect to homeomorphisms [35]. Since the Lyapunov dimension is used as an upper estimate of Hausdorff dimension, the question arises whether the Lyapunov dimension is invariant under diffeomorphisms (see, e.g. [44, 80]).

Consider the dynamical system $\big(\{\varphi^t\}_{t\geq 0}, (\mathcal{U}\subset\mathbb{R}^n, |\cdot|)\big)$ under the change of coordinates $w = h(u)$, where $h:\mathcal{U}\subset\mathbb{R}^n\to\mathbb{R}^n$ is a diffeomorphism. In this case the semi-orbit $\gamma^+(u) = \{\varphi^t(u), t\geq 0\}$ is mapped to the semi-orbit defined by $\varphi_h^t(w) = \varphi_h^t(h(u)) = h(\varphi^t(u))$, the dynamical system $\big(\{\varphi^t\}_{t\geq 0}, (\mathcal{U}\subset\mathbb{R}^n, |\cdot|)\big)$ is transformed to the dynamical system $\big(\{\varphi_h^t\}_{t\geq 0}, (h(\mathcal{U})\subset\mathbb{R}^n, |\cdot|)\big)$, and a compact set $\mathcal{K}\subset\mathcal{U}$ invariant with respect to $\{\varphi^t\}_{t\geq 0}$ is mapped to the compact set $h(\mathcal{K})\subset h(\mathcal{U})$ invariant with respect to $\{\varphi_h^t\}_{t\geq 0}$. Here we have

$$D_w\varphi_h^t(w) = D_w\big(h(\varphi^t(h^{-1}(w)))\big) = D_u h(\varphi^t(h^{-1}(w)))D_u\varphi^t(h^{-1}(w))D_w h^{-1}(w),$$

$$D_u\big(\varphi_h^t(h(u))\big) = D_w\varphi_h^t(h(u))D_u h(u) = D_u\big(h(\varphi^t(u))\big) = D_u h(\varphi^t(u))D_u\varphi^t(u).$$

Therefore it takes place

$$D_w h^{-1}(w) = \big(D_u h(u)\big)^{-1}$$

and

$$D\varphi_h^t(w) = Dh(\varphi^t(u))D\varphi^t(u)\big(Dh(u)\big)^{-1}. \tag{6.62}$$

If $u\in\mathcal{K}$, then $\varphi^t(u)$ and $\varphi_h^t(h(u))$ define bounded semi-orbits. Remark that Dh and $(Dh)^{-1}$ are continuous and, thus, $Dh(\varphi^t(u))$ and $(Dh(\varphi^t(u)))^{-1}$ are bounded in t. From (6.9) it follows that for any $d\in[0,n]$ there is a constant $c = c(d)\geq 1$ such that for any $t\geq 0$

$$\max_{u\in\mathcal{K}}\omega_d\big(Dh(u)\big)\leq c,\quad \max_{u\in\mathcal{K}}\omega_d\big((Dh(u))^{-1}\big)\leq c,\quad t\geq 0. \tag{6.63}$$

Lemma 6.8 *If for a fixed $t > 0$ there exist a diffeomorphism $h : \mathcal{U} \subset \mathbb{R}^n \to \mathbb{R}^n$ and a $d \in [0, n]$ such that the estimate*

$$\max_{w \in h(\mathcal{K})} \omega_d\big(D\varphi_h^t(w)\big) = \max_{u \in \mathcal{K}} \omega_d\left(Dh(\varphi^t(u))D\varphi^t(u)\big(Dh(u)\big)^{-1}\right) < 1 \tag{6.64}$$

is valid,[12] *then for $u \in \mathcal{K}$*

$$\liminf_{t \to +\infty} \left(\omega_d\big(D\varphi^t(u)\big) - \omega_d\big(D\varphi_h^t(h(u))\big)\right) = 0$$

and

$$\liminf_{t \to +\infty} \omega_d\big(D\varphi_h^t(h(u))\big) = \liminf_{t \to +\infty} \omega_d\big(D\varphi^t(u)\big) = 0.$$

Proof Applying (6.10) to (6.62), we get

$$\omega_d\big(D\varphi_h^t(h(u))\big) \le \omega_d\big(Dh(\varphi^t(u))\big)\omega_d\big(D\varphi^t(u)\big)\omega_d\big((Dh(u))^{-1}\big).$$

By (6.63) we obtain

$$\omega_d\big(D\varphi_h^t(h(u))\big) \le c^2 \omega_d\big(D\varphi^t(u)\big).$$

Similarly we have

$$\omega_d\big(D\varphi^t(u)\big) \le \omega_d\big((Dh(\varphi^t(u)))^{-1}\big)\omega_d\big(D\varphi_h^t(h(u))\big)\omega_d\big(Dh(u)\big)$$

and

$$\omega_d\big(D\varphi^t(u)\big) \le c^2 \omega_d\big(D\varphi_h^t(h(u))\big).$$

Therefore for any $d \in [0, n]$, $t \ge 0$, and $u \in \mathcal{K}$

$$c^{-2}\omega_d\big(D\varphi_h^t(h(u))\big) \le \omega_d\big(D\varphi^t(u)\big) \le c^2\omega_d\big(D\varphi_h^t(h(u))\big) \tag{6.65}$$

and

$$(c^{-2} - 1)\omega_d\big(D\varphi_h^t(h(u))\big) \le \omega_d\big(D\varphi^t(u)\big) - \omega_d\big(D\varphi_h^t(h(u))\big) \le (c^2 - 1)\omega_d\big(D\varphi_h^t(h(u))\big).$$

If for a fixed $t \ge 0$ there is a $d \in [0, n]$ such that $\sup_{u \in \mathcal{K}} \omega_d\big(D\varphi_h^t(h(u))\big) < 1$, then by Corollary 6.3 we have

$$\liminf_{t \to +\infty} \omega_d\big(D\varphi_h^t(h(u))\big) = 0$$

[12] The expression in (6.64) corresponds to the expressions considered in [54] for $p(u) = Dh(u)$, [55] and [56] for $Q(u) = Dh(u)$.

and

$$0 \le \liminf_{t\to+\infty} \left(\omega_d\big(D\varphi^t(u)\big) - \omega_d\big(D\varphi_h^t(h(u))\big)\right) \le 0.$$

□

Corollary 6.9 (see, e.g. [46]) *For $u \in \mathcal{K}$ we have*

$$\lim_{t\to+\infty} \left(\nu_i\left(D\varphi_h^t(h(u))\right) - \nu_i\left(D\varphi^t(u)\right)\right) = 0, \qquad i = 1, 2, .., n$$

and, therefore,

$$\limsup_{t\to+\infty} \nu_i\left(D\varphi_h^t(h(u))\right) = \limsup_{t\to+\infty} \nu_i\left(D\varphi^t(u)\right), \qquad i = 1, 2, .., n.$$

Proof For $t > 0$ from (6.65) we get

$$\frac{1}{t}\log c^{-2} + \frac{1}{t}\log\omega_d\big(D\varphi_h^t(h(u))\big) \le \frac{1}{t}\log\omega_d\big(D\varphi^t(u)\big) \le \frac{1}{t}\log c^2 + \frac{1}{t}\log\omega_d\big(D\varphi_h^t(h(u))\big). \tag{6.66}$$

Thus for the integer $d = m$ we have

$$\lim_{t\to+\infty}\left(\frac{1}{t}\log\omega_m\big(D\varphi^t(u)\big) - \frac{1}{t}\log\omega_m\big(D\varphi_h^t(h(u))\big)\right)$$
$$= \lim_{t\to+\infty}\left(\sum_{i=1}^m \nu_i\left(D\varphi^t(u)\right) - \sum_{i=1}^m \nu_i\left(D\varphi_h^t(h(u))\right)\right) = 0.$$

□

The above statements are rigorous reformulations from [46, 58] and imply the following

Proposition 6.12 ([44]) *The Lyapunov dimension of the dynamical system $\{\varphi^t\}_{t\ge0}$ with respect to the compact invariant set $\mathcal{K}$ is invariant with respect to any diffeomorphism $h : \mathcal{U} \subset \mathbb{R}^n \to \mathbb{R}^n$, i.e.*

$$\dim_L(\{\varphi^t\}_{t\ge0}, \mathcal{K}) = \dim_L(\{\varphi_h^t\}_{t\ge0}, h(\mathcal{K})). \tag{6.67}$$

Proof Lemma 6.8 implies that if $\max_{w\in h(\mathcal{K})}\omega_d\big(D\varphi_h^t(w)\big) < 1$ for a fixed $t > 0$ and $d \in [0, n]$, then there exists $T > t$ such that

$$\max_{u\in\mathcal{K}}\omega_d\big(D\varphi^T(u)\big) < 1 \tag{6.68}$$

and vice versa. Thus the set of d, over which $\inf_{t>0}$ is taken in (6.31), is the same for $D\varphi^t(u)$ and $D\varphi_h^t(w)$ and, therefore,

$$\inf_{t>0} \inf\{d \in [0, n] : \max_{u\in\mathcal{K}} \omega_d(D\varphi^t(u)) < 1\} = \inf_{t>0} \inf\{d \in [0, n] : \max_{w\in h(\mathcal{K})} \omega_d(D\varphi_h^t(w)) < 1\}.$$

□

Corollary 6.10 *Suppose $H(u)$ is a $n \times n$ matrix, the elements of which are scalar continuous functions of u and* $\det H(u) \neq 0$ *for $u \in \mathcal{K}$. If for a fixed $t > 0$ there is $d \in (0, n]$ such that*

$$\max_{w\in h(\mathcal{K})} \omega_d\big(D\varphi_h^t(w)\big) = \max_{u\in\mathcal{K}} \omega_d\left(H(\varphi^t(u))D\varphi^t(u)\big(H(u)\big)^{-1}\right) < 1, \tag{6.69}$$

then by (6.64) *with $H(u)$ instead of $Dh(u)$,* (6.67) *and* (6.68) *for sufficiently large $t = T > 0$ we have*

$$\dim_H \mathcal{K} \le \dim_L(\{\varphi^t\}_{t\ge0}, \mathcal{K}) \le \dim_L(\varphi^T, \mathcal{K}) \le d.$$

If it is assumed that $H(u) = \varkappa(u)S$, where $\varkappa(\cdot) : \mathcal{U} \subset \mathbb{R}^n \to \mathbb{R}^1$ is a continuous positive scalar function and S is a non-singular $n \times n$ matrix, then condition (6.69) *takes the form*

$$\begin{aligned} &\sup_{u\in\mathcal{K}} \omega_d\left(H(\varphi^t(u))D\varphi^t(u)\big(H(u)\big)^{-1}\right) \\ &= \sup_{u\in\mathcal{K}} \left(\big(\varkappa(\varphi^t(u))(\varkappa(u))^{-1}\big)^d \omega_d\big(SD\varphi^t(u)S^{-1}\big)\right) < 1. \end{aligned} \tag{6.70}$$

Consider now the Leonov method of analytical estimation of the Lyapunov dimension and its relation with the invariance of Lyapunov dimension with respect to diffeomorphisms. As it is shown below the multiplier of the type $\varkappa(\varphi^t(u))(\varkappa(u))^{-1}$ in (6.70) plays the role of a Lyapunov-like function.[13]

Let us apply the linear change of variables $w = h(u) = Su$ with a non-singular $n \times n$ matrix S. Then $\varphi^t(p) = u(t, p)$ is transformed into $\varphi_S^t(w_0)$:

$$\varphi_S^t(w_0) = w(t, w_0) = S\varphi^t(p) = Su(t, S^{-1}w_0).$$

Consider the transformed systems (6.1) and (6.3)

$$\dot{w} = Sf(S^{-1}w) \text{ or } w_{t+1} = S\varphi(S^{-1}w_t)$$

[13] In [78] and in Chap. 7 it is interpreted as changes of Riemannian metrics.

and their linearizations along the solution $\varphi_S^t(w_0) = w(t, w_0) = S\varphi^t(p)$:

$$\begin{aligned} \dot{\upsilon} &= J_S(w(t, w_0))\upsilon \ \ or \ \ \upsilon_{t+1} = J_S(w(t, w_0))\upsilon_t, \\ J_S(w(t, w_0)) &= S\,J(S^{-1}w(t, w_0))\,S^{-1} = S\,J(u(t, p))\,S^{-1}. \end{aligned} \tag{6.71}$$

For the corresponding fundamental matrices we have $D\varphi_S^t(w) = SD\varphi^t(u)S^{-1}$.

First we consider a continuous time dynamical system. Let $\lambda_i(p, S) = \lambda_i(S\varphi^t(p))$, $i = 1, 2, ..., n$, be the eigenvalues of the symmetrized Jacobian matrix

$$\frac{1}{2}\left(SJ(u(t, p))S^{-1} + (SJ(u(t, p))S^{-1})^*\right) = \frac{1}{2}\left(J_S(w(t, w_0)) + J_S(w(t, w_0))^*\right), \tag{6.72}$$

ordered so that $\lambda_1(p, S) \geq \cdots \geq \lambda_n(p, S)$ for any $p \in \mathcal{U}$. The following theorem is rigorous reformulation of results from [55–57].

Theorem 6.4 *Let be $d = d_0 + s \in [1, n]$ with integer $d_0 = \lfloor d \rfloor \in \{1, \ldots, n\}$ and real $s = (d - \lfloor d \rfloor) \in [0, 1)$. If there are a differentiable scalar function $V(\cdot) : \mathcal{U} \subset \mathbb{R}^n \to \mathbb{R}^1$ and a non-singular $n \times n$ matrix S such that*

$$\sup_{u \in \mathcal{K}} \left(\lambda_1(u, S) + \cdots + \lambda_{d_0}(u, S) + s\lambda_{d_0+1}(u, S) + \dot{V}(u)\right) < 0, \tag{6.73}$$

where $\dot{V}(u) = (\text{grad}(V(u)))^* f(u)$, then

$$\dim_H \mathcal{K} \leq \dim_L(\{\varphi^t\}_{t \geq 0}, \mathcal{K}) \leq \dim_L(\varphi^T, \mathcal{K}) \leq d_0 + s$$

for sufficiently large $T > 0$.

Proof From the following relations (see Liouville's formula, Chap. 2, and, e.g., [56])

$$\begin{aligned} &\omega_{d_0+s}\left(SD\varphi^t(u)S^{-1}\right) \\ &= \exp\left(\int_0^t \lambda_1(S\varphi^\tau(u)) + \cdots + \lambda_{d_0}(S\varphi^\tau(u)) + s\lambda_{d_0+1}(S\varphi^\tau(u))d\tau\right) \end{aligned} \tag{6.74}$$

and

$$\left(\varkappa(\varphi^t(u))(\varkappa(u))^{-1}\right)^{d_0+s} = \exp\left(V(\varphi^t(u)) - V(u)\right) = \exp\left(\int_0^t \dot{V}(\varphi^\tau(u))d\tau\right)$$

we get

$$\begin{aligned} &\left(\varkappa(\varphi^t(u))(\varkappa(u))^{-1}\right)^{d_0+s}\omega_{d_0+s}\left(SD\varphi^t(u)S^{-1}\right) \\ &\leq \exp\left(\int_0^t \left(\lambda_1(S\varphi^\tau(u)) + \cdots + \lambda_{d_0}(S\varphi^\tau(u)) + s\lambda_{d_0+1}(S\varphi^\tau(u)) + \dot{V}(\varphi^\tau(u))\right)d\tau\right). \end{aligned} \tag{6.75}$$

Since $\varphi^t(u) \in \mathcal{K}$ for any $u \in \mathcal{K}$, for $t > 0$ by (6.73) we have

$$\max_{u\in\mathcal{K}} \left(\left(\varkappa(\varphi^t(u))\varkappa(u)^{-1}\right)^{d_0+s} \omega_{d_0+s}\big(SD\varphi^t(u)S^{-1}\big) \right) < 1, \quad \forall t > 0.$$

Therefore by Corollary 6.10 with $H(u) = \varkappa(u)S$, where $\varkappa(u) = \left(e^{V(u)}\right)^{\frac{1}{d}}$, we get the assertion of the theorem. □

Now consider a discrete time dynamical system. Let $\lambda_i(p, S) = \lambda_i(S\varphi^t(p))$, $i = 1, 2, ..., n$, be the positive square roots of the eigenvalues of the symmetrized Jacobian matrix

$$\left((SJ(u(t, p))S^{-1})^* SJ(u(t, p))S^{-1}\right) = \left(J_S(w(t, w_0))^* J_S(w(t, w_0))\right), \tag{6.76}$$

ordered so that $\lambda_1(p, S) \geq \cdots \geq \lambda_n(p, S)$ for any $p \in \mathcal{U}$.

Theorem 6.5 ([44]) *Let be $d = (d_0 + s) \in [1, n]$ with integer $d_0 = \lfloor d \rfloor \in \{1, \ldots, n\}$ and real $s = (d - \lfloor d \rfloor) \in [0, 1)$. If there are a scalar continuous function $V(\cdot) : \mathcal{U} \subset \mathbb{R}^n \to \mathbb{R}^1$ and a non-singular $n \times n$ matrix S such that*

$$\sup_{u\in\mathcal{K}} \left(\log \lambda_1(u, S) + \cdots + \log \lambda_{d_0}(u, S) + s \log \lambda_{d_0+1}(u, S) + \big(V(\varphi(u)) - V(u)\big) \right) < 0, \tag{6.77}$$

then

$$\dim_H \mathcal{K} \leq \dim_L(\{\varphi^t\}_{t\geq 0}, \mathcal{K}) \leq \dim_L(\varphi^T, \mathcal{K}) \leq d_0 + s$$

for sufficiently large $T > 0$.

Proof By (2) for $D\varphi_S^t(w) = SD\varphi^t(u)S^{-1} = \prod_{\tau=0}^{t-1} \left(S\, J(u(\tau, p))\, S^{-1}\right)$ we have

$$\begin{aligned} &\omega_{d_0+s}\big(SD\varphi^t(u)S^{-1}\big) \\ &\leq \prod_{\tau=0}^{t-1} \omega_{d_0+s}\big(S\, J(u(\tau, p))\, S^{-1}\big). \end{aligned} \tag{6.78}$$

Therefore by the discrete analog of (6.74) we have

$$\omega_{d_0+s}\big(SD\varphi^t(u)S^{-1}\big) \leq \prod_{\tau=0}^{t-1} \lambda_1(S\varphi^\tau(u)) \cdots \lambda_{d_0}(S\varphi^\tau(u))\big(\lambda_{d_0+1}(S\varphi^\tau(u))\big)^s. \tag{6.79}$$

By the relation

$$\left(\varkappa(\varphi^t(u))\varkappa(u)^{-1}\right)^{d_0+s} = \exp\left(V(\varphi^t(u)) - V(u)\right)$$
$$= \exp\left(\sum_{\tau=0}^{t-1} V(\varphi^{\tau+1}(u)) - V(\varphi^{\tau}(u))\right)$$

and we get

$$\log\left(\varkappa(\varphi^t(u))\varkappa(u)^{-1}\right)^{d_0+s} + \log \omega_{d_0+s}\left(SD\varphi^t(u)S^{-1}\right)$$
$$\leq \sum_{\tau=0}^{t-1}\Bigg(\log \lambda_1(S\varphi^{\tau}(u)) + \cdots + \log \lambda_{d_0}(S\varphi^{\tau}(u))$$
$$+ s \log \lambda_{d_0+1}(S\varphi^{\tau}(u)) + V(\varphi(\varphi^{\tau}(u))) - V(\varphi^{\tau}(u))\Bigg).$$

Since $\varphi^t(u) \in \mathcal{K}$ for any $u \in \mathcal{K}$, by (6.77) and Corollary 6.10 with $H(u) = \varkappa(u)S$, where $\varkappa(u) = \left(e^{V(u)}\right)^{\frac{1}{d}}$, we get the assertion of the theorem. □

From (6.38) we have

Corollary 6.11 *If at an equilibrium point* $u_{eq}^{cr} \equiv \varphi^t(u_{eq}^{cr})$ *for a certain* $t > 0$ *the relation*

$$\dim_L(\varphi^t, u_{eq}^{cr}) = d_0 + s$$

holds, then for any invariant set $\mathcal{K} \ni u_{eq}^{cr}$ *we get the analytical formula of exact Lyapunov dimension*

$$\dim_H \mathcal{K} \leq \dim_L(\{\varphi^t\}_{t\geq 0}, \mathcal{K}) = \dim_L(\{\varphi^t\}_{t\geq 0}, u_{eq}^{cr}) = d_0 + s.$$

Remark that in the above approach we need only the Douady-Oesterlé theorem (see Theorem 6.2) and do not use the results on the Lyapunov dimension developed by Eden, Constantin, Foias, Temam in [15, 26] (see (6.40),(6.44), Propositions 6.6 and 6.7).

In [7, 85] and Chapter 5 it is demonstrated, how a technique similar to the above can be effectively applied to derive constructive upper bounds of the topological entropy of dynamical systems.

6.5 Analytical Formulas of Exact Lyapunov Dimension for Well-Known Dynamical Systems

Next we consider examples smooth dynamical systems generated by difference and differential equations (for an example of PDE, see e.g. [21]) in which the critical point, corresponding to the maximum of the local Lyapunov dimension, is one of the equilibrium points (see (6.47)). In these examples we assume the existence of an invariant set $\mathcal{K}$ in which the corresponding dynamical system $\{\varphi^t\}_{t\geq 0}$ is defined, and use the compact notation $\dim_L(\mathcal{K})$ for the Lyapunov dimension instead of (6.33).

6.5.1 *Hénon Map*

Let us consider again (see also Sect. 5.1.3) the Hénon map $\varphi : \mathbb{R}^2 \to \mathbb{R}^2$ given by

$$(x, y) \in \mathbb{R}^2 \longmapsto (a + by - x^2, x), \tag{6.80}$$

where $a > 0$ and $0 < b < 1$ are parameters. The fixed points $(x_\pm, x_\pm)$ of this map are defined by $(x_\pm, x_\pm)$ where $x_\pm = \frac{1}{2}\left(b - 1 \pm \sqrt{(b-1)^2 + 4a}\right)$. Following [55] we introduce the matrix $S = \begin{pmatrix} 1 & 0 \\ 0 & \sqrt{b} \end{pmatrix}$ and parameter $\gamma = \frac{1}{(b-1-2x_-)\sqrt{x_-^2+b}}$, $s \in [0, 1)$. It follows that for $(x, y) \in \mathbb{R}^2$ we have

$$SJ\big((x, y)\big)S^{-1} = \begin{pmatrix} -2x & \sqrt{b} \\ \sqrt{b} & 0 \end{pmatrix} \text{ and } \lambda_1((x, y), S) = \left(\sqrt{x^2 + b} + |x|\right),\ \lambda_2((x, y), S) = \frac{b}{\lambda_1((x, y), S)}.$$

If we take now $V((x, y)) = \gamma(1 - s)(x + by)$, the condition (6.77) with $d_0 = 1$ and

$$s > s^* = \frac{\log |\lambda_1((x_-, x_-), S)|}{|\log b - \log |\lambda_1((x_-, x_-), S)||}$$

is satisfied for all $(x, y) \in \mathbb{R}^2$ and we need not localize the invariant set $\mathcal{K}$ in the phase space. By Corollary 6.11 and (6.56), at the equilibrium point $u_{eq}^{cr} = (x_-, x_-)$ we get

$$\dim_L((x_-, x_-)) = \dim_L^{\text{KY}}(\{\log \lambda_i(x_-, x_-)\}_1^2) = 1 + s^*.$$

Therefore, for a bounded invariant set $\mathcal{K} \ni (x_-, x_-)$ we have [55]

$$\dim_L(\mathcal{K}) = \dim_L((x_-, x_-)) = 1 + \frac{\log |\lambda_1((x_-, x_-), S)|}{|\log b - \log |\lambda_1((x_-, x_-), S)||}.$$

Here for $a = 1.4$ and $b = 0.3$ we have $\dim_L(\mathcal{K}) = 1.495 \ldots$.

6.5.2 Lorenz System

Consider the Lorenz system [74]:

$$\begin{cases} \dot{x} = \sigma(y - x), \\ \dot{y} = rx - y - xz, \\ \dot{z} = -bz + xy, \end{cases} \tag{6.81}$$

with parameters $\sigma > 0,\ r > 0,\ b \in (0, 4]$.

The following estimates are shown in Chap. 1 (see also [53, 59]):

Lemma 6.9 *For any solution $(x(t), y(t), z(t))$ of system* (6.81) *we have the estimates:*

$$\limsup_{t\to\infty} \left(y^2(t) + (z(t) - r)^2\right) \le l^2 r^2, \tag{6.82}$$

$$\limsup_{t\to\infty} \left(x(t)^2 + y(t)^2 + (z(t) - r - \sigma)^2\right) \le \tfrac{b(\sigma+r)^2}{2c}, \tag{6.83}$$

$$\liminf_{t\to+\infty} \left[z(t) - \tfrac{x(t)^2}{2\sigma}\right] \ge 0 \ \ \text{if } 2\sigma \ge b. \tag{6.84}$$

Here $l = \begin{cases} 1, & \text{for } b \le 2, \\ \frac{b}{2\sqrt{b-1}}, & \text{for } b > 2, \end{cases}$ *(see also (1.20), Chap. 1),* $c = \min(\sigma, 1, \frac{b}{2})$.

Corollary 6.12 *For system* (6.81) *we have the absorbing set*

$$\mathcal{Z} = \left\{(x, y, z) \in \mathbb{R}^3 \mid (x^2 + y^2 + (z - r - \sigma)^2) < \tfrac{b(\sigma+r)^2}{2c}\right\}, \tag{6.85}$$

where c is defined in Lemma 6.9.

Corollary 6.13 *For $r > 1$, $2\sigma \ge b$ and some $C > 0$, $D > 4C\sigma$ on the global attractor $\mathcal{A}$ of system* (6.81) *the following estimate*

$$-Cx^4 + Dx^2 z \le (2D - 4C\sigma)\sigma z^2 \tag{6.86}$$

is valid.

Theorem 6.6 *Suppose $\mathcal{A}$ is the global attractor of system* (6.81). *If*

$$2\sigma > b \tag{6.87}$$

and

$$r \ge (b + 1)(\tfrac{b}{\sigma} + 1), \tag{6.88}$$

then

$$\dim_L \mathcal{A} = 3 - \frac{2(\sigma+b+1)}{\sigma+1+\sqrt{(\sigma-1)^2+4\sigma r}}, \tag{6.89}$$

otherwise any positive semi-orbit tends to the equilibria set as $t \to +\infty$.

Proof Following [53] let us present the sketch of the proof.

Case (1): $r \le 1$. In this case system (6.81) has the unique equilibrium $u_1 = (0, 0, 0)$. Using Theorem 1.1 it was shown in Example 1.10, Chap. 1 that the equilibrium u_1 is globally asymptotically stable.

Case (2): $r > 1$ and $2\sigma \le b$. Now system (6.81) has three equilibrium points and the assumptions of Corollary (2.15), Chap. 2 are satisfied. It follows that any positive semi-orbit of system (6.81) converges to an equilibrium.

Case (3): $1 < r < (b+1)(\frac{b}{\sigma}+1)$, $2\sigma > b$. In this case we apply Corollary 5.9, Chap. 5. To do this we specify a linear change of coordinates by a nonsingular (3×3)-matrix S and differentiable scalar function $V : U \subset \mathbb{R}^3 \to \mathbb{R}^1$ to satisfy the condition

$$\lambda_1(u, S) + \lambda_2(u, S) + \dot{V}(u) < 0, \quad \forall u \in \mathcal{Z}, \tag{6.90}$$

where $\lambda_1(u, S) > \lambda_2(u, S) > \lambda_3(u, S)$ are the eigenvalues of the symmetrized matrix (see (6.72))

$$Q = \frac{1}{2}\left(SJ(u)S^{-1} + (SJ(u)S^{-1})^*\right)$$

and

$$J = J(x, y, z) = \begin{pmatrix} -\sigma & \sigma & 0 \\ r-z & -1 & -x \\ y & x & -b \end{pmatrix} \tag{6.91}$$

is the 3×3 Jacobian matrix of system (6.81).

For that we consider the two cases

$$\left[\begin{array}{ll} 1 < r < (b+1)(\frac{b}{\sigma}+1), & b \in (0, 2], \\ 1 < r < r_0, & b \in [2, 4], \end{array}\right. \quad \text{and} \quad r_0 \le r < (b+1)(\tfrac{b}{\sigma}+1), b \in [2, 4], \tag{6.92}$$

where $r_0 = \frac{2}{(1+l^2)}(b+1)(\frac{b}{\sigma}+1)$ and constant l is defined according to (6.82).

3.1) For $1 < r < (b+1)(\frac{b}{\sigma}+1)$, $b \in (0, 2]$ or $1 < r < r_0$, $b \in [2, 4]$ we consider $S = \begin{pmatrix} \sqrt{\frac{r}{\sigma}} & 0 & 0 \\ 0 & 1 & 0 \\ 0 & 0 & 1 \end{pmatrix}$ and $V \equiv 0$. Then condition (6.90) is equivalent to the relation

$$Q + \mu I > 0, \tag{6.93}$$

where $\mu = -\text{tr}J = \sigma + b + 1 > 0$. Condition (6.93) means that all leading principal minors $\Delta_{1,2,3}$ of the corresponding matrix are positive. For the chosen matrix S we have $\Delta_1 = -\sigma + \mu > 0$, $\Delta_2 = \frac{\Delta_3}{(-b+\mu)} + \left(\Delta_3^{''}\right)^2$ and relation (6.93) can be expressed as

$$\Delta_3 = \begin{vmatrix} -\sigma + \mu & \Delta_3^{'} & \Delta_3^{''} \\ \Delta_3^{'} & -1 + \mu & 0 \\ \Delta_3^{''} & 0 & -b + \mu \end{vmatrix} > 0, \tag{6.94}$$

where $\Delta_3^{'} = \sqrt{r\sigma} - \frac{z}{2}\sqrt{\frac{\sigma}{r}}$, $\Delta_3^{''} = \frac{y}{2}\sqrt{\frac{\sigma}{r}}$. Condition (6.94) can be rewritten as

$$\frac{r}{\sigma}\left((\mu - \sigma)(\mu - 1) - r\sigma\right) - \frac{z^2}{4} - \frac{y^2}{4}\frac{(\mu - 1)}{(\mu - b)} + rz > 0. \tag{6.95}$$

From (6.95) and (6.82) we conclude that inequality (6.90) is true.

3.2) For $r_0 \leq r < (b+1)(\frac{b}{\sigma} + 1)$ and $b \in [2, 4]$ we consider the matrix $S = \begin{pmatrix} -A^{-1} & 0 & 0 \\ -\frac{(b-1)}{\sigma} & 1 & 0 \\ 0 & 0 & 1 \end{pmatrix}$ and the Lyapunov function $V(x, y, z) = \frac{V_1(x,y,z)}{\sqrt{(\sigma-1)^2+4\sigma r}}$, where $A = \frac{\sigma}{\sqrt{\sigma r + (\sigma - b)(b-1)}}$ and

$$V_1(x, y, z) = \gamma_1(y^2 + z^2) + \gamma_2 x^2 + \gamma_3\left(\frac{x^4}{4\sigma^2} - \frac{x^2 z}{\sigma} - y^2 - 2xy\right) - \frac{\sigma}{b}z \tag{6.96}$$

with parameters γ_1, γ_2, and γ_3.

Using (6.84) one can for all $z \geq \frac{x^2}{2\sigma}$ obtain the estimate

$$\begin{aligned} \dot{V}_1 \leq\ & [2rxy - 2dy^2 - 2bz^2]\gamma_1 + [2\sigma xy - 2\sigma x^2]\gamma_2 + \\ & + [(4\sigma + 2b)z^2 + 2(\sigma - r + d)xy - 2(\sigma - d)y^2 - 2rx^2]\gamma_3 + \sigma z - \frac{\sigma}{b}xy. \end{aligned} \tag{6.97}$$

It is easy to see that the matrix Q has the eigenvalues

$$\lambda_2(u, S) = -b, \quad \lambda_{1,3}(u, S) = -\frac{(\sigma+1)}{2} \pm \frac{1}{2}\sqrt{(\sigma + 1 - 2b)^2 + \left(Az - \frac{2\sigma}{A}\right)^2 + A^2\left(y + \frac{b-1}{\sigma}x\right)^2},$$

for which the relations $\lambda_1(u, S) > \lambda_2(u, S) > \lambda_3(u, S)$ hold.

With these eigenvalues condition (6.90) takes the form

$$\begin{aligned} 2(\lambda_1(u, S) + \lambda_2(u, S) + \dot{V}(u)) \leq & -(\sigma + 1 + 2b) + \sqrt{(\sigma - 1)^2 + 4\sigma r} + \\ & + \frac{2}{\sqrt{(\sigma-1)^2 + 4\sigma r}}\left(-\sigma z + \frac{A^2 z^2}{4} + \frac{A^2}{4}\left(y + \frac{b-1}{\sigma}x\right)^2 + \dot{V}_2\right). \end{aligned}$$

Supposing $\gamma_1 \geq \frac{A^2}{8b} + \gamma_3(\frac{2\sigma}{b} + 1)$ and using (6.97), we get the estimate

$$-\sigma z + \frac{A^2 z^2}{4} + \frac{A^2}{4}(y + \frac{b-1}{\sigma}x^2) + \dot{V}_2 \leq B_1 x^2 + B_2 xy + B_3 y^2, \quad \forall z \geq \frac{x^2}{2\sigma},$$

where

$$B_1 = -2\gamma_2\sigma + \tfrac{A^2(b-1)^2}{4\sigma^2} - 2\gamma_3 r, \quad B_3 = -2\gamma_1 + \tfrac{A^2}{4} - 2(\sigma-1)\gamma_3,$$
$$B_2 = 2\big(r\gamma_1 + \tfrac{A^2(b-1)}{4\sigma} - \frac{\sigma}{2b} + \sigma\gamma_2 + (\sigma+1-r)\gamma_3\big).$$

The inequality $B_1x^2 + B_2xy + B_3y^2 \le 0$ holds for any $x, y \in \mathbb{R}$ if

$$B_1 \le 0, \quad B_3 \le 0, \quad 4B_1B_3 - B_2^2 \ge 0. \tag{6.98}$$

If we take the coefficients

$$\gamma_1 = A^2\tfrac{3\sigma+b-1}{8\sigma(b+2)}, \quad \gamma_3 = A^2\tfrac{b-1}{8\sigma+2}, \quad \gamma_2 = \tfrac{r-2}{\sigma}(\gamma_3-\gamma_1) - A^2\tfrac{(b+\sigma-1)}{4\sigma^2} + \tfrac{1}{2b},$$

conditions (6.98) are equivalent to the condition

$$b \ge 1, \quad r\sigma + (\sigma-b)(b-1) > 0, \quad \tfrac{(2r-1)\sigma(8\sigma+b\sigma+b^2-b)}{b(b+1)} > -3\sigma^2 + 6\sigma(b-1) + (b-1)^2,$$

which are satisfied for $b \in [2,4]$ and $r \ge r_0$.

Thus, for these values condition (6.90) takes the form

$$2(\lambda_1(u,S) + \lambda_2(u,S) + \dot{V}(u)) \le -(\sigma+1+2b) + \sqrt{(\sigma-1)^2 + 4\sigma r} < 0.$$

Case (4): $r \ge (b+1)(\frac{b}{\sigma}+1)$ and $2\sigma > b$. In this case, following the ideas from Case (3) we estimate (see condition (6.73))

$$\lambda_1(u,S) + \lambda_2(u,S) + s\lambda_3(u,S) + \dot{V}(u,S) < 0$$

for $b \in (0,2]$ and $V = 0$ or $b \in [2,4]$ and $V = \frac{V_1}{\sqrt{(\sigma-1)^2+4\sigma r}}$ and get

$$2(\lambda_1(u,S) + \lambda_2(u,S) + s\lambda_3(u,S) + \dot{V}(u)) \le$$
$$-(\sigma+1+2b) - s(\sigma+1) + (1-s)\sqrt{(\sigma-1)^2 + 4\sigma r}, \quad \forall z \ge x^2/2\sigma. \tag{6.99}$$

Therefore it follows from Theorem 6.4, that

$$\dim_L \mathcal{A} \le 3 - \frac{2(\sigma+b+1)}{\sigma+1+\sqrt{(\sigma-1)^2+4\sigma r}}.$$

Since the Jacobi matrix $J(u)$ at the equilibrium u_1 has the simple real eigenvalues

$$\lambda_1 = -b, \ \lambda_{1,3} = \tfrac{1}{2}\left(-(\sigma+1) \pm \sqrt{(\sigma-1)^2 + 4\sigma r}\right) \tag{6.100}$$

we get

$$\dim_L u_1 = 3 - \frac{2(\sigma + b + 1)}{\sigma + 1 + \sqrt{(\sigma - 1)^2 + 4\sigma r}}.$$

Therefore, finally by Corollary 6.11 we obtain $\dim_L \mathcal{A} = \dim_L u_1$.

The existence of an analytical formula for the exact Lyapunov dimension of the Lorenz system with classical parameters is known (see, e.g. [70]) as the Eden conjecture on the Lorenz system (see [23–25]).

6.5.3 Glukhovsky-Dolzhansky System

Consider a system, suggested by Glukhovsky and Dolzhansky [31] (see also Sect. 5.3.3., Chapt. 5)

$$\begin{cases} \dot{x} = -\sigma x + z + a_0 yz, \\ \dot{y} = R - y - xz, \\ \dot{z} = -z + xy, \end{cases} \tag{6.101}$$

where σ, R, a_0 are positive numbers (here $u = (x, y, x)$). By the change of variables

$$(x, y, z) \to (x, R - \frac{\sigma}{a_0 R + 1} z, \frac{\sigma}{a_0 R + 1} y) \tag{6.102}$$

system (6.101) becomes

$$\begin{cases} \dot{x} = -\sigma x + \sigma y - \frac{a_0 \sigma^2}{(a_0 R + 1)^2} yz, \\ \dot{y} = \frac{R}{\sigma}(a_0 R + 1)x - y - xz, \\ \dot{z} = -z + xy. \end{cases} \tag{6.103}$$

System (6.103) is a generalization of the Lorenz system (6.81) and can be written as

$$\begin{cases} \dot{x} = \sigma(y - x) - Ayz, \\ \dot{y} = rx - y - xz, \\ \dot{z} = -bz + xy, \end{cases} \tag{6.104}$$

where

$$A = \frac{a_0 \sigma^2}{(a_0 R + 1)^2}, \quad r = \frac{R}{\sigma}(a_0 R + 1), \quad b = 1. \tag{6.105}$$

Theorem 6.7 ([67]) *If*

1. $\sigma = Ar$, $4\sigma r > (b + 1)(b + \sigma)$
   or

2. *$b = 1$, $r > 2$, and*

$$\begin{cases} \sigma > \frac{-3+2\sqrt{3}}{3} Ar, & \text{if } \ 2 < r \leq 4, \\ \sigma \in \left(\frac{-3+2\sqrt{3}}{3} Ar, \ \frac{3r+2\sqrt{r(2r+1)}}{r-4} Ar \right), & \text{if } \ r > 4, \end{cases}$$

then for a bounded invariant set $\mathcal{K} \ni (0, 0, 0)$ of system (6.104) *with $b = 1$ or $\sigma = Ar$ we have*

$$\dim_{L} \mathcal{K} = 3 - \frac{2(\sigma + 2)}{\sigma + 1 + \sqrt{(\sigma - 1)^2 + 4\sigma r}}. \tag{6.106}$$

If $(0, 0, 0) \notin \mathcal{K}$, *then the right-hand side of the above relation is an upper bound of* $\dim_{L}(\{\varphi^t\}_{t \geq 0}, \mathcal{K})$.

Note that this formula coincides with the dimension formula for the Lorenz system (6.89) for $b = 1$. Remark that system (6.101) is dissipative and possesses a global attractor (see, e.g. [66]).

6.5.4 Yang and Tigan Systems

Consider the Yang system [101]

$$\begin{cases} \dot{x} = \sigma(y - x), \\ \dot{y} = rx - xz, \\ \dot{z} = -bz + xy, \end{cases} \tag{6.107}$$

where $\sigma > 0, b > 0$, and r is a real number. Consider also the T-system (Tigan system) [99]

$$\begin{cases} \dot{x} = a(y - x), \\ \dot{y} = (c - a)x - axz, \\ \dot{z} = -bz + xy. \end{cases} \tag{6.108}$$

By the transformation $(x, y, z) \to (\frac{x}{\sqrt{a}}, \frac{y}{\sqrt{a}}, \frac{z}{a})$ the Tigan system takes the form of the Yang system with parameters $\sigma = a, r = c - a$.

Theorem 6.8 ([63])

1. *Assume that $r = 0$ and the following inequalities $b(\sigma - b) > 0$, $\sigma - \frac{(\sigma+b)^2}{4(\sigma-b)} \geq 0$ are satisfied. Then any bounded on $[0, +\infty)$ solution of system* (6.107) *tends to a certain equilibrium as $t \to +\infty$.*
2. *Assume that $r < 0$ and $r\sigma + b(\sigma - b) > 0$. Then any bounded on $[0, +\infty)$ solution of system* (6.107) *tends to a certain equilibrium as $t \to +\infty$.*
3. *Assume that $r > 0$ and there are two distinct real roots $\gamma^{(II)} > \gamma^{(I)}$ of equation*

$$4br\sigma^2(\gamma + 2\sigma - b)^2 + 16\sigma b\gamma(r\sigma^2 + b(\sigma + b)^2 - 4\sigma(\sigma r + \sigma b - b^2)) = 0 \tag{6.109}$$

such that $\gamma^{(II)} > 0$.
In this case

a. *if*

$$b(b - \sigma) < r\sigma < b(\sigma + b),$$

then any bounded on $[0, +\infty)$ *solution of system* (6.107) *tends to a certain equilibrium as* $t \to +\infty$.

b. *if*

$$r\sigma > b(\sigma + b), \tag{6.110}$$

then

$$\dim_L \mathcal{K} = 3 - \frac{2(\sigma + b)}{\sigma + \sqrt{\sigma^2 + 4\sigma r}}, \tag{6.111}$$

where $\mathcal{K} \ni (0, 0, 0)$ *is a bounded invariant set of system* (6.107). *If* $(0, 0, 0) \notin \mathcal{K}$, *then the right-hand side of the above relation is an upper bound of* $\dim_L(\{\varphi^t\}_{t\geq 0}, \mathcal{K})$.

6.5.5 Shimizu-Morioka System

Consider the Shimizu-Morioka system [92] of the form

$$\begin{cases} \dot{x} = y, \\ \dot{y} = x - by - xz, \\ \dot{z} = -az + x^2, \end{cases} \tag{6.112}$$

where a, b are positive parameters.

Using the diffeomorphism of $\mathbb{R}^3$

$$\begin{pmatrix} x \\ y \\ z \end{pmatrix} \to \begin{pmatrix} x \\ y \\ z - \frac{x^2}{2} \end{pmatrix}, \tag{6.113}$$

system (6.112) can be reduced to the following system

$$\begin{cases} \dot{x} = y, \\ \dot{y} = x - by - xz + \dfrac{x^3}{2}, \\ \dot{z} = -az + xy + \left(1 + \dfrac{a}{2}\right)x^2, \end{cases} \tag{6.114}$$

where α, λ are the positive parameters of system (6.112). We say that system (6.114) is the *transformed* Shimizu-Morioka system.

Theorem 6.9 ([58]) *Suppose, $\mathcal{K}$ is a bounded invariant set of system* (6.114)*: $(0, 0, 0) \in \mathcal{K}$, and the following relations*

$$b - 4 \leq \sqrt{10 + \frac{3}{a} - 13a}, \; b < \frac{1}{a} - a, \; 4 - b \leq \sqrt{\frac{8 + 15a - 8a^2 - 24a^3}{2a(a+1)}} \tag{6.115}$$

are satisfied. Then

$$\dim_L \mathcal{K} = 3 - \frac{2(b+a)}{b + \sqrt{4 + b^2}}. \tag{6.116}$$

If $(0, 0, 0) \notin \mathcal{K}$, then the right-hand side of relation (6.116) *is an upper bound of* $\dim_L(\{\varphi^t\}_{t\geq 0}, \mathcal{K})$.

In the proof there are used the Lyapunov function of the form

$$V(x, y, z) = \frac{1 - s}{4\sqrt{4 + b^2}}\theta,$$

where

$$\begin{aligned}\theta = {} & \mu_1(2y^2 - 2xy - x^4 + 2x^2z) + \mu_2x^2 - \frac{4}{a}z \\ & + \mu_3(z^2 - x^2z + \frac{x^4}{4} + xy) + \mu_4(z^2 + y^2 - \frac{x^4}{4} - x^2),\end{aligned}$$

and the non-singular matrix

$$S = \begin{pmatrix} -\frac{1}{k} & 0 & 0 \\ b - a & 1 & 0 \\ 0 & 0 & 1 \end{pmatrix}.$$

6.6 Attractors of Dynamical Systems

6.6.1 Computation of Attractors and Lyapunov Dimension

The study of a dynamical system typically begins with an analysis of the equilibria, which are easily found numerically or analytically. Therefore, from a computational perspective, it is natural to suggest the following classification of attractors, which is based on the simplicity of finding their basins of attraction in the phase space:

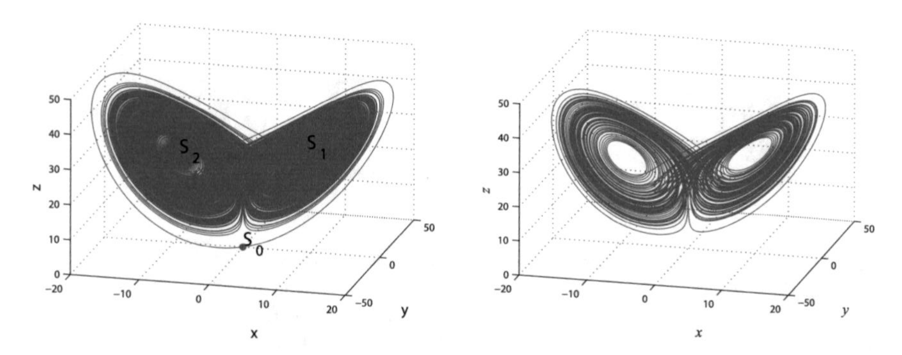

Fig. 6.1 Numerical visualization of self-excited chaotic attractor in the Lorenz system. Global $\mathcal{B}$-attractor (left subfigure), $\dim_L(\mathcal{K}) = \sup_{u\in\mathcal{K}} \dim_L(u) = \dim_L(S_0) = 2.4013$ according to (93) global attractor (right subfigure), $\dim_L(\mathcal{K}) \approx 2.0565$ by numerical computation. Parameters: $r = 28, \sigma = 10, b = 8/3$

Definition 6.12 [61, 66, 68, 69] An attractor is called a *self-excited attractor* if its basin of attraction intersects with any open neighborhood of a stationary state (an equilibrium), otherwise it is called a *hidden attractor* .

Self-excited attractor in a system can be found using the standard computational procedure, i.e., by constructing a solution using initial data from a small neighborhood of the equilibrium, observing how it is attracted and, thus, visualizes the attractor. For example, in the Lorenz system (6.81) with classical parameters $\sigma = 10$, $b = 8/3$, $r = 28$ there is a chaotic attractor, which is self-excited with respect to all three equilibria and could have been found using the standard computational procedure with initial data in vicinity of any of the equilibria (see Fig. 6.1).

Here it is possible to check numerically that for the considered parameters the local attractor is a global attractor (i.e., there are no other attractors in the phase space). In this case the global $\mathcal{B}$-attractor involves the chaotic local attractor, three unstable equilibria and their unstable manifolds attracted to the chaotic local attractor.

However it is known that for other values of parameters, e.g. $\sigma = 10$, $b = 8/3$, $r = 24.5$ [95], the chaotic local attractor in the Lorenz system may be self-excited with respect to the zero unstable equilibrium only. In this case there are three coexisting minimal local attractors (see Fig. 6.2): chaotic local attractor and two trivial local attractors—stable equilibria $S_{1,2}$.

Self-excited attractors in a multistable system can be found using the standard computational procedure, whereas there is no standard way of predicting the existence of hidden attractors in a system.

While the *multistability* is a property of the system, the *self-excited* and *hidden* properties are properties of the attractor and its basin. For example, hidden attractors are attractors in systems with no equilibria or with only one stable equilibrium (a special case of multistability and coexistence of attractors).

In general, there is no straightforward way of predicting the existence or coexistence of hidden attractors in a system (see, e.g. [18, 45, 48–50, 61, 65, 66, 68, 69]). A

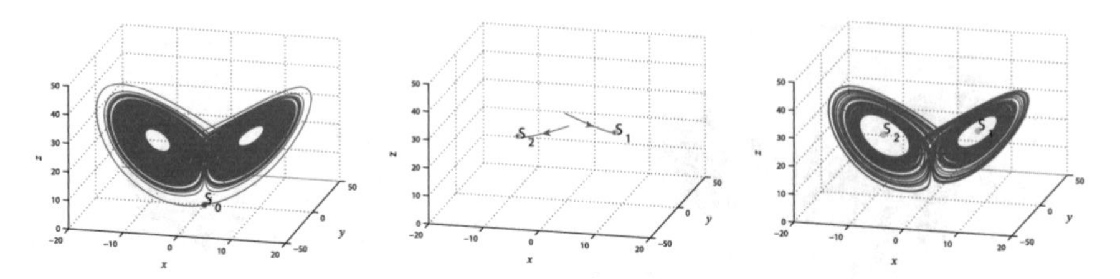

Fig. 6.2 Numerical visualization of self-excited chaotic local attractor in the Lorenz system. Local $\mathcal{B}$-attractor involves self-excited chaotic local attractor, unstable zero equilibrium, and its unstable manifold attracted to the chaotic local attractor (left subfigure), $\dim_L(\mathcal{K}) = \sup_{u\in\mathcal{K}} \dim_L(u) = \dim_L(S_0) = 2.3727$ according to (93) Trajectories with the initial data $(\pm 1.3276, \mp 9.7014, 28.7491)$ tend to trivial local attractors—equilibria $S_{1,2}$ (middle subfigure), $\dim_L(S_{1,2}) = 1.9989$. Global attractor is the union of three coexisting local attractors: self-excited chaotic local attractor and two trivial local attractors (right subfigure), $\dim_L(\mathcal{K}) \approx 2.0489$ by numerical computation. Parameters: $r = 24.5$, $\sigma = 10$, $b = 8/3$

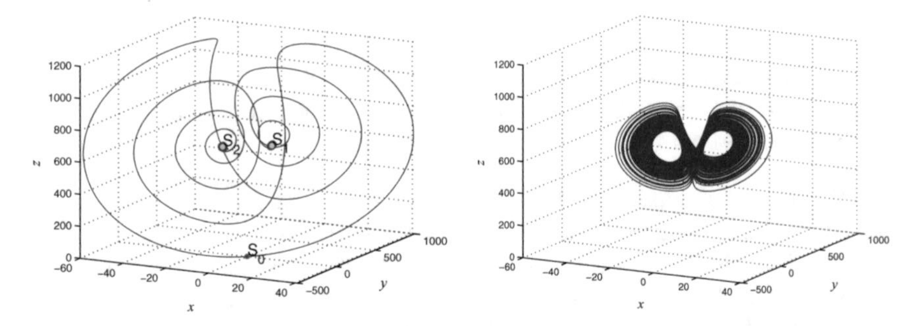

Fig. 6.3 Numerical visualization of local $\mathcal{B}$-attractor and hidden local attractor in the Glukhovsky-Dolzhansky system. Local $\mathcal{B}$-attractor involves outgoing separatrix (blue) of the saddle S_0 (red) attracted to the stable equilibria $S_{1,2}$ (green) (left subfigure). Hidden local attractor (magenta, $\dim_L(\mathcal{K}) \approx 2.1322$ by numerical computation) coexists with local $\mathcal{B}$-attractor ($\dim_L(\mathcal{K}) = \sup_{u\in\mathcal{K}} \dim_L(u) = \dim_L(S_0) = 2.8917$ by (6.106)). Global $\mathcal{B}$-attractor involves the local $\mathcal{B}$-attractor and the hidden local attractor

numerical search of hidden attractors by evolutionary algorithms is discussed in [104, 105]. Recent examples of hidden attractors can be found in *The European Physical Journal Special Topics: Multistability: Uncovering Hidden Attractors*, 2015 (see [9, 27, 28, 38, 73, 83, 86, 89–91, 96, 100, 106]).

For example, in the Glukhovsky-Dolzhansky system and the corresponding generalized Lorenz system (6.104) with parameters $r = 700$, $A = 0.0052$, $\sigma = rA$, $b = 1$ a hidden chaotic local attractor can be found [65, 66] (see Fig. 6.3).

Remark that if a system is proved to an absorbing set, then all self-excited or hidden local attractors of the system are inside this absorbing bounded domain and can be found numerically. However, in general, *the determination of the number and mutual disposition of chaotic minimal local attractors in the phase space for a system* may be a challenging problem [62] (see, e.g. the corresponding well-known problem for two-dimensional polynomial systems — the second part of the 16th Hilbert problem on the number and mutual disposition of limit cycles [33]).[14] Thus the advantage of the analytical method for the Lyapunov dimension estimation, suggested in Theorem 6.4, is that it is useful not only for dissipative systems (see, e.g. estimation of the Lyapunov dimension for one of the Rössler systems [59]) but also allows one to estimate the Lyapunov dimension of invariant set without localization of the set in the phase space.

Remark that, from a computational perspective, it is not feasible to check numerically the attractivity property for all initial states of the phase space of a dynamical system. A natural generalization of the notion of an attractor is the consideration of the weaker attraction requirements: almost everywhere or on a set of positive measure (see, e.g. [77] and Chap. 1). See also *trajectory attractors* [11, 13, 88]. In numerical computations, to distinguish an artificial computer generated chaos from a real behavior of the system, one can consider the shadowing property of the system (see, e.g., the survey in [84]).

We can typically see an attractor (or global attractor) in numerical experiments. The notion of a $\mathcal{B}$-attractor is mostly used in the theory of dimensions, where we consider invariant sets covered by balls. The uniform attraction requirement of the attractor implies that a global $\mathcal{B}$-attractor involves a set of stationary points $\mathcal{S}$ and the corresponding unstable manifolds W^u (see Chap. 1). The same is true for $\mathcal{B}$-attractor if the considered neighborhood $\mathcal{K}_\varepsilon$ of the attractor contains some of the stationary points from $\mathcal{S}$. This allows one to get analytical estimations of the Lyapunov dimension for $\mathcal{B}$-attractors and even formulas since the local Lyapunov dimension at a stationary point can be easily obtained analytically (but this does not help for chaotic minimal local attractors, hidden $\mathcal{B}$-attractors since they do not involve any stationary points).

From a computational perspective, numerically check the attractivity property of an attractor is also difficult. Therefore, if the basin of attraction involves unstable manifolds of equilibria, then computing the minimal attractor and the unstable manifolds that are attracted to it may be regarded as an approximation of the minimal $\mathcal{B}$-attractor. For example, consider the visualization of the Lorenz attractor from the neighborhood of the zero saddle equilibria. Note that a minimal global attractor involves the set $\mathcal{S}$ and its basin of attraction involves the set $W^u(\mathcal{S})$.

For the computation of the Lyapunov dimension of an attractor $\mathcal{A}$ we consider a sufficiently large time T and a sufficiently dense grid of points $\mathcal{A}_{\text{grid}}$ on the attractor, compute the local Lyapunov dimensions by the corresponding Kaplan-Yorke formula $\dim_L^{\text{KY}}(\{\nu_i(T,u)\}_1^n)$, and take the maximum on the grid: $\max_{u\in\mathcal{A}_{\text{grid}}} \dim_L^{\text{KY}}(\{\nu_i(T,u)\}_1^n)$.

---

[14]The numerical search of hidden attractors can be complicated by the small size of the basin of attraction with respect to the considered set of parameters $p \in P$ and subset of the phase space $\mathcal{U}_0 \subset \mathcal{U}$: following [9, 103], the attractor may be called a *rare attractor* if the measure μ of the basin of attractors $\beta(\mathcal{K}_p)$ for the considered set of parameters $p \in P$ is small with respect to the considered part of the phase space $\mathcal{U}_0 \subset \mathcal{U}$, i.e. $\frac{\int_{p\in P} \mu(\beta(\mathcal{K}_p)\cap\mathcal{U}_0)}{\mu(\mathcal{U}_0)} << 1$. Also computational difficulties may be caused by the shape of basin of attraction, e.g. by Wada and riddled basins.

Since numerically we can check only that all points of the grid belong to the basin of attraction, the following remark is useful. Let a point u_0 belong to the basin of attraction of the attractor $\mathcal{A}$. Consider the union of the semi-orbit $\gamma^+(u_0) = \{\varphi^t(u_0), t \geq 0\}$ and the attractor $\mathcal{A}$: $\mathcal{K}(u_0) = \mathcal{A} \cup \gamma^+(u_0)$. According to the definition of the basin of attraction, the ω-limit set of $\varphi^t(u_0)$ belong to $\mathcal{A}$, thus the set $\mathcal{K}(u_0)$ is compact and invariant. Since $\mathcal{A} \supset \mathcal{K}(\varphi^t(u_0)) \supset \mathcal{K}(u_0)$, we have

$$\dim_L(\varphi^t, \mathcal{A}) = \max_{u \in \mathcal{A}} \dim_L(\varphi^t, u) \leq \max_{u \in \mathcal{K}(\varphi^t(u_0))} \dim_L(\varphi^t, u) \leq \max_{u \in \mathcal{K}(u_0)} \dim_L(\varphi^t, u).$$

Since $\rho\big(\mathcal{K}(u_0), \mathcal{K}(\varphi^t(u_0))\big) \to 0$ for $t \to +\infty$, from the property (6.32) and the continuity (Lemma 6.2), it follows that

$$\dim_L = \liminf_{t \to +\infty} \max_{u \in \mathcal{K}(\varphi^t(u_0))} \dim_L(\varphi^t, u).$$

6.7 Computation of the Finite-Time Lyapunov Exponents and Dimension in MATLAB

The singular value decomposition (*SVD*) of a fundamental matrix $D\varphi^t(u_0)$ has the form

$$D\varphi^t(u_0) = \mathrm{U}(t, u_0)\text{˝}(t, u_0)\mathrm{V}^T(t, u_0): \quad \mathrm{U}(t, u_0)^T\mathrm{U}(t, u_0) \equiv I \equiv \mathrm{V}(t, u_0)^T\mathrm{V}(t, u_0),$$

where $\text{˝}(t) = \mathrm{diag}\{\alpha_1(t, u_0), ..., \alpha_n(t, u_0)\}$ is a diagonal matrix with positive real diagonal entries consistingof singular values. We now give a MATLAB implementation [66] of the discrete SVD method for computing the finite-time Lyapunov exponents $\{\nu_i(t, u_0)\}_1^n$ based on the product SVD algorithm (see, e.g., [20, 97]).

Listing 6.1 productSVD.m – product SVD algorithm

```
1  function [U, R, V] = productSVD(initFactorization, nIterations)
2  % Parameters:
3  %   initFactorization - the array contains factor matrices of the
4  %                       fundamental matrix X, such that:
5  %           X = initFactorization(:,:,1) * ... * initFactorization(:,:,end);
6  %   nIterations - the number of iterations in the product SVD algorithm.
7
8  % dimOde - dimension of the ODEs, nFactors - the number of factor matrices
9  [~, dimOde, nFactors] = size(initFactorization);
10
11 % A - 2d array of matrices storing the factor matrices at each iteration
12 A = zeros(dimOde, dimOde, nFactors, nIterations); A(:, :, :, 1) =
13   initFactorization;
14
15 % Q - array of matrices storing orhogonal matrices of the QR decomposition
16 Q = zeros(dimOde, dimOde, nFactors+1);
17
18 % U, V - orthogonal matrices in the SVD decomposition
19  U = eye(dimOde); V = eye(dimOde);
20
21 % R - array of upper triangular factor matrices, such that after
22 % the last iteration \Sigma = R(:,:,1) * ... * R(:,:,end)
23  R = zeros(dimOde, dimOde, nFactors);
24
25 % Main loop
26 for iIteration = 1 : nIterations
```

```
27        Q(:, :, nFactors + 1) = eye(dimOde, dimOde);
28        for jFactor = nFactors : -1 : 1
29            C = A(:, :, jFactor, iIteration) * Q(:, :, jFactor+1);
30            [Q(:, :, jFactor), R(:, :, jFactor)] = qr(C);
31            for kCoord = 1 : dimOde
32                if R(kCoord, kCoord, jFactor) < 0
33                    R(kCoord, :, jFactor) = -1 * R(kCoord, :, jFactor);
34                    Q(:, kCoord, jFactor) = -1 * Q(:, kCoord, jFactor);
35                end;
36            end;
37        end;
38
39        if mod(iIteration, 2) == 1
40            U = U * Q(:, :, 1);
41        else
42            V = V * Q(:, :, 1);
43        end
44
45        for jFactor = 1 : nFactors
46            A(:, :, jFactor, iIteration + 1) = R(:, :, nFactors-jFactor+1)';
47        end
48    end
49
50 end
```

Listing 6.2 computeLEs.m – computation of the Lyapunov exponents

```
1  function LEs = computeLEs(extOde, initPoint, tStep, ...
2                                   nFactors, nSvdIterations, odeSolverOptions)
3  % Parameters:
4  %   extOde - extended ODE system (system of ODEs + var. eq.);
5  %   initPoint -  initial point;
6  %   tStep - time-step in the factorization procedure;
7  %   nFactors - number of factor matrices in the factorization procedure;
8  %   nSvdIterations - number of iterations in the product SVD algoritm;
9  %   odeSolverOptions - solver options (sover = ode45);
10
11 % Dimension of the ODE :
12  dimOde = length(initPoint);
13
14 % Dimension of the extended ODE (ODE + Var. Eq.):
15  dimExtOde = dimOde * (dimOde + 1);
16
17   tBegin = 0; tEnd = tStep; tSpan = [tBegin, tEnd]; initFundMatrix =
18   eye(dimOde); initCond = [initPoint(:); initFundMatrix(:)];
19
20 X = zeros(dimOde, dimOde, nFactors);
21
22 % Main loop : factorization of the fundamental matrix
23 for iFactor = 1 : nFactors
24     [~, extOdeSolution] = ode45(extOde, tSpan, initCond, odeSolverOptions);
25
26     X(:, :, iFactor) = reshape(...
27                           extOdeSolution(end, (dimOde + 1) : dimExtOde), ...
28                                                             dimOde, dimOde);
29     currInitPoint = extOdeSolution(end, 1 : dimOde);
30     currInitFundMatrix = eye(dimOde);
31
32     tBegin = tBegin + tStep;
33     tEnd = tEnd + tStep;
34     tSpan = [tBegin, tEnd];
35     initCond = [currInitPoint(:); currInitFundMatrix(:)];
36  end
37
38 % Product SVD of factorization X of the fundamental matrix
39   [~, R, ~] = productSVD(X, nSvdIterations);
40
41 % Computation of the Lyapunov exponents
42   LEs = zeros(1, dimOde); for jFactor = 1 : nFactors
43     LEs = LEs + log(diag(R(:, :, jFactor))');
44    end; finalTime = tStep * nFactors; LEs = LEs / finalTime;
45
46   end
```

Listing 6.3 lyapunovDim.m – computation of the Lyapunov dimension

```
function LD = lyapunovDim( LEs )
% For the given array of finite-time Lyapunov exponents at a point the function
% computes the local Lyapunov dimension by the Kaplan-Yorke formula.

% Parameters:
%   LEs - array of the finite-time Lyapunov exponents.

% Initialization of the local Lyapunov dimension:
  LD = 0;

% Number of LEs :
 nLEs = length(LEs);

% Sorted LEs :
  sortedLEs = sort(LEs, 'descend');

% Main loop :
  leSum = sortedLEs(1); if ( sortedLEs(1) > 0 )
    for i = 1 : nLEs-1
       if sortedLEs(i+1) ~= 0
          LD = i + leSum / abs( sortedLEs(i+1) );
          leSum = leSum + sortedLEs(i+1);
          if leSum < 0
             break;
          end
       end
    end
 end end
```

Listing 6.4 genLorenzSyst.m – generalized Lorenz system (6.104) along with the variational equation

```
function OUT = genLorenzSyst(t, x, r, sigma, b, a)

% Generalized Lorenz system with
% parameters: r sigma b a

  OUT(1) = sigma*(x(2) - x(1)) - a*x(2)*x(3); OUT(2) = r*x(1) - x(2)
  - x(1)*x(3); OUT(3) = -b*x(3) + x(1)*x(2);

% Jacobian at the point [x(1), x(2), x(3)]
J = [-sigma, sigma-a*x(3),  -a*x(2);
    r-x(3),      -1,      -x(1);
    x(2),      x(1), -b];

X = [x(4), x(7), x(10);
    x(5), x(8), x(11);
    x(6), x(9), x(12)];

% Variational equation
 OUT(4:12) = J*X;
```

Listing 6.5 main.m – computation of the Lyapunov exponents and local Lyapunov dimension for the hidden attractor of generalized Lorenz system (6.104)

```
function main

% Parameters of generalized Lorenz system
% that correspond to the hidden attractor
 r = 700; sigma = 4; b = 1; a = 0.0052;

% Initial point for the trajectory which visualizes the hidden attractor
x0 = [-14.551336132013954 -173.86811769236883 718.92035664071227];

 tStep = 0.1; nFactors = 10000; nSvdIterations = 3;

% ODE solver parameters
 acc = 1e-8; RelTol = acc; AbsTol = acc; InitialStep = acc/10;
 odeSolverOptions = odeset('RelTol', RelTol, 'AbsTol', AbsTol, ...
                   'InitialStep', InitialStep, 'NormControl', 'on');

 LEs = computeLEs(@(t, x) genLorenzSyst(t, x, r, sigma, b, a), ...
                x0, tStep, nFactors, nSvdIterations, odeSolverOptions);

```

```
20     fprintf('Lyapunov exponents: %6.4f, %6.4f, %6.4f\n', LEs);
21
22   LD = lyapunovDim(LEs);
23
24   fprintf('Lyapunov dimension: %6.4f\n', LD);
25
26   end
```

References

1. Abarbanel, H., Brown, R., Kennel, M.: Variation of Lyapunov exponents on a strange attractor. J. Nonl. Sci. **1**(2), 175–199 (1991)
2. Adrianova, L.Y.: Introduction to Linear systems of Differential Equations. Amer. Math. Soc, Providence, Rhode Island (1998)
3. Barabanov, E.: Singular exponents and properness criteria for linear differential systems. J. Diff. Equ. **41**, 151–162 (2005)
4. Barreira, L., Gelfert, K.: Dimension estimates in smooth dynamics: a survey of recent results. Ergodic Theory Dyn. Syst. **31**, 641–671 (2011)
5. Barreira, L., Schmeling, J.: Sets of "Non-typical" points have full topological entropy and full Hausdorff dimension. Israel J. of Math. **116**(1), 29–70 (2000)
6. Bogoliubov, N., Krylov, N.: La theorie generalie de la mesure dans son application a l'etude de systemes dynamiques de la mecanique non-lineaire. Ann. Math. II (French) (Annals of Mathematics) **38** (1), 65–113 (1937)
7. Boichenko, V.A., Leonov, G.A.: Lyapunov's direct method in estimates of topological entropy. Zap. Nauchn. Sem. POMI **231**, 62–75 (1995) (Russian); English transl. J. Math. Sci. **91**(6), 3370–3379 (1998)
8. Boichenko, V.A., Leonov, G.A., Reitmann, V.: Dimension Theory for Ordinary Differential Equations. Teubner, Stuttgart (2005)
9. Brezetskyi, S., Dudkowski, D., Kapitaniak, T.: Rare and hidden attractors in van der Pol-Duffing oscillators. Eur. Phys. J. Spec. Topics **224**(8), 1459–1467 (2015)
10. Bylov, B.E., Vinograd, R.E., Grobman, D.M., Nemytskii, V.V.: Theory of Characteristic Exponents and its Applications to Problems of Stability. Nauka, Moscow (1966). (Russian)
11. Chepyzhov, V., Vishik, M.: Attractors for Equations of Mathematical Physics. Amer. Math. Soc, Providence, Rhode Island (2002)
12. Choquet, G., Foias, C.: Solution d'un probleme sur les iteres d'un operateur positif sur $C(K)$ et proprietes de moyennes associees. Annales de l'institut Fourier **25**(3/4), 109–129 (1975) (French)
13. Chueshov, I., Siegmund, S.: On dimension and metric properties of trajectory attractors. J. Dynam. Diff. Equ. **17**(4), 621–641 (2005)
14. Constantin, P., Foias, C.: Global Lyapunov exponents, Kaplan-Yorke formulas and the dimension of the attractors for 2D Navier-Stokes equations. Commun. Pure Appl. Math. **38**(1), 1–27 (1985)
15. Constantin, P., Foias, C., Temam, R.: Attractors representing turbulent flows. Amer. Math. Soc. Memoirs. Providence, Rhode Island **53**(314) (1985)
16. Cvitanović, P., Artuso, R., Mainieri, R., Tanner, G., Vattay, G.: Chaos: Classical and Quantum. Niels Bohr Institute, Copenhagen. http://www.ChaosBook.org (2012)
17. Czornik, A., Nawrat, A., Niezabitowski, M.: Lyapunov exponents for discrete time-varying systems. Stud. Comput. Intell. **440**, 29–44 (2013)
18. Danca, M.-F., Feckan, M., Kuznetsov, N.V., Chen, G.: Looking more closely at the Rabinovich-Fabrikant system. Intern. J. of Bifurcation Chaos **26**(2), art. num. 1650038 (2016)
19. Dellnitz, M., Junge, O.: Set oriented numerical methods for dynamical systems. In: Handbook of Dynamical Systems, vol. 2, 221–264, Elsevier Science (2002)

20. Dieci, L., Elia, C.: SVD algorithms to approximate spectra of dynamical systems. Math. Comput. Simul. **79**(4), 1235–1254 (2008)
21. Doering, C., Gibbon, J., Holm, D., Nicolaenko, B.: Exact Lyapunov dimension of the universal attractor for the complex Ginzburg-Landau equation. Phys. Rev. Lett. **59**, 2911–2914 (1987)
22. Douady, A., Oesterlé, J.: Dimension de Hausdorff des attracteurs. C. R. Acad. Sci. Paris, Ser. A 290, 1135–1138 (1980)
23. Eden, A.: An abstract theory of L-exponents with applications to dimension analysis (Ph.D. thesis). Indiana University (1989)
24. Eden, A.: Local Lyapunov exponents and a local estimate of Hausdorff dimension. ESAIM: Math. Modell. Numer. Anal. Modelisation Mathematique et Analyse Numerique **23**(3), 405–413 (1989)
25. Eden, A.: Local estimates for the Hausdorff dimension of an attractor. J. Math. Anal. Appl. **150**(1), 100–119 (1990)
26. Eden, A., Foias, C., Temam, R.: Local and global Lyapunov exponents. J. Dynam. Diff. Equ. **3**, 133–177 (1991) [Preprint No. 8804, The Institute for Applied Mathematics and Scientific Computing, Indiana University, 1988]
27. Feng, Y., Pu, J., Wei, Z.: Switched generalized function projective synchronization of two hyperchaotic systems with hidden attractors. Eur. Phys. J.: Spec. Topics **224**(8), 1593–1604 (2015)
28. Feng, Y., Wei, Z.: Delayed feedback control and bifurcation analysis of the generalized Sprott B system with hidden attractors. Eur. Phys. J.: Spec. Topics **224**(8), 1619–1636 (2015)
29. Frederickson, P., Kaplan, J., Yorke, E., Yorke, J.: The Liapunov dimension of strange attractors. J. Diff. Equ. **49**(2), 185–207 (1983)
30. Gelfert, K.: Maximum local Lyapunov dimension bounds the box dimension. Direct proof for invariant sets on Riemannian manifolds. Zeitschrift für Analysis und ihre Anwendungen (ZAA) **22**(3), 553–568 (2003)
31. Glukhovsky, A.B., Dolzhanskii, F.V: Three-component geostrophic model of convection in rotating fluid. Izv. Akad. Nauk SSSR, Fiz. Atmos. i Okeana, **16**, 451–462 (1980) (Russian)
32. Gundlach, V., Steinkamp, O.: Products of random rectangular matrices. Mathematische Nachrichten **212**(1), 51–76 (2000)
33. Hilbert, D.: Mathematical problems. Bull. Amer. Math. Soc. **8**, 437–479 (1901–1902)
34. Horn, R.A., Johnson, C.R.: Topics in Matrix Analysis. Cambridge University Press, Cambridge (1991)
35. Hurewicz, W., Wallman, H.: Dimension Theory. Princeton University Press, Princeton (1948)
36. Il'yashenko, Y.S., Weigu, L.: Nonlocal Bifurcations. Amer. Math. Soc (1999)
37. Izobov, N.A.: Lyapunov Exponents and Stability. Cambridge Scientific Publishers, Cambridge (2012)
38. Jafari, S., Sprott, J., Nazarimehr, F.: Recent new examples of hidden attractors. Eur. Phys. J.: Spec. Topics **224**(8), 1469–1476 (2015)
39. Kaplan, J.L., Yorke, J.A.: Chaotic behavior of multidimensional difference equations. In: Functional Differential Equations and Approximations of Fixed Points, pp. 204–227, Springer, Berlin (1979)
40. Kolmogorov, A.: On entropy per unit time as a metric invariant of automorphisms. Dokl. Akad. Nauk SSSR **124**(4), 754–755 (1959) (Russian)
41. Kuczma, M., Gilányi, A.: An Introduction to the Theory of Functional Equations and Inequalities: Cauchy's Equation and Jensen's Inequality. Birkhäuser Basel (2009)
42. Kunze, M., Kupper, T.: Non-smooth dynamical systems: An overview. In: Fiedler, B. (ed.) Ergodic Theory, Analysis, and Efficient Simulation of Dynamical Systems, pp. 431–452. Springer, New York, Berlin (2001)
43. Kuratowski, K.: Topology. Academic Press, New York (1966)
44. Kuznetsov, N.V.: The Lyapunov dimension and its estimation via the Leonov method. Phys. Lett. A **380**(25–26), 2142–2149 (2016)
45. Kuznetsov, N.V.: Hidden attractors in fundamental problems and engineering models. A short survey. Lecture Notes in Electrical Engineering, vol. 371, 13–25, (plenary lecture at AETA 2015: Recent Advances in Electrical Engineering and Related Sciences) (2016)

46. Kuznetsov, N.V., Alexeeva, T.A., Leonov, G.A.: Invariance of Lyapunov exponents and Lyapunov dimension for regular and irregular linearizations. Nonl. Dyn. **85**(1), 195–201 (2016)
47. Kuznetsov, N.V., Leonov, G.A.: On stability by the first approximation for discrete systems. In: 2005 International Conference on Physics and Control, PhysCon 2005, Proc. Vol. 2005, pp. 596–599. IEEE (2005)
48. Kuznetsov, N.V., Leonov, G.A.: Hidden attractors in dynamical systems: systems with no equilibria, multistability and coexisting attractors. IFAC Proc. Vol. (IFAC-PapersOnline) **19**, 5445–5454 (2014)
49. Kuznetsov, N.V., Leonov, G.A., Mokaev, T.N.: Hidden attractor in the Rabinovich system. (2015) Available via arXiv:1504.04723v1
50. Kuznetsov, N.V., Leonov, G.A., Vagaitsev, V.I.: Analytical-numerical method for attractor localization of generalized Chua's system. IFAC Proc. Vol. (IFAC-PapersOnline) **4**(1), 29–33 (2010)
51. Kuznetsov, N.V., Mokaev, T.N., Vasilyev, P.A.: Numerical justification of Leonov conjecture on Lyapunov dimension of Rossler attractor. Commun. Nonlinear Sci. Numer. Simul. **19**, 1027–1034 (2014)
52. Ledrappier, F.: Some relations between dimension and Lyapunov exponents. Commun. Math. Phys. **81**, 229–238 (1981)
53. Kuznetsov, N.V., Mokaev, T.N., Kuznetsova, O.A., Kudryashova, E.V., Leonov, G.A.: The Lorenz system: hidden boundary of practical stability and the Lyapunov dimension (2020) Available via arXiv. http://arxiv.org
54. Leonov, G.A.: On estimations of the Hausdorff dimension of attractors. Vestn. Leningrad Gos. Univ. Ser. 1, **15**, 41–44 (1991) (Russian); English transl. Vestn. Leningrad Univ. Math. **24**(3), 38–41 (1991)
55. Leonov, G.A.: Lyapunov dimensions formulas for Hénon and Lorenz attractors. Alg. Anal. **13**, 155–170 (2001) (Russian); English transl. St. Petersburg Math. J. **13**(3), 453–464 (2002)
56. Leonov, G.A.: Strange Attractors and Classical Stability Theory. St. Petersburg State Univ. Press, St.Petersburg (2008)
57. Leonov, G.A.: Lyapunov functions in the attractors dimension theory. J. Appl. Math. Mech. **76**(2), 129–141 (2012)
58. Leonov, G.A., Alexeeva, T.A., Kuznetsov, N.V.: Analytic exact upper bound for the Lyapunov dimension of the Shimizu-Morioka system. Entropy **17**(7), 5101 (2015)
59. Leonov, G.A., Boichenko, V.A.: Lyapunov's direct method in the estimation of the Hausdorff dimension of attractors. Acta Appl. Math. **26**, 1–60 (1992)
60. Leonov, G.A., Kuznetsov, N.V.: Time-varying linearization and the Perron effects. Intern. J. Bifurcation Chaos **17**(4), 1079–1107 (2007)
61. Leonov, G.A., Kuznetsov, N.V.: Hidden attractors in dynamical systems. From hidden oscillations in Hilbert-Kolmogorov, Aizerman, and Kalman problems to hidden chaotic attractors in Chua circuits. Intern. J. Bifurcation Chaos **23**(1), Art. no. 1330002 (2013)
62. Leonov, G.A., Kuznetsov, N.V.: On differences and similarities in the analysis of Lorenz, Chen, and Lu systems. Appl. Math. Comput. **25**(6), 334–343 (2015)
63. Leonov, G., Kuznetsov, N., Korzhemanova, N., Kusakin, D.: Lyapunov dimension formula of attractors in the Tigan and Yang systems (2015) Available via arXiv:1510.01492v1
64. Leonov, G.A., Kuznetsov, N.V., Korzhemanova, N.A., Kusakin, D.V.: Lyapunov dimension formula for the global attractor of the Lorenz system. Commun. Nonlinear Sci. Numer. Simul. **41**, 84–103 (2016)
65. Leonov, G.A., Kuznetsov, N.V., Mokaev, T.N.: Hidden attractor and homoclinic orbit in Lorenz-like system describing convective fluid motion in rotating cavity. Commun. Nonlinear Sci. Numer. Simul. **28**, 166–174 (2015)
66. Leonov, G.A., Kuznetsov, N.V., Mokaev, T.: Homoclinic orbits, and self-excited and hidden attractors in a Lorenz-like system describing convective fluid motion. Eur. Phys. J. Spec. Topics **224**(8), 1421–1458 (2015)
67. Leonov, G.A., Kuznetsov, N.V., Mokaev, T.N.: The Lyapunov dimension formula of self-excited and hidden attractors in the Glukhovsky-Dolzhansky system (2015) Available via arXiv:1509.09161

68. Leonov, G.A., Kuznetsov, N.V., Vagaitsev, V.I.: Localization of hidden Chua's attractors. Phys. Lett. A **375**(23), 2230–2233 (2011)
69. Leonov, G.A., Kuznetsov, N.V., Vagaitsev, V.I.: Hidden attractor in smooth Chua systems. Phys. D: Nonlin. Phenomena **241**(18), 1482–1486 (2012)
70. Leonov, G.A., Lyashko, S.: Eden's hypothesis for a Lorenz system. Vestn. S. Peterburg Gos. Univ., Matematika, **26**(3), 15–18 (1993) (Russian); English transl. Vestn. St. Petersburg Univ. Math. Ser. 1 **26**(3), 14–16 (1993)
71. Leonov, G.A., Pogromsky, A.Yu., Starkov, K.E.: Erratum to "The dimension formula for the Lorenz attractor". Phys. Lett. A **375**(8), 1179 (2011), Phys. Lett. A **376**(45), 3472–3474 (2012)
72. Leonov, G.A., Poltinnikova, M.S.: On the Lyapunov dimension of the attractor of Chirikov dissipative mapping. AMS Transl. Proc. St.Petersburg Math. Soc., Vol. X **224**, 15–28 (2005)
73. Li, C., Hu, W., Sprott, J., Wang, X.: Multistability in symmetric chaotic systems. Eur. Phys. J.: Spec. Topics **224**(8), 1493–1506 (2015)
74. Lorenz, E.N.: Deterministic nonperiodic flow. J. Atmos. Sci. **20**, 130–141 (1963)
75. Lyapunov, A.M.: The general problem of the stability of motion. Kharkov (1892) (Russian); Engl. transl. Intern. J. Control (Centenary Issue) **55**, 531–572 (1992)
76. Millionschikov, V.M.: A formula for the entropy of smooth dynamical systems. Diff. Urav. (Russian) **12**(12), 2188–2192, 2300 (1976)
77. Milnor, J.W.: Attractor. Scholarpedia **1**, 11 (2006). https://doi.org/10.4249/scholarpedia.1815
78. Noack, A., Reitmann, V.: Hausdorff dimension estimates for invariant sets of time-dependent vector fields. Zeitschrift für Analysis und ihre Anwendungen (ZAA) **15**(2), 457–473 (1996)
79. Oseledec, V.I.: A multiplicative ergodic theorem: Lyapunov characteristic numbers for dynamical systems. Trans. Moscow Math. Soc. **19**, 197–231 (1968)
80. Ott, E., Withers, W., Yorke, J.: Is the dimension of chaotic attractors invariant under coordinate changes? J. Stat. Phys. **36**(5–6), 687–697 (1984)
81. Ott, W., Yorke, J.: When Lyapunov exponents fail to exist. Phys. Rev. E **78** (2008)
82. Pesin, Ya.B.: Characteristic Lyapunov exponents and smooth ergodic theory. Uspekhi Mat. Nauk **43**, 55–112 (1977) (Russian); English transl. Russ. Math. Surveys **32**, 55–114 (1977)
83. Pham, V., Vaidyanathan, S., Volos, C., Jafari, S.: Hidden attractors in a chaotic system with an exponential nonlinear term. Eur. Phys. J.: Spec. Topics **224**(8), 1507–1517 (2015)
84. Pilyugin, S.: Theory of pseudo-orbit shadowing in dynamical systems. J. Diff. Equ. **47**(13), 1929–1938 (2011)
85. Pogromsky, A.Y., Matveev, A.S.: Estimation of topological entropy via the direct Lyapunov method. Nonlinearity **24**(7), 1937 (2011)
86. Saha, P., Saha, D., Ray, A., Chowdhury, A.: Memristive non-linear system and hidden attractor. Eur. Phys. J.: Spec. Topics **224**(8), 1563–1574 (2015)
87. Schmeling, J.: A dimension formula for endomorphisms—the Belykh family. Ergodic Theory Dyn. Syst. **18**, 1283–1309 (1998)
88. Sell, G.R.: Global attractors for the three-dimensional Navier-Stokes equations. J. Dyn. Diff. Equ. **8**(1), 1–33 (1996)
89. Semenov, V., Korneev, I., Arinushkin, P., Strelkova, G., Vadivasova, T., Anishchenko, V.: Numerical and experimental studies of attractors in memristor-based Chua's oscillator with a line of equilibria. Noise-induced effects. Eur. Phys. J.: Spec. Topics **224**(8), 1553–1561 (2015)
90. Shahzad, M., Pham, V.-T., Ahmad, M., Jafari, S., Hadaeghi, F.: Synchronization and circuit design of a chaotic system with coexisting hidden attractors. Eur. Phys. J.: Spec. Topics **224**(8), 1637–1652 (2015)
91. Sharma, P.R., Shrimali, M.D., Prasad, A., Kuznetsov, N.V., Leonov, G.A.: Control of multistability in hidden attractors. Eur. Phys. J.: Spec. Topics **224**(8), 1485–1491 (2015)
92. Shimizu, T., Morioka, N.: On the bifurcation of a symmetric limit cycle to an asymmetric one in a simple model. Phys. Lett. A **76**(3–4), 201–204 (1980)
93. Sinai, Ya.G.: On the concept of entropy of a dynamical system. Dokl. Akad. Nauk, SSSR **124**, 768–771 (1959) (Russian)

94. Smith, R.A.: Some applications of Hausdorff dimension inequalities for ordinary differential equations. Proc. Roy. Soc. Edinburgh **104A**, 235–259 (1986)
95. Sparrow, C.: The Lorenz Equations, Bifurcations, Chaos, and Strange Attractors. Springer, New York (1982)
96. Sprott, J.: Strange attractors with various equilibrium types. Eur. Phys. J.: Spec. Topics **224**(8), 1409–1419 (2015)
97. Stewart, D.E.: A new algorithm for the SVD of a long product of matrices and the stability of products. Electron. Trans. Numer. Anal. **5**, 29–47 (1997)
98. Temam, R.: Infinite-Dimensional Dynamical Systems in Mechanics and Physics, 2nd edn. Springer, New York (1997)
99. Tigan, G., Opris, D.: Analysis of a 3d chaotic system. Chaos Solitons and Fractals **36**(5), 1315–1319 (2008)
100. Vaidyanathan, S., Pham, V.-T., Volos, C.: A 5-D hyperchaotic Rikitake dynamo system with hidden attractors. Eur. Phys. J.: Spec. Topics **224**(8), 1575–1592 (2015)
101. Yang, Q., Chen, G.: A chaotic system with one saddle and two stable node-foci. Intern. J. Bifurcation Chaos **18**, 1393–1414 (2008)
102. Young, L.-S.: Mathematical theory of Lyapunov exponents. J. Phys. A: Math. Theor. **46**(25), 254001 (2013)
103. Zakrzhevsky, M., Schukin, I., Yevstignejev, V.: Scientific Proc. Riga Technical Univ. Transp. Engin. **6**, 79 (2007)
104. Zelinka, I.: A survey on evolutionary algorithms dynamics and its complexity—mutual relations, past, present and future. Swarm Evolut. Comput. **25**, 2–14 (2015)
105. Zelinka, I.: Evolutionary identification of hidden chaotic attractors. Eng. Appl. Artif. Intell. **50**, 159–167 (2016). https://doi.org/10.1016/j.engappai.2015.12.002
106. Zhusubaliyev, Z., Mosekilde, E., Churilov, A., Medvedev, A.: Multistability and hidden attractors in an impulsive Goodwin oscillator with time delay. Eur. Phys. J.: Spec. Topics **224**(8), 1519–1539 (2015)

Part III
Dimension Estimates on Riemannian Manifolds

Chapter 7
Basic Concepts for Dimension Estimation on Manifolds

Abstract In this chapter we present some auxiliary results from the linear operator theory and stability theory which are used in the sequel for dimension estimation. In Sect. 7.1 some elements of the exterior calculus of linear operators in linear spaces are introduced. Section 7.2 is concerned with orbital stability results for vector fields on Riemannian manifolds.

7.1 Exterior Calculus in Linear Spaces, Singular Values of an Operator and Covering Lemmas

7.1.1 Multiplicative and Additive Compounds of Operators

In this subsection we introduce for operators acting between finite dimensional linear spaces, their multiplicative and additive compounds. It will be shown that in special bases the matrices of these operators coincide with the multiplicative and additive compound matrices, respectively, introduced in Sect. 2.3, Chap. 2.

Our representation in Sect. 7.1.1 is based on the results of [10]. Similar problems in Hilbert space are treated in [13, 31, 34].

Definition 7.1 Suppose that $\mathbb{V}$ is an n-dimensional linear space over the field $\mathbb{K}$ and $k \in \{1, 2, \ldots, n\}$ is a natural number. With $\mathbb{V}^{\wedge k}$ or $\Lambda^k \mathbb{V}$ we denote the *k-th exterior power of* $\mathbb{V}$, i.e. the linear space consisting of all linear combinations of the formal expressions $\xi_1 \wedge \xi_2 \wedge \cdots \wedge \xi_k$ with $\xi_j \in \mathbb{V},\ j = 1, 2, \ldots, k$, such that the following rules hold:

(1) For any permutation $\pi = (j_1, \ldots, j_k)$ of $(1, \ldots, k)$ we have
$\xi_{j_1} \wedge \cdots \wedge \xi_{j_k} = \operatorname{sign} \pi \cdot \xi_1 \wedge \cdots \wedge \xi_k\ ,\ \forall\ \xi_i \in \mathbb{V},\ i = 1, \ldots, k\,;$

(2) $\xi_1 \wedge \xi_2 \wedge \cdots \wedge \xi_k + \eta_1 \wedge \xi_2 \wedge \cdots \wedge \xi_k = (\xi_1 + \eta_1) \wedge \xi_2 \wedge \cdots \wedge \xi_k\ ,\forall\ \xi_i, \eta_1 \in \mathbb{V}, i = 1, \ldots, k\,;$

(3) $(\lambda\, \xi_1) \wedge \xi_2 \wedge \cdots \wedge \xi_k = \lambda(\xi_1 \wedge \cdots \wedge \xi_k)\ ,\quad \forall \lambda \in \mathbb{K},\ \forall \xi_i \in \mathbb{V},\ i = 1, 2, \ldots, k\,;$

N. Kuznetsov and V. Reitmann, *Attractor Dimension Estimates for Dynamical Systems: Theory and Computation*, Emergence, Complexity and Computation 38,
https://doi.org/10.1007/978-3-030-50987-3_7

(4) There exist n linearly independent vectors $e_1, \dots, e_n \in \mathbb{V}$ such that the $\binom{n}{k}$ vectors $e_{i_1} \wedge \cdots \wedge e_{i_k}$, $1 \le i_1 < i_2 < \cdots < i_k \le n$, are linearly independent in $\mathbb{V}^{\wedge k}$.

Proposition 7.1 *Suppose $\mathbb{V}$ is an n-dimensional linear space over $\mathbb{K}$ and $k \in \{1, 2, \dots, n\}$ is arbitrary. Then:*

(1) $\mathbb{V}^{\wedge k}$ has the dimension $\binom{n}{k}$ and for any basis $e_1, \dots, e_n$ of $\mathbb{V}$ the vectors $e_{i_1} \wedge \cdots \wedge e_{i_k}$, $1 \le i_1 < \cdots < i_k \le n$, form a basis of $\mathbb{V}^{\wedge k}$.
(2) $\xi_1 \wedge \cdots \wedge \xi_k = 0$ for $\xi_i \in \mathbb{V}$, $i = 1, \dots, k$, if and only if $\xi_1, \dots, \xi_k$ are linearly dependent in $\mathbb{V}$.

Proof See a standard book on linear algebra. □

Remark 7.1 (a) If $k > n$ then any vectors $\xi_1, \dots, \xi_k$ from $\mathbb{V}$ are linearly dependent. According to Proposition 7.1 it is naturally to define in this case $\xi_1 \wedge \cdots \wedge \xi_k = 0$. It follows that $\mathbb{V}^{\wedge k} = \{0\}$ for $k = n+1, n+2, \dots$. Let us define $\mathbb{V}^{\wedge 1} := \mathbb{V}$ and $\mathbb{V}^{\wedge 0} := \mathbb{K}$. Then $\mathbb{V}^{\wedge k}$ exists for all $k = 0, 1, \dots$.

(b) Again from Proposition 7.1 we have $\dim \mathbb{V}^{\wedge k} = \binom{n}{k} = \binom{n}{n-k} = \dim \mathbb{V}^{\wedge (n-k)}$ for $k = 0, 1, \dots, n$.

(c) For $k = 1, 2, \dots$ we have $\mathbb{V}^{\wedge k} = \text{span}\{\xi_1 \wedge \cdots \wedge \xi_k \mid \xi_i \in \mathbb{V} , \ i = 1, 2, \dots, k\}$. Elements of $\mathbb{V}^{\wedge k}$ which have the form $\xi_1 \wedge \cdots \wedge \xi_k$ are called *decomposable* or *simple k-vectors*.

(d) $\mathbb{V}^{\wedge} := \bigoplus_{k=0}^{n} \mathbb{V}^{\wedge k}$ is the *exterior* or *Grassmann algebra* of $\mathbb{V}$.

Definition 7.2 Suppose that $\mathbb{V}$ and $\mathbb{W}$ are linear spaces of dimension n and m, respectively, over $\mathbb{K}$, $k \in \{1, 2, \dots, n\}$ is a number and $S : \mathbb{V} \to \mathbb{W}$ is a linear map. The *k-th multiplicative compound* or *k-th exterior power* of S is the linear map $S^{\wedge k} : \mathbb{V}^{\wedge k} \to \mathbb{W}^{\wedge k}$, defined for decomposable vectors of $\mathbb{V}^{\wedge k}$ by

$$S^{\wedge k}(\xi_1 \wedge \cdots \wedge \xi_k) = S\xi_1 \wedge \cdots \wedge S\xi_k , \quad \xi_i \in \mathbb{V}, \ i = 1, \dots, k ,$$

and for non-decomposable vectors by linearity, i.e.

$$\begin{aligned} S^{\wedge k}(\beta(\xi_1 \wedge \cdots \wedge \xi_k) + \beta'(\eta_1 \wedge \cdots \wedge \eta_k)) \\ = \beta S^{\wedge k}(\xi_1 \wedge \cdots \wedge \xi_k) + \beta' S^{\wedge k}(\eta_1 \wedge \cdots \wedge \eta_k) , \ \xi_i, \eta_j \in \mathbb{V}, \\ i, j = 1, \dots, k, \beta, \beta' \in \mathbb{K} . \end{aligned}$$

Remark 7.2 It is easy to see that Definition 7.2 is correct, i.e. depends in fact only on the products $\xi_1 \wedge \cdots \wedge \xi_k$, $\xi_i \in \mathbb{V}, i = 1, 2, \dots, k$.

Proposition 7.2 *Suppose that $\mathbb{V}, \mathbb{W}$ and $\mathbb{W}'$ are linear spaces of dimension n, m and r, respectively, over $\mathbb{K}$, and $S : \mathbb{V} \to \mathbb{W}$ and $T : \mathbb{W} \to \mathbb{W}'$ are linear maps. Then we have:*

(a) *If* $k \in \{1, 2, \dots, \min(m, n, r)\}$ *then* $(TS)^{\wedge k} = T^{\wedge k} S^{\wedge k}$;
(b) *If* $k \in \{1, 2, \dots, n\}$ *then* $(\mathrm{id}_{\mathbb{V}})^{\wedge k} = \mathrm{id}_{\mathbb{V}^{\wedge k}}$;
(c) *Suppose that* $n = m$, *the map* $S : \mathbb{V} \to \mathbb{W}$ *is invertible and* $k \in \{1, 2, \dots, n\}$. *Then* $S^{\wedge k} : \mathbb{V}^{\wedge k} \to \mathbb{W}^{\wedge k}$ *is invertible and* $(S^{\wedge k})^{-1} = (S^{-1})^{\wedge k}$;
(d) *Suppose that* $\mathbb{V} = \mathbb{W}$ *and* $S : \mathbb{V} \to \mathbb{V}$ *has the eigenvalues* $\lambda_1, \dots, \lambda_n$ *(each eigenvalue* λ_i *repeated with respect to the algebraic multiplicity) and the associated eigenvectors* $e_1, \dots, e_n$. *Then, for any* $k \in \{1, \dots, n\}$ *the map* $S^{\wedge k}$ *has the* $\binom{n}{k}$ *eigenvalues* $\lambda_{i_1}, \dots, \lambda_{i_k}$ *and the associated eigenvectors* $e_{i_1} \wedge \dots \wedge e_{i_k}$, $1 \le i_1 < \dots < i_k \le n$;
(e) *Suppose* $\mathbb{V} = \mathbb{W}$ *and* $S : \mathbb{V} \to \mathbb{V}$ *is a linear operator. Then* $S^{\wedge n} = \det S$.

Proof (a) We have by definition for arbitrary $\xi_i \in \mathbb{V},\ i = 1, \dots, k$, and decomposable vectors

$$\begin{aligned}(T^{\wedge k} S^{\wedge k})\, (\xi_1 \wedge \dots \wedge \xi_k) &= T^{\wedge k}(S\,\xi_1 \wedge \dots \wedge S\,\xi_k) \\ &= (TS\,\xi_1 \wedge \dots \wedge TS\,\xi_k) = (TS)^{\wedge k}(\xi_1 \wedge \dots \wedge \xi_k)\ .\end{aligned}$$

For non-decomposable vectors the same formula holds by linearity.

(b) Again we have $(\mathrm{id}_{\mathbb{V}})^{\wedge k}(\xi_1 \wedge \dots \wedge \xi_k) = \xi_1 \wedge \dots \wedge \xi_k = \mathrm{id}_{\mathbb{V}^{\wedge k}}\, \xi_1 \wedge \dots \wedge \xi_k$ for all decomposable vectors. The remaining part follows by linearity.

(c) By (a) and (b) we conclude from $SS^{-1} = \mathrm{id}_{\mathbb{W}}$ that $(S^{\wedge k})\,(S^{-1})^{\wedge k} = (\mathrm{id}_{\mathbb{W}})^{\wedge k} = \mathrm{id}_{\mathbb{W}^{\wedge k}}$. It follows that $(S^{\wedge k})^{-1} = (S^{-1})^{\wedge k}$.

(d) and (e) can be shown as for matrices (see Proposition 2.9, Chap. 2). □

Definition 7.3 Suppose that $\mathbb{E}$ is an n-dimensional Euclidean space over $\mathbb{K}$ with scalar product $(\cdot, \cdot)_{\mathbb{E}}$ and suppose that $k \in \{1, \dots, n\}$ is arbitrary. A *scalar product* $(\cdot, \cdot)_{\mathbb{E}^{\wedge k}}$ in $\mathbb{E}^{\wedge k}$ is defined for decomposable vectors $\xi_1 \wedge \dots \wedge \xi_k$, $\eta_1 \wedge \dots \wedge \eta_k \in \mathbb{E}^{\wedge k}$ by

$$(\xi_1 \wedge \dots \wedge \xi_k, \eta_1 \wedge \dots \wedge \eta_k)_{\mathbb{E}^{\wedge k}} = \det\,[(\xi_i, \eta_j)_{\mathbb{E}}\,|_{i,j=1}^{k}]$$

and for non-decomposable vectors from $\mathbb{E}^{\wedge k}$ by linearity, i.e.

$$\begin{aligned}&(\beta(\xi_1 \wedge \dots \wedge \xi_k) + \beta'(\xi_1' \wedge \dots \wedge \xi_k')\,,\ \eta_1 \wedge \dots \wedge \eta_k)_{\mathbb{E}^{\wedge k}} \\ &= \beta(\xi_1 \wedge \dots \wedge \xi_k, \eta_1 \wedge \dots \wedge \eta_k)_{\mathbb{E}^{\wedge k}} + \beta'(\xi_1' \wedge \dots \wedge \xi_k', \eta_1 \wedge \dots \wedge \eta_k)_{\mathbb{E}^{\wedge k}}\ , \\ &\forall\, \xi_i, \xi_j',\ \eta_l \in \mathbb{E},\ i, j, l = 1, \dots, k, \quad \forall\, \beta, \beta' \in \mathbb{K}\,.\end{aligned}$$

Definition 7.4 Two bases $\{\xi_1, \dots, \xi_k\}$ and $\{\eta_1, \dots, \eta_k\}$ of a k-dimensional subspace of the n-dimensional Euclidean space $(\mathbb{E}, (\cdot, \cdot)_{\mathbb{E}})$ over $\mathbb{R}$ have the same *orientation* if in the representation $\eta_i = \sum_{j=1}^{k} c_{ij}\xi_j$ with $(c_{ij}) \in M_k(\mathbb{R})$ we have $\det(c_{ij}) > 0$.

Proposition 7.3 *Suppose that* $\{\xi_1, \dots, \xi_k\}$ *is a system of linearly independent vectors in the* n*-dimensional Euclidean space* $(\mathbb{E}, (\cdot, \cdot)_{\mathbb{E}})$ *over* $\mathbb{R}$. *If* $\{\eta_1, \dots, \eta_k\}$ *is some other linearly independent system in* $\mathbb{E}$, *then* $\xi_1 \wedge \dots \wedge \xi_k = \eta_1 \wedge \dots \wedge \eta_k$ *if and only if the following conditions are satisfied:*

(a) $\text{span}\{\xi_1, \ldots, \xi_k\} = \text{span}\{\eta_1, \ldots, \eta_k\}$;
(b) *Both systems have the same orientation;*
(c) $\det [(\xi_i, \xi_j)_{\mathbb{E}}|_{i,j=1}^k] = \det [(\eta_i, \eta_j)_{\mathbb{E}}|_{i,j=1}^k]$.

Proof See [9] or any other text book on linear algebra. □

Remark 7.3 The properties of the determinant show that $(\cdot, \cdot)_{\mathbb{E}^{\wedge k}}$ is really a scalar product. The norm in $\mathbb{E}^{\wedge k}$ is as usual $|\upsilon|_{\mathbb{E}^{\wedge k}} = \sqrt{(\upsilon, \upsilon)_{\mathbb{E}^{\wedge k}}}$, $\forall \upsilon \in \mathbb{E}^{\wedge k}$.

Proposition 7.4 *If* $\{e_i\}_{i=1}^n$ *is an orthonormal basis of the* n*-dimensional Euclidean space* $(\mathbb{E}, (\cdot, \cdot)_{\mathbb{E}})$*, then for any* $k \in \{1, \ldots, n\}$ *the family of vectors*
$\{e_{i_1} \wedge \cdots \wedge e_{i_k}\}_{1 \le i_1 < \cdots < i_k \le n}$ *forms an orthonormal basis in* $(\mathbb{E}^{\wedge k}, (\cdot, \cdot)_{\mathbb{E}^{\wedge k}})$.

Proof Follows immediately from the definition of $(\cdot, \cdot)_{\mathbb{E}^{\wedge k}}$. □

Definition 7.5 Suppose that $(\mathbb{E}, (\cdot, \cdot)_{\mathbb{E}})$ and $(\mathbb{F}, (\cdot, \cdot)_{\mathbb{F}})$ are two n- resp. m-dimensional Euclidean spaces over $\mathbb{R}$ and $S : \mathbb{E} \to \mathbb{F}$ is a linear operator. The linear operator $S^{[*]} : \mathbb{F} \to \mathbb{E}$ defined (uniquely) by

$$(S\,\xi, \eta)_{\mathbb{F}} = (\xi, S^{[*]}\eta)_{\mathbb{E}} \ , \quad \forall \xi \in \mathbb{E}, \ \forall \eta \in \mathbb{F} \ ,$$

is the *adjoint operator* of S.

Example 7.1 (a) Suppose that $(\mathbb{E}, (\cdot, \cdot)_{\mathbb{E}})$ is an n-dimensional Euclidean space, $S : \mathbb{E} \to \mathbb{E}$ is a linear operator. Then $S^{[*]} = S^*$ is the usual adjoint operator.

(b) Consider $\mathbb{R}^n$ as Euclidean space $\mathbb{E}$ with the scalar product $(\xi, \eta)_{\mathbb{E}} := (\xi, G\eta)_n, \forall \xi, \eta \in \mathbb{R}^n$, where $G = G^* > 0$ is a given positive definite matrix. The Euclidean space $\mathbb{F}$ is defined as $\mathbb{R}^n$ equipped with the scalar product $(\xi, \eta)_{\mathbb{F}} := (\xi, G'\eta), \forall \xi, \eta \in \mathbb{R}^n$, where again $G' = G'^* > 0$ is an $n \times n$ matrix.

Assume that a linear operator is in the canonical basis $e_1, \ldots, e_n$ of $\mathbb{R}^n$ given by the matrix S. By definition the adjoint operator $S^{[*]}$ is characterized by the equation

$$(S\,\xi, \eta)_{\mathbb{F}} = (\xi, S^{[*]}\eta)_{\mathbb{E}} \ , \quad \forall \xi, \eta \in \mathbb{R}^n \ , \text{ i.e. by}$$
$$(S\xi, G'\eta)_n = (\xi, GS^{[*]}\eta)_n \ , \quad \forall \xi, \eta \in \mathbb{R}^n \ .$$

It follows that $S^{[*]} = G^{-1}S^*G'$, where S^* is the usual transposed matrix to S.

Proposition 7.5 *Suppose that* $(\mathbb{E}, (\cdot, \cdot)_{\mathbb{E}})$ *and* $(\mathbb{F}, (\cdot, \cdot)_{\mathbb{F}})$ *are* n*-resp.* m*-dimensional Euclidean spaces over* $\mathbb{R}$*,* $S : \mathbb{E} \to \mathbb{F}$ *is a linear operator, and suppose that* $k \in \{1, \ldots, n\}$ *is arbitrary. Then*

$$(S^{[*]})^{\wedge k} = (S^{\wedge k})^{[*]} \ .$$

Proof It is sufficient to show the assertion for decomposable vectors. Consider for arbitrary $\xi_i \in \mathbb{E}, \eta_j \in \mathbb{F}, \ i, j = 1, \ldots, k$,

$$
\begin{aligned}
&(\xi_1\wedge\cdots\wedge\xi_k, (S^{[*]})^{\wedge k}(\eta_1\wedge\cdots\wedge\eta_k))_{\mathbb{E}^{\wedge k}} = (\xi_1\wedge\cdots\wedge\xi_k, S^{[*]}\eta_1\wedge\cdots\wedge S^{[*]}\eta_k)_{\mathbb{E}^{\wedge k}} \\
&= \det\left[(\xi_i, S^{[*]}\eta_j)_{\mathbb{E}}\Big|_{i,j=1}^{k}\right] = \det\left[(S\,\xi_i, \eta_j)_{\mathbb{F}}\Big|_{i,j=1}^{k}\right] \\
&= (S\,\xi_1\wedge\cdots\wedge S\,\xi_k, \eta_1\wedge\cdots\wedge\eta_k)_{\mathbb{F}^{\wedge k}} = (S^{\wedge k}(\xi_1\wedge\cdots\wedge\xi_k)\,,\ \eta_1\wedge\cdots\wedge\eta_k)_{\mathbb{F}^{\wedge k}}\ .
\end{aligned}
$$

It follows that $(S^{[*]})^{\wedge k} = (S^{\wedge k})^{[*]}$. □

Definition 7.6 Suppose that $(\mathbb{E}, (\cdot,\cdot)_{\mathbb{E}})$ and $(\mathbb{F}, (\cdot,\cdot)_{\mathbb{F}})$ are n-dimensional Euclidean spaces over $\mathbb{R}$ and $S:\mathbb{E}\to\mathbb{F}$ is an invertible linear operator. The map S is called *orthogonal* if $S^{-1} = S^{[*]}$.

Proposition 7.6 *Suppose that $(\mathbb{E}, (\cdot,\cdot)_{\mathbb{E}})$ and $(\mathbb{F}, (\cdot,\cdot)_{\mathbb{F}})$ are n-dimensional Euclidean spaces over $\mathbb{R}$ and $S:\mathbb{E}\to\mathbb{F}$ is orthogonal. Then for any $k\in\{1,\dots,n\}$ the operator $S^{\wedge k}:\mathbb{E}^{\wedge k}\to\mathbb{F}^{\wedge k}$ is orthogonal and*

$$(S^{\wedge k})^{-1} = (S^{\wedge k})^{[*]} = (S^{[*]})^{\wedge k}\ .$$

Proof By Proposition 7.2 we have, using the orthogonality,

$$(S^{\wedge k})\,(S^{[*]})^{\wedge k} = (SS^{[*]})^{\wedge k} = (\mathrm{id}_{\mathbb{F}})^{\wedge k} = \mathrm{id}_{\mathbb{F}^{\wedge k}}\ .$$

From this and by Proposition 7.5 it follows that

$$(S^{\wedge k})^{-1} = (S^{[*]})^{\wedge k} = (S^{\wedge k})^{[*]}\ .$$

□

Definition 7.7 Suppose that $\mathbb{V}$ is an n-dimensional linear space over $\mathbb{K}$, $S:\mathbb{V}\to\mathbb{V}$ is a linear map and $k\in\{1,\dots,n\}$ is arbitrary. The *k-th additive compound* or *k-th derivation operator* S_k is the linear operator $S_k:\mathbb{V}^{\wedge k}\to\mathbb{V}^{\wedge k}$, defined for all $\xi_1,\dots,\xi_k\in\mathbb{V}$ by

$$S_k(\xi_1\wedge\cdots\wedge\xi_k) = S\xi_1\wedge\cdots\wedge\xi_k + \xi_1\wedge S\,\xi_2\wedge\cdots\wedge\xi_k + \cdots + \xi_1\wedge\cdots\wedge S\,\xi_k$$

and extended to $\mathbb{V}^{\wedge k}$ by linearity.

Some frequently used notions and their symbols are presented in Table 7.1.

Proposition 7.7 *Definition 7.7 is correct, i.e. for any $k\in\{1,\dots,n\}$ and any $\xi_1,\dots,\xi_k\in\mathbb{V}$ the value $S_k(\xi_1\wedge\cdots\wedge\xi_k)$ depends only on the product $\xi_1\wedge\cdots\wedge\xi_k$.*

Proof Case 1: $\xi_1\wedge\cdots\wedge\xi_k = 0$. Then, w.l.o.g., $\xi_k = \sum\limits_{j=1}^{k-1}\beta_j\xi_j$ with some $\beta_j\in\mathbb{K}$. It follows that $\xi_1\wedge\cdots\wedge\left(\sum\limits_{j=1}^{k-1}\beta_j\xi_j\right) = \sum\limits_{j=1}^{k-1}\beta_j(\xi_1\wedge\cdots\wedge\xi_{k-1}\wedge\xi_j) = 0$. On the other hand we have

Table 7.1 Frequently used notions and their symbols

Symbol	Notion	Sections
$\mathbb{V}^{\wedge k}$	k-th exterior power of a linear space	7.1.1
$S^{\wedge k}$	k-th multiplicative compound or exterior power of an operator	7.1.1
S_k	k-th additive compound of a linear operator	7.1.1

$$S_k\Big(\xi_1\wedge\cdots\wedge\sum_{j=1}^{k-1}\beta_j\xi_j\Big)=\sum_{j=1}^{k-1}\beta_j S_k(\xi_1\wedge\cdots\wedge\xi_{k-1}\wedge\xi_j)$$

$$=\sum_{j=1}^{k-1}\beta_j\,[\xi_1\wedge\cdots\wedge S\,\xi_j\wedge\cdots\wedge\xi_{k-1}\wedge\xi_j+\xi_1\wedge\cdots\wedge\xi_j\wedge\cdots\wedge S\,\xi_j]=0\,.$$

<u>*Case*2</u>: $\xi_1\wedge\cdots\wedge\xi_k\neq 0$. Suppose that $\xi_1\wedge\cdots\wedge\xi_k=\eta_1\wedge\cdots\wedge\eta_k$ for some $\eta_i\in\mathbb{V}$. By Proposition 7.3 it follows that $\eta_j=\sum_{i=1}^{n}a_{ij}\xi_i$ with some $a_{ij}\in\mathbb{K}$ and $\eta_1\wedge\cdots\wedge\eta_k=\det(a_{ij})\,(\xi_1\wedge\cdots\wedge\xi_k)$ with $\det(a_{ij})=1$. Consequently we have

$$S_k(\eta_1\wedge\cdots\wedge\eta_k)=S\eta_1\wedge\cdots\wedge\eta_k+\cdots+\eta_1\wedge\cdots\wedge S\eta_k$$

$$=\ \Big(\sum a_{i_1 1}\ldots a_{i_k k}(-1)^{\mathrm{sign}(i_1,\ldots,i_k)}\Big)\,(S\xi_1\wedge\cdots\wedge\xi_k+\cdots+\xi_1\wedge\cdots\wedge S\,\xi_k)\,.$$

But the first factor of the last expression is $\det(a_{ij})=1$ and the second gives $S_k(\xi_1\wedge\cdots\wedge\xi_k)$. □

Proposition 7.8 *Suppose* $\mathbb{V}$ *is an n-dimensional linear space over* $\mathbb{K}$, $S:\mathbb{V}\to\mathbb{V}$ *is a linear operator, and* $k\in\{1,\ldots,n\}$ *is arbitrary. Then*

$$S_k=\frac{d}{dh}\big[\mathrm{id}_{\mathbb{V}}+hS\big]^{\wedge k}_{|h=0}\,.$$

Proof For arbitrary $\xi_1,\ldots,\xi_k\in\mathbb{V}$ and $h\in\mathbb{R}$ we have

$$[\mathrm{id}_{\mathbb{V}}+hS]^{\wedge k}(\xi_1\wedge\cdots\wedge\xi_k)=(\mathrm{id}_{\mathbb{V}}+hS)\xi_1\wedge\cdots\wedge(\mathrm{id}_{\mathbb{V}}+hS)\xi_k$$

$$=\xi_1\wedge\cdots\wedge\xi_k+h\big[S\xi_1\wedge\cdots\wedge\xi_k+\xi_1\wedge S\,\xi_2\wedge\cdots\wedge\xi_k+\cdots+\xi_1\wedge\cdots\wedge S\,\xi_k\big]+o\,(h)$$

$$=\big[\mathrm{id}_{\mathbb{V}}+hS\big]^{\wedge k}_{|h=0}(\xi_1\wedge\cdots\wedge\xi_k)+hS_k(\xi_1\wedge\cdots\wedge\xi_k)+o\,(h)\,.$$

□

Proposition 7.9 *Suppose that $(\mathbb{E}, (\cdot,\cdot)_{\mathbb{E}})$ is an n-dimensional Euclidean space over $\mathbb{R}$, $k \in \{1,\ldots,n\}$ is arbitrary and $S, T : \mathbb{E} \to \mathbb{E}$ are two linear operators. Then we have:*

(a) $[S+T]_k = S_k + T_k$; *(b)* $S_1 = S,\ S_n = \operatorname{tr} S$;
(c) $(\mathrm{id}_{\mathbb{E}})_k = k\,\mathrm{id}_{\mathbb{E}^{\wedge k}}$; *(d)* $(S^{[*]})_k = (S_k)^{[*]}$;
(e) If $\lambda_1,\ldots,\lambda_n$ is the complete system of eigenvalues of the operator S then $\lambda_{i_1} + \cdots + \lambda_{i_k}$, $1 \le i_1 < \cdots < i_k \le n$, is the complete system of eigenvalues for S_k.

Proof (a) Follows by definition.
(b) $S_1 = \frac{d}{dh}\big[\mathrm{id}_{\mathbb{E}} + h\,S\big]^{\wedge 1}_{|h=0} = S$; $S_n = \frac{d}{dh}[\mathrm{id}_{\mathbb{E}} + hS]^{\wedge n}_{|h=0}$
$= \frac{d}{dh}\big(\det\big[\mathrm{id}_{\mathbb{E}} + h\,S\big]\big)_{|h=0} = \frac{d}{dh}\big(1 + h \operatorname{tr} S + o\,(h)\big)_{|h=0} = \operatorname{tr} S$.
(c) For any decomposable vectors $\xi_1 \wedge \cdots \wedge \xi_k$ with $\xi_1,\ldots,\xi_k \in \mathbb{E}$ we have $[\mathrm{id}_{\mathbb{E}}]_k(\xi_1 \wedge \cdots \wedge \xi_k) = k \cdot \xi_1 \wedge \cdots \wedge \xi_k$. The general case follows by linearity.
(d) By Proposition 7.8 we have

$$[S^{[*]}]_k = \frac{d}{dh}[\mathrm{id}_{\mathbb{E}} + hS^{[*]}]^{\wedge k}_{|h=0} = \frac{d}{dh}\{[\mathrm{id}_{\mathbb{E}} + hS]^{\wedge k}\}^{[*]}_{|h=0} = (S_k)^{[*]} .$$

(e) If $e_{i_1},\ldots,e_{i_k}$ are the associated to $\lambda_{i_1},\ldots,\lambda_{i_k}$ eigenvectors, we have

$$\begin{aligned} S_k(e_{i_1} \wedge \cdots \wedge e_{i_k}) &= Se_{i_1} \wedge \cdots \wedge e_{i_k} + \cdots + e_{i_1} \wedge \cdots \wedge Se_{i_k} \\ &= [\lambda_{i_1} + \cdots + \lambda_{i_k}]e_{i_1} \wedge \cdots \wedge e_{i_k} \text{ for arbitrary } 1 \le i_1 < \cdots < i_k \le n. \end{aligned}$$

□

Proposition 7.10 *Suppose that $\mathbb{V}$ and $\mathbb{W}$ are n-resp. m-dimensional vector spaces over $\mathbb{K}$, $k \in \{1,\ldots,\min(n,m)\}$, $\{e_i\}_{i=1}^n$ and $\{f_j\}_{j=1}^m$ are bases of $\mathbb{V}$ and $\mathbb{W}$, respectively, $\{e_{i_1} \wedge \cdots \wedge e_{i_k}\}$ and $\{f_{j_1} \wedge \cdots \wedge f_{j_k}\}$ are the lexicographically ordered bases of $\mathbb{V}^{\wedge k}$ and $\mathbb{W}^{\wedge k}$, respectively, and $S : \mathbb{V} \to \mathbb{W}$ is a linear operator. If $[S]$ and $[S^{\wedge k}]$ denote the matrices of the operators S and $S^{\wedge k}$, respectively, with the respect to the above bases then*

$$[S^{\wedge k}] = [S]^{(k)} ,$$

i. e. the k-th multiplicative compound matrix of the matrix $[S]$.

Proof Introduce the lexicographic ordering of the bases of $\mathbb{V}^{\wedge k}$ and $\mathbb{W}^{\wedge k}$ by

$$\widetilde{e}_i = e_{i_1} \wedge \cdots \wedge e_{i_k} ,\ i = 1,\ldots,\tbinom{n}{k} ,\ (i) = (i_1,\ldots,i_k) ,$$

and

$$\widetilde{f}_j = f_{j_1} \wedge \cdots \wedge f_{j_k} ,\ j = 1,\ldots,\tbinom{m}{k} ,\ (j) = (j_1,\ldots,j_k) .$$

Denote $[S] = (s_{ij}) = \mathcal{S}$, i.e. $Se_{i_p} = \sum_{l=1}^{m} s_{li_p} f_l$, $p = 1,\ldots,k$. It follows that

$$S^{\wedge k}\widetilde{e}_i = Se_{i_1} \wedge \cdots \wedge Se_{i_k} = \left(\sum_{p_1=1}^{m} s_{p_1 i_1} f_{p_1} \wedge \cdots \wedge \sum_{p_k=1}^{m} s_{p_k j_k} f_{p_k} \right)$$

$$= \sum_{p_1,\ldots,p_k=1}^{m} (-1)^{\mathrm{sign}\{p_1,\ldots,p_k\}} s_{p_1 i_1} \cdot s_{p_2 i_2} \ldots s_{p_k i_k} (f_{p_1} \wedge \cdots \wedge f_{p_k})$$

$$= \sum_{p=1}^{\binom{m}{k}} \det \mathcal{S}\left[(p)\big|(i)\right] \widetilde{f}_p \,.$$

From this we see that

$$[S^{\wedge k}] = \begin{pmatrix} \det S\left[(1)\big|(1)\right] & \ldots & \det S\left[(1)\big|(\binom{n}{k})\right] \\ \det S\left[(2)\big|(1)\right] & \ldots & \det S\left[(2)\big|(\binom{n}{k})\right] \\ \ldots & \ldots & \ldots \\ \det S\left[(\binom{m}{k})\big|(1)\right] & \ldots & \det S\left[(\binom{m}{k})\big|(\binom{n}{k})\right] \end{pmatrix} = [S]^{(k)} \,.$$

□

Remark 7.4 Proposition 7.10, can be used to give a short proof of the Binet-Cauchy theorem (Proposition 2.3, Chap. 2). Consider two given matrices $A \in M_{n,m}(\mathbb{K})$ and $B \in M_{m,p}(\mathbb{K})$ as matrix realization of the two linear operators $S : \mathbb{W} \to \mathbb{W}'$ and $T : \mathbb{V} \to \mathbb{W}$, where $\mathbb{V}, \mathbb{W}$ and $\mathbb{W}'$ are linear spaces of dimensions p, m and n, respectively.

Proposition 7.11 *Suppose that $\mathbb{V} = \mathbb{K}^n$ is the n-dimensional vector space of n-columns over $\mathbb{K}$, $k \in \{1, \ldots, n\}$ is a natural number and $u_1, \ldots, u_k \in \mathbb{K}^n$ are arbitrary vector columns. Then we have $u_1 \wedge \cdots \wedge u_k = C^{(k)}$, where C is the $n \times k$ matrix having as columns the vectors $u_1, u_2, \ldots, u_k$.*

Proof Let us consider for simplicity $k = 2$ and the $\binom{n}{2}$-dimensional vector space $(\mathbb{K}^n)^{\wedge 2}$. Suppose that

$u = \begin{pmatrix} \xi_1 \\ \vdots \\ \xi_n \end{pmatrix}$ and $v = \begin{pmatrix} \eta_1 \\ \vdots \\ \eta_n \end{pmatrix}$ are two vectors of $\mathbb{K}^n$written in components with respect to the canonical basis$\{e_1, \ldots, e_n\}$ of $\mathbb{K}^n$ with $= \begin{pmatrix} 0 \\ \vdots \\ 0 \\ 1 \\ 0 \\ \vdots \\ 0 \end{pmatrix} \Big\} i$.

This means that $u = \sum_{i=1}^{n} \xi_i e_i$ and $v = \sum_{j}^{n} \eta_j e_j$. Using the properties of the exterior product (see Definition 7.1), the lexicographic ordering (see Sect. 2.3.1, Chap. 2) and Definition 2.2, Chap. 2 of the multiplicative compound matrix we can write

$$u \wedge v = \sum_{\substack{i,\\ j=i}}^{n} \xi_i \eta_j (e_i \wedge e_j) = \sum_{i=1}^{\binom{n}{2}} \begin{vmatrix} \xi_{i_1} & \eta_{i_1} \\ \xi_{i_2} & \eta_{i_2} \end{vmatrix} \widetilde{e}_{(i)}$$

with $(i) = (i_1, i_2)$ and

$$e_{(i)} = e_{i_1} \wedge e_{i_2} = \begin{pmatrix} 0 & 0 \\ \vdots & \vdots \\ 1 & 0 \\ \vdots & \vdots \\ 0 & 1 \\ \vdots & \vdots \\ 0 & 0 \end{pmatrix}^{(2)} \begin{matrix} \leftarrow i_1 \\ \leftarrow i_2 \end{matrix} = \begin{pmatrix} 0 \\ \vdots \\ 1 \\ \vdots \\ 0 \end{pmatrix} \}^{i} = \widetilde{e}_i \in \mathbb{K}^{\binom{n}{2}} .$$

□

The next proposition shows the connection between additive compounds of an operator and additive compound matrices.

Proposition 7.12 *Suppose that $\mathbb{V}$ is an n-dimensional vector space over $\mathbb{K}$, $k \in \{1, \ldots, n\}$ is a natural number, $\{e_1, \ldots, e_n\}$ is a basis of $\mathbb{V}$, $\{\widetilde{e}_1, \ldots, \widetilde{e}_{\binom{n}{k}}\}$ is the lexicographically ordered basis of $\mathbb{V}^{\wedge k}$ given by $\widetilde{e}_j = e_{j_1} \wedge \cdots \wedge e_{j_k}$ with $(j) = (j_1, \ldots, j_k)$. If $S : \mathbb{V} \to \mathbb{V}$ is a linear operator, $S_k : \mathbb{V}^{\wedge k} \to \mathbb{V}^{\wedge k}$ is the k-th additive compound operator of S, and $[S]$ and $[S_k]$ are the matrices of these operators in the above bases, then $[S_k] = [S]^{[k]}$.*

Proof For any k-tuple $1 \le j_1 < \cdots < j_k \le n$ we can write by definition

$$\begin{aligned} S_k(e_{j_1} \wedge \cdots \wedge e_{j_k}) &= Se_{j_1} \wedge \cdots \wedge e_{j_k} + \cdots + e_{j_1} \wedge \cdots \wedge Se_{j_k} \\ &= \Big(\sum_{p_1=1}^{n} s_{p_1 j_1} e_{p_1}\Big) \wedge \cdots \wedge e_{j_k} + \cdots + e_{j_1} \wedge \cdots \wedge \Big(\sum_{p_k=1}^{n} s_{p_k j_k} e_{p_k}\Big) \\ &= \sum_{p_1=1}^{n} s_{p_1 j_1} (e_{p_1} \wedge \cdots \wedge e_{j_k}) + \cdots + \sum_{p_k=1}^{n} s_{p_k j_k} (e_{j_1} \wedge \cdots \wedge e_{p_k}) , \end{aligned} \tag{7.1}$$

where the matrix $[S] = (s_{ij})$ is defined by $Se_j = \sum_{i=1}^{n} s_{ij} e_i$, $j = 1, 2, \ldots, n$. In order to determine the element $[S_k]_{ij}$ of the matrix $[S_k]$ we consider three cases.

<u>*Case*1</u>: $(i) = (j) = (j_1, \ldots, j_k)$. It follows from (7.1) that

$$[S_k]_{jj} = s_{j_1 j_1} + s_{j_2 j_2} + \cdots + s_{j_k j_k} .$$

<u>*Case*2</u>: (i) is different from (j) in more than two symbols. Then we get from (7.1) $[S_k]_{ij} = 0$.

<u>*Case*3</u>: Exactly one symbol i_r from (i) is not included in (j) and exactly one symbol j_p from (j) is not included in (i). Thus $(i) \cap (j) =: \{q_1, \ldots, q_{k-1}\}$ with

$1 \leq q_1 < \cdots < q_{k-1} \leq n$, where $(i) = \{q_1 \ldots, q_v, i_r, q_{v+1}, \ldots, q_{k-1}\}$ and $(j) = \{q_1, \ldots, q_w, j_p, q_{w+1}, \ldots, q_{k-1}\}$.

This means that the element $[S_k]_{ij}$ depends only on the terms

$$e_{q_1} \wedge \cdots \wedge e_{q_w} \wedge Se_{j_p} \wedge e_{q_{w+1}} \wedge \cdots \wedge e_{q_{k-1}} .$$

Since $Se_{j_p} = s_{i_r j_p} e_{i_r} + \ldots$, we see that $[S_k]_{ij}$ is the coefficient of

$$\begin{aligned} & s_{i_r j_p}(e_{q_1} \wedge \cdots \wedge e_{q_w} \wedge e_{i_r} \wedge e_{q_{w+1}} \wedge \cdots \wedge e_{q_{k-1}}) \\ & = (-1)^\sigma s_{i_r j_p}(e_{q_1} \wedge \cdots \wedge e_{q_v} \wedge e_{i_r} \wedge e_{q_{r+1}} \wedge \cdots \wedge e_{q_{k-1}}), \\ & \text{i.e. } [S_k]_{ij} = (-1)^\sigma s_{i_r j_p} \text{ , where } \sigma = \begin{cases} p - r \text{ if } p \geq r , \\ r - p \text{ if } p < r . \end{cases} \end{aligned}$$

In both cases we have $(-1)^\sigma = (-1)^{p+r}$.

Thus in all three cases the element $[S_k]_{ij}$ of the matrix $[S_k]$ is computed from the elements of the matrix $[S]$ according to the rules of computation for the matrix $[S]^{[k]}$ (see Proposition 2.12, Chap. 2). This shows that $[S_k] = [S]^{[k]}$. □

7.1.2 Singular Values of an Operator Acting Between Euclidean Spaces

Suppose that $\mathbb{E}$ and $\mathbb{F}$ are n-dimensional Euclidean spaces with scalar products $(\cdot, \cdot)_\mathbb{E}$ and $(\cdot, \cdot)_\mathbb{F}$, respectively, and $S : \mathbb{E} \to \mathbb{F}$ is a linear operator. Recall that the adjoint operator $S^{[*]} : \mathbb{F} \to \mathbb{E}$ is defined uniquely by the relation

$$(S\xi, \eta)_\mathbb{F} = (\xi, S^{[*]}\eta)_\mathbb{E} , \quad \forall \xi \in \mathbb{E}, \ \forall \eta \in \mathbb{F} .$$

It follows that the operator $S^{[*]}S : \mathbb{E} \to \mathbb{E}$ is selfadjoint, i.e. $(S^{[*]}S)^{[*]} = S^{[*]}S$ and non-negative, i.e. $(S^{[*]}S\xi, \xi)_\mathbb{E} \geq 0, \quad \forall \xi \in \mathbb{E}$.

It follows also that the eigenvalues of $S^{[*]}S$ are non-negative. Denote the complete and ordered system of eigenvalues of $S^{[*]}S$ by

$$\lambda_1(S^{[*]}S) \geq \cdots \geq \lambda_n(S^{[*]}S) .$$

Definition 7.8 The numbers $\alpha_1(S) \geq \alpha_2(S) \geq \cdots \geq \alpha_n(S)$ defined by $\alpha_i(S) := \sqrt{\lambda_i(S^{[*]}S)}$, $i = 1, \ldots, n$, are the *singular values* of S.

Example 7.2 (a) Suppose that $(\mathbb{E}, (\cdot, \cdot)_\mathbb{E})$ is an n-dimensional Euclidean space over $\mathbb{R}$ and $S : \mathbb{E} \to \mathbb{E}$ is a self-adjoint and non-negative linear operator. In this case $S^{[*]}S = S^2$ and the singular values of S are the eigenvlues of S, i.e. $\alpha_i(S) = \lambda_i(S), i = 1, \ldots, n$.

(b) Let us consider the linear operator given by the $n \times n$ matrix S from Example 7.1 (b). It was shown that in the notion of this example $S^{[*]} = G^{-1}S^*G'$. This means that the singular values $\alpha_i(S)$ of the matrix S are the non-negative roots of the eigenvalues of the matrix $G^{-1}S^*G'S$. Note that the last matrix is similar to the matrix

$$G^{1/2}G^{-1}S^*G'SG^{-1/2} = G^{-1/2}S^*(G')^{1/2}(G')^{1/2}SG^{-1/2} = A^*A$$

with $A := (G')^{1/2}SG^{-1/2}$.

The geometric properties of the singular values of an operator acting between Euclidean spaces are described in the next proposition.

Proposition 7.13 *Suppose that* $(\mathbb{E}, (\cdot,\cdot)_{\mathbb{E}})$ *and* $(\mathbb{F}, (\cdot,\cdot)_{\mathbb{F}})$ *are two n-dimensional Euclidean spaces over* $\mathbb{R}$, *and* $S : \mathbb{E} \to \mathbb{F}$ *is a linear operator with singular values* $\alpha_1 \geq \alpha_2 \geq \cdots \geq \alpha_n > 0$. *Let* $\mathcal{B}_r(0)$ *be the closed ball in* $\mathbb{E}$ *of radius* $r > 0$ *and with center at* 0. *Then the image of* $\mathcal{B}_r(0)$ *under S is an ellipsoid* $\mathcal{E}$ *in* $\mathbb{F}$ *whose semi-axes have the length* $\alpha_j r$, $j = 1, 2, \ldots, n$.

Proof Since $S^{[*]}S$ is self-adjoint there exists an orthonormal basis of $\mathbb{E}$, $e_1, \ldots, e_n$, which consists of eigenvalues of $S^{[*]}S$ associated with the eigenvalues $\alpha_1^2 \geq \alpha_2^2 \geq \cdots \geq \alpha_n^2 > 0$. For arbitrary $i, j \in \{1, 2, \ldots, n\}$ we have

$$\left(\frac{Se_i}{\alpha_i}, \frac{Se_j}{\alpha_j}\right)_{\mathbb{F}} = \frac{1}{\alpha_i\alpha_j}(e_i, S^{[*]}Se_j)_{\mathbb{E}} = \frac{\alpha_j}{\alpha_i}(e_i, e_j)_{\mathbb{E}} = \begin{cases} 1 \text{ for } i = j\,, \\ 0 \text{ for } i \neq j\,. \end{cases}$$

It follows that the vectors $\{f_i\}_{i=1}^n$ with $f_i = \frac{Se_i}{\alpha_i}$, $i = 1, 2, \ldots, n$, form an orthonormal basis of $\mathbb{F}$. Let us write the given ball as

$$\mathcal{B}_r(0) = \left\{ \sum_{i=1}^n \xi_i e_i \;\middle|\; \xi_i \in \mathbb{R}, \sum_{i=1}^n \xi_i^2 \leq r^2 \right\}.$$

Then we have $S\mathcal{B}_r(0) =$

$$= \left\{ \sum_{i=1}^n \xi_i Se_i \;\middle|\; \xi_i \in \mathbb{R}, \sum_{i=1}^n \xi_i^2 \leq r^2 \right\} = \left\{ \sum_{i=1}^n \xi_i \alpha_i \frac{Se_i}{\alpha_i} \;\middle|\; \xi_i \in \mathbb{R}, \sum_{i=1}^n \left(\frac{\xi_i}{r}\right)^2 \leq 1 \right\}$$
$$= \left\{ \sum_{i=1}^n \eta_i f_i \;\middle|\; \eta_i \in \mathbb{R}, \sum_{i=1}^n \left(\frac{\eta_i}{\alpha_i r}\right)^2 \leq 1 \right\} = \mathcal{E}\,.$$

□

As in the matrix case the singular values of an operator can be characterized by a min-max property.

Theorem 7.1 (Fischer-Courant) *Suppose that* $(\mathbb{E}, (\cdot,\cdot)_{\mathbb{E}})$ *and* $(\mathbb{F}, (\cdot,\cdot)_{\mathbb{F}})$ *are two n-dimensional Euclidean spaces with associated vector norms* $|\cdot|_{\mathbb{E}}$ *and* $|\cdot|_{\mathbb{F}}$, *respectively, and* $T : \mathbb{E} \to \mathbb{F}$ *is a linear operator. Then the singular values of* T, $\alpha_1(T) \geq \alpha_2(T) \geq \cdots \geq \alpha_n(T)$, *can be computed by*

$$a) \quad \alpha_1(T) = \max_{|u|_{\mathbb{E}}=1} |Tu|_{\mathbb{F}}\,, \tag{7.2}$$

$$\alpha_k(T) = \min_{\substack{\mathbb{L}\subset\mathbb{E} \\ \dim \mathbb{L}=k-1}} \max_{\substack{|u|_{\mathbb{E}}=1 \\ u\in\mathbb{L}^{\perp}}} |Tu|_{\mathbb{F}}\,, k = 2, 3, \ldots, n\,, \tag{7.3}$$

where in (7.3) the minimum is taken over all linear subspaces $\mathbb{L}$ *of* $\mathbb{E}$ *with* $\dim \mathbb{L} = k-1$ *and* $\mathbb{L}^{\perp} = \{u \in \mathbb{E} \,|\, (u, v)_{\mathbb{E}} = 0, \forall v \in \mathbb{L}\}$.

$$b) \quad \alpha_k(T) = \max_{\substack{\mathbb{L}\subset\mathbb{E} \\ \dim \mathbb{L}=k}} \min_{\substack{u\in\mathbb{L} \\ |u|_{\mathbb{E}}=1}} |\, Tu\,|_{\mathbb{F}}\,, \quad k = 1, 2, \ldots, n\,. \tag{7.4}$$

Proof We omit the proof which can be done similarly as the proof of Theorem 2.1, Chap. 2. For a full proof see [5]. □

Consider a linear operator $T : \mathbb{E} \to \mathbb{F}$, where $(\mathbb{E}, (\cdot, \cdot)_{\mathbb{E}})$ and $(\mathbb{F}, (\cdot, \cdot)_{\mathbb{F}})$ are Euclidean spaces of dimension n. The singular values of T, i.e., the eigenvalues of the positive operator $(T^{[*]}T)^{1/2} : \mathbb{E} \to \mathbb{E}$, ordered with respect to its size and multiplicity we denote by $\alpha_1(T) \geq \alpha_2(T) \geq \cdots \geq \alpha_n(T)$.

For an arbitrary $k \in \{0, 1, \ldots, n\}$ we define $\omega_k(T)$ as for a matrix in (2.11), Chap. 2 by

$$\omega_k(T) = \begin{cases} \alpha_1(T) \cdots \alpha_k(T)\,, & \text{for } k > 0\,, \\ 1 & \text{for } k = 0\,. \end{cases} \tag{7.5}$$

More generally, for an arbitrary real number $d \in [0, n]$ written in the form $d = d_0 + s$ with $d_0 \in \{0, 1, \ldots, n-1\}$ and $s \in (0, 1]$ we introduce the *singular value function of order d of the operator* $T : \mathbb{E} \to \mathbb{F}$ by

$$\omega_d(T) = \begin{cases} \omega_{d_0}(T)^{1-s} \cdots \omega_{d_0+1}(T)^s\,, & \text{for } d \in (0, n]\,, \\ 1 & \text{for } d = 0\,. \end{cases} \tag{7.6}$$

Obviously this can also be interpreted as $\omega_d(T) = |\bigwedge^{d_0} T|^{1-s} |\bigwedge^{d_0+1} T|^s$, where $|\bigwedge^k T|$ stands for the norm of the k-th exterior power of T, i.e., the norm of the linear operator $\bigwedge^k T : \bigwedge^k \mathbb{E} \to \bigwedge^k \mathbb{F}$.

If $\mathbb{H}$ is a Hilbert space with scalar product and norm $(\cdot, \cdot)$ and $|\cdot|$, respectively, and $T \in \mathcal{L}(\mathbb{H}, \mathbb{H})$ is a linear bounded operator, the singular values $\alpha_1(T) \geq \alpha_2(T) \geq \cdots$ are defined by

$$\alpha_k(T) := \sup_{\substack{\mathbb{L}\subset\mathbb{H} \\ \dim \mathbb{L}=k}} \inf_{\substack{u\in\mathbb{L} \\ |u|=1}} |\, Tu\,|\,, \quad k = 1, 2, \ldots\,. \tag{7.7}$$

It is shown in [34] that in case that T is a compact operator the singular values of T coincide with the eigenvalues of $(T^*T)^{1/2}$, where T^* denotes the Hilbert adjoint operator. For each real $d \geq 0$ the singular value function $\omega_d(T)$ *of order* d is defined as above by the formulas (7.5) and (7.6).

7.1.3 Lemmas on Covering of Ellipsoids in an Euclidean Space

In this subsection we continue the investigation of some covering techniques for ellipsoids from Sect. 2.1.3, Chap. 2. Let $\mathcal{E}$ be an ellipsoid in the n-dimensional Euclidean space $\mathbb{E}$ over $\mathbb{R}$ and let $a_1(\mathcal{E}) \geq \cdots \geq a_n(\mathcal{E})$ denote the length of its semi-axes. For an arbitrary number $d \in [0, n]$ written as $d = d_0 + s$ with
$d_0 \in \{0, 1, \ldots, n-1\}$ and $s \in (0, 1]$ we introduce the *d-dimensional ellipsoid measure* $\omega_d(\mathcal{E})$ by the formula (2.5), Chap. 2.

Assume now that $\mathbb{F}$ is another n-dimensional Euclidean space over $\mathbb{R}$. For the invertible linear operator $T : \mathbb{E} \to \mathbb{F}$ and the ball $\mathcal{B}_r(0)$ in $\mathbb{E}$ of radius r around the origin 0 of $\mathbb{E}$ the image $T\mathcal{B}_r(0)$ is by Proposition 7.13 an ellipsoid in $\mathbb{F}$ with length of semi-axes $\alpha_i(T)r$, $i = 1, 2, \ldots, n$. In addition, for $d \in [0, n]$ it holds

$$\omega_d(T\mathcal{B}_r(0)) = \omega_d(T)\, r^d. \tag{7.8}$$

The next lemma is similar to Lemma 2.1, Chap. 2 and is stated without proof.

Lemma 7.1 *Let us consider numbers $d \in (0, n]$ (written as above), $\kappa > 0$, $\delta > 0$ and $\eta > 0$ and assume $\kappa \leq \delta^d$. Let $\mathcal{E}$ be an ellipsoid in $\mathbb{E}$ such that $a_1(\mathcal{E}) \leq \delta$ and $\omega_d(\mathcal{E}) \leq \kappa$. Further, we take a ball $\mathcal{B}_\eta(0)$ of radius η in $\mathbb{E}$. Then the set $\mathcal{E} + \mathcal{B}_\eta(0)$ is contained in an ellipsoid $\mathcal{E}'$ which satisfies*

$$\omega_d(\mathcal{E}') \leq \left(1 + \left(\frac{\delta^{d_0}}{\kappa}\right)^{\frac{1}{s}} \eta\right)^d \kappa.$$

We call an ellipsoid $\mathcal{E} \subset \mathbb{E}$ *degenerated* if $a_i(\mathcal{E}) = 0$ for some $i \in \{1, \ldots, n\}$. We reformulate Lemma 7.1, which is true for a non-degenerated ellipsoid $\mathcal{E}$ only. The following two lemmas where obtained in [12].

Lemma 7.2 *Let $\mathcal{E}$ be an ellipsoid in an n-dimensional Euclidean space $\mathbb{E}$. Let $\delta, \kappa, \varsigma$ be positive numbers, $d \in (0, n]$ written as $d = d_0 + s$ with $d_0 \in \{0, 1, \ldots, n-1\}$ and $s \in (0, 1]$ and $\kappa \leq \delta^d$. Suppose that $\omega_i(\mathcal{E})\varsigma^{d-i} \leq \kappa$ for any $i = 0, \ldots, d_0$, $\omega_d(\mathcal{E}) \leq \kappa$ and $a_1(\mathcal{E}) \leq \delta$. Then for any $\eta > 0$ the sum $\mathcal{E} + \mathcal{B}_\eta(0)$ is contained in an ellipsoid $\mathcal{E}' \subset \mathbb{E}$ which satisfies*

$$\omega_d(\mathcal{E}') \leq \left(1 + \left(\frac{\delta^{d_0}}{\kappa}\right)^{\frac{1}{s}} \eta\right)^d \kappa\,,$$

$$a_{d_0+1}(\mathcal{E}') \leq \left(1 + \left(\frac{\delta^{d_0}}{\kappa}\right)^{\frac{1}{s}} \eta\right) \max\left\{\varsigma, a_{d_0+1}(\mathcal{E})\right\}. \tag{7.9}$$

Proof We enlarge the ellipsoid $\mathcal{E}$ as follows: If $a_{d_0+1}(\mathcal{E}) < \varsigma$, we replace the lengths $a_{d_0+1}(\mathcal{E}), \ldots, a_n(\mathcal{E})$ by ς. These values determine a non-degenerate ellipsoid which contains $\mathcal{E}$ and for which Lemma 7.1 can be applied. □

With Lemma 7.2 and with methods developed in [34] and Sect. 2.1, Chap. 2, we obtain:

Lemma 7.3 *We keep the assumptions of Lemma 7.2. Then for any number* $\eta > 0$ *the set* $\mathcal{E} + \mathcal{B}_\eta(0)$ *can be covered by* $\left[\frac{2^d \kappa}{\widetilde{\varsigma}^d}\right]$ *balls with radius*

$$\left(1 + \left(\frac{\delta^{d_0}}{\kappa}\right)^{\frac{1}{s}} \eta\right) \widetilde{\varsigma}\sqrt{d_0 + 1}$$

where we set $\widetilde{\varsigma} := \max\left\{\varsigma, a_{d_0+1}(\mathcal{E})\right\}$.

The next two lemmas are similar to Lemma 2.2 and 2.3, Chap. 2, and are stated without proof.

Lemma 7.4 *Let* $(\mathbb{E}, (\cdot,\cdot)_{\mathbb{E}})$ *be an n-dimensional Euclidean space,* $u_1, \ldots, u_n$ *an orthonormal basis and*

$$\mathcal{E} = \left\{a_1 u_1 + \ldots + a_n u_n \in \mathbb{E} \,\middle|\, (a_1, \ldots, a_n) \in \mathbb{R}^n,\ \left(\frac{a_1}{\alpha_1(\mathcal{E})}\right)^2 + \ldots + \left(\frac{a_n}{\alpha_n(\mathcal{E})}\right)^2 \leq 1\right\}$$

an ellipsoid with $\alpha_1(\mathcal{E}) \geq \ldots \geq \alpha_n(\mathcal{E}) > 0$. *Then for any* $\eta > 0$, *the set* $\mathcal{E} + \mathcal{B}_\eta(0)$, *where* $\mathcal{B}_\eta(0)$ *denotes the ball with radius* η *centered at the origin, is contained in the ellipsoid* $\mathcal{E}' = \left(1 + \frac{\eta}{\alpha_n(\mathcal{E})}\right)\mathcal{E}$.

Lemma 7.5 *Let* $(\mathbb{E}, (\cdot,\cdot)_{\mathbb{E}})$ *be an n-dimensional Euclidean space,* $u_1, \ldots, u_n$ *an orthonormal basis,*

$$\mathcal{E} = \left\{a_1 u_1 + \ldots + a_n u_n \in \mathbb{E} \,\middle|\, \left(\frac{a_1}{\alpha_1(\mathcal{E})}\right)^2 + \ldots + \left(\frac{a_n}{\alpha_n(\mathcal{E})}\right)^2 \leq 1\right\}$$

an ellipsoid with $\alpha_1(\mathcal{E}) \geq \ldots \geq \alpha_n(\mathcal{E}) > 0$ *and* $0 < r < \alpha_n(\mathcal{E})$. *Then the relation* $N_{\sqrt{n}r}(\mathcal{E}) \leq \frac{2^n \omega_n(\mathcal{E})}{r^n}$ *holds, where* $\omega_n(\mathcal{E})$ *is defined in Chap. 2.*

7.1.4 Singular Value Inequalities for Operators

In this subsection we derive some inequalities for the products of singular values of an operator which are useful in dimension estimations. In contrast to similar inequalities in Sect. 2.2, Chap. 2 we consider here the general case of an operator acting between Euclidean spaces.

The main references for the first two propositions are [17, 25, 34].

Proposition 7.14 *Suppose $\mathbb{E}$, $\mathbb{F}$ and $\mathbb{F}'$ are n-dimensional Euclidean spaces with scalar product $(\cdot,\cdot)_{\mathbb{E}}$, $(\cdot,\cdot)_{\mathbb{F}}$ and $(\cdot,\cdot)_{\mathbb{F}'}$, respectively, $T:\mathbb{E}\to\mathbb{F}$ and $S:\mathbb{F}\to\mathbb{F}'$ are linear operators and $k\in\{1,\dots,n\}$ is arbitrary. Then*

$$\alpha_k(ST)\le |\,S\,|\,\alpha_k(T)\,,$$

where $|\,S\,|$ is the operator norm of $S:\mathbb{F}\to\mathbb{F}'$ and $\alpha_k(\cdot)$ denotes the singular values of the operator.

Proof By definition we have for $k=1,2,\dots,n$ the following equalities $\alpha_k(ST)^2=\lambda_k((ST)^{[*]}ST)$ and $\alpha_k(T)^2=\lambda_k(T^{[*]}T)$, where $\lambda_k(\cdot)$ denotes the ordered eigenvalues $\lambda_1(\cdot)\ge\cdots\ge\lambda_n(\cdot)$ of the given positive operator. For any $\xi\in\mathbb{E}$ we can write

$$\begin{aligned}((ST)^{[*]}ST\xi,\xi)_{\mathbb{E}} &= ((ST)\xi,(ST)\xi)_{\mathbb{F}'}\\ &= |\,(ST)\xi\,|^2_{\mathbb{F}'}\le |\,S\,|^2|\,T\xi\,|^2_{\mathbb{F}}=|\,S\,|^2((T^{[*]}T)\xi,\xi)_{\mathbb{E}}\,.\end{aligned}$$

It follows that

$$\lambda_k\big((ST)^{[*]}(ST)\big)\le\lambda_k\big(|\,S\,|^2(T^{[*]}T)\big)=|\,S\,|^2\lambda_k(T^{[*]}T),\ k=1,2,\dots,n.$$

But this implies that $\alpha_k(ST)\le|\,S\,|\alpha_k(T)\,k=1,2,\dots,n$. □

Proposition 7.15 *Suppose that $\mathbb{E}$ and $\mathbb{F}$ are n-dimensional Euclidean spaces, $T:\mathbb{E}\to\mathbb{F}$ is a linear operator and $k\in\{1,2,\dots,n\}$ is arbitrary.*

Then it holds:

(a) $\alpha_1(T)\alpha_2(T)\cdots\alpha_k(T)=\sqrt{\lambda_1((T^{\wedge k})^{[*]}T^{\wedge k})}=|\,T^{\wedge k}\,|$, where $|\,T^{\wedge k}\,|$ is the operator norm of $T^{\wedge k}:\mathbb{E}^{\wedge k}\to\mathbb{F}^{\wedge k}$;

(b) $\alpha_{n-k+1}(T)\alpha_{n-k+2}(T)\cdots\alpha_n(T)=\sqrt{\lambda_{\binom{n}{k}}((T^{\wedge k})^{[*]}T^{\wedge k})}\,.$

Proof (a) Using the definition of a singular value and the properties of eigenvalues, we can write

$$\lambda_1\big((T^{\wedge k})^{[*]}T^{\wedge k}\big)=\lambda_1\big(T^{[*]}T\big)\,\lambda_2\big(T^{[*]}T\big)\cdots\lambda_k\big(T^{[*]}T\big)=\alpha_1(T)^2\cdots\alpha_k(T)^2\,.$$

Again by definition

$$|T^{\wedge k}|^2 = \sup_{\substack{\xi\in\mathbb{E}^{\wedge k}\\ |\xi|=1}} \left(T^{\wedge k}\xi,\, T^{\wedge k}\xi\right)_{\mathbb{F}^{\wedge k}} = \max_{\substack{\xi\in\mathbb{E}^{\wedge k}\\ |\xi|=1}} \left((T^{\wedge k})^{[*]}T^{\wedge k}\xi, \xi\right)_{\mathbb{E}^{\wedge k}} = \lambda_1\left((T^{\wedge k})^{[*]}T^{\wedge k}\right)$$

by the Fischer-Courant theorem (Theorem 7.1). Part (b) can be shown similarly. □

The following proposition is due to [17, 28].

Proposition 7.16 *(Generalized Horn's inequality) Suppose that $\mathbb{E}$, $\mathbb{F}$ and $\mathbb{F}'$ are n-dimensional Euclidean spaces over $\mathbb{R}$, $T : \mathbb{E} \to \mathbb{F}$ and $S : \mathbb{F} \to \mathbb{F}'$ are linear operators. Then for $k \in \{1, \dots, n-1\}$ we have $\omega_k(ST) \le \omega_k(S)\omega_k(T)$ and $\omega_n(ST) = \omega_n(S)\omega_n(T)$.*

Proof The first part follows immediately from the operator norm properties and Propositions 7.2, 7.15:

$$\omega_k(ST) = |(ST)^{\wedge k}| = |S^{\wedge k}|\,|T^{\wedge k}| = \omega_k(S)\omega_k(T)\,.$$

The second part follows from these propositions by

$$\omega_n(ST) = |(ST)^{\wedge n}| = |\det(ST)| = |\det S|\,|\det T| = \omega_n(S)\omega_n(T)\,. \qquad \square$$

We finish this subsection with two statements from [21].

Lemma 7.6 *Let $T : \mathbb{E} \to \mathbb{E}$ be a self-adjoint, linear operator on an n-dimensional Euclidean space $\mathbb{E}$ with scalar product $(\cdot,\cdot)_{\mathbb{E}}$ and let $\alpha_1(T) \ge \dots \ge \alpha_n(T)$ denote the singular values of T ordered with respect to size and multiplicity. Suppose that for a certain k-dimensional subspace $\mathbb{L}^k$ of $\mathbb{E}$ $(1 \le k \le n)$ and for some number $\kappa \in \mathbb{R}$ the relation*

$$(\upsilon, T\upsilon)_{\mathbb{E}} \ge \kappa\,|\upsilon|^2 \quad \textit{for all} \quad \upsilon \in \mathbb{L}^k \tag{7.10}$$

is satisfied. Then

$$\alpha_k(T) \ge \kappa.$$

Proof Let $\upsilon_1, \dots, \upsilon_n$ denote an orthonormal system of eigenvectors of T belonging to the eigenvalues $\lambda_1(T) = \alpha_1(T), \dots, \lambda_n(T) = \alpha_n(T)$ of T. If $k > 1$ then choose $\upsilon \in \mathbb{L}^k$ such that $\upsilon \ne 0$ and

$$(\upsilon, \upsilon_i)_{\mathbb{E}} = 0 \qquad \text{for all} \quad i = 1, \dots, k-1$$

and if $k = 1$ then take any $\upsilon \in \mathbb{L}^1$ with $\upsilon \ne 0$. In both cases υ can be written as $\upsilon = \sum_{i=k}^{n} a_i\upsilon_i$ with $a_i \in \mathbb{R}$. Using (7.10) and taking into account the ordering of the singular values we obtain

$$\kappa\,|\upsilon|^2 \le (\upsilon, T\upsilon)_{\mathbb{E}} = \sum_{i=k}^{n} a_i^2\alpha_i(T) \le \alpha_k(T)|\upsilon|^2,$$

which completes the proof. □

Lemma 7.7 *Let $T : \mathbb{E} \to \mathbb{E}$ be as in Lemma 7.6 and furthermore invertible. Suppose that for a certain k-dimensional subspace $\mathbb{L}^k$ of $\mathbb{E}$ $(1 \le k \le n)$ and some number $\kappa \in \mathbb{R}$ the relation*

$$(\upsilon, T\upsilon)_{\mathbb{E}} \le \kappa \,|\, \upsilon \,|^2 \qquad \textit{for all} \quad \upsilon \in \mathbb{L}^k \tag{7.11}$$

is satisfied. Then

$$\alpha_{n-k+1}(T) \le \kappa.$$

Proof One proceeds analogously as in the proof of Lemma 7.6 considering the inverse operator T^{-1} instead of T. □

7.2 Orbital Stability for Flows on Manifolds

The results from [21] represented in Sects. 7.2.1–7.2.7 of this section develop and generalize various approaches and methods for vector fields on manifolds that go back to [18, 22]. These methods can be used to derive orbital stability criteria for nonlinear feedback systems in terms of the frequency-domain characteristics and transfer function of the linear part together with conditions on the nonlinear part of the dynamical system.

7.2.1 *The Andronov-Vitt Theorem*

Let $(\mathcal{M}, g)$ be a Riemannian manifold of dimension n and smoothness C^m $(m > 3)$. We consider on $\mathcal{M}$ the differential equation

$$\dot{u} = F(u), \tag{7.12}$$

where $F : \mathcal{M} \to T\mathcal{M}$ is a vector field of class C^l $(2 < l \le m - 1)$. For simplicity we assume that (7.12) has a flow $u : \mathbb{R} \times \mathcal{M} \to \mathcal{M}$ (which is C^l) and put $\varphi^t(\cdot) := u(t, \cdot)$ for every $t \in \mathbb{R}$. It follows that (see Sect. A.6, Appendix A) the curve $t \mapsto \varphi^t(p)$, $t \in \mathbb{R}$, is the unique solution of (7.12) with the initial condition $\varphi^0(p) = p$.

Let $\gamma(p) := \{\varphi^t(p) \mid t \in \mathbb{R}\}$ denote the orbit through p of (7.12) and let $\gamma_+(p) := \{\varphi^t(p) \mid t \ge 0\}$ be the corresponding positive semi-orbit.

By $\rho(\cdot, \cdot)$ we denote the geodesic distance on the Riemannian manifold $(\mathcal{M}, g)$ and define by $\mathrm{dist}(p, \mathcal{U}) := \inf\{\rho(p, \tilde{p}) \mid \tilde{p} \in \mathcal{U}\}$ the distance between the point p and the set $\mathcal{U}$ for an arbitrary $p \in \mathcal{M}$ and an arbitrary subset $\mathcal{U} \subset \mathcal{M}$.

An orbit of (7.12) through p is called *(positive) Lagrange stable* if $\overline{\gamma}_+(p)$ (i.e. the closure of $\gamma_+(p)$) is compact. A solution $\varphi^{(\cdot)}(p)$ of (7.12) is called *(positive) orbitally stable* if for each $\varepsilon > 0$ there exists a $\delta > 0$ such that for all $\tilde{p} \in \mathcal{M}$ satisfying $\rho(p, \tilde{p}) < \delta$ and for all $t \geq 0$

$$\text{dist}(\varphi^t(\tilde{p}), \gamma_+(p)) < \varepsilon$$

holds. We say that $\varphi^{(\cdot)}(p)$ is *asymptotically orbitally stable* if $\varphi^{(\cdot)}(p)$ is orbitally stable and if there exists a $\Delta > 0$ such that

$$\lim_{t \to \infty} \text{dist}(\varphi^t(\tilde{p}), \gamma_+(p)) = 0$$

holds for each $\tilde{p} \in \mathcal{M}$ satisfying $\rho(p, \tilde{p}) < \Delta$. A solution is called *orbitally unstable* if it is not orbitally stable. It is also common to speak of the stability or instability of the orbit $\gamma(p)$ instead of the orbital stability or instability of the solution $\varphi^{(\cdot)}(p)$.

Note that in case $\gamma(p)$ is an equilibrium point of (7.12) then the orbital stability and asymptotic orbital stability coincide with the stability and asymptotic stability in the sense of Lyapunov, respectively, of the equilibrium.

For periodic orbits the well-known Andronov-Vitt theorem (e.g. [1, 2]), offers criteria for stability and instability in terms of the characteristic exponents or multipliers of the periodic orbit. The *multipliers* of a T-periodic orbit $\gamma(p) = \{\varphi^t(p) \mid t \in [0, T]\}$ of (7.12) are the eigenvalues $\rho_1(p), \rho_2(p), \ldots, \rho_n(p)$ of the differential $d_p\varphi^T : T_p\mathcal{M} \to T_p\mathcal{M}$ ordered with respect to their modulus and algebraic multiplicity by $|\rho_1(p)| \geq |\rho_2(p)| \geq \cdots \geq |\rho_n(p)|$. One can show that the spectrum of eigenvalues of $d_p\varphi^T$ consists of 1 and of the eigenvalues of $d_u P : T_u\mathcal{S} \to T_u\mathcal{S}$, where $P : \mathcal{S} \to \mathcal{S}$ is the Poincaré map at an arbitrary chosen point $u \in \gamma(p)$ with respect to a local transversal section $\mathcal{S} \subset \mathcal{M}$. The Andronov-Vitt theorem says that if for the second multiplier $|\rho_2(p)| < 1$ is satisfied (Andronov-Vitt condition), then the periodic orbit $\gamma(p)$ is asymptotically stable and the solution paths near $\gamma(p)$ possess asymptotic phases [1].

7.2.2 Various Types of Variational Equations

The behaviour of the flow of system (7.12) near a given solution $\varphi^{(\cdot)}(p)$ is described by the standard *variational equation* of (7.12)

$$\frac{Dy}{dt} = \nabla F(\varphi^t(p))y. \tag{7.13}$$

Here $\nabla F(p) : T_p\mathcal{M} \to T_p\mathcal{M}$ is the covariant derivative of F at $p \in \mathcal{M}$. The absolute derivative $\frac{Dy}{dt}$ is taken along the integral curve $t \mapsto \varphi^t(p)$. In local coordinates of a chart x around $\varphi^t(p)$ this derivative of $y = y^i(t)\partial_i(\varphi^t(p))$ has the form

$$\frac{Dy}{dt} = \left(\frac{dy^i}{dt} + \Gamma^k_{ij} y^j \dot{x}^i\right) \partial_k(\varphi^t(p)),$$

where $x^i(t)$ are the local coordinates of $\varphi^t(p)$ in the chart x.

Remark 7.5 The variational equation (7.13) describes a special linear flow on vector bundles. Suppose that $(\mathcal{M}, g)$ is an n-dimensional Riemannian C^3-manifold, $\mathbb{V}$ is an m-dimensional real vector space, $\pi : \mathcal{B} \to \mathcal{M}$ is a C^k-vector bundle over $\mathcal{M}$ with typical fiber $\mathbb{V}$. Suppose that on $\mathcal{B}$ is defined the C^{k-1}-bundle metric $g \in \mathcal{E}^{k-1}(T_2^0(\mathcal{B}))$ and a connection with covariant derivative ∇ (see Sect. A.10, Appendix A). Let us assume that the connection is metrical.

A *linear flow* Φ on the vector bundle $\pi : \mathcal{B} \to \mathcal{M}$ is a flow on $\mathcal{B}$ preserving fibers such that

$$\Phi(t, e_1 + e_2) = \Phi(t, e_1) + \Phi(t, e_2)\,, \quad \forall\, t \in \mathbb{R}, \quad \forall\, e_1, e_2 \in \mathcal{E}_p\,,$$
$$\Phi(t, \alpha\, e) = \alpha\, \Phi(t, e)\,, \quad \forall\, t \in \mathbb{R}, \quad \forall\, \alpha \in \mathbb{R}, \quad \forall\, e \in \mathcal{E}_p\,.$$

Note that Φ includes a flow $\pi\,\Phi : (\mathrm{id}, \mathbb{T})\,(\mathbb{R} \times \mathcal{B}) \to \mathcal{M}$ on the base space $\mathcal{M}$, which we denote by $p \cdot t$ for $t \in \mathbb{R},\ p \in \mathcal{M}$.

Suppose that $A : \mathcal{M} \to \mathcal{L}(\mathcal{B}, \mathcal{B})$ is a section in the vector bundle $\pi_{\mathcal{L}} : \mathcal{L}(\mathcal{B}, \mathcal{B}) \to \mathcal{M}$ consisting of all bundle maps whose fiber over $p \in \mathcal{M}$ is the vector space of all linear maps $L : \mathcal{B}_p \to \mathcal{B}_p$. This means that A associates to any point $p \in \mathcal{M}$ a linear map $A(p) : \mathcal{B}_p \to \mathcal{B}_p$. Let us assume that the linear flow Φ on the vector bundle $\pi : \mathcal{B} \to \mathcal{M}$ is C^1, i.e. there exists a section A as described above such that for any $p \in \mathcal{M}$ and $\upsilon \in \mathcal{B}_p$ the curve $t \mapsto \Phi(t, (p, \upsilon))$ satisfies the differential equation

$$\frac{D}{dt}\Phi(t, (p, \upsilon)) = A(p \cdot t)\Phi(t, (p, \upsilon))\,. \tag{7.14}$$

Here $\frac{D}{dt}$ denotes the covariant derivative along the curve $t \mapsto p \cdot t$.

The linearization of the flow φ, generated by the vector field (7.12), i.e. the map $\Phi : \mathbb{R} \times T\mathcal{M} \to T\mathcal{M}$ with $\Phi(t, (p, \upsilon)) := d_p\varphi^t(\upsilon)$ for $p \in \mathcal{M}$, $(p, \upsilon) \in T\mathcal{M}$ and $t \in \mathbb{R}$, is a linear flow on the tangent bundle $T\mathcal{M}$. For fixed $p \in \mathcal{M}$ the behaviour of the flow lines of φ in a neighborhood of $\varphi^{(\cdot)}(p)$ is characterized by the curve $t \mapsto \Phi(t, (p, \upsilon))$ with $\upsilon \in T_p\mathcal{M}$, which satisfies for fixed υ the variational equation (7.14).

Let $Y(\cdot, p)$ be the fundamental operator solution of (7.13) satisfying the initial condition $Y(0, p) = \mathrm{id}_{T_p\mathcal{M}}$. Thus $Y(t, p) = d_p\varphi^t$ for all $t \in \mathbb{R}$. Note that, in particular, $F(\varphi^{(\cdot)}(p))$ is a (vector) solution of (7.13).

A short calculation shows that the relation

$$\frac{d}{dt}(y(t), y(t)) = 2\Big(\frac{Dy(t)}{dt}, y(t)\Big) \tag{7.15}$$

holds for any C^1-curve $y(\cdot)$ satisfying $y(t) \in T_{\varphi^t(p)}\mathcal{M}$ for all $t \in \mathbb{R}$. Here for every $t \in \mathbb{R}$ the term $(\cdot, \cdot)$ stands for the scalar product in $T_{\varphi^t(p)}\mathcal{M}$ introduced by the Riemannian metric.

For every $u \in \mathcal{M}$ with $F(u) \neq O_u$, where O_u denotes the origin of the tangent space $T_u\mathcal{M}$, we introduce the linear subspace

$$T^{\perp}(u) := \{z \in T_u\mathcal{M} \mid (z, F(u)) = 0\}$$

of the tangent space $T_u\mathcal{M}$. To describe how the orthogonal deviation of a perturbation in the initial conditions evolves we split a solution $y(\cdot)$ of (7.13) into orthogonal components

$$y(t) = z(t) + \mu(t)F(\varphi^t(p)), \tag{7.16}$$

with $z(t) \in T^{\perp}(\varphi^t(p))$ and $\mu(t)$ a time-dependent factor which will be specified below. Then $z(\cdot)$ is a solution of the *system in normal variations* with respect to the solution $\varphi^{(\cdot)}(p)$ of (7.12) which is of the form

$$\frac{Dz}{dt} = A(\varphi^t(p))z \tag{7.17}$$

with a map $A(u) : T_u\mathcal{M} \to T_u\mathcal{M}$ given by

$$A(u)\upsilon := \nabla F(u)\upsilon - \frac{2(F(u), S\nabla F(u)\upsilon)}{|F(u)|^2}F(u) \tag{7.18}$$

for all $\upsilon \in T_u\mathcal{M}$. In local coordinates of an arbitrary chart x around u the operator A is defined by

$$A_i^k := \nabla_i f^k - \frac{2}{g_{rs} f^r f^s} f^k g_{jl} f^l S_i^j,$$

where f^k, g_{jl} are the coordinates of the vector field F and the Riemannian metric tensor g in the chart x. Here $|\cdot|$ stands for the norm in the tangent space derived from the scalar product in the space and $S_i^j := \frac{1}{2}(g^{jk}\nabla_k f^l g_{li} + \nabla_i f^j)$ is the representation in coordinates of the symmetric part $S\nabla F$ of the covariant derivative of the vector field F. For $\mathcal{M} = \mathbb{R}^n$ and $(g_{ij}) = I$ the standard metric the operator (7.18) is given in Sect. 3.3. Let $Z(\cdot, p)$ denote the fundamental operator solution of (7.17) with $Z(0, p) = \mathrm{id}_{T^{\perp}(p)}$. From the definition (7.18) of A we see that for any $t \in \mathbb{R}$ and $p \in \mathcal{M}$ the linear operator $Z(t, p)$ acts between the orthogonal subspaces $T^{\perp}(p)$ and $T^{\perp}(\varphi^t(p))$ of $T_p\mathcal{M}$ and of $T_{\varphi^t(p)}\mathcal{M}$, respectively.

Now let $y(\cdot)$ be a solution of (7.13). We want to consider the splitting (7.16) into orthogonal components in more detail. The factor $\mu(\cdot)$ is a scalar valued C^{l-1}-function determined by

$$\mu(t) := \frac{(y(t), F(\varphi^t(p)))}{|F(\varphi^t(p))|^2}. \tag{7.19}$$

Differentiating the expressions in formula (7.16) and applying formula (7.16) again gives $\nabla F(\varphi^t(p))z(t) = A(\varphi^t(p))z(t) + \dot{\mu}(t)F(\varphi^t(p))$. If we take now the scalar product of this term with $z(t)$ we obtain

$$(A(\varphi^t(p))z(t), z(t)) = (\nabla F(\varphi^t(p))z(t), z(t)). \tag{7.20}$$

For an arbitrary number $k \in \{1, \dots n\}$ we consider the *k-th compound equation* of (7.13)

$$\frac{Dy}{dt} = \left(\nabla F(\varphi^t(p))\right)_k y, \tag{7.21}$$

where $(\nabla F(u))_k : \bigwedge^k T_u\mathcal{M} \to \bigwedge^k T_u\mathcal{M}$ is the k-th additive compound operator of $\nabla F(u)$ (see Sect. 7.1.1). The absolute derivative $\frac{D}{dt}[y_1(\cdot) \wedge \cdots \wedge y_k(\cdot)]$ of $[y_1(\cdot) \wedge \cdots \wedge y_k(\cdot)]$ along $t \mapsto \varphi^t(p)$ is defined by

$$\frac{D}{dt}[y_1(\cdot) \wedge \cdots \wedge y_k(\cdot)] = \frac{Dy_1(\cdot)}{dt} \wedge \cdots \wedge y_k(\cdot) + \cdots + y_1(\cdot) \wedge \cdots \wedge \frac{Dy_k(\cdot)}{dt},$$

where $\frac{Dy_i(\cdot)}{dt}$ is the absolute derivative of y_i along $t \mapsto \varphi^t(p)$. We remark that relation (7.15) can be generalized to

$$\frac{d}{dt}(y(t), y(t))_{\bigwedge^k T_{\varphi^t(p)}\mathcal{M}} = 2\left(\frac{Dy(t)}{dt}, y(t)\right)_{\bigwedge^k T_{\varphi^t(p)}\mathcal{M}},$$

which is valid for every $k \in \{1, \dots, n\}$ and every C^1-curve $y(\cdot)$ with $y(t) \in \bigwedge^k T_{\varphi^t(p)}\mathcal{M}$ for all $t \in \mathbb{R}$. If $Y(\cdot, p)$ is the fundamental operator solution of (7.13) satisfying $Y(0, p) = \mathrm{id}_{T_p\mathcal{M}}$ then $Y^{\wedge k}(\cdot, p)$ is the fundamental operator solution of (7.21) satisfying $Y^{\wedge k}(0, p) = \mathrm{id}_{\Lambda^k T_p\mathcal{M}}$.

Finally, the trivial solution $y \equiv 0$ of (7.21) is called *exponentially stable* if there exist constants $C > 0$ and $a > 0$ such that for any solution $y(\cdot)$ of (7.21) and any $s \geq 0$ the inequality

$$|y(t)| \leq C|y(s)|e^{-a(t-s)} \tag{7.22}$$

holds for all $t \geq s$.

7.2.3 Asymptotic Orbital Stability Conditions

The following theorem from [21] provides a result on orbital stability of solutions of (7.12) requiring exponential stability properties of the compound variational equation (7.21) for $k = 2$. The theorem generalizes similar results for systems in $\mathbb{R}^n$ (see [3, 6, 14, 22]).

We say that an orbit $\gamma(p)$ of (7.12) belongs to BO_+ if $\gamma_+(p)$ lies in some open bounded set $\mathcal{U} = \mathcal{U}(p) \subset \mathcal{M}$ and any equilibrium of (7.12) which is contained in the closure $\overline{\mathcal{U}}$ is asymptotically stable.

Theorem 7.2 *Consider the equation (7.12) and suppose $\gamma(p) \in BO_+$. If the zero solution of the second compound equation*

$$\frac{Dw}{dt} = \left(\nabla F(\varphi^t(p))\right)_2 w \tag{7.23}$$

is exponentially stable, then the solution $\varphi^{(\cdot)}(p)$ *of (7.12) is asymptotically orbitally stable.*

To investigate the stability of a fixed orbit of (7.12) it suffices to consider stability properties of the projected flow of system (7.12) orthogonally to the perturbed flow line. Keeping this in mind the following result from [15] on the reparametrization of solutions of (7.12) in the neighborhood of a fixed solution is of great importance. We formulate this result under slightly weaker assumptions (see [21, 26]).

Lemma 7.8 *Suppose* $\varphi^{(\cdot)}(p)$ *with* $p \in \mathcal{M}$ *to be a non-constant solution of (7.12) with bounded positive semi-orbit* $\gamma_+(p)$ *such that for a certain* $C_0 > 0$ *the inequality* $|F(u)| > C_0$ *is satisfied for all* $u \in \gamma_+(p)$. *Then it holds:*

(a) For any finite time $T_0 > 0$ *there exists a* $\delta = \delta(T_0) > 0$ *such that for any pair* $(r, \upsilon) \in [0, \delta] \times (T^\perp(p) \cap \mathcal{B}(O_p, 1))$ *we may find a* C^{l-1}*-diffeomorphism (for* $l = 1$ *a homeomorphism)* $s(\cdot, r, \upsilon) : \mathbb{R}_+ \to \mathbb{R}_+$ *with* $s(t, 0, \upsilon) = t$ *for all* $t \in \mathbb{R}_+$ *and such that near* $\gamma_+(p)$ *the reparameterized flow* $\phi(t, r, \upsilon) := \varphi^{s(t,r,\upsilon)}(\exp_p(r\upsilon))$ *of (7.12) satisfies the condition*

$$(D_2\phi(t, r, \upsilon), F(\phi(t, r, \upsilon))) = 0 \tag{7.24}$$

for all $t \in [0, T_0]$. *(Here* $D_2\phi(t, r, \upsilon)$ *stands for the derivative of* ϕ *with respect to the second argument.)*

(b) Suppose that the zero solution of the equation in normal variations (7.17) (with respect to the solution $\varphi^{(\cdot)}(p)$*) is exponentially stable. Then there exist numbers* $\delta > 0$ *and* $C > 0$ *such that for every* $u = u(r, \upsilon) \in \mathcal{B}^\perp(p, \delta) := \exp_p(\mathcal{B}(O_p, \delta) \cap T^\perp(p))$, *parameterized in the above sense, a homeomorphism* $s(\cdot, r, \upsilon) : \mathbb{R}_+ \to \mathbb{R}_+$ *can be found which satisfies*

$$\rho(\varphi^{s(t,r,\upsilon)}(u(r, \upsilon)), \varphi^t(p)) \le C\rho\,(u(r, \upsilon), p) \quad \text{for all } t \ge 0. \tag{7.25}$$

Proof (a) Let $\upsilon \in B^\perp(p, 1)$ be arbitrary. We seek a familiy of parametrizations $s(\cdot, \cdot, \upsilon)$ such that the derivative w.r.t. r of $\phi(t, r, \upsilon)$ in an arbitrary point $\varphi^{s(t,r,\upsilon)}$ $(u(r, \upsilon))$ is a vector which belongs to $T^\perp(\varphi^{s(t,r,\upsilon)}(u(r, \upsilon)))$. This means that (7.24) has to be satisfied. Suppose that f^i and ϕ^j are the local coordinates of F and ϕ, respectively, in a chart x around the point $\phi(t, r, \upsilon)$. Then $D_2\phi(t, r, \upsilon)$ has the local representation

$$\frac{d\phi^i}{dr} = f^i \frac{\partial s}{\partial r} + \frac{\partial \phi^i}{\partial r} \,.$$

It follows that

$$D_2\phi = F(\varphi^{s(t,r,\upsilon)}(u(r, \upsilon))\frac{\partial s}{\partial r} + d_u\varphi^{s(t,r,\upsilon)}\tau_p^u(\upsilon) \,. \tag{7.26}$$

Thus for any $v \in \mathcal{B}^{\perp}(p, 1)$ we get the Cauchy problem for $s(\cdot, \cdot, v)$

$$\frac{\partial s}{\partial r} = -\frac{\left(y(s, \tau_p^{u(r,v)}(v)), F(\varphi^s(u(r, v)))\right)}{|F(\varphi^s(u(r, v)))|^2}, \quad s(t, 0, v) = t \quad \text{for all } t \geq 0 . \tag{7.27}$$

Here $y(\cdot, \tau_p^u v)$ is the solution of the variational equation (7.13), computed with respect to the curve $t \mapsto \varphi^t(u(r, v))$ with initial state $y(0, \tau_p^u v) = \tau_p^u v$.

If we compare the right-hand side of (7.27) with the definition (7.19) of the function $\mu(\cdot)$ computed also with respect to $\varphi^{(\cdot)}(u(r, v))$, we conclude that $\mu = -\frac{\partial s}{\partial r}$. Together with (7.16) and (7.26) we get

$$D_2\phi(t, r, v) = z(s(t, r, v)\tau_p^u v) \tag{7.28}$$

for all $t \geq 0, r \in [0, \varepsilon]$ and $v \in \mathcal{B}^{\perp}(p, 1)$. Here $z(\cdot, \tau_p^u v)$ denotes the solution of (7.17), computed w.r.t. the curve $t \mapsto \varphi^{(t)}(u(r, v))$, and satisfying $z(0, \tau_p^u(v)) = \tau_p^u(v)$. Since the right-hand side of (7.27) is C^l there exist the continuous second-order derivatives $\frac{\partial}{\partial t}\frac{\partial s}{\partial r} = \frac{\partial}{\partial r}\frac{\partial s}{\partial t}$. Furthermore, there is a $\delta > 0$ such that for all $t \in [0, T_0]$ the solution $s(t, \cdot, v)$ exists on $[0, \delta]$ and is C^{l-1} w.r.t. all arguments. Additionally we have $\frac{\partial s}{\partial r} \in C^{l-1}$. It was shown in [15] that $s(\cdot, r, v)$ is strongly monotone increasing.

(b) Since $\gamma_+(p)$ belongs to an open bounded set $\mathcal{U} \subset \mathcal{M}$ there exists an $\varepsilon_0 > 0$ such that $\exp_u^{-1}$ is for any $u \in \gamma_+(u)$ a diffeomorphism on $\mathcal{B}(u, \varepsilon_0)$. From part (a) and from the boundedness of $\gamma_+(p)$ it follows that for any finite but fixed $T_0 > 0$ there exists a $\delta > 0$ such that the parametrization $s(\cdot, \cdot, v)$ is defined on $[0, T_0] \times [0, \delta]$ for all $u \in \gamma_+(p)$ and $v \in \mathcal{B}^{\perp}(u, 1)$. At the next step we extend the existence interval $[0, T_0]$ for the parametrization. Since the right-hand side of (7.27) is C^{k-1}-smooth we see that $\frac{\partial s}{\partial r}(\cdot, \cdot, \cdot)$ is also C^{k-1} w.r.t. all arguments. It follows that

$$\sup_{\substack{t \in [0, T_0] \\ u(r,v) \in \mathcal{B}^{\perp}(q,\delta),\, q \in \gamma_+(p)}} \left| \frac{\partial s}{\partial r}(t, r, v) \right| < \infty . \tag{7.29}$$

For an arbitrary $u(r, v) \in \mathcal{B}^{\perp}(p, \delta_1)$ with arbitrary $r \in [0, \delta]$ and $v \in S^{\perp}(0)$ we consider the solution $z(\cdot, \tau_p^u(v))$ of (7.17) w.r.t. the integral curve $t \mapsto \varphi^t(u(r, v))$. The Taylor expansion results in

$$z(s(t, r, v), \tau_p^u(v)) = \tau_{\varphi_u^t}^{\varphi_u^s}\left\{z(t) + A(\varphi^t(p))z(t)\frac{\partial s}{\partial r}(t, 0, v)r + 0\left(|z(t)|r^2\right\}\right., \tag{7.30}$$

where $t \in [0, T_0]$ is arbitrary, and $z(\cdot)$ is the solution of (7.17) w.r.t. the curve $t \leftrightarrow \varphi^t(p)$ and with initial state $z(0) = v$. Because of $\inf |F(u)| > 0\, u \in \gamma_+(u)$ and the definition of the linear operator A we have $\sup_{u \in \overline{\mathcal{U}}} |A(u)| < \infty$. Together with (7.29) and (7.30) this gives

$$|z(s(t,r,\upsilon)\,,\,\tau_p^u(\upsilon)| \;=\; |z(t)|\,[1+0(r)] \tag{7.31}$$

for all $t \in [0, T_0]$ and $u\,(r, \upsilon) \in \mathcal{B}^{\perp}(p, \delta)$. (Note that norms are computed in different tangent spaces.) Using (7.28) and (7.31) we get for the length of the curve $c : \widetilde{r} \to \phi(t, \widetilde{r}, \upsilon)$ on $[0, r]$ the formula

$$l(c) = \int_0^r |D_2\phi(t,\widetilde{r},\upsilon)|d\widetilde{r} = \int_0^r |z(\widetilde{s}, \tau_p^{\widetilde{u}}(\upsilon))|d\widetilde{r} = |z(t)|r(1+0(r))\,. \tag{7.32}$$

Here $u(r, \upsilon) \in \mathcal{B}^{\perp}(p, \delta)$ and $t \in [0, T_0]$ are arbitrary and $\widetilde{s} := s(t, \widetilde{r}, \upsilon), \widetilde{u} := u(\widetilde{r}, \upsilon)$. By assumption the trivial solution of (7.17), is uniformly Lyapunov stable. It follows that there exists a constant $C_0 > 0$ with $|z(t)| \leq C_0$ for all $t \geq 0$. By definition of the geodesic distance we have $\rho(\phi(t, r, \upsilon), \varphi^t(p)) \leq l(c)$. Thus (7.32) implies that

$$\rho(\phi(t,r,\upsilon), \varphi^t(p)) \leq C_0 r(1+0(r))\,. \tag{7.33}$$

Therefore, we can find a $\delta_0 \in [0, \delta]$ such that

$$\rho(\phi(t,r,\upsilon), \varphi^t(p)) < \delta \tag{7.34}$$

for all $t \in [0, T_0]$ and all $r \in [0, \delta_0]$. After choosing T_0 and δ we can continue the parametrization onto the interval $[0, 2\,T_0]$. Since the bound in (7.34) does not depend on the time, we can derive the existence of a parametrization $s(t, r, \upsilon)$ for all $t \geq 0$. With $r = \rho(u(r, \upsilon), p)$ and $\rho(\phi(t, r, \upsilon), \varphi^t(p)) \leq l(c)$ we get directly from (7.33) that with an appropriate constant $C > C_0$

$$d(\phi(t,r,\upsilon),\, \varphi^t(p)) \leq C\rho(u(r,\upsilon), p)$$

for all $t \geq 0$ and all $u(r, \upsilon) \in \mathcal{B}^{\perp}(p, \delta_0)$. □

Remark 7.6 ([11]) It follows from the construction in Lemma 7.8 that under the condition $\inf_{u\in\mathcal{M}}|F(u)| > 0$ there exists (see Sect. A.10, Appendix A) a unique foliation $\mathcal{F}$ of $\mathcal{M}$ of the codimension 1 for which F is a section in the normal bundle $T(\mathcal{F})^{\perp}$. The leaf $\mathcal{L}$ of this foliation which contains the point p is a local transversal section of f in p. In order to see this we consider the $(n-1)$-dimensional submanifold of $\mathcal{M}$ given by $\mathcal{W} := \{\varphi^{s(0,r,\upsilon)}(u(r, \upsilon)) \,|\, u(r, \upsilon) \in \mathcal{B}^{\perp}(p, \delta)\}$.

The normal bundle of $\mathcal{W}$ contains the vectors of the vector field $F_{|\mathcal{W}}$. It follows that $\mathcal{W}$ is part of the leaf $\mathcal{L}$ of the foliation $\mathcal{F}$ which contains the point p.

The differential equation (7.13) is a special local version of the system (7.14). The reparametrization generates a linear local flow $\phi : \mathcal{U} \to \mathcal{E}$, which is defined on an open neighborhood $\mathcal{U}$ of $[0, T] \times (\mathcal{E} \cap T_{\gamma(p)}\mathcal{M})$ in $\mathbb{R} \times \mathcal{E}$. For any pairs $(t, (w, z))$ with $t \in [0, T]$, $w \in \mathcal{W}$ and $z \in \mathcal{E}_w$ this flow is defined as follows: Any point $w \in \mathcal{W}$ we can associate uniquely to a point $u_w \in \mathcal{B}^{\perp}(p, \delta)$ which is defined by $u_w = \exp_p(r_w \upsilon_w)$ with $r_w \in [0, \delta)$ and $\upsilon_w \in \mathcal{E}_p$, $|\upsilon_w|_{\mathcal{E}_p} = 1$ such that $w = \phi(0, r_w, \upsilon_w)$. For this point the reparametrization through u_w is defined. Now we introduce the map

$h : \mathcal{W} \to [0, \delta) \times \partial\mathcal{B}^{\perp}(0_p, 1)$ that for any $w \in \mathcal{W}$ is given by $h(w) := (r_w, \upsilon_w)$. For any $t \in [0, T]$ we define the map $P_t : \mathcal{W} \to \mathcal{M}$ through

$$P_t(w) := \phi(t, h(w)), \quad \forall\, w \in \mathcal{W}.$$

Now the local flow ϕ is given for all $t \in [0, T]$, $w \in \mathcal{W}$, $z \in \mathcal{E}_w$ by

$$\phi(t, (w, z)) := d_w P_t z\,.$$

For any time $t \in [0, T]$ we consider the $(n-1)$- dimensional submanifold of $\mathcal{M}$ defined by

$$\mathcal{W}_t := \{\varphi^{s(t,r,\upsilon)}(u(r, \upsilon)) \mid u(r, \upsilon) \in \mathcal{B}^{\perp}(p, \varepsilon)\}.$$

Because of (7.24) this submanifold is part of a leaf $\mathcal{F}$ of the foliation. For any point $w \in \mathcal{W}$ we have the relation

$$d_w P_t T_p \mathcal{W} = T_{P_t(w)} \mathcal{W}_t\,.$$

In particular this means that $\phi(t, (w, \cdot))$ maps the vectors $z \in \mathcal{E}_w$ into $\mathcal{E}_{P_t(w)}$. The result of the reparametrization is indeed a linear local flow on the subbundle $\mathcal{E}$.

Note that relation (7.25) establishes the orbital stability of $\varphi^{(\cdot)}(p)$ (Fig. 7.1). It remains to prove the asymptotic properties.

Remark 7.7 If under the assumptions of Theorem 7.2 the ω-limit set $\omega(p)$ of p does not contain equilibrium points of the vector field (7.12) one can show (see Theorem 7.10), using the asymptotic orbital stability properties, that $\omega(p)$ is a periodic orbit which is asymptotically stable and nearby solution paths have asymptotic phases. For $\mathcal{M} = \mathbb{R}^n$ this is done in [15, 24]; for the cylinder it is demonstrated in [20, 23].

Proof of Theorem 7.2. For a given solution $\varphi^{(\cdot)}(p)$ of (7.12) which is contained in the set $\mathcal{U} = \mathcal{U}(p)$ we consider separately the following two cases.

Case1: There exists a sequence $\{t_i\}_{i\in\mathbb{N}}$, $t_i \geq 0$ and $\lim_{i\to+\infty} t_i = \infty$ such that

$$\lim_{i\to+\infty} |F(\varphi^{t_i}(p))| = 0 \qquad \text{holds.}$$

Case2:

$$\inf_{u\in\overline{\mathcal{U}}} |F(u)| > 0. \tag{7.35}$$

First we will discuss Case 1. Since $\gamma_+(p) \subset \overline{\mathcal{U}}$ is bounded the sequence $\{\varphi^{t_i}(p)\}$ has a convergent subsequence.

Without loss of generality we can assume that there exists a point $u^* \in \overline{\mathcal{U}}$ such that

$$\lim_{i\to+\infty} \rho\,(\varphi^{t_i}(p), u^*) = 0. \tag{7.36}$$

It follows that $F(u^*) = O_{u^*}$ and by the assumption on $\mathcal{U}$ we know that u^* is an asymptotically stable equilibrium point of system (7.12). Thus we find an open neighborhood $\mathcal{V}(u^*)$ of u^* such that any positive semi-orbit through $u \in \mathcal{V}(u^*)$ is attracted by u^*. On account of (7.36) there exists $k \in \mathbb{N}$ such that $\varphi^{t_k}(p) \in \mathcal{V}(u^*)$, so we obtain $\rho(\varphi^t(p), u^*) \to 0$ for $t \to +\infty$ and also that points of a sufficiently small neighborhood of p are attracted by u^* as well. But this implies the asymptotic stability of $\gamma(p)$.

The proof in the Case 2 needs more work. For the sake of clearness we divide this proof into four steps.

Step 1. Using the exponential stability of the zero solution of the second compound equation (7.23) we deduce the exponential stability of the zero solution of (7.17). By the assumption of the theorem there exist positive numbers δ_0, a and C such that for any $s \geq 0$

$$|w(t)| \leq C|w(s)|e^{-a(t-s)} \tag{7.37}$$

holds for an arbitrary solution $w(\cdot)$ of (7.23) satisfying $|w(s)| < \delta_0$ and all $t \geq s$. We consider an arbitrary solution $y(\cdot)$ of the variational equation (7.13) with $y(0) \neq O_p$ and $|y(0) \wedge F(p)| < \delta_0$. Let $z(\cdot)$ be the solution of the system (7.17) coupled with $y(\cdot)$ in the sense that $y(t) = z(t) + \mu(t)F(\varphi^t(p))$, where the factor $\mu(\cdot)$ is determined by (7.19). In addition, suppose $y(0)$ to be such that $z(0) \neq O_p$. As mentioned above $F(\varphi^{(\cdot)}(p))$ is a solution of (7.13). Thus, $w(\cdot) := y(\cdot) \wedge F(\varphi^{(\cdot)}(p))$ is a solution of (7.23) and by means of the fundamental operator solution $Y^{\wedge 2}$ of

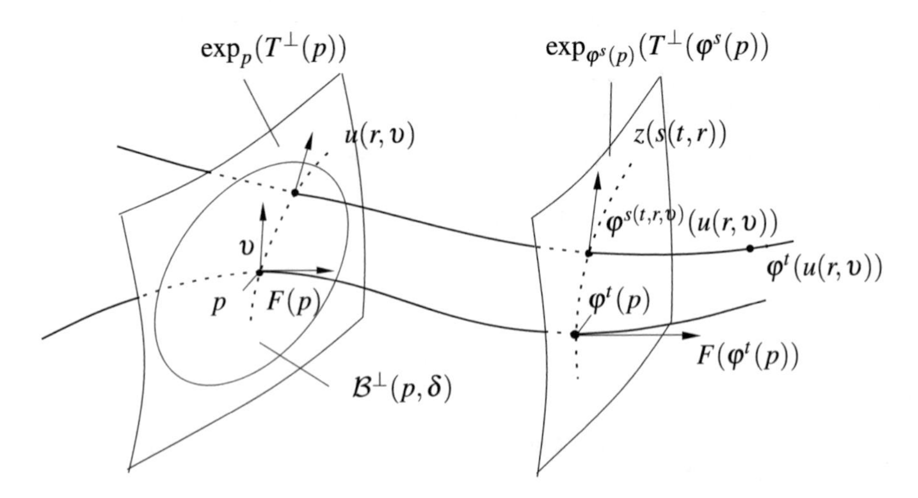

Fig. 7.1 Reparametrization of the flow

(7.23) can be written in the form $w(t) = Y^{\wedge 2}(t)[y(0) \wedge F(p)]$. This gives

$$|w(t)| = |z(t)| \cdot |F(\varphi^t(p))| \tag{7.38}$$

for all $t \geq 0$ and $|w(0)| \neq 0$.

The assumed boundedness of $\gamma_+(p)$ and the inequality (7.35) imply that there is a constant $C_1 > 0$ such that

$$\sup_{u,u' \in \gamma_+(p)} \frac{|F(u)|}{|F(u')|} < C_1, \tag{7.39}$$

where the norms are taken in the corresponding tangent spaces. Using (7.37) – (7.39) we derive

$$|z(t)| \leq CC_1 |z(s)| e^{-a(t-s)} \tag{7.40}$$

for an arbitrary solution $z(\cdot)$ of (7.17) with $0 < |z(s)| < \delta_0/|F(p)|$ and for all $t \geq s$.

Step 2. Utilizing the exponential stability of the zero solution of the system in normal variations we deduce the orbital stability of the reference orbit by a reparametrization of the semi-flow near this orbit. Further we investigate some properties of this reparametrization, which will be used in Step 3. Consider again $\gamma_+(p)$ and an arbitrary point $u \in \mathcal{M}$ in the $(n-1)$-dimensional submanifold $\mathcal{B}^\perp(p, \Delta)$ through p. As indicated above, u can be uniquely represented by a pair (r, v) due to $u(r, v) = \exp_p(rv)$. Under the assumptions of the theorem and with (7.40) all requirements of the part (b) of the Lemma 7.8 are fulfilled. This yields the existence of a number $\delta_p \in (0, \Delta)$ such that for all solutions $\varphi^{(\cdot)}(u(r, v))$ with $u(r, v) = \exp_p(rv) \in \mathcal{B}^\perp(p, \delta_p)$ there is a reparametrization $s(\cdot, r, v) : \mathbb{R}_+ \to \mathbb{R}_+$ for which (7.25) is satisfied. As it was mentioned above from (7.25) the orbital stability follows.

Step 3. From (7.25) we see that $\gamma(p)$ is stable. Let $\varepsilon > 0$ be fixed and choose $\delta \in (0, \delta_p)$ so that for every $u(r, v) \in \mathcal{B}^\perp(p, \delta)$ the corresponding solution stays in the ε-neighbourhood $\gamma_+(p)$ for $t \geq 0$. Then, from the construction of the parametrization, we have for arbitrary $u(r, v) \in \mathcal{B}^\perp(p, \delta)$ and arbitrary $t \geq 0$ that there exist $\widetilde{r} = \widetilde{r}(t) \in [0, \varepsilon)$ and $\widetilde{v} = \widetilde{v}(t) \in T^\perp(\varphi^t(p))$ such that

$$s(t, r, v) = s(0, \widetilde{r}, \widetilde{v}). \tag{7.41}$$

By a generalization of Liouville's formula (see Proposition 2.18, Chap. 2) for any $\tau > 0$ and any $u \in \mathcal{M}$ the inequality

$$|Y(\tau, u)| \leq \exp\left\{ \int_0^\tau \lambda_1(\varphi^t(u)) dt \right\}$$

is valid, where $Y(\tau, u)$ is, as above, the fundamental operator solution of (7.13) with respect to $\varphi^{(\cdot)}(u)$ and λ_1 is the largest eigenvalue of the symmetric part of ∇F. Put $\mathcal{V} := \mathcal{U} \cup \bigcup_{u \in \partial \mathcal{U}} \mathcal{B}^{\perp}(u, \varepsilon)$, where $\mathcal{U}$ is the open bounded set introduced above for which $\gamma_+(p) \subset \overline{\mathcal{U}}$. Then the smoothness of $F(\varphi^{(\cdot)}(\cdot))$ guarantees $\sup_{u \in \overline{\mathcal{V}}} \lambda_1(u) < \infty$. Hence we obtain

$$\sup_{\substack{\tau \in [0,\varepsilon] \\ u \in \overline{\mathcal{V}}}} \int_0^{\tau} \lambda_1(\varphi^t(u))dt < \infty,$$

which implies

$$\sup_{\substack{\tau \in [0,\varepsilon] \\ u \in \overline{\mathcal{V}}}} |Y(\tau, u)| < \infty.$$

This, together with the assumption $F(u) \neq O_u$ for all $u \in \overline{\mathcal{U}}$ and relations (7.27) and (7.44), give the existence of a constant $C_s > 0$ such that

$$\sup_{\substack{t \geq 0 \\ u(r,\upsilon) \in \mathcal{B}^{\perp}(p,\delta)}} \left| \frac{\partial s}{\partial r}(t, r, \upsilon) \right| \leq \sup_{\substack{\widetilde{p} \in \gamma(p) \\ \widetilde{u}(\widetilde{r},\widetilde{\upsilon}) \in \mathcal{B}^{\perp}(\widetilde{p},\varepsilon)}} \left| \frac{\partial s}{\partial r}(0, \widetilde{r}, \widetilde{\upsilon}) \right| < C_s \tag{7.42}$$

is satisfied.

For an arbitrary $u(r, \upsilon) \in \mathcal{B}^{\perp}(p, \delta)$ we consider the solution $z(\cdot, \tau_p^u(\upsilon))$ of (7.17) with respect to $\varphi^{(\cdot)}(u)$ and initial condition $z(0, \tau_p^u(\upsilon)) = \tau_p^u(\upsilon)$. Since $\gamma(p)$ is stable, we can expand $z(s(t, r, \upsilon), \tau_p^u(\upsilon))$ by Taylor's formula to obtain

$$z(s(t, r, \upsilon), \tau_p^u(\upsilon)) = \tau_{\varphi^t(p)}^{\varphi^s(u)} \left\{ z(t) + A(\varphi^t(p))z(t)\frac{\partial s}{\partial r}(t, 0, \upsilon)r + O(|z(t)|r^2) \right\}, \tag{7.43}$$

where $z(t)$ is the solution of (7.17) with respect to $\varphi^{(\cdot)}(p)$ having $z(0) = \upsilon$. By (7.35) and the definition of the operator A the inequality $\sup_{u \in \overline{U}} |A(u)| < \infty$ holds. This fact, (7.42) and (7.43) imply that

$$|z(s(t, r, \upsilon), \tau_p^u(\upsilon))| = |z(t)|[1 + O(r)] \tag{7.44}$$

holds for any $u(r, \upsilon) \in \mathcal{B}^{\perp}(p, \delta)$ and all $t \geq 0$ in tangent space $T_{\varphi^s(u)}\mathcal{M}$.

Step 4. Let us fix now a time $t \geq 0$ and $\upsilon \in T^{\perp}(p)$ with $|\upsilon| = 1$ and consider the C^l-curve $\beta(\cdot) := \phi(t, \cdot, \upsilon)$ on $\mathcal{M}$ and the lifted curve $w(\cdot) := \exp_{\varphi^t(p)}^{-1}(\phi(t, \cdot, \upsilon))$ in the tangent bundle. By the properties of the exponential map for any $r \in [0, \delta)$ the tangent vectors $\beta'(r) = D_2\phi(t, r, \upsilon) \in T_{\beta(r)}\mathcal{M}$ and $w'(r) \in T_{\varphi^t(p)}\mathcal{M}$ are related by parallel transport between the tangent spaces, i.e. $w'(r) = \tau_{\beta(r)}^{\varphi^t(p)}(\beta'(r))$. Thus, relations (7.28), (7.43), (7.44) and the fact that $\exp_{\varphi^t(p)}^{-1}(\phi(t, 0, p)) = O_{\varphi^t(p)}$ yield

$$
\begin{aligned}
|\exp^{-1}_{\varphi^t(p)}(\phi(t,r,v))| &= \left| \int_0^r \tau^{\varphi^t(p)}_{\varphi^{\widetilde{s}}(\widetilde{u})}(D_2\phi(t,\widetilde{r},v))d\widetilde{r} \right| \\
&= \left| \int_0^r \tau^{\varphi^t(p)}_{\varphi^{\widetilde{s}}(\widetilde{u})}(z(\widetilde{s},\tau^{\widetilde{u}}_p(v)))d\widetilde{r} \right| |z(t)| r\,(1+O(r))
\end{aligned}
\tag{7.45}
$$

for all $u(r,v) \in \mathcal{B}^{\perp}(p,\delta)$ and all $t \geq 0$, where for brevity we have written $\widetilde{s} = s(t,\widetilde{r},v)$ and $\widetilde{u} = u(\widetilde{r},v)$. From (7.45) and (7.40) it follows immediately $|\exp^{-1}_{\varphi^t(p)}(\phi(t,r,v))| \to 0$ for $t \to +\infty$ for arbitrary $u(r,v) \in \mathcal{B}^{\perp}(p,\delta)$. Since by definition of ϕ the inequality

$$
\operatorname{dist}\left(\varphi^{s(t,r,v)}(u(r,v)),\gamma(p)\right) \leq |\exp^{-1}_{\varphi^t(p)}(\phi(t,r,v))|
$$

holds for all $t \geq 0$ and since from the validity of relation (7.25) we have the stability of the orbit $\gamma(p)$, the convergence of $|\exp^{-1}_{\varphi^t(p)}(\phi(t,r,v))|$ to zero establishes the asymptotic stability of $\gamma(p)$.

Using the property of Lyapunov instability we can now formulate the following.

Corollary 7.1 *Consider equation (7.12) and $p \in \mathcal{M}$. Then holds:*

(a) If $\gamma(p) \in BO_+$ and the zero solution of the system (7.17) with A from (7.18) is exponentially stable, then the orbit $\gamma(p)$ is asymptotically stable.

(b) If $\varphi^{(\cdot)}(p)$ is a periodic solution of (7.12) and the zero solution of (7.17) is Lyapunov unstable, then the orbit $\gamma(p)$ is unstable.

Proof From the proof of Theorem 7.2 the part (a) of this corollary follows directly. Let us prove the part (b). Using assertion (a) of Lemma 7.8 we find a number $\varepsilon > 0$ such that the bundle of parametrizations $s(\cdot,\cdot,\cdot) : \mathbb{R}_+ \times [0,\varepsilon] \times T^{\perp}(p) \cap \mathcal{B}(O_p,1)) \to \mathbb{R}_+$ exists. Let $\delta > 0$ be chosen as in the proof of Theorem 7.2 and let $\mathcal{S}_0 := \left\{\varphi(s(0,r,v),u) | u(r,v) \in \mathcal{B}^{\perp}(p,\delta)\right\}$ be the orthogonal section of $\gamma(p)$ through p. For $0 < \varepsilon_0 < \min(\varepsilon,\delta)$ we consider $(n-1)$-dimensional submanifolds $\mathcal{U} := \{\varphi(s(0,r,v),u) | u = u(r,v) \in \mathcal{B}^{\perp}(p,\varepsilon_0)\}$ and $\mathcal{V} := \{\varphi(s(T,r,v),u) | u = u(r,v) \in \mathcal{B}^{\perp}(p,\varepsilon_0)\}$ of M. Obviously $p \in \mathcal{U}$ and $\mathcal{U} \subset \mathcal{S}_0$. Using the periodicity of $\varphi^{(\cdot)}(p)$ we further have $p \in \mathcal{V}$ and with the relationship (7.24) for the reparametrized flow $\phi(t,r,v)$ as defined in the proof of Theorem 7.2 $\phi(T,r,v) \in \mathcal{S}_0$ for all $u \in \mathcal{U}$. This implies $\mathcal{V} \subset \mathcal{S}_0$.

We now introduce the injective map $h : \mathcal{U} \to \mathcal{B}^{\perp}(0,\varepsilon_0)$ given by $h(u) = (r,v)$ and define a Poincaré map $P : \mathcal{U} \to \mathcal{V}$ for $\gamma(p)$ with respect to the orthogonal section $\mathcal{S}_0$ by

$$
P(u) = \phi(T,h(u)).
$$

The properties of the parametrization $s(\cdot,\cdot,\cdot)$ guarantee that P is indeed a Poincaré map. By construction it follows for the differential of P at p

$$
d_p P = Z(T,p),
\tag{7.46}
$$

where $Z(T, p) : T^{\perp}(p) \to T^{\perp}(p)$ denotes the fundamental operator solution of (7.17) at time $t = T$. The assumption of the corollary is equivalent to the existence of an unbounded solution of system (7.17), say, $\overline{z}(\cdot)$. We start from the contrary and suppose that $\gamma(p)$ is stable. Then we find an open set $\mathcal{U}_0 \subset \mathcal{U} \cap \mathcal{V}$ such that all iterates $P^k, k \in \mathbb{N}$, of the Poincaré map $P : \mathcal{U} \to \mathcal{V}$ exist on $\mathcal{U}_0$. For sufficiently small $\delta_0 > 0$ we introduce the map $h_0 : \mathcal{B}^{\perp}(p, \delta) \to \mathcal{U}_0$ given by $h_0(u(r, \upsilon)) := \phi(0, r, \upsilon)$. Then by the properties of the parametrization $s(\cdot, \cdot, \cdot)$ this map is a diffeomorphism. We introduce further a map $q_k : [0, \delta) \to T_p\mathcal{S}_0 = T^{\perp}(p)$ given by

$$q_k(r) := \exp_p^{-1}(P^k(h_0(u(r, \upsilon)))),$$

where $r \in [0, \delta)$ and $\upsilon \in T^{\perp}(p), |\upsilon| = 1$, are such that $u = u(r, \upsilon) = \exp_p(r\upsilon)$. Using Taylor's formula and $q_k(0) = 0_{T_p\mathcal{S}}$ we obtain

$$|q_k(r)| = |q_k'(0) + o(r)|r = (|q_k'(0)| + o(r))r.$$

By definition it holds $q_k'(0) = d_p P^k \frac{\partial}{\partial r} h_0(u(r, \upsilon))_{|r=0}$. With $\frac{\partial}{\partial r} h_0(u(r, \upsilon))_{|r=0} = D_2\phi(0, 0, \upsilon) = \upsilon$ we have $q_k'(0) = T_p P^k \upsilon$. Thus

$$|q_k(r)| = (|d_p P^k \upsilon| + o(r))r,$$

for arbitrary $k \in \mathbb{N}$. Take now $r \in [0, \delta)$ fixed and $\upsilon := \overline{z}(0)$. Using the abbreviation $\widetilde{u} := h_0(u(r, \upsilon))$ we get with (7.46)

$$|q_k(r)| = (|\overline{z}(kT)| + o(r))r.$$

But the solution $\overline{z}(\cdot)$ of (7.17) is unbounded, therefore the last contradicts the assumed stability of $\gamma(p)$. □

Remark 7.8 For a T-periodic orbit $\gamma(p)$ the $n - 1$ multipliers different from the one which is the eigenvalue corresponding to $F(p)$ coincide with the eigenvalues of the fundamental operator solution $Z(T, p) : T_p^{\perp}\mathcal{M} \to T_p^{\perp}\mathcal{M}$. Obviously, in this case the assumption of the first part of Corollary 7.1 is an equivalent formulation of the Andronov-Vitt condition [2, 6].

We now propose a condition [21] sufficient for the exponential stability of the zero solution of (7.17). For any $u \in \mathcal{M}$ denote by $\lambda_1(u) \geq \cdots \geq \lambda_n(u)$ the eigenvalues of the symmetric part of the covariant derivative $S\nabla F(u)$ ordered with respect to their size and multiplicity.

Theorem 7.3 *If for an orbit $\gamma(p)$ of (7.12) with $\gamma(p) \in BO_+$ there exists a number $\kappa > 0$ and a sequence $\{t_j\}_{j\in\mathbb{N}}$ of positive numbers satisfying $\lim_{j\to+\infty} t_j = \infty$, $0 < t_{j+1} - t_j \leq \kappa$ for $j = 1, 2, \ldots$ and such that*

$$\sup_{j\in\mathbb{N}} \frac{1}{j} \int_0^{t_j} \left[\lambda_1(\varphi^t(p)) + \lambda_2(\varphi^t(p))\right] dt < 0, \tag{7.47}$$

then the orbit $\gamma(p)$ is asymptotically stable.

Proof We distinguish between the case where the orbit is attracted by an equilibrium point in $\overline{\mathcal{U}}$ and the case where it is not, and handle the first case completely analogously as in the proof of Theorem 7.2. In the other case, where relation (7.35) is valid, the first steps are also similar to the proof of Theorem 7.2. Consider the solution $w(\cdot) := y(\cdot) \wedge F(\varphi^{(\cdot)}(p))$ of (7.23), where $y(\cdot)$ is a solution of (7.13) with $|y(0) \wedge F(p)| \neq 0$ and with properties as in the proof of Theorem 7.2. Let us couple $y(\cdot)$ with a solution $z(\cdot)$ of (7.17) by means of $z(t) = y(t) + \mu(t)F(\varphi^t(p))$. Then we have already shown that $|w(t)| = |z(t)| \cdot |F(\varphi^t(p))| > 0$ for $t \geq 0$. A straightforward calculation shows that for all $t \geq 0$

$$\frac{d}{dt}|w(t)|^2 = 2((S\nabla F(\varphi^t(p)))_2 \upsilon(t), \upsilon(t))|w(t)|^2,$$

where $\upsilon(t) = w(t)/|w(t)|$. Integrating both sides of the last equality on $[0, t]$ leads to

$$|w(t)| = |w(0)| \exp\left\{ \int_0^t ((S\nabla F(\varphi^\tau(p)))_2 \upsilon(\tau), \upsilon(\tau)) d\tau \right\}.$$

This gives

$$|z(t)| \leq |z(0)| \exp\left\{ \int_0^t \left[\lambda_1(\varphi^\tau(p)) + \lambda_2(\varphi^\tau(p))\right] d\tau \right\} \frac{|F(p)|}{|F(\varphi^t(p))|} \tag{7.48}$$

for arbitrary $t \geq 0$. Since this holds for an arbitrary non-zero solution of (7.17) and $\sup_{u,u'\in\gamma_+(p)} \frac{|F(u)|}{|F(u')|} < C_1$ is satisfied with the constant C_1, we conclude that the operator norm of $Z(t, p) : T^\perp(p) \to T^\perp(\varphi^t(p))$, i.e. of the fundamental operator solution of system (7.17), satisfies

$$|Z(t, p)| \leq C_1 \exp\left\{ \int_0^t \left[\lambda_1(\varphi^\tau(p)) + \lambda_2(\varphi^\tau(p))\right] d\tau \right\}. \tag{7.49}$$

The assumption of the theorem implies that there exist an index $j_0 \in \mathbb{N}$ and some number $\varepsilon > 0$ such that for all times t_j with $j \geq j_0$

$$\int_0^{t_j} \left[\lambda_1(\varphi^\tau(p)) + \lambda_2(\varphi^\tau(p))\right] d\tau < -j\varepsilon. \tag{7.50}$$

Let $t \geq t_{j_0}$ be arbitrary. Then there is an index $j \geq j_0$ such that $t \in [t_j, t_{j+1}]$. Since F is C^l-smooth, and fulfills the above condition and since $\gamma_+(p)$ is bounded and $|t_{j+1} - t_j| < \kappa$ we have $\sup_{u \in \gamma_+(p)} \int_0^\kappa \left|\lambda_1(\varphi^t(u)) + \lambda_2(\varphi^t(u))\right| dt < +\infty$. Using (7.49), (7.50), the relations $t_j \leq j\kappa$ and $t < t_j + \kappa$, finally we find a constant $C > 0$ such that

$$|Z(t, p)| \leq Ce^{-\frac{\varepsilon}{\kappa}t} \tag{7.51}$$

is valid for $t \geq 0$. Thus, the zero solution of (7.17) is exponentially stable. Applying Corollary 7.1 provides the assertion of Theorem 7.3. □

Remark 7.9 If $\gamma(p) \in BO_+$ is a T-periodic orbit of (7.12) then condition (7.47) takes the form

$$\int_0^T [\lambda_1(\varphi^t(p)) + \lambda_2(\varphi^t(p))]dt < 0,$$

which guarantees the asymptotic stability of the orbit. If we introduce on $\mathcal{M}$ a new metric tensor by $\widetilde{g}_{|u} := \rho^2(u) g_{|u}$ by means of $\rho(u) := e^{\frac{V(u)}{2}}$, where V is a smooth function on $\mathcal{M}$, the eigenvalues $\widetilde{\lambda}_i$ of the symmetric part of $\widetilde{\nabla} F(u)$ (the covariant derivative with respect to the new metric tensor) are $\widetilde{\lambda}_i = \lambda_i + \frac{\dot{V}}{2}, i = 1, 2, \ldots, n$, where $\dot{V} \equiv \mathcal{L}_F V$ is the Lie derivative of V in direction of the vector field F. Thus $\widetilde{\lambda}_1 + \widetilde{\lambda}_2 = \lambda_1 + \lambda_2 + \dot{V}$. It follows that the condition

$$\int_0^T [\lambda_1(\varphi^t(p)) + \lambda_2(\varphi^t(p)) + \dot{V}(\varphi^t(p))]dt < 0$$

is sufficient for the asymptotic stability of the T-periodic orbit $\gamma(p)$. The last condition is often used for $\mathcal{M} = \mathbb{R}^n$ in investigations with Lyapunov functions of orbital or Zhukovskii stability of periodic solutions (see [19, 32]).

Example 7.3 Consider the equation

$$\dot{u} = f(u) \tag{7.52}$$

with a C^l-vector field $f : \mathbb{R}^n \to \mathbb{R}^n (l > 1)$. Suppose that the flow $\varphi^{(\cdot)}(\cdot) : \mathbb{R} \times \mathbb{R}^n \to \mathbb{R}^n$ of (7.52) exists. We introduce the subgroup of $\mathbb{R}^n$

$$\Gamma := \left\{ \sum_{i=1}^m k_i d_i | k_i \in \mathbb{Z} \right\}, \tag{7.53}$$

where $m \leq n$, $\{d_i\}_{i=1}^m$ are linearly independent vectors in $\mathbb{R}^n$. The vector field (7.52) is supposed to have the property $f(u + \vartheta) = f(u)$ for every $\vartheta \in \Gamma$ and every $u \in \mathbb{R}^n$. In this case system (7.52) is called *pendulum-like* with respect to Γ [23].

Note that a broad class of differential equations with angular coordinates in mechanics, phase-synchronization and other fields can be considered as pendulum-like systems. For the flow of (7.52) we have then the *equivariance propery* $\varphi^t(u + \vartheta) = \varphi^t(u) + \vartheta$ for every $u \in \mathbb{R}^n$, $t \in \mathbb{R}$ and $\vartheta \in \Gamma$. Therefore, system (7.52) can be interpreted as vector field F on the flat cylinder $\mathbb{R}^n/\Gamma (m < n)$ or on the flat torus $(m = n)$, respectively (see Sect. A.1, Appendix A).

A solution $\varphi^{(\cdot)}(p)$ of (7.52) is called a *cycle of the second kind* if there exists a number $T > 0$ and some $\vartheta \in \Gamma\backslash\{0\}$, such that $\varphi^T(p) = p + \vartheta$. We call the minimal positive number T with this property the *period* of $\varphi^{(\cdot)}(p)$. Note that any cycle of the second kind of (7.52) becomes closed for the vector field $\widetilde{F}$ on the cylinder or torus.

We will now consider a system (7.52) in $\mathbb{R}^2$ of the form

$$\dot{x}_1 = a \ \sin x_1 + b, \qquad \dot{x}_2 = -x_2 \tag{7.54}$$

with parameters $b > a > 1$. Since the right-hand side f of (7.54) is 2π-periodic with respect to the coordinate x_1, the system can be interpreted as vector field on the flat cylinder $\mathbb{R}^2/\Gamma$ with respect to the subgroup $\Gamma := \{k(2\pi, 0)|k \in \mathbb{Z}\}$ of $\mathbb{R}^2$.

On account of the assumption $b > a$ the circle $\{(x_1, 0)|x_1 \mod 2\pi\}$ coincides with a closed orbit of system (7.54) on $\mathbb{R}^2/\Gamma$, which we denote by $\overline{\gamma}$. Consider a solution $\overline{\varphi}(\cdot)$ of (7.54) which corresponds to $\overline{\gamma}$. Then $\overline{\varphi}(\cdot)$ is of the form $\overline{\varphi}(t) = (\overline{x}_1(t), 0)$. The variational system (7.13) with respect to $\overline{\varphi}(\cdot)$ is given by

$$\dot{y}_1 = a \ \cos \overline{x}_1 \cdot y_1, \qquad \dot{y}_2 = -y_2. \tag{7.55}$$

Since $f(\overline{\varphi}(t)) = (a \sin \overline{x}_1(t) + b, 0)$, the vectors in $T^{\perp}((\overline{x}_1, 0))$ are of the form $z = (0, x)$ with $x \in \mathbb{R}$. Observe that the differential equations in system (7.54) for the tangent and the orthogonal directions are not coupled along $\overline{\gamma}$. Thus, the restriction of system (7.17) to $T^{\perp}((\overline{x}_1(t), 0))$ can simply be written as differential equation for x, given by

$$\dot{x} = -x\,. \tag{7.56}$$

It is obvious that the trivial solution of the system in normal variations (7.56) is exponentially stable and Corollary 7.1 provides that the orbit $\overline{\gamma}$ is asymptotically stable.

Example 7.4 Consider the geodesic flow (see Sect. A.8, Appendix A) on a compact Riemannian manifold $(\mathcal{M}, g)$ of dimension n and of class C^m $(m \geq 3)$. The geodesics $c_{p,\upsilon}(\cdot)$ are obtained as solutions of the second order differential equation

$$\frac{D\dot{c}_{p,\upsilon}}{dt}(t) = 0 \tag{7.57}$$

with initial conditions $c_{p,\upsilon}(0) = p \in \mathcal{M}$ and $\dot{c}_{p,\upsilon}(0) = \upsilon \in T_p\mathcal{M}$, which is given in local coordinates of a chart x around some initial p by

$$\ddot{x}^k + \Gamma^k_{ij}\dot{x}^i\dot{x}^j = 0. \tag{7.58}$$

The geodesic flow is given by the map $\phi : \mathbb{R} \times T\mathcal{M} \to T\mathcal{M}$ with $\phi(t, (p, \upsilon)) = \dot{c}_{p,\upsilon}(t)$. We use the notation $\phi^t(p, \upsilon) \equiv \phi(t, (p, \upsilon))$. System (7.58) is equivalent to the first order system (7.12) on the product manifold $\mathcal{M} \times T\mathcal{M}$ given by

$$\dot{u} = \upsilon \ , \quad \dot{\upsilon} = G\,(u, \upsilon), \tag{7.59}$$

where $G : \mathcal{M} \times T\mathcal{M} \to TT\mathcal{M}$ is determined in local coordinates of the chart x through (7.58) by $\dot{\upsilon}^i = -\Gamma^i_{jk}\upsilon^j\upsilon^k$ and υ^j are the coordinates of υ.

The tangent space $T_\upsilon T\mathcal{M}$ at any $\upsilon \in T_p\mathcal{M}$ is isomorphic to $T_p\mathcal{M} \times T_p\mathcal{M}$, so that $w \in T_\upsilon T\mathcal{M}$ can be identified with a pair $(w_1, w_2) \in T_p\mathcal{M} \times T_p\mathcal{M}$. Now we consider the splitting of $w \in T_\upsilon T\mathcal{M}$ into parts $d\pi(w)$ and Cw, where $d\pi : T_\upsilon T\mathcal{M} \to T_p\mathcal{M}$ denotes the differential of the projection map $\pi : T\mathcal{M} \to \mathcal{M}, \pi((p, \upsilon)) = p$, defined by $d\pi(w) = w_1$ and C denotes the connection map given in local coordinates of some chart x around p by $(Cw)^i = w_2^i + \Gamma^i_{jk} w_1^j \upsilon^k$.

A *Jacobi field* $y(\cdot) : t \mapsto T_{c(t)}\mathcal{M}$ along a geodesic $c(t)$ is a solution of the Jacobi equation

$$\frac{D^2 y}{dt^2} + K(t)y = 0, \tag{7.60}$$

where $K(t)\upsilon = R(\upsilon, \dot{c}(t))\dot{c}(t)$ and R denotes the Riemannian curvature tensor (see Sect. A.9, Appendix A). If we identify the differential $d_{p,\upsilon}\phi^t : T_\upsilon T\mathcal{M} \to T_{\phi^t(p,\upsilon)}T\mathcal{M}$ at any $w \in T_\upsilon T\mathcal{M}$ with the pair $(d\pi(d_{p,\upsilon}\phi^t w), Cd_{p,\upsilon}\phi^t w)$ then the Jacobi equation, written as a first order system on $T\mathcal{M} \times T\mathcal{M}$, is the variational system (7.13) with respect to an integral curve $(c_{p,\upsilon}(\cdot), \phi^{(\cdot)}(p, \upsilon))$ of system (7.60) (see [7, 16, 30]).

Let $y(\cdot)$ be a Jacobi field along some geodesic $c(\cdot)$. Then $y(\cdot)$ can be split into a tangent part y^T in direction of $\dot{c}$ and a part $y^\perp = y - y^T \in T^\perp(c(t))$. Both $y^T(\cdot)$ and $y^\perp(\cdot)$ are Jacobi fields along $c(\cdot)$ and further $\frac{D^2 y^T}{dt^2} = 0$ (see [16]) and $(\frac{Dy^\perp}{dt}, \dot{c}) = 0$. This gives that the integral curve

$$z(t) = (z_1(t), z_2(t)) := (y^\perp(t), \frac{Dy^\perp}{dt}(t))$$

is the solution of the system in normal variations (7.17) of system (7.60) with respect to $(c_{p,\upsilon}(\cdot), \phi^{(\cdot)}(p, \upsilon))$

$$\frac{Dz_1}{dt} = z_2(t), \quad \frac{Dz_2}{dt} = -K(t)z_1. \tag{7.61}$$

Now we want to investigate the stability behaviour of closed geodesics using Corollary 7.1. On the base of our general criterion we show the orbital instability of closed geodesics on negatively curved manifolds, which is of course not surprising because of the hyperbolic and expansive character of such a geodesic flow [7, 30].

Since the geodesics are curves of constant velocity we can restrict our investigation to the geodesic flow on the unit tangent bundle $S\mathcal{M} := \{(u, \upsilon) \in T\mathcal{M} \,|\, u \in \mathcal{M}, \upsilon \in$

$T_u\mathcal{M}, |v| = 1\}$. Let $S_u\mathcal{M}$ denote the unit tangent space at the point $u \in \mathcal{M}$. Fix an arbitrary geodesic $c_{p,v}(\cdot)$ with initial velocity $v \in S_p$.

Let us indicate that an appropriate way to write system (7.61) in the bundle of transversal sections is to use Fermi coordinates along $c_{p,v}(\cdot)$. This means to take an orthonormal base $\{e_i(0)\}_{i=1}^{n-1}$ of $T^{\perp}(c_{p,v}(0))$ and to generate an orthonormal base $\{e_i(t)\}_{i=1}^{n-1}$ in $T^{\perp}(c_{p,v}(t))$ by parallel transport along $c_{p,v}(\cdot)$. To define a base in $T_{c_{p,v}(t)}\mathcal{M}$ the tupel $\{e_i(t)\}_{i=1}^{n-1}$ is accomplished by $\dot{c}_{p,v}(t)$ for any time $t \in \mathbb{R}$. The construction implies that $\frac{De_i}{dt} = 0$ for $i = 1, \ldots, n-1$.

An arbitrary Jacobi field with initial condition in $T^{\perp}(v) \times T^{\perp}(v)$ can be written as $y(t) = \nu^i(t)e_i(t)$ with C^{m-1}-functions $\nu^i : \mathbb{R} \to \mathbb{R}$ and we obtain $\frac{Dy}{dt}(t) = \dot{\nu}^i(t)e_i(t)$ and $\frac{D^2y}{dt^2}(t) = \ddot{\nu}^i e_i(t)$. Hence, replacing the expressions in (7.61) leads to

$$\ddot{\nu}^i + K^i_j \nu^j = 0, \tag{7.62}$$

where $K^i_j := (R(e_i, \dot{c})\dot{c}, e_j)$. We define in $T_vT\mathcal{M}$ the norm

$$|w| = \sqrt{|d\pi(w)|_{T_pM} + |Cw|_{T_pM}}.$$

Suppose now that the sectional curvature on $\mathcal{M}$ is negative, say, bounded by a constant $-k^2 < 0$. Note, that for $v \in S_p\mathcal{M}$ and any Jacobi field $y(\cdot)$ along $c_{p,v}(\cdot)$ holds

$$(K(t)y(t), y(t)) \leq -k^2(y(t), y(t))$$

for all $t \geq 0$. Let $\mathcal{K}_\delta$ denote the cone, generated by all vectors $z \in TT\mathcal{M}$ written as $z = (z_1, z_2) = (d\pi(z), Cz)$ which satisfy $\frac{(z_1, z_2)_{T_pM}}{|z|} \geq \delta$. A result from [16] says, that in this situation any such cone $\mathcal{K}_\delta$ is invariant if $\delta > 0$ is sufficiently small and further, that there exists a constant $\varepsilon_0 > 0$, for which each solution $z(\cdot)$ of (7.61) with $z(0) \in \mathcal{K}_\delta$ satisfies

$$\frac{d}{dt}|z(t)| > \varepsilon_0 |z(t)|.$$

This implies $|z(t)| \geq |z(0)|e^{\varepsilon_0 t}$ for $t \geq 0$ and establishes the Lyapunov instability of the trivial solution $z \equiv 0_{T_vT\mathcal{M}}$ of (7.61). Finally, the application of the second part of Corollary 7.1, (b) provides that all periodic solutions $(c_{p,v}(\cdot), \phi^{(\cdot)}(p, v))$ are orbitally unstable and, therefore, all closed geodesics with unit speed are orbitally unstable.

7.2.4 *Characteristic Exponents*

Let us consider again on the Riemannian manifold $(\mathcal{M}, g)$ the vector field $f : \mathcal{M} \to T\mathcal{M}$ generating (7.12) and the associated variational equation (7.13) along a solution

$\varphi^{(\cdot)}(p)^{(\cdot)}(u)$ with the operator solution $Y(\cdot, u)$ normed at $t = 0$. For such a point $u \in \mathcal{M}$ and an arbitrary vector $v \in T_u\mathcal{M}$, $v \neq 0_{T_u\mathcal{M}}$, the number

$$\chi(u, v) = \limsup_{t\to\infty} \frac{1}{t} \ln |Y(t, u)v| \tag{7.63}$$

is called the *characteristic exponent of* (7.12) (or *of the flow* $\{\varphi^{(\cdot)}(p)^t\}$) *at* $u \in \mathcal{M}$ *in direction* v (see e.g. [30]). Furthermore, we need for $k = 1, \dots, n$ the singular values of $Y^{\wedge k}(t, u)$, *i.e.*, the square roots of the eigenvalues of the positive definite linear operator $(Y^{\wedge k}(t, u))^* Y^{\wedge k}(t, u) : \bigwedge^k T_u\mathcal{M} \to \bigwedge^k T_u\mathcal{M}$. These singular values we denote by $\alpha_1^{(k)}(t, u) \geq \dots \geq \alpha_{\binom{n}{k}}^{(k)}(t, u)$, whereas for the singular values of $Y(t, u)$ we simply write $\alpha_1(t, u) \geq \dots \geq \alpha_n(t, u)$.

We continue by listing some well-known properties of the characteristic exponents (see [29, 30] and, for the case of $\mathbb{R}^n$, [4]).

Lemma 7.9 *Let $u \in \mathcal{M}$ be some point such that the corresponding semi-orbit $\gamma_+(u)$ of (7.12) is bounded. Then the following holds:*

(1) $-\infty < \chi(u, v) < \infty$, *for all $v \in T_u\mathcal{M}$ with $v \neq 0_{T_u\mathcal{M}}$;*
(2) *$\chi(u, \cdot)$ possesses at most n different values, which we denote by $\widetilde{\chi}_1(u) > \dots > \widetilde{\chi}_{N(u)}(u)$ according to size, where $0 < N(u) \leq n$;*
(3) *provided that in a neighborhood of $\gamma_+(u)$ no equilibrium point of (7.12) is contained, then at least one value $\widetilde{\chi}_i(u)$ vanishes.*

Proof Here we only give a proof of 3) in order to show briefly in which direction the characteristic exponent is equal to zero. It appears that in the vector field direction $f(u)$ at $u \in \mathcal{M}$ the characteristic exponent always vanishes. Indeed, by the assumption that $\gamma_+(u) \subset \overline{\mathcal{U}}$ for some open bounded set $\mathcal{U} \subset \mathcal{M}$ there exists a constant $C > 0$ such that $\sup_{p\in\overline{\mathcal{U}}} |f(p)| < C$ holds and since $f(\varphi^{(\cdot)}(p)(\cdot, u))$ is a solution of (7.13) by definition (7.63) we have

$$\chi(u, f(u)) = \limsup_{t\to\infty} \frac{1}{t} \ln |f(\varphi^{(\cdot)}(p)^t(u))| \leq \lim_{t\to\infty} \frac{1}{t} \ln C = 0.$$

Since the additional assumption of 3) implies $\inf_{t\geq 0} |f(\varphi^{(\cdot)}(p)^t(u))| > 0$, we obtain by an analogous argumentation that $\chi(u, f(u)) \geq 0$. □

We call $\widetilde{\chi}_1(u) > \dots > \widetilde{\chi}_{N(u)}(u)$ the *characteristic exponents of* (7.12) *in u* (or *of the semi-orbit* $\gamma_+(u)$.)

More in general, for an arbitrary $u \in \mathcal{M}, k = 1, \dots, n$ and arbitrary $w \in \bigwedge^k T_u\mathcal{M}$ we define the *characteristic exponents of* (7.12) *of order k at $u \in \mathcal{M}$ in direction* $w \neq 0_{\bigwedge^k T_u\mathcal{M}}$ (see [29]) by

$$\chi^{(k)}(u, w) = \limsup_{t\to\infty} \frac{1}{t} \ln |Y^{\wedge k}(t, u)w|. \tag{7.64}$$

Obviously, the above definition (7.63) coincides with (7.64) for the case $k = 1$, so $\chi_i(u, v) = \chi_i^{(1)}(u, v)$ are, in fact, the characteristic exponents of first order. From Lemma 7.9 we obtain analogous properties 1) and 2) for the characteristic exponents of k-th order. This is due to the fact that we only replace the operator solution $Y(t, u)$ of (7.13) normed at $t = 0$ operating in the n-dimensional linear space $T_{\varphi^{(\cdot)}(p)^t(u)}\mathcal{M}$ by the operator solution $Y^{\wedge k}(t, u)$ of (7.21) normed at $t = 0$ operating in the linear space $\bigwedge^k T_{\varphi^{(\cdot)}(p)^t(u)}\mathcal{M}$ of dimension $\binom{n}{k}$. Hence there are at most $\binom{n}{k}$ different values of $\chi^{(k)}(u, \cdot)$ for every $k = 1, \ldots, n$. We denote them by $\widetilde{\chi}_1^{(k)}(u) > \ldots > \widetilde{\chi}_{N(k,u)}^{(k)}$ ordered with respect to size, where $0 < N(k, u) \le \binom{n}{k}$, and call them *characteristic exponents of* (7.12) *of order* k *at* u *(or of the semi-orbit* $\gamma_+(u)$*)*. In particular, for $k = n$, there is exactly one value $\chi^{(n)}(u, w) = \widetilde{\chi}_1^{(n)}(u)$ for any nonzero $w \in \bigwedge^n T_u\mathcal{M}$ and by definition (7.64) we have

$$\widetilde{\chi}_1^{(n)}(u) = \limsup_{t\to\infty} \frac{1}{t} \ln |\det Y(t, u)|. \tag{7.65}$$

The next lemma indicates the relationship between the characteristic exponents of order k and the singular values of $Y^{\wedge k}(t, u)$, which will prove later to be of great importance in analyzing the stability behavior of the underlying dynamical system.

Lemma 7.10 *For arbitrary* $u \in \mathcal{M}$ *and* $k \in \{1, \ldots, n\}$

(1) $\widetilde{\chi}_1^{(k)}(u) = \limsup_{t\to\infty} \frac{1}{t} \ln \alpha_1^{(k)}(t, u)$ *and*

(2) for all nonzero $w = w_1 \wedge \ldots \wedge w_k \in \bigwedge^k T_u\mathcal{M}$ *holds*

$$\chi^{(k)}(u, w) \le \chi(u, w_1) + \cdots + \chi(u, w_k). \tag{7.66}$$

Proof We prove assertion (1) for $k = 1$. (For arbitrary k one proceeds analogously.) Let $\{y_i\}$ be an orthonormal base with respect to $(\cdot, \cdot)_{T_u\mathcal{M}}$ of $T_u\mathcal{M}$. Since $\widetilde{\chi}_1(u)$ is the largest characteristic exponent at u in any direction the relation

$$\chi(u, y_i) \le \widetilde{\chi}_1(u) \tag{7.67}$$

is valid for each $i \in \{1, \ldots, n\}$. Let us for any $t \ge 0$ denote by $v_1(t)$ an eigenvector of $Y(t, u)^*Y(t, u)$ corresponding to the eigenvalue $\alpha_1^2(t, u)$, normalized by $|v_1(t)| = 1$. Then there exist functions $a_i(t)$ such that

$$v_1(t) = \sum_{i=1}^{n} a_i(t) y_i$$

and, consequently, $|a_i(t)| \le 1$ for all $t \ge 0$. Since

$$\alpha_1(t, u) = |Y(t, u)v_1(t)| \le \sum_{i=1}^{n} |a_i(t)||Y(t, u)y_i|,$$

we find for any $t \geq 0$ an index $k = k(t) \in \{1, \ldots, n\}$ such that

$$|Y(t, u)y_k| \geq \frac{1}{n}\alpha_1(t, u). \tag{7.68}$$

It follows that for any sequence $t_i \to \infty$ there is a subsequence which we also denote by $\{t_i\}_i$ and a fixed index $j \in \{1, \ldots, n\}$ such that

$$|Y(t_i, u)y_j| \geq \frac{1}{n}\alpha_1(t_i, u) \quad \text{for } i = 1, 2, \ldots \tag{7.69}$$

is satisfied. From this we obtain for all i

$$\frac{1}{t_i} \ln |Y(t_i)y_j| \geq \frac{1}{t_i} \ln \alpha_1(t_i, u) - \frac{1}{t_i} \ln n,$$

so according to (7.63) and (7.67) we have

$$\widetilde{\chi}_1(u) \geq \chi(u, y_j) \geq \limsup_{t\to\infty} \frac{1}{t} \ln \alpha_1(t, u).$$

The opposite inequality is obvious since $|Y(t, u)v| \leq |Y(t, u)| = \alpha_1(t, u)$ holds for all $v \in T_u\mathcal{M}$ and all times $t \geq 0$. The statement 2) follows immediately from the fact that the supremum over a sum is always less than or equal to the sum over the suprema. □

Now we want to derive some estimates for the characteristic exponents of (7.12) at $u \in \mathcal{M}$ in terms of the right hand-side of the variational system (7.21), which will be needed in the sequel. Let $k \in \{1, \ldots, n\}$ be arbitrarily chosen and let $w(\cdot)$ be an arbitrary nonzero solution of (7.21) considered with respect to $\varphi^{(\cdot)}(p)^{(\cdot)}(u)$. Remembering that $|w(t)|^2 = (w(t), w(t)))_{\bigwedge^k T_{\varphi^{(\cdot)}(p)^t(u)}\mathcal{M}}$ we obtain

$$\frac{d}{dt} \ln |w(t)|^2 = 2\Big((S\nabla f(\varphi^{(\cdot)}(p)^t(u))_k \frac{w(t)}{|w(t)|}, \frac{w(t)}{|w(t)|}\Big)$$

and

$$\ln |w(t)| = \ln |w(0)| + \int_0^t \Big((S\nabla f(\varphi^{(\cdot)}(p)^\tau(u))_k \frac{w(\tau)}{|w(\tau)|}, \frac{w(\tau)}{|w(\tau)|}\Big)d\tau \tag{7.70}$$

for all $t \geq 0$, where $S\nabla f(\cdot)$ is as above the symmetric part of the covariant derivative of f. Then by (7.64) and (7.70) we have

$$\chi^{(k)}(u, w_0) = \limsup_{t\to\infty} \frac{1}{t} \int_0^t \Big((S\nabla f(\varphi^{(\cdot)}(p)^\tau(u)))_k v(\tau), v(\tau)\Big)d\tau, \tag{7.71}$$

where $v(t) := w(t)/|w(t)|$ and $w_0 := w(0)$. In terms of the eigenvalues $\alpha_i(\varphi^{(\cdot)}(p)^t(u))$ of $S\nabla f(\varphi^{(\cdot)}(p)^t(u))$ we derive from (7.71) the estimate

$$\widetilde{\chi}_1^{(k)}(u) \leq \limsup_{t\to\infty} \frac{1}{t} \int_0^t (\alpha_1(\varphi^{(\cdot)}(p)^\tau(u)) + \cdots + \alpha_k(\varphi^{(\cdot)}(p)^\tau(u)))d\tau. \quad (7.72)$$

In particular, for $k = n$, we already mentioned that for any $w_0 \neq 0_{\bigwedge^n T_u\mathcal{M}}$ exists exactly one value $\widetilde{\chi}_1^{(n)}(u) = \chi^{(n)}(u, w_0)$. Then relation (7.71) gives

$$\begin{aligned} \widetilde{\chi}_1^{(n)}(u) &= \limsup_{t\to\infty} \frac{1}{t} \int_0^t \operatorname{tr} S\nabla f(\varphi^{(\cdot)}(p)^\tau(u))d\tau \\ &= \limsup_{t\to\infty} \frac{1}{t} \int_0^t \operatorname{div} f(\varphi^{(\cdot)}(p)^\tau(u))d\tau. \quad (7.73) \\ &= \limsup_{t\to\infty} \frac{1}{t} \int_0^t (\alpha_1(\varphi^{(\cdot)}(p)^\tau(u)) + \cdots + \alpha_n(\varphi^{(\cdot)}(p)^\tau(u)))d\tau. \end{aligned}$$

7.2.5 Orbital Stability Conditions in Terms of Exponents

We come back now to the orbital stability investigation beginning with a criterion for orbital stability of arbitrary bounded solutions written in terms of characteristic exponents.

Theorem 7.4 *Suppose that for an orbit $\gamma(u)$ of (7.12) with $u \in W_+$ the largest characteristic exponent of order 2 satisfies*

$$\widetilde{\chi}_1^{(2)}(u) < 0.$$

Then the orbit $\gamma(u)$ is asymptotically stable.

Proof The assumptions of the theorem and statement 1) of Lemma 7.10 provide that for sufficiently large $t > 0$

$$\frac{1}{t} \ln \alpha_1^{(2)}(t, u) \leq -\kappa < 0, \quad (7.74)$$

where for brevity $\kappa := -\widetilde{\chi}_1^{(2)}(u)$. Therefore

$$|Y^{\wedge 2}(t, u)| = \alpha_1^{(2)}(t, u) \leq e^{-\kappa t}$$

holds for sufficiently large $t > 0$. This yields that the trivial solution of (7.23) is asymptotically stable in the sense of Lyapunov and Theorem 7.2 can be applied. □

Remark 7.10 Consider now a T-periodic orbit $\gamma(u)$ of system (7.12). It is well-known that the Lyapunov exponents ν of an invariant measure concentrated an $\gamma(u)$ are related to the multipliers ρ_i by (see e.g. [8])

$$\nu(u) = \frac{1}{T} \ln |\rho_i(u)|, \quad i = 1, \ldots, n. \tag{7.75}$$

This confirms incidently that one of them is equal to zero. On account of the Andronov-Witt theorem we conclude that if $\lambda_1(u) = 0$ and $\lambda_2(u) < 0$ then the orbit is asymptotically stable.

For completeness we formulate an orbital stability result that was proved in [27]. One easily checks that the second statement of this theorem is simply the appropriate formulation of Theorem 7.3 for the case of a periodic orbit.

Theorem 7.5 *Let $\gamma(u)$ be a T-periodic orbit of system (7.12). Let $\lambda_1(\cdot) \geq \lambda_2(\cdot)$ be the largest two eigenvalues of $S\nabla f(\cdot)$. Suppose that one of the conditions*

(1) $\rho_2(u) < 1$,

(2) $\int\limits_0^T \left[\lambda_1(\varphi^{(\cdot)}(p)^t(u) + \lambda_2(\varphi^{(\cdot)}(p)^t(u))\right] dt < 0$

is satisfied. Then $\gamma(u)$ is asymptotically stable.

7.2.6 Estimating the Singular Values and Orbital Stability

In this subsection we continue studying the fundamental operator solution $Y(t, p)$ of the variational system (7.13) along a solution $\varphi^t(p)$ of (7.12) on the manifold $(\mathcal{M}, g)$. In order to use Lyapunov-type functions for upper and lower estimates of the singular values of the fundamental operator solution of (7.13) we introduce some formal assumptions. In the next subsection it will be shown that frequency-domain conditions for feedback-control systems on the cylinder may effectively generate such type of assumptions.

Let H be a map from an open and bounded set $\widetilde{\mathcal{U}} \subset \mathcal{M}$ into the space of linear operators in tangent space such that

(H1) $H(u) : T_u\mathcal{M} \to T_u\mathcal{M}$ is a linear operator depending C^{m-1}-smoothly on u and satisfying $H^*(u) = H(u)$ for every $u \in \widetilde{\mathcal{U}}$.

Further we will suppose that H has one of the following properties:

(H2a) $H(u)$ has for any $u \in \widetilde{\mathcal{U}}$ at least k $(1 \leq k \leq n)$ negative eigenvalues.
(H2b) $H(u)$ has for any $u \in \widetilde{\mathcal{U}}$ at least k $(1 \leq k \leq n)$ positive eigenvalues.

We put $\Omega_H(u) := \{v \in T_u\mathcal{M} \mid (v, H(u)v) < 0\}$ and
$\Omega^H(u) := \{v \in T_u\mathcal{M} \mid (v, H(u)v) > 0\}$ for any $u \in \widetilde{\mathcal{U}}$.

Consider a solution $\varphi^t(p)$ of (7.12) and denote the singular values of the fundamental operator solution $Y(t, p)$ of the variational system (7.13) by $\alpha_1(t, p) \geq \cdots \geq \alpha_n(t, p)$.

Proposition 7.17 *Let H be a map on the open and bounded set $\widetilde{\mathcal{U}} \subset \mathcal{M}$ with properties* **(H1)** *and* **(H2a)**. *Let $\mathcal{U} \subset \mathcal{M}$ be also open with $\overline{\mathcal{U}} \subset \widetilde{\mathcal{U}}$ and suppose that for the orbit through p of (7.12) with $\gamma_+(p) \subset \mathcal{U}$ there exists a real function $\theta : \mathbb{R}_+ \to \mathbb{R}$ such that*

$$(y(t, v), H(\varphi^t(p))\nabla f(\varphi^t(p))y(t, v)) + (y(t, v), \tfrac{D}{dt}(H(\varphi^t(p))y(t, v)))$$

$$+ 2\theta(t)(y(t, v), H(\varphi^t(p))y(t, v)) \leq 0 \tag{7.76}$$

for all solutions $y(\cdot, v)$ of (7.13) with $y(0, v) = v \in \Omega_H(p)$ and for all such times $t \geq 0$ for which $y(t, v) \in \Omega_H(\varphi^t(p))$. Then

$$\alpha_k(t, p) \geq \beta e^{-\int_0^t \theta(\tau)d\tau} \qquad \text{for all } t \geq 0$$

with some constant $\beta > 0$.

Proof Fix an arbitrary $v \in \Omega_H(p)$ and consider the solution $y(\cdot, v)$ of the variational system (7.13) with $y(0, v) = v$ and the auxiliary function $V : \mathbb{R}_+ \to \mathbb{R}$ given by $V(t) := (y(t, v), H(\varphi^t(p))y(t, v))$. Obviously, $V(0) < 0$ holds. By the continuity of the solution of (7.12) and (7.13) we can argue that there is a time $t_0 = t_0(v) > 0$ such that $y(t, v) \in \Omega_H(\varphi^t(p))$ for all $t \in [0, t_0]$. Therefore inequality (7.76) reads as

$$\dot{V}(t) + 2\theta(t)V(t) \leq 0 \qquad \text{for all } t \in [0, t_0]. \tag{7.77}$$

Since (7.77) can be written as

$$\frac{d}{dt}\left(V(t)\, e^{2\int_0^t \theta(\tau)d\tau} \right) \leq 0 \qquad \text{for all } t \in [0, t_0],$$

we conclude that

$$V(t) \leq V(0)\, e^{-2\int_0^t \theta(\tau)d\tau} \qquad \text{for all } t \in [0, t_0]. \tag{7.78}$$

From $V(0) < 0$ and (7.78) we obtain that the solution $y(t, v)$ stays inside $\Omega_{\mathcal{H}}(\varphi^t(p))$ for all finite times $t \geq 0$. This implies that (7.77) as well as (7.78) hold for all $t \geq 0$. Hence by definition of V we have

$$(y(t,\upsilon), H(\varphi^t(p))y(t,\upsilon)) \leq (\upsilon, H(p)\upsilon)e^{-2\int_0^t \theta(\tau)d\tau} \tag{7.79}$$

for all $t \geq 0$ and $\upsilon \in \Omega_H(p)$. From property **(H2a)** we know that for every $u \in \widetilde{U}$ the linear operator $H(u)$ has at least k negative eigenvalues $\alpha_{n-k+1}(H(u)) \geq \ldots \geq \alpha_n(H(u))$. Thus, on account of the symmetry of $H(u)$ we can construct a subspace $\mathbb{L}^j(u)$ of $T_u\mathcal{M}$ of dimension $j \geq k$ generated by a system of eigenvectors which correspond to these k eigenvectors of $H(u)$. Obviously, $\mathbb{L}^j(u) \subset \Omega_H(u)$ holds. Now put $\beta := \frac{\alpha_k(H(p))}{\sup_{u\in\overline{\mathcal{U}}}\alpha_n(H(u))}$. From the symmetry of $H(u)$ for every $u \in \overline{\mathcal{U}}$ and with (7.79) we derive

$$|Y(t,p)\upsilon| \geq \beta|\upsilon|e^{-\int_0^t \theta(\tau)d\tau}$$

for all $\upsilon \in \mathbb{L}^j(p)$ and $t \geq 0$. Applying Lemma 7.6, finishes the proof. □

Remark 7.11 The condition of Proposition 7.17 generalizes the eigenvalue condition for the covariant derivative ∇F given with $H(u) \equiv \mathrm{id}_{T_u\mathcal{M}}$ in the form $(y(t,\upsilon), \nabla F(\varphi^t(p))y(t,\upsilon)) \leq -\theta(t)|y(t,\upsilon)|^2$ for $t \geq 0$ with respect to the solutions of (7.12) and (7.13). Note, that in case $\widetilde{\mathcal{U}} = \mathcal{M}$ for a compact manifold $\mathcal{M}$ property **(H1)** guarantees that the family of bilinear forms $\widetilde{g}(u;\upsilon,w) = g(u;\upsilon,Hw)$ generates a pseudo Riemannian metric on $\mathcal{M}$ and the subspaces $\mathbb{L}^j(u)$ appearing in the proof of Proposition 7.17 are constructed in such a way that these bilinear forms are positive when restricted to $\mathbb{L}^j(u)$.

Consider now the adjoint to (7.13) variational system

$$\frac{Dy}{dt} = -\nabla F^*(\varphi^t(p))y. \tag{7.80}$$

Notice that the fundamental operator solution of (7.80) having the identity at $t = 0$ is given by $Y^*(-t,u)$ where $Y(t,u)$ denotes the fundamental operator solution of the variational system (7.13) satisfying $Y(0,p) = \mathrm{id}_{T_p\mathcal{M}}$.

Proposition 7.18 *Let H be a map on an open and bounded set $\widetilde{\mathcal{U}} \subset \mathcal{M}$ with properties* **(H1)** and **(H2b)**. Let further $\mathcal{U} \subset \mathcal{M}$ be open with $\overline{\mathcal{U}} \subset \widetilde{\mathcal{U}}$ and suppose that for the orbit through p of (7.12) with $\gamma_+(p) \subset \mathcal{U}$ there exists a real function $\theta : \mathbb{R} \to \mathbb{R}$ such that

$$\begin{aligned}(y(t,\upsilon), H(\varphi^t(p))\nabla F^*(\varphi^t(p))y(t,\upsilon)) + (y(t,\upsilon), \tfrac{D}{dt}(H(\varphi^t(p))y(t,\upsilon)))\\ + 2\theta(t)(y(t,\upsilon), H(\varphi^t(p))y(t,\upsilon)) \leq 0\end{aligned} \tag{7.81}$$

for all solutions $y(\cdot,\upsilon)$ of (7.80) with $y(0,\upsilon) = \upsilon \in \Omega^H(p)$ and for all such times $t \geq 0$ for which $y(t,\upsilon) \in \Omega^H(\varphi^t(p))$. Then

$$\alpha_{n-k+1}(t, p) \le \beta e^{-\int_0^t \theta(\tau)d\tau} \quad \text{for all } t \ge 0 \tag{7.82}$$

with constant $\beta > 0$.

Proof The argumentation is similar to that of the proof of Proposition 7.17 since the solutions of (7.80) satisfy an inequality (7.76) which results from inequality (7.81) if ∇F is replaced by $-\nabla F^*$, H by $-H$ and θ by $-\theta$. □

In the next lemma we deduce sufficient conditions for the boundedness on $\mathbb{R}_+$ of the largest singular value of the fundamental solution operator of (7.13).

Lemma 7.11 *Assume that for an orbit of (7.12) through p the following conditions are satisfied:*

(a)

$$\alpha_2(t, p) \to 0 \quad \text{as} \quad t \to +\infty. \tag{7.83}$$

(b) There exists a solution $y_1(t, p)$ of (7.13) and constants C_1 and C_2 such that

$$0 < C_1 \le |y_1(t, p)| \le C_2 \quad \text{for all} \quad t \ge 0. \tag{7.84}$$

Then $\alpha_1(t, p)$ is bounded on $[0, +\infty)$.

Proof Suppose the opposite, i.e., suppose that there exists a sequence $t_k \to +\infty$ for $k \to +\infty$ such that $\alpha_1(t_k, p) \to +\infty$ for $k \to +\infty$. It follows that $|Y(t_k, p)| \to +\infty$ as $k \to +\infty$ and there exists a solution $y_2(t, p)$ of (7.13), satisfying

$$|y_2(t_k, p)| \to +\infty \quad \text{as} \quad k \to +\infty. \tag{7.85}$$

Take now the vectors $y_1(0, p)$ and $y_2(0, p)$ and numbers $\varepsilon_1 > 0,\ \varepsilon_2 > 0$ to define the two-dimensional annular region

$$\begin{aligned}\big\{\alpha y_1(0, p) + \beta y_2(0, p) : 0 < \varepsilon_1 \le |\alpha y_1(0, p) + \beta y_2(0, p)|\\ \le \varepsilon_2,\ \alpha, \beta \in \mathbb{R}\big\} \subset T_p\mathcal{M}.\end{aligned} \tag{7.86}$$

We consider a point $\alpha y_1(0, p) + \beta y_2(0, p)$ from this region (7.86). We find a $\delta = \delta(\varepsilon_1) > 0$ such that with respect to the above sequence $\{t_k\}$ (or possibly with respect to some subsequence)

$$|Y(t_k, p)(\alpha y_1(0, p) + \beta y_2(0, p)| = |\alpha y_1(t_k, p) + \beta y_2(t_k, p)| \ge \delta \tag{7.87}$$

for all $k = 1, 2, \ldots$ holds. Indeed, suppose on the contrary that such a δ does not exist. Then we can assume the existence of bounded sequences of numbers $\{\alpha_k\}$ and $\{\beta_k\}$ generating points in the annular region (7.86) and such that

$$\alpha_k y_1(t_k, p) + \beta_k y_2(t_k, p) \to 0 \quad \text{as} \quad k \to +\infty. \tag{7.88}$$

Take now convergent subsequences (we use the previous notations) $\alpha_k \to \alpha$ and $\beta_k \to \beta$ as $k \to +\infty$ and consider two cases.

Case1: $\beta = 0$. Then it follows from (7.86) that $\alpha \neq 0$. But, because of (7.84), this implies that property (7.88) is impossible.

Case2: $\beta \neq 0$. Passing possibly to a subsequence with (7.85) we conclude that for some $\delta > 0$ the inequalities

$$|\alpha y_1(t_k, p) + \beta y_2(t_k, p)| \geq |\beta y_2(t_k, p)| - |\alpha y_1(t_k, p)| \geq \delta$$

hold for all $k \in \mathbb{N}$. But this contradicts (7.88). This shows that the property (7.87) is satisfied. For an arbitrary $t \geq 0$ let us take an orthonormalized system of vectors $\{u_j(t, p)\}_{j=1}^n$ in $T_p\mathcal{M}$ such that

$$Y(t, p)^* Y(t, p) u_j(t, p) = \alpha_j^2(t, p) u_j(t, p) \quad (j = 1, 2, \ldots, n).$$

Then the vectors $Y(t, p)u_j(t, p), \ j = 1, \ldots, n$, are orthogonal in $T_{\varphi^t(p)}\mathcal{M}$ and $|Y(t, p)u_j(t, p)| = \alpha_j(t, p)$. Obviously, for any $t \geq 0$ we have the unique representation

$$\alpha y_1(0, p) + \beta y_2(0, p) = \sum_{j=1}^{n} \tau_j(t, p, \alpha, \beta) u_j(t, p)$$

with $\tau_j(t, p, \alpha, \beta) = (\alpha y_1(0, p) + \beta y_2(0, p), u_j(t, p)), \ j = 1, 2, \ldots, n$. Thus, by (7.86)

$$|\tau_j(t, p, \alpha, \beta)| \leq |\alpha y_1(0, p) + \beta y_2(0, p)| \leq \varepsilon_2 \quad \text{for } j = 1, 2, \ldots, n \text{ and } t \geq 0. \tag{7.89}$$

It follows $Y(t, p)\,(\alpha y_1(0, p) + \beta y_2(0, p))$

$$= Y(t, p) \left(\sum_{j=1}^{n} \tau_j(t, p, \alpha, \beta) u_j(t, p) \right) = \sum_{j=1}^{n} \tau_j(t, p, \alpha, \beta) Y(t, p) u_j(t, p).$$

Using (7.83) and (7.89) we get

$$\left| \sum_{j=2}^{n} \tau_j(t, p, \alpha, \beta) Y(t, p) u_j(t, p) \right| \leq \sum_{j=2}^{n} |\tau_j(t, p, \alpha, \beta)|\, |Y(t, p) u_j(t, p)| \leq \sum_{j=2}^{n} \varepsilon_2 \alpha_j(t, p)$$

with $\sum_{j=2}^{n} \varepsilon_2 \alpha_j(t, p) \to 0 \quad \text{as} \quad t \to +\infty$.

From this we conclude that with $z(t) := \tau_1(t, p, \alpha, \beta)\, Y(t, p) u_1(t, p)$ the estimate

$$|\alpha y_1(t, p) + \beta y_2(t, p) - z(t)| \to 0 \quad \text{as} \quad t \to +\infty$$

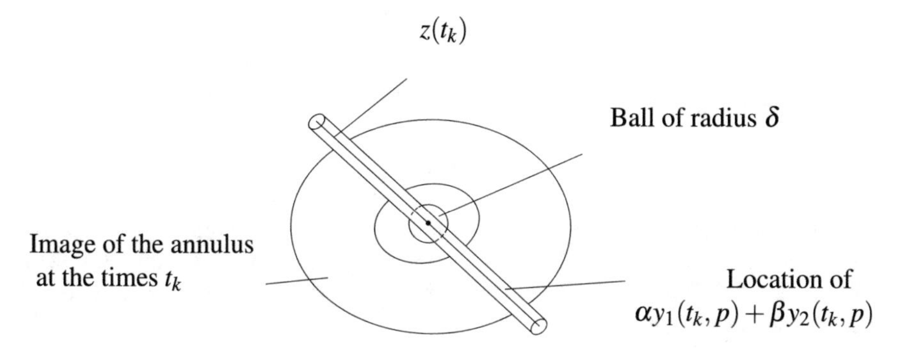

Fig. 7.2 The annular region

holds. But this contradicts the fact that the image under the fundamental operator $Y(t, p)$ of the annular region (7.86) has property (7.87) for times t_k. Further, on the other side its image under this non-singular linear map is again an annular region as shown in Fig. 7.2. □

The next theorem contains the main stability result of this section. In Sect. 7.2.7 we will reduce the stability investigation for feedback-control systems on the cylinder to a situation described in this theorem.

Recall that BO_+ denotes the set of all orbits of (7.12) for which there exists an open bounded set $\mathcal{U} \subset \mathcal{M}$ containing $\gamma_+(p)$ and such that any equilibrium of system (7.12) lying in $\overline{\mathcal{U}}$ is asymptotically stable.

Theorem 7.6 *Let $\gamma(p)$ be an orbit of (7.12) with $\gamma(p) \in BO_+$ and $\mathcal{U}$ the set containing $\gamma_+(p)$. Suppose that the assumptions of Proposition 7.18 are satisfied with respect to $\mathcal{U}$, $k = n - 1$ and a function $\theta(\cdot)$ which satisfies $\inf_{t \in \mathbb{R}_+} \theta(t) > 0$. Then the orbit $\gamma(p)$ is asymptotically stable.*

Proof Since the assumptions of Proposition 7.18 are satisfied for $k = n - 1$, we find a constant $B > 0$ such that

$$\alpha_2(t, p) \leq B e^{-\int\limits_0^t \theta(\tau) d\tau} \tag{7.90}$$

for all $t \in \mathbb{R}_+$ is valid. From this it follows that $\alpha_2(t, p) \to 0$ for $t \to +\infty$ and thus condition (a) of Lemma 7.11 is satisfied. The function $y_1(t, p) = F(\varphi^t(p))$ is a solution of the variational equation (7.13) and the assumption $\gamma(p) \in BO_+$ guarantees that $y_1(\cdot, p)$ satisfies the condition (b) of Lemma 7.11. We conclude with this lemma that $\alpha_1(t, p)$ is bounded on $\mathbb{R}_+$. Denote by $\alpha_1^{(2)}(t, p)$ the largest singular value of the second multiplicative compound operator $Y^{\wedge 2}(t, p)$ which is the fundamental operator solution of the second compound variational equation (7.23). Since $\alpha_1^{(2)}(t, p) = \alpha_1(t, p)\alpha_2(t, p)$ for all $t \geq 0$, $\alpha_1(\cdot, p)$ is bounded and $\alpha_2(\cdot, p)$ satisfies (7.90), there is a constant $C > 0$ such that for sufficiently large $t > 0$

$$\alpha_1^{(2)}(t, p) \leq Ce^{-\int_0^t \theta(\tau)d\tau}.$$

Using the assumptions on $\theta(\cdot)$ we find a constant $a > 0$ such that

$$\alpha_1^{(2)}(t, p) \leq Ce^{-at}$$

for sufficiently large $t > 0$. Since $\alpha_1^{(2)}(t, p) = |Y^{\wedge 2}(t, p)|$, where $|\cdot|$ denotes the associated operator norm, it follows by the last inequality that the trivial solution of the equation (7.23) is exponentially stable. Applying now the statement of Theorem 7.2 we obtain the desired result. □

7.2.7 *Frequency-Domain Conditions for Orbital Stability in Feedback Control Equations on the Cylinder*

We now consider feedback control systems with one scalar nonlinearity and a linear part written in the form

$$\dot{x} = Px + q\xi, \quad \xi = \phi(w), \quad w = (r, x), \tag{7.91}$$

where P is a real $n \times n$ matrix, q and r are real n-vectors and $\phi : \mathbb{R} \to \mathbb{R}$ is a continuously differentiable function. We assume that ϕ is periodic with some period $\Delta > 0$, for example a sine-type function $\phi(w) = \sin w - \rho$ with a constant ρ. In order to define with (7.91) a vector field on the cylinder we suppose also that there exists a vector $d \in \mathbb{R}^n$, $d \neq 0$, with $Pd = 0$ and $(r, d) = \Delta$. Hence we obtain a system (7.91) which is pendulum-like with respect to the subgroup Γ generated by the vector d. (See Example 7.3.) It follows that (7.91) can be considered as vector field on the cylinder $\mathbb{R}^n/\Gamma$. To explore the stability behaviour of solutions of system (7.91) we use results from the preceeding subsection as well as the solvability of special matrix inequalities (see Sect. 2.5, Chap. 2). It will not be difficult to verify conditions **(H1)** and **(H2)** for system (7.91).

Here we consider only solutions $\varphi(\cdot, x_0)$ of (7.91) with $\varphi(0, x_0) = x_0$ and $\gamma(x_0) \in BO_+$ on $\mathbb{R}^n/\Gamma$. One class of such solutions are the *circular solutions*, i.e., solutions of (7.91) for which there exist $\varepsilon > 0$ and $\tau \geq 0$ such that for $w(t) = (r, \varphi(t, x_0))$ the inequality $\dot{w}(t) \geq 0$ (or $\dot{w}(t) \leq -\varepsilon$) is satisfied on $[\tau, \infty)$. We say that a circular solution is *asymptotically orbitally stable* if the corresponding bounded orbit on $\mathbb{R}^n/\Gamma$ is asymptotically stable.

The variational system (7.13) with respect to a solution $\varphi(\cdot, x_0)$ of (7.91) is given by

$$\dot{y} = \left(P + \phi'((r, \varphi(t, x_0)))qr^*\right) y. \tag{7.92}$$

For brevity we put $u(t, x_0) := \phi'((r, \varphi(t, x_0)))$. Then $u(\cdot, x_0)$ is a continuous n-vector function for fixed x_0 and system (7.92) becomes

$$\dot{y} = (P + u(t, x_0)qr^*)y,$$

while the adjoint to (7.92) system becomes

$$\dot{y} = -(P^* + u(t, x_0)rq^*)y. \tag{7.93}$$

Let now $\varphi(\cdot, x_0)$ be a solution of (7.91) with $\gamma(x_0) \in BO_+$. We assume that there are given a Hermitian form $\mathcal{F}(\cdot, \cdot) : \mathbb{C} \times \mathbb{C}^n \to \mathbb{R}$, a real n-vector c and a number $\varepsilon > 0$ such that

$$\mathcal{F}((c, z), z) \geq \varepsilon(c, z)^2 \tag{7.94}$$

for all $z \in \mathbb{R}^n$ and

$$\mathcal{F}(u(t, x_0)(q, z), z) \geq 0 \tag{7.95}$$

for all $t \geq 0$ and $z \in \mathbb{R}^n$. We can establish the following result [21].

Theorem 7.7 *Consider (7.91) and suppose that the pair (P^*, r) is controllable and that c is an n-vector such that the pair (P^*, c) is observable. Let $\gamma(x_0) \in BO_+$ and suppose that $\mathcal{F}(\xi, z)$ is a Hermitian form such that with respect to c and $u(t, x_0) := \phi'((r, \varphi(t, x_0)))$ the inequalities (7.94) and (7.95) are satisfied. Let $\kappa > 0$ be some number such that the following conditions hold:*

(1) The matrix $P^* + rc^* + \kappa I$ has at least $n - 1$ eigenvalues with negative real part.
(2) $\mathcal{F}(\xi, [(i\omega - \kappa)I - P^*]^{-1} r\xi) \leq 0$ for all $\xi \in \mathbb{C}$, $\omega \in \mathbb{R}$ with $\det[(i\omega - \kappa)I - P^*] \neq 0$.

Then the solution $\varphi(\cdot, x_0)$ is asymptotically orbitally stable.

Proof Because of the controllability of (P^*, r) and the assumption 2) of the theorem the Yakubovich-Kalman theorem (Theorem 2.7, Chap. 2) guarantees the existence of a real $n \times n$ matrix $H = H^*$ satisfying

$$2(H[P^* + \kappa I]y + r\xi, y) + \mathcal{F}(\xi, y) \leq 0 \quad \text{for all } y \in \mathbb{R}^n \text{ and } \xi \in \mathbb{R}. \tag{7.96}$$

We take, in particular, $\xi = (c, y)$ in (7.96) and use (7.94) to obtain

$$2(H[P^* + rc^* + \kappa I]y, y) \leq -\varepsilon(c, y)^2 \quad \text{for all } y \in \mathbb{R}^n. \tag{7.97}$$

Applying Lemma 2.7, Chap. 2, we see that the observability of the pair (P^*, c), assumption 1) of the theorem, and (7.97) provide that H has at least $n - 1$ positive eigenvalues.

If we put $\xi = u(t, x_0)(q, y)$ in (7.96) and use the inequality (7.95) we derive from (7.96)

$$(H[P^* + u(t, x_0)rq^*]y, y) + \kappa(Hy, y) \leq 0 \quad \text{for all } y \in \mathbb{R}^n \text{ and } t \geq 0.$$

Taking now $\mathcal{M} = \mathbb{R}^n$, $\widetilde{\mathcal{U}} = \mathcal{U}$ (the bounded set containing $\gamma(x_0)$) and $H(u) \equiv H$ on $\mathcal{U}$, we see that for H the assumptions **(H1)** and **(H2b)** and the inequality (7.81) with $\theta(t) \equiv \frac{\kappa}{2}$ are satisfied. Thus, we can apply Theorem 7.6 to conclude that the orbit $\gamma(x_0)$ of (7.91) is asymptotically stable. □

Consider now the transfer function of the linear part of (7.91)

$$W(z) = r^*(P - zI)^{-1}q, \tag{7.98}$$

which is defined for all $z \in \mathbb{C}$ with $\det(P - zI) \neq 0$. Notice that under our assumption $W(\cdot)$ is a rational function of the form $W(z) = R(z)/N(z)$, where $N(z) = \det(zI - P)$ and $R(z)$ is a polynomial of degree less than n. We say that $W(\cdot)$ is *non-degenerate* if the polynomials $R(\cdot)$ and $N(\cdot)$ are coprime. It is well-known that $W(\cdot)$ is non-degenerate if and only if the pair (P, q) is controllable and the pair (P, r) is observable (see [35]).

Notice further that W is non-degenerate if and only if W^* is non-degenerate.

Corollary 7.2 *Consider system (7.91) and let $\gamma(x_0) \in BO_+$. Suppose that the transfer function $W(\cdot)$ of (7.91) is non-degenerate and that there exists a number $\kappa > 0$ such that the following conditions are fulfilled:*

(1) The matrix $P^* + \kappa I$ has at least $n - 1$ eigenvalues with negative real part;
(2) There are numbers $\kappa_1 < 0 < \kappa_2$ such that for $u(t, x_0) := \phi'((r, \varphi(t, x_0)))$ we have

$$\kappa_1 \le u(t, x_0) \le \kappa_2 \quad \textit{for all} \quad t \ge 0 \quad \textit{and}$$

$$|W(i\omega - \kappa)|^2 + \left(\kappa_1^{-1} + \kappa_2^{-1}\right) \operatorname{Re} W(i\omega - \kappa) \le -\kappa_1^{-1}\kappa_2^{-1} \tag{7.99}$$

for all $\omega \in \mathbb{R}$.

Then the solution $\varphi(\cdot, x_0)$ is asymptotically orbitally stable.

Proof We consider in $\mathbb{C} \times \mathbb{C}^n$ the Hermitian form

$$\mathcal{F}(\xi, y) := 2\operatorname{Re}\left[(q^*y - \kappa_1^{-1}\xi)^*(q^*y - \kappa_2^{-1}\xi)\right].$$

Then condition (2) of the present corollary implies condition (2) of Theorem 7.7 for this particular form $\mathcal{F}$. We define the vector $c = \delta q$, where $\delta \neq 0$ is a sufficiently small number such that the matrix $P^* + \delta r q^* + \kappa I$, as in condition (1), has also at least $n - 1$ eigenvalues with negative real part and the inequality $\varepsilon := \frac{1}{\delta^2}\left(1 - \kappa_1^{-1}\delta\right)\left(1 - \kappa_2^{-1}\delta\right) > 0$ holds. It follows immediately that the inequalities (7.94), (7.95) and condition 1) of Theorem 7.7 are satisfied and this theorem is applicable. □

A large class of pendulum-like systems can be written in the *second canonical form* (see [20, 23]) as

$$\dot{x} = Ax + b\phi(w), \quad \dot{w} = (c, x) + b'\phi(w), \tag{7.100}$$

where A is a real $m \times m$ matrix, b and c are real m-vectors, $\phi(\cdot)$ is a Δ-periodic function on $\mathbb{R}$ and b' is a parameter. System (7.100) can easily be brought into feedback form (7.91) by choosing $n = m + 1$ and

$$P = \begin{bmatrix} A & 0 \\ c^* & 0 \end{bmatrix}, \quad q = \begin{bmatrix} b \\ b' \end{bmatrix}, \quad r = \begin{bmatrix} 0 \\ \vdots \\ 0 \\ 1 \end{bmatrix}.$$

We conclude that system (7.100) is pendulum-like with respect to the subgroup Γ which is generated by the vector $d = \Delta r$. The transfer function of (7.100) is given by $W(z) = 1/z[(c, (A - zI)^{-1}b) - b']$. In analyzing the stability behaviour of circular solutions of (7.100), the flat cylinder interpretation makes it possible to apply the above results. In particular, Corollary 7.2 requires that for a certain $\kappa > 0$ the matrix $P^* + \kappa I$ possesses n eigenvalues with negative real part and that the transfer function $W(\cdot)$ satisfies an appropriate frequency-domain condition.

In the sequel we suppose that the nonlinearity $\phi(w)$ in system (7.100) does not have zeros, i.e., $\phi(w) \neq 0$ for all $w \in \mathbb{R}$, that the matrix A has only eigenvalues with negative real part and $(c, A^{-1}(b) \neq b'$. In this case it is well-known ([23], p.128) that there exists a cycle of the second kind and all solutions of system (7.100) are circular ones. In addition system (7.100) is dissipative in the cylindrical phase space (see Sect. 1.2, Chap. 1), i.e., there exists a number D such that for any solution $(x(t), w(t))$ of (7.100) we have $\limsup_{t \to +\infty} |x(t)| \leq D$. From this and from Corollary 7.2 we obtain the following:

Proposition 7.19 *Suppose that the Δ-periodic nonlinearity ϕ in (7.100) does not have zeros and that there exist numbers $\kappa_1 < 0 < \kappa_2$ such that*

$$\kappa_1 \leq \phi'(w) \leq \kappa_2 \quad \textit{for all} \quad w \in \mathbb{R}. \tag{7.101}$$

Suppose further that the transfer function $W(\cdot)$ of (7.100) is non-degenerate, that there exists a number $\kappa > 0$ such that inequality (7.99) is satisfied and that the matrix $A + \kappa I$ has only eigenvalues with negative real part. Then system (7.100) has a unique cycle of the second kind, say, $\varphi(\cdot, x_0)$ which is asymptotically orbitally stable and which attracts any other orbit of the system for $t \to +\infty$, i.e., $\lim_{t \to +\infty} dist(\varphi(t, u), \gamma_+(x_0)) = 0$ for any solutions $\varphi(\cdot, u)$ of (7.100).

Remark 7.12 The frequency-domain condition (7.99) coincides with the frequency-domain condition of the circle criterion for Lagrange stability of system (7.100) in the phase space $\mathbb{R}^n$ [20, 23].

In contrast to the Theorem 7.7, in the circle criterion it is supposed that the nonlinearity ϕ has zeros. Thus in case that the nonlinearity ϕ has zeros by preservation of the rest of the assumptions on ϕ and the transfer function W the circular globally asymptotically stable cycle disappears and system (7.100) becomes stable in the sense of Lagrange in the phase space $\mathbb{R}^n$.

Example 7.5 ([21]) Consider system (7.100) with the nonlinearity $\phi(w) = \sin w - \gamma$ and the transfer function

$$W(z) = \frac{1}{z(z+\beta)^j}, \tag{7.102}$$

where β is a positive number and j is a natural number. For $j = 1$ we have the pendulum equation with constant torque $\gamma \geq 0$ and a viscous resistance $\beta \geq 0$

$$\ddot{w} + \beta\dot{w} + \sin w = \gamma.$$

If j is an arbitrary natural number the transfer function (7.102) generates a system (7.100) in the $j+1$-dimensional cylindrical phase space. It is easy to see that if we choose $-\kappa_1 = \kappa_2 = 1$ the inequality (7.101) is satisfied and condition (7.99) becomes

$$|W(i\omega - \kappa)|^2 \leq 1 \quad \text{for all} \quad \omega \in \mathbb{R}. \tag{7.103}$$

Obviously, for (7.103) it suffices to require that

$$\kappa^2(\beta - \kappa)^{2j} \geq 1. \tag{7.104}$$

Since for system (7.100) with $W(\cdot)$ from (7.102) we have $\det(zI - A) = (z+\beta)^j$, the assumptions of Proposition 7.19 on $A + \kappa I$ are satisfied, if $\kappa < \beta$. Define $\kappa = \frac{\beta}{j+1}$ and write condition (7.104) as

$$a^j \left(\frac{j}{j+1}\right)^j \geq 1. \tag{7.105}$$

Thus on account of Proposition 7.19, if condition (7.105) is satisfied, then for $|\gamma| > 1$ the system (7.100) with the nonlinearity $\sin w - \gamma$ and the transfer function (7.102) has an globally asymptotically orbitally stable cycle of the second kind. If $|\gamma| \leq 1$ this cycle disappears and in the covering space $\mathbb{R}^{j+1}$ all positive semi-orbits of (7.100) are bounded.

7.2.8 Dynamical Systems with a Local Contraction Property

In this subsection we describe general properties of dynamical systems on manifolds which have a certain contraction property transversal to a considered orbit. It will be shown that this local property characterizes the global behaviour of the system. All results in this subsection go back to Stenström [33].

Suppose that $(\mathcal{M}, g)$ is an n-dimensional C^k-manifold with Riemannian metric. Let $\mathcal{U}$ be an open connected subset of $\mathcal{M}$ such that its closure $\overline{\mathcal{U}}$ in $\mathcal{M}$ is compact and its boundary $\partial\mathcal{U}$ is an $(n-1)$-dimensional C^1-submanifold of $\mathcal{M}$. Let F be a C^1-vector field on some open set $\mathcal{U}_1 \supset \overline{\mathcal{U}}$, i.e.

$$\dot{u} = F(u) \tag{7.106}$$

with $F : \mathcal{U}_1 \to T\mathcal{U}_1$. Assume the following conditions:

(A1) F penetrates $\partial\mathcal{U}$ inwards, that is, $(F(p), n(p)) > 0$ for every $p \in \partial\mathcal{U}$, where $n(p) \in T_p\mathcal{M}$ is the inner normal to $\partial\mathcal{U}$ at p and $(\cdot,\cdot)$ stands for the scalar product in the tangent space induced by the Riemannian metric; $|\cdot|$ is the corresponding norm.

(A2) For each $p \in \mathcal{U}$ we have $(\nabla_\upsilon F(p), \upsilon) < 0$ for every vector $\upsilon \in T_p\mathcal{M}$ with $(F(p), \upsilon) = 0$. Here $\nabla F(p) : T_p\mathcal{M} \to T_p\mathcal{M}$ is the covariant derivative of F at $p \in \mathcal{M}$. Note that in local coordinates **(A2)** has the following form:

(A3) At each point $p \in \mathcal{U}$ we have $\left(\frac{\partial f^i}{\partial x^j} + \Gamma^i_{jk} f^k\right)\xi^j \xi^m g_{im} < 0$ for every $\upsilon = \xi^i \partial_i(p) \in T_p\mathcal{M}$ with $f^i \xi^j g_{ij} = 0$.

Here $\upsilon = \xi^i \partial_i(p)$, $F = f^i \partial_i(p)$ and g_{ij} are the representations of υ, F and g, respectively, in an arbitrary chart x around p.

Consider a solution $\varphi^{(\cdot)}(p)$ of (7.106) starting at $p \in \overline{\mathcal{U}}$ at $t = 0$. The maps $p \mapsto \varphi^t(p)$ form a semi-group on $\overline{\mathcal{U}}$.

Definition 7.9 For a given $\varepsilon > 0$ the (closed) ε-neighborhood $\mathcal{N}_\varepsilon(p)$ around a semi-orbit $\gamma_+(p)$ of (7.106) is defined as

$$\mathcal{N}_\varepsilon(p) := \{q \in \overline{\mathcal{U}} | \rho(q, \gamma_+(p)) \le \varepsilon\} .$$

(Here $\rho(\cdot,\cdot)$ denotes the metric on $\mathcal{M}$ generated by g.) If $F(p) \ne 0$ we call $\mathcal{N}_\varepsilon(p)$ an ε-tube; if $F(p) = 0$ the set $\mathcal{N}_\varepsilon(p)$ is called a (closed) ε-ball.

The ε-neighborhood $\mathcal{N}_\varepsilon(p)$ is called *normal* if the ε-neighborhood $\mathcal{V}$ of every point of $\gamma_+(p)$ has the property that any two points in $\mathcal{V}$ can be joined by a unique geodesic in $\mathcal{V}$. The *section* at $u \in \gamma_+(p)$ of a normal ε-tube $\mathcal{N}_\varepsilon(p)$ consists of those q in $\mathcal{N}_\varepsilon(p)$ that can be reached from u along a geodesic in $\mathcal{N}_\varepsilon(p)$ of length $\le \varepsilon$ perpendicular to $F(u)$. Normal ε-neighborhoods always exist when $\overline{\mathcal{U}}$ is compact.

In case $\gamma_+(p)$ does not contain an equilibrium point, we require $\mathcal{N}_\varepsilon(p)$ to satisfy also the following condition:

(A3) $(F(u), \tau^u_q F(q)) > 0$ for all $u, q \in \mathcal{U}$ with $u \in \gamma_+(p)$ and $\rho(\mathcal{U}, q) \le \varepsilon$. (Here τ^u_q denotes the parallel transport in $\mathcal{M}$; see Sect. A.8, Appendix A.)

Note that **(A3)** can be required since the inequality $(F(u), \tau_q^u F(q)) > 0$ is satisfied on the diagonal of the compact set $\overline{\gamma_+(p)} \times \overline{\gamma_+(p)}$ of $\overline{\mathcal{U}} \times \overline{\mathcal{U}}$, and therefore is satisfied in some neighborhood of the diagonal.

The next theorem is due to [33].

Theorem 7.8 *(a) Let $\mathcal{N}_\varepsilon(p)$ be a normal ε-ball around the equilibrium point p. Then a solution of (7.106) starting in $\mathcal{N}_\varepsilon(p)$ approaches p with monotonously decreasing distance from p.*

(b) Let $\mathcal{N}_\varepsilon(p)$ be a normal ε-tube around $\gamma_+(p)$, p not an equilibrium of (7.106), and suppose $u \in \mathcal{N}_\varepsilon(p)$. Then it holds:

(1) If $\varphi^{(\cdot)}(p)$ tends to an equilibrium q of (7.106), then also $\varphi^{(\cdot)}(u)$ tends to q ;

(2) If $\varphi^{(\cdot)}(p)$ does not tend to any equilibrium point of (7.106), then $\varphi^{(\cdot)}(u)$ approaches $\gamma_+(p)$ with monotonously decreasing distance.

Proof Let $u \in \gamma_+(p)$ and consider any geodesic Γ starting at u. For each point $q \in \Gamma$, let υ_q be the tangent vector of unit length to Γ at q, showing in the direction away from u. We will show that if $(F(u), \upsilon_u) = 0$ then $(F(q), \upsilon_q) < 0$ for all $q \in \Gamma,\ q \neq u$. Assume the opposite, i.e. suppose that $(F(q_1), \upsilon_{q_1}) \geq 0$ for some $q_1 \in \Gamma$. The function $q \mapsto (F(q), \upsilon_q)$ is a C^1-function on Γ and

$$\nabla_\upsilon (F(q), \upsilon_q) = (\nabla_\upsilon F(q), \upsilon_q) + (F(q), \nabla_\upsilon \upsilon_q) .$$

But $\nabla_\upsilon \upsilon_q = 0$, so by condition **(A2)** we have $(F(q), \upsilon_q) < 0$ in a neighborhood of $u \neq q$ on Γ. By continuity there exists a nearest q_2 to u on Γ where $(F(q_2), \upsilon_{q_2}) = 0$ and then $(F(q), \upsilon_q) \leq 0$ between u and q_2 on Γ. By the same argument for q_2 instead of u, we find that $(F(q), -\upsilon_q) < 0$ in a neighborhood of $q_2 \neq q$ on Γ, which gives a contradiction.

Now let $s(t)$ be the distance of $\varphi^t(q)$, $q \in \mathcal{N}_\varepsilon$, from $u \in \gamma_+(p)$. When $\varphi^t(q)$ is in the ε-sphere around p, we find, using normal local coordinates x^i around u, that

$$\frac{ds}{dt} = \frac{\partial s}{\partial x^i}\frac{dx^i}{dt} = g_{ij}\, \xi^j f^i = (F, \upsilon) .$$

It follows that no solution can leave the ε-tube (resp. the ε-sphere) $\mathcal{N}_\varepsilon(p)$. □

Using Theorem 7.8 we can show the following two results from [33].

Theorem 7.9 *If Eq. (7.106) has an equilibrium point, then this point is the limit set of (7.106).*

Proof The proof follows from the fact that the limit set is connected and from Theorem 7.8. □

Theorem 7.10 *If Eq. (7.106) has no equilibrium point, then it has a periodic orbit which is the limit set of (7.106).*

Proof Let p be an arbitrary point in the limit set of (7.106). Choose a normal ε-tube $\mathcal{N}_\varepsilon$ around $\gamma_+(p)$ and let $\overline{\mathcal{B}_\varepsilon(p)}$ be the ε-ball around p. If ε is small enough, every solution of (7.106) starting in $\overline{\mathcal{B}_\varepsilon(p)}$ stays in $\mathcal{N}_\varepsilon$ and we can find $\varepsilon_1 < \varepsilon$ such that after some time all the solutions are in the ε_1-tube around $\gamma_+(p)$. Since p is in the limit set, there is a point u with $\rho(u, p) < \frac{1}{3}(\varepsilon - \varepsilon_1)$ such that $\varphi^{(\cdot)}(u)$ returns to the $\frac{1}{3}(\varepsilon - \varepsilon_1)$-neighborhood of p for arbitrary large t. Since $\varphi^t(u)$ is for $t \geq 0$ in the $\frac{1}{3}(\varepsilon - \varepsilon_1)$-tube around $\gamma_+(p)$, this means that $\varphi^T(p)$ is in the $\frac{2}{3}(\varepsilon - \varepsilon_1)$- neighborhood of p for some arbitrary large T. Then the section at $\varphi^T(p)$ of the ε_1- tube around $\gamma_+(p)$ is contained in $\overline{\mathcal{B}_\varepsilon(p)}$. Every solution of (7.106) starting at $t = 0$ in $\overline{\mathcal{B}_\varepsilon(p)}$ reaches this section at a time nearly to T, which gives a continuous map of $\overline{\mathcal{B}_\varepsilon(p)}$ into the section, i.e. into $\overline{\mathcal{B}_\varepsilon(p)}$. By the Brouwer fixed point theorem (Theorem B.3, Appendix B) this map has a fixed point. Thus every neighborhood of p contains an initial point for a periodic solution of (7.106). It follows then from Theorem 7.8 that $\varphi^{(\cdot)}(p)$ is the unique periodic solution of (7.106). □

Recall for the next theorem from [33] that the solid n-torus (see Sect. A.10, Appendix A) is the set $\mathbb{R}^{n-1} \times S^1$, while the solid Klein bottle is the non-trivial fiber bundle of $\mathbb{R}^{n-1}$ over S^1.

Theorem 7.11 *(a) System (7.106) has an equilibrium point if and only if $\mathcal{U}$ is homeomorphic with $\mathbb{R}^n$;*

(b) System (7.106) has a periodic solution if and only if $\mathcal{U}$ is homeomorphic with either a solid torus or a solid Klein bottle.

Proof It follows immediately from Theorem 7.10 and the Euler Poincaré formula for the Euler characteristic (see Sect. B.4, Appendix B). □

In the Euclidean case the obtained theorems lead to the following result. Suppose that

$$\dot{x} = f(x) \tag{7.107}$$

is a C^1-vector field $f : \mathcal{G} \to \mathbb{R}^n$ on an open subset $\mathcal{G}$ of $\mathbb{R}^n$. Let us assume that $\mathcal{U}$ is an open, connected and bounded subset of $\mathcal{G}$ with $\overline{\mathcal{U}} \subset \mathcal{G}$ and such that f penetrates $\partial\mathcal{U}$ inwards. The scalar product and the norm in $\mathbb{R}^n$ are denoted by $(\cdot, \cdot)$ and $|\cdot|$, respectively. Then the following theorem [33] is true.

Theorem 7.12 *Suppose that there exists a constant, positive definite, symmetric $n \times n$ matrix P such that for each $x \in \mathcal{U}$,*

$$(Df(x)y, Py) < 0 \quad \textit{for all} \quad y \neq 0, \ (f(x), Py) = 0.$$

Then it holds:

(a) $\mathcal{U}$ is homeomorphic with either $\mathbb{R}^n$ or a solid torus ;

(b) If $\mathcal{U}$ is homeomorphic with $\mathbb{R}^n$ then system (7.107) has an equilibrium point as limit set $\omega(\mathcal{U})$. If $\mathcal{U}$ is homeomorphic with a solid torus then system (7.107) has a limit cycle as limit set $\omega(\mathcal{U})$.

Proof Apply the previous Theorem 7.11 with the flat metric on $\mathbb{R}^n$ defined by the matrix P. □

Remark 7.13 Suppose that $f(y) \neq 0, \forall y \in \mathcal{U}$, for the vector field (7.107), where $\mathcal{U} \subset \mathbb{R}^n$ is again a bounded, connected and open set in $\mathcal{G}$. Suppose that $\varphi : \mathbb{R}_+ \to \mathcal{U}$ is a solution of (7.107) with dist $(\varphi(t), \partial\mathcal{U}) > \zeta > 0,\ \forall t \geq 0$. Define $\vartheta(y) := \max(Df(y)x, x)$ for all x, $|x| = 1$, such that $(x, f(y)) = 0$, and suppose that $\vartheta(y) \leq -c < 0,\ \forall y \in \mathcal{U}$.

Then the ω-limit set of φ is a periodic solution φ_0 of (7.106) which has $n-1$ characteristic exponents with negative real part and, consequently, is asymptotically orbitally stable.

References

1. Abraham, R., Marsden, J.E., Ratiu, T.: Manifolds, Tensor-Analysis, and Applications. Springer, New York (1988)
2. Andronov, A.A., Vitt, A.A., Khaikin, S.E.: Theory of Oscillations. Pergamon Press, Oxford (1966)
3. Borg, G.: A condition for existence of orbitally stable solutions of dynamical systems. Kungl. Tekn. Högsk. Handl. Stockholm **153**, 3–12 (1960)
4. Cesari, L.: Asymptotic Behavior and Stability Problems in Ordinary Differential Equations. Springer, Berlin (1959)
5. Courant, R., Hilbert, D.: Methods of Mathematical Physics, v. I., Wiley (Interscience), New York (1953)
6. Demidovich, B.P.: Lectures on Mathematical Stability Theory. Nauka, Moscow (1967). (Russian)
7. Eberlein, P.: When is a geodesic flow of Anosov type? I., J. Diff. Geom. **8**, 437–463 (1973)
8. Eckmann, J.-P., Ruelle, D.: Ergodic theory of chaos and strange attractors. Rev. Mod. Phys. **57**, 617 (1985)
9. Faddeev, D.K.: Lectures on Algebra. Nauka, Moscow (1984). (Russian)
10. Fiedler, M.: Additive compound matrices and an inequality for eigenvalues of symmetric stochastic matrices. Czechoslovak Math. J. **24**, 392–402 (1974)
11. Gelfert, K.: Estimates of the box dimension and of the topological entropy for volume-contracting and partially volume-expanding dynamical systems on manifolds. Doctoral Thesis, University of Technology Dresden (2001) (German)
12. Gelfert, K.: Maximum local Lyapunov dimension bounds the box dimension. Direct proof for invariant sets on Riemannian manifolds. Zeitschrift für Analysis und ihre Anwendungen (ZAA) **22** (3), 553–568 (2003)
13. Ghidaglia, J.M., Temam, R.: Attractors for damped nonlinear hyperbolic equations. J. Math. Pures et Appl. **66**, 273–319 (1987)
14. Giesl, P.: Converse theorems on contraction metrics for an equilibrium. J. Math. Anal. Appl. **424**(2), 1380–1403 (2015)
15. Hartman, P., Olech, C.: On global asymptotic stability of solutions of ordinary differential equations. Trans. Amer. Math. Soc. **104**, 154–178 (1962)
16. Katok, A., Hasselblatt, B.: Introduction to the Modern Theory of Dynamical Systems. (Encyclopedia of Mathematics and its Applications), **54**, Cambridge University Press, Cambridge (1995)
17. Ledrappier, F.: Some relations between dimension and Lyapunov exponents. Commun. Math. Phys. **81**, 229–238 (1981)

18. Leonov, G.A.: Lyapunov Exponents and Problems of Linearization From Stability to Chaos. St. Petersburg State Univ. Press, St. Petersburg (1997)
19. Leonov, G.A.: Generalization of the Andronov-Vitt theorem. Regul. Chaotic Dyn. **11** (2), 281–289 (2006) (Russian)
20. Leonov, G.A., Burkin, I.M., Shepelyavyi, A.I.: Frequency Methods in Oscillation Theory. Kluwer Academic Publishers, Ser. Mathematics and its Applications, vol. 357, Dordrecht, Boston, London (1996)
21. Leonov, G.A., Noack, A., Reitmann, V.: Asymptotic orbital stability conditions for flows by estimates of singular values of the linearization. Nonl. Anal. Theory Methods Appl. **44**, 1057–1085 (2001)
22. Leonov, G.A., Ponomarenko, D.V., Smirnova, V.B.: Local instability and localization of attractors. From stochastic generators to Chua's system. Acta Appl. Math. **40**, 179–243 (1995)
23. Leonov, G.A., Reitmann, V., Smirnova, V.B.: Non-local Methods for Pendulum-like Feedback Systems. Teubner-Texte zur Mathematik, Bd. 132, B. G. Teubner Stuttgart- Leipzig (1992)
24. Li, M.Y., Muldowney, J.S.: Phase asymptotic semiflows, Poincaré's condition, and the existence of stable limit cycles. J. Diff. Equ. **124**(2), 425–448 (1996)
25. Lidskii, V.B.: On the characteristic numbers of the sum and product of symmetric matrices. Dokl. Akad. Nauk, SSSR **75**, 769–772 (1950) (Russian)
26. Noack, A.: Dimension and entropy estimates and stability investigations for nonlinear systems on manifolds. Doctoral Thesis, University of Technology Dresden (1998) (German)
27. Noack, A.: On Orbital Stability of Solutions of Differential Equations on Riemannian Manifolds. University of Technology Dresden, Preprint (1997)
28. Noack, A., Reitmann, V.: Hausdorff dimension estimates for invariant sets of time-dependent vector fields. Zeitschrift für Analysis und ihre Anwendungen (ZAA) **15**(2), 457–473 (1996)
29. Oseledec, V.I.: A multiplicative ergodic theorem: Lyapunov characteristic numbers for dynamical systems. Trans. Moscow Math. Soc. **19**, 197–231 (1968)
30. Pesin, Ya.B.: Characteristic Lyapunov exponents and smooth ergodic theory. Uspekhi Mat. Nauk **43**, 55–112 (1977) (Russian); English Transl. Russ. Math. Surv. **32**, 55–114 (1977)
31. Reitmann, V.: Applied Theory of Partial Differential Equations. St. Petersburg State Univ. Press, St. Petersburg (2019). (Russian)
32. Shiryaev, A.S., Khusainov, R.R., Mamedov, Sh.N., Gusev, S.V., Kuznetsov, N.V.: On Leonov's method for computing the linearization of transverse dynamics and analyzing Zhukovsky stability. Vestn. St. Petersburg Univ. Math., **52** (4), 344–341 (2019)
33. Stenström, B.: Dynamical systems with a certain local contraction property. Math. Scand. **11**, 151–155 (1962)
34. Temam, R.: Infinite-Dimensional Dynamical Systems in Mechanics and Physics. Springer, New York, Berlin (1988)
35. Yakubovich, V.A., Leonov, G.A., Gelig, A.K.: Stability of Stationary Sets in Control Systems with Discontinuous Nonlinearities. World Scientific, Singapore (2004)

Chapter 8
Dimension Estimates on Manifolds

Abstract In this chapter generalizations of the Douady-Oesterlé theorem (Theorem 5.1, Chap. 5) are obtained for maps and vector fields on Riemannian manifolds. The proof of the generalized Douady-Oesterlé theorem on manifolds is given in Sect. 8.1. In Sect. 8.2 it is shown that the Lyapunov dimension is an upper bound for the Hausdorff dimension. A tubular Carathéodory structure is used in Sect. 8.3 for the estimation of the Hausdorff dimension of invariant sets.

8.1 Hausdorff Dimension Estimates for Invariant Sets of Vector Fields

8.1.1 Introduction

In this section a version of the Douady-Oesterlé theorem [5] is obtained by requiring weaker conditions for the map under consideration. Lyapunov functions are introduced to modify the Jacobian matrix of the tangent map. That includes naturally a change of the singular values of the Jacobian matrix.

In this section, we also show that these results follow directly from a generalization of the Douady-Oesterlé theorem to Riemannian manifolds. With slightly stronger assumptions, such kind of generalization is quoted in [15].

In the last part of this section, the number of closed orbits for autonomous differential equations will be estimated from above in the form of a Bendixson-Dulac-criterion for Riemannian manifolds.

N. Kuznetsov and V. Reitmann, *Attractor Dimension Estimates for Dynamical Systems: Theory and Computation*, Emergence, Complexity and Computation 38, https://doi.org/10.1007/978-3-030-50987-3_8

8.1.2 Hausdorff Dimension Bounds for Invariant Sets of Maps on Manifolds

Let $(\mathcal{M}, g)$ be a Riemannian manifold without boundary, of dimension n and class C^1. Let $\mathcal{U} \subset \mathcal{M}$ be an open subset and let us consider a map $\varphi : \mathcal{U} \to \mathcal{M}$ of class C^1. The tangent map of φ at a point $u \in \mathcal{M}$ is denoted by $d_u\varphi : T_u\mathcal{M} \to T_{\varphi(u)}\mathcal{M}$.

Remark 8.1 Let $u \in \mathcal{U}$ be an arbitrary point and consider charts x and x' at u and $\varphi(u)$, respectively. We introduce the matrices $G := (g_{ij}(u))$ and $G' = (g'_{ij}(u))$ that realize the metric fundamental tensor g in the canonical bases of $T_u\mathcal{M}$ and $T_{\varphi(u)}\mathcal{M}$, respectively. The tangent map of φ at u written in coordinates of the charts x and x' is given by the matrix $\Phi := D(x' \circ \varphi \circ x^{-1})(x(u))$. From Example 7.1, Chap. 7, it follows that the singular values of the tangent map $d_u\varphi : T_u\mathcal{M} \to T_{\varphi(u)}\mathcal{M}$ coincide with the singular values of the matrix $\sqrt{G'}\Phi\sqrt{G^{-1}}$.

The following theorem from [24] generalizes the results of Sect. 5.1, Chap. 5 to Riemannian manifolds.

Theorem 8.1 *Let $d \in (0, n)$ be a real number and $\mathcal{K} \subset \mathcal{U}$ be a compact set which is negatively invariant with respect to φ, i.e. $\varphi(\mathcal{K}) \supset \mathcal{K}$. If the inequality*

$$\sup_{u \in \mathcal{K}} \omega_d(d_u\varphi) < 1 \tag{8.1}$$

holds, then $\dim_H \mathcal{K} < d$.

The proof of Theorem 8.1 is postponed to the end of this subsection. Now we proceed with some corollaries. The first one concerns a result which has been formulated in Sect. 5.1, Chap. 5 for the case $\mathcal{M} = \mathbb{R}^n$ using slightly stronger conditions for the map.

Corollary 8.1 *Let $\mathcal{K} \subset \mathcal{U} \subset \mathcal{M}$ be a compact set satisfying $\varphi(\mathcal{K}) \supset \mathcal{K}$. If for some continuous function $\varkappa : \mathcal{U} \to \mathbb{R}_+$ and for some number $d \in (0, n]$ the inequality*

$$\sup_{u \in \mathcal{K}} \left(\frac{\varkappa(\varphi(u))}{\varkappa(u)} \omega_d(d_u\varphi) \right) < 1 \tag{8.2}$$

holds, then $\dim_H \mathcal{K} < d$.

Proof On the open set $\mathcal{U} \subset \mathcal{M}$ we introduce a new metric tensor $\widetilde{g}$ by $\widetilde{g}_{|_u} := \varkappa^2(u) g_{|_u}$. It is easy to show that this is really a Riemannian metric equivalent to the given one on compact subsets of $\mathcal{M}$. Since $\mathcal{K}$ is compact, the new equivalent metric does not alter the value of the Hausdorff dimension of $\mathcal{K}$.

Let us consider an arbitrary point $u \in \mathcal{K}$ and two charts x and x' around u and $\varphi(u)$, respectively. Suppose that $G := (g_{ij}(u))$ and $G' := (g'_{ij}(\varphi(u)))$ are the realizations with respect to the canonical bases in $T_u\mathcal{M}$ and $T_{\varphi(u)}\mathcal{M}$ of the metric tensors g and

g', respectively. As indicated in Remark 8.1 the singular values of the tangent map $d_u\varphi$ in the new metric are the singular values of the matrix

$$\sqrt{\widetilde{G}'}\Phi\sqrt{\widetilde{G}^{-1}} = \varkappa(\varphi(u))\sqrt{G'}\Phi\varkappa(u)^{-1}\sqrt{G^{-1}}.$$

Thus, condition (8.2) guarantees, that in the new metric the inequality (8.1) is valid and Theorem 8.1 can be applied. □

The next corollary [24] provides a method for estimating the Hausdorff dimension without explicit computation of the singular values.

Corollary 8.2 *Let $\mathcal{K} \subset \mathcal{U} \subset \mathcal{M}$ be a compact set which is assumed to be negatively invariant with respect to φ. Let $\theta : \mathcal{U} \to \mathbb{R}_+$ be a continuous function and let $d \in (0, n]$ be a real number such that the conditions*

(a) $\quad \left([(d_u\varphi)^{[*]}d_u\varphi]v, v\right) \geq \theta(u)^2|v|^2\,, \quad \forall u \in \mathcal{K}, \quad \forall v \in T_u\mathcal{M},$ *and*

(b) $\quad \dfrac{|\det d_u\varphi|}{\theta(u)^{n-d}} < 1\,, \quad \forall u \in \mathcal{K}\,,$

are satisfied. Then $\dim_H \mathcal{K} < d$.

Proof From condition a) for the singular values $\alpha_i(u)$ of the tangent map $d_u\varphi$ we obtain the inequalities

$$\alpha_i(u) \geq \theta(u) \quad , \quad \forall u \in \mathcal{K}, \quad i = 1, 2, \dots, n.$$

Thus, for any $k \in \{0, 1, \dots, n\}$ and $u \in \mathcal{K}$ with $\alpha_0(u) := 1$ it follows that

$$\alpha_1(u)\cdots\alpha_k(u)\theta(u)^{n-k} \leq \alpha_1(u)\cdots\alpha_n(u) = |\det d_u\varphi|.$$

For $d = d_0 + s$ with $d_0 \in \{0, 1, \dots, n-1\}$ and $s \in (0, 1]$ the last relation together with condition b) of the corollary leads to

$$\begin{aligned}\alpha_1(u)\cdots\alpha_{d_0}(u)\alpha_{d_0+1}^s(u) &= \left(\alpha_1(u)\cdots\alpha_{d_0}(u)\right)^{1-s}\left(\alpha_1(u)\cdots\alpha_{d_0}(u)\alpha_{d_0+1}(u)\right)^s \\ &\leq \frac{|\det d_u\varphi|^{1-s}}{\theta(u)^{(n-d_0)(1-s)}}\frac{|\det d_u\varphi|^s}{\theta(u)^{(n-d_0-1)s}} = \frac{|\det d_u\varphi|}{\theta(u)^{n-d}} < 1.\end{aligned}$$

Now, the only thing left to do is to apply Theorem 8.1. □

Let us come back now to the Riemannian manifold $(\mathcal{M}, g)$ and consider the exponential map (see Sect. A.8, Appendix A) $\exp_u : T_u\mathcal{M} \to \mathcal{M}$ at an arbitrary point $u \in \mathcal{M}$. Then the set $\exp_u(\mathcal{E})$ is the image of an ellipsoid $\mathcal{E}$ in the tangent space $T_u\mathcal{M}$ centered at 0 under the map $\exp_u$.

Let $\mathcal{K} \subset \mathcal{U}$ be a compact set, let $\varepsilon > 0$ be a sufficiently small number and let us fix a number $d \in (0, n]$. The *ellipsoid premeasure at level* ε *and of order* d of $\mathcal{K}$ is given by

$$\widetilde{\mu}_H(\mathcal{K}, d, \varepsilon) := \inf \sum_i \omega_d(\mathcal{E}_i)$$

where the infimum is taken over all finite covers $\bigcup_i \exp_{u_i}(\mathcal{E}_i) \supset \mathcal{K}$, where $u_i \in \mathcal{M}$, and $\mathcal{E}_i \subset T_{u_i}\mathcal{M}$ are ellipsoids satisfying $\omega_d(\mathcal{E}_i)^{1/d} \leq \varepsilon$. Now we show the equivalence of both the Hausdorff premeasure and the ellipsoid premeasure in a similar way as is done in Sect. 5.1, Chap. 5.

Lemma 8.1 *For an arbitrary number $d \in (0, n]$ written in the form $d = d_0 + s$ with $d_0 \in \{0, 1, \dots, n-1\}$ and $s \in (0, 1]$ we define the numbers $C_d := 2^{d_0}(d_0 + 1)^{d/2}$ and $\lambda_d := \sqrt{d_0 + 1}$. Then for a compact set $\mathcal{K} \subset \mathcal{U}$ and for every sufficiently small $\varepsilon > 0$ the inequalities*

$$\mu_H(\mathcal{K}, d, \varepsilon) \geq \widetilde{\mu}_H(\mathcal{K}, d, \varepsilon) \geq C_d^{-1} \mu_H(\mathcal{K}, d, \lambda_d \varepsilon) \tag{8.3}$$

hold.

Proof In a number of technical details the proof differs from the one given in Sect. 5.1, Chap. 5.

In analogous manner as in Sect. 5.1, Chap. 5, it is established that for sufficiently small $\varepsilon > 0$, for an arbitrary $u \in \mathcal{K}$ and any ellipsoid $\mathcal{E} \subset T_u\mathcal{M}$ satisfying $\omega_d(\mathcal{E})^{1/d} \leq \varepsilon$ the relation

$$\mu_H(\exp_u(\mathcal{E}), d, \lambda_d \varepsilon) \leq C_d \, \omega_d(\mathcal{E}) \tag{8.4}$$

is valid.

Let us now fix a finite cover of $\mathcal{K}$ consisting of sets $\{\exp_{u_i}(\mathcal{E}_i)\}$, where $\omega_d(\mathcal{E}_i)^{1/d} \leq \varepsilon$ holds for all indices i. The properties of the premeasures then guarantee the following relation

$$\mu_H(\mathcal{K}, d, \lambda_d \varepsilon) \leq \mu_H\left(\bigcup_i \exp_{u_i}(\mathcal{E}_i), d, \lambda_d \varepsilon\right) \leq \sum_i \mu_H(\exp_{u_i}(\mathcal{E}_i), d, \lambda_d \varepsilon).$$

Using (8.4) we obtain $\mu_H(\mathcal{K}, d, \lambda_d \varepsilon) \leq C_d \sum_i \omega_d(\mathcal{E}_i)$ and since the cover was arbitrary among all those satisfying the restriction for ω_d we have

$$\mu_H(\mathcal{K}, d, \lambda_d \varepsilon) \leq C_d \, \widetilde{\mu}_H(\mathcal{K}, d, \varepsilon).$$

□

Lemma 8.2 *Let $\mathcal{K} \subset \mathcal{U}$ be a compact set and consider a map $\varphi : \mathcal{U} \to \mathcal{M}$ of class C^1. For a number $d \in (0, n]$ we assume that $\sup_{u \in \mathcal{K}} \omega_d(d_u\varphi) \leq k$. Then, for every $l > k$ there exists a number $\varepsilon_0 > 0$ such that for every $\varepsilon \in (0, \varepsilon_0]$*

$$\mu_H(\varphi(\mathcal{K}), d, \lambda_d \, l^{\frac{1}{d}} \varepsilon) \leq C_d \, l \, \mu_H(\mathcal{K}, d, \varepsilon)$$

holds, where C_d and λ_d are defined as in Lemma 8.1.

Proof In a first step we show for sufficiently small numbers $\varepsilon > 0$ the inequality

$$\widetilde{\mu}_H(\varphi(\mathcal{K}), d, l^{\frac{1}{d}}\varepsilon) \le l\mu_H(k, d, \varepsilon). \tag{8.5}$$

Obviously it is always possible to find an open set $\mathcal{V} \subset \mathcal{U}$ containing $\mathcal{K}$ which itself lies inside a compact set $\widetilde{\mathcal{K}} \subset \mathcal{U}$ with the property

$$k' := \sup_{u \in \widetilde{\mathcal{K}}} \omega_d(d_u\varphi) < l.$$

We choose a number $\delta > 0$ such that $k' < \delta^d$ and

$$\sup_{u \in \widetilde{\mathcal{K}}} |d_u\varphi| \le \delta \tag{8.6}$$

hold. Further we can find a number $\eta > 0$ satisfying

$$\left[1 + \left(\frac{\delta^{d_0}}{k'}\right)^{\frac{1}{s}} \eta\right]^d k' = l\,. \tag{8.7}$$

We take $\varepsilon > 0$ small enough such that

$$\varepsilon \le \frac{1}{2} \inf_{\substack{u \in \mathcal{K} \\ u' \in \mathcal{U} \setminus \mathcal{V}}} \rho(u, u') \quad \text{and} \quad |\tau_{\varphi(v)}^{\varphi(u)} d_v\varphi\tau_u^v - d_u\varphi| \le \eta$$

for all $u, v \in \mathcal{V}$ with $\rho(u, v) \le \varepsilon$. By $\rho(\cdot, \cdot)$ we mean the geodesic distance between the points of $\mathcal{M}$ and by τ_u^v we denote the isometry between $T_u\mathcal{M}$ and $T_v\mathcal{M}$ defined by parallel transport along the geodesic for points lying sufficiently near to each other (see Sect. A.8, Appendix A).

Let us fix a finite cover with balls $\{\mathcal{B}(u_i, r_i)\}_i$ of radius $r_i \le \varepsilon$ of $\mathcal{K}$. Then every ball $\mathcal{B}(u_i, r_i)$ satisfying $\mathcal{B}(u_i, r_i) \cap \mathcal{K} \ne \emptyset$ is entirely contained in the open set $\mathcal{V}$. The Taylor formula for differentiable maps provides that for every $v \in \mathcal{B}(u_i, r_i)$

$$|\exp_{\varphi(u_i)}^{-1} \varphi(v) - d_{u_i}\varphi(\exp_{u_i}^{-1}(v))| \le \sup_{w \in \mathcal{B}(u_i, r_i)} |\tau_{\varphi(w)}^{\varphi(u_i)} d_w\varphi\tau_{u_i}^w - d_{u_i}\varphi| \cdot |\exp_{u_i}^{-1}(w)| \tag{8.8}$$

holds. Thus, for every ball $\mathcal{B}(u_i, r_i)$ of the cover with $\mathcal{B}(u_i, r_i) \cap \mathcal{K} \ne \emptyset$ the image under the map φ is included in the following set

$$\varphi(\mathcal{B}(u_i, r_i)) \subset \exp_{\varphi(u_i)}\left(d_{u_i}\varphi(\mathcal{B}_{T_{u_i}\mathcal{M}}(0, r_i)) + \mathcal{B}_{T_{\varphi(u_i)}\mathcal{M}}(0, \eta r_i)\right)\,. \tag{8.9}$$

The notions $\mathcal{B}_{T_{u_i}\mathcal{M}}(0, r_i)$ and $\mathcal{B}_{T_{\varphi(u_i)}\mathcal{M}}(0, \eta r_i)$ stand for balls in the tangent spaces $T_{u_i}\mathcal{M}$ and $T_{\varphi(u_i)}\mathcal{M}$, respectively. Obviously the set $d_{u_i}\varphi(\mathcal{B}_{T_{u_i}\mathcal{M}}(0, r_i))$ is an ellipsoid with half-axes of length $r_i\alpha_j(u_i)$ $(j = 1, \ldots, n)$, where $\alpha_j(u_i)$ $(j = 1, \ldots, n)$ denote the singular values of the linear operator $d_{u_i}\varphi$. Concerning the definition of k' we may conclude

$$\omega_d\left(d_{u_i}\varphi(\mathcal{B}_{T_{u_i}\mathcal{M}}(0,r_i))\right) \le r_i^d k'. \tag{8.10}$$

With (8.6) it follows

$$\alpha_i\left(d_{u_i}\varphi(\mathcal{B}_{T_{u_i}\mathcal{M}}(0,r_i))\right) \le \delta r_i\ . \tag{8.11}$$

Further by using (8.7), (8.10) and (8.11) and by Lemma 7.1, Chap. 7 there can be found an ellipsoid $\mathcal{E}_i'$ containing $\exp_{u_i}^{-1}(\varphi(\mathcal{B}(u_i,r_i)))$ and satisfying $\omega_d(\mathcal{E}_i') \le l r_i^d$.

We can summarize that every finite cover of the compact set $\mathcal{K}$ with balls $\{\mathcal{B}(u_i,r_i)\}_i$ of radius $r_i \le \varepsilon$ such that $\mathcal{B}(u_i,r_i) \cap \mathcal{K}$ is non-empty, generates a cover $\{\exp_{u_i}(\mathcal{E}_i')\}_i$ of $\varphi(\mathcal{K})$, where $\mathcal{E}_i'$ denotes an ellipsoid in $T_{u_i}\mathcal{M}$ satisfying $\omega_d(\mathcal{E}_i') \le l r_i^d$. Therefore, we have

$$\sum_i \omega_d(\mathcal{E}_i') \le l \sum_i r_i^d.$$

Since the result is valid for every such cover it must be true for the infimum as well. So we have

$$\sum_i \omega_d(\mathcal{E}_i') \le l\mu_H(\mathcal{K},d,\varepsilon).$$

If at the left-hand side we pass to the infimum, then the last inequality becomes (8.5). Lemma 8.1 and Eq. (8.5) finally guarantee the inequalities

$$l\mu_H(\mathcal{K},d,\varepsilon) \ge \widetilde{\mu}_H(\varphi(\mathcal{K}),d,l^{\frac{1}{d}}\varepsilon) \ge C_d^{-1}\mu_H(\varphi(\mathcal{K}),d,\lambda_d l^{\frac{1}{d}}\varepsilon).$$

But this is exactly what we wanted to prove. □

Proof of Theorem 8.1 The essence of the proof of Theorem 8.1 is contributed by Lemma 8.2. The lemma claims that the Hausdorff premeasure defined on a Riemannian manifold exhibits the same properties concerning the effect of a map φ of class C^1 as it has in $\mathbb{R}^n$. When applying Lemma 8.2 the statement of Theorem 8.1 follows directly from arguments that agree with the last steps of the proof of Theorem 5.1 in Sect.5.1, Chap. 5. □

The next theorem provides another important result, the proof of which follows directly from Lemma 8.2 using the same arguments as in [19].

Theorem 8.2 *Let the manifold $\mathcal{M}$ be compact and let $\varphi : \mathcal{M} \to \mathcal{M}$ be a map of class C^1. Suppose that $\sup_{u\in\mathcal{M}} \omega_d(d_u\varphi) < 1$ holds for a number $d \in (0,n]$. If for a compact set $\mathcal{K} \subset \mathcal{M}$ the condition $\mu_H(\mathcal{K},d) < \infty$ is satisfied, then*

$$\lim_{m\to\infty} \mu_H(\varphi^m(\mathcal{K}),d) = 0.$$

Remark 8.2 A version of Theorem 8.1 for differential equations in Hilbert spaces is given in [3]. For vector fields on Hilbert manifolds a similar theorem is shown in [13, 14].

8.1.3 Time-Dependent Vector Fields on Manifolds

Let $(\mathcal{M}, g)$ be a Riemannian manifold without boundary of dimension n and of class C^2, let $\mathcal{U} \subset \mathcal{M}$ be an open subset and $\mathcal{I}_1 \subset \mathbb{R}$ an open interval with $0 \in \mathcal{I}_1$. We consider a time-dependent vector field $F : \mathcal{I}_1 \times \mathcal{U} \to T\mathcal{U}$ of class C^1 and the corresponding differential equation

$$\dot{u} = F(t, u) . \tag{8.12}$$

Suppose, that for a point $(t, u) \in \mathcal{I}_1 \times \mathcal{U}$ the covariant derivative of the vector field F is $\nabla F(t, u) : T_u\mathcal{M} \to T_u\mathcal{M}$. We assume for (8.12) that there can be found an open set $\mathcal{D} \subset \mathcal{U}$ and an open interval $\mathcal{I} \subset \mathcal{I}_1$ such that the solution $\varphi(\cdot, u)$ of (8.12) starting at $u \in \mathcal{D}$ for $t = 0$ exists everywhere on $\mathcal{I}$. For every $t \in \mathcal{I}$ we can define the t-shift operator $\varphi^t : \mathcal{D} \to \mathcal{U}$ by $\varphi^t(u) := \varphi(t, u)$. In case the differential equation (8.12) is autonomous, the family $\{\varphi^t\}_{t \in \mathcal{I}}$ of all those t-shifts is a local flow.

Since the vector field F is continuously differentiable, the same holds for the time-t-shift operators φ^t $(t \in \mathcal{I})$. For an arbitrary point $u \in \mathcal{D}$, the tangent map $d_u\varphi^t$ solves the variation equation

$$y' = \nabla F(t, \varphi^t(u))y \tag{8.13}$$

with initial condition $d_u\varphi^t_{|t=0} = \mathrm{id}_{T_u\mathcal{M}}$. Here the absolute derivative y' is taken along the integral curve $t \mapsto \varphi^t(u)$ in the direction of the vector field F. Let us denote the eigenvalues of the symmetric part of the covariant derivative $\nabla F(t, u)$, i.e., of the operator

$$S(t, u) := \frac{1}{2}\left[\nabla F(t, u) + \nabla F(t, u)^{[*]}\right]$$

by $\lambda_i(t, u)$ $(i = 1, \ldots, n)$ and order them with respect to its size and multiplicity, i.e., $\lambda_1(t, u) \geq \cdots \geq \lambda_n(t, u)$. The divergence $\operatorname{div} F(t, u)$ of the vector field F at $(t, u) \in \mathcal{I}_1 \times \mathcal{U}$ is the trace of the linear operator $\nabla F(t, u) : T_u\mathcal{M} \to T_u\mathcal{M}$ and therefore $\operatorname{div} F(t, u) = \operatorname{tr} \nabla F(t, u) = \lambda_1(t, u) + \cdots + \lambda_n(t, u)$ holds. The next theorem is the main result of this section and extends a result of [27] to Riemannian manifolds. The proof is given at the end of the section.

Theorem 8.3 *Let $d \in (0, n]$ be a real number written in the form $d = d_0 + s$ with $d_0 \in \{0, 1, \ldots, n-1\}$ and $s \in (0, 1]$ and let $\mathcal{K} \subset \mathcal{D}$ be a compact set satisfying $\varphi^\tau(\mathcal{K}) \supset \mathcal{K}$ for a certain $\tau \in \mathcal{I} \cap \mathbb{R}_+$. If the condition*

$$\sup_{u \in \mathcal{K}} \int_0^\tau \left[\lambda_1(t, \varphi^t(u)) + \cdots + \lambda_{d_0}(t, \varphi^t(u)) + s\lambda_{d_0+1}(t, \varphi^t(u))\right] dt < 0 \tag{8.14}$$

holds, then $\dim_H \mathcal{K} < d$.

Remark 8.3 We shall now consider an arbitrary $u \in \mathcal{U}$ and a chart x around u. In local coordinates of x and in the canonical basis $\partial_1(u), \dots, \partial_n(u)$ of the tangent space $T_u\mathcal{M}$, the vector field of (8.12) then becomes $F(t, u) = f^i(t, u)\partial_i(u)$ and the covariant derivative $\nabla F(t, u) : T_u\mathcal{M} \to T_u\mathcal{M} : v \mapsto \nabla_v F(t, u)$ is given by $\nabla_v F(t, u) = \nabla_i f^k(t, u)v^i \partial_k(u)$, where $v = v^i\partial_i(u)$ is an arbitrary vector in $T_u\mathcal{M}$ and

$$\nabla_i f^k = \frac{\partial f^k}{\partial x^i} + \Gamma^k_{ij} f^j .$$

Here by Γ^k_{ij} the Christoffel symbols in the chart x corresponding to the metric tensor g are denoted (see Sect. A.5, Appendix A). The symmetric part $S(t, u)$ of $\nabla F(t, u)$ in the canonical basis of $T_u\mathcal{M}$ is realized by the matrix

$$\frac{1}{2}\left[G^{-1}\Phi^T G + \Phi\right] , \tag{8.15}$$

where G is as in Remark 8.1 and $\Phi = (\nabla_i f^k)$. The expression for the variational equation (8.13) in the chart x is

$$y^{k\,\prime} = \dot{y}^k + \Gamma^k_{ij} f^j y^i = \nabla_i f^k y^i .$$

Let us define $f_{s,i} = g_{st}\nabla_i f^t$ and consider the quadratic form $e_{si} = \frac{1}{2}\left[f_{s,i} + f_{i,s}\right]$. Then e_{si} is related to (8.15) in the sense that the eigenvalues of this quadratic form, i.e., the solutions of $\det\left[e_{si} - \lambda g_{si}\right] = 0$, coincide with the eigenvalues of the matrix (8.15).

Let us now, by means of the notion $c_{jk} = g_{ik}\Gamma^i_{js} f^s$, introduce the derivative of the metric tensor g in the direction of the vector field f^i by $\dot{g}_{jk} = \frac{1}{2}\left[c_{jk} + c_{kj}\right]$. Then the quadratic form can be written as

$$e_{si} = \frac{1}{2}\left[g_{sk}\frac{\partial f^k}{\partial x^i} + \frac{\partial f^k}{\partial x^s} g_{ik}\right] + \dot{g}_{si} .$$

Before we devote ourselves to the proof of Theorem 8.3, we shall add some corollaries. The first one generalizes results from Chap. 5 formulated there for the case $\mathcal{M} = \mathbb{R}^n$ and under slightly stronger conditions. Keeping in mind the second method of Lyapunov which is often used in stability theory, that result is in Sect. 5.1, Chap. 5 referred to as "introduction of a Lyapunov function in Hausdorff dimension estimates".

From the point of view of the present section this approach can be treated as the introduction of a new metric tensor on the manifold. Note that a special case of this theorem for vector fields on a cylinder is considered in [17].

For a differentiable function $V : \mathcal{U} \to \mathbb{R}$ the map $\dot{V} : \mathcal{I}_1 \times \mathcal{U} \to \mathbb{R}$ defined by $\dot{V}(t, u) = (d_u V, F(t, u))$ is the derivative of V in the direction of the vector field F.

Corollary 8.3 *Let $\mathcal{K} \subset \mathcal{D}$ be a compact set such that $\varphi^\tau(\mathcal{K}) \supset \mathcal{K}$ is true for some $\tau \in \mathcal{I} \cap \mathbb{R}_+$. Let $V : \mathcal{U} \to \mathbb{R}$ be a differentiable function and denote by $\lambda_1(t,u) \geq \cdots \geq \lambda_n(t,u)$ the eigenvalues of $S(t,u)$. If for a real number $d \in (0,n]$ written in the form $d = d_0 + s$ with $d_0 \in \{0,1,\ldots,n-1\}$ and $s \in (0,1]$ the condition*

$$\sup_{u\in\mathcal{K}} \int_0^\tau \left[\lambda_1(t,\varphi^t(u)) + \cdots + \lambda_{d_0}(t,\varphi^t(u)) + s\lambda_{d_0+1}(t,\varphi^t(u)) + \dot{V}(t,\varphi^t(u))\right] dt < 0\,, \tag{8.16}$$

holds, then $\dim_H \mathcal{K} < d$.

Proof We shall introduce on $\mathcal{U}$ a new metric tensor $\widetilde{g}_{|u} = \varkappa^2(u) g_{|u}$ by means of some function $\varkappa : \mathcal{U} \to \mathbb{R}_+$ of class C^1. Let us fix a point $u \in \mathcal{U}$ and consider a chart x around u. Further, let the metric tensor g and the vector field F be expressed in the canonical basis of $T_u\mathcal{M}$ by g_{ij} and f^i, respectively. The symmetric part of the covariant derivative $\widetilde{\nabla} F(t,u)$ at $u \in \mathcal{U}$ with respect to the new metric is then determined according to Remark 8.3 by the matrix representation

$$\frac{1}{2}\left[G^{-1}\Phi^T G + \Phi\right] + \frac{\dot{\varkappa}}{\varkappa}\mathrm{Id}\,. \tag{8.17}$$

If, in particular, if we choose

$$\varkappa(u) := e^{\frac{V(u)}{d}} \quad (u \in \mathcal{U})$$

then $\dot{\varkappa}(u) := \varkappa(u)\frac{\dot{V}(u)}{d}$ implies that the eigenvalues $\widetilde{\lambda}_i$ of (8.17) are related to the eigenvalues with respect to the original metric g by the formula $\widetilde{\lambda}_i = \lambda_i + \frac{\dot{V}}{d}$. Finally

$$\widetilde{\lambda}_1 + \cdots + \widetilde{\lambda}_{d_0} + s\widetilde{\lambda}_{d_0+1} = \lambda_1 + \cdots + \lambda_{d_0} + s\lambda_{d_0+1} + \dot{V}$$

guarantees (8.16) and therefore (8.14) of Theorem 8.3. □

Suppose that Λ is a logarithmic norm on the space of linear operators $L_u : T_u\mathcal{M} \to T_u\mathcal{M}$. For a number $d = d_0 + s$ with integer $d_0 \in [0, n-1]$ and $s \in [0,1]$ we introduce the partial d-trace w.r.t. Λ of $\nabla F(t,u) : T_u\mathcal{M} \to T_u\mathcal{M}$

$$\mathrm{tr}_{d,\Lambda}\, \nabla F(t,u) = s\Lambda(\nabla F(t,u)^{[d_0+1]} + (1-s)\Lambda(\nabla F(t,u)^{[d_0]}.$$

Then the following Corollary 8.4 is true.

Corollary 8.4 *Let $\mathcal{K} \subset \mathcal{D}$ be a compact set such that $\varphi^\tau(\mathcal{K}) \supset \mathcal{K}$ is true for some $\tau \in \mathcal{I} \cap \mathbb{R}_+$. Let Λ be a logarithmic norm and a continuously differentiable on $\mathcal{K}$ function Λ satisfying with $d = d_{0+s}$ the inequality*

$$\sup_{u\in\mathcal{K}} \int_0^\tau \left[\mathrm{tr}_{d,\Lambda}(\nabla F(t,u) + \dot{V}(t,\varphi^t(u))\right] dt < 0 \tag{8.18}$$

holds, then $\dim_H \mathcal{K} < d$.

To verify the conditions of Theorem 8.3 we need to compute the eigenvalues of the symmetric part of the covariant derivative. The next two corollaries are variations of this theorem using conditions on the divergence of the vector field of (8.12). Similar results for the case $\mathcal{M} = \mathbb{R}^n$ can be found in Sect. 5.2, Chap. 5.

Corollary 8.5 *Let* $\mathcal{K} \subset \mathcal{D}$ *be a compact set such that* $\varphi^\tau(\mathcal{K}) \supset \mathcal{K}$ *holds for some* $\tau \in \mathcal{I} \cap \mathbb{R}_+$. *Assume that for a continuous function* $\theta : \mathcal{I} \times \mathcal{U} \to \mathbb{R}$ *and for some* $d \in (0, n]$ *the conditions*

(a) $\big(S(t,u)v, v\big) \geq -\theta(t,u)|v|^2 \quad \forall t \in [0,\tau], u \in \mathcal{U}, v \in T_u\mathcal{M}$ *and*

(b) $\sup_{u\in\mathcal{K}} \int\limits_0^\tau \big[\operatorname{div} F(t,\varphi^t(u)) + (n-d)\theta(t,\varphi^t(u))\big]\,dt < 0$

are satisfied. Then $\dim_H \mathcal{K} < d$.

Proof From condition (a) we have for the eigenvalues λ_i of $S(t,u)$,

$$\lambda_i(t,u) \geq -\theta(t,u) \qquad \forall (t,u) \in [0,\tau] \times \mathcal{U}, i = 1, \ldots, n\,. \tag{8.19}$$

Thus, if $k \in \{0, 1, \ldots, n\}$, $t \in [0,\tau]$ and $u \in \mathcal{U}$ are arbitrary then

$$\lambda_1(t,u) + \cdots + \lambda_k(t,u) - (n-k)\theta(t,u) \leq \operatorname{tr} S(t,u) = \operatorname{div} F(t,u).$$

This implies

$$\lambda_1(t,u) + \cdots + \lambda_{d_0}(t,u) + s\lambda_{d_0+1}(t,u) \leq \operatorname{div} F(t,u) + (n-d)\theta(t,u).$$

By using condition (b) and Theorem 8.3 the proof is complete. □

Corollary 8.6 *Consider (8.12) on an open set* $\mathcal{U} \subset \mathbb{R}^n$. *Suppose that system (8.12) possesses a compact set* $\mathcal{K}$ *satisfying* $\varphi^\tau(\mathcal{K}) \supset \mathcal{K}$ *for some* $\tau \in \mathcal{I} \cap \mathbb{R}_+$. *Further, assume that there exist a number* $d \in (0, n]$, *an* $n \times n$ *matrix* $H = H^T > 0$ *and a continuous function* $\theta : \mathcal{I} \times \mathcal{U} \to \mathbb{R}$ *such that the condition b) of Corollary 8.5 and the inequality*

$$\frac{1}{2}\big[HD_2F(t,u) + D_2F(t,u)^T H\big] \geq -\theta(t,u)H \quad \forall (t,u) \in [0,\tau] \times \mathcal{U} \tag{8.20}$$

are satisfied. Then $\dim_H \mathcal{K} < d$.

Proof Let us introduce in $\mathcal{U}$ a new metric by means of the matrix $(g_{ij}) \equiv H$. From Remark 8.3 it follows that the eigenvalues $\lambda_i(t,u)$ of $S(t,u)$ with respect to the new metric, agree with the eigenvalues of the quadratic form corresponding to (8.15). Therefore, they satisfy the relation

$$\frac{1}{2}\left[H D_2 F(t,u) + D_2 F(t,u)^T H\right] = \lambda_i(t,u) H.$$

Using this and (8.20) we obtain (8.19). All further steps can be analogously be carried out as in the proof of Corollary 8.5. □

Proof of Theorem 8.2 Let us fix an arbitrary $u \in \mathcal{K}$, a number $k \in \{1, \dots, n\}$ and arbitrary $\upsilon_1, \dots, \upsilon_k \in T_u\mathcal{M}$. For every $t \in [0, \tau]$ we introduce

$$w(t) := |d_u\varphi^t \upsilon_1 \wedge \cdots \wedge d_u\varphi^t \upsilon_k|^2_{\wedge^k T_{\varphi^t(u)}\mathcal{M}} .$$

Applying the variational equation (8.13) and Definition 7.7, Chap. 7, we achieve for every t in $[0, \tau]$ the equation

$$\dot{w}(t) = 2\left(\left[S(t,\varphi^t(u))\right]_k (d_u\varphi^t \upsilon_1 \wedge \cdots \wedge d_u\varphi^t \upsilon_k), d_u\varphi^t \upsilon_1 \wedge \cdots \wedge d_u\varphi^t \upsilon_k\right)_{\wedge^k T_{\varphi^t(u)}\mathcal{M}} .$$

With Proposition 7.9, Chap. 7, for every $t \in [0, \tau]$ this leads to

$$\begin{aligned} &2\left[\lambda_{n-k+1}(t,\varphi^t(u)) + \cdots + \lambda_n(t,\varphi^t(u))\right] w(t) \\ \le\ &\dot{w}(t) \le 2\left[\lambda_1(t,\varphi^t(u)) + \cdots + \lambda_k(t,\varphi^t(u))\right] w(t). \end{aligned} \tag{8.21}$$

Therefore we conclude

$$\begin{aligned} &|d_u\varphi^\tau \upsilon_1 \wedge \cdots \wedge d_u\varphi^\tau \upsilon_k|_{\wedge^k T_{\varphi^t(u)}\mathcal{M}} \\ &\quad \le |\upsilon_1 \wedge \cdots \wedge \upsilon_k|_{\wedge^k T_u\mathcal{M}} \exp\left\{\int_0^\tau \left[\lambda_1(t,\varphi^t(u)) + \cdots + \lambda_k(t,\varphi^t(u))\right] dt\right\}. \end{aligned} \tag{8.22}$$

Let us apply the Fischer-Courant Theorem (Theorem 7.1, Chap. 7) to the product of the squares of the first k singular values of $d_u\varphi^\tau$ and use (8.22) in order to receive the following result

$$\begin{aligned} \alpha_1(\tau,u)^2 \cdots \alpha_k(\tau,u)^2 &= \lambda_1\left(\bigwedge^k \left[(d_u\varphi^\tau)^* d_u\varphi^\tau\right]\right) \\ &= \sup_{\substack{\upsilon \in \wedge^k T_u\mathcal{M} \\ |\upsilon|_{\wedge^k T_u\mathcal{M}}=1}} \left|\bigwedge^k d_u\varphi^\tau \upsilon\right|^2_{\wedge^k T_{\varphi^\tau(u)}\mathcal{M}} \\ &= \sup_{\substack{\upsilon_1,\dots,\upsilon_k \in T_u\mathcal{M} \\ |\upsilon_i|_{T_u\mathcal{M}}=1}} |d_u\varphi^\tau \upsilon_1 \wedge \cdots \wedge d_u\varphi^\tau \upsilon_k|^2_{\wedge^k T_{\varphi^\tau(u)}\mathcal{M}} \\ &\le \exp\left\{2\int_0^\tau \left[\lambda_1(t,\varphi^t(u)) + \cdots + \lambda_k(t,\varphi^t(u))\right] dt\right\}. \end{aligned}$$

This last inequality and the assumptions of Theorem 8.3 finally guarantee that

$$\begin{aligned}
&\sup_{u\in\mathcal{K}} \omega_d(d_u\varphi^\tau) \\
&= \sup_{u\in\mathcal{K}} \left[\alpha_1(\tau,u)\cdots\alpha_{d_0}(\tau,u)\right]^{1-s}\left[\alpha_1(\tau,u)\cdots\alpha_{d_0+1}(\tau,u)\right]^s \\
&\le \sup_{u\in\mathcal{K}} \exp\left\{\int_0^\tau \left[\lambda_1(t,\varphi^t(u))+\cdots+\lambda_{d_0}(t,\varphi^t(u))+s\lambda_{d_0+1}(t,\varphi^t(u))\right]dt\right\} < 1\,.
\end{aligned} \tag{8.23}$$

This shows that for the map φ^τ, the assumptions of Theorem 8.1 are valid.

□

Remark 8.4 The inequality (8.21) can be interpreted as a generalization of Liouville's truncated trace formula for linear differential equations in Euclidean spaces (Sect. 2.4, Chap. 2). In particular, from (8.21), by setting $k = n$ and indicating that

$$\lambda_1(t,\varphi^t(u))+\cdots+\lambda_n(t,\varphi^t(u)) = \mathrm{tr}\nabla F(t,\varphi^t(u)) = \mathrm{div}F(t,\varphi^t(u)) \tag{8.24}$$

we obtain Liouville's trace formula in the form

$$\left(|\det d_u\varphi^t|\right)^{\cdot} = \mathrm{div}F(t,\varphi^t(u))|\det d_u\varphi^t| \quad \forall t\in[0,\tau]\,, u\in\mathcal{K}\,. \tag{8.25}$$

For a point u and a chart x in the neighborhood of this point let again g_{ij}, f^i and ξ^i represent g, F and φ^t in coordinates of the chart x. Then, formula (8.25) agrees locally with

$$\left(\sqrt{\gamma}\left|\det\left(\frac{\partial\xi^i}{\partial x^j}\right)\right|\right)^{\cdot} = \nabla_k\xi^k\sqrt{\gamma}\left|\det\left(\frac{\partial\xi^i}{\partial x^j}\right)\right|$$

where γ stands for $\det(g_{ij})$. For a Lebesgue measurable set $\Omega\subset\mathcal{D}$ of finite volume, we denote the volume of $\varphi^t(\Omega)$ by V_t. Then the formula (8.25) provides the transport lemma ([1]; see also Sect. A.6, Appendix A) for Riemannian manifolds in the form

$$\dot V_t = \int_{\varphi^t(\Omega)} \mathrm{div}F dV.$$

8.1.4 *Convergence for Autonomous Vector Fields*

We shall now consider compact Riemannian manifolds $(\mathcal{M}, g)$ without boundary of dimension n and of class C^2. Let on $\mathcal{M}$ be given a vector field $F : \mathcal{M}\to T\mathcal{M}$ of class C^1 and the corresponding differential equation

$$\dot u = F(u)\,. \tag{8.26}$$

We assume that the global flow $\varphi : \mathbb{R} \times \mathcal{M} \to \mathcal{M}$ of (8.26) exists. As in the previous subsection we denote by $\lambda_1(u) \geq \cdots \geq \lambda_n(u)$ the eigenvalues of the symmetric part $S(u) = \frac{1}{2}\left[\nabla F(u)^{[*]} + \nabla F(u)\right]$ of the covariant derivative $\nabla F(u) : T_u\mathcal{M} \to T_u\mathcal{M}$ of F at a point $u \in \mathcal{M}$.

The main result of this subsection is Theorem 8.4 from [23] which can be considered as a general formulation of the Bendixson-Dulac-criterion for Riemannian manifolds of dimension n. Certain generalizations of the original Bendixson-Dulac-criterion for differential equations in $\mathbb{R}^n$ (see Chap. 5) can be derived from that theorem when adapting it to the particular situation.

In the following, the dimension of the 1-homology group $H_1(\mathcal{M})$ of $\mathcal{M}$ is denoted by b_1, i.e., the first Betti-number of $\mathcal{M}$ (see Sect. B.4, Appendix B).

Theorem 8.4 *Let the manifold $\mathcal{M}$ with Betti-number b_1 be compact and suppose that for the eigenvalues of the symmetric part S of ∇F one of the inequalities*

(a) $\lambda_1(u) + \lambda_2(u) < 0$ *or*
(b) $\lambda_{n-1}(u) + \lambda_n(u) > 0$

is valid on $\mathcal{M}$. Then the system (8.26) possesses on $\mathcal{M}$ at most b_1 non-trivial periodic orbits.

Proof We shall only consider the case of condition (a), the other one can be performed in a similar fashion by considering the negative time evolution of the flow of (8.26). We first take $b_1 = 0$ and show that every closed orbit is constant. Suppose that (8.26) has a non-trivial closed orbit γ. Let $\mathcal{S}$ be a generalized surface in $\mathcal{M}$ of minimal two-dimensional Hausdorff measure $0 < \mu_H(\mathcal{S}, 2) < \infty$ with boundary γ. Notice, that such a surface as solution of the Plateau problem for Riemannian manifolds [8, 22] in our situation exists. The properties of the flow ensure that for arbitrary $t \geq 0$ the set $\varphi^t(\mathcal{S})$ is again a generalized surface in $\mathcal{M}$ with boundary γ. Due to Theorem 8.2 and condition (a), for sufficiently large $t > 0$ we have

$$\mu_H(\varphi^t(\mathcal{S}), 2) < \mu_H(\mathcal{S}, 2).$$

But this is in contradiction to the fact that the surface $\mathcal{S}$ was taken to be of minimal two-dimensional Hausdorff measure.

Now, consider the case $b_1 \geq 1$. Suppose that (8.26) possesses more than b_1 closed non-trivial orbits in $\mathcal{M}$. Then there are at least two orbits γ_1 and γ_2 among them which are homologous to each other. Let $\mathcal{S}$ be a surface of minimal two-dimensional Hausdorff measure with boundary $\gamma_1 \cup \gamma_2$. Then again, for arbitrary $t > 0$ the set $\varphi^t(\mathcal{S})$ is a surface in $\mathcal{M}$ with $\gamma_1 \cup \gamma_2$ as boundary. For sufficiently large $t > 0$ the same argument as above together with Theorem 8.2 leads to a contradiction. □

Let us now add a version of Theorem 8.4 for the case $n = 2$ that in principle agrees with the classical negative Bendixson-Dulac-criterion. The difference to Theorem (8.4) is actually that here a modification of the vector field is allowed in the way that products ςF are considered with a C^1-smooth function $\varsigma : \mathcal{M} \to \mathbb{R}$. In the proof [23] it is confirmed that the introduction of the function ς can be interpreted, similar to the methods of the previous sections, as a transition to an equivalent Riemannian metric on $\mathcal{M}$.

Corollary 8.7 *Let the manifold $\mathcal{M}$ with Betti-number b_1 be two-dimensional and compact. Assume that a function $\varsigma : \mathcal{M} \to \mathbb{R}$ of class C^1 exists such that the divergence* $\operatorname{div}(\varsigma F)$ *does not vanish on $\mathcal{M}$. Then the system (8.26) possesses on $\mathcal{M}$ at most b_1 non-trivial periodic orbits.*

Proof We pass on $\mathcal{M}$ to the new Riemannian metric $\widetilde{g}_{|u} := \varkappa(u) g_{|u}$ for $u \in \mathcal{M}$. Consider the two eigenvalues $\widetilde{\lambda}_1(u) \geq \widetilde{\lambda}_2(u)$ of the symmetric part of the covariant derivative of F at $u \in \mathcal{M}$ in the new metric $\widetilde{g}$. Then we have $\widetilde{\operatorname{div}} F(u) = \widetilde{\lambda}_1(u) + \widetilde{\lambda}_2(u)$ for the divergence of F with respect to $\widetilde{g}$. On the other hand, a straightforward calculation using representation (8.17) gives

$$\widetilde{\operatorname{div}} F = \operatorname{div} F + \frac{\dot{\varkappa}}{\varkappa} = \frac{1}{\varkappa} \operatorname{div}(\varkappa F).$$

When combining relation (8.24) with the previous result, it becomes clear that one of the conditions (a) or (b) of Theorem 8.4 are satisfied if the inequality $\operatorname{div} F > 0$ or $\operatorname{div} F < 0$ holds on $\mathcal{M}$, respectively. □

Corollary 8.8 *Let the manifold $\mathcal{M}$ with Betti-number b_1 be compact and suppose that there exists a continuous function $\theta : \mathcal{M} \to \mathbb{R}$ such that for the symmetric part S of ∇F one of the following conditions is satisfied:*

(a) $\big(S(u)v, v\big)_{T_u\mathcal{M}} \geq -\theta(u)(v, v)_{T_u\mathcal{M}}$ *and* $\operatorname{div} F(u) + (n-2)\theta(u) < 0$
$(u \in \mathcal{M}, v \in T_u\mathcal{M})$ *or*
(b) $\big(S(u)v, v\big)_{T_u\mathcal{M}} \leq \theta(u)(v, v)_{T_u\mathcal{M}}$ *and* $\operatorname{div} F(u) - (n-2)\theta(u) > 0$
$u \in \mathcal{M}, v \in T_u\mathcal{M})$.

Then the system (8.26) possesses on $\mathcal{M}$ at most b_1 non-trivial periodic orbits.

Proof Again we only prove the case of condition (a). For the second case the same method can be applied. Analogously to Corollary (8.7) we obtain

$$\lambda_1(u) + \lambda_2(u) \leq (n-2)\theta + \operatorname{div} F < 0$$

and this coincides with the condition (a) of Theorem 8.4. □

We want to add a further corollary for the special case that the manifold has the Betti-number $b_1 = 0$. It demonstrates what kind of convergence a system (8.26) satisfying the assumptions of Theorem 8.4 in this situation necessarily exhibits.

Corollary 8.9 *Let the manifold $\mathcal{M}$ be compact and $b_1 = 0$. Let the set of equilibria of system (8.26) consists of isolated points only. If one of the conditions (a) or (b) of Theorem 8.4 holds, then every orbit of (8.26) converges both for $t \to \infty$ and for $t \to -\infty$ to an equilibrium point.*

Proof Again we shall restrict ourselves to the case of condition a). Consider an arbitrary integral curve $\varphi(\cdot, q)$ of (8.26) for $t \to \infty$. The manifold is compact and therefore the ω-limit set $\omega(q)$ is not empty. If we assume that there exists an element $p \in \omega(q)$ such that p is not an equilibrium point for (8.26), then by Pugh's closing lemma (Theorem A.4, Appendix A) in every small neighborhood of F in $\mathcal{X}^1(\mathcal{M})$ we can find a vector field $\widetilde{F}$ such that the corresponding differential equation possesses a non-trivial periodic orbit through p.

The compactness of $\mathcal{M}$ implies that $\widetilde{F}$ can be chosen such that for the first two eigenvalues $\widetilde{\lambda}_1(u) \geq \widetilde{\lambda}_2(u)$ of the symmetric part of the covariant derivative of $\widetilde{F}$ the property $\widetilde{\lambda}_1(u) + \widetilde{\lambda}_2(u) < 0$ is maintained on $\mathcal{M}$. But this contradicts the statement of Theorem 8.4. Thus, we can conclude that p has to be an equilibrium point of the original system (8.26). Remember that the set of all equilibria of (8.26) was assumed to consist of isolated points only. This finally gives $\omega(q) = \{p\}$, or in other words, the considered integral curve converges for $t \to \infty$ to p.

It is clear, that the convergence for $t \to -\infty$ follows in analogous manner when investigating the α-limit set instead. □

8.2 The Lyapunov Dimension as Upper Bound of the Fractal Dimension

8.2.1 Statement of the Results

In this section we shall show that the maximum local Lyapunov dimension of a set is an upper bound of the fractal dimension. Our representation follows the paper of Gelfert [10]. Suppose that $(\mathcal{M}, g)$ is an n-dimensional Riemannian C^3-manifold $\mathcal{M}$. Let $\mathcal{U} \subset \mathcal{M}$ be an open set and let $\varphi\colon \mathcal{U} \to \mathcal{M}$ be a C^1-map. Given $u \in \mathcal{U}$, we consider the singular values $\alpha_1(d_u\varphi) \geq \cdots \geq \alpha_n(d_u\varphi) \geq 0$ of the tangent map $d_u\varphi\colon\ T_u\mathcal{M} \to T_{\varphi(u)}\mathcal{M}$. We denote by $\dim_L(\varphi, u)$ the *local Lyapunov dimension* of φ at $u \in \mathcal{U}$ which is defined to be the largest number $d \in (0, n]$ for which $\omega_d(d_u\varphi) \geq 1$. If $\alpha_1(d_u\varphi) < 1$, we set $\dim_L(\varphi, u) = 0$. Note that the functions $u \mapsto \alpha_i(d_u\varphi)$, $i = 1, \ldots, n$, are continuous on $\mathcal{U}$. The function $u \mapsto \dim_L(\varphi, u)$ is continuous on $\mathcal{U}$ except at a point u which satisfies $\alpha_1(d_u\varphi) = 1$, where it is still upper semi-continuous. For a compact set $\mathcal{K} \subset \mathcal{M}$ we introduce the notation

$$\dim_L(\varphi, \widetilde{\mathcal{K}}) := \sup_{u \in \widetilde{\mathcal{K}}} \dim_L(\varphi, u)$$

and call this value as in the $\mathbb{R}^n$ case *Lyapunov dimension of* φ *on* $\widetilde{\mathcal{K}}$.

We can now state the theorem ([10]).

Theorem 8.5 *Let $(\mathcal{M}, g)$ be a Riemannian C^3-manifold. Let further $\mathcal{U} \subset \mathcal{M}$ be an open set and let $\varphi: \mathcal{U} \to \mathcal{M}$ be a C^1-map. Then for compact sets $\mathcal{K}$, $\widetilde{\mathcal{K}} \subset \mathcal{U}$ which satisfy $\mathcal{K} \subset \varphi^t(\mathcal{K}) \subset \widetilde{\mathcal{K}}$ for all $t \in \mathbb{N}$ we have* $\dim_F \mathcal{K} \le \dim_L(\varphi, \widetilde{\mathcal{K}})$.

From Theorem 8.5 we deduce the following result [12].

Corollary 8.10 *Let $\mathcal{U} \subset \mathbb{R}^n$ be an open set, let $\varphi : \mathcal{U} \to \mathbb{R}^n$ be a C^1-map, and let $\mathcal{K} \subset \mathcal{U}$ be a compact invariant set of φ (that is, $\varphi(\mathcal{K}) = \mathcal{K}$). Then*

$$\dim_F \mathcal{K} \le \dim_L(\varphi, \mathcal{K}).$$

Remark 8.5 As in Corollary 8.1 it can be shown that Theorem 5.17, Chap. 5 is a special case of Theorem 8.5.

Theorem 8.5 will be proved in Sect. 8.2.2.

8.2.2 Proof of Theorem 8.5

Let us start with a dimension estimate from [9]. Recall (Sect. 3.2.2, Chap. 3) that $\mu_B(\cdot, d, \varepsilon)$ denotes the capacitive d-measure at level ε.

Lemma 8.3 *Let $(\mathcal{M}, \rho)$ be a metric space. If for a compact set $\mathcal{K} \subset \mathcal{M}$ and for numbers $d \ge 0$, $\eta' > 0$ and $0 < D < 1$ we have $\mu_B(\mathcal{K}, d, D\eta) \le \mu_B(\mathcal{K}, d, \eta)$ for every $\eta \in (0, \eta']$, then $\dim_F \mathcal{K} \le d$.*

Proof Let $r \in (0, \eta')$ be chosen arbitrarily. Because of $D < 1$ there exists a number $j \in \mathbb{N}$ for which $D^j\eta' \le r < D^{j-1}\eta'$. Therefore,

$$\mu_B(\mathcal{K}, d, r) = N_r(\mathcal{K})r^d < N_{D^j\eta'}(\mathcal{K})\,(D^{j-1}\eta')^d = D^{-d}\mu_B(\mathcal{K}, d, D^j\eta'). \quad (8.27)$$

Setting $\eta = D^{j-1}\eta', \dots, \eta = \eta'$ we obtain $\mu_B(\mathcal{K}, d, r) < D^{-d}\mu_B(\mathcal{K}, d, \eta')$. Since $\mathcal{K}$ is compact, $\mu_B(\mathcal{K}, d, \eta')$ is finite. Thus, $\mu_B(\mathcal{K}, d, r)$ is uniformly bounded from above for all $r < \eta'$ which implies $\dim_F \mathcal{K} \le d$. □

Next we shall prove the following lemma in which we set $\widehat{\mathcal{K}} := \overline{\cup_{t\ge 0}\varphi^t(\mathcal{K})}$.

Lemma 8.4 *Let $d \in (0, n)$ written as $d = d_0 + s$ with $d_0 \in \{0, 1, \dots, n-1\}$ and $s \in (0, 1]$. Assume that*

$$2(8\sqrt{d_0+1})^d\omega_d(d_p\varphi) < 1 \quad \forall\, p \in \widehat{\mathcal{K}}.$$

Then there exist numbers $\eta_0 > 0$ and $a, b \in (0, 1)$ and a uniformly continuous function $\zeta: \widehat{\mathcal{K}} \to (a, b)$ such that for any $l \in \mathbb{N}$ the following holds:

For every ball $\mathcal{B}(q,\eta)$ with $\eta \in (0,\eta_0)$ and $q \in \varphi^l(\mathcal{K})$ there exists a family of filial balls $\mathcal{F}^{(1)}(\mathcal{B}(q,\eta))$, each with radius $\zeta(q)\eta$ and center in $\varphi^{l+1}(\mathcal{K})$, whose union covers $\varphi(\mathcal{B}(q,\eta)) \cap \varphi^{l+1}(\mathcal{K})$. For the minimal number $N(q)$ of balls in $\mathcal{F}^{(1)}(\mathcal{B}(q,\eta))$ we have $N(q) \le \frac{1}{2(\zeta(q))^d}$.

Proof We introduce the notation $\omega_d(\varphi,\mathcal{K}) := \max_{p\in\mathcal{K}} \omega_d(d_p\varphi)$. Choose a number $h > \omega_d(\varphi,\widehat{\mathcal{K}})$ satisfying

$$\left(8\sqrt{d_0+1}\right)^d h < \frac{1}{2} \tag{8.28}$$

and an open set $\mathcal{Z} \subset \mathcal{U}$ containing $\widehat{\mathcal{K}}$ and which is itself contained within a compact set $\mathcal{A} \subset \mathcal{U}$ satisfying $\omega_d(\varphi,\mathcal{A}) < h$. Further, choose a number $\kappa < 1$ satisfying $\omega_d(\varphi,\mathcal{A}) \le \kappa < h$ and a number $\delta > 0$ for which $\kappa \le \delta^d$ and $\omega_1(\varphi,\mathcal{A}) \le \delta$ hold and set $C := \left(\frac{\delta^{d_0}}{\kappa}\right)^{\frac{1}{s}}$. At last, choose $\varsigma > 0$ satisfying

$$\omega_m(\varphi,\widehat{\mathcal{K}})\varsigma^{d-l} \le \kappa\,, \qquad m = 0,\dots,d_0. \tag{8.29}$$

The equation

$$(1+C\eta)^d\kappa = h \tag{8.30}$$

uniquely determines a number $\eta > 0$. Since

$$\sup_{p\in\mathcal{A}} \alpha_{d_0+1}(d_p\varphi) \le \omega_d(\varphi,\mathcal{A})^{\frac{1}{d}} \le \kappa^{\frac{1}{d}}$$

we have

$$(1+C\eta)\sup_{p\in\mathcal{A}} \alpha_{d_0+1}(d_p\varphi) \le h^{\frac{1}{d}}. \tag{8.31}$$

Denote by $\exp_q : T_q\mathcal{M} \to \mathcal{M}$ the exponential map at a point $q \in \widetilde{\mathcal{K}}$. Since $\exp_q$ is a smooth map which satisfies $|d_{O_q}\exp_q| = 1$, for every point $q \in \mathcal{M}$ there is a number $\delta_q > 0$ such that $|d_v \exp_q| \le 2$ for any $v \in \mathcal{B}(O_q,\delta_q)$. Further, since $\widetilde{\mathcal{K}}$ is compact, there is a number $\delta_0 = \min_{q\in\widetilde{\mathcal{K}}} \delta_q > 0$ and, consequently,

$$\rho\,(\exp_q v_1, \exp_q v_2) \le 2|v_1 - v_2|_{T_q\mathcal{M}}$$

for any $q \in \widetilde{\mathcal{K}}$ and any $v_1, v_2 \in \mathcal{B}(O_q,\delta_0)$. Here $\rho(\cdot,\cdot)$ denotes the geodesic distance on $\mathcal{M}$ and $\tau_u^q : T_u\mathcal{M} \to T_q\mathcal{M}$ denotes the isometric operator defined by the parallel transport (see Sect. A.7, Appendix A) along the geodesic for points lying sufficiently close to each other. Let $\eta_0 > 0$ be so small such that:

(1) For every $q \in \widehat{\mathcal{K}}$, the set $\mathcal{B}(q,2\eta_0)$ is contained in $\mathcal{Z}$.

(2) $\eta_0(1+\theta+\eta) \le \delta_0$.

(3) $|\tau_{\varphi(w)}^{\varphi(q)} \circ d_w\varphi \circ \tau_q^w - d_q\varphi| \le \eta$ for all $w, q \in \mathcal{Z}$ with $\rho\,(w,q) \le \eta_0$.

Then every ball $\mathcal{B}(q,\eta)$ $\;(\eta \le \eta_0)$ which intersects $\widehat{\mathcal{K}}$ is contained in $\mathcal{Z}$. Taylor's formula for the differentiable map φ gives for any $u \in \mathcal{B}(q,\eta)$ the estimate

$$\begin{aligned} &\left| \exp_{\varphi(q)}^{-1} \varphi(u) - d_q\varphi(\exp_q^{-1} u)\right| \\ &\quad \le \sup_{w\in\mathcal{B}(q,\eta)} \left|\tau_{\varphi(w)}^{\varphi(q)} \circ d_w\varphi \circ \tau_q^w - d_q\varphi\right| \left|\exp_q^{-1} u\right| \end{aligned} \tag{8.32}$$

which together with property (3) implies the relation

$$\varphi(\mathcal{B}(q,\eta)) \subset \exp_{\varphi(q)}\left(d_q\varphi\, \mathcal{B}(O_q,\eta) + \mathcal{B}(O_{\varphi(q)},\eta\eta_0)\right). \tag{8.33}$$

Because of the choice of ς in (8.29) and because of Lemma 7.2, Chap. 7, for every point $q \in \varphi^l(\mathcal{K})$ the set $d_q\varphi\, \mathcal{B}(q,\eta) + \mathcal{B}(O_{\varphi(q)},\eta\eta_0)$ can be covered by $\left[\frac{2^d\kappa}{\widetilde{\varsigma}^d}\right]$ balls of radius $\sqrt{d_0+1}(1+C\eta)\widetilde{\varsigma}\,\eta$ where $\widetilde{\varsigma} := \max\{\varsigma, \alpha_{d_0+1}(d_q\varphi)\}$. Here the cover can be evidently chosen in such a way that any ball is contained in a ball of radius $(1+\delta+\eta)\eta$ centered at $O_{\varphi(q)}$, which follows from $\omega_1(d_q\varphi) < \theta$ and from (8.18) and (8.28). Hence, by (8.29) and property 2), the set $\varphi(\mathcal{B}(q,\eta))$ can be covered by $\left[\frac{2^d\kappa}{\widetilde{\varsigma}^d}\right]$ balls of radius $2\sqrt{d_0+1}(1+C\eta)\widetilde{\varsigma}\eta$. For this cover any ball intersecting $\varphi^{l+1}(\mathcal{K})$ can be replaced by a ball which is centered at a point in $\varphi^{l+1}(\mathcal{K})$ and with twice the radius.

For $u \in \mathcal{U}$ we put

$$\kappa(u) := 4\sqrt{d_0+1}\,(1+C\eta)\cdot \max\left\{\varsigma, \alpha_{d_0+1}(d_u\varphi)\right\}.$$

Thus, for any $q \in \varphi^l(\mathcal{K})$ the set $\varphi(\mathcal{B}(q,\eta)) \cap \varphi^{l+1}(\mathcal{K})$ can be covered by $N(q)$ balls $\mathcal{B}(q_j,\kappa(q)\eta)$, $j=1,\dots,N(q)$, which are centered at $q_j \in \varphi^{l+1}(\mathcal{K})$. Here we have

$$N(q) \le \left[\frac{2^d\kappa}{\kappa(q)^d}\left(4\sqrt{d_0+1}(1+C\eta)\right)^d\right] \le \frac{1}{2\kappa(q)^d}$$

where for the second inequality we have used (8.30) and (8.28). The function $\kappa: \mathcal{U} \to \mathbb{R}$, $u \mapsto \kappa(u)$ is uniformly continuous on the compact set $\widehat{\mathcal{K}}$ because of smooth dependence of the singular values of $d_u\varphi$ on u. Because of (8.31) and (8.28) there exist numbers $a, b \in (0,1)$ for which

$$a < \kappa(u) < b \;, \qquad \forall\, u \in \widehat{\mathcal{K}}. \tag{8.34}$$

This proves the lemma. □

Proof of Theorem 8.3 Let us assume that $\dim_L(\varphi,\widetilde{\mathcal{K}}) < n$ and let us choose an arbitrary number $d \in (\dim_L(\varphi,\widetilde{\mathcal{K}}), n)$. Recall that

$$\sup\left\{\omega_d(d_p\varphi^t) \mid p \in \bigcup_{\tau\ge 0}\varphi^\tau(\mathcal{K})\right\} \le \max_{p\in\widetilde{\mathcal{K}}} \omega_d(d_p\varphi)^t$$

for any natural number t. Hence, for sufficiently large number $t \in \mathbb{N}$, we have $2(8\sqrt{d_0+1})^d \omega_d(\varphi^t, \widehat{\mathcal{K}}) < 1$ and thus the assumption of Lemma 8.4 is satisfied for the map $\psi := \varphi^t$.

Let us choose an arbitrary finite cover $\mathfrak{U} := \cup_{j=1}^{J} \mathcal{B}(p_j, \eta)$ of $\mathcal{K}$ with $p_j \in \mathcal{K}$. We construct a family of filial covers. By Lemma 8.4, for any ball $\mathcal{B}(p_j, \eta)$ we find a family of balls

$$\mathcal{F}^{(1)}(\mathcal{B}(p_j, \eta)) = \left\{\mathcal{B}(q_i, \varkappa(p)\eta)\right\}_{i=1}^{N(p)}$$

which cover the set $\psi(\mathcal{B}(p_j, \eta)) \cap \psi(\mathcal{K})$. We call $\mathcal{F}^{(1)}(\mathcal{B}(p_j, \eta))$ in accordance with [2] a *family of filial balls* for $\mathcal{B}(p_j, \eta)$ *of order* 1. Further, we define a sequence *families of filial balls of order t* recursively by setting

$$\mathcal{F}^{(t)}(\mathcal{B}(p_j, \eta)) = \bigcup \left\{\mathcal{F}^{(1)}(\mathcal{B}) \mid \mathcal{B} \in \mathcal{F}^{(t-1)}(\mathcal{B}(p_j, \eta))\right\}, \qquad t \geq 2, t \in \mathbb{N}.$$

Let us denote by $r(\mathcal{B})$ the radius of a ball $\mathcal{B}$. For each family of filial balls of $\mathcal{B}(p_j, \eta)$, $p_j \in \mathcal{K}$, of order t we therefore obtain the estimates

$$\begin{aligned}\sum_{\mathcal{B} \in \mathcal{F}^{(t)}(\mathcal{B}(p_j,\eta))} r(\mathcal{B})^d &= \sum_{\mathcal{B} \in \mathcal{F}^{(t-1)}(\mathcal{B}(p_j,\eta))} \; \sum_{\mathcal{B}' \in \mathcal{F}^{(1)}(\mathcal{B})} r(\mathcal{B}')^d \\ &\leq \sum_{\mathcal{B} \in \mathcal{F}^{(t-1)}(\mathcal{B}(p_j,\eta))} \frac{r(\mathcal{B})^d}{2} \leq \frac{\eta^d}{2^t}. \end{aligned} \tag{8.35}$$

We shall now assign certain iteration depths. For every point $p \in \mathcal{K}$ we fix a prehistory $\{s_0(p), s_1(p), \dots\}$ with respect to ψ as follows:

$$s_0(p) := p\,, \quad s_i(p) := q\,, \quad i = 1, 2, \dots\,, \quad \text{for some } q \in \left\{u \in \mathcal{K} : \; \psi(u) = s_i(p)\right\}. \tag{8.36}$$

Further, we shall choose some number $c \in (0, \frac{1}{2})$ satisfying

$$\frac{2^{-\frac{\log c}{\log a}}}{a^d} < 2^{-(d+2)}. \tag{8.37}$$

Because of (8.34), to any point $p \in \mathcal{K}$ we can assign a prehistory of finite length $I(p)$ for which the inequalities

$$ac < \kappa(s_1(p)) \cdots \kappa(s_{I(p)}(p)) \leq c \tag{8.38}$$

hold. Because of (8.34) and (8.38) we obtain $ac < b^{I(p)}$ and $a^{I(p)} < c$, and therefore

$$\frac{\log ac}{\log b} > I(p) > \frac{\log c}{\log a} \tag{8.39}$$

for any $p \in \mathcal{K}$. Without loss of generality we can assume that c has been chosen small enough such that $I(p) > 1$ for all $p \in \mathcal{K}$. We set $I := \sup_{p \in \mathcal{K}} I(p)$ which is finite because of (8.39).

We now construct the homogeneous cover $\mathfrak{G}$ of $\mathcal{K}$. First, for each point $p \in \mathcal{K}$ we construct a ball of radius approximately $c\eta$ containing p as follows: We take the history $\{s_0(p), s_1(p), \ldots\}$ of p and choose some ball in $\mathfrak{U}$ which contains the point $s_{I(p)}(p)$ and denote it by $\mathcal{B}_{p,I(p)}$. Along the orbit

$$s_{I(p)}(p) \mapsto s_{I(p)-1}(p) \mapsto \cdots \mapsto s_1(p) \mapsto s_0(p) = p$$

of length $I(p)$ we construct balls $\mathcal{B}_{p,I(p)-(i+1)}$, $i = 0, \ldots, I(p) - 1$, which are defined recursively as follows. The union of filial balls of the family $\mathcal{F}^{(1)}(\mathcal{B}_{p,I(p)-i})$ covers the set $\psi(\mathcal{B}_{p,I(p)-i}) \cap \psi^i(\mathcal{K})$. Choose $\mathcal{B}_{p,I(p)-(i+1)}$ as a ball from this cover which contains the point $s_{I(p)-(i+1)}(p)$. We obtain

$$s_{I(p)}(p) \in \mathcal{B}_{p,I(p)}, \ldots, s_0(p) = p \in \mathcal{B}_{p,0}.$$

We denote by $\widetilde{s}_i(p)$ the center point of the corresponding ball $\mathcal{B}_{p,i}$, $i = 0, \ldots, I(p)$. By construction in the proof of Lemma 8.4, $\widetilde{s}_i(p) \in \psi^{I(p)-i}(\mathcal{K})$. Since $\mathcal{B}_{p,0}$ is an element of a family of filial balls for $\mathcal{B}_{p,I(p)}$ of order $I(p)$ we have

$$r(\mathcal{B}_{p,0}) = \kappa(\widetilde{s}_{I(p)}(p)) \cdots \kappa(\widetilde{s}_1(p))\, r(\mathcal{B}_{p,I(p)}).$$

Since $\mathcal{B}_{p,I(p)} \in \mathcal{U}$, we have $r(\mathcal{B}_{p,I(p)}) = \eta$ and therefore

$$r(\mathcal{B}_{p,0}) = \kappa(\widetilde{s}_{I(p)}(p)) \cdots \kappa(\widetilde{s}_1(p))\eta. \tag{8.40}$$

Further,

$$\rho\big(s_{I(p)-i}(p), \widetilde{s}_{I(p)-i}(p)\big) \leq \eta, \qquad i = 0, \ldots, I(p). \tag{8.41}$$

Now we choose a sub-family $\widetilde{\mathfrak{G}} := \{\mathcal{B}_{p_l,0}\}_{l=1}^{L}$ of the family $\{\mathcal{B}_{p,0}\}_{p \in \mathcal{K}}$ such that the union $\cup_{l=1}^{L} \mathcal{B}_{p_l,0}$ covers the compact set $\mathcal{K}$ and set

$$R := \max_{l=1,\ldots,L} r(\mathcal{B}_{p_l,0}). \tag{8.42}$$

Each ball $\mathcal{B}_{p_l,0}$ in $\widetilde{\mathfrak{G}}$ with radius $r(\mathcal{B}_{p_l,0})$ and center point $p_l = \widetilde{s}_0(p_l)$ can be replaced by the concentric ball with radius R. This gives us a cover

$$\mathfrak{G} := \{\mathcal{B}(\widetilde{s}_0(p_l), R)\}_{l=1}^{L}$$

of $\mathcal{K}$ with balls of equal radius R, where $R \in (ac\eta, c\eta]$ because of (8.38).

We shall now study the oscillation of the radii of balls within the cover $\widetilde{\mathfrak{G}}$. For this, choose some number $\Delta > 1$ satisfying

$$\Delta^{2dI} < 2. \tag{8.43}$$

From Lemma 8.4 we obtain the uniform continuity of the function κ on $\widehat{\mathcal{K}}$. Further, by (8.35) this function is on $\widehat{\mathcal{K}}$ uniformly bounded from below by a positive number. Thus, there exists $\eta_1 > 0$ such that

$$\frac{\kappa(p)}{\kappa(q)} \le \Delta, \qquad \forall\, p, q \in \widehat{\mathcal{K}}, \rho(p, q) < \eta_1. \tag{8.44}$$

W.l.o.g. we can take $\eta_1 = \eta_0$. From (8.41), (8.44) and (8.38), for every $l = 1, \ldots, L$ we conclude

$$\frac{r(\mathcal{B}_{p_l,0})}{\eta} = \kappa(\widetilde{s}_1(p_l)) \cdots \kappa(\widetilde{s}_{I(p_l)}(p_l)) \le \Delta^{I(p_l)} c.$$

Analogously we obtain

$$\frac{\eta}{r(\mathcal{B}_{p_l,0})} \le \frac{\Delta^{I(p_l)}}{\kappa(s_1(p_l)) \cdots \kappa(s_{I(p_l)}(p_l))} < \frac{\Delta^{I(p_l)}}{ac}.$$

Thus, for any two balls $\mathcal{B}_{p_l,0}$ and $\mathcal{B}_{p_k,0}$ from the cover $\widetilde{\mathfrak{G}}$ we have

$$\frac{r(\mathcal{B}_{p_l,0})}{r(\mathcal{B}_{p_k,0})} < \frac{\Delta^{I(p_k)}}{ac} \Delta^{I(p_l)} c \le \frac{\Delta^{2I}}{a}.$$

Finally, for the radius R, from (8.42)

$$R \le \frac{\Delta^{2I}}{a} r(\mathcal{B}_{p_l,0}) \tag{8.45}$$

follows.

We shall now estimate the capacitive d-measure at level R of $\mathcal{K}$, i.e., the outer measure $\mu_B(\cdot, d, R)$ on M. Since $\mathcal{K}$ has been covered by balls from $\mathfrak{G}$ with equal radius R, we obtain by (8.45)

$$\mu_B(\mathcal{K}, d, R) \le \sum_{l=1}^{L} \mu_B\big(\mathcal{B}(\widetilde{s}_0(p_l), R), d, R\big) = \sum_{l=1}^{L} R^d \le \frac{\Delta^{2dI}}{a^d} \sum_{l=1}^{L} r(\mathcal{B}_{p_l,0})^d. \tag{8.46}$$

To each ball $\mathcal{B}_{p_l,0}$, $l = 1, \ldots, L$, we assigned a ball $\mathcal{B}(p_j, \eta) \in \mathcal{U}$ such that $\mathcal{B}_{p_l,0}$ belongs to the family of filial balls of $\mathcal{B}(p_j, \eta)$ of order $I(p_j)$. Consequently, each term in sum in the right-hand term in (8.46) occurs at most once as a term in the sum

$$\sum_{j=1}^{J}\sum_{i=I^*}^{I}\sum_{\mathcal{B}\in\mathcal{F}^{(i)}(\mathcal{B}(p_j,\eta))} r(\mathcal{B})^d$$

where we have set $I^* := \min_{p\in\mathcal{K}} I(p)$. Thus, we obtain

$$\mu_B(\mathcal{K},d,R) \le \frac{\Delta^{2dI}}{a^d}\sum_{j=1}^{J}\sum_{i=I^*}^{I}\sum_{\mathcal{B}\in\mathcal{F}^{(i)}(\mathcal{B}(p_j,\eta))} r(\mathcal{B})^d \le \frac{\Delta^{2dI}}{a^d} J \sum_{i=I^*}^{\infty}\frac{1}{2^i}\eta^d \tag{8.47}$$

where we have used (8.35). By (8.39) and by definition of the number I there holds

$$\frac{\log c}{\log a} < I^* \le I.$$

From this we deduce for the capacitive d-measure at level R of $\mathcal{K}$

$$\mu_B(\mathcal{K},d,R) < \mathcal{J}\, 2^{-\frac{\log c}{\log a}} 2\eta^d \frac{\Delta^{2dI}}{a^d}.$$

Now (8.43) and (8.37) imply

$$\mu_B(\mathcal{K},d,R) < 4\frac{2^{-\frac{\log c}{\log a}}}{a^d}\mathcal{J}\,\eta^d < 2^{-d}\mathcal{J}\eta^d. \tag{8.48}$$

We shall now apply Lemma 8.3. The initial cover $\mathfrak{U}$ of $\mathcal{K}$ of balls of radius η centered at a point in $\mathcal{K}$ has been chosen arbitrarily. Any ball intersecting $\mathcal{K}$ of radius $\frac{\eta}{2}$ can be replaced by one which is centered in $\mathcal{K}$ and with radius η. Thus, we can replace the right-hand side in (8.43) by $\mu_B(\mathcal{K},d,\frac{\eta}{2})$. All assumptions of Lemma 8.3 are thus satisfied. From $\mu_B(\mathcal{K},d,R) < \mu_B(\mathcal{K},d,\frac{\eta}{2})$ and $R \le c\eta < \frac{\eta}{2}$ the estimate $\dim_F \mathcal{K} \le d$ follows. This holds for arbitrary $d > \dim_L(\psi,\widetilde{\mathcal{K}})$ which proves Theorem 8.5. □

Remark 8.6 For a C^1-map in $\mathbb{R}^n$ and a compact invariant set Theorem 8.5 has been shown by Hunt [12], where the author uses the on $\mathbb{R}^n$ equivalent definition of the fractal dimension via a grid covering. For twice continuously Frechét-differentiable maps in a separable Hilbert space, Blinchevskaya and Ilyashenko [2] extended this result by showing that a compact invariant set has fractal dimension not exciting k if the map contracts k-volumina.

8.2.3 Global Lyapunov Exponents and Upper Lyapunov Dimension

Let us consider the long-term behaviour of the dynamical system $\varphi\colon \mathbb{N} \times \mathcal{K} \to \mathcal{K}$ generated by the iterates φ^t $(t \in \mathbb{N})$ on an invariant set $\mathcal{K} \subset \mathcal{U}$. We introduce the *global Lyapunov exponents* $\nu_1^u \geq \cdots \geq \nu_n^u$ of φ on $\mathcal{K}$ which are recursively defined by

$$\nu_1^u + \cdots + \nu_j^u := \lim_{t\to+\infty} \frac{1}{t} \log \max_{p\in\mathcal{K}} \omega_j(d_p\varphi^t)\,, \qquad j = 1, \ldots, n.$$

The *upper Lyapunov dimension* of φ on $\mathcal{K}$ with respect to the global Lyapunov exponents is

$$\dim_L^u(\varphi, \mathcal{K}) := k + \frac{\nu_1^u + \cdots + \nu_k^u}{|\nu_{k+1}^u|}$$

where $k \in \{0, \ldots, n-1\}$ denotes the smallest number satisfying $\nu_1^u + \cdots + \nu_{k+1}^u < 0$.

Using Theorem 8.5 and the method of proof of Theorem 3.3 in [28], we obtain the following result [10].

Theorem 8.6 *Let (M, g) be a Riemannian C^3-manifold. Let $\mathcal{U} \subset \mathcal{M}$ be an open set, and let $\varphi\colon \mathcal{U} \to \mathcal{M}$ be a C^1-map. For a compact and invariant set $\mathcal{K} \subset \mathcal{U}$ we have* $\dim_F \mathcal{K} \leq \dim_L^u(\varphi, \mathcal{K})$.

Remark 8.7 Since for invariant sets $\mathcal{K}$ the function $t \mapsto \max_{p\in\mathcal{K}} \omega_d(d_p\varphi^t)$ is sub-exponential (cp. [28]), we have

$$\dim_L^u(\varphi, \mathcal{K}) \leq \inf_{t\in\mathbb{N}} \dim_L(\varphi^t, \mathcal{K}) = \lim_{t\to+\infty} \dim_L(\varphi^t, \mathcal{K}). \tag{8.49}$$

Further, recall [15] that

$$\inf_{t\to+\infty} \dim_L(\varphi^t, \mathcal{K}) = \sup_{\mu} \dim_L(\varphi, \mu)$$

where $\dim_L(\varphi, \mu)$ denotes the Lyapunov dimension with respect to the Lyapunov exponents $\nu_i(\mu)$ of μ and where the supremum is taken over all invariant ergodic probability measures supported on $\mathcal{K}$.

In correspondence to the global Lyapunov exponents the *local Lyapunov exponents* $\nu_1(p) \geq \cdots \geq \nu_n(p)$ *of* φ *at a point* $p \in \mathcal{K}$ (see also Chap. 6) are defined recursively by

$$\nu_1(p) + \cdots + \nu_j(p) = \limsup_{t\to+\infty} \frac{1}{t} \log \omega_j(d_p\varphi^t)\,, \qquad j = 1, \ldots, n.$$

The *local upper Lyapunov dimension of φ at p with respect to the local Lyapunov exponents* is then given by

$$\dim_L(\varphi, p) := k(p) + \frac{\nu_1(p) + \cdots + \nu_{k(p)}(p)}{|\nu_{k(p)+1}(p)|}$$

where $k(p) \in \{0, \ldots, n-1\}$ denotes the smallest number satisfying $\nu_1(p) + \cdots + \nu_{k(p)+1}(p) < 0$.

Remark 8.8 For an invariant set $\mathcal{K}$ the inequality

$$\sup_{p \in \mathcal{K}} \dim_L(\varphi, p) \leq \dim_L^u(\varphi, \mathcal{K})$$

has been proven by Eden ([6]). He presumed that for a typical system ([6]) there always exists a point p satisfying $\dim_L(\varphi, p) = \dim_L^u(\varphi, \mathcal{K})$. He refers also to examples for which strict inequality holds.

8.2.4 Application to the Lorenz System

To handle systems on Riemannian manifolds gives us the freedom also to construct adapted metrics. Recall that, within a class of equivalent metrics, the box dimension of a compact set is the same. However, notice that the local Lyapunov dimensions of a differentiable map strongly depends on the Riemannian metric. This fact can be used to optimize dimension estimates (see, for instance, [4, 19, 23] for a related approach using adapted Lyapunov functions).

As an example we consider the flow $\varphi\colon \mathbb{R} \times \mathbb{R}^3 \to \mathbb{R}^3$ of the Lorenz system

$$\dot{x} = -\sigma x + \sigma y \quad \dot{y} = rx - y - xz \quad \dot{z} = -bz + xy \tag{8.50}$$

with given parameters $b = \frac{8}{3}$, $\sigma = 10$ and $r = 28$. The flow is dissipative and has a global attractor $\mathcal{A} \subset \mathbb{R}^3$. Since the divergence of the vector field f given by (8.50) equals $\operatorname{div} f = -(\sigma + 1 + b) < 0$, the Lorenz system is volume-contracting, hence the Lorenz attractor has Lebesgue measure zero. It was shown in Sect. 6.5.2, Chap. 6 that the maximal local Lyapunov dimension of the time-1-map φ^1 on $\mathcal{A}$ equals the local Lyapunov dimension of φ^1 at the equilibrium point $p_0 = (0, 0, 0)$. There were used certain linear transformations and Lyapunov-type functions. We put this approach into the framework of adaption of metrics. For this we introduce the family of matrices

$$S(p) := \exp\left(\frac{V(p)}{d}\right)A\,, \quad p = (x, y, z) \in \mathbb{R}^3\,,$$

with

$$A := \begin{pmatrix} a & 0 & 0 \\ -\frac{1}{\sigma}(b-1) & 1 & 0 \\ 0 & 0 & 1 \end{pmatrix}$$

where

$$a := \frac{1}{\sigma}\sqrt{r\sigma + (b-1)(\sigma - b)}$$

and

$$V(p) = \frac{1}{2a\theta}(3-d)\Big(\kappa_1 x^2 + \kappa_2\Big(y^2 + z^2 - x^2\frac{(b-1)^2}{\sigma^2}\Big) + \kappa_3 z\Big)$$

with

$$\theta := 2\sqrt{(\sigma + 1 - 2b)^2 + (2\sigma b)^2}$$

and

$$\begin{aligned} \kappa_2 &:= \frac{1}{2a},\ \kappa_3 := -\frac{4\sigma a}{b}, \\ \kappa_1 &:= -\frac{1}{2\sigma}\Big[2\kappa_2\frac{r\sigma - (b-1)^2}{\sigma} + \kappa_3 + \frac{2(b-1)}{a\sigma}\Big]. \end{aligned} \tag{8.51}$$

We consider the metric tensor g on $\mathbb{R}^3$ given at a point $p \in \mathbb{R}^3$ by

$$g(p)(\upsilon, w) := \exp\Big(2\frac{V(p)}{d}\Big)(A^T A\upsilon, w)_{\mathbb{R}^3},\ \upsilon, w \in \mathbb{R}^3$$

where $(\cdot,\cdot)_{\mathbb{R}^3}$ is induced by the Euclidean metric. Here d is a positive parameter which will be specified below. Note that with respect to this metric structure the singular value function of order $2+s$ $(s \in (0,1))$ of the map φ^1 can be estimated as

$$\begin{aligned} \omega_{2+s}(d_p\varphi^1) \le \exp \int_0^1 \Big[\lambda_1(\varphi^\tau(p)) + \lambda_2(\varphi^\tau(p)) + s\lambda_3(\varphi^\tau(p))\Big]d\tau \\ + V(\varphi^1(p)) - V(p) =: \Upsilon_{2+s}(\varphi^1, p) \end{aligned} \tag{8.52}$$

where $\lambda_1(u) \ge \lambda_2(u) \ge \lambda_3(u)$ denote the eigenvalues of the matrix

$$\frac{1}{2}\big(A\,Df(u)\,A^{-1} + (A\,Df(u)\,A^{-1})^T\big)$$

with $Df(u)$ being the Jacobian of $f(u)$. Note that

$$\lambda_2(p_0) = -b\,,\ \ \lambda_{1/3}(p_0) = -\frac{\sigma+1}{2} \pm \frac{1}{2}\sqrt{(\sigma-1)^2+4\sigma r}$$

which implies

$$\dim_L(\varphi^1, p_0) = 2 + s_0, \quad s_0 = -1 + \frac{2(\sigma+1+b)}{\sigma+1+\sqrt{(\sigma-1)^2+4\sigma r}}$$

and $\Upsilon_{2+s}(\varphi^1, p_0) \le 1$ for any $s \ge s_0$. Moreover,

$$\Upsilon_{2+s}(\varphi^1, p) \le \Upsilon_{2+s_0}(\varphi^1, p_0) \qquad (p \in \mathbb{R}^3, s \ge s_0)$$

and thus

$$\dim_L(\varphi^1, p_0) = \sup_{p\in\mathbb{R}^3} \dim_L(\varphi^1, p).$$

Since p_0 is an equilibrium point, $\dim_L(\varphi^1, p_0) = \dim_L^u(\varphi^1, p_0)$. Hence, the dynamical system generated by (8.50) on $\mathcal{A}$ is typical in the sense of Remark 8.8. Moreover,

$$\dim_F \mathcal{A} \le \dim_L(\varphi^1, p_0) = \dim_L^u(\varphi^1, \mathcal{A}) = \inf_{t\in\mathbb{N}} \dim_L(\varphi^t, \mathcal{A}) \approx 2.401\,. \tag{8.53}$$

The Lorenz attractor $\mathcal{A}$ observed numerically has been the object of several analytic estimates of the box dimension (e.g., Eden et al. [7] obtained $\dim_F \mathcal{A} \le 2.408$). Since the local Lyapunov dimension of an equilibrium point is invariant under changes of metrics, estimate (8.53) is optimal in terms of methods developed in this section and in [4, 7, 12, 28]. However, numerical investigations suggest $\dim_B \mathcal{A} \approx 2.05$.

8.3 Hausdorff Dimension Estimates by Use of a Tubular Carathéodory Structure and their Application to Stability Theory

8.3.1 The System in Normal Variation

An important class of invariant sets of dynamical systems are strange attractors which locally have the structure of the product of a smooth (often one-dimensional) submanifold directed "along the attractor" and a Cantor-like set, "transversal" to the attractor [25]. Thus, it is natural to investigate the stability and dimension properties of such attractors considering the intersection of the attractors with surfaces which are locally transversal to the attractor [11, 16]. The use of transverse intersections is well-known in stability theory.

Consider now a Riemannian manifold $(\mathcal{M}, g)$ of dimension $n (n \geq 2)$ and, for simplicity, of class C^∞. Let $F : \mathcal{M} \to T\mathcal{M}$ be a vector field of class C^2 on $\mathcal{M}$ and let us consider the corresponding differential equation

$$\dot{u} = F(u). \tag{8.54}$$

For simplicity, let us assume that the global flow $\varphi : \mathbb{R} \times \mathcal{M} \to \mathcal{M}$ of (8.54) exists. This flow φ can also be written as a one-parameter family of C^2-diffeomorphisms $\{\varphi^t\}_{t \in \mathbb{R}}$ with $\varphi^t(\cdot) = \varphi(t, \cdot)$.

The behaviour of system (8.54) near a given solution $\varphi^{(\cdot)}(p)$ is described by the variational equation

$$\frac{Dy}{dt} = \nabla F(\varphi^t(p))y\,. \tag{8.55}$$

All points $p \in \mathcal{M}$ with $F(p) \neq O_p$ $(F(p) = O_p)$, where O_p denotes the origin of the tangent space $T_p\mathcal{M}$, we call *regular* (*singular*) points of the vector field F. If p is a regular point we may consider the *system in normal variations* with respect to the solution $\varphi^{(\cdot)}(p)$ of (8.55)

$$\frac{Dz}{dt} = A(\varphi^t(p))z, \tag{8.56}$$

where the linear operator $A(p) : T_p\mathcal{M} \to T_p\mathcal{M}$ is given by

$$\begin{aligned} A(p) &= \nabla F(p) - B(p), \quad \text{where} \\ B(p)v &= 2\frac{F(p)}{|F(p)|^2}\,(F(p), S\nabla F(p)v) \quad \text{for all } v \in T_p\mathcal{M}. \end{aligned} \tag{8.57}$$

The scalar product $(\cdot,\cdot)$ and the associated norm $|\cdot|$ are taken in the tangent space $T_p\mathcal{M}$. In coordinates of an arbitrary chart $x : \mathcal{D}(x) \to \mathcal{R}(x)$ around the regular point p, the linear operator $A(p)$ is given by

$$A_i^k = \nabla_i f^k - \frac{2}{g_{mn} f^m f^n} f^k g_{jl} f^l S_i^j, \quad k, i = 1, \dots, n, \tag{8.58}$$

where f^k and g_{jk} are the coordinates of the vector field F and the Riemannian metric tensor g in the chart x and $S_i^j = \frac{1}{2}\left[g^{jk}\nabla_k f^p g_{pi} + \nabla_i f^j\right]$ is the representation in coordinates of the symmetric part $S\nabla F(p)$ of the covariant derivative of the vector field F in this chart. Note that for ODE's in $\mathbb{R}^n$ with standard metric, the system in normal variations (8.56) coincides with the system in modified variations. Suppose that $p \in \mathcal{M}$ is a regular point of F and $y(\cdot)$ is a solution of (8.56) along $\varphi^{(\cdot)}(p)$. This solution can be split for any $t \in \mathbb{R}$ into two orthogonal components as

$$y(t) = z(t) + \mu(t)F(\varphi^t(p)), \tag{8.59}$$

where $z(\cdot)$ is the solution of (8.56) with respect to $\varphi^{(\cdot)}(p)$ with initial condition $z(0) = y(0)$ and $\mu(\cdot)$ is a scalar valued C^1-function given by $\mu(t) = (y(t), F(\varphi^t(p))) / |F(\varphi^t(p))|^2$.

For every regular point $p \in \mathcal{M}$ of F we introduce the $(n-1)$-dimensional linear subspace

$$T^{\perp}(p) = \{v \in T_p\mathcal{M} : (v, F(p)) = 0\}$$

of the tangent space $T_p\mathcal{M}$. Denote by $SA(p) := \frac{1}{2}[A(p) + A(p)^{[*]}]$ the symmetric part of the operator $A(p)$. A straight forward calculation shows that for all $v \in T^{\perp}(p)$ the following two relations

$$(F(p), SA(p)v) = 0 \quad \text{and} \quad (v, A(p)v) = (v, \nabla F(p)v) \tag{8.60}$$

are satisfied. Hence, we have $SA(p) : T^{\perp}(p) \to T^{\perp}(p)$. Using this fact one can easily prove the first part of the following lemma [21].

Lemma 8.5 *For an arbitrary regular point $p \in \mathcal{M}$ of the vector field (8.55), the eigenvalues of the operator $SA(p) : T_p\mathcal{M} \to T_p\mathcal{M}$ are the eigenvalues of the operator $SA(p)$ which is restricted to the linear subspace $T^{\perp}(p)$ and the value $-(\nabla F(p)F(p), F(p))/|F(p)|^2$. Further we have*

$$S\nabla F(p)z - \frac{F(p)}{|F(p)|^2}(F(p), S\nabla F(p)z) = SA(p)z \quad \text{for all } z \in T^{\perp}(p).$$

In the following, we denote at any regular point p of (8.54) the eigenvalues of the operator $SA(p)$ restricted to the subspace $T^{\perp}(p)$ by $\lambda_1^{\perp}(p) \geq \cdots \geq \lambda_{n-1}^{\perp}(p)$, which are ordered with respect to size and multiplicity. By $Z(t, p)$ we denote the operator solution of (8.56) with initial condition $Z(0, p) = \mathrm{id}_{T^{\perp}(p)}$. For every $t \in \mathbb{R}$, the linear operator $Z(t, p) : T^{\perp}(p) \to T^{\perp}(\varphi^t(p))$ maps between the subspaces $T^{\perp}(p)$ and $T^{\perp}(\varphi^t(p))$ being orthogonal to the vector field in p and $\varphi^t(p)$, respectively. The next lemma can be proved analogously to Theorem 8.3.

Lemma 8.6 *Suppose that $p \in \mathcal{M}$ is a regular point of the vector field (8.54) and $Z(\cdot, p)$ is the operator solution of (8.56). Let $d \in (0, n-1]$ be written in the form $d = d_0 + s$ with $d_0 \in \{0, 1, \ldots, n-2\}$ and $s \in (0, 1]$. Then for all $t \geq 0$ it holds*

$$\omega_d(Z(t, p)) \leq \exp\left\{\int_0^t \left[\lambda_1^{\perp}(\varphi^\tau(p)) + \cdots + \lambda_{d_0}^{\perp}(\varphi^\tau(p)) + s\lambda_{d_0+1}^{\perp}(\varphi^\tau(p))\right] d\tau\right\}.$$

Let $\mathcal{B}(O_p, r)$ denote the ball of radius r around the origin O_p of $T_p\mathcal{M}$. For a regular point $p \in \mathcal{M}$ of F let $\mathcal{B}^{\perp}(O_p, r) := \mathcal{B}(O_p, r) \cap T^{\perp}(p)$ be the ball in the subspace $T^{\perp}(p)$ centered in the origin O_p of $T_p\mathcal{M}$ with radius r. Fix p and r and consider for any $t \geq 0$ the ellipsoid $\mathcal{E}(t) := Z(t, p)\mathcal{B}^{\perp}(O_p, r)$ in the subspace

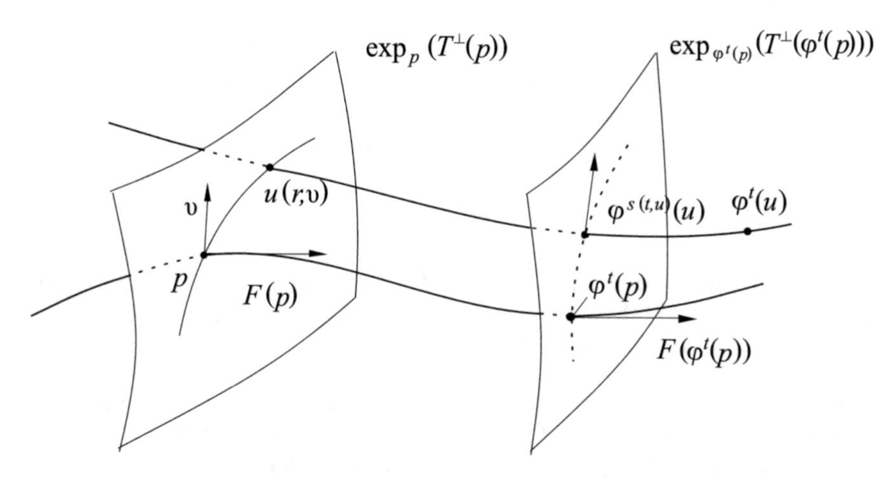

Fig. 8.1 Reparametrization of the flow

$T^{\perp}(\varphi^t(p))$. If $a_1(\mathcal{E}(t)) \geq \cdots \geq a_{n-1}(\mathcal{E}(t))$ are the lengths of the semi-axes of $\mathcal{E}(t)$ and d is an arbitrary number in $(0, n-1]$ we have by Eq. (7.8), Chap. 7,

$$\omega_d(\mathcal{E}(t)) = \omega_d(Z(t, p))\, r^d\,. \tag{8.61}$$

Our aim is to describe the variation of time translated pieces of $(n-1)$-dimensional submanifolds, orthogonal to a considered orbit of (8.54). For this purpose we use the methods from Sect. 7.2, Chap. 7, developed there for stability investigations of flows on manifolds. Considering a non-equilibrium solution $\varphi^{(\cdot)}(p)$ of (8.54) with $p \in \mathcal{M}$ the local transformation of small pieces of $(n-1)$-dimensional submanifolds can be described by a *reparametrized local flow*. For $\delta > 0$, so small that $\exp_p$ is defined on $\mathcal{B}(O_p, \delta)$, we shall consider the $(n-1)$-dimensional submanifold

$$\mathcal{B}^{\perp}(p, \delta) := \exp_p(\mathcal{B}^{\perp}(O_p, \delta))$$

of $\mathcal{M}$ through p which is local transversal at the point p to the trajectory of the vector field passing through the point p. Every point $u \in \mathcal{B}^{\perp}(p, \delta)$ can be uniquely written in the form $u = \exp_p(r\upsilon)$, where $\upsilon \in T^{\perp}(p)$ is a vector of length $|\upsilon| = 1$ and $r \in [0, \delta)$ measures the arc length of the geodesic $c_{p,\upsilon}$ connecting p and u. This defines for us a unique representation $u = u(r, \upsilon)$ of a point $u \in \mathcal{B}^{\perp}(p, \delta)$.

The main properties of the reparametrization (see Fig. 8.1) are summarized in the following lemmata [21] whose proofs are similar ommited to Lemma 7.8, Chap. 7.

Lemma 8.7 *Suppose that $\varphi^{(\cdot)}(p)$ is a non-equilibrium solution of the C^2-vector field (8.54). Then for any finite number $T_0 > 0$ there exists a number $\varepsilon_1 > 0$ such that for every $u \in \mathcal{B}^{\perp}(p, \varepsilon_1)$ there is a monotonously increasing differentiable function $s(\cdot, u) : \mathbb{R}_+ \to \mathbb{R}_+$ satisfying $s(\cdot, p) = \mathrm{id}|_{[0,\mathrm{T}_0]}$ and*

$$\left(\exp^{-1}_{\varphi^t(p)}\left(\varphi^{s(t,u)}(u)\right), F(\varphi^t(p))\right) = 0 \quad \textit{for all } t \in [0, T_0]. \tag{8.62}$$

The next lemma states that for any regular point $p \in \mathcal{M}$ of F for the locally defined reparametrized flow $\phi^t(\cdot) \equiv \phi(t, \cdot) := \varphi(s(t, \cdot), \cdot)$ the differential $d_p\phi^t$ of ϕ^t restricted to $T^{\perp}(p)$ satisfies (8.56).

Lemma 8.8 *Suppose that $\varphi^{(\cdot)}(p)$ is a non-equilibrium solution of (8.54) and the function $s(\cdot, \cdot) : [0, T_0] \times \mathcal{B}^{\perp}(p, \varepsilon_1) \to \mathbb{R}_+$ as given in Lemma 8.7 defines a reparametrized local flow $\phi^t(u) := \varphi^{s(t,u)}(u)$. Then for all $t \in [0, T_0]$ there holds*

$$d_p\phi^t|_{T^{\perp}(p)} = Z(t, p),$$

where $Z(t, p)$ denotes the operator solution of (8.56) with $Z(0, p) = id_{T^{\perp}(p)}$.

8.3.2 Tubular Carathéodory Structure

In this subsection we define a special Carathéodory structure in the sense of Sect. 3.4.1, Chap. 3, for flow negatively invariant sets on Riemannian manifolds. The outer measures which arise from this structure will majorize the Hausdorff measures and will be applied to obtain Hausdorff dimension estimates of flow-invariant sets on the manifold.

Let $(\mathcal{M}, g)$ be a smooth n-dimensional Riemannian manifold and ρ the metric induced by g. For a piecewise smooth curve $c : \mathcal{I} \to \mathcal{M}$ ($\mathcal{I} \subset \mathbb{R}$ an interval) of finite length and arbitrary $\varepsilon > 0$ we define the *ε-tubular neighborhood* $\Omega(c, \varepsilon)$ of c by

$$\Omega(c, \varepsilon) = \bigcup_{u \in c(\mathcal{I})} \mathcal{B}(u, \varepsilon),$$

where $\mathcal{B}(u, \varepsilon) = \{p \in \mathcal{M} | \rho(u, p) < \varepsilon\}$ is a metric ε-ball on $\mathcal{M}$ centered at the point u. For simplicity we call the ε-tubular neighborhood $\Omega(c, \varepsilon)$ around the curve c of length l shortly *tube of length l*.

For a given compact set $\mathcal{K} \subset \mathcal{M}$ and a given number $l_0 > 0$ we denote by $\Gamma(l_0) = \{c\}$ a family of piecewise smooth curves of a finite length $l(c) = l_0$ such that for any $\varepsilon > 0$ the following condition is satisfied:

(H1) $\mathcal{K}$ is contained in the union of ε-tubular neighborhoods $\Omega(c, \varepsilon)$ with $c \in \Gamma(l_0)$.

For a family $\Gamma(l_0)$ satisfying **(H1)** we define a family of subsets $\mathfrak{F}$, a parameter set $\mathbb{P}$, and the functions $\xi : \mathfrak{F} \times \mathbb{P} \to [0, \infty)$, $\eta : \mathfrak{F} \times \mathbb{R} \to [0, \infty)$, and $\psi : \mathfrak{F} \to [0, \infty)$ by

$$
\begin{gathered}
\mathfrak{F} = \{\Omega(c, \varepsilon) \cap \mathcal{K} \mid c \in \Gamma(l_0), \varepsilon > 0\} \cup \{\emptyset\}, \qquad \mathbb{P} = [1, +\infty), \\
\xi(\Omega(c, \varepsilon) \cap \mathcal{K}, d) := \varepsilon^{d-1}, \quad \eta(\Omega(c, \varepsilon) \cap \mathcal{K}, s) := \varepsilon^s, \quad \psi(\Omega(c, \varepsilon) \cap \mathcal{K}) := \varepsilon
\end{gathered}
\tag{8.63}
$$

for $c \in \Gamma(l_0)$, $\varepsilon > 0$ with $\Omega(c, \varepsilon) \cap \mathcal{K} \neq \emptyset$, $\xi(\emptyset, d) = \psi(\emptyset) = 0$, and $\eta(\emptyset, s) = 1$ for all $d \in \mathbb{P}$, $s \in \mathbb{R}$.

Straight forwardly, one can verify that the collection $(\mathfrak{F}, \mathbb{P}, \xi, \eta, \psi)$ defined via (8.63) with $\Gamma(l_0)$ satisfying **(H1)** is a Carathéodory structure on $\mathcal{K}$ in the sense of Sect. 3.4, Chap. 3. In the sequel we will call such a structure simply *Carathéodory structure with tubes of length* l_0 on $\mathcal{K}$. The next proposition ([21]) shows the relations between the Carathéodory measures and the Hausdorff measures, as well as between the Carathéodory dimension, generated by this structure, and the Hausdorff dimension. For the $\mathbb{R}^n$-case we refer to [18].

Proposition 8.1 *Suppose that $\mathcal{K}$ is a compact set on the smooth n-dimensional Riemannian manifold $(\mathcal{M}, g)$. Suppose that $(\mathfrak{F}, \mathbb{P}, \xi, \eta, \psi)$ is a tubular Carathéodory structure on $\mathcal{K}$ with tubes of length l_0 defined by (8.63) and with respect to this structure $\mu_C(\cdot, d, \varepsilon)$, $\mu_C(\cdot, d)$, and $\dim_C$ are the Carathéodory d-measure at level ε, the Carathéodory d-measure, and the Carathéodory dimension, respectively. Then there exist two numbers $k > 0$ and $\varepsilon_0 > 0$ depending only on $\mathcal{K}$ such that for any set $\mathcal{Y} \subset \mathcal{K}$ and any $d \geq 1$ the inequality*

$$
\mu_H(\mathcal{Y}, d, \varepsilon) \leq l_0 k \mu_C(\mathcal{Y}, d, \varepsilon) \tag{8.64}
$$

holds for all $\varepsilon \in (0, \varepsilon_0]$. Therefore, we have

$$
\mu_H(\mathcal{Y}, d) \leq l_0 k \mu_C(\mathcal{Y}, d) \quad \textit{and thus} \quad \dim_H \mathcal{Y} \leq \dim_C \mathcal{Y}.
$$

As in the previous subsection we shall consider the complete C^2-vector field $F : \mathcal{M} \to T\mathcal{M}$ on a smooth n-dimensional Riemannian manifold $\mathcal{M}$ and the corresponding differential equation (8.54) with the global flow $\{\varphi^t\}_{t \in \mathbb{R}}$. Let $\mathcal{K}$ and $\widetilde{\mathcal{K}}$ be two compact sets in $\mathcal{M}$ satisfying

$$
\mathcal{K} \subset \varphi^t(\mathcal{K}) \subset \widetilde{\mathcal{K}} \quad \text{for all } t \geq 0. \tag{8.65}
$$

At first we suppose that the set $\mathcal{K}$ does not contain equilibrium points of (8.54).

To construct the family $\Gamma(l_0)$ we can denote by $\mathcal{S}$ the set of all equilibrium points of (8.54) in $\widetilde{\mathcal{K}}$ and set $e_1 = \frac{1}{2}\operatorname{dist}(\mathcal{Z}, \mathcal{K})$, where $\operatorname{dist}(\mathcal{Z}, \mathcal{K}) = \inf\limits_{\substack{u \in \mathcal{Z}, \\ p \in \mathcal{K}}} \rho(u, p)$, and we can define

$$\mathcal{Z} := \widetilde{\mathcal{K}} \cap \bigcup_{p \in \mathcal{K}} \mathcal{B}(p, e_1). \tag{8.66}$$

With respect to the vector field F, the compact set $\widetilde{\mathcal{K}}$, and the set $\mathcal{Z}$ define the coefficient

$$\kappa(F, \widetilde{\mathcal{K}}, \mathcal{Z}) := \frac{\max_{u \in \widetilde{\mathcal{K}}} |F(u)|_{T_u \mathcal{M}}}{\min_{u \in \mathcal{Z}} |F(u)|_{T_u \mathcal{M}}}. \tag{8.67}$$

For any $p \in \mathcal{K}$ we take a time $b_p > 0$ such that $\varphi^t(p) \in \mathcal{Z}$ for all $t \in [0, b_p]$. Further, since $d_p \varphi^t|_{t=0} = \mathrm{id}_{\mathrm{T_p}\mathcal{M}}$ we can suppose that $|d_p \varphi^t| \leq 2$ holds for all $t \in [0, b_p]$. Since $\mathcal{K}$ is compact and contains no equilibrium points of F there exists a number $e_2 > 0$ such that for the length of the integral curve pieces it holds $l(\varphi(\cdot, p)|_{[0, b_p]}) \geq e_2$ for any $p \in \mathcal{K}$. We put

$$l_0 := \frac{1}{2} \min\{e_1, e_2\},$$

introduce for any $q \in \mathcal{K}$ the number $\tau(q) > 0$ satisfying $l(\varphi(\cdot, q)|_{[0, \tau(q)]}) = l_0$, and define the set

$$\Gamma = \{\, \varphi(\cdot, q)|_{[0, \tau(q)]} \;\; | \; q \in \mathcal{K}\}. \tag{8.68}$$

Obviously $\Gamma(l_0)$ satisfies condition **(H1)** and $(\mathfrak{F}, \mathbb{P}, \xi, \eta, \psi)$ defined by (8.63) is a Carathéodory structure on $\mathcal{K}$.

8.3.3 *Dimension Estimates for Sets Which are Negatively Invariant for a Flow*

In the present subsection we derive upper bounds for the Hausdorff dimension of compact sets being negatively invariant with respect to the flow of (8.54).

Our main result is the following theorem [21].

Theorem 8.7 *Let F be the C^2-vector field (8.54) on the smooth n-dimensional ($n \geq 2$) Riemannian manifold $(\mathcal{M}, g)$ satisfying the following conditions:*

(a) The flow $\{\varphi^t\}_{t \in \mathbb{R}}$ of (8.54) satisfies (8.65) with respect to the compact sets $\mathcal{K}$ and $\widetilde{\mathcal{K}}$ in $\mathcal{M}$, where $\mathcal{K}$ does not contain equilibrium points of (8.54).

(b) For any $p \in \widetilde{\mathcal{K}}$ let $\lambda_1^{\perp}(p) \geq \cdots \geq \lambda_{n-1}^{\perp}(p)$ be the eigenvalues of the symmetric part $SA(p)$ restricted to the subspace $T^{\perp}(p)$ where $A(p)$ is the operator from (8.57). There exists a number $d \in (0, n-1]$, written as $d = d_0 + s$ with $d_0 \in \{0, 1, \ldots, n-2\}$ and $s \in (0, 1]$, a number $\Theta > 0$, and a time $T_0 > 0$ such that

$$\int_0^{T_0} \left[\lambda_1^{\perp}(\varphi^\tau(p)) + \cdots + \lambda_{d_0}^{\perp}(\varphi^\tau(p)) + s\lambda_{d_0+1}^{\perp}(\varphi^\tau(p))\right] d\tau \le -\Theta \tag{8.69}$$

is satisfied for all regular points $p \in \widetilde{\mathcal{K}}$.

Then it holds $\dim_H \mathcal{K} < d+1$. *If* $d = 1$ *we have* $\dim_H \mathcal{K} \le 1$.

Before proving Theorem 8.7 let us formulate some lemmata from [21]. For an arbitrary piecewise smooth curve $c : [t_1, t_2] \to \mathcal{M}$ we denote its length by $l(c)$.

Lemma 8.9 *Suppose that* $\{\varphi^t\}_{t\in\mathbb{R}}$ *is the flow of (8.54),* $\mathcal{Z}$ *and* $\widetilde{\mathcal{K}}$ *are compact sets in* $\mathcal{M}$*,* $\mathcal{Z}$ *does not contain any equilibrium of (8.54), and* $\kappa(F, \widetilde{\mathcal{K}}, \mathcal{Z})$ *is the coefficient from (8.67) and let* $c^t : [t_1, t_2] \to \mathcal{M}$ *be the restriction of* $\varphi(\cdot, p)$ *on* $[t_1, t_2]$ *given by* $c^t(\cdot) := \varphi(t + \cdot, p)|_{[t_1,t_2]}$ *and satisfying* $c^0([t_1, t_2]) \subset \mathcal{Z}$ *and* $c^t([t_1, t_2]) \subseteq \widetilde{\mathcal{K}}$ *for all* $t > 0$. *Then the length* $l(c^t)$ *of such a restriction satisfies* $l(c^t) \le \kappa(F, \widetilde{\mathcal{K}}, \mathcal{Z})\, l(c^0)$ *for all* $t \ge 0$.

Proof The statement follows immediately from

$$l(c^t) = \int_{t_1}^{t_2} |\dot\varphi(\tau, \varphi^t(p))| d\tau = \int_{t_1}^{t_2} \frac{|\dot\varphi(\tau + t, p)|}{|\dot\varphi(\tau, p)|} |\dot\varphi(\tau, p)| d\tau \le \kappa(F, \widetilde{\mathcal{K}}, \mathcal{Z}) l(c^0).$$

□

Lemma 8.10 *Suppose* $\{\varphi^t\}_{t\in\mathbb{R}}$ *is the flow of (8.54) satisfying (8.65) with respect to the compact sets* $\mathcal{K}$ *and* $\widetilde{\mathcal{K}}$ *in* $\mathcal{M}$*, where* $\mathcal{K}$ *does not contain equilibrium points of (8.54). Suppose also that* $\mathcal{Z}$*,* $\kappa(F, \widetilde{\mathcal{K}}, \mathcal{Z})$*, and* l_0 *are given by (8.66), (8.67), and (8.68), respectively. For* $p \in \widetilde{\mathcal{K}}$ *let* $\lambda_1(p)$ *be the largest eigenvalue of* $S\nabla F(p)$*, and for a regular point* $p \in \widetilde{\mathcal{K}}$ *let* $\lambda_1^{\perp}(p) \ge \cdots \ge \lambda_{n-1}^{\perp}(p)$ *be the eigenvalues of* $SA(p)|_{T^{\perp}(p)}$ *where* $A(p)$ *is the operator from (8.57). Define for a number* $d \in (0, n-1]$ *written as* $d = d_0 + s$ *with* $d_0 \in \{0, 1, \ldots, n-2\}$ *and* $s \in (0, 1]$*, and a time* $T_0 > 0$ *the values*

$$k := \max_{p\in\mathcal{K}} \exp\left\{\int_0^{T_0} \left[\lambda_1^{\perp}(\varphi^\tau(p)) + \cdots + \lambda_{d_0}^{\perp}(\varphi^\tau(p)) + (d-d_0)\lambda_{d_0+1}^{\perp}(\varphi^\tau(p))\right] d\tau\right\},$$

$$\begin{aligned} a &:= \exp\left[3l_0 \max_{p\in\widetilde{\mathcal{K}}} \alpha_1(p) \frac{\kappa(F, \widetilde{\mathcal{K}}, \mathcal{Z})}{\min_{p\in\mathcal{Z}} |F(p)|_{T_p\mathcal{M}}}\right], \\ \beta &:= 2^6\sqrt{d_0+1}\, a, \quad \text{and} \quad C := \left(3\kappa(F, \widetilde{\mathcal{K}}, \mathcal{Z}) + 1\right) 2^{d_0} \beta^d. \end{aligned} \tag{8.70}$$

Then for any $l > k$ *there exists an* $\varepsilon_0 > 0$ *such that for all* $\varepsilon \in (0, \varepsilon_0]$ *the Carathéodory* $(d+1)$*-measure* $\mu_C(\cdot, d+1, \varepsilon)$ *at level* ε*, generated with respect to the Carathéodory structure on* $\mathcal{K}$ *with tubes of length* l_0 *from (8.68), satisfies the inequality*

$$\mu_C(\varphi^{T_0}(\mathcal{K}) \cap \mathcal{K}, d+1, \beta l^{1/d}\varepsilon) \leq C l \mu_C(\mathcal{K}, d+1, \varepsilon). \tag{8.71}$$

Proof Fix some $c \in \Gamma(l_0)$. For arbitrary $l > k$ we can choose an $\varepsilon > 0$ such that the set $\mathcal{V} := \bigcup_{p \in \mathcal{K}} \mathcal{B}(p, \varepsilon)$ contains no equilibrium points of (8.54) and the inequality

$$k' := \max_{u \in \overline{\mathcal{V}}} \exp \left\{ \int_0^{T_0} \left[\lambda_1^{\perp}(\varphi^{\tau}(u)) + \cdots + \lambda_{d_0}^{\perp}(\varphi^{\tau}(u)) \right. \right.$$
$$\left. \left. + (d - d_0)\lambda_{d_0+1}^{\perp}(\varphi^{\tau}(u)) \right] d\tau \right\} < l \tag{8.72}$$

is satisfied. We set

$$\sigma := \max_{p \in \overline{\mathcal{V}}} \exp \left\{ \int_0^{T_0} \lambda_1^{\perp}(\varphi^{\tau}(p)) d\tau \right\} \tag{8.73}$$

and take a number $m > 0$ such that $k' < m^d$ and $\sigma \leq m$ are satisfied. Since $l > k'$ the equation

$$\left[1 + \left(\frac{m^{d_0}}{k'} \right)^{1/(1-d_0)} \eta \right]^d k' = l$$

uniquely defines a number $\eta > 0$.

Choose $\delta > 0$ such that for any $u \in \widetilde{\mathcal{K}}$ the map $\exp_u$ maps the ball $\mathcal{B}(O_u, \delta) \subset T_u\mathcal{M}$ diffeomorphically onto the geodesic ball $\mathcal{B}(u, \delta) \subset \mathcal{M}$. Further with $|d_{O_u} \exp_p| = 1$ we can suppose that $|d_v \exp_p| \leq 2$ and therefore $\rho(\exp_u v_1, \exp_u v_2) \leq 2\rho(v_1, v_2)$ holds for all u and $v_1, v_2 \in \mathcal{B}(O_u, \delta)$.

To simplify the use of the reparametrized local flow we cover $\Omega(\gamma, \varepsilon)$ by a set $\mathcal{T}(\gamma_p, \varepsilon)$ as follows. Let for some $p \in \mathcal{K}$ and the associated time $t(p) > 0$ be $\gamma_p(\cdot) = \varphi(\cdot, p)|_{[0, t(p)]}$ the integral curve of length $2l_0$ such that $\gamma_p \supset \gamma$ and for any $\varepsilon > 0$ the inclusion $\Omega(\gamma, \varepsilon) \subset \mathcal{T}(\gamma_p, \varepsilon)$ holds, where

$$\mathcal{T}(\gamma_p, \varepsilon) := \bigcup_{u \in \gamma_p} \mathcal{B}^{\perp}(u, \varepsilon).$$

Let p and $t(p)$ be fixed. We take now $\varepsilon_0(\gamma) < \frac{1}{4} \min\{\varepsilon, \delta, \text{dist}(\mathcal{K}, \mathcal{M} \backslash \mathcal{V})\}$ small enough such that the following conditions are satisfied :

(1) The function $s : [0, \max\{T_0, t(p)\}] \times \mathcal{B}^{\perp}(p, 4\varepsilon_0(\gamma)) \to \mathbb{R}_+$ as characterized in the Lemma 8.7 defines a local reparametrization of the flow φ by $\phi : [0, \max\{T_0, t(p)\}] \times \mathcal{B}^{\perp}(p, 4\varepsilon_0(\gamma)) \to \mathcal{M}$ with $\phi(t, \cdot) \equiv \phi^t(\cdot) := \varphi^{s(t, \cdot)}(\cdot)$ for $t \in [0, \max\{T_0, t(p)\}]$.

(2) $\phi^{T_0}(\mathcal{B}^{\perp}(p, 4\varepsilon_0(\gamma))) \subset \mathcal{B}(\varphi^{T_0}(p), \delta)$

(3) The distance between the points $\phi^t(u)$ on an integral curve starting in $u =$

$\exp_p(r\upsilon) \in \mathcal{B}^{\perp}(p, \varepsilon_0(\gamma))$ and the reference orbit through p for a fixed $t \in [0, t(p)]$ is of the size

$$\rho\,(\varphi^t(p), \phi^t(u)) = |d_p\phi^t| \cdot r(1 + O(r))$$

as $r \to 0$. It holds $|d_p\phi^t| \le |d_p\varphi^t|$ and $|d_p\varphi^t| \le 2$ for any $t > 0$ such that $l(\varphi([0, t], p)) \le 2l_0$. Thus, for any $u \in \mathcal{B}^{\perp}(p, \varepsilon_0(\gamma))$ it is $\rho\,(\varphi^t(p), \phi^t(u)) \le 4\rho\,(p, u)$ for any such t. We can assume analogous assumptions for the flow in reverse time-direction. Let for $\varepsilon_0(\gamma) > 0$ the following be satisfied : If $\gamma' = \phi([0, t(p)], u)$ is some arc of trajectory intersecting $\mathcal{T}(\gamma_p, \varepsilon_0(\gamma))$ then γ' is completely contained in $\mathcal{T}(\gamma_p, 4\,\varepsilon_0(\gamma))$ and satisfies $l(\gamma') \le 3\,l_0$.
(4) For any $u \in \widetilde{\mathcal{K}}$ and for the associated time $\iota(u) > 0$ such that the integral curve $\varphi([0, \iota(u)], u)$ is of length $3\,l_0\kappa(F, \widetilde{\mathcal{K}}, \Phi)$ it holds

$$\sup_{q \in \mathcal{B}(u, 16\sigma\,\varepsilon_0(\gamma))} |\tau_{\varphi^t(q)}^{\varphi^t(u)} d_q\varphi^t \tau_q^u - d_u\varphi^t| \le a, \quad \text{for all } t \in (0, \iota(u)). \tag{8.74}$$

Suppose that it holds

$$\sup_{q \in \mathcal{B}^{\perp}(p, 4\varepsilon_0(\gamma))} |\tau_{\phi^{T_0}(q)}^{\phi^{T_0}(p)} d_q\phi^{T_0} \tau_p^q - d_p\phi^{T_0}| \le \eta. \tag{8.75}$$

(5) For any $u = u(r, \upsilon) \in \mathcal{B}^{\perp}(p, 4\varepsilon_0(\gamma))$ the deviation arising from the local reparametrization of the flow is of the form $s(T_0, u(r, \upsilon)) - T_0 = O(r)$ as $r \to 0$ which gives for the point $\phi^{T_0}(u) = \varphi^{s(T_0, u) - T_0}(\varphi^{T_0}(u))$ the representation

$$\exp^{-1}_{\varphi^{T_0}(u)}(\phi^{T_0}(u)) = O_{\varphi^{T_0}(u)} + F(\varphi^{T_0}(u))O(r) + o(r)$$

as $r \to 0$. The vector field C^2-varies on $\mathcal{M}$. So we can suppose that for any point $u \in \mathcal{B}^{\perp}(\varphi^{T_0}(p), \delta)$ for $\nu < 2^4\sqrt{d_0 + 1}\sigma\,\varepsilon_0(\gamma)$ any set $(\phi^{T_0} \circ \varphi^{-T_0})\mathcal{B}(u, \nu)$ is contained in a 2ν-tubular neighborhood of a curve $\varphi(\cdot, \varphi^{T_0}(u))|_{(-\tau,\tau)}$ of some finite length, say of length l_0.

Now let $r \le \varepsilon_0(\gamma)$. Suppose $\varphi^{T_0}(\Omega(\gamma, r)) \cap \mathcal{K} \ne \emptyset$. The set $\mathcal{B}(p, 4r)$ is contained in the open set $\mathcal{V}$. Taylor's formula for the differentiable map ϕ^{T_0} provides that for every $u \in \mathcal{B}^{\perp}(p, 4r)$

$$\begin{aligned} &|\exp^{-1}_{\phi^{T_0}(p)} \phi^{T_0}(u) - d_p\phi^{T_0}(\exp_p^{-1}(u))| \\ &\qquad \le \sup_{q \in \mathcal{B}(p, 4r)} |\tau_{\phi^{T_0}(q)}^{\phi^{T_0}(p)} d_q\phi^{T_0} \tau_p^q - d_p\phi^{T_0}| \cdot |\exp_p^{-1}(q)| \end{aligned} \tag{8.76}$$

holds. Considering the image of $\mathcal{B}^{\perp}(p, 4r)$ under ϕ^{T_0} with (8.75) we obtain the inclusion

$$\exp^{-1}_{\phi^{T_0}(p)}\left(\phi^{T_0}\left(\mathcal{B}^{\perp}(p, 4r)\right)\right) \subset d_p\phi^{T_0}\left(\mathcal{B}^{\perp}(O_p, 4r)\right) + \mathcal{B}^{\perp}(O_{\varphi^{T_0}(p)}, \eta 4r).$$

The set $d_p\phi^{T_0}\left(\mathcal{B}^\perp(O_p, 4r)\right)$ is an ellipsoid with half-axes of length $4r\alpha_k(p)$, where $\alpha_k(p)$ $(k = 1, \dots, n-1)$ denote the singular values of the linear operator $d_p\phi^{T_0}$: $T^\perp(p) \to T^\perp(\varphi^{T_0}(p))$. Using the definition of k', Lemma 8.6 and Eq. (8.61) we conclude

$$\omega_d\left(d_p\phi^{T_0}(\mathcal{B}^\perp(O_p, 4r))\right) \le (4r)^d\, k'. \tag{8.77}$$

By standard covering results (see e.g. Sect. 8.1.2 or [24]) an ellipsoid $\mathcal{E} \subset T^\perp(\varphi^{T_0}(p))$ can be found containing $d_p\phi^{T_0}\left(\mathcal{B}^\perp(O_p, 4r)\right) + \mathcal{B}(O_{\varphi^{T_0}(p)}, \eta 4r)$ and satisfying $\omega_d(\mathcal{E}) \le l(4r)^d$. Any set $\mathcal{E}$ can be covered by N balls of radius $R = \sqrt{d_0+1}\alpha_{d_0+1}(\mathcal{E})$. The number N can be estimated from above by

$$N \le \frac{2^{d_0}\omega_d(\mathcal{E})}{\alpha_{d_0+1}(\mathcal{E})^d}\,.$$

Thus, any set $\exp_{\varphi^{T_0}(p)}(\mathcal{E})$ and therefore $\phi^{T_0}(\mathcal{B}^\perp(p, 4r))$ can be covered by N geodesic balls in $\mathcal{M}$ of radius $2R$. Fixing such a cover $\{\mathcal{B}(\widetilde{u}_j, 2R)\}_{j\ge1}$ where $\widetilde{u}_j \in \mathcal{M}$ $(j \ge 1)$ we choose in every set $\mathcal{K} \cap \mathcal{B}(\widetilde{u}_j, 2R) \cap \mathcal{B}^\perp(\varphi^{T_0}(p), \delta)$ a point u_j and obtain the cover $\{\mathcal{B}_j\}_{j\ge1}$ of the set $\phi^{T_0}(\mathcal{B}^\perp(p, 4r)) \cap \mathcal{K}$ with $\mathcal{B}_j = \mathcal{B}(u_j, 4R) \cap \mathcal{B}^\perp(\varphi^{T_0}(p), \delta)$.

Now we consider the deviation arising from the reparametrization. By the property (8.76) any set $(\phi^{T_0} \circ \varphi^{-T_0})(\mathcal{B}_j)$ is with precision $o(r)$ $(r \le \varepsilon_0(\gamma))$ contained in a $4R$-neighborhood of the orbit trough $\varphi^{T_0}(u)$, or more precise, in an $8R$-neighborhood of a trajectory piece $\varphi(\cdot, (\varphi^{T_0} \circ \phi^{-T_0})(u_j))|_{(-\tau,\tau)}$ of length l_0.

By the choice of $\varepsilon_0(\gamma)$ any trajectory piece in $\mathcal{T}(\gamma, 4r)$ which intersects $\mathcal{T}(\gamma, r)$ is of maximal length $3l_0$. We shift the balls $\mathcal{B}((\phi^{T_0} \circ \varphi^{-T_0})(u_j), 8R)$ along the flow lines. Thus, with the above and (8.74) the set $\varphi^{T_0}(\mathcal{T}(\gamma, r))$ can be covered by N tubes of length $3l_0\kappa(F, \widetilde{\mathcal{K}}, \mathcal{Z}) + l_0$ and diameter $2a \cdot 8R$.

Covering each curve arc by curve arcs of length l_0 we conclude

$$\begin{aligned}
\mu_C(\varphi^{T_0}(\Omega(\gamma, r)) \cap \mathcal{K}, d+1, 2^6\sqrt{d_0+1}l^{1/d}a\varepsilon_0(\gamma))& \\
\le N(3\kappa(F, \widetilde{\mathcal{K}}, \Phi) + 1)\left(2^6 a\sqrt{d_0+1}\alpha_{d_0+1}(\mathcal{E})\right)^d& \\
\le Cl\varepsilon_0(\gamma)^d.&
\end{aligned} \tag{8.78}$$

Since Γ is the set of trajectory pieces starting at a point p in the compact set $\mathcal{K}$ we can pass to $\varepsilon_0 = \inf_{\gamma\in\Gamma} \varepsilon_0(\gamma) > 0$ such that the (8.78) holds for any $\Omega(\gamma, \varepsilon)$ with $\gamma \in \Gamma$ and $\varepsilon \le \varepsilon_0$. Let $\varepsilon \le \varepsilon_0$. For any $\nu > 0$ there exists a finite family $\{\Omega(\gamma_i, r_i)\}_{i\ge1}$ with $\gamma_i \in \Gamma$, $r_i \le \varepsilon$ having the property that $\bigcup_i \Omega(\gamma_i, r_i) \supset \mathcal{K}$ and $\sum_i r_i^d \le \mu_C(\varphi^{T_0}(\mathcal{K}) \cap \mathcal{K}, d+1, \varepsilon) + \nu$. We obtain $\mu_C(\varphi^{T_0}(\mathcal{K}) \cap \mathcal{K}, d+1, \beta l^{1/d}\varepsilon) \le \sum_i \mu_C(\varphi^{T_0}(\Omega(\gamma_i, r_i)) \cap \mathcal{K}, d+1, \beta l^{1/d}\varepsilon) \le Cl\sum_i r_i^d \le Cl$ $(\mu_C(\mathcal{K}, d+1, \varepsilon) + \nu)$ where β and C are defined by (8.70). Since ν has been chosen arbitrarily, we obtain that (8.71) holds for any $\varepsilon \in (0, \varepsilon_0]$. $\square$

Although we are mainly interested in upper estimates of the Hausdorff dimension of flow negatively invariant sets, we can deduce upper bounds of its Carathéodory dimension with respect to the chosen tubular Carathéodory structure.

Proposition 8.2 *Let the differential equation (8.54) satisfy the conditions of Theorem 8.7 with the number $d \in (0, n-1]$ in (8.69) and the negatively invariant set $\mathcal{K}$. Then the Carathéodory dimension of $\mathcal{K}$, determined with respect to the Carathéodory structure (8.63) on $\mathcal{K}$ consisting of tubes with length l_0 determined in (8.68), satisfies*

$$\dim_C \mathcal{K} < d + 1.$$

Proof It follows from (8.69) that for an arbitrarily small number $\kappa \in (0, 1)$ there exists some number $m = m(\kappa) > 0$ such that

$$k := \sup_{p \in \mathcal{K}} \exp\left\{ \int_0^{mT_0} [\lambda_1^{\perp}(\varphi^t(p)) + \cdots + \lambda_{d_0}^{\perp}(\varphi^t(p)) + s\lambda_{d_0+1}^{\perp}(\varphi^t(p))]dt \right\}$$
$$\leq \exp(-m\Theta) < \kappa\,. \quad (8.79)$$

Without loss of generality we can assume that this number k satisfies $\beta k^{1/d} < 1$ and $Ck < 1$, where β and C are the constants given in (8.70). We choose $l > k$ with $\beta l^{1/d} < 1$ and $Cl < 1$. Lemma 8.10, applied to the map φ^{mT_0}, guarantees that for the chosen number l there exists a number $\varepsilon_0 > 0$ such that for all $\varepsilon \in (0, \varepsilon_0]$ the inequality

$$\mu_C(\varphi^{mT_0}(\mathcal{K}) \cap \mathcal{K}, d+1, \beta l^{1/d}\varepsilon) \leq Cl\mu_C(\mathcal{K}, d+1, \varepsilon) \quad (8.80)$$

holds. Let $\varepsilon \in (0, \varepsilon_0]$ be arbitrarily small. Since $\mathcal{K}$ is compact the value $\mu_C(\mathcal{K}, d+1, \varepsilon)$ is finite. Since $\mathcal{K}$ is negatively invariant with respect to φ^{mT_0} we have $\mathcal{K} = \varphi^{mT_0}(\mathcal{K}) \cap \mathcal{K}$. Using the inequality $Cl < 1$ we conclude $\mu_C(\mathcal{K}, d+1, \beta l^{1/d}\varepsilon) < \mu_C(\mathcal{K}, d+1, \varepsilon)$. By $\beta l^{1/d} < 1$ and the fact that $\mu_C(\mathcal{K}, d+1, \varepsilon)$ is monotonously increasing as $\varepsilon \to 0+0$, we get $\mu_C(\mathcal{K}, d+1, \varepsilon) = 0$, which contradicts our assumption. Thus, the equality $\mu_C(\mathcal{K}, d+1, \varepsilon) = 0$ holds for every $\varepsilon \in (0, \varepsilon_0]$. We see that $\mu_C(\mathcal{K}, d+1) = 0$. This implies $\dim_C \mathcal{K} \leq d+1$. Since (8.79) holds true if we slightly reduce d we conclude $\dim_C \mathcal{K} < d+1$. s □

Proof of Theorem 8.4 Applying Propositions 8.1 and 8.2 we obtain $\dim_H \mathcal{K} < d + 1$. If condition (8.69) is also satisfied for $d = 1$ it is satisfied for all $d \in (0, n-1]$. Thus, $\dim_H \mathcal{K} < d+1$ for all $d \in (0, n-1]$ and we obtain $\dim_H \mathcal{K} \leq 1$. This proves the Theorem. □

Let us again consider compact sets $\mathcal{K}$ and $\widetilde{\mathcal{K}}$ in $\mathcal{M}$ satisfying (8.65) with respect to the flow of (8.54). We may now assume that the set $\mathcal{K}$ possesses equilibrium points and satisfies the following condition:

(H2) The set $\mathcal{K}$ contains at most a finite number of equilibrium points of (8.54). Every such equilibrium point possesses a local stable manifold with a dimension at least $n - 1$. Trajectories starting in local unstable manifolds or local center manifolds of such an equilibrium point in $\mathcal{K}$, converge for $t \to +\infty$ to an asymptotically stable equilibrium point of (8.54) in $\widetilde{K}$.

The special structure of equilibrium points satisfying **(H2)** allows us to obtain the following theorem.

Theorem 8.8 *Let F be a C^2-vector field (8.54) on the smooth n-dimensional Riemannian manifold $(\mathcal{M}, g)$. Suppose that the flow $\{\varphi^t\}_{t\in\mathbb{R}}$ of (8.54) satisfies (8.65) and condition* **(H2)** *with respect to compact sets $\mathcal{K}$ and $\widetilde{\mathcal{K}}$ in $\mathcal{M}$. Suppose also that condition b) of Theorem 8.7 is satisfied. Then the conclusion of Theorem 8.7 holds.*

In the following statement from [21] we denote for a differentiable function $V : \mathcal{U} \subset \mathcal{M} \to \mathbb{R}, \mathcal{U}$ an open subset, by $L_F V(p)$ the Lie derivative of V in p in direction of the vector field F (Sect. A.3, Appendix A).

Corollary 8.11 *Suppose that the flow $\{\varphi^t\}_{t\in\mathbb{R}}$ of (8.54) satisfies (8.65) and condition* **(H2)** *with respect to compact sets $\mathcal{K}$ and $\widetilde{\mathcal{K}}$ in $\mathcal{M}$.*

Denote by $\mathcal{S}$ the set of equilibrium points of (8.54) in $\mathcal{M}$. For $p \in \mathcal{M}\backslash\mathcal{S}$ let $\lambda_1^{\perp}(p) \geq \cdots \geq \lambda_{n-1}^{\perp}(p)$ be the eigenvalues of the symmetric part $SA(p)$ restricted to the subspace $T^{\perp}(p)$, where $A(p)$ is the operator from (8.57), and let $V : \mathcal{M}\backslash\mathcal{S} \to \mathbb{R}$ be a C^1-function. Suppose also that for a number $d \in (0, n-1]$, written as $d = d_0 + s$ with $d_0 \in \{0, 1, \ldots, n-1\}$ and $s \in (0, 1]$, and a time $T_0 > 0$ such that

$$\int_0^{T_0} \left[\lambda_1^{\perp}(\varphi^t(p)) + \cdots + \lambda_{d_0}^{\perp}(\varphi^t(p)) + s\lambda_{d_0+1}^{\perp}(\varphi^t(p)) + L_F V(\varphi^t(p))\right] dt \leq -\Theta \tag{8.81}$$

holds for all regular points $p \in \widetilde{\mathcal{K}}$. Then the conclusion of Theorem 8.7 holds.

Proof On open and flow positively invariant neighborhoods of equilibrium points of (8.54) which satisfy **(H2)**, the flow preserves its contracting properties with respect to the Hausdorff measure. So it remains to show that for any compact, flow negatively invariant set $\mathcal{K}_1 \subset \mathcal{K}$ which does not contain equilibrium points of (8.54) it holds $\dim_H \mathcal{K}_1 < d + 1$. On $\mathcal{M}\backslash\mathcal{S}$ we introduce a new metric tensor by $\widehat{g}(p) := \exp\left(\frac{2V(p)}{d}\right) g(p)$ for $p \in \mathcal{M}\backslash\mathcal{S}$. On $\mathcal{K}_1$ the Riemannian metric $\widehat{g}$ is equivalent to g. Changing to the metric $\widehat{g}$ does not alter the Hausdorff dimension of the compact set $\mathcal{K}_1$. Consider the operator $\widehat{A}(p)$ from (8.57), the symmetric part $S\widehat{A}(p)$ of $\widehat{A}(p)$, the operator $\widehat{\nabla} F(p)$, and $S\widehat{\nabla} F(p)$, which are defined with regard to the scalar product in $T_p\mathcal{M}$ induced by the metric $\widehat{g}$. As in Sect. 7.2, Chap. 7, one shows that $S\widehat{\nabla} F(p) = S\nabla F(p) + \frac{L_F V(p)}{d} \mathrm{id}_{T_p\mathcal{M}}$. Using (8.60) we obtain that for a regular point $p \in \mathcal{M}$ the eigenvalues $\widehat{\lambda}_i^{\perp}(p)$ of the operator $S\widehat{A}(p)|_{T^{\perp}(p)}$ are related to the eigenvalues $\lambda_i^{\perp}(p), i = 1, \ldots, n-1$ with respect to the original metric g by $\widehat{\lambda}_i^{\perp}(p) = \lambda_i^{\perp}(p) + \frac{L_F V(p)}{d}$. Therefore,

$$\widehat{\lambda}_1^{\perp}(p) + \cdots + \widehat{\lambda}_{d_0}^{\perp}(p) + s\widehat{\lambda}_{d_0+1}^{\perp}(p) = \lambda_1^{\perp}(p) + \cdots + \lambda_{d_0}^{\perp} + s\lambda_{d_0+1}^{\perp}(p) + L_F V(p)$$

guarantees (8.81) and thus (8.69) of Theorem 8.7. Hence $\dim_H \mathcal{K}_1 < d + 1$. □

Corollary 8.12 *Consider a 2-dimensional smooth Riemannian manifold $\mathcal{M}$. Suppose that the flow $\{\varphi^t\}_{t\in\mathbb{R}}$ of (8.54) satisfies (8.65) and condition* **(H2)** *with respect to compact sets $\mathcal{K}$ and $\widetilde{\mathcal{K}}$ in $\mathcal{M}$. If* div $F(p) < 0$ *holds for any regular points $p \in \widetilde{\mathcal{K}}$ then* $\dim_H \mathcal{K} \leq 1$.

Proof For the operator $A(p)$ from (8.57) it holds that $\mathrm{tr}(SA(p)|_{T^{\perp}(p)}) = \mathrm{tr}\nabla F(p) - (\nabla F(p)F(p), F(p))/|F(p)|^2$. We define the C^1-function V on the set of all regular points p in $\mathcal{M}$ by $V(p) := \frac{1}{2}\log|F(p)|^2$. The statement follows with Corollary 8.11. □

8.3.4 *Flow Invariant Sets with an Equivariant Tangent Bundle Splitting*

The considered outer measures, defined via tube covers, show in many cases a better contraction behaviour under the flow operator of a vector field in positive time direction, than conventional outer measures defined via a covering of balls. Using such an approach for a class of generalized hyperbolic flows on n-dimensional Riemannian manifolds, we may improve upper Hausdorff dimension estimates which are obtained with methods from Sect. 5.2, Chap. 5 and Sect. 8.1.

Consider again the vector field (8.54) on the smooth n-dimensional manifold $(\mathcal{M}, g)$. Let us assume that a flow-invariant compact set $\mathcal{K} \subset \mathcal{M}$ possesses an *equivariant tangent bundle splitting* $T_{\mathcal{K}}\mathcal{M} = \mathbb{E}^1 \oplus \mathbb{E}^2$ with respect to the flow $\{\varphi^t\}_{t\in\mathbb{R}}$, i.e. (see Sect. A.10, Appendix A) for any $p \in \mathcal{K}$ and $i = 1, 2$ the space $\mathbb{E}^i_p = \mathbb{E}^i \cap T_p\mathcal{M}$ is an n_i-dimensional subspace of $T_p\mathcal{M}$ such that $n_1 + n_2 = n$ and $d_p\varphi^t(\mathbb{E}^i_p) = \mathbb{E}^i_{\varphi^t(p)}$ hold for any $p \in \mathcal{K}$ and $t \in \mathbb{R}$.

For $d \in (0, n - n_2]$ and $t \in \mathbb{R}$ we introduce the *singular value function of order d of φ^t on $\mathcal{K}$ with respect to the splitting* $\mathbb{E}^1 \oplus \mathbb{E}^2$ which is defined by

$$\omega_{d,\mathcal{K}}^{\mathbb{E}^1,\mathbb{E}^2}(\varphi^t) = \sup_{p\in\mathcal{K}} \omega_d(d_p\varphi^t|_{\mathbb{E}^1(p)}).$$

Since $\omega_{d,\mathcal{K}}^{\mathbb{E}^1,\mathbb{E}^2}(\varphi^t)$ is a sub-exponential function, the limit

$$\nu_d = \lim_{t\to+\infty} \frac{1}{t} \log \omega_{d,\mathcal{K}}^{\mathbb{E}^1,\mathbb{E}^2}(\varphi^t)$$

exists [28] for any $d \in (0, n - n_2]$. We call the numbers

$$\nu_1^u := \nu_1, \quad \nu_i^u := \nu_i - \nu_{i-1}, \quad i = 1, 2, \dots, n - n_2$$

the *uniform Lyapunov exponents of the flow* $\{\varphi^t\}_{t\in\mathbb{R}}$ *with respect to the splitting* $\mathbb{E}^1 \oplus \mathbb{E}^2$. Let us investigate the splitting $T_{\mathcal{K}}\mathcal{M} = \mathbb{E}^2 \oplus \mathbb{E}^2$ such that $\mathbb{E}^1 = T^{\perp}$ with $\mathbb{E}_p^1 := T^{\perp}(p)$ and $\mathbb{E}^2 = T^{\parallel}$ with $\mathbb{E}_p^2 := T^{\parallel}(p) = \mathrm{span}\{F(p)\}$.

With the help of Lemma 8.5 one shows that for any regular point $p \in \mathcal{M}$ satisfying

$$(S\nabla F(p)z, F(p)) = 0 \quad \text{for all } z \in T^{\perp}(p) \tag{8.82}$$

the $n-1$ eigenvalues $\alpha_1(p), \dots, \alpha_{n-1}(p)$ of $SA(p)|_{T^{\perp}(p)}$, with the operator $A(p)$ from (8.57), coincide with $n-1$ eigenvalues of $S\nabla F(p)$. The subspace $T^{\parallel}(p)$ is the eigenspace of the remaining nth eigenvalue $\overline{\lambda}(p) = (\nabla F(p)F(p), F(p))/|F(p)|^2$ oF $S\nabla F(p)$.

We consider now two compact sets $\mathcal{K}$ and $\widetilde{\mathcal{K}}$ of $\mathcal{M}$ without equilibrium points of (8.54) satisfying (8.65) and suppose that (8.82) is satisfied for any $p \in \widetilde{\mathcal{K}}$. By $\lambda_1(p) \geq \dots \geq \lambda_n(p)$, denote the eigenvalues of $S\nabla F(p)$. For that case Theorem 8.3 states that if for some $d = d_0 + s$ with $d_0 \in \{0, \dots, n-1\}$ and $s \in (0, 1]$ the inequality

$$\lambda_1(p) + \dots + \lambda_{d_0}(p) + s\lambda_{d_0+1}(p) < 0$$

holds for all $p \in \widetilde{\mathcal{K}}$, the estimate $\dim_H \mathcal{K} < d$ is true. For the C^1-function $V : \widetilde{\mathcal{K}} \to \mathbb{R}$ given by $V(p) := \frac{1}{2}\log|F(p)|^2$ we have $L_F V(p) = (\nabla F(p)F(p), F(p))/|F(p)|^2 = \overline{\lambda}(p)$ for each $p \in \widetilde{\mathcal{K}}$. If $\overline{\lambda}(p) \geq 0$ holds for all $p \in \widetilde{\mathcal{K}}$ then

$$\lambda_1(p) + \dots + \lambda_{d_0}(p) + s\lambda_{d_0+1}(p) = \lambda_1^{\perp}(p) + \dots + \lambda_{d_0-1}^{\perp}(p) + s\lambda_{d_0}^{\perp}(p) + L_F V(p).$$

With this Corollary 8.11 gives an upper bound of $\dim_H \mathcal{K}$ which is less than or equal to the upper bound we would get applying Theorem 8.3. If $d = 1$ then Corollary 8.11 gives the better estimate $\dim_H \mathcal{K} \leq 1$.

One easily shows that a compact, flow-invariant set $\mathcal{K}$ without equilibrium points possesses an equivariant tangent bundle splitting $T^{\perp} \oplus T^{\parallel}$ if and only if (8.82) holds for any $p \in \mathcal{K}$. Obviously the flow $\{\varphi^t\}_{t\in\mathbb{R}}$ on $\mathcal{K}$ then is already reparametrized globally if one considers the reparametrization described in Lemma 8.7. For that case the assumptions of Theorem 8.7 can be weakened if we consider the long-time behaviour.

Proposition 8.3 *Let F be the C^2-vector field from (8.54) on the n-dimensional Riemannian manifold $\mathcal{M}$. Suppose that $\mathcal{K} \subset \mathcal{M}$ is a compact and flow-invariant set without equilibrium points of (8.54) and that $\mathcal{K}$ possesses an equivariant tangent bundle splitting $T_{\mathcal{K}}\mathcal{M} = T^{\perp} \oplus T^{\parallel}$ with respect to the flow. Let $D \in \{0, \dots, n-1\}$ be the smallest number such that $\nu_1^u + \cdots + \nu_D^u + \nu_{D+1}^u < 0$. Then it holds*

$$\dim_H \mathcal{K} \leq D + \frac{\nu_1^u + \cdots + \nu_D^u}{|\nu_{D+1}^u|} + 1 .$$

Proof Take an arbitrary number $d \in \left(D + \frac{\nu_1^u + \dots + \nu_D^u}{|\nu_{D+1}^u|}, n-1\right]$. Then it holds $\nu_d = \nu_1^d + \cdots + \nu_{d_0}^u + s\nu_{d_0+1}^u < 0$. Fix some $\varepsilon \in (0, \nu_d)$. By definition of ν_d there is a finite number $T_0 > 0$ such that $\frac{1}{T_0} \log \omega_{d,\mathcal{K}}^{T^{\perp}, T^{\parallel}}(\varphi^{T_0}) < \nu_d + \varepsilon$, i.e. $\omega_{d,\mathcal{K}}^{T^{\perp}, T^{\parallel}}(\varphi^{T_0}) < \exp(T_0(\nu_d + \varepsilon)) < 1$. Theorem 8.7 basically uses properties of the singular value function. Thus, the proposition can be proved applying arguments and using the property $\omega_{d,\mathcal{K}}^{T^{\perp}, T^{\parallel}}(\varphi^{T_0}) = \sup_{p \in \mathcal{K}} \omega_d(d_p \varphi^{T_0}|_{T^{\perp}(p)})$. □

Example 8.1 Consider the vector field in $\mathbb{R}^2$ given by

$$\dot{x}_1 = a \sin x_1 , \quad \dot{x}_2 = -x_2 + b , \tag{8.83}$$

where $a \geq 1, b \neq 0$ are parameters. The arising dynamical system can be interpreted as a dynamical system on the flat cylinder Z (see Sect. A.1, Appendix A), i.e. on a 2-dimensional Riemannian manifold with the standard metric for factor manifolds. Every solution of (8.83) is bounded in the second coordinate. Obviously, the set $\mathcal{K} := \{z \in Z | z = [u], u = (x_1, 0), x_1 \in \mathbb{R}\}$ is compact and flow-invariant with respect to (8.83). The variational system (8.55) and the system in normal variations (8.56) with respect to any solution $(x_1(t), 0)$ in $\mathcal{K}$ are given by

$$\dot{y}_1 = a \cos x_1(t) \cdot y_1 , \qquad \dot{y}_2 = -\dot{y}_2 ,$$

and

$$\dot{z}_1 = -a \cos x_1(t) \cdot z_1 , \qquad \dot{z}_2 = -\dot{z}_2 ,$$

respectively. Thus $\lambda_1^{\perp}(z) = -1$ for any $z = (z_1, z_2) \in \mathcal{K}$ and condition (8.69) is satisfied with $d = 1$ and $\Theta = T = 1$. By Theorem 8.7 we conclude that $\dim_H \mathcal{K} \leq 1$. Note that in the present situation Theorem 8.1 is not applicable since the divergence of the right-hand side of (8.83) is $a \cos x_1 - 1$ which is, in contrast to the assumptions of Theorem 8.1, not always negative.

8.3.5 Generalizations of the Theorems of Hartman-Olech and Borg

Consider an arbitrary C^2-vector field F in $\mathbb{R}^3$ with the standard Euclidean metric, i.e., the differential equation

$$\dot{u} = F(u). \tag{8.84}$$

Suppose that for (8.84) the global flow $\{\varphi^t\}_{t\in\mathbb{R}}$ exists. Let $\mathcal{K}$ and $\widetilde{\mathcal{K}}$ be two compact sets in $\mathbb{R}^3$ satisfying (8.65). For any $u \in \mathbb{R}^3$ the covariant derivative $\nabla F(u)$ can be identified with the Jacobi matrix $DF(u)$ of F in u. Suppose that (8.84) possesses in $\widetilde{\mathcal{K}}$ a finite number of equilibrium points and for any such equilibrium point u all eigenvalues of $DF(u)$ have negative real part.
Consider the symmetric part $SDF(u) = \frac{1}{2}(DF(u) + DF(u)^*)$ of $DF(u)$. For any regular point u of F define the hyperplane $T^\perp(u) := \{w \in \mathbb{R}^3 | \, (w, F(u)) = 0\}$. Then the linear operator $SA(u) : T^\perp(u) \to T^\perp(u)$ is given for $w \in T^\perp(u)$ by

$$SA(u)w := SDF(u)w - \frac{F(u)}{|F(u)|^2}(SDF(u)w, F(u)).$$

Denote the eigenvalue of $SDF(u)$, ordered with respect to size and multiplicity, by $\lambda_1(u) \geq \lambda_2(u) \geq \lambda_3(u)$. Suppose that $\lambda_1^\perp(u) \geq \lambda_2^\perp(u)$ are the eigenvalues of $SA(u)$ restricted to $T^\perp(u)$ and suppose further that $\lambda_1^\perp(u)$ and $\lambda_2^\perp(u)$ are not eigenvalues of $SDF(u)$. It is easy to see that $\lambda_1^\perp(u)$ and $\lambda_2^\perp(u)$ are the zeros of the equation

$$\big((\lambda I - SDF(u))^{-1}F(u), F(u)\big) = 0.$$

We introduce the polynomial

$$\det(\lambda I - DF(u)) \equiv \lambda^3 + \delta_2(u)\lambda^2 + \delta_1(u)\lambda + \delta_0(u). \tag{8.85}$$

Let $u \in \widetilde{\mathcal{K}}$. Note that we have $\delta_2(u) := -(\lambda_1(u) + \lambda_2(u) + \lambda_3(u))$, $\delta_1(u) := \lambda_1(u)\lambda_2(u) + \lambda_2(u)\lambda_3(u) + \lambda_1(u)\lambda_3(u)$ and $\delta_0(u) := -\lambda_1(u)\lambda_2(u)\lambda_3(u)$. From this it follows that the eigenvalues $\lambda_i^\perp(u)$, $i = 1, 2$, of $SA(u)$ are the zeros of the equation

$$\lambda^2 + [\delta_2(u) + \Delta_1(u)]\,\lambda + [\delta_1(u) + \delta_2(u)\Delta_1(u) + \Delta_2(u)] = 0\,,$$

where

$$\Delta_1(u) := (Df(u)F(u), F(u)) \quad \text{and} \quad \Delta_2(u) := \big(DF(u)^2F(u), F(u)\big)\,. \tag{8.86}$$

Using this one sees immediately that the assumptions of Corollary 8.11 are satisfied for (8.84) if we suppose for the function $V(u) := \frac{1}{2}\log|F(u)|^2$, defined for all regular points of (8.84), the following condition:

(H3) There exists a continuous function $\zeta : \widetilde{\mathcal{K}} \to [0, s]$ with $s \in (0, 1]$ such that for any regular point $u \in \widetilde{\mathcal{K}}$ of (8.84) with $h(u) := \frac{1-\zeta(u)}{1+\zeta(u)}$ the inequalities

$$\delta_2(u) - h(u)\Delta_1(u) > 0 \quad \text{and}$$
$$\frac{1}{4h(u)^2}(\delta_2(u) - h(u)\Delta_1(u))^2 > \frac{1}{4}(\delta_2(u) - \Delta_1(u))^2 - \delta_1(u) - \Delta_2(u)$$

hold.

As a corollary of **(H3)** we get that if the inequalities

$$\delta_2(u) - \Delta_1(u) > 0 \quad \text{and} \quad \delta_1(u) + \Delta_2(u) > 0 \tag{8.87}$$

are satisfied for all regular points u of (8.84) on $\widetilde{\mathcal{K}}$ then by Corollary 8.11 it holds that $\dim_H \mathcal{K} \leq 1$. Further, the set $\mathcal{K}$ consists of a finite number of equilibrium points and closed trajectories of (8.84).

The Hartman-Olech condition ([11]) requires that $\lambda_1(u) + \lambda_2(u) < 0$ for all regular points $u \in \widetilde{\mathcal{K}}$. Note that this is always sufficient for the condition (8.87).

Let us formulate a further corollary from Theorem 8.8 for the case $\mathcal{M} = \mathbb{R}^3$.

(H4) Suppose that $\delta_2(u) > 0$ for all regular points $u \in \widetilde{\mathcal{K}}$ of (8.84) and that there exists a continuous function $\zeta : \widetilde{\mathcal{K}} \to [0, s)$ with $s \in (0, 1]$ such that the inequalities

$$\frac{1+\zeta(u)}{1-\zeta(u)}\delta_2(u) - \Delta_1(u) \geq 0 \quad \text{and}$$
$$\frac{\zeta(u)}{(1-\zeta(u))^2}\delta_2(u)^2 - \frac{\zeta(u)}{1-\zeta(u)}\delta_2(u)\Delta_1(u) + \delta_1(u) + \Delta_2(u) \geq 0 \tag{8.88}$$

hold for all regular $u \in \widetilde{\mathcal{K}}$ of (8.84).

Under the condition **(H4)** it follows from Corollary 8.11 that $\dim_H \mathcal{K} < 2 + s$. From Sect. 8.1 it follows that a sufficient condition for the dimension estimate $\dim_H \mathcal{K} < 2 + s$ is the inequality

$$\lambda_1(u) + \lambda_2(u) + s\lambda_3(u) < 0 \quad \text{for all } u \in \widetilde{\mathcal{K}}. \tag{8.89}$$

It is easy to show that condition (8.88) is always satisfied, supposing that (8.89) is satisfied.

References

1. Abraham, R., Marsden, J.E., Ratiu, T.: Manifolds, Tensor-Analysis, and Applications. Springer, New York (1988)
2. Blinchevskaya, M.A., Ilyashenko, Yu.S.: Estimates for the entropy dimension of the maximal attractor for k-contracting systems in an infinite-dimensional space. Russ. J. Math. Phys. **6**, 20–26 (1999)
3. Boichenko, V.A. Leonov, G.A.: On Lyapunov functions in estimates of the Hausdorff dimension of attractors. Leningrad, Dep. at VINITI 28.10.91, No. 4 123–B 91 (1991). (Russian)
4. Boichenko, V.A., Leonov, G.A., Franz, A., Reitmann, V.: Hausdorff and fractal dimension estimates of invariant sets of non-injective maps. Zeitschrift für Analysis und ihre Anwendungen (ZAA) **17**(1), 207–223 (1998)
5. Douady, A., Oesterlé, J.: Dimension de Hausdorff des attracteurs. C. R. Acad. Sci. Paris, Ser. A **290**, 1135–1138 (1980)
6. Eden, A.: Local Lyapunov exponents and a local estimate of Hausdorff dimension. ESAIM: Mathematical Modelling and Numerical Analysis - Modelisation Mathematique et Analyse Numerique, **23**(3), 405–413 (1989)
7. Eden, A., Foias, C., Temam, R.: Local and global Lyapunov exponents. J. Dynam. Diff. Equ. **3**, 133–177 (1991) [Preprint No. 8804, The Institute for Applied Mathematics and Scientific Computing, Indiana University, 1988]
8. Federer, H.: Geometric Measure Theory. Springer, New York (1969)
9. Gelfert, K.: Estimates of the box dimension and of the topological entropy for volume-contracting and partially volume-expanding dynamical systems on manifolds. Doctoral Thesis, University of Technology Dresden, (2001) (German)
10. Gelfert, K.: Maximum local Lyapunov dimension bounds the box dimension. Direct proof for invariant sets on Riemannian manifolds. Zeitschrift für Analysis und ihre Anwendungen (ZAA) **22**(3), 553–568 (2003)
11. Hartman, P., Olech, C.: On global asymptotic stability of solutions of ordinary differential equations. Trans. Amer. Math. Soc. **104**, 154–178 (1962)
12. Hunt, B.: Maximum local Lyapunov dimension bounds the box dimension of chaotic attractors. Nonlinearity **9**, 845–852 (1996)
13. Kruk, A.V., Reitmann, V.: Upper Hausdorff dimension estimates for invariant sets of evolutionary systems on Hilbert manifolds. In: Proceedings of Equadiff, pp. 247–254. Bratislava (2017)
14. Kruk, A.V., Malykh, A.E., Reitmann, V.: Upper bounds for the Hausdorff dimension and stratification of an invariant set of an evolution system on a Hilbert manifold. J. Diff. Equ. **53**(13), 1715–1733 (2017)
15. Ledrappier, F.: Some relations between dimension and Lyapunov exponents. Commun. Math. Phys. **81**, 229–238 (1981)
16. Leonov, G.A.: Estimation of the Hausdorff dimension of attractors of dynamical systems. Diff. Urav., **27**(5), 767–771 (1991) (Russian); English transation J. Diff. Equ., **27**, 520–524 (1991)
17. Leonov, W.G.: Estimate of Hausdorff dimension of invariant sets in cylindric phase space. J. Diff. Equ. **30**(7), 1274–1276 (1994) (Russian)
18. Leonov, G.A.: Construction of a special outer Carathéodory measure for the estimation of the Hausdorff dimension of attractors. Vestnik St. Petersburg University, Matematika, **1**(22), 24–31 (1995) (Russian); English translation Vestnik St. Petersburg University Math. Ser. 1, **28**(4), 24–30 (1995)
19. Leonov, G.A., Boichenko, V.A.: Lyapunov's direct method in the estimation of the Hausdorff dimension of attractors. Acta Appl. Math. **26**, 1–60 (1992)
20. Leonov, G.A., , Poltinnikova, M.S.: On the Lyapunov dimension of the attractor of the Chirikov dissipative mapping. Amer. Math. Soc. transl. In: Proceedings of the St. Petersburg Math. Soc. **214**(2), 15–28 (2005)

21. Leonov, G.A., Gelfert, K., Reitmann, V.: Hausdorff dimension estimates by use of a tubular Carathéodory structure and their application to stability theory. Nonlinear Dyn. Syst. Theory **1**(2), 169–192 (2001)
22. Morrey, C.: The problem of Plateau on a Riemannian manifold. Ann. Math. **49**, 807–851 (1948)
23. Noack, A.: Dimension and entropy estimates and stability investigations for nonlinear systems on manifolds. Doctoral Thesis, University of Technology Dresden (1998) (German)
24. Noack, A., Reitmann, V.: Hausdorff dimension estimates for invariant sets of time-dependent vector fields. Zeitschrift für Analysis und ihre Anwendungen (ZAA) **15**(2), 457–473 (1996)
25. Pesin, Ya.B.: Dimension type characteristics for invariant sets of dynamical systems. Uspekhi Mat. Nauk **43**(4), 95–128 (1988) (Russian); English translation. Russian Math. Surveys, **43**(4), 111–151 (1988)
26. Reitmann, V.: Dimension estimates for invariant sets of dynamical systems. In: Fiedler, B. (ed.) Ergodic Theory, Analysis, and Efficient Simulation of Dynamical Systems, pp. 585–615. Springer, New York-Berlin (2001)
27. Smith, R.A.: An index theorem and Bendixson's negative criterion for certain differential equations of higher dimensions. Proc. Roy. Soc. Edinburgh **91A**, 63–777 (1981)
28. Temam, R.: Infinite-Dimensional Dynamical Systems in Mechanics and Physics. Springer, New York - Berlin (1988)
29. Thieullen, P.: Entropy and the Hausdorff dimension for infinite-dimensional dynamical systems. J. Dyn. Diff. Equ. **4**(1), 127–159 (1992)

Chapter 9
Dimension and Entropy Estimates for Global Attractors of Cocycles

Abstract In this chapter we derive dimension and entropy estimates for invariant sets and global $\mathcal{B}$-attractors of cocycles in non-fibered and fibered spaces. A version of the Douady-Oesterlé theorem will be proven for local cocycles in an Euclidean space and for cocycles on Riemannian manifolds. As examples we consider cocycles, generated by the Rössler system with variable coefficients. We also introduce time-discrete cocycles on fibered spaces and define the topological entropy of such cocycles. Upper estimates of the topological entropy along an orbit of the base system are given which include the Lipschitz constants of the evolution system and the fractal dimension of the parameter dependent phase space.

9.1 Basic Facts from Cocycle Theory in Non-fibered Spaces

9.1.1 Definition of a Cocycle

Studying nonautonomous differential equations leads to considering the theory of cocycles and their attractors ([5, 14–16, 31]). Using the concept of a cocycle it proves possible to examine random dynamical systems and the corresponding random attractors. Elements of the theory of estimates of the Hausdorff dimension of random attractors were developed in [6, 7, 12]. Cocycles generated by PDE's were considered, e.g. in [10, 11, 30]. Dimension-like properties of cocycles given by variational inequalities with delay are reported in [27].

Suppose $\mathbb{T} \in \{\mathbb{R}, \mathbb{Z}, \}$ is a time set, $(\mathcal{M}, \rho_{\mathcal{M}})$ is a metric space and $(\{\varphi^t\}_{t\in\mathbb{T}}, (\mathcal{M}, \rho_{\mathcal{M}}))$ is the *base dynamical system.*

Definition 9.1 Suppose $(\mathcal{N}, \rho_{\mathcal{N}})$ is a metric space. The pair

$$(\{\psi^t(u, \cdot)\}_{t\in\mathbb{T}_+, u\in\mathcal{M}}, (\mathcal{N}, \rho_{\mathcal{N}}))\,, \tag{9.1}$$

N. Kuznetsov and V. Reitmann, *Attractor Dimension Estimates for Dynamical Systems: Theory and Computation*, Emergence, Complexity and Computation 38, https://doi.org/10.1007/978-3-030-50987-3_9

where $\psi^t(u,\cdot):\mathcal{N}\to\mathcal{N},\quad \forall t\in\mathbb{T}_+,\ \forall u\in\mathcal{M}$ is a continuous map is called a *cocycle over the base system*

$$(\{\varphi^t\}_{t\in\mathbb{T}},(\mathcal{M},\rho_{\mathcal{M}}))\,, \tag{9.2}$$

if

(1) $\psi^0(u,\cdot)=\mathrm{id}_{\mathcal{N}},\quad \forall u\in\mathcal{M}$;
(2) $\psi^{t+s}(u,\cdot)=\psi^t(\varphi^s(u),\psi^s(u,\cdot)),\quad \forall t,s\in\mathbb{T}_+,\ \forall u\in\mathcal{M}$.

Shortly we denote such a system by (φ,ψ). Let us introduce the product space $\widehat{\mathcal{M}}:=\mathcal{M}\times\mathcal{N}$ with the metric

$$\rho_{\widehat{\mathcal{M}}}((u,v'),(u',v')):=\sqrt{\rho^2_{\mathcal{M}}(u,u')+\rho^2_{\mathcal{N}}(v,v')}\quad\text{or}$$
$$\rho_{\widehat{\mathcal{M}}}((u,v),(u',v')):=\max\{\rho_{\mathcal{M}}(u,u'),\rho_{\mathcal{N}}(v,v')\},\ \forall(u,v),(u',v')\in\mathcal{M}\times\mathcal{N}.$$

The associated dynamical system (*skew product dynamical system*)

$$(\{\widehat{\varphi}^t\}_{t\in\mathbb{T}},(\widehat{\mathcal{M}},\rho_{\widehat{\mathcal{M}}})) \tag{9.3}$$

is defined by $\widehat{u}=(u,v)\in\mathcal{M}\times\mathcal{N}\mapsto\widehat{\varphi}^t(\widehat{u}):=(\varphi^t(u),\psi^t(u,v))$.

Remark 9.1 Let us show that (9.3) is really a dynamical system:

(1) $\widehat{\varphi}^0(u,v)=(\varphi^0(u),\psi^0(u,v))=(u,v),\quad\forall(u,v)\in\mathcal{M}\times\mathcal{N}$;
(2) $\widehat{\varphi}^{t+s}(u,v)=(\varphi^{t+s}(u),\psi^{t+s}(u,v))=(\varphi^t(\varphi^s(u)),\psi^t(\varphi^s(u),\psi^s(u,v)))$
$=\widehat{\varphi}^t(\varphi^s(u),\psi^s(u,v))=\widehat{\varphi}^t(\widehat{\varphi}^s(u,v)),\quad\forall t,s\in\mathbb{T}_+,(u,v)\in\mathcal{M}\times\mathcal{N}$.

Example 9.1 Consider the autonomous differential equation

$$\dot{\varphi}=f(\varphi) \tag{9.4}$$

where $f:\mathbb{R}^n\to\mathbb{R}^n$ is a smooth vector field. Suppose that for any $u\in\mathbb{R}^n$ there exists on $\mathbb{R}$ the solution $\varphi(\cdot,u)$ satisfying $\varphi(0,u)=u$. Put $\varphi^t(u):=\varphi(t,u)$ and $\mathcal{M}:=\mathbb{R}^n$ with the Euclidean norm $|\cdot|$. It follows that $(\{\varphi^t\}_{t\in\mathbb{R}},(\mathbb{R}^n,|\cdot|))$ is a dynamical system which will be considered as base system. For any $u\in\mathcal{M}$ the function $\Psi(t):=D\varphi^t(u)$ is the solution of the matrix differential equation $\dot{\Psi}=Df(\varphi^t(u))\Psi$, $\Psi(0)=I$. Furthermore, for any $v\in\mathcal{N}:=\mathbb{R}^n$ the function $\psi(t,v):=\Psi(t)v$ is the solution of the vector differential equation

$$\dot{\psi}=Df(\varphi^t(u))\psi,\ \psi(0)=v.$$

Put $\psi^t(u,v):=D\varphi^t(u)v$ for $u\in\mathcal{M}$, $v\in\mathcal{N}$ and $t\in\mathbb{R}_+$. It is easy to see that the pair (φ,ψ) is a cocycle in the sense of Definition 9.1:

(1) $\psi^0(u,v)=D\varphi^0(u)v=v,\quad\forall u\in\mathcal{M},\ v\in\mathcal{N}$;
(2) $\psi^{t+s}(u,v)=D\varphi^{t+s}(u)v=D\varphi^t(\varphi^s(u))D\varphi^s(u)v=\psi^t(\varphi^s(u),D\varphi^s(u)v)$
$=\psi^t(\varphi^s(u),\psi^s(u,v)),\quad\forall u\in\mathcal{M},v\in\mathcal{N},\ t,s\in\mathbb{T}_+=\mathbb{R}_+$.

Example 9.2 Consider the non-autonomous differential equation

$$\dot{\psi} = g(t, \psi), \tag{9.5}$$

where $g : \mathbb{R} \times \mathbb{R}^n \to \mathbb{R}^n$ is a smooth map. Suppose that for arbitrary $u \in \mathcal{M} := \mathbb{R}$ and $v \in \mathcal{N} := \mathbb{R}^n$ there exists a unique solution $\psi(t, u, v)$ of (9.5) satisfying $\psi(u, u, v) = v$. Let us show that $\psi^t(u, v) := \psi(t + u, u, v)$, $\forall u \in \mathbb{R}$, $v \in \mathbb{R}^n$ defines a cocycle over the base flow $(\{\varphi^t\}_{t\in\mathbb{R}} (\mathbb{R}, |\cdot|))$ with $\varphi^t(u) := t + u$, $\forall t \in \mathbb{R}$, $\forall u \in \mathcal{M}$:

(1) $\psi^0(u, v) = \psi(u, u, v) = v, \quad \forall u \in \mathbb{R}, \forall v \in \mathbb{R}^n$, i.e.
$\psi^0(u, \cdot) = \mathrm{id}_{\mathcal{N}}, \quad \forall u \in \mathbb{R}$;

(2) $\psi^{t+s}(u, v) = \underbrace{\psi(t + s + u, u, v)}_{(*)} \stackrel{?}{=} \psi^t(\underbrace{\varphi^s(u)}_{s+u}, \underbrace{\psi^s(u, v)}_{\psi(s+u,u,v)})$
$= \underbrace{\psi(t + s + u, s + u, \psi(s + u, u, v))}_{(**)}.$

Consider $(**)$ for $t = 0 : \psi(s + u, \psi(s + u, u, v)) = \psi(s + u, u, v)$. It follows that $(*)$ and $(**)$ are solutions of (9.5) with the same initial point at $t = 0$. Using the uniqueness property of solutions this shows that $(*)$ and $(**)$ coincide.

9.1.2 *Invariant Sets*

Definition 9.2 (Refs. [14, 15]) Suppose that there is a continuous map $\mathcal{M} \ni u \mapsto \mathcal{Z}(u) \subset \mathcal{N}$ and $\{\mathcal{Z}(u)\}_{u\in\mathcal{M}}$ is a family of sets depending on u. The family $\widetilde{\mathcal{Z}} := \{\mathcal{Z}(u)\}_{u\in\mathcal{M}}$ is called *closed* (*compact* or *bounded*) if for every $u \in \mathcal{M}$ the set $\mathcal{Z}(u) \subset \mathcal{N}$ is closed (compact or bounded). The family $\widetilde{\mathcal{Z}} = \{\mathcal{Z}(u)\}_{u\in\mathcal{M}}$ is called with respect to the cocycle (φ, ψ) *positively invariant* if $\psi^t(u, \mathcal{Z}(u)) \subset \mathcal{Z}(\varphi^t(u))$, $\forall t \in \mathbb{T}_+$, $\forall u \in \mathcal{M}$ (Fig. 9.1) *negatively invariant*, if $\psi^t(u, \mathcal{Z}(u)) \supset \mathcal{Z}(\varphi^t(u))$, $\forall t \in \mathbb{T}_+$, $\forall u \in \mathcal{M}$ and *invariant* if $\psi^t(u, \mathcal{Z}(u)) = \mathcal{Z}(\varphi^t(u))$, $\forall t \in \mathbb{T}_+$, $\forall u \in \mathcal{M}$.

Remark 9.2 Consider in $\widehat{\mathcal{M}} = \mathcal{M} \times \mathcal{N}$ the map $\mathcal{M} \ni u \mapsto \mathcal{Z}(u) \subset \mathcal{N}$, i.e., the set $\widehat{\mathcal{Z}} = \{\widehat{u} = (u, v) \in \widehat{\mathcal{M}} | u \in \mathcal{M}, v \in \mathcal{Z}(u)\}$.
The set $\widehat{\mathcal{Z}} \subset \widehat{\mathcal{M}}$ is positively invariant w.r.t. the skew product flow $(\{\widehat{\varphi}^t\}_{t\in\mathbb{T}}, (\widehat{\mathcal{M}}, \rho_{\widehat{\mathcal{M}}}))$ if $\widehat{\varphi}^t(\widehat{\mathcal{Z}}) \subset \widehat{\mathcal{Z}}, \quad \forall t \in \mathbb{T}_+$, i.e., $\widehat{\varphi}^t(\widehat{\mathcal{Z}}) = \{(\varphi^t(u), \psi^t(u, v)) \mid u \in \mathcal{M}, v \in \mathcal{Z}(u)\} \subset \{(u, v) \in \widehat{\mathcal{M}} | u \in \mathcal{M}, v \in \mathcal{Z}(u)\} \Leftrightarrow \psi^t(u, v) \in \mathcal{Z}(\varphi^t(u))$, $u \in \mathcal{M}, v \in \mathcal{Z}(u), t \in \mathbb{T}_+$.

9.1.3 *Global $\mathcal{B}$-Attractors of Cocycles*

Let us consider the cocycle (φ, ψ). Denote by $\mathfrak{B}(\mathcal{N})$ the family of non-empty bounded subsets of the metric space $(\mathcal{N}, \rho_{\mathcal{N}})$.

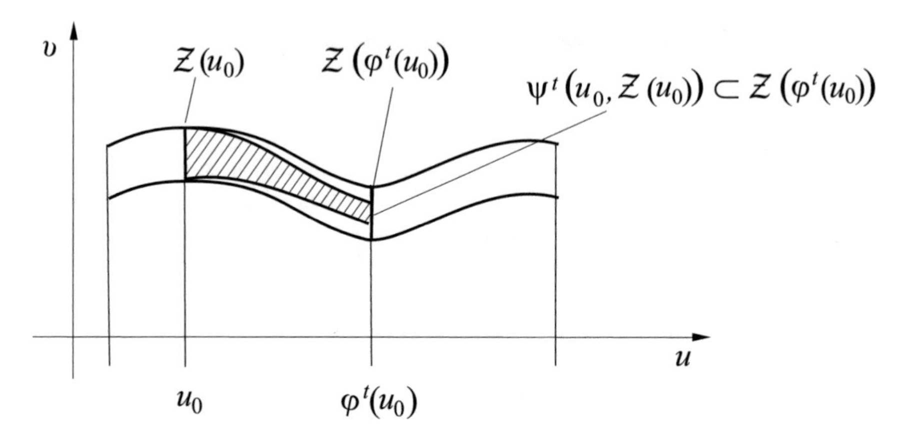

Fig. 9.1 Positive invariance

Definition 9.3 (Refs. [14, 15]) Suppose that (φ, ψ) is a cocycle and $\widetilde{\mathcal{Z}} = \{\mathcal{Z}(u)\}_{u\in\mathcal{M}}$ is a family of subsets of $\mathcal{N}$. The family $\widetilde{\mathcal{Z}}$ is called

(a) *globally $\mathcal{B}$-forward absorbing* for (φ, ψ) if
$\forall u \in \mathcal{M}\ \forall\ \mathcal{B} \in \mathfrak{B}(\mathcal{N})\ \exists T = T(u, \mathcal{B}) : \psi^t(u, \mathcal{B}) \subset \mathcal{Z}(\varphi^t(u)),\ \forall t \geq T(u, \mathcal{B}),$
$t \in \mathbb{T}_+$;

(b) *globally $\mathcal{B}$-pullback absorbing* if
$\forall u \in \mathcal{M}\ \forall\ \mathcal{B} \in \mathfrak{B}(\mathcal{N})\ \exists T = T(u, \mathcal{B}) : \psi^t(\varphi^{-t}(u), \mathcal{B}) \subset \mathcal{Z}(u), \forall t \geq T(u, \mathcal{B}),$
$t \in \mathbb{T}_+$;

(c) *globally $\mathcal{B}$-forward attracting* if
$\forall u \in \mathcal{M}\ \forall\ \mathcal{B} \in \mathfrak{B}(\mathcal{N}) : \lim_{t\to\infty} \operatorname{dist}(\psi^t(u, \mathcal{B}),\ \mathcal{Z}(\varphi^t(u))) = 0;$

(d) *globally $\mathcal{B}$-pullback attracting* if
$\forall u \in \mathcal{M}\ \forall\ \mathcal{B} \in \mathfrak{B}(\mathcal{N}) : \lim_{t\to\infty} \operatorname{dist}(\psi^t(\varphi^{-t}(u), \mathcal{B}),\ \mathcal{Z}(u)) = 0;$

(e) *globally $\mathcal{B}$-forward attractor (globally $\mathcal{B}$-pullback attractor)* if
$\widetilde{\mathcal{Z}}$ is compact, invariant and globally $\mathcal{B}$-forward attracting (globally $\mathcal{B}$-pullback attracting) (Fig. 9.2 and 9.3).

Theorem 9.1 (Kloeden–Schmalfuss [15]).

(a) *Consider the cocycle (9.1), (9.2) where $(\mathcal{N}, \rho_{\mathcal{N}})$ is a complete metric space. Suppose that the cocycle (9.1), (9.2) has a compact globally $\mathcal{B}$-pullback attracting set $\widetilde{\mathcal{Z}} = \{\mathcal{Z}(u)\}_{u\in\mathcal{M}}$. Then the cocycle (9.1), (9.2) has a unique global $\mathcal{B}$-pullback attractor $\widetilde{\mathcal{A}} = \{\mathcal{A}(u)\}_{u\in\mathcal{M}}$ where*

$$\mathcal{A}(u) = \bigcap_{t\in\mathbb{T}_+} \overline{\bigcup_{\substack{s\geq t\\ s\in\mathbb{T}_+}} \psi^s(\varphi^{-s}(u), \mathcal{Z}(\varphi^{-s}(u)))}, \quad \forall u \in \mathcal{M};$$

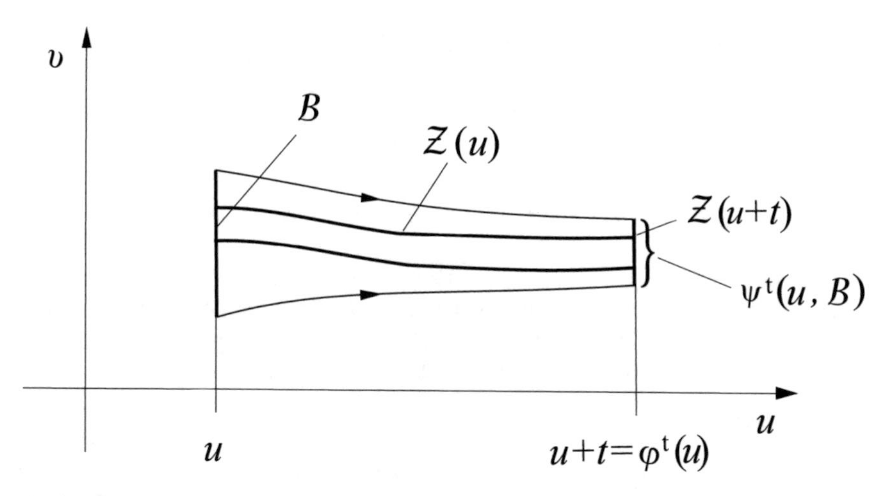

Fig. 9.2 Global $\mathcal{B}$-forward attractor

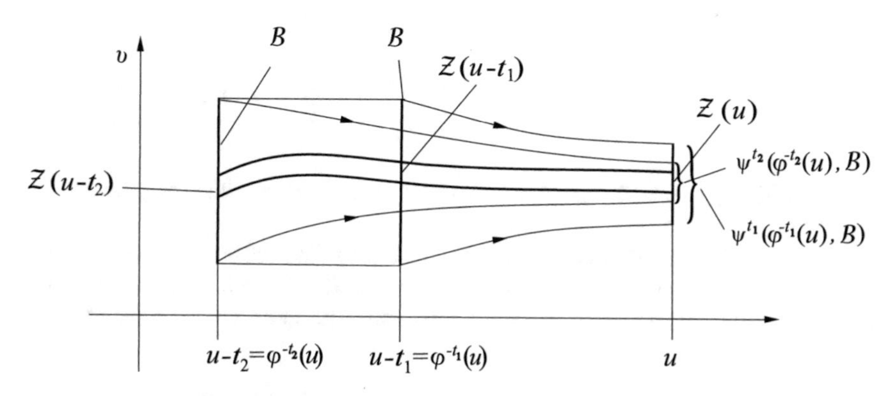

Fig. 9.3 Global $\mathcal{B}$-pullback attractor

(b) *Suppose that the cocycle (9.1), (9.2) has a compact globally $\mathcal{B}$-forward attracting set $\mathcal{Z}$.*
Then the cocycle (9.1), (9.2) has a unique global $\mathcal{B}$-pullback attractor $\widetilde{\mathcal{A}} = \{\mathcal{A}(u)\}_{u\in\mathcal{M}}$ where

$$\mathcal{A}(u) = \bigcap_{t\in\mathbb{T}_+} \overline{\bigcup_{\substack{s\geq t \\ s\in\mathbb{T}_+}} \psi^s(\varphi^{-s}(u), \mathcal{Z})}, \quad \forall u \in \mathcal{M}.$$

9.1.4 Extension System Over the Bebutov Flow on a Hull

Let us consider again equation (9.5). Introduce the shift map for the right-hand side of (9.5) by

$$\mathbb{R} \times \mathcal{M} \ni (s, u) \mapsto \varphi^s(u) := g(\cdot + s, \cdot), \quad \text{where} \quad u = g(t, \cdot).$$

Suppose that $\mathcal{M} \equiv \mathcal{H}(g) := \overline{\{g(\cdot + s, \cdot) : s \in \mathbb{R}\}}$ is the *hull* of g. The closure is taken in the compact-open topology . In this topology we have the property $h_m \to h$ as $m \to \infty (h_m, h : \mathbb{R} \times \mathbb{R}^n \to \mathbb{R}^n)$ if and only if for arbitrary compact sets $\mathcal{J} \subset \mathbb{R}$ and $\mathcal{K} \subset \mathbb{R}^n$ we have $\sup_{(t,\upsilon) \in \mathcal{J} \times \mathcal{K}} |h_m(t, \upsilon) - h(t, \upsilon)| \to 0$ as $m \to \infty$.

Some properties of the hull:

(1) $\mathcal{H}(g)$ is metrizable. Indeed, suppose $\{\widetilde{\mathcal{K}}_m\}_{m=1}^{\infty}$ is a sequence of compact sets in $\mathbb{R} \times \mathbb{R}^n$ such that $\widetilde{\mathcal{K}}_m \subset \widetilde{\mathcal{K}}_{m+1}$, $m = 1, 2, \ldots$ and $\bigcup_{m=1}^{\infty} \widetilde{\mathcal{K}}_m = \mathbb{R} \times \mathbb{R}^n$. Then the metric defined by

$$\rho_{\mathcal{H}(g)}(h_1, h_2) := \sum_{m=1}^{\infty} 2^{-m} \frac{\sup\{|h_1(t, \upsilon) - h_2(t, \upsilon)| : (t, \upsilon) \in \widetilde{\mathcal{K}}_m\}}{1 + \sup\{|h_1(t, \upsilon) - h_2(t, \upsilon)| : (t, \upsilon) \in \widetilde{\mathcal{K}}_m\}},$$

for arbitrary $h_1, h_2 \in \mathcal{H}(g)$ generates a topology in $\mathcal{H}(g)$ which is equivalent to the previous one.

(2) The set $\mathcal{H}(g)$ is compact if and only if the map $\mathbb{R} \times \mathbb{R}^n \ni (t, \upsilon) \mapsto g(t, \upsilon)$ is bounded and equicontinuous on every set $\mathbb{R} \times \mathcal{K}$, where $\mathcal{K} \subset \mathbb{R}^n$ is compact. A function $g(t, \upsilon)$ which is continuous and T-periodic w.r. to t has the last properties.

On the set $\mathcal{M} \times \mathbb{R}^n$ one considers the *evaluation map* $\widetilde{g} : \mathcal{M} \times \mathbb{R}^n \to \mathbb{R}^n$ given by $\mathcal{M} \times \mathbb{R}^n \ni (u, \upsilon) \mapsto u(0, \upsilon)$, i.e., $\widetilde{g}(\varphi^t(u), \upsilon)$. Suppose $u = g \in \mathcal{H}(g)$. Then $\widetilde{g}(\varphi^t(g), \upsilon) = g(t, \upsilon)$, i.e., $\widetilde{g}$ is the *extension* of g. Instead of the single equation (9.5) we now consider the family of systems (*Bebutov flow*)

$$\dot{\psi} = \widetilde{g}(\varphi^t(u), \psi), \quad u \in \mathcal{M} = \mathcal{H}(g). \tag{9.6}$$

Theorem 9.2 (Wakeman [30]) *Suppose that the following conditions are satisfied for equation (9.5):*

(1) *The map $g : \mathbb{R} \times \mathbb{R}^n \to \mathbb{R}^n$ is continuous;*

(2) *The map $\mathbb{R} \times \mathbb{R}^n \ni (t, \upsilon) \mapsto g(t, \upsilon)$ is locally Lipschitz according to the second argument and there exist measurable integrable functions $\alpha(t)$ and $\beta(t)$ such that*

$$|g(t, \upsilon)| \le \alpha(t) \cdot |\upsilon| + \beta(t), \quad \forall (t, \upsilon) \in \mathbb{R} \times \mathbb{R}^n.$$

Then equation (9.5) generates a cocycle ψ over the Bebutov flow $\{\varphi^t\}_{t\in\mathbb{R}}$ on the hull $\mathcal{H}(g)$. This cocycle can be written as

$$\psi^t(u,\upsilon) = \upsilon + \int_0^t \widetilde{g}(\varphi^s(u), \psi^s(u,\upsilon))ds\,, \quad \forall(u,\upsilon) \in \mathcal{M} \times \mathbb{R}^n\,.$$

The cocycle map $\psi^{(\cdot)}(\cdot,\cdot) : \mathbb{R} \times \mathcal{M} \times \mathbb{R}^n \to \mathbb{R}^n$ is continuous.

Example 9.3 Consider the system

$$\dot{\psi} = A(t)\psi + h(t,\psi) =: g(t,\psi), \tag{9.7}$$

where $A(t)$ is a continuous $n \times n$-matrix function and $h : \mathbb{R} \times \mathbb{R}^n \to \mathbb{R}^n$ is a continuous function which satisfies the conditions of Theorem 9.2. Then there exists a cocycle ψ generated by (9.7) and given over the Bebutov base flow $\{\varphi^t\}_{t\in\mathbb{R}}$ on $\mathcal{M} = \mathcal{H}(g)$, i.e.,

$$\dot{\psi} = \widetilde{A}(\varphi^t(u))\psi + \widetilde{h}(\varphi^t(u),\psi) =: A_u(t)\psi + h_u(t,\psi)\,. \tag{9.8}$$

Suppose that there exist on $\mathbb{R}$ continuous scalar functions $c_{1,u}, c_{2,u}, c_{3,u}$ and constants $c_4 > 0$ and $c_0 > 0$ such that the following conditions are satisfied:

(A1) $(A_u(t)\upsilon,\upsilon) \le -c_{1,u}(t)|\upsilon|^2, \quad \forall t \in \mathbb{R},\ \forall \upsilon \in \mathbb{R}^n,\ \forall u \in \mathcal{M}$;
(A2) $(h_u(t,\upsilon),\upsilon) \le c_{2,u}(t)|\upsilon|^2 + c_{3,u}(t), \quad \forall t \in \mathbb{R},\ \forall \upsilon \in \mathbb{R}^n,\ \forall u \in \mathcal{M}$;
(A3) $-c_{1,u}(t) + c_{2,u}(t) \le -c_0 < 0, \quad \forall t \in \mathbb{R},\ \forall u \in \mathcal{M}$;
(A4) $c_{3,u}(t) \le c_4 < \infty, \quad \forall t \in \mathbb{R},\ \forall u \in \mathcal{M}$.

Let us show that for (9.8) there exists a global $\mathcal{B}$-pullback attractor. Introduce the Lyapunov function

$$V(\upsilon) := \frac{1}{2}(\upsilon,\upsilon) = \frac{1}{2}|\upsilon|^2, \quad \forall \upsilon \in \mathbb{R}^n.$$

Suppose that $\psi(t)$ is an arbitrary solution of (9.8) with fixed parameter u. Then we have

$$\begin{aligned}
&\frac{d}{dt}V(\psi(t)) = (\psi(t),\dot{\psi}(t)) = (\psi(t), A_u(t)\psi(t) + h_u(t,\psi(t)))\\
&= (\psi(t), A_u(t)\psi(t)) + (\psi(t), h_u(t,\psi(t)))\\
&\overset{(A1),(A2)}{\le} -c_{1,u}(t)|\psi(t)|^2 + c_{2,u}(t)|\psi(t)|^2 + c_{3,u}(t)\\
&\overset{(A3),(A4)}{\le} -c_0|\psi(t)|^2 + c_4 = -2c_0V(\psi(t)) + c_4\,.
\end{aligned}$$

It follows that

$$\frac{d}{dt}V(\psi(t)) \le -2c_0V(\psi(t)) + c_4, \quad \forall t \in \mathbb{R}\,.$$

Introduce the function $W : \mathbb{R}^n \to \mathbb{R}$ by $V(v) = W(v) + \frac{c_4}{2c_0}$. Then we have

$$\dot{V}(\psi(t)) = \dot{W}(\psi(t)) \text{ and } \dot{W}(\psi(t)) + 2c_0 W(\psi(t)) \leq 0\,, \quad \forall t \in \mathbb{R},$$

and, consequently,

$$\dot{W}(\psi(t))e^{2c_0 t} + 2c_0 e^{2c_0 t} W(\psi(t)) \leq 0, \quad \forall t \in \mathbb{R}.$$

From this it follows that for arbitrary $t_0 \leq t$ we have

$$\int_{t_0}^{t} \frac{d}{dt}(W(\psi(t))e^{2c_0 t}) \leq 0\,.$$

This gives the estimate

$$W(\psi(t))e^{2c_0 t} \leq W(\psi(t_0))e^{2c_0 t} \text{ or } W(\psi(t)) \leq e^{2c_0(t_0 - t)} W(\psi(t_0))\,, \quad \forall t \geq t_0\,.$$

For the first function V this means that

$$V(\psi(t)) - \frac{c_4}{2c_0} \leq e^{-2c_0(t-t_0)} \left(V(\psi(t)) - \frac{c_4}{4c_0} \right)\,, \quad \forall t \geq t_0\,.$$

Finally we get the inequality $\limsup_{t\to\infty} V(\psi(t)) \leq \frac{c_4}{2c_0}$. From this it follows that the set $\mathcal{Z} := \{v \in \mathbb{R}^n |v|^2 \leq \frac{c_4}{c_0}\}$ is a globally $\mathcal{B}$-forward attracting set for (9.8). Using Theorem 9.1 we conclude that there exists a unique global $\mathcal{B}$-pullback attractor $\widetilde{\mathcal{A}} = \{\mathcal{A}(u)\}_{u\in\mathcal{M}}$ where $\mathcal{A}(u) \quad \forall u \in \mathcal{M}$, is given by this theorem.

9.2 Local Cocycles Over the Base Flow in Non-fibered Spaces

9.2.1 Definition of a Local Cocycle

A *local cocycle* ([19, 23]) on $\mathbb{R}_+$ over a *base flow* $(\{\varphi^t\}_{t\in\mathbb{R}}, (\mathcal{M}, \rho_{\mathcal{M}}))$ is a pair $(\{\psi^t(u, \cdot)\}_{u\in\mathcal{M},\, t\in[0,\beta(u,\cdot))}, (\mathbb{R}^n, |\cdot|))$, where $\psi^{(\cdot)}(\cdot, \cdot)$ is continuous on the set

$$\mathcal{D} = \{(t, u, v) \mid (u, v) \in \mathcal{M} \times \mathbb{R}^n,\ t \in [0, \beta(u, v))\},$$

and $[0, \beta(u, v))$ is the non-negative part of the maximal interval of existence of the mapping ψ^t that passes through the point (u, v). Here ψ satisfies the following conditions:

(1) $\psi^0(u, \cdot) = \mathrm{id}_{\mathbb{R}^n}, \quad \forall\, u \in \mathcal{M}$,
(2) $\psi^{t+s}(u, v) = \psi^t(\varphi^s(u), \psi^s(u, v)), \quad \forall\, (u, v) \in \mathcal{M} \times \mathbb{R}^n, \forall\, s \in [0, \beta(u, v)),$
$\forall\, t \in \left[0, \beta(\varphi^s(u), \psi^s(u, v))\right), t + s < \beta(u, v).$

In the following, for brevity, (φ, ψ) denotes the local cocycle

$(\{\psi^t(u, \cdot)\}_{u \in \mathcal{M},\, t \in [0, \beta(u, \cdot))}, (\mathbb{R}^n, |\cdot|))$ on $\mathbb{R}_+$ over the base flow $(\{\varphi^t\}_{t \in \mathbb{R}}, (\mathcal{M}, \rho_{\mathcal{M}})$.

Given a mapping $\mathcal{M} \ni u \mapsto \mathcal{Z}(u) \subset \mathbb{R}^n$, the set $\widetilde{\mathcal{Z}} = \{\mathcal{Z}(u)\}_{u \in \mathcal{M}}$ is called *parametrized*. A parametrized set $\widetilde{\mathcal{Z}} = \{\mathcal{Z}(u)\}_{u \in \mathcal{M}}$ is said to be *compact* if the set $\mathcal{Z}(u) \subset \mathbb{R}^n$ is compact for any $u \in \mathcal{M}$.

Definition 9.4 A set $\widetilde{\mathcal{Z}}$ is called negatively invariant for a local cocycle (φ, ψ) if there exists $0 < \tau < \min\limits_{\substack{u \in \mathcal{M} \\ v \in \mathcal{Z}(u)}} \beta(u, v)$, such that

$$\psi^\tau(u, \mathcal{Z}(u)) \supset \mathcal{Z}(\varphi^\tau(u)), \quad \forall\, u \in \mathcal{M}.$$

9.2.2 *Upper Bounds of the Hausdorff Dimension for Local Cocycles*

Now we can formulate the main result of this section. Suppose that there is given a local cocycle (φ, ψ) with C^1-smooth maps $\psi^t(u, \cdot) : \mathbb{R}^n \to \mathbb{R}^n$ for all $u \in \mathcal{M}$ and $t \in [0, \beta(u, \cdot))$.
For the subsequent presentation, we need the following assumptions.

(A5) The parametrized set $\widetilde{\mathcal{Z}} = \{\mathcal{Z}(u)\}_{u \in \mathcal{M}}$ is compact and negatively invariant for the local cocycle (φ, ψ) with some $\tau > 0$ in the sense of Definition 9.4 and $\mathcal{Z}(u) \subset \mathcal{Z}(\varphi^\tau(u))$, $u \in \mathcal{M}$.
(A6) Given arbitrary points $(u, v) \in \mathcal{M} \times \mathbb{R}^n$ and $t \in [0, \beta(u, \cdot))$, the differential of the function $\psi^t(u, v)$ with respect to v is denoted by $\partial_2 \psi^t(u, v) : \mathbb{R}^n \to \mathbb{R}^n$ and satisfies the following conditions:

(a) For any $\varepsilon > 0$ and $0 < t < \min\limits_{\substack{u \in \mathcal{M}, \\ v \in \mathcal{Z}(u)}} \beta(u, v)$ the function

$$\eta_\varepsilon(t, u) := \sup_{\substack{w, v \in \mathcal{Z}(0), \\ 0 < |w - v| \le \varepsilon}} \frac{|\psi^t(u, w) - \psi^t(u, v) - \partial_2 \psi^t(u, v)(w - v)|}{|w - v|}$$

is bounded on $\mathcal{M}$ and tends to zero as $\varepsilon \to 0$ for each fixed t;
(b) For any $0 < t < \min\limits_{\substack{u \in \mathcal{M}, \\ v \in \mathcal{Z}(u)}} \beta(u, v)$ we have

$$\sup_{u\in\mathcal{M}}\ \sup_{v\in\mathcal{Z}(u)} |\partial_2\psi^t(u,v)|_{op} < \infty,$$

where $|L|_{op}$ denotes the operator norm of an $n \times n$-matrix L.

Theorem 9.3 (Refs. [19, 28]) *Under assumptions* **(A5)** *and* **(A6)** *suppose that:*

(1) *There exists a compact set* $\widetilde{\mathcal{K}} \subset \mathbb{R}^n$ *such that*

$$\overline{\bigcup_{u\in\mathcal{M}} \mathcal{Z}(u)} \subset \widetilde{\mathcal{K}};$$

(2) *There exists a bounded continuous function* $\varkappa : \mathcal{M} \times \mathbb{R}^n \to \mathbb{R}_+$ *and a number* $d \in (0, n]$ *such that*

$$\sup_{(u,v)\in\mathcal{M}\times\widetilde{\mathcal{K}}} \frac{\varkappa(\varphi^\tau(u), \psi^\tau(u,v))}{\varkappa(u,v)}\, \omega_d(\partial_2\psi^\tau(u,v)) < 1\,,$$

where $\tau > 0$ *is the number mentioned in assumption* **(A5)** *and* ω_d *is the singular value function. Then* $\dim_H \mathcal{Z}(u) \le d$ *for any* $u \in \mathcal{M}$.

Proof From assumption (2) it follows that there exists a number $0 < \kappa_1 < 1$ such that

$$\sup_{v\in\widetilde{\mathcal{K}}} \frac{\varkappa(\varphi^\tau(u), \psi^\tau(u,v))}{\varkappa(u,v)}\, \omega_d(\partial_2\psi^\tau(u,v)) \le \kappa_1\,, \quad \forall u \in \mathcal{M}\,. \tag{9.9}$$

Given an arbitrary $m \in \mathbb{N}$, we define

$$\kappa(m,u) := \kappa_1^m \sup_{v\in\mathcal{Z}(0)} \frac{\varkappa(\varphi^\tau(u), \psi^\tau(u,v))}{\varkappa(u,v)}. \tag{9.10}$$

Clearly, $\kappa(m,u)$ can be made arbitrary small for any $u \in \mathcal{M}$ by taking m sufficiently large. In other words, for any $l > 0$ there exists a sufficiently large $m_0 \in \mathbb{N}$ such that

$$0 < \kappa(m,u) < l, \quad \forall m \ge m_0, \quad \forall u \in \mathcal{M}. \tag{9.11}$$

Let (a sufficiently large) $m \ge m_0$ be fixed. In what follows, the construction of the mapping ψ is based on solving a system of ordinary differential equations. Since this solution is given on a compact set $\mathcal{M} \times \widetilde{\mathcal{K}}$, it can be extended to the right in t. Hence, in subsequent considerations it will be assumed that ψ is defined on the interval $[0, m\tau]$ (for sufficiently large m). Hence, by the chain rule for composition, we get:

$$\begin{aligned}\partial_2\psi^{m\tau}(u,v) &= \partial_2\psi^\tau(\varphi^{(m-1)\tau}(u), \psi^{(m-1)\tau}(u,v)) \cdot \\ &\quad \cdot \partial_2\psi^\tau(\varphi^{(m-2)\tau}(u), \psi^{(m-2)\tau}(u,v)) \cdot \ldots \cdot \partial_2\psi^\tau(u,v).\end{aligned}$$

Using Horn's inequality (Proposition 2.4, Chap. 2), we get

$$\omega_d(\partial_2\psi^{m\tau}(u,\upsilon)) \leqslant \prod_{j=1}^{m}\omega_d(\partial_2\psi^{\tau}(\varphi^{(m-j)\tau}(u),\psi^{(m-j)\tau}(u,\upsilon))). \tag{9.12}$$

Further, taking into account (9.9) for arguments of the form $(\varphi^{(m-j)\tau}(u),\psi^{(m-j)\tau}(u,\upsilon))$, $j=1,\dots,m$ and applying the estimate (9.12), we have

$$\begin{aligned}\omega_d\left(\partial_2\psi^{m\tau}(u,\upsilon)\right) &\leqslant \prod_{j=1}^{m}\kappa_1\cdot\frac{\varkappa(\varphi^{(m-j)\tau}(u),\psi^{(m-j)\tau}(u,\upsilon))}{\varkappa(\varphi^{(m-j+1)\tau}(u),\psi^{(m-j+1)\tau}(u,\upsilon))}\\ &=\kappa_1^m\cdot\frac{\varkappa(u,\upsilon)}{\varkappa\left(\varphi^{m\tau}(u),\psi^{m\tau}(u,\upsilon)\right)}\leqslant\kappa(m,u).\end{aligned} \tag{9.13}$$

From assumption **(A6)** it follows that, for any $\varepsilon>0,\ u\in\mathcal{M}$, and any fixed $t=m\tau$,

$$\left|\psi^{mt}(u,w)-\psi^{mt}(u,\upsilon)-\partial_2\psi^{mt}(u,\upsilon)(w-\upsilon)\right|\leqslant\eta_\varepsilon(m\tau,u)|w-\upsilon|, \tag{9.14}$$

$\forall u\in\mathcal{M},\ \forall w\in\mathcal{B}_r(\upsilon),\ r\leqslant\varepsilon,\ w,\upsilon\in\mathcal{Z}(u)$.
Also, since $\eta_\varepsilon(t,u)$ is bounded, there exists ζ such that, for any $\varepsilon>0,\ u\in\mathcal{M}$ and fixed $0<t<\min\limits_{\substack{u\in\mathcal{M},\\ \upsilon\in\mathcal{Z}(u)}}\beta(u,\upsilon)$ we have $\eta_\varepsilon(t,u)\leq\zeta$. Then (9.14) can be replaced by

$$\left|\psi^{mt}(u,w)-\psi^{mt}(u,\upsilon)-\partial_2\psi^{mt}(u,\upsilon)(w-\upsilon)\right|\leqslant\zeta|w-\upsilon|, \tag{9.15}$$

$\forall u\in\mathcal{M},\ \forall w\in\mathcal{B}_r(\upsilon),\ r\leqslant\varepsilon,\ w,\upsilon\in\mathcal{Z}(u)$.
Since $\eta_\varepsilon(t,u)\underset{\varepsilon\to 0}{\longrightarrow}0$, it follows that ζ can be made arbitrarily small by taking sufficiently small ε. Further, by (9.15), we have

$$\psi^{m\tau}(u,\mathcal{B}_r(\upsilon))\subset\psi^{m\tau}(u,\upsilon)+\partial_2\psi^{m\tau}(u,\upsilon)\mathcal{B}_r(\upsilon)+\mathcal{B}_{r\zeta}(\upsilon). \tag{9.16}$$

Put $\mathcal{E}:=\partial_2\psi^{m\tau}(u,\upsilon)\mathcal{B}_r(\upsilon)$. It is easily verified (Proposition 2.2, Chap. 2) that $\mathcal{E}$ is an ellipsoid with semiaxes of length $a_i(\mathcal{E})=r\alpha_i(\partial_2\psi^{m\tau}(u,\upsilon)),\ i=1,\dots,n$.
According to (9.11), $\kappa(m,u)$ can be made arbitrarily small by taking sufficiently large m. Hence there exists a $\kappa_2(m)$ such that $\kappa(m,u)\leqslant\kappa_2(m),\ \forall u\in\mathcal{M}$. Clearly, $\kappa_2(m)$ can also be made arbitrarily small by taking large m.
We apply Lemma 2.1, Chap. 2 with $\kappa=\kappa_2(m)$, and take arbitrary δ so that we have $\sup\limits_{\upsilon\in\mathcal{Z}(u)}|\partial_2\psi^{m\tau}(u,\upsilon)|\leqslant\delta$ and $\kappa_2(m)\leqslant\delta^d$ for a sufficiently large m. Also, we choose a small ζ so that

$$\left(1+\left(\frac{\delta^{d_0}}{\kappa_2(m)}\right)^{\frac{1}{s}}\zeta\right)^d\kappa_2(m)<l \tag{9.17}$$

for fixed $\kappa_2(m)$ and δ satisfying the above conditions. Here l is the same as in (9.11). It is then easily verified that the parameters $\kappa':=r^d\kappa_2(m)$, $\delta':=r\delta,\ \eta':=r\zeta$ satisfy the hypotheses of Lemma 2.1, Chap. 2. Then, by this lemma we have

$$\omega_d(\mathcal{E}') \leqslant \left(1+\left(\frac{\delta'^{d_0}}{\kappa'}\right)^{\frac{1}{s}}\eta'\right)^d\kappa' = \left(1+\frac{\delta^{d_0/s}}{\kappa_2^{1/s}(m)}\zeta\right)^d\kappa_2(m)r^d < lr^d. \tag{9.18}$$

Here $\omega_d(\mathcal{E})$ denotes the d-dimensional ellipsoidal measure of $\mathcal{E}$ defined as

$$\omega_d(\mathcal{E}) := \begin{cases} a_1(\mathcal{E})a_2(\mathcal{E})\dots a_{d_0}(\mathcal{E})a^s_{d_0+1}(\mathcal{E}), & \text{for} \quad d>0, \\ 1, & \text{for} \quad d=0, \end{cases}$$

where $d = d_0 + s$, $d_0 \in \{0, 1, \dots, n-1\}$, $s \in (0, 1]$, and $a_i(\mathcal{E})$ are the lengths of the semiaxes of $\mathcal{E}$ with the ordering $a_1 \geqslant a_2 \geqslant \dots \geqslant a_n > 0$. Further, if $\{\mathcal{B}_{r_j}(u_j)\}$ is a countable covering of $\mathcal{Z}(u)$ by balls of radii $r_j \leqslant \varepsilon$, then we can build a countable covering of $\psi^{m\tau}(u, \mathcal{Z}(u))$ by ellipsoids $\mathcal{E}'_j$, for which $\omega_d(\mathcal{E}'_j) \leqslant lr_j^d$.
Let us introduce some new notation: for any compact set $\mathcal{K} \subset \mathbb{R}^n$ we define $\widetilde{\mu}_H(\mathcal{K}, d, \varepsilon) := \inf \sum_j \omega_d(\mathcal{E}'_j)$, the infimum being taken over all countable coverings of $\mathcal{K}$ by ellipsoids $\mathcal{E}'_j$, for which $\omega_d(\mathcal{E}'_j) \leqslant \varepsilon^d$.
From the definitions of $\widetilde{\mu}_H$, μ_H and inequality (9.18), we get

$$\widetilde{\mu}_H\left(\psi^{m\tau}(u, \mathcal{Z}(u)), d, l^{1/d}\varepsilon\right) \leqslant l\mu_H(\mathcal{Z}(u), d, \varepsilon). \tag{9.19}$$

Using Lemma 2.1. of Chap. 2, once again, for a countable covering of the compact set $\mathcal{K}$ by ellipsoids $\mathcal{E}_j$, for which $\left(\omega_d(\mathcal{E}_j)\right)^{1/d} \leqslant \varepsilon$, we obtain the following estimate: $\mu_H\left(\mathcal{E}_j, d, \sqrt{d_0+1}\,\varepsilon\right) \leqslant 2^{d_0}(d_0+1)^{\frac{d}{2}}\omega_d(\mathcal{E}_j)$. Consequently,

$$\mu_H\left(\mathcal{K}, d, \sqrt{d_0+1}\,\varepsilon\right) \leqslant 2^{d_0}(d_0+1)^{\frac{d}{2}}\sum_j \omega_d(\mathcal{E}_j).$$

Hence, taking the infimum, we obtain

$$\mu_H\left(\mathcal{K}, d, \sqrt{d_0+1}\,\varepsilon\right) \leqslant 2^{d_0}(d_0+1)^{\frac{d}{2}}\widetilde{\mu}_H(\mathcal{K}, d, \varepsilon). \tag{9.20}$$

We apply (9.20) and then (9.19) to the set $\mathcal{K} = \psi^{m\tau}(u, \mathcal{Z}(u))$. This gives

$$\begin{aligned} \mu_H\left(\psi^{m\tau}(u, \mathcal{Z}(u)), d, \sqrt{d_0+1}\,l^{1/d}\varepsilon\right) &\leqslant 2^{d_0}(d_0+1)^{\frac{d}{2}}\widetilde{\mu}_H\left(\psi^{m\tau}(u, \mathcal{Z}(u)), d, l^{1/d}\varepsilon\right) \\ &\leqslant 2^{d_0}(d_0+1)^{\frac{d}{2}}l\mu_H(\mathcal{Z}(u), d, \varepsilon). \end{aligned} \tag{9.21}$$

Assume that $\mu_H(\mathcal{Z}(u), d) \leqslant \mu_0 < \infty$ and take $\varepsilon \to 0$ in (9.21). It follows that

$$\mu_H\left(\psi^{m\tau}(u, \mathcal{Z}(u)), d\right) \leqslant 2^{d_0}(d_0+1)^{\frac{d}{2}}l\mu_H(\mathcal{Z}(u), d) \leqslant 2^{d_0}(d_0+1)^{\frac{d}{2}}l\mu_0. \tag{9.22}$$

Recall that l is an (arbitrarily small) positive number, for which a sufficiently large m was chosen so that we were able to apply the above estimates. Hence, for a sufficiently large m, the right-hand side of (9.22) can be made arbitrarily small. Therefore, for each $u \in \mathcal{M}$, we have shown that if $\mu_H(\mathcal{Z}(u), d) < \infty$, then

$$\lim_{\varepsilon \to 0} \mu_H(\psi^{m\tau}(u, \mathcal{Z}(u)), d, \varepsilon) = 0. \tag{9.23}$$

Further, since l is an arbitrary number, it may be assumed that it satisfies the following restrictions: $\sqrt{d_0 + 1}\, l^{1/d} < 1$ and $2^{d_0}(d_0 + 1)^{\frac{d}{2}} l < 1$. In this case, we have

$$\mu_H(\mathcal{Z}(u), d, \varepsilon) \leqslant \mu_H\left(\mathcal{Z}(u), d, \sqrt{d_0 + 1}\, l^{1/d} \varepsilon\right). \tag{9.24}$$

By assumption **(A5)** of the theorem, $\mathcal{Z}(u) \subset \mathcal{Z}(\varphi^{\tau}(u)) \subset \ldots \subset \mathcal{Z}(\varphi^{m\tau}(u))$, and so

$$\mu_H(\mathcal{Z}(u), d, \sqrt{d_0 + 1}\, l^{1/d} \varepsilon) \leqslant \mu_H(\mathcal{Z}(\varphi^{m\tau}(u)), d, \sqrt{d_0 + 1}\, l^{1/d} \varepsilon). \tag{9.25}$$

Further, $\widetilde{\mathcal{Z}}$ is negatively invariant by assumption **(A5)**. In other words, $\mathcal{Z}(\varphi^{m\tau}(u)) \subset \psi^{m\tau}(u, \mathcal{Z}(u))$. Hence,

$$\mu_H(\mathcal{Z}(\varphi^{m\tau}(u)), d, \sqrt{d_0 + 1}\, l^{1/d} \varepsilon) \leqslant \mu_H(\psi^{m\tau}(u, \mathcal{Z}(u)), d, \sqrt{d_0 + 1}\, l^{1/d} \varepsilon). \tag{9.26}$$

Also, it follows from (9.21) that

$$\mu_H(\psi^{m\tau}(u, \mathcal{Z}(u)), d, \sqrt{d_0 + 1}\, l^{1/d} \varepsilon) \leqslant 2^{d_0}(d_0 + 1)^{\frac{d}{2}} l \mu_H(\mathcal{Z}(u), d, \varepsilon). \tag{9.27}$$

Combining together (9.25), (9.26),(9.27), and (9.24), we obtain

$$\mu_H(\mathcal{Z}(u), d, \sqrt{d_0 + 1}\, l^{1/d} \varepsilon) \leqslant 2^{d_0}(d_0 + 1)^{\frac{d}{2}} l \mu_H(\mathcal{Z}(u), d, \sqrt{d_0 + 1}\, l^{1/d} \varepsilon), \tag{9.28}$$

where the factor $2^{d_0}(d_0 + 1)^{\frac{d}{2}} l$ is strictly less than 1 by the choice of l.
It follows that $\mu_H(\mathcal{Z}(u), d, \sqrt{d_0 + 1}\, l^{1/d} \varepsilon)$ can only be 0 for all $u \in \mathcal{M}$. Making $\varepsilon \to 0$, we get $\mu_H(\mathcal{Z}(u), d) = 0, \ \forall u \in \mathcal{M}$. Hence, by the definition of the Hausdorff dimension, $\dim_H \mathcal{Z}(u) \leqslant d, \ \forall u \in \mathcal{M}$. □

Remark 9.3 Using the results of Sect. 5.4, Chap. 5 for fractal dimension estimates of negatively invariant sets of dynamical systems together with the above technique it is possible to get a similar estimate for the fractal dimension of negatively invariant sets of cocycles ([9]). A different approach for fractal dimension estimates is developed in [17].

9.2.3 Upper Estimates for the Hausdorff Dimension of Local Cocycles Generated by Differential Equations

Let us consider the non-autonomous ordinary differential equation

$$\dot{\psi} = g(t, \psi), \tag{9.29}$$

where $g : \mathbb{R} \times \mathbb{R}^n \to \mathbb{R}^n$ is a C^k-smooth ($k \geqslant 2$) vector field. Relative to the vector field (9.29) we introduce the hull of g defined by

$$\mathcal{H}(g) = \overline{\{g(\cdot + t, \cdot), t \in \mathbb{R}\}},$$

where the closure is taken in the compact open topology.
We assume that $\mathcal{H}(g)$ is compact. For this, it suffices to require that $g(t, \upsilon)$ in (9.29) is not only smooth in υ but also almost periodic in t.
Using the map $\widetilde{g}$, introduced in Subsect. 9.1.4, we can associate with system (9.29) the following family of vector fields

$$\dot{\psi} = \widetilde{g}(\varphi^t(u), \psi), \tag{9.30}$$

where $u \in \mathcal{H}(g)$ is arbitrary. The initial system (9.29) is contained in (9.30) as a special case.
Using, for example, almost periodicity in t of the map $(t, \upsilon) \mapsto g(t, \upsilon)$ and considering the above assumptions, one may show that for (9.30) there exists a local cocycle $(\{\psi^t(u, \cdot)\}_{u\in\mathcal{H}(g),\, t\in[0,\beta(u,\cdot))}, (\mathbb{R}^n, |\cdot|))$ over the base flow $(\{\varphi^t\}_{t\in\mathbb{R}}, (\mathcal{H}(g), \rho))$ (see [30]), where ψ^t is given in terms of the solution operator of system (9.30), and $[0, \beta(u, \upsilon))$ is the non-negative part of the maximal interval of existence of the motion passing through the point $(u, \upsilon) \in \mathcal{M} \times \mathbb{R}^n$. For a point $(u, \upsilon) \in \mathcal{M} \times \mathbb{R}^n$ let $\psi(t, \upsilon)$ be the solution of the variational equation along the trajectory of the cocycle through the point (u, υ); this is the solution of the equation

$$\dot{\psi} = \partial_2 \widetilde{g}(\varphi^t(u), \psi^t(u, \upsilon))\psi \tag{9.31}$$

with the initial condition $\psi(0, \psi_0) = \psi_0 \in \mathbb{R}^n$. Hence $\partial_2 \psi^t(u, \upsilon) w = \psi(t, w)$ for $0 \leqslant t < \beta(u, \upsilon)$. Let $\lambda_1(u, \upsilon) \geqslant \lambda_2(u, \upsilon) \geqslant \cdots \geqslant \lambda_n(u, \upsilon)$ be the ordered eigenvalues of the matrix $\frac{1}{2}\left[\partial_2 \widetilde{g}(u, \upsilon) + \partial_2 \widetilde{g}(u, \upsilon)^T\right]$.

Theorem 9.4 (Refs. [19, 28]) *Suppose that the local cocycle (ψ, φ) generated by the differential equation (9.29) over the Bebutov flow satisfies assumption* **(A5)** *and the following conditions:*

(1) *Condition (1) of Theorem 9.3 is satisfied with the set $\widetilde{\mathcal{K}}$;*
(2) *There exists a continuous function $V : \mathcal{M} \times \mathbb{R}^n \to \mathbb{R}$ with derivative $\frac{d}{dt} V(\varphi^t(u), \psi^t(u, \upsilon))$ along a given trajectory, and there exists a number $d \in (0, n]$ written as $d = d_0 + s$, where $d_0 \in \{0, 1, \ldots, n-1\}$ and $s \in (0, 1]$, such that*

$$\int_0^\tau [\lambda_1(\varphi^t(u), \psi^t(u, \upsilon) + \ldots + \lambda_{d_0}(\varphi^t(u), \psi^t(u, \upsilon)) + s\lambda_{d_0+1}(\varphi^t(u), \psi^t(u, \upsilon)) + \frac{d}{dt} V(\varphi^t(u), \psi^t(u, \upsilon))]dt < 0$$

for all $u \in \mathcal{M}$ *and* $\upsilon \in \widetilde{\mathcal{K}}$*, where* $\tau > 0$ *is the number from assumption* **(A5)***.*
Then $\dim_H \mathcal{Z}(u) \leqslant d$ *for all* $u \in \mathcal{M}$*.*

The proof of this theorem and the following one goes parallel to the proof of Theorem 5.5 and Corollary 5.4, Chap. 5, for dynamical systems and is omitted here.

For a number $d = d_0 + s$ with $d_0 \in [0, n-1]$ integer and $s \in (0, 1]$ and logarithmic norm Λ on the space of $n \times n$-matrices we introduce as in Sect. 2.4, Chap. 2, the partial d-trace w.r.t Λ of the map $\partial_2 \widetilde{g} : \mathcal{M} \times \mathbb{R}^n \to \mathbb{R}^n$ by

$$\mathrm{tr}_{d,\Lambda}\, \partial_2 \widetilde{g}(u, \upsilon) := s\Lambda\big(\partial_2 \widetilde{g}(u, \upsilon)^{[d_0+1]}\big) + (1+s)\Lambda\big(\partial_2 \widetilde{g}(u, \upsilon)^{[d_0]}\big),$$

for $u \in \mathcal{M}, \upsilon \in \mathbb{R}^n$.

Theorem 9.5 (Refs. [19, 28]) *Suppose that there exist an integer number* $d_0 \in [0, n-1]$*, a real* $s \in (0, 1]$*, a logarithmic norm* Λ *and a continuously differentiable on* $\widetilde{\mathcal{K}}$ *function* V *satisfying with* $d = d_0 + s$ *the inequality*

$$\int_0^\tau \left[\mathrm{tr}_{d,\Lambda}\, \partial_2 \widetilde{g}\left(\varphi^t(u), \psi^t(u, \upsilon)\right) + \frac{d}{dt} V\left(\varphi^t(u), \psi^t(u, \upsilon)\right) \right] dt < 0$$

for all $u \in \mathcal{M}$ *and* $\upsilon \in \widetilde{\mathcal{K}}$*, where* $\tau > 0$ *is the number from assumption* **(A5)***.*
Then $\dim_H \mathcal{Z}(u) \leq d$ *for all* $u \in \mathcal{M}$*.*

Remark 9.4 In a similar way as in Chap. 6 we could introduce different types of Lyapunov exponents for cocycles and consider the concept of Lyapunov dimension and Lyapunov dimension formula ([18]).

9.2.4 Upper Estimates for the Hausdorff Dimension of a Negatively Invariant Set of the Non-autonomous Rössler System

Consider the non-autonomous Rössler system (see [19, 29])

$$\begin{cases} \dot{y}_1 = -y_2 - y_3, \\ \dot{y}_2 = y_1, \\ \dot{y}_3 = -b(t)y_3 + a(t)(y_2 - y_2^2), \end{cases} \tag{9.32}$$

where $a, b : \mathbb{R} \to \mathbb{R}_+$ are functions defined by

$$a(t) = a_0 + a_1(t), \quad b(t) = b_0 + b_1(t).$$

Here a_0 and b_0 are positive constants, while $a_1(\cdot)$ and $b_1(\cdot)$ are C^1-smooth functions satisfying the inequalities

$$|a_1(t)| \leqslant \varepsilon a_0, \quad |b_1(t)| \leqslant \varepsilon b_0, \quad \forall\, t \in \mathbb{R}, \tag{9.33}$$

where $\varepsilon \in (0, 1)$ is a small parameter. Assume also that there exists $l > 0$ such that

$$|\dot{b}(t)| \leqslant \varepsilon l, \quad \forall\, t \in \mathbb{R} \tag{9.34}$$

and that the hull $\mathcal{H}(g)$ with g equal to the right-hand side of (9.32) is compact. For this purpose, it is sufficient that a and b are almost periodic.
Instead of (9.32), we consider the family of systems of type (9.30):

$$\begin{cases} \dot{y}_1 = -y_2 - y_3, \\ \dot{y}_2 = y_1, \\ \dot{y}_3 = -b_u(t) y_3 + a_u(t)(y_2 - y_2^2). \end{cases} \tag{9.35}$$

Here, for brevity,

$$a_u(t) \equiv \widetilde{a}(\varphi^t(u)), \quad b_u(t) \equiv \widetilde{b}(\varphi^t(u)).$$

Since system (9.32) has all the properties of system (9.29), it generates a local cocycle $(\{\psi^t(u, \cdot)\}_{u\in\mathcal{H}(g), t\in[0,\beta(u,\upsilon))}, (\mathbb{R}^n, |\cdot|))$ over the base flow $(\{\varphi^t\}_{t\in\mathbb{R}}, (\mathcal{H}(g), \rho_{\mathcal{H}(g)}))$, where $[0, \beta(u, \upsilon))$ is the non-negative part of the maximal interval on which there exits a solution of (9.35) passing through the point $(u, \upsilon) \in \mathcal{M} \times \mathbb{R}^n$. Assume that, for this cocycle, there exists a compact set $\widetilde{\mathcal{Z}} = \{\mathcal{Z}(u)\}_{u\in\mathcal{H}(g)}$, satisfying assumption (1) of Theorem 9.3 with the compact set $\widetilde{\mathcal{K}}$ and there exists a time $0 < \tau < \min\limits_{\substack{u\in\mathcal{H}(g),\\ \upsilon\in\mathcal{Z}(u)}} \beta(u, \upsilon)$, such that $\widetilde{\mathcal{Z}}$ is negatively invariant for the local cocycle in the sense of Definition 9.4 and (**A5**).
To estimate the Hausdorff dimension of $\widetilde{\mathcal{Z}}$ from above with the help of Theorem 9.4, we need to verify the inequality

$$\lambda_{1,u}(t, y_1, y_2, y_3) + \lambda_{2,u}(t, y_1, y_2, y_3) + s\lambda_{3,u}(t, y_1, y_2, y_3) + \frac{d}{dt} V_u(t, y_1, y_2, y_3) < 0, \tag{9.36}$$

for all $t \in [0, \tau]$, $(y_1, y_2, y_3) \in \widetilde{\mathcal{K}}$ and $u \in \mathcal{H}(g)$. Here

$$\lambda_{k,u}(t, y_1, y_2, y_3) \equiv \lambda_k(\varphi^t(u), \psi^t(u, y_1, y_2, y_3)), \quad k = 1, 2, 3$$

are the eigenvalues of the symmetrized Jacobi matrix of the right-hand side of (9.35) arranged in non-increasing order $\lambda_{1,u} \geqslant \lambda_{2,u} \geqslant \lambda_{3,u}$, and

$$V_u(t, y_1, y_2, y_3) \equiv V(\varphi^t(u), \psi^t(u, y_1, y_2, y_3))$$

is a Lyapunov-type function defined for all $(y_1, y_2, y_3) \in \widetilde{\mathcal{K}},\ u \in \mathcal{H}(g)$ and $t \in [0, \tau]$ by the relation

$$V(\varphi^t(u), y_1, y_3) := \frac{1}{2}(1-s)\xi(y_3 - b_u(t)y_1), \tag{9.37}$$

where ξ is a variable parameter. We calculate the eigenvalues $\lambda_{k,u}$ and the derivative $\frac{d}{dt}V_u$ and substitute them into (9.36).
It is easy to see that

$$\begin{aligned}\lambda_{1,u} &= \frac{1}{2}\left(-b_u(t) + \sqrt{b_u^2(t) + 1 + a_u^2(t)(1-2y_2)^2}\right), \\ \lambda_{2,u} &= 0\,, \\ \lambda_{3,u} &= \frac{1}{2}\left(-b_u(t) - \sqrt{b_u^2(t) + 1 + a_u^2(t)(1-2y_2)^2}\right).\end{aligned} \tag{9.38}$$

A direct calculation shows that

$$\dot{V}_u = \frac{1}{2}(1-s)\xi\left((a_u(t)+b_u(t))y_2 - b_u(t)y_1 - a_u(t)y_2^2\right). \tag{9.39}$$

It follows that the inequality (9.36) is satisfied if

$$\begin{aligned}&-b_u(t)(1+s) + (1-s)\,h_u\,(t, y_1, y_2; \xi) < 0\,, \\ &\forall\, t \in [0, \tau],\, u \in \mathcal{H}(g) \text{ and } (y_1, y_2) \in \mathrm{pr}_{y_1,y_2}\widetilde{\mathcal{K}},\end{aligned} \tag{9.40}$$

where

$$\begin{aligned}&h_u(t, y_1, y_2; \xi) \\ &= \sqrt{b_u^2(t) + 1 + a_u^2(t)(1-2y_2)^2} + \xi\left((a_u(t)+b_u(t))y_2 - b_u(t)y_1 - a_u(t)y_2^2\right)\end{aligned} \tag{9.41}$$

and $\mathrm{pr}_{y_1,y_2}\widetilde{\mathcal{K}}$ is the projection of $\widetilde{\mathcal{K}}$ on the subspace of y_1 and y_2.
Let us estimate $h(t, y_1, y_2; \xi)$ from above. We can write this expression as

$$\begin{aligned}h_u(t, y_1, y_2; \xi) = &-\left(\eta\sqrt{b_u^2(t) + 1 + a_u^2(t)(1-2y_2)^2} - \frac{1}{2\eta}\right)^2 \\ &+ \eta^2(b_u^2(t) + 1 + a_u^2(t)(1-2y_2)^2) + \frac{1}{4\eta^2} \\ &+ \xi\left((a_u(t)+b_u(t))y_2 - b_u(t)y_1 - a_u(t)y_2^2\right),\end{aligned} \tag{9.42}$$

where $\eta \neq 0$ is another varying parameter.
After some transformations we get for all arguments the inequality

$$
\begin{aligned}
h_u(t, y_1, y_2; \xi) \leq\ & \eta^2(a_u^2(t) + b_u^2(t) + 1) + \frac{1}{4\eta^2} - \xi b_u(t) y_1 \\
& - (\xi a_u(t) - 4\eta^2 a_u^2(t)) \left(y_2 + \frac{4\eta^2 a_u^2(t) - \xi(a_u(t) + b_u(t))}{\xi a_u(t) - 4\eta^2 a_u^2(t)} \right)^2 \\
& + \frac{(4\eta^2 a_u^2(t) - \xi(a_u(t) + b_u(t)))}{4(\xi a_u(t) - 4\eta^2 a_u^2(t))} .
\end{aligned}
\tag{9.43}
$$

Let us take ξ and η so that

$$
\xi a_u(t) - 4\eta^2 a_u^2(t) > 0, \quad \forall\, t \in [0, \tau],\ \forall\, u \in \mathcal{H}(g) . \tag{9.44}
$$

This is possible under our conditions for sufficiently small $\varepsilon > 0$. Using (9.43) and (9.44) we get

$$
\begin{aligned}
h_u(t, y_1, y_2; \xi) \leq \eta^2(a_u^2(t) + b_u^2(t) + 1) + \frac{1}{4\eta^2} & - \xi b_u(t) y_1 \\
+ \frac{(4\eta^2 a_u^2(t) - \xi(a_u(t) + b_u(t)))^2}{4(\xi a_u(t) - 4\eta^2 a_u^2(t))} & .
\end{aligned}
\tag{9.45}
$$

Since $\mathrm{pr}_{y_1} \widetilde{\mathcal{K}}$ is compact there exists an $m > 0$ such that

$$
|y_1| \leq m \quad \text{for all} \quad y_1 \in \mathrm{pr}_{y_1} \widetilde{\mathcal{K}} . \tag{9.46}
$$

Let us choose now the parameters as

$$
\xi := 4\eta^2 a_0 \frac{a_0 + 2b_0}{a_0 + b_0} \quad \text{and} \quad \eta^2 := \frac{1}{2\sqrt{(a_0 + 2b_0)^2 + b_0^2 + 1}} . \tag{9.47}
$$

Substituting these values into (9.45), taking a number of direct calculations and using the estimates (9.44) and (9.46) we finally get the estimate

$$
\begin{aligned}
& h_u(t, y_1, y_2; \xi) \leq \sqrt{(a_0 + 2b_0)^2 + b_0^2 + 1} + \varepsilon \cdot C \\
& \text{for all} \quad t \in [0, \tau],\ u \in \mathcal{H}(g) \text{ and } (y_1, y_2) \in \mathrm{pr}_{y_1, y_2} \widetilde{\mathcal{K}} ,
\end{aligned}
\tag{9.48}
$$

where C is a term which can be directly calculated by means of the parameters a_0, b_0, ε, l and m of the system and which is bounded from above for all small $\varepsilon > 0$. In order to use Theorem 9.3 effectively we need to find the minimal s for which the inequality (9.40) still holds. Thus, from (9.40), (9.48) and Theorem 9.3 it follows that

$$
\begin{aligned}
\dim_H \mathcal{Z}(u) &\leq 3 - \frac{2b_u(t)}{b_u(t) + h_u(t, y_1, y_2; \xi)} \\
&\leq 3 - \frac{2(1-\varepsilon)b_0}{(1+\varepsilon)b_0 + \sqrt{(a_0 + 2b_0)^2 + b_0^2 + 1 + \varepsilon \cdot C}}.
\end{aligned}
\tag{9.49}
$$

Direct computations with the use of (9.33), (9.34) and Theorem 9.4 finally yield the estimate

$$
\dim_H \mathcal{Z}(u) \leqslant 3 - \frac{2(1-\varepsilon)b_0}{(1+\varepsilon)b_0 + \sqrt{(a_0 + 2b_0)^2 + b_0^2 + 1} + \varepsilon \cdot C}
\tag{9.50}
$$

for all $u \in \mathcal{H}(g)$, where C is a positive number, which can be obtained from the parameters a_0, b_0, ε, l, of our system and which is bounded for all small $\varepsilon > 0$. Returning to the Rössler autonomous system (making $\varepsilon \to 0$), we arrive at the already known estimate for the Hausdorff dimension of a compact negatively invariant set $\widetilde{\mathcal{K}}$ of the Rössler system,

$$
\dim_H \widetilde{\mathcal{K}} \leqslant 3 - \frac{2b_0}{b_0 + \sqrt{(a_0 + 2b_0)^2 + b_0^2 + 1}},
\tag{9.51}
$$

(see Theorem 5.19, Chap. 5).

Remark 9.5 Similar Hausdorff dimension estimates for the Lorenz system are derived in [3, 19]. Dimension properties of cocycles generated by partial differential equations are considered in [10, 11]. Borg's criterion for almost periodic differential equations is shown in [13].

9.3 Dimension Estimates for Cocycles on Manifolds (Non-fibered Case)

9.3.1 *The Douady-Oesterlé Theorem for Cocycles on a Finite Dimensional Riemannian Manifold*

Suppose that $(\mathcal{N}, \rho_{\mathcal{N}})$ is a smooth m-dimensional Riemannian manifold, $(\mathcal{M}, \rho_{\mathcal{M}})$ is a complete metric space, $\psi^t(u, \cdot) : \mathcal{N} \to \mathcal{N}, u \in \mathcal{M}, t \in \mathbb{R}$ is a family of smooth maps and $\varphi^t : \mathcal{M} \to \mathcal{M}$ is a continuous base system.
To formulate the next theorem, we need the following assumptions.

(A7) The set $\widetilde{\mathcal{Z}} = \{\mathcal{Z}(u)\}_{u \in \mathcal{M}}$ is compact and negatively invariant w.r.t the cocycle.
(A8) For all $t > 0$ we have

$$\sup_{u\in\mathcal{M}} \sup_{v\in\widetilde{\mathcal{Z}}(u)} |\partial_2\psi^t(u,v)|_{op} < \infty,$$

where $|\cdot|_{op}$ is the operator norm.

Theorem 9.6 (Ref. [23]) *Under the assumptions* **(A7)**, **(A8)** *suppose additionally that:*

(1) *There exists a set* $\widetilde{\mathcal{K}} \subset \mathcal{N}$ *such that*

$$\overline{\bigcup_{u\in\mathcal{M}} \mathcal{Z}(u)} \subset \widetilde{\mathcal{K}};$$

(2) *There exists a bounded function* $\varkappa : \mathcal{M}\times\mathcal{N} \to \mathbb{R}_+ \setminus \{0\}$, *numbers* $\tau > 0$ *and* $d \in (0, n]$ *such that*

$$\sup \frac{\varkappa(\varphi^\tau(u), \psi^\tau(u,v))}{\varkappa(u,v)} \omega_d(\partial_2\psi^\tau(u,v)) < 1 \tag{9.52}$$

and $\mathcal{Z}(u) \subset \mathcal{Z}(\varphi^\tau(u))$, $u \in \mathcal{M}$. *Then* $\dim_H \mathcal{Z}(u) \le d$ *for all* $u \in \mathcal{M}$.

Proof From assumption (2) of Theorem 9.6 it follows that there exist a bounded continuous function $\varkappa : \mathcal{M}\times\mathcal{N} \to \mathbb{R}_+ \setminus \{0\}$, numbers $\tau > 0$ and $d \in (0, n]$ such that

$$\sup_{(u,v)\in\mathcal{M}\times\widetilde{\mathcal{K}}} \frac{\varkappa(\varphi^\tau(u), \psi^\tau(u,v))}{\varkappa(u,v)} \omega_d(\partial_2\psi^\tau(u,v)) < 1.$$

Thus, for all $u \in \mathcal{M}$

$$\sup_{v\in\mathcal{Z}(u)} \frac{\varkappa(\varphi^\tau(u), \psi^\tau(u,v))}{\varkappa(u,v)} \omega_d(\partial_2\psi^\tau(u,v)) < 1.$$

Then there exists a number $0 < \kappa_1 < 1$, such that for all $u \in \mathcal{M}$

$$\sup_{v\in\mathcal{Z}(u)} \frac{\varkappa(\varphi^\tau(u), \psi^\tau(u,v))}{\varkappa(u,v)} \omega_d(\partial_2\psi^\tau(u,v)) < \kappa_1. \tag{9.53}$$

Given an arbitrary $m \in \mathbb{N}$, we define

$$\kappa(m,u) := \kappa_1^m \sup_{v\in\mathcal{Z}(u)} \frac{\varkappa(\varphi^\tau(u), \psi^\tau(u,v))}{\varkappa(u,v)}. \tag{9.54}$$

By assumption, $\varkappa$ is bounded. Hence, $\sup\limits_{v\in\mathcal{Z}(u)} \frac{\varkappa(\varphi^\tau(u),\psi^\tau(u,v))}{\varkappa(u,v)}$ is finite and not dependent on m. Furthermore, for all $u \in \mathcal{M}$ we can make $\kappa(m,u)$ arbitrary small by taking sufficiently large m.

In other words, for all $l > 0$ there exists $m_0 \in \mathbb{N}$ such that

$$0 \le \kappa(m, u) \le l, \quad \forall m \ge m_0, u \in \mathcal{M}. \tag{9.55}$$

Let $m \ge m_0$ be a sufficiently large fixed number. By the chain rule, we get

$$\begin{aligned}\partial_2\psi^{m\tau}(u, \upsilon) &= \partial_2\psi^{\tau}\left(\varphi^{(m-1)\tau}(u), \psi^{(m-1)\tau}(u, \upsilon)\right)\\ &\circ \partial_2\psi^{\tau}\left(\varphi^{(m-2)\tau}(u), \psi^{(m-2)\tau}(u, \upsilon)\right) \circ \ldots \circ \partial_2\psi^{\tau}(u, \upsilon).\end{aligned}$$

Using Horn's inequality, we get

$$\begin{aligned}\omega_d(\partial_2\psi^{m\tau}(u, \upsilon)) &\le \omega_d\left(\partial_2\psi^{\tau}(\varphi^{(m-1)\tau}(u), \psi^{(m-1)\tau}(u, \upsilon))\right)\\ &\cdot \omega_d\left(\partial_2\psi^{\tau}(\varphi^{(m-2)\tau}(u), \psi^{(m-2)\tau}(u, \upsilon))\right) \cdot \ldots \cdot \partial_2\psi^{\tau}(u, \upsilon)\\ &= \prod_{j=1}^{m} \omega_d\left(\partial_2\psi^{\tau}(\varphi^{(m-j)\tau}(u), \psi^{(m-j)\tau}(u, \upsilon))\right).\end{aligned}$$

Further, with the use of (9.53) for $(\varphi^{(m-j)\tau}(u), \psi^{(m-j)\tau}(u, \upsilon))$, $j = 1, \ldots, m-1$, we have the following estimates

$$\begin{aligned}&\omega_d(\partial_2\psi^{\tau}(u, \upsilon)) \le \kappa_1 \cdot \frac{\varkappa(u, \upsilon)}{\varkappa(\varphi^{\tau}(u), \psi^{\tau}(u, \upsilon))},\\ &\quad \ldots\\ &\omega_d(\partial_2\psi^{\tau}(\varphi^{(m-1)\tau}(u), \psi^{(m-1)\tau}(u, \upsilon))) \le \kappa_1 \cdot \frac{\varkappa(\varphi^{(m-1)\tau}(u), \psi^{(m-1)\tau}(u, \upsilon))}{\varkappa(\varphi^{m\tau}(u), \psi^{m\tau}(u, \upsilon))}.\end{aligned} \tag{9.56}$$

Using (9.53) and (9.56), we can deduce that

$$\omega_d(\partial_2\psi^{m\tau}(u, \upsilon)) \le \prod_{j=1}^{m} \kappa_1 \cdot \frac{\varkappa(\varphi^{(m-j)\tau}(u), \psi^{(m-j)\tau}(u, \upsilon))}{\varkappa(\varphi^{(m-j+1)\tau}(u), \psi^{(m-j+1)\tau}(u, \upsilon))}.$$

The function $\varkappa$ is positive and bounded. Thus, after simplifying, we get

$$\omega_d(\partial_2\psi^{m\tau}(u, \upsilon)) \le \kappa_1^m \cdot \frac{\varkappa(u, \upsilon)}{\varkappa(\varphi^{m\tau}(u), \psi^{m\tau}(u, \upsilon))} \le \kappa(m, u).$$

Then for all $u \in \mathcal{M}$, $\upsilon \in \mathcal{Z}(u)$ and sufficiently large fixed $m \ge m_0$

$$\omega_d(\partial_2\psi^{m\tau}(u, \upsilon)) \le \kappa(m, u). \tag{9.57}$$

Let $\varepsilon > 0$ be sufficiently small to fulfill

$$\left|\iota_{\psi^{m\tau}(u,w)}^{\psi^{m\tau}(u,\upsilon)} \circ \partial_2\psi^{m\tau}(u, w) \circ \iota_{\upsilon}^{w} - \partial_2\psi^{m\tau}(u, \upsilon)\right| \le \zeta,$$

for all υ, w such that $\rho(\upsilon, w) \le \varepsilon$, where $\rho(\cdot,\cdot)$ is the geodesic distance on $\mathcal{N}$ and ι_υ^w is the isometry between $T_\upsilon\mathcal{N}$ and $T_w\mathcal{N}$.

Using Taylor's formula, we get

$$\left|\exp^{-1}_{\psi^{m\tau}(u,\upsilon)}\psi^{m\tau}(u,w) - \partial_2\psi^{m\tau}(u,\upsilon)(\exp^{-1}_\upsilon(w))\right| \le$$
$$\sup_{z\in\mathcal{B}(\upsilon,r)}\left|\iota^{\psi^{m\tau}(u,\upsilon)}_{\psi^{m\tau}(u,z)}\circ\partial_2\psi^{m\tau}(u,z)\circ\iota^z_\upsilon - \partial_2\psi^{m\tau}(u,\upsilon)\right|\cdot\left|\exp^{-1}_\upsilon(z)\right|,$$

for all $w \in \mathcal{B}(\upsilon, r)$ with $r < \varepsilon$. It follows that the image of $\mathcal{B}(\upsilon, r)$ under the map ψ is contained in

$$\psi^{m\tau}(u, \mathcal{B}(\upsilon, r)) \subset \exp_{\psi^{m\tau}(u,\upsilon)}\left(\partial_2\psi^{m\tau}(u,\upsilon)(\mathcal{B}_{T_u\mathcal{N}}(0,r)) + \mathcal{B}_{T_{\psi^{m\tau}(u,\upsilon)}\mathcal{N}}(0, r\zeta)\right). \tag{9.58}$$

By taking ε sufficiently small we can make ζ arbitrary small. Let $\mathcal{E} := \partial_2\psi^{m\tau}(u,\upsilon)(\mathcal{B}_{T_\upsilon\mathcal{N}}(0,r))$ and $\partial_2\psi^{m\tau}(u,\upsilon)$ be a linear operator; hence, by Proposition 7.13, Chap. 7, the set $\mathcal{E}$ is an ellipsoid with semiaxes length $a_i(\mathcal{E}) = r\alpha_i(\partial_2\psi^{m\tau}(u,\upsilon))$, $i = 1, \dots, n$. We have

$$\sup_{\upsilon\in\mathcal{Z}(u)} |\partial_2\psi^{m\tau}(u,\upsilon)| \le \delta, \quad \text{and} \quad \kappa(m) \le \delta^d,$$

for all sufficiently large m and

$$\left(1 + \left(\frac{\delta^{d_0}}{\kappa(m)}\right)^{1/s}\zeta\right)\kappa(m) < l. \tag{9.59}$$

Now we need to verify the conditions of Lemma 7.1, Chap. 7. Recall that

$$a_i(\mathcal{E}) = r\alpha_i(\partial_2\psi^{m\tau}(u,\upsilon)).$$

By the Fischer-Courant theorem (Theorem 7.1, Chap. 7) we have

$$\alpha_1(\partial_2\psi^{m\tau}(u,\upsilon)) = \sup_{\upsilon\in\mathcal{Z}(u),\ |\upsilon|=1} |\partial_2\psi^{m\tau}(u,\upsilon)|_{op} \le \delta$$

which leads to $a_1(\mathcal{E}) \le r\delta$ and

$$\begin{aligned}
\omega_d(\mathcal{E}) &:= a_1(\mathcal{E})\cdot\ldots\cdot a_{d_0}(\mathcal{E})a^s_{d_0+1}(\mathcal{E})\\
&= r\alpha_1(\partial_2\psi^{m\tau}(u,\upsilon))\ldots r\alpha_{d_0}(\partial_2\psi^{m\tau}(u,\upsilon))r^s\alpha^s_{d_0+1}(\partial_2\psi^{m\tau}(u,\upsilon))\\
&= r^{d_0+s}\alpha_1(\partial_2\psi^{m\tau}(u,\upsilon))\ldots\alpha_{d_0}(\partial_2\psi^{m\tau}(u,\upsilon))\alpha^s_{d_0+1}(\partial_2\psi^{m\tau}(u,\upsilon))\\
&= r^d\omega_d(\partial_2\psi^{m\tau}(u,\upsilon)).
\end{aligned}$$

From (9.57) it follows that

$$\omega_d(\partial_2\psi^{m\tau}(u,v)) \le \kappa(m,u) \le \kappa(m),$$

hence,

$$\omega_d(\mathcal{E}) \le r^d\kappa(m).$$

Furthermore, $\kappa(m) \le \delta^d$ and $r^d\kappa(m) \le (r\delta)^d$.
The conditions of Lemma 7.1, Chap. 7, hold for $\kappa' := r^d\kappa(m)$, $\delta' := r\delta$, $\eta' := r\zeta$. It follows that the set $\mathcal{E} + \mathcal{B}(0, r\zeta)$ is contained in the ellipsoid $\mathcal{E}'$ with

$$\omega_d(\mathcal{E}') \le \left(1 + \left(\frac{\delta'^{d_0}}{\kappa'}\right)^{1/s}\eta'\right)^d \kappa' = \left(1 + \left(\frac{r^{d_0}\delta^{d_0}}{r^d\kappa(m)}\right)^{1/s} r\eta\right)^d \kappa(m)$$
$$= \left(1 + \frac{\delta^{d_0/s}}{\kappa^{1/s}(m)}\eta\right)^d \kappa(m)r^d.$$

Moreover, with the use of (9.59) we have

$$\left(1 + \frac{\delta^{d_0/s}}{\kappa^{1/s}(m)}\zeta\right)^d \kappa(m) < l$$

and then

$$\omega_d(\mathcal{E}') < lr^d, \tag{9.60}$$

where $l > 0$ is an arbitrary small number.
If $\{\mathcal{B}(v_j, r_j)\}$ is a countable covering of $\mathcal{Z}(u)$ by balls of radius $r_j \le \varepsilon$ we can construct a countable cover of $\psi^{m\tau}(u, \mathcal{Z}(u))$ by ellipsoids $\mathcal{E}'_j$ with $\omega_d(\mathcal{E}'_j) \le lr_j^d$.
For any compact set $\mathcal{K} \subset \mathcal{N}$ define

$$\widetilde{\mu}_H(\mathcal{K}, d, \varepsilon) := \inf \sum_j \omega_d(\mathcal{E}'_j),$$

where the infimum is taken over all countable coverings of $\mathcal{K}$ by ellipsoids $\mathcal{E}'_j$ with $\omega_d(\mathcal{E}'_j) \le lr_j^d$.
From the definitions of $\widetilde{\mu}_H$, μ_H and (9.60), we get

$$\mu_H(\mathcal{Z}(u), d, \varepsilon) = \inf \left\{ \sum_j r_j^d \mid r_j^d \le \varepsilon \right\} \quad \text{and}$$

$$\widetilde{\mu}_H(\psi^{m\tau}(u, \mathcal{Z}(u)), d, l^{1/d}\varepsilon) = \inf \left\{ \sum_j \omega_d(\mathcal{E}'_j) \mid \omega_d(\mathcal{E}'_j) \le l\varepsilon^d \right\},$$

where $\sum_j \omega_d(\mathcal{E}'_j) \le l \sum_j r_j^d$ and $\omega_d(\mathcal{E}'_j) \le l r_j^d \le l\varepsilon^d$. Thus,

$$\widetilde{\mu}_H(\psi^{m\tau}(u, \mathcal{Z}(u)), d, l^{1/d}\varepsilon) \le l\mu_H(\mathcal{Z}(u), d, \varepsilon).$$

We have $(\omega_d(\mathcal{E}_j))^{1/d} \le \varepsilon$ for each ellipsoid $\mathcal{E}_j$ from the covering of $\mathcal{K}$, and

$$\mu_H(\mathcal{E}_j, d, \sqrt{d_0+1}\varepsilon) \le 2^{d_0}(d_0+1)^{d/2}\omega_d(\mathcal{E}_j).$$

Consequently, we get

$$\begin{aligned} \mu_H(\mathcal{K}, d, \sqrt{d_0+1}\varepsilon) &\le \mu_H\Big(\bigcup_j \mathcal{E}_j, d, \sqrt{d_0+1}\varepsilon\Big) \\ &\le \sum_j \mu_H(\mathcal{E}_j, d, \sqrt{d_0+1}\varepsilon) \le 2^{d_0}(d_0+1)^{d/2}\sum_j \omega_d(\mathcal{E}_j). \end{aligned}$$

Taking the infimum, we obtain

$$\mu_H(\mathcal{K}, d, \sqrt{d_0+1}\varepsilon) \le 2^{d_0}(d_0+1)^{d/2}\widetilde{\mu}_H(\mathcal{K}, d, \varepsilon). \tag{9.61}$$

Applying (9.61) to the set $\mathcal{K} = \psi^{m\tau}(u, \mathcal{Z}(u))$, we get

$$\begin{aligned} &\mu_H(\psi^{m\tau}(u, \mathcal{Z}(u)), d, \sqrt{d_0+1}\,l^{1/d}\varepsilon) \le \\ &2^{d_0}(d_0+1)^{d/2}\widetilde{\mu}_H(\psi^{m\tau}(u, \mathcal{Z}(u)), d, l^{1/d}\varepsilon) \le m2^{d_0}(d_0+1)^{d/2} l\mu_H(\mathcal{Z}(u), d, \varepsilon). \end{aligned} \tag{9.62}$$

Assuming that $\mu_H(\mathcal{Z}(u), d) \le \mu_0 < \infty$ and taking $\varepsilon \to 0$, we obtain

$$\mu_H(\psi^{m\tau}(u, \mathcal{Z}(u)), d) \le 2^{d_0}(d_0+1)^{d/2} l\mu_H(\mathcal{Z}(u), d) \le 2^{d_0}(d_0+1)^{d/2} l\mu_0. \tag{9.63}$$

Recall that l is an arbitrary small positive number, for which we chose a sufficiently large number m so that we were able to apply the above estimates. Hence, for a sufficiently large m the right-hand side of (9.63) can be made arbitrary small. Therefore, for all $u \in \mathcal{M}$, we have shown that if $\mu_H(\mathcal{Z}(u), d) < \infty$, then

$$\lim_{\varepsilon \to 0} \mu_H(\psi^{m\tau}(u, \mathcal{Z}(u)), d) = 0.$$

As l is an arbitrary number we can assume that it satisfies

$$\sqrt{d_0+1}\,l^{1/d} < 1, \quad \text{and} \quad 2^{d_0}(d_0+1)^{d/2} l < 1.$$

In that case

$$\mu_H(\mathcal{Z}(u), d, \varepsilon) \le \mu_H(\mathcal{Z}(u), d, \sqrt{d_0+1}\,l^{1/d}\varepsilon). \tag{9.64}$$

By assumption of the theorem, $\mathcal{Z}(u) \subset \mathcal{Z}(\varphi^{\tau}(u)) \subset \ldots \subset \mathcal{Z}(\varphi^{m\tau}(u))$. This leads to the inequality

$$\mu_H(\mathcal{Z}(u), d, \sqrt{d_0+1}l^{1/d}\varepsilon) \le \mu_H(\mathcal{Z}(\varphi^{m\tau}(u)), d, \sqrt{d_0+1}l^{1/d}\varepsilon). \tag{9.65}$$

By assumption **(A7)**, $\widetilde{\mathcal{Z}}$ is negatively invariant, which means that $\mathcal{Z}(\varphi^{m\tau}(u)) \subset \psi^{m\tau}(u, \mathcal{Z}(u))$. Then

$$\mu_H(\mathcal{Z}(\varphi^{m\tau}(u)), d, \sqrt{d_0+1}l^{1/d}\varepsilon) \le \mu_H(\psi^{m\tau}(u, \mathcal{Z}(u)), d, \sqrt{d_0+1}l^{1/d}\varepsilon). \tag{9.66}$$

From (9.62), we have

$$\mu_H(\psi^{m\tau}(u, \mathcal{Z}), d, \sqrt{d_0+1}l^{1/d}\varepsilon) \le 2^{d_0}(d_0+1)^{d/2} l \mu_H(\mathcal{Z}(u), d, \sqrt{d_0+1}l^{1/d}\varepsilon). \tag{9.67}$$

Combining (9.64), (9.65), (9.66) and (9.67), we obtain

$$\mu_H(\mathcal{Z}(u), d, \sqrt{d_0+1}, l^{1/d}\varepsilon) \le 2^{d_0}(d_0+1)^{d/2} l \mu_H(\mathcal{Z}(u), d, \sqrt{d_0+1}l^{1/d}\varepsilon), \tag{9.68}$$

where $2^{d_0}(d_0+1)^{d/2}l$ is strictly less that 1 by the choice of l.
From here we get that $\mu_H(\mathcal{Z}(u), d, \sqrt{d_0+1}l^{1/d}\varepsilon)$ can only be 0 for all $u \in \mathcal{M}$. Taking $\varepsilon \to 0$ we have $\mu_H(\mathcal{Z}(u), d) = 0$ for all $u \in \mathcal{M}$. Hence, by the definition of the Hausdorff dimension, $\dim_H \mathcal{Z}(u) \le d$, for all $u \in \mathcal{M}$. □

9.3.2 *Upper Bounds for the Haussdorff Dimension of Negatively Invariant Sets of Discrete-Time Cocycles*

Let $\mathcal{N}$ be an n-dimensional smooth Riemannian manifold equipped with a discrete-time cocycle $(\{\psi^k(u,\cdot)\}_{\substack{k\in\mathbb{Z}_+\\ u\in\mathcal{M}}}, (\mathcal{N}, g))$ over the base flow $(\{\varphi^k\}_{k\in\mathbb{Z}}, (\mathcal{M}, \rho_{\mathcal{M}}))$, where $\psi^k(u,\cdot): \mathcal{N} \to \mathcal{N}$. We make the following assumptions:

(A9) The parametrized set $\widetilde{\mathcal{Z}} = \{\mathcal{Z}(u)\}_{u\in\mathcal{M}}$ is compact, negatively invariant for the cocycle (φ, ψ) and satisfies $\mathcal{Z}(u) \subset \mathcal{Z}(\varphi(u))$, $\forall u \in \mathcal{M}$;

(A10) For any $k > 0$, one has:

$$\sup_{u\in\mathcal{M}} \sup_{v\in\mathcal{Z}(u)} |\partial_2 \psi^k(u, v)|_{op} < \infty,$$

where $|\partial_2 \psi^k(u, v)|_{op}$ is the operator norm of the linear mapping

$$\partial_2 \psi^k(u, v): T_v\mathcal{N} \to T_{\psi^k(u,v)}\mathcal{N}, \quad u \in \mathcal{M}, \quad v \in \mathcal{Z}(u).$$

Let us give a generalization of the Douady-Oesterlé theorem ([8]) to the case of discrete time cocycles on a finite-dimensional Riemannian manifold. The proof of the theorem is similar to the proof of Theorem 9.6 and omitted here.

Theorem 9.7 (Ref. [22]) *Let assumptions* **(A9)**, **(A10)** *and the following conditions hold for the cocycle* (φ, ψ)*:*

(1) *There exists a compact set* $\widetilde{\mathcal{K}} \in \mathcal{N}$ *such that*

$$\overline{\bigcup_{u \in \mathcal{M}} \mathcal{Z}(u)} \subset \widetilde{\mathcal{K}};$$

(2) *There exists a continuous bounded function* $\varkappa : \mathcal{M} \times \mathcal{N} \to \mathbb{R}_+ \backslash \{0\}$, *a time* $j > 0$, *and a number* $d \in (0, n]$ *such that*

$$\sup_{(u,v) \in \mathcal{M} \times \widetilde{\mathcal{K}}} \frac{\varkappa(\varphi^j(u), \psi^j(u, v))}{\varkappa(u, v)} \omega_d(\partial_2 \psi^j(u, v)) < 1. \tag{9.69}$$

Then $\dim_H \mathcal{Z}(u) \leqslant d$ *for each* $u \in \mathcal{M}$.

The following theorem generalizes the result obtained in [24] to the case of discrete-time cocycles.

Theorem 9.8 (Ref. [22]) *Assume that the set* $\{\mathcal{K}(u)\}_{u \in \mathcal{M}}$ *is a negatively invariant for a discrete-time cocycle* (φ, ψ), $\mathcal{K}(u) \subset \mathcal{K}(\varphi(u))$, $u \in \mathcal{M}$, *and let* $\mathcal{D} \subset \mathcal{N}$ *be an open set such that* $\bigcup_{u \in \mathcal{M}} \mathcal{K}(u) \subset \mathcal{D}$.
Assume that there exists a continuous function $\kappa : \mathcal{D} \to \mathbb{R}_+$, *satisfying the following conditions:*

(1) $(w, (\partial_2 \psi(u, v))^* \partial_2 \psi(u, v) w)_{T_v \mathcal{N}} \geqslant \kappa^2(u, v)(w, w)_{T_v \mathcal{N}}, \quad \forall u \in \mathcal{M},$
$\forall v \in \mathcal{K}(u), \quad \forall w \in T_v \mathcal{N};$

(2) $\sup_{u \in \mathcal{M}, v \in \mathcal{K}(u)} \frac{|\det \partial_2 \psi(u, v)|}{\kappa(u, v)^{n-1}} < 1.$

Then $\dim_H(\mathcal{K}) < d$.

Proof Consider the tangent mapping $\partial_2 \psi(u, \cdot) : T_v \mathcal{N} \to T_{\psi(u,v)} \mathcal{N}$. For each square of singular number $\alpha_i^2 := \alpha_i^2(u, v)$, $i = 1, 2, \ldots, n$ of the linear operator $\partial_2 \psi$, there exists an eigenvector $w_i = w_i(u, v) \in T_v \mathcal{N}$ such that

$$(\partial_2 \psi(u, v))^* \partial_2 \psi(u, v) w_i = \alpha_i^2 w_i.$$

It follows from assumption (1) that

$$\alpha_i^2(u, v) \geqslant \kappa^2(u, v), \quad i = 1, 2, \ldots, n, \quad u \in \mathcal{M}, \quad v \in \mathcal{K}(u), \tag{9.70}$$

whence

$$\alpha_1(u, \upsilon) \cdot \ldots \cdot \alpha_k(u, \upsilon)\kappa(u, \upsilon)^{n-k} \leqslant |\det \partial_2\psi(u, \upsilon)| \tag{9.71}$$

for any $k \in \{1, \ldots, n\}$. From this we find that $\kappa(u, \upsilon)^n \leqslant |\det \partial_2\psi(u, \upsilon)|$. Using assumption (2) of the theorem, we obtain the inequality

$$\omega_d(\partial_2\psi(u, \upsilon)) \leqslant \frac{|\det \partial_2\psi(u, \upsilon)|}{\kappa(u, \upsilon)^{n-d}} < 1.$$

The assertion of Theorem 9.8 follows now from Theorem 9.7. □

Let us present an example demonstrating the application of Theorem 9.8 to the Hénon system for the case of parameters depending on some base systems on a metric space. A similar example for constant parameters can be found in [24].

Example 9.4 Consider the time-varying Hénon system

$$\begin{cases} x_{k+1} = 1 - a_k x_k^2 + y_k, \\ y_{k+1} = b_k x_k, \quad k \in \mathbb{Z}_+, \end{cases} \tag{9.72}$$

where $\{a_k\}_{k=0}^{\infty}$ and $\{b_k\}_{k=0}^{\infty}$ are sequences of the form $a_k = a + \widetilde{a}_k$ and $b_k = b + \widetilde{b}_k$. Here a and b are positive parameters and $\{\widetilde{a}_k\}$ and $\{\widetilde{b}_k\}$ are bounded sequences satisfying the inequalities

$$|\widetilde{a}_k| \leqslant \varepsilon a \quad \text{and} \quad |\widetilde{b}_k| \leqslant \varepsilon b, \quad k \in \mathbb{Z}_+,$$

where $\varepsilon \in (0, 1)$ is a small parameter.

Together with system (9.72), consider the family of systems

$$\begin{cases} x_{k+1} = 1 - a_u(k) x_k^2 + y_k, \\ y_{k+1} = b_u(k) x_k, \quad k \in \mathbb{Z}_+, \end{cases} \tag{9.73}$$

where we write $a_u(k) = \widetilde{a}(\varphi^k(u))$ and $b_u(k) = \widetilde{b}(\varphi^k(u))$ for brevity. Here

$$(\{\varphi^k\}_{k\in\mathbb{Z}}, (\mathcal{M}, \rho_{\mathcal{M}})) \tag{9.74}$$

is a base system on a compact metric space $(\mathcal{M}, \rho_{\mathcal{M}})$ and $\widetilde{a}, \widetilde{b} : \mathcal{M} \to \mathbb{R}_+$ are continuous functions.

Let (φ, ψ) be the cocycle generated by systems (9.72) and (9.74). Assume that there exists a compact invariant set $\widetilde{\mathcal{K}} = \{\mathcal{K}(u)\}_{u\in\mathcal{M}}$ for (φ, ψ). Using Theorem 9.8, we estimate the Hausdorff dimension from above. Let $\upsilon = (x, y)$; then

$$\partial_2\psi(u, \upsilon) = \begin{bmatrix} -2a_u x & 1 \\ b_u & 0 \end{bmatrix}.$$

Hence $|\det \partial_2\psi(u, \upsilon)| = b_u$ for all $u \in \mathcal{M}$, $\upsilon \in \mathcal{K}$. For $d = 1 + s$, $s \in [0, 1]$, consider the function of the singular numbers $\alpha_1(u, x)$ and $\alpha_2(u, x)$ given by

$$\omega_d(\partial_2\psi) = \alpha_1(u,x)\alpha_2^s(u,x) = \alpha_1(u,x)^{1-s}b_u^s.$$

The maximum singular values $\alpha_1(u,x)$ can be computed by the formula

$$\alpha_1^2(u,x) = \frac{4a_u^2x + b_u^2 + 1}{2} + \sqrt{\frac{(4a_u^2x + b_u^2 + 1)^2}{4} - b_u^2}.$$

Assume that $\sup_{\substack{x\in \mathrm{pr}\,\mathcal{K}(u),\\ u\in\mathcal{M}}} \alpha_1(u,x)^{1-s}bu < 1$, for some $s \in [0,1]$; then $\dim_H \mathcal{K}(u) < 1+s$ for any $u \in \mathcal{M}$. Here $\mathrm{pr}\,\mathcal{K}(u)$ is the projection of $\mathcal{K}(u)$ onto the first coordinate.

Remark 9.6 Hausdorff dimension estimates for the time-varying Hénon system with a cellular automaton as driven system are derived in [9].

9.3.3 Frequency Conditions for Dimension Estimates for Discrete Cocycles

Consider the system ([1, 2, 20–22, 26])

$$v_{k+1} = Av_k + B\phi(k, w_k), \quad w_k = C^*v_k, \quad k = 0, 1, 2, \dots, \tag{9.75}$$

where A is a constant $n \times n$ matrix and B and C are constant $n \times m$ matrices. Further, let $\mathcal{D} \subset \mathbb{R}^m$ be an open pathwise connected set and let $\phi : \mathbb{Z} \times \mathcal{D} \to \mathbb{R}^m$ be a nonlinear mapping smooth in the second argument.
Together with (9.75), consider the cocycle (φ, ψ). Thus, we have a cocycle that is described via the family of systems

$$v_{k+1} = Av_k + B\widetilde{\phi}(\varphi^k(u), w_k), \quad w_k = C^*v_k, \quad u \in \mathcal{M} \quad k = 0, 1, 2, \dots. \tag{9.76}$$

Let $W(z) = C^*(A - zI)^{-1}B$, $z \in \mathbb{C}$: $\det(A - zI) \neq 0$, be the transfer function of the linear part of system (9.75).
The following theorem generalizes the result from [24] to the case of discrete cocycles.

Theorem 9.9 (Refs. [21, 22]) *Let $\{\mathcal{K}(u)\}_{u\in\mathcal{M}}$ be a compact negatively invariant set for the cocycle (φ, ψ) such that $\bigcup_{u\in\mathcal{M}} \mathcal{K}(u) \supset \mathcal{K}$ and $\mathcal{K}(u) \subset \mathcal{K}(\varphi(u))$, $u \in \mathcal{M}$. Assume that the pair (A, B) is controllable and the pair (A, C) is observable. Suppose that there exist numbers $d \in (0, n]$, $\lambda > 0$, and $\delta > 0$ such that the following conditions hold.*

(1) *All eigenvalues of $\frac{1}{\lambda}A$ lie outside the unit circle.*

(2) $\frac{1}{2}\eta^*[(\partial_2\widetilde{\phi}(u, C^*v))^* + \partial_2\widetilde{\phi}(u, C^*v)]\eta \leqslant \delta|\eta|^2$, $\forall u \in \mathcal{M}, \forall v \in \mathcal{D}, \forall \eta \in \mathbb{R}^n$;

(3) $\operatorname{Re} W(\lambda z) + \delta W^*(\lambda z) W(\lambda z) \leqslant 0, \ \forall z \in \mathbb{C}, |z| = 1$;

(4) $\sup\limits_{\substack{u \in \mathcal{M}, \\ \upsilon \in \mathcal{K}(u)}} \frac{|\det \partial_2 \psi(u,\upsilon)|}{\kappa(u,\upsilon)^{n-1}} < 1.$

Then $\dim_H(\mathcal{K}) < d$ *for each* $u \in \mathcal{M}$.

To prove this theorem, we need the following lemma.

Lemma 9.1 *Let* $\delta > 0$ *and* $\lambda > 0$ *be real numbers with respect to which the pair* (A, B) *of system (9.76) is controllable and the pair* (A, C) *is observable, and let conditions* (1) $--$(3) *of Theorem 9.9 be satisfied. Then there exists a real negative definite* $n \times n$ *matrix* $P = P^*$ *such that*

$$\partial_2 \psi(u, \upsilon)^* P \partial_2 \psi(u, \upsilon) \leqslant \lambda^2 P, \quad \forall\, (u, \upsilon) \in \mathcal{M} \times \mathcal{D}.$$

Proof Consider the quadratic form

$$\mathcal{F}(\upsilon, \xi) := -\xi^* C^* \upsilon + \delta |C^* \upsilon|^2, \quad \xi \in \mathbb{R}^m, \quad \upsilon \in \mathbb{R}^n. \tag{9.77}$$

It follows from the assumptions of the lemma that the pair $(\frac{1}{\lambda} A, \sqrt{\delta} C)$ is observable and the inverse matrix $(zI - \frac{1}{\lambda} A)$ exists for all $z \in \mathbb{C}$ with $|z| = 1$. Consider the following Hermitian extension of the quadratic form $\mathcal{F}$:

$$\mathcal{F}_{\mathbb{C}}(\upsilon, \xi) = -\operatorname{Re}(\xi^* C^* \upsilon) + \delta |C^* \upsilon|^2, \quad \xi \in \mathbb{R}^m, \quad \upsilon \in \mathbb{R}^n.$$

By assumption 3) of the theorem we have

$$\operatorname{Re}\left[\xi^* C^* \left(zI - \frac{1}{\lambda} A\right)^{-1} \frac{1}{\lambda} B \xi\right] + \delta \left[C^* \left(zI - \frac{1}{\lambda} A\right)^{-1} \frac{1}{\lambda} B \xi\right]^2 \leqslant 0$$

for all $z \in \mathbb{C}$ with $|z| = 1$ and all $\xi \in \mathbb{C}^m$. It follows from the Kalman–Szegö frequency theorem (Theorem 2.10, Chap. 2) that there exists a matrix $P = P^*$ satisfying the inequality

$$\frac{1}{\lambda^2}(A\upsilon + B\xi)^* P (A\upsilon + B\xi) - \upsilon^* P \upsilon - \xi^* C^* \upsilon + \delta |C^* \upsilon|^2 \leqslant 0, \tag{9.78}$$

for all $(\upsilon, \xi) \in \mathbb{R}^n \times \mathbb{R}^m$. Let $\xi = 0$, then (9.78) acquires the form

$$\upsilon^* \left(\frac{1}{\lambda} A\right)^* P \frac{1}{\lambda} A \upsilon - \upsilon^* P \upsilon \leqslant -\delta |C^* \upsilon|^2, \quad \forall \upsilon \in \mathbb{R}^n.$$

Consequently, the matrix P is negative definite by the Lyapunov lemma.
For arbitrary $u \in \mathcal{M}$ and $\upsilon_1, \upsilon_2 \in \mathcal{D}$, set $\upsilon = \upsilon_1 - \upsilon_2$ and $\xi = \widetilde{\phi}(u, C^* \upsilon_1) - \widetilde{\phi}(u, C^* \upsilon_2)$. Then we obtain

$$\xi = \int_0^1 \partial_2 \widetilde{\phi}(u, C^* \upsilon_1 \tau + C^* \upsilon_2 (1-\tau)) C^* \upsilon \mathrm{d}\tau$$

and further

$$\upsilon^* C \xi = \int_0^1 \upsilon^* C \partial_2 \widetilde{\phi}(u, C^* \upsilon_1 \tau + C^* \upsilon_2 (1-\tau)) C^* \upsilon \mathrm{d}\tau,$$

$$= \int_0^1 \tfrac{1}{2} \upsilon^* C \left[\partial_2^* \widetilde{\phi}(u, C^* \upsilon_1 \tau + C^* \upsilon_2 (1-\tau)) + \partial_2 \widetilde{\phi}(u, C^* \upsilon_1 \tau + C^* \upsilon_2 (1-\tau)) \right] C^* \upsilon \mathrm{d}\tau.$$

It follows from the last relation and assumption 2) of the theorem that for $\eta := C^* \upsilon$ we have $\upsilon^* C \xi \leqslant \delta |C^* \upsilon|^2$ for all $\upsilon \in \mathbb{R}^n$. Thus, it follows from (9.78) that

$$\frac{1}{\lambda^2} (A\upsilon + B\xi)^* P (A\upsilon + B\xi) - \upsilon^* P \upsilon \leqslant 0, \quad \forall \upsilon \in \mathbb{R}^n. \tag{9.79}$$

Take $\upsilon \in \mathcal{D}$, $\upsilon_2 = \upsilon$, and $\upsilon_1 = \upsilon + h\bar{\upsilon}$, $h \in \mathbb{R}$, $\bar{\upsilon} \neq 0$ such that $\upsilon_1 \in \mathcal{D}$. We substitute this into inequality (9.79) and obtain

$$\begin{aligned} &\frac{1}{\lambda} \left(A\bar{\upsilon} + B \frac{\widetilde{\phi}(u, C^*(\upsilon + h\bar{\upsilon})) - \widetilde{\phi}(u, C^* \upsilon)}{h} \right)^* P \\ &\times \left(A\bar{\upsilon} + B \frac{\widetilde{\phi}(u, C^*(\upsilon + h\bar{\upsilon})) - \widetilde{\phi}(u, C^* \upsilon)}{h} \right) - \bar{\upsilon}^* P \bar{\upsilon} \leqslant 0. \end{aligned} \tag{9.80}$$

We pass to the limit as $h \to 0$ and use the relation $\partial_2 \psi(u, \upsilon) = A + B \partial_2 \widetilde{\phi}(u, C^* \upsilon) C^*$ to obtain

$$\bar{\upsilon}^* \partial_2 \psi(u, \upsilon)^* P \partial_2 \psi(u, \upsilon) \bar{\upsilon} \leqslant \lambda^2 \bar{\upsilon}^* P \bar{\upsilon}, \ \forall \bar{\upsilon} \in \mathbb{R}^n, \forall \upsilon \in \mathcal{D}, \forall u \in \mathcal{M}.$$

□

Proof of Theorem 9.9 . The proof follows from Theorem 9.8 and the last lemma. □

9.3.4 Upper Bound for Hausdorff Dimension of Invariant Sets and $\mathcal{B}$-attractors of Cocycles Generated by Ordinary Differential Equations on Manifolds

Let $(\mathcal{N}, g)$ be a finite dimensional Riemannian manifold. Consider the non-autonomous ordinary differential equation

$$\dot{\psi} = g(t, \psi), \tag{9.81}$$

where $g : \mathbb{R} \times \mathcal{N} \to T\mathcal{N}$ is a C^k-smooth ($k \geq 2$) non-autonomous vector field. We extend this system to a Bebutov flow. Define for this the hull $\mathcal{H}(g)$ of g as

$$\mathcal{H}(g) = \overline{\{g(\cdot + t, \cdot),\ t \in \mathbb{R}\}}$$

and the evaluation map $\widetilde{g}$ as

$$\dot{\psi} = \widetilde{g}(\varphi^t(u), \psi), \quad u \in \mathcal{M}, \tag{9.82}$$

where $\mathcal{M} := \mathcal{H}(g)$. The initial equation is contained in the extended system as $u = g$. Assume that equation (9.82) with initial condition $t_0 \in \mathbb{R},\ \upsilon_0 \in \mathcal{N}$ has a unique continuous solution $\upsilon(\cdot, t_0, \upsilon_0)$ defined on $\mathbb{R}_+$ and $\upsilon(t_0, \upsilon_0, \upsilon_0) = \upsilon_0$. Under this assumption equation (9.82) generates the cocycle $\big(\{\psi^t(u, \cdot)_{u\in\mathcal{M}, t\in\mathbb{R}_+}\}, (\mathcal{N}, \rho_{\mathcal{N}})\big)$ over the base flow $(\{\varphi^t\}_{t\in\mathbb{R}}, (\mathcal{M}, \rho_{\mathcal{M}}))$.
Now we define the linearization of the cocycle. Let $w(t, u_0, \upsilon_0)$ be a solution of the variation equation along the trajectory of the cocycle through the point $(u_0, \upsilon_0) \in \mathcal{M} \times \mathcal{N}$. This variation equation has the form

$$\begin{aligned} &\dot{w} = \nabla_2\widetilde{g}(\varphi^t(u_0), \psi^t(u_0, \upsilon_0))w, \\ &w(0, u_0, w_0) = w_0 \in T_{\upsilon_0}\mathcal{N} \end{aligned} \tag{9.83}$$

where $\nabla_2\widetilde{g}(\cdot, \cdot) : T\mathcal{N} \to T\mathcal{N}$ is the covariant derivative with respect to the second argument. Then

$$\partial_2\psi^t(u_0, \upsilon_0)(w_0) = w(t, u_0, w_0)$$

for all $t \in \mathbb{R}_+$. Thus, $\partial_2\psi^t(u_0, \upsilon_0)(w_0)$ is a fundamental matrix of the (9.83).
Let $\lambda_1(u, \upsilon) \geq \lambda_2(u, \upsilon) \geq \ldots \geq \lambda_n(u, \upsilon)$ be the eigenvalues of

$$\frac{1}{2}\left[\nabla_2\widetilde{g}(u, \upsilon) + \nabla_2\widetilde{g}(u, \upsilon)^*\right],$$

each eigenvalue appears p_i times, where p_i is the eigenvalue's algebraic multiplicity.

Theorem 9.10 (Ref. [23]) *Assume that*

(1) **(A7), (A8)** *with* $\mathcal{M} = \mathcal{H}(g)$ *holds;*
(2) *there exists a compact set* $\widetilde{\mathcal{K}} \subset \mathcal{N}$, *such that*

$$\overline{\bigcup_{u\in\mathcal{H}(g)} \mathcal{Z}(u)} \subset \widetilde{\mathcal{K}};$$

(3) *there are a continuous function* $V : \mathcal{H}(g) \times \mathcal{N} \to \mathbb{R}$, *such that the derivative along the trajectory* $\frac{d}{dt}V(\varphi^t(u), \psi^t(u, \upsilon)$ *exists, a number* $\tau > 0$ *and a number* $d \in (0, n],\ d = d_0 + s$, *where* $d_0 \in \{0, 1, \ldots, n-1\}, s \in (0, 1]$, *such that*

$$\mathcal{Z}(u) \subset \mathcal{Z}(\varphi^{\tau}(u));$$

(4) $$\int_0^{\tau} [\lambda_1(\varphi^t(u), \psi^t(u, \upsilon)) + \ldots + \lambda_{d_0}(\varphi^t(u), \psi^t(u, \upsilon)) + s\lambda_{d_0+1}(\varphi^t(u), \psi^t(u, \upsilon))$$

$$+\frac{d}{dt} V(\varphi^t(u), \psi^t(u, \upsilon))]dt < 0 \quad \textit{for all} \quad u \in \mathcal{H}(g), \upsilon \in \widetilde{\mathcal{K}}.$$

Then $\dim_H \mathcal{Z}(u) \le d$ *for all* $u \in \mathcal{H}(g)$.

Proof To prove Theorem 9.10, we need to show the existence of a function $\varkappa\colon \mathcal{H}(g) \times \mathcal{N} \to \mathbb{R}_+$, such that (9.52) holds.
Recall that $\partial_2 \psi^t(u, \upsilon)$ is a solution of the variation equation

$$\dot{w} = \nabla_2 \widetilde{g}(\varphi^t(u), \psi^t(u, \upsilon))w, \tag{9.84}$$

Lets fix $u \in \mathcal{H}(g)$ $k \in \mathbb{N}$. For all t we have

$$w(t) = \left|\partial_2 \psi^t(u, \upsilon)\upsilon_1 \wedge \ldots \wedge \partial_2 \psi^t(u, \upsilon)\upsilon_k\right|_{\Lambda^k T_{\psi^t(u,\upsilon)}\mathcal{N}} \cdot$$

With the use of the variation equation and Definition 7.7, Chap. 7 we get

$$\begin{aligned} \dot{w} = 2\langle\big[S(\varphi^t(u), \psi^t(u, \upsilon))\big]_k (\partial_2 \psi^t(u, \upsilon)\upsilon_1 \wedge \ldots \wedge \partial_2 \psi^t(u, \upsilon)\upsilon_k), \\ \partial_2 \psi^t(u, \upsilon)\upsilon_1 \wedge \ldots \wedge \partial_2 \psi^t(u, \upsilon)\upsilon_k\rangle_{\Lambda^k T_{\psi^t(u,\upsilon)}\mathcal{N}}, \quad \forall t \in [0, \tau]. \end{aligned} \tag{9.85}$$

From Proposition 7.9, Chap. 7, it follows that

$$\dot{w} \le 2\left(\lambda_1(\varphi^t(u), \psi^t(u, \upsilon)) + \ldots + \lambda_k(\varphi^t(u), \psi^t(u, \upsilon))\right) w(t), \tag{9.86}$$

for all $t \in [0, \tau]$. Thus

$$\begin{aligned} &|\partial_2 \psi^{\tau}(u, \upsilon)\upsilon_1 \wedge \ldots \wedge \partial_2 \psi^{\tau}(u, \upsilon)\upsilon_k|_{\bigwedge^k T_{\psi^{\tau}(u,\upsilon)}\mathcal{N}} \\ &\le |\upsilon_1 \wedge \ldots \wedge \upsilon_k|_{\bigwedge^k T_{\upsilon}\mathcal{N}} \cdot \exp\left\{ \int_0^{\tau} [\lambda_1(\varphi^t(u), \psi^t(u, \upsilon)) + \ldots + \right. \\ &\left. + \lambda_k(\varphi^t(u), \psi^t(u, \upsilon))]\, dt \right\}. \end{aligned} \tag{9.87}$$

By the Fischer-Courant theorem (Theorem 7.1, Chap. 7) we have

$$\begin{aligned}
&\alpha_1(\tau)^2+\ldots+\alpha_k(\tau)^2=\lambda_1\left(\wedge^k\left[(\partial_2\psi^\tau(u,\upsilon))^*\partial_2\psi^\tau(u,\upsilon)\right]\right) \\
&=\sup_{w\in\wedge^k T_\upsilon\mathcal{N},\ |w|_{\wedge^k T_\upsilon \mathcal{N}}=1} |\wedge^k \partial_2\psi^\tau(u,\upsilon)w|^2_{\wedge^k T_{\psi^\tau(u,\upsilon)}\mathcal{N}} \\
&=\sup_{w_1,\ldots,w_k\in T_w\mathcal{N},\ |w_i|_{T_\upsilon\mathcal{N}}=1} |\partial_2\psi^\tau(u,\upsilon)w_1\wedge\ldots\wedge\partial_2\psi^\tau(u,\upsilon)w_k|^2_{\wedge^k T_{\psi^\tau(u,\upsilon)}\mathcal{N}} \\
&\le \exp\left\{2\int_0^\tau[\lambda_1(\varphi^t(u),\psi^t(u,\upsilon))+\ldots+\lambda_k(\varphi^t(u),\psi^t(u,\upsilon))]\,dt\right\}.
\end{aligned} \tag{9.88}$$

By definition, for all $d\in(0,n],\ d=d_0+s,\ d_0\in\{0,1,\ldots,n-1\},\ s\in(0,1]$ we have

$$\omega_d(\partial_2\psi^\tau(u,\upsilon))=\omega_{d_0}^{1-s}(\partial_2\psi^\tau(u,\upsilon))\omega^s_{d_0+1}(\partial_2\psi^\tau(u,\upsilon)).$$

Thus

$$\begin{aligned}
\omega_d(\partial_2\psi^\tau(u,\upsilon)) &\le \exp\left\{(1-s)\int_0^\tau[\lambda_{1,u}(t,\upsilon)+\ldots+\lambda_{d_0,u}(t,\upsilon)]\,dt\right. \\
&\quad \left.+s\int_0^\tau[\lambda_{1,u}(t,\upsilon)+\ldots+\lambda_{d_0,u}(t,\upsilon)+\lambda_{d_0+1,u}(t,\upsilon)]\,dt\right\} \\
&=\exp\left\{\int_0^\tau[\lambda_{1,u}(t,\upsilon)+\ldots+\lambda_{d_0,u}(t,\upsilon)+\lambda_{d_0+1,u}(t,\upsilon)]\,dt\right\}
\end{aligned} \tag{9.89}$$

We choose $\varkappa\colon \mathcal{H}(g)\times\mathcal{N}\to\mathbb{R}_+$ as $\varkappa(u,\upsilon):=\exp\{V(u,\upsilon)\}$ for all $u\in\mathcal{H}(g)$, $\upsilon\in\mathcal{N}$. Then

$$\begin{aligned}
\frac{\varkappa(\varphi^\tau(u),\psi^\tau(u,\upsilon))}{\varkappa(u,\upsilon)} &= \exp\{V(\varphi^\tau(u),\psi^\tau(u,\upsilon))\} \\
&= \exp\left\{\int_0^\tau \frac{d}{dt}V(\varphi^t(u),\psi^t(u,\upsilon))\,dt.\right\}
\end{aligned} \tag{9.90}$$

From (9.89) and (9.90) we get

$$\begin{aligned}
&\frac{\varkappa(\varphi^\tau(u),\psi^\tau(u,\upsilon))}{\varkappa(u,\upsilon)}\omega_d(\partial_2\psi^\tau(u,\upsilon)) \\
&\le \exp\left\{\int_0^\tau[\lambda_1(\varphi^t(u),\psi^t(u,\upsilon))+\ldots+\lambda_{d_0}(\varphi^t(u),\psi^t(u,\upsilon))\right. \\
&\left.+s\lambda_{d_0+1}(\varphi^t(u),\psi^t(u,\upsilon))+\frac{d}{dt}V(\varphi^t(u),\psi^t(u,\upsilon))]\,dt\right\}<\exp(0)=1
\end{aligned}$$

and by Theorem 9.6 we have $\dim_H \mathcal{Z}(u)\le d$ for all $u\in\mathcal{H}(g)$. □

9.3.5 Upper Bounds for the Hausdorff Dimension of Attractors of Cocycles Generated by Differential Equations on the Cylinder

Suppose that $\phi(\cdot,\cdot):\mathbb{R}\times\mathbb{R}\to\mathbb{R}$ is a bounded, smooth and 2π-periodic with respect to the second argument function, A is a stable $(n-1)\times(n-1)$-matrix (i.e. for all eigenvalues of A we have $\operatorname{Re}\lambda<0$); c and b are $(n-1)$-vectors.
Let us consider the differential equation ([1, 2, 20])

$$\dot{\theta}=c^{*}y,\quad \dot{y}=Ay+b\phi(t,\theta). \tag{9.91}$$

It is well-known that instead of (9.91) we can consider a differential equation on the cylinder $(\mathcal{C},\rho_{\mathcal{C}}))$. With $\upsilon=(\theta,y)$ and $g(t,\upsilon)=(c^{*}y,Ay+b\phi(t,\theta))^{T}$ equation (9.91) can be represented as

$$\dot{\upsilon}=g(t,\upsilon),\ \ \upsilon\in\mathcal{C}. \tag{9.92}$$

From Subsect. 9.3.4 we can consider the evaluation map $\widetilde{g}$ and the equation

$$\dot{\upsilon}=\widetilde{g}(\varphi^{t}(u),\upsilon), \tag{9.93}$$

which generates a cocycle $(\{\psi^{t}(u,\cdot)\}_{u\in\mathcal{H}(g)},(\mathcal{C},\rho_{\mathcal{C}}))$ over the Bebutov flow $(\{\varphi^{t}(\cdot)\}_{t\in\mathbb{R}},(\mathcal{H}(g),\rho_{\mathcal{H}(g)}))$, where $\mathcal{H}(g)$ is the hull of g.
For a fixed $u_0\in\mathcal{H}(g)$ let $\upsilon(\cdot,t_0,u_0,\upsilon_0)=(\theta(\cdot,t_0,u_0,\theta_0,y_0),y(\cdot,t_0,u_0,\theta_0,y_0))$ be the solution of (9.93) such that $\upsilon(t_0,t_0,u_0,\upsilon_0)=u_0$, where $\upsilon_0=(y_0,\theta_0)$.
We have for $t\ge t_0$ the estimates

$$\begin{aligned}|\upsilon(t,t_0,u_0,\upsilon_0)|_{C}&\le|\theta(t,t_0,u_0,\theta_0,y_0)|+|y(t,t_0,u_0,\theta_0,y_0)|\\ &\le 2\pi+|y(t,t_0,q_0,\theta_0,y_0)|,\end{aligned}$$

where $|\cdot|$ is the Euclidean norm.
Since A is a stable matrix, there exist $\varepsilon>0$, $C_2>0$ such that $|e^{At}|\le C_2e^{-\varepsilon t}$ for all $t>0$. By the Cauchy formula we have

$$|y(t,t_0,u_0,\theta_0,y_0)|\le|e^{(t-t_0)A}y_0|+\left|\int_{t_0}^{t}e^{(t-t_0-\tau)A}b\phi(\tau,\theta)d\tau\right|.$$

From the stability of A we deduce that $|e^{(t-t_0)A}y_0|\to 0$ as $t\to+\infty$. From the boundedness of $\phi(t,\theta)$ we get that there is a $C_1>0$ such that $|\phi(t,\theta)|\le C_1$ for all t,θ. Then for all $t>t_0$ we have

$$\left|\int_{t_0}^{t} e^{(t-t_0-\tau)A} b\phi(\tau,\theta)d\tau\right| \leq \int_{t_0}^{t} |e^{(t-t_0-\tau)A}|\,|b|\,|\phi(\tau,\theta)|d\tau$$

$$\leq |b|C_1 \int_{t_0}^{t} |e^{(t-t_0-\tau)A}|d\tau \leq |b|C_1C_2 \int_{t_0}^{t} e^{(t-t_0-\tau)}d\tau \leq \frac{1}{\varepsilon}|b|C_1C_2.$$

For all $u_0 \in \mathcal{H}(g)$ we get

$$\lim_{t\to+\infty} \sup |\upsilon(t,t_0,u_0,\upsilon_0)|_C \leq \frac{1}{\varepsilon}|b|C_1C_2 2\pi. \tag{9.94}$$

Thus, the cocycle $(\{\varphi^t(\cdot)\}_{t\in\mathbb{R}}, (\mathcal{H}(g), \rho_{\mathcal{H}(g)}))$ has a unique global $\mathcal{B}$-forward attractor $\widetilde{\mathcal{A}}$.

Example 9.5 Consider the differential equation

$$\dot{\theta} = cy, \quad \dot{y} = ay + b\phi(t,\theta), \tag{9.95}$$

where $a < 0$, $b \in \mathbb{R}$, $c > 0$ are some parameters, $\phi(\cdot,\cdot) : \mathbb{R}\times\mathbb{R} \to \mathbb{R}$ is a bounded and 2π-periodic in the second argument smooth function. Instead of (9.95) we consider the family of systems

$$\dot{\theta} = cy\,, \quad \dot{y} = ay + b\widetilde{\phi}(\varphi^t(u),\theta) \tag{9.96}$$

where $\widetilde{\phi} : \mathbb{R}\times S^1 \to \mathbb{R}$ is the extension of ϕ to the hull $\mathcal{H}(\phi)$. Let us write system (9.96) in the form

$$\dot{\theta} = cy\,, \quad \dot{y} = ay + b\phi_u(t,\theta) \tag{9.97}$$

From Theorem 9.2 it follows that equation (9.96) generates a cocycle (φ, ψ) on the phase space $\mathbb{R}\times S^1$. Using Theorem 9.1 we see that this cocycle has a global $\mathcal{B}$-attractor $\widetilde{\mathcal{A}}$. In order to get a dimension estimate we use Theorem 9.10 with $V \equiv 1$. Suppose that $\lambda_{1,u}(t,y^1,y^2) \geq \lambda_{2,u}(t,y^1,y^2)$ are the eigenvalues of the symmetrized matrix $\frac{1}{2}\left[\nabla_2 g_u(t,\upsilon) + \nabla_2 g_u^*(t,\upsilon)\right]$ where

$$g_u(t,\upsilon) = \begin{pmatrix} cy \\ ay + b\phi_u(t,\theta) \end{pmatrix}$$ with $\upsilon = (y^1,y^2) = (\theta,y)$ in some chart.

Define $\eta_u(t,y^1) := \frac{\partial}{\partial y^1}\phi_u(t,y^1)$.

(a) The Case of a Trivial Metric Tensor

Here we consider the trivial metric tensor $(g_{ij}) = \begin{pmatrix} 1 & 0 \\ 0 & 1 \end{pmatrix}$. Then the symmetrized Jacobian matrix has the form

$$\frac{1}{2}\begin{pmatrix} 0 & c + b\eta_u(t, y^1) \\ c + b\eta_u(t, y^1) & 2a \end{pmatrix}.$$

Its eigenvalues are $\lambda_{1,2;u}(t, y^1) = \frac{1}{2}a \pm \sqrt{a^2 + c + (b\eta_u(t, y^1))^2}$. From Theorem 9.10 it follows that in order to show that the Hausdorff dimension of the attractor is smaller than $1 + s$ it is sufficient to show that

$$\lambda_{1,u}(t, y^1, y^2) + s\lambda_{2,u}(t, y^1, y^2) < 0 \quad \text{for all} \quad t > 0$$
$$\text{and} \quad (y^1, y^2) \in \mathcal{R}(\widetilde{\mathcal{A}}), u \in \mathcal{H}(g).$$

Here $\mathcal{R}(\widetilde{\mathcal{A}})$ denotes the (y^1, y^2)-components of the attractor $\widetilde{\mathcal{A}}$. The last inequality is satisfied if

$$s > \sup \frac{a + \sqrt{a^2 + (c + b)\eta_u(t, y^1)^2}}{\sqrt{a^2 + (c + b\eta_u(t, y^1))^2} - a},$$

where the supremum is taken over $(y^1, y^2) \in \mathcal{R}(\widetilde{\mathcal{A}})$ and $u \in \mathcal{H}(g)$.
Assume in the following that $\eta_u(t, y^1) = \xi_u(t)\sin(y^1)$ and there exist $0 < \kappa_1 < \kappa_2$ such that

$$\kappa_1 < |b\xi_u(t)| < \kappa_2, \quad \forall u \in \mathcal{H}(g), \ \forall t \geq 0. \tag{9.98}$$

Then if $c > \frac{\kappa_1 + \kappa_2}{2}$ we have

$$s \geq \frac{a + \sqrt{a^2 + (c + \kappa_2)^2}}{\sqrt{a^2 + (c - \kappa_2)^2} - a}.$$

This estimate holds only if

$$\frac{a + \sqrt{a^2 + (c + \kappa_2)^2}}{\sqrt{a^2 + (c + \kappa_2)^2} - a} < 1$$

or if

$$-2a > \sqrt{a^2 + (c + \kappa_2)^2} - \sqrt{a^2 + (c + \kappa_2)^2}.$$

After taking the square of the left and right sides and dividing by 2 we get

$$\sqrt{a^2 + (c + \kappa_2)^2} \cdot \sqrt{a^2 + (c + \kappa_2)^2} > c^2 + \kappa_2^2 - a^2.$$

Hence $a^2 > \frac{c^2\kappa_2^2}{c^2 + \kappa_2^2}$. It follows that if $c \leq \frac{\kappa_1 + \kappa_2}{2}$ the inequality

$$s \geq \frac{a + \sqrt{a^2 + (c + \kappa_2)^2}}{\sqrt{a^2 + (c - \kappa_2)^2} - a}$$

is satisfied for $a^2 > \frac{c^2\kappa_2^2}{c^2+\kappa_2^2}$. This means that for

$$a^2 > \frac{((c+\kappa_2)^2 - (c-\kappa_1)^2)^2}{8((c+\kappa_2)^2 + (c-\kappa_1)^2)}$$

the estimate

$$s \geq \frac{a + \sqrt{a^2 + (c+\kappa_2)^2}}{\sqrt{a^2 + (c-\kappa_2)^2} - a}$$

holds. From this it follows that

$$\dim_H \widetilde{\mathcal{A}} \leq 1 + s\,. \tag{9.99}$$

(b) The Case of a Non-trivial Metric Tensor

Now we consider the non-trivial metric tensor

$$(g_{ij}(\theta)) = \begin{pmatrix} 2+\sin\theta & 0 \\ 0 & r \end{pmatrix}, \theta \in S^1,$$

where $r > 0$. The symmetrized Jacobian matrix has the form

$$\frac{1}{2}\begin{pmatrix} cy^2\cos(y^1) & c + b\eta_u(t, y^1) \\ c + b\eta_u(t, y^1) & 2a \end{pmatrix}$$

and its eigenvalues are

$$\lambda_{1,2;u}(t, y^1, y^2) = \frac{1}{2}\Bigg[a + \frac{cy^2\cos(y^1)}{2(2+\sin(y^1))} \pm \sqrt{\left(a + \frac{cy^2\cos(y^1)}{2(2+\sin(y^1))}\right)^2 + (c + b\eta_u(t, y^1))^2 - \frac{2cy^2\cos(y^1)}{2+\sin(y^1)}}\,.\Bigg]$$

Again we assume that $\eta_u(t, y^1) = \xi_u(t)\sin(y^1)$ and $\xi_u(\cdot)$ satisfies (9.98). From Theorem 9.10 it follows that in order to show that (9.99) is satisfied for the attractor $\widetilde{\mathcal{A}}$ it is sufficient to show that

$$\lambda_{1,u}(t, y^1, y^2) + s\lambda_{2,u}(t, y^1, y^2) < 0 \quad \text{for all} \quad t \geq 0 \quad \forall (y^1, y^2) \in \mathcal{R}(\widetilde{\mathcal{A}}), \forall u \in \mathcal{H}(g).$$

Hence we have to show that for these arguments

$$s > \sup \frac{a + \frac{cy^2\cos(y^1)}{2(2+\sin(y^1))} + \sqrt{\left(a + \frac{cy^2\cos(y^1)}{2(2+\sin(y^1))}\right)^2 + (c + b\eta_u(t, y^1, y^2))^2 - \frac{2cy^2\cos(y^1)}{2+\sin(y^1)}}}{\sqrt{\left(a + \frac{cy^2\cos(y^1)}{2(2+\sin(y^1))}\right)^2 + (c + b\eta_u(t, y^1))^2 - \frac{2cy^2\cos(y^1)}{2+\sin(y^1)}} - a - \frac{cy^2\cos(y^1)}{2(2+\sin(y^1))}},$$

where the supremum is taken over $(y^1, y^2) \in \mathcal{R}(\widetilde{\mathcal{A}})$ and $u \in \mathcal{H}(g)$. Since the attractor is compact there exist a $\kappa_3 > 0$ such that $|y^2| < \kappa_3$ for all $y^2 \in \mathrm{pr}_2(\mathcal{R}(\mathcal{A}))$. Clearly, we have $\left|\frac{\cos(y^1)}{2+\sin(y^1)}\right| \leq \frac{1}{\sqrt{3}}$ for all $y^1 \in \mathrm{pr}_1(\mathcal{R}(\widetilde{\mathcal{A}}))$. Suppose that $a < \frac{-\kappa_3 c}{2\sqrt{3}}$ and assume the inequality (9.98).
It follows that for $c > \frac{\kappa_1+\kappa_2}{2}$ we have

$$s > \frac{a + \frac{\kappa_3 c}{2\sqrt{3}} + \sqrt{a - \frac{\kappa_3 c}{2\sqrt{3}} + (c-\kappa_2)^2 - \frac{2\kappa_3 c}{2\sqrt{3}}}}{\sqrt{\left(a - \frac{\kappa_3 c}{2\sqrt{3}}\right)^2 + (c+\kappa_2)^2 - \frac{2\kappa_3 c}{2\sqrt{3}}} - a - \frac{\kappa_3 c}{2\sqrt{3}}}.$$

This estimate holds only if

$$\frac{a + \frac{\kappa_3 c}{2\sqrt{3}} + \sqrt{\left(a - \frac{\kappa_3 c}{2\sqrt{3}}\right)^2 + (c+\kappa_2)^2 - \frac{2\kappa_3 c}{2\sqrt{3}}}}{\sqrt{\left(a - \frac{\kappa_3 c}{2\sqrt{3}}\right)^2 + (c-\kappa_2)^2 - \frac{2\kappa_3 c}{2\sqrt{3}}} - a - \frac{\kappa_3 c}{2\sqrt{3}}} < 1$$

or

$$-2\left(a + \frac{\kappa_3 c}{2\sqrt{3}}\right) > \sqrt{\left(a - \frac{\kappa_3 c}{2\sqrt{3}}\right)^2 + (c+\kappa_2)^2 - \frac{2a\kappa_3 c}{\sqrt{3}}} - \sqrt{\left(a + \frac{\kappa_3 c}{2\sqrt{3}}\right)^2 + (c-\kappa_2)^2 - \frac{2a\kappa_3 c}{\sqrt{3}}}.$$

After taking squares and dividing by 2, we get

$$\sqrt{\left(a - \frac{\kappa_3 c}{2\sqrt{3}}\right)^2 + (c+\kappa_2)^2 - \frac{2a\kappa_3 c}{\sqrt{3}}} \cdot \sqrt{\left(a + \frac{\kappa_3 c}{2\sqrt{3}}\right)^2 + (c-\kappa_2)^2 - \frac{2a\kappa_3 c}{\sqrt{3}}} < c^2 + \kappa_2^2 - (a + \frac{\kappa_3 c}{2\sqrt{3}})^2 - \frac{a\kappa_3 c}{\sqrt{3}},$$

which is satisfied for $a < \frac{1}{2}\left(\frac{-2\kappa_3 c}{\sqrt{3}} - \sqrt{\kappa_3^2 c^2 + 4(c^2+\kappa_2^2)}\right)$.

Thus if $c > \frac{\kappa_1+\kappa_2}{2}$ and $a < \frac{1}{2}(\frac{-2\kappa_3 c}{\sqrt{3}} - \sqrt{\kappa_3^2 c^2 + 4(c^2+\kappa_2^2)})$, we have the estimate (9.99) provided that

$$s > \frac{a + \frac{\kappa_3 c}{2\sqrt{3}} + \sqrt{\left(a - \frac{\kappa_3 c}{2\sqrt{3}}\right)^2 + (c+\kappa_2)^2 - \frac{2\kappa_3 c}{2\sqrt{3}}}}{\sqrt{\left(a - \frac{\kappa_3 c}{2\sqrt{3}}\right)^2 + (c-\kappa_2)^2 - \frac{2\kappa_3 c}{2\sqrt{3}}} - a - \frac{\kappa_3 c}{2\sqrt{3}}}. \tag{9.100}$$

If $c \le \frac{\kappa_1+\kappa_2}{2}$ we have (9.100) provided that

$$s > \frac{a + \frac{\kappa_3 c}{2\sqrt{3}} + \sqrt{\left(a - \frac{\kappa_3 c}{2\sqrt{3}}\right)^2 + (c+\kappa_2)^2 - \frac{2\kappa_3 c}{2\sqrt{3}}}}{\sqrt{\left(a - \frac{\kappa_3 c}{2\sqrt{3}}\right)^2 + (c-b_1)^2 - \frac{2\kappa_3 c}{2\sqrt{3}}} - a - \frac{\kappa_3 c}{2\sqrt{3}}}.$$

This holds if

$$\frac{a + \frac{\kappa_3 c}{2\sqrt{3}} + \sqrt{\left(a - \frac{\kappa_3 c}{2\sqrt{3}}\right)^2 + (c+\kappa_2)^2 - \frac{2\kappa_3 c}{2\sqrt{3}}}}{\sqrt{\left(a - \frac{\kappa_3 c}{2\sqrt{3}}\right)^2 + (c-b_1)^2 - \frac{2\kappa_3 c}{2\sqrt{3}}} - a - \frac{\kappa_3 c}{2\sqrt{3}}} < 1$$

or if

$$-2\left(a + \frac{\kappa_3 c}{2\sqrt{3}}\right) > \sqrt{\left(a - \frac{\kappa_3 c}{2\sqrt{3}}\right)^2 + (c+\kappa_2)^2 - \frac{2a\kappa_3 c}{\sqrt{3}}} - \sqrt{\left(a + \frac{\kappa_3 c}{2\sqrt{3}}\right)^2 + (c-b_1)^2 + \frac{2a\kappa_3 c}{\sqrt{3}}}.$$

After taking squares we get

$$2\sqrt{\left(a - \frac{\kappa_3 c}{2\sqrt{3}}\right)^2 + (c+\kappa_2)^2 - \frac{2a\kappa_3 c}{\sqrt{3}}} \cdot \sqrt{\left(a + \frac{\kappa_3 c}{2\sqrt{3}}\right)^2 + (c-\kappa_1)^2 + \frac{2a\kappa_3 c}{\sqrt{3}}} < (c+\kappa_2)^2 + (c-\kappa_1)^2 - 2\left(a + \frac{\kappa_3 c}{2\sqrt{3}}\right)^2 - \frac{2a\kappa_3 c}{\sqrt{3}}.$$

Obviously the last inequality holds if

$$a < \frac{1}{4}\left(-\frac{4\kappa_3 c}{\sqrt{3}} - \sqrt{\frac{14}{3}\kappa_3^2 c^2 + 8((c+\kappa_2)^2 + (c-\kappa_1)^2}\right). \tag{9.101}$$

Thus if $c \le \frac{\kappa_1+\kappa_2}{2}$ and (9.101) are satisfied we have the estimate (9.99), provided that

$$s > \frac{a + \frac{\kappa_3 c}{2\sqrt{3}} + \sqrt{\left(a - \frac{\kappa_3 c}{2\sqrt{3}}\right)^2 + (c+\kappa_2)^2 - \frac{2\kappa_3 c}{2\sqrt{3}}}}{\sqrt{\left(a - \frac{\kappa_3 c}{2\sqrt{3}}\right)^2 + (c-\kappa_1)^2 - \frac{2\kappa_3 c}{2\sqrt{3}}} - a - \frac{\kappa_3 c}{2\sqrt{3}}}.$$

9.4 Discrete-Time Cocycles on Fibered Spaces

9.4.1 Definition of Cocycles on Fibered Spaces

Assume that $\big(\{\varphi^k\}_{k\in\mathbb{Z}}, (\mathcal{M}, \rho_{\mathcal{M}})\big)$ is a dynamical system on the complete metric space $(\mathcal{M}, \rho_{\mathcal{M}})$ which is called again *base system*. Let $\{(\mathcal{N}(u), \rho_u)\}_{u\in\mathcal{M}}$ be a family of complete metric spaces. The family of maps $\psi^k(u, \cdot) : \mathcal{N}(u) \to \mathcal{N}(\varphi^k(u))$ is called discrete-time *cocycle over the base system* ([12, 15]) if the following conditions are satisfied:

(1) $\psi^0(u, \cdot) = \mathrm{id}_{\mathcal{N}(u)}$;
(2) $\psi^{k+j}(u, \cdot) = \psi^k(\varphi^j(u), \psi^j(u, \cdot)), \quad \forall k, j \in \mathbb{Z}_+\,,\ u \in \mathcal{M}\,;$
(3) For any $k \in \mathbb{Z}_+$, $u \in \mathcal{M}$ the map $\psi^k(u, \cdot) : \mathcal{N}(u) \to \mathcal{N}(\varphi^k(u))$ is continuous.

Example 9.6 Suppose $\mathcal{M}$ is a C^s-smooth m-dimensional Riemannian manifold, $\rho_{\mathcal{M}}$ is the metric, $\varphi : \mathcal{M} \to \mathcal{M}$ is a C^r-diffeomorphisms, $1 \le r \le s$. Introduce the base system $\big(\{\varphi^k\}_{k\in\mathbb{Z}}, (\mathcal{M}, \rho_{\mathcal{M}})\big)$ by

$$\varphi^k = \begin{cases} \underbrace{\varphi \circ \cdots \circ \varphi}_{k\text{-times}}, & \text{if } k \in \mathbb{N}\,, \\ \mathrm{id}_{\mathcal{M}}\,, & \text{if } k = 0\,, \\ \underbrace{\varphi^{-1} \circ \cdots \circ \varphi^{-1}}_{-k\text{-times}}, & \text{if } -k \in \mathbb{N}\,. \end{cases}$$

Let us consider the differential $d_u\varphi^k : T_u\mathcal{M} \to T_{\varphi^k(u)}\mathcal{M}$ given by (see Appendix A)

$$d_u\varphi^k([u, x, \xi]) := \Big[\varphi^k(u), y, (y \circ \varphi^k \circ x^{-1})'(x(u))\xi\Big]$$

where x is a chart around u, y is a chart around $\varphi^k(u)$ and $\xi \in \mathbb{R}^m$ is an arbitrary vector. Introduce the map $\psi^k(u, v) := d_u\varphi^k(v)$, where $v = [u, x, \xi] \in T_u\mathcal{N}$, and the spaces $\mathcal{N}(u) := T_uQ$. It follows that

$$\psi^k(u, \cdot) : \mathcal{N}(u) \to \mathcal{N}(\varphi^k(u)) \qquad \forall k \in \mathbb{Z}_+, \forall u \in \mathcal{M}\,. \tag{9.102}$$

As it is easy to see we have the cocycle property

$$\begin{aligned} &\psi^{k+j}(u, v) = d_u\varphi^{k+j}(v) = d_{\varphi^j(u)}\varphi^k(d_u\varphi^j(v)) \\ =\ &\psi^k(\varphi^j(u), d_u\varphi^j(v)) = \psi^k(\varphi^j(u), \psi^j(u, v)), \qquad \forall k, j \in \mathbb{Z}_+. \end{aligned}$$

9.4.2 Global Pullback Attractors

Let $\widetilde{\mathcal{D}}$ be a family of parametrized subsets $\{\mathcal{D}(u)\}_{u\in\mathcal{M}}, \mathcal{D}(u) \subset \mathcal{N}(u)$. We call a system $\widetilde{\mathcal{D}}$ *inclusion closed* ([12]) if it fulfills the properties.

(1) If $\mathcal{D} \in \widetilde{\mathcal{D}}$ then for any $u \in \mathcal{M}$ the set $\mathcal{D}(u) \subset \mathcal{N}(u)$ is non-empty.
(2) If $\mathcal{D} \in \widetilde{\mathcal{D}}$ and $\emptyset \neq \mathcal{D}'(u) \subset \mathcal{D}(u)$ for any $u \in \mathcal{M}$ then $\widetilde{\mathcal{D}'} \in \widetilde{\mathcal{D}}$.

A parametrized set $\widetilde{\mathcal{Z}} = \{\mathcal{Z}(u)\}_{u\in\mathcal{M}} \in \widetilde{\mathcal{D}}$ is called *$\mathcal{D}$-absorbing* if for any $\mathcal{D} \in \widetilde{\mathcal{D}}, u \in \mathcal{M}$, there exists $k_0 = k(u, \mathcal{D})$ such that $\psi^k(\varphi^k(u), \mathcal{D}(\varphi^k(u)) \subset \mathcal{Z}(u)$ if $k \geq k_0$. It follows from this definition that the domain of the cocycle operators in (9.102) in $\mathcal{N}(\varphi^{-k}(u)), k \geq 0$, is depending on k. But the image set is always contained in $\mathcal{N}(u)$. Thus, we can study ω-limit sets contained in a fixed fibre $\mathcal{N}(u)$. A parametrized family $\widetilde{\mathcal{A}} = \{\mathcal{A}(u)\}_{u\in\mathcal{M}} \in \widetilde{\mathcal{D}}$ is called *global $\mathcal{D}$- pullback attractor* if for any $u \in \mathcal{M}$

$$\psi^k(u, \mathcal{A}(u)) = \mathcal{A}(\varphi^k(u)) \quad \text{for} \quad k \in \mathbb{Z}_+ \qquad \text{and}$$

$$\lim_{k\to\infty} \text{dist}_u(\overline{\psi^k(\varphi^{-k}(u), \mathcal{D}(\varphi^{-k}(u))}^{\rho_u}, \quad \mathcal{A}(u)) = \{0\}$$

for any $\mathcal{D} \in \widetilde{\mathcal{D}}$ where $\text{dist}_u(\mathcal{Z}_1, \mathcal{Z}_2) = \sup_{v\in\mathcal{Z}_1} \inf_{w\in\mathcal{Z}_2} \rho_u(v, w)$.

Theorem 9.11 (Ref. [12]) *Let $\{(\mathcal{N}(u), \rho_u)\}_{u\in\mathcal{M}}$ be a family of complete metric spaces. The family of operators $\{\psi^k(u, \cdot)\}_{u\in\mathcal{M}}$ is defined to be a cocycle over the flow $\{\varphi^k\}_{k\in\mathbb{Z}}$ fulfilling (9.102). The maps $\psi^k(u, \cdot)$, $u \in \mathcal{M}$, $k \in \mathbb{Z}_+$, are assumed to be continuous. Moreover, we assume the existence of a $\mathcal{D}$-absorbing set $\widetilde{\mathcal{Z}} = \{\mathcal{Z}(u)\}_{u\in\mathcal{M}}$. Each of these sets $\mathcal{Z}(u)$, $u \in \mathcal{M}$, is supposed to be compact. Then the cocycle $\{\psi^k(u, \cdot)\}_{u\in\mathcal{M}, k\in\mathbb{Z}_+}$ has a unique global $\mathcal{D}$-pullback attractor*

$$\mathcal{A}(u) = \bigcap_{j\geq k_0(u,\mathcal{Z})} \overline{\bigcup_{k\geq j} \psi^k(\varphi^{-k}(u), \mathcal{Z}(\varphi^{-k}(u))}^{\rho_u}$$

where $k_0(u, \mathcal{Z})$ is given in the above definition.

9.4.3 The Topological Entropy of Fibered Cocycles

The basic properties of topological entropy for cocycles are considered in [9, 16, 25]. Topological entropy estimates for systems with multiple time are considered in [4].

(a) The Characterization by Open Covers

Suppose that $(\mathcal{N}, \rho_{\mathcal{N}})$ and $(\mathcal{N}', \rho_{\mathcal{N}'})$ are compact metric spaces, $\psi : \mathcal{N} \to \mathcal{N}'$ is a continuous map. Suppose also that $\mathfrak{U}$ and $\mathfrak{U}'$ are open covers of $\mathcal{N}$ and $\mathcal{N}'$ respectively.

Denote by $N(\mathfrak{U})$(resp. $N(\mathfrak{U}')$) the minimal number of elements $\mathfrak{U}$(resp. $\mathfrak{U}'$) necessary for the covering of $\mathcal{N}$(resp. $\mathcal{N}'$) and define

$$H(\mathfrak{U}) := \log N(\mathfrak{U}) \qquad (9.103)$$
$$(\text{resp.} H(\mathfrak{U}') := \log N(\mathfrak{U}')) .$$

The next lemma is an analogon of Lemma 3.6, Chap. 3.

Lemma 9.2 *Suppose $\psi : \mathcal{N} \to \mathcal{N}'$ is continuous, $(\mathcal{N}, \rho_{\mathcal{N}})$ and $(\mathcal{N}', \rho'_{\mathcal{N}})$ are compact metric spaces, $\mathfrak{U}'$ and $\mathfrak{V}'$ are open covers of $\mathcal{N}'$. Then the following holds:*

(1) $\psi^{-1}(\mathfrak{U}' \vee \mathfrak{V}') = \psi^{-1}\mathfrak{U}' \vee \psi^{-1}\mathfrak{V}'$;
(2) $H(\mathfrak{U}') \le H(\mathfrak{U}' \vee \mathfrak{V}') \le H(\mathfrak{U}') + H(\mathfrak{V}')$;
(3) $H(\psi^{-1}\mathfrak{U}') \le H(\mathfrak{U}')$;
(4) $H(\psi^{-1}\mathfrak{U}') = H(\mathfrak{U}')$ *if ψ is surjective.*

Proof The proof of the lemma follows directly from the definitions. □

Let us assume that the parametrized metric spaces $\{(\mathcal{N}(u), \rho_u)\}_{u\in\mathcal{M}}$ are compact. Suppose that $\gamma(u) = \{\varphi^k(u), k \in \mathbb{Z}\}$ is the orbit of φ through $u \in \mathcal{M}$. Let us fix a point $u \in \mathcal{M}$ and consider the family of open covers $\mathfrak{U}_u := \{\mathfrak{U}(u')\}_{u'\in\gamma(u)}$ such that $\mathfrak{U}(u')$ is an open cover of $\mathcal{N}(u')$, where $u' \in \gamma(u)$. We define the *topological entropy of the cocycle* (φ, ψ) *along the orbit through u with respect to the family of open covers* $\mathfrak{U}_u$ as

$$h_{\text{top}}((\varphi, \psi), u, \mathfrak{U}_u) := \limsup_{m\to\infty} \log N\left(\bigvee_{k=0}^{m-1} \psi^{-k}(u, \mathfrak{U}(\varphi^k(u))\right).$$

It is easy to see that the following lemma is true.

Lemma 9.3 $h_{\text{top}}((\varphi, \psi), u, \mathfrak{U}_u) = h_{\text{top}}((\varphi, \psi), \varphi^1(u), \mathfrak{U}_u), \forall u \in \mathcal{M}.$

From this we get immediately the next result:

Proposition 9.1 *For any $u \in \mathcal{M}$ and $k \in \mathbb{Z}$ we have*

$$h_{\text{top}}((\varphi, \psi), u, \mathfrak{U}_u) = h_{\text{top}}((\varphi, \psi), \varphi^k(u), \mathfrak{U}_u)$$

Let us take new covers $\mathfrak{U}_u = \{\mathfrak{U}(u')\}_{u'\in\gamma(u)}$ for which the maximal diameter of $\mathfrak{U}(\varphi^k(u))$ converges to zero for $k \to \infty$. Then we can get arbitrary large values of $h_{\text{top}}((\varphi, \psi), u, \mathfrak{U}_u))$. To avoid this effect we introduce the following definition.

Definition 9.5 The *topological entropy of the cocycle* (φ, ψ) *along the orbit through* u is defined by

$$h_{\text{top}}((\varphi, \psi), u) := \sup h_{\text{top}}((\varphi, \psi), u, \mathfrak{U}_u)$$

where the supremum is taken over all families of finite covers $\mathfrak{U}_u = \{\mathcal{U}(u')\}_{u'\in\gamma(u)}$ which have a positive Lebesgue number.

(b) The Bowen-Dinaburg-type Definition

Suppose $u \in \mathcal{M}$ is arbitrary. On the space $\mathcal{N}(u)$ we consider for $m \in \mathbb{N}$ the family of metrics given through

$$\rho_{u,m}(v, w) := \max_{0 \le k \le m-1} \rho_{\varphi^k(u)}(\psi^k(u, v), \psi^k(u, w)).$$

A set $\mathcal{P} \subset \mathcal{N}(u)$ we call (m, ε)-*spanning* if for any $v \in \mathcal{N}(u)$ there is a point $w \in \mathcal{P}$ such that $\rho_{u,m}(v, w) < \varepsilon$. A set $\mathcal{R} \subset \mathcal{N}(u)$ is said to be (m, ε)-*separated* if for any $v, w \in \mathcal{R}$, $v \neq w$, we have $\rho_{u,m}(v, w) > \varepsilon$. Suppose $N_\varepsilon(\mathcal{N}(u), m)$ is the smallest cardinality of an (m, ε)-spanning set in $\mathcal{N}(u)$ and $S_\varepsilon(\mathcal{N}(u), m)$ is the largest cardinality of an (m, ε)-separated set in $\mathcal{N}(u)$. Repeating the proof of Lemma 3.5, Chap. 3, for the fibre $\mathcal{N}(u)$ we get the following result which is an analogon of Proposition 3.26, Chap. 3.

Proposition 9.2 *Suppose $\{(\mathcal{N}(u), \rho_u)\}_{u \in \mathcal{M}}$ is a family of compact metric spaces and $\psi^k(u, \cdot) : \mathcal{N}(u) \to \mathcal{N}(\varphi^k(u))$, $u \in \mathcal{M}, k \in \mathbb{Z}$ is a family of continuous maps. Then for any $u \in \mathcal{M}$ we have*

$$\begin{aligned} h_{\text{top}}((\psi, \varphi), u) &= \lim_{\varepsilon \to 0+} \limsup_{m \to \infty} \frac{1}{m} \log N_\varepsilon(\mathcal{N}(u), m) \\ &= \lim_{\varepsilon \to 0+} \limsup_{m \to \infty} \frac{1}{m} \log S_\varepsilon(\mathcal{N}(u), m) . \end{aligned}$$

(c) Upper Estimates for the Topological Entropy

Let us derive an analogon of Ito's entropy estimate (Theorem 5.28, Chap. 5) for fibered cocycle systems.

We say that the cocycle (φ, ψ) *satisfies* a *Lipschitz condition along the orbit* $\gamma(u)$ of φ if there exist positive numbers $\lambda_k = \lambda_k(u)$, $k = 0, 1, 2, \ldots$, such that the following inequalities are satisfied

$$\begin{aligned} &\rho_{\varphi^{k+1}(u)}\big(\psi^1(\varphi^k(u), v), \psi^1(\varphi^k(u), w)\big) \\ &\le \lambda_k \rho_{\varphi^k(u)}(v, w), \quad \forall v, w \in \mathcal{N}(\varphi^k(u)), \qquad k = 0, 1, \ldots . \end{aligned}$$

Theorem 9.12 *Suppose $u \in \mathcal{M}$ is arbitrary,* $\dim_F(\mathcal{N}(u)) < +\infty$ *and the cocycle (φ, ψ) satisfies along the orbit $\gamma(u)$ of φ a Lipschitz condition with constants $\lambda_k = \lambda_k(u)$, $k = 0, 1, \ldots$. Suppose that* $\lambda := \limsup_{m \to \infty} \frac{1}{m} \log \prod_{k=0}^{m-1} \max\{\lambda_k, 1\}$.
Then $h_{\text{top}}((\varphi, \psi), u) \le \lambda \dim_F(\mathcal{N}(u))$, $\forall u \in \mathcal{M}$.

Proof Take a number $\zeta > \dim_F(\mathcal{N}(u))$ such that

$$\frac{\log N_\delta(\mathcal{N}(u))}{\log 1/\delta} < \zeta$$

is satisfied for all $\delta > 0$ sufficiently small. Recall that the number $N_\varepsilon(\mathcal{N}(u), m)$ characterizes in the metric $\rho_{u,m}$ the minimal number of balls with radius $\varepsilon > 0$ necessary for covering of $\mathcal{N}(u)$. From this it follows that

$$\rho_{\mathcal{N}(\varphi^k(u))}(\psi^k(u, v), \psi^k(u, w)) \le \lambda_{k-1} \cdot \ldots \cdot \lambda_0 \rho_{\mathcal{N}(u)}(v, w), \quad \forall v, w \in \mathcal{N}(u).$$

Suppose that m is a positive integer and $\lambda'_m := \prod_{k=0}^{m-1} \max\{\lambda_k, 1\} + 2^{-m}$. It follows that for all sufficiently small $\varepsilon > 0$ we have $\mathcal{B}_\delta \subset \mathcal{B}_\varepsilon(u, m)$, where $\mathcal{B}_\delta$ and $\mathcal{B}_\varepsilon(u, m)$ are balls in spaces with metric ρ_p and $\rho_{m,\rho}$, respectively, and $\delta = \frac{\varepsilon}{\lambda'_m}$. From this we get the inequality

$$\frac{\log N_\varepsilon(u, m)}{m} \le \frac{\log N_\delta(\mathcal{N}(u))}{m} = \frac{\log N_\delta(\mathcal{N}(u))}{\log 1/\delta} \cdot \frac{\log 1/\delta}{m}.$$

Using the smallness of $\varepsilon > 0$ (and δ) we get the inequality $\frac{\log N_\delta(\mathcal{N}(u))}{\log 1/\delta} < \zeta$, and, consequently, $\frac{\log N_\varepsilon(\mathcal{N}(u),m)}{m} \le \zeta \left(\frac{\log \lambda'_m}{m} + \frac{\log 1/\varepsilon}{m} \right)$. From this it follows that

$$\limsup_{m \to \infty} \frac{\log N'_\varepsilon(\mathcal{N}(u), m)}{m} < \zeta \limsup_{m \to \infty} \frac{1}{m} \log \prod_{k=0}^{m-1} \max\{\lambda_k, 1\}.$$

Now, taking the limit for $\varepsilon \to 0+$ and using the fact that ζ is arbitrarily close to $\dim_F \mathcal{N}_u$, we get that

$$h_{\text{top}}((\varphi, \psi), u) \le \lambda \dim_F \mathcal{N}(u).$$

□

Remark 9.7 Theorem 9.12 is used in [9] to derive upper estimates of the topological entropy for discrete time cocycles, see also Remark 9.3.

References

1. Abramovich, S., Koryakin, Yu., Leonov, G., Reitmann, V.: Frequency-domain conditions for oscillations in discrete systems. I., Oscillations in the sense of Yakubovich in discrete systems. Wiss. Zeitschr. Techn. Univ. Dresden, **25**(5/6), 1153 – 1163 (1977) (German)
2. Abramovich, S., Koryakin, Yu., Leonov, G., Reitmann, V.: Frequency-domain conditions for oscillations in discrete systems. II., Oscillations in discrete phase systems. Wiss. Zeitschr. Techn. Univ. Dresden, **26**(1), 115–122 (1977) (German)
3. Anguiano, M., Caraballo, T.: Asymptotic behaviour of a non-autonomous Lorenz-84 system. Discrete Contin. Dynam. Syst. **34**(10), 3901–3920 (2014)
4. Anikushin, M.M., Reitmann, V.: Development of the topological entropy conception for dynamical systems with multiple time. Electr. J. Diff. Equ. and Contr. Process. **4** (2016) (Russian); English Trans. J. Diff. Equ. **52**(13), 1655 – 1670 (2016)

5. Cheban, D.N.: Nonautonomous Dyn. Springer Monographs in Mathematics, Berlin (2020)
6. Crauel, H., Flandoli, F.: Hausdorff dimension of invariant sets for random dynamical systems. J. Dyn. Diff. Equ. **10**(3), 449–474 (1998)
7. Debussche, A.: Hausdorff dimension of a random invariant set. Math. Pures Appl. **77**, 967–988 (1998)
8. Douady, A., Oesterlé, J.: Dimension de Hausdorff des attracteurs. C. R. Acad. Sci. Paris, Ser. A 290, 1135–1138 (1980)
9. Egorova, V.E., Reitmann, V.: Estimation of topological entropy for cocycles with cellular automaton as a base system. Electron. J. Diff. Equ. Contr. Process. **3**, 102–122 (2018) (Russian)
10. Ermakov, I.V., Kalinin, Yu.N., Reitmann, V.: Determining modes and almost periodic integrals for cocycles. Electron. J. Diff. Equ. Contr. Process. **4** (2011) (Russian); English Trans. J. Diff. Equ. **47**(13), 1837–1852 (2011)
11. Ermakov, I.V., Reitmann, V., Skopinov, S.: Determining functionals for cocycles and application to the microwave heating problem. Abstracts, Equadiff, Loughborough, UK, 135 (2011)
12. Flandoli, F., Schmalfuss, B.: Random attractors for the 3D stochastic Navier-Stokes equation with multiplicative white noise. Stoch. Stoch. Rep. **59**(1–2), 21–45 (1996)
13. Giesl, P., Rasmussen, M.: Borg's criterion for almost periodic differential equations. Nonlinear Anal. **69**(11), 3722–3733 (2008)
14. Kloeden, P.E., Rasmussen, M.: Nonautonomous Dynamical Systems. Mathematical Surveys and Monographs, Amer. Math. Soc. **176** (2011)
15. Kloeden, P.E., Schmalfuß, B.: Nonautonomous systems, cocycle attractors and variable time-step discretization. Numer. Algorithms **14**, 141–152 (1997)
16. Kolyada, S., Snoha, L.: Topological entropy of nonautonomous dynamical systems. Random Comput. Dyn. **4**(2/3), 205–233 (1996)
17. Langa, J.A., Schmalfuss, B.: Finite dimensionality of attractors for non-autonomous dynamical systems given by partial differential equations. Stoch. Dyn. **4**(3), 385–404 (2004)
18. Ledrappier, F., Young, L.S.: Dimension formula for random transformations. Commun. Math. Phys. **117**(4), 529–548 (1988)
19. Leonov, G.A., Reitmann, V., Slepuchin, A.S.: Upper estimates for the Hausdorff dimension of negatively invariant sets of local cocycles. Dokl. Akad. Nauk, T. **439**, 6 (2011) (Russian); English Trans. Dokl. Mathematics, **84**(1), 551–554 (2011)
20. Leonov, G., Tschschigowa, T., Reitmann, V.: A frequency-domain variant of the Belykh-Nekorkin comparison method in phase synchronisation theory. Wiss. Zeitschr. Techn. Univ. Dresden, **32**(1), 51–59 (1983) (German)
21. Maltseva, A.A., Reitmann, V.: Global stability and bifurcations of invariant measures for the discrete cocycles of the cardiac conduction system's equations. J. Diff. Equ. **50**(13), 1718–1732 (2014)
22. Maltseva, A.A., Reitmann, V.: Existence and dimension properties of a global $\mathcal{B}$-pullback attractor for a cocycle generated by a discrete control system. J. Diff. Equ. **53**(13), 1703–1714 (2017)
23. Maricheva, A.V.: Hausdorff dimension bounds for cocycle attractors on a finite dimensional Riemannian manifold. Diploma Thesis, St. Petersburg State University (2015) (Russian)
24. Noack, A.: Hausdorff dimension estimates for time-discrete feedback control systems. ZAMM **77**(12), 891–899 (1997)
25. Pogromsky, A.Y., Matveev, A.S.: Estimation of topological entropy via the direct Lyapunov method. Nonlinearity **24**(7), 1937 (2011)
26. Reitmann, V.: About bounded and periodic trajectories in nonlinear impulse systems. Wiss. Zeitschr. Techn. Univ. Dresden, **27**(2), 355–357 (1978) (German)
27. Reitmann, V., Anikushin, M.M., Romanov, A.O.: Dimension-like properties and almost periodicity for cocycles generated by variational inequalities with delay. Abstracts, Equadiff, Leiden, The Netherlands, 90 (2019)
28. Reitmann, V., Slepuchin, A.V.: On upper estimates for the Hausdorff dimension of negatively invariant sets of local cocycles. Vestn. St. Petersburg Univ. Math. **44**(4), 292–300 (2011)

29. Rössler, O. E.: Different types of chaos in two simple differential equations. Z. Naturforsch. 31 a, 1664–1670 (1976)
30. Wakeman, D.R.: An application of topological dynamics to obtain a new invariance property for nonautonomous ordinary differential equations. J. Diff. Equ. **17**(2), 259–295 (1975)
31. Wang, Y., Zhong, C., Zhou, S.: Pullback attractors of nonautonomous dynamical systems. Discrete and Cont. Dynam. Syst. **16**(3), 587–614 (2006)

Chapter 10
Dimension Estimates for Dynamical Systems with Some Degree of Non-injectivity and Nonsmoothness

Abstract In this chapter dimension estimates for maps and dynamical systems with specific properties are derived. In Sect. 10.1 a class of non-injective smooth maps is considered. Dimension estimates for piecewise non-injective maps are given in Sect. 10.2. For piecewise smooth maps with a special singularity set upper Hausdorff dimension estimates are shown in Sect. 10.3. Lower dimension estimates are shown in Sect. 10.4.

10.1 Dimension Estimates for Non-injective Smooth Maps

10.1.1 Hausdorff Dimension Estimates

In Chap. 5 Hausdorff dimension estimates for compact sets $\mathcal{K} \subset \mathbb{R}^n$ that are invariant under C^1-maps φ are given. The main idea consists in showing that for a number $j \in \mathbb{N}$ the Hausdorff outer measure of $\varphi^j(\mathcal{K})$ is by a certain factor smaller than the outer measure of $\mathcal{K}$, i.e. the iterated map is contracting with respect to the Hausdorff outer measure on $\mathcal{K}$. The contraction constant can be estimated by means of a singular value function of the tangent map, i.e. if the singular value function is less than 1, then the map is contracting. In Chap. 5 the condition for the contraction of the Hausdorff outer measure in $\mathbb{R}^n$ is weakened using Lyapunov-type functions. The latter results are generalized in Chap. 8 to maps on Riemannian manifolds (see also [17]). Using a technique similar to that of Douady and Oesterlé, Temam gave in [38] (see also [39]) upper bounds for the Hausdorff and fractal dimensions of semiflow invariant sets in a Hilbert space. Analogously fractal dimension estimates are derived in [13] for semiflows on Riemannian manifolds.

In practice the maps and vector fields describing concrete physical or technical systems are often non-injective (see for instance [4]). For such non-injective maps it may be possible to use information about the "degree of non-injectivity" in order to get Hausdorff and fractal outer measure and dimension estimates under weakened conditions compared with the theorems mentioned above. For the first

N. Kuznetsov and V. Reitmann, *Attractor Dimension Estimates for Dynamical Systems: Theory and Computation*, Emergence, Complexity and Computation 38, https://doi.org/10.1007/978-3-030-50987-3_10

time such Douady-Oesterlé-type Hausdorff dimension estimates using the "degree of non-injectivity" are considered in [25]. There a class of k-1-endomorphisms is described, where the given invariant set can be split into k compact subsets and where each of those subsets is mapped onto the whole invariant set. The factor $\frac{1}{k}$ can be used to compensate the missing contraction property for the Hausdorff outer measure. Note that another class of modified systems is given by multivalued differential and difference equations [34].

In the present section we consider a class of maps satisfying even a weaker non-injectivity condition than the k-1-property. In general, such a class may be described as follows. Let φ be a C^1-map on a smooth (for simplicity C^∞) n-dimensional Riemannian manifold $(\mathcal{M}, g)$ and $\mathcal{K} \subset \mathcal{M}$ a compact set. (A class of maps that are only piecewise C^1 is considered in [33]. For these maps many results of this section are also true.) Suppose that for a given outer measure $\mathfrak{m}(\cdot, d)$ on $\mathcal{M}$ (d-dimensional Hausdorff or fractal outer measure of the given set or of a covering class of this set) there exist a number $0 < a < 1$ and a family $\{\mathcal{K}_j\}_{j \geq j_0}$ of subsets of $\mathcal{K}$ such that $\mathfrak{m}(\varphi^j(\mathcal{K}_j), d) = \mathfrak{m}(\varphi^j(\mathcal{K}), d)$ and $\mathfrak{m}(\mathcal{K}_j, d) \leq a^j \mathfrak{m}(\mathcal{K}, d)$ for all $j \geq j_0$. A map φ with such properties can be considered as piecewise $\mathfrak{m}(\cdot, d)$-expansive on $\mathcal{K}$ ($\frac{1}{a}$ is the expansion parameter and also describes the "degree of non-injectivity"). It follows that for such a map and any set $A \subset \mathcal{K}$ there exists a $j \geq j_0$ such that $\mathfrak{m}(\mathcal{K}_j, d) \leq a^j \mathfrak{m}(\mathcal{K}, d) \leq \mathfrak{m}(A, d)$ and $\mathfrak{m}(\varphi^j(\mathcal{K}_j), d) = \mathfrak{m}(\varphi^j(\mathcal{K}), d)$, i.e. the semidynamical system $\{\varphi^j\}_{j \geq 0}$ generated by a piecewise $\mathfrak{m}(\cdot, d)$-expansive map has a certain transitive Markov-type property on $\mathcal{K}$. It will be shown that for negatively invariant sets $\mathcal{K}$ of piecewise $\mathfrak{m}(\cdot, d)$-expansive maps, where $m(\cdot, d)$ is the d-dimensional Hausdorff ($\mu_H(\cdot, d)$) or fractal outer measure, the parameter d is an upper bound of the associated dimension. Our presentation in Sect. 10.1 is based on the results of [5].

Theorem 10.1 *Let $(\mathcal{M}, g)$ be a smooth n-dimensional Riemannian manifold, $\mathcal{U} \subset \mathcal{M}$ be an open set and $\varphi : \mathcal{U} \to \mathcal{M}$ be a C^1-map. Suppose $\mathcal{K}$ and $\widetilde{\mathcal{K}}$ are compact sets satisfying the relations $\mathcal{K} \subset \widetilde{\mathcal{K}} \subset \mathcal{U}$ and $\varphi^j(\mathcal{K}) \subset \widetilde{\mathcal{K}}$ for any $j = 1, 2, \ldots$. Suppose that for some numbers $d \in (0, n]$ and $a > 0$ the following conditions are satisfied:*

(1) $\omega_{d,\widetilde{\mathcal{K}}}(\varphi) < \frac{1}{a}$.
(2) There is a number $j_0 \in \mathbb{N}$ such that for any natural number $j \geq j_0$ there exists a set $\mathcal{K}_j \subset \mathcal{K}$ such that $\mu_H\left(\varphi^j(\mathcal{K}_j), d\right) = \mu_H(\varphi^j(\mathcal{K}), d)$ and $\mu_H\left(\mathcal{K}_j, d\right) \leq a^j \mu_H(\mathcal{K}, d)$.
(3) $\mu_H(\mathcal{K}, d) < \infty$.

Then $\lim_{j \to \infty} \mu_H(\varphi^j(\mathcal{K}), d) = 0$.

Proof It follows from Chap. 7 that the singular value function satisfies the relation

$$\omega_{d,\mathcal{K}}(\varphi^j) \leq \omega^j_{d,\widetilde{\mathcal{K}}}(\varphi)$$

for any $j \in \mathbb{N}$. Further for any number $\delta > 0$ using condition (1) we find a number $j_\delta > j_0$ so that for d written as $d = d_0 + s$ with $d_0 \in \{0, \ldots, n-1\}$ and $s \in (0, 1]$

the inequality

$$2^{d_0}(d_0+1)^{\frac{d}{2}}(\omega_{d,\widetilde{\mathcal{K}}}(\varphi)\cdot a)^j \leq \delta$$

will be true for any $j > j_\delta$. Using additionally condition (2) and Lemma 8.2,Chap. 8 we get for any $j > j_\delta$ the relations

$$\begin{aligned}\mu_H(\varphi^j(\mathcal{K}), d) &= \mu_H\left(\varphi^j(\mathcal{K}_j), d\right) \leq 2^{d_0}(d_0+1)^{\frac{d}{2}}\omega^j_{d,\widetilde{\mathcal{K}}}(\varphi)\mu_H(\mathcal{K}_j, d)\\ &\leq 2^{d_0}(d_0+1)^{\frac{d}{2}}(\omega_{d,\widetilde{\mathcal{K}}}(\varphi)\cdot a)^j\mu_H(\mathcal{K}, d) \leq \delta\mu_H(\mathcal{K}, d).\end{aligned}$$

Since δ can be chosen arbitrarily small, by condition (3) we get

$$\lim_{j\to\infty}\mu_H(\varphi^j(\mathcal{K}), d) = 0.$$

□

Corollary 10.1 *If the conditions (1) and (2) of Theorem 10.1 are satisfied for certain numbers $a > 0$ and $d \in (0, n]$ and furthermore $\varphi(\mathcal{K}) \supset \mathcal{K}$ holds, then either $\mu_H(\mathcal{K}, d) = 0$, or $\mu_H(\mathcal{K}, d) = \infty$.*

Corollary 10.2 *Let the conditions (2) and (3) of Theorem 10.1 be satisfied for certain numbers $a > 0$ and $d \in (0, n]$. Furthermore let $\varkappa : \widetilde{\mathcal{K}} \to \mathbb{R}_+ = \{x \in \mathbb{R} \mid x > 0\}$ be a continuous function such that the condition*

$$\sup_{u\in\widetilde{\mathcal{K}}}\left(\frac{\varkappa(\varphi(u))}{\varkappa(u)}\omega_d(d_u\varphi)\right) < \frac{1}{a} \tag{10.1}$$

is satisfied. Then $\lim_{j\to\infty}\mu_H(\varphi^j(\mathcal{K}), d) = 0$.

Proof It follows from condition (10.1) that there is a number $0 < \kappa < 1$ with $a\frac{\varkappa(\varphi(u))}{\varkappa(u)}\omega_d(d_u\varphi) < \kappa$ for any $u \in \widetilde{\mathcal{K}}$. Therefore by the chain rule and by applying (10.1) we get for any $u \in \widetilde{\mathcal{K}}$ and arbitrary $j \in \mathbb{N}$

$$\begin{aligned}a^j\omega_d(d_u\varphi^j) &\leq a^j\omega_d(d_{\varphi^{j-1}(u)}\varphi)\dots\omega_d(d_u\varphi) \leq a^j\frac{\kappa}{a}\frac{\varkappa(\varphi^{j-1}(u))}{\varkappa(\varphi^j(u))}\cdots\frac{\kappa}{a}\frac{\varkappa(u)}{\varkappa(\varphi(u))}\\ &= \kappa^j\frac{\varkappa(u)}{\varkappa(\varphi^j(u))} \leq \kappa^j\frac{\sup_{u\in\widetilde{\mathcal{K}}}\varkappa(u)}{\inf_{u\in\widetilde{\mathcal{K}}}\varkappa(u)}.\end{aligned}$$

For any $\delta > 0$ we find a number $j_\delta > j_0$ such that for d ($d = d_0 + s$ with $d_0 \in \{0, \dots, n-1\}$ and $s \in (0, 1]$) the relation

$$2^{d_0}(d_0+1)^{\frac{d}{2}}a^j\omega_{d,\widetilde{\mathcal{K}}}(\varphi^j) \leq \delta$$

will be true for any $j > j_\delta$. For these numbers j, similarly as in the proof of Theorem 10.1, we get

$$\begin{aligned}\mu_H(\varphi^j(\mathcal{K}), d) &= \mu_H\left(\varphi^j(\mathcal{K}_j), d\right) \le 2^{d_0}(d_0+1)^{\frac{d}{2}}\omega_{d,\tilde{\mathcal{K}}}(\varphi^j)\mu_H(\mathcal{K}_j, d) \\ &\le 2^{d_0}(d_0+1)^{\frac{d}{2}}\omega_{d,\tilde{\mathcal{K}}}(\varphi^j)a^j\mu_H(\mathcal{K}, d) \le \delta\mu_H(\mathcal{K}, d)\end{aligned}$$

and therefore, $\lim_{j\to\infty} \mu_H(\varphi^j(\mathcal{K}), d) = 0.$ □

Example 10.1 Consider the modified horseshoe map φ introduced in Example 1.6, Chap. 1. Let us choose $\mathcal{K}_1 := \mathcal{K} \cap [0, 1] \times [0, \frac{51}{65}]$ and $\mathcal{K}_j := \mathcal{K}_{j-1} \cap \varphi^{-1}(\mathcal{K}_{j-1})$, $j = 2, 3, \ldots$. Then these sets satisfy $\varphi^j(\mathcal{K}_j) = \mathcal{K} = \varphi^j(\mathcal{K})$, so the first part of condition (2) of Theorem 10.1 is true. Furthermore the set $\mathcal{K}_j$ is contained in 4^j rectangles with edges $\frac{1}{3^j}$ out of the 6^j rectangles forming $\mathcal{K}^j$ (see Fig. 10.1). Let ε_j denote half of the minimal distance between two of the different 6^j rectangles. If we cover $\mathcal{K}$ by balls of radii smaller than ε_j, then every ball can contain points of only one of the 6^j rectangles. If we consider only the part of $\mathcal{K}_{j-1}$ which is contained in one of the 6^{j-1} rectangles of $\mathcal{K}^{j-1}$, then the part of $\mathcal{K}_j$ which is contained in the same rectangle consists of four linear copies of the part of $\mathcal{K}_{j-1}$ with a factor $\frac{1}{3}$. Therefore for $\varepsilon < \varepsilon_j$ we have

$$\mu_H(\mathcal{K}_j, d, \varepsilon) = \frac{4}{3^d}\mu_H(\mathcal{K}_{j-1}, d, 3\varepsilon)$$

for any $d \in [0, 2]$, in the limit $\varepsilon \to 0+0$ we get

$$\mu_H(\mathcal{K}_j, d) = \left(\frac{4}{3^d}\right)^j \mu_H(\mathcal{K}, d).$$

So the second part of condition (2) is satisfied with $a = \frac{4}{3^d}$. Now we have to check condition (1), i.e. we have to find a number $d \in [1, 2]$ such that $\omega_{d,\mathcal{K}}(\varphi)a < 1$ is satisfied. The singular value function here has the form

$$\omega_d(d_u\varphi) = \begin{cases} 3\left(\frac{1}{3}\right)^{d-1}, & \text{for } u \in \mathcal{K}_1, \\ 5\left(\frac{1}{3}\right)^{d-1}, & \text{for } u \in \mathcal{K}\setminus\mathcal{K}_1, \end{cases}$$

so we have $\omega_{d,\mathcal{K}}(\varphi) = 5\left(\frac{1}{3}\right)^{d-1}$. It is easy to see that the inequality $\frac{4}{3^d}5\left(\frac{1}{3}\right)^{d-1} < 1$ is equivalent to $d > \frac{1}{2}\left(\frac{\ln 20}{\ln 3} + 1\right) \approx 1.863$. For such numbers d the conditions (1) and (2) of Theorem 10.1 are satisfied, and Corollary 10.1 yields to $\mu_H(\mathcal{K}, d) = 0$ or $\mu_H(\mathcal{K}, d) = \infty$.

Now we want to use the same method as above to find an upper estimate of the Hausdorff dimension of the set $\mathcal{K}$ considered in Theorem 10.1. Let us additionally assume that $\mathcal{K}$ is negatively invariant under φ, i. e. $\mathcal{K} \subset \varphi(\mathcal{K})$ as in Corollary 10.1.

Fig. 10.1 Construction of the invariant set $\mathcal{K}$

In order to find an upper bound for the Hausdorff dimension of $\mathcal{K}$ we can not assume $\mu_H(\mathcal{K}, d) < \infty$ as in Theorem 10.1. However it is possible to consider the Hausdorff outer measure of the class of finite covers of $\mathcal{K}$ by balls of radii at most ε instead of the Hausdorff outer measure of $\mathcal{K}$ itself, because the outer measure of a finite cover is always finite.

Theorem 10.2 *Let $(\mathcal{M}, g)$ be a smooth n-dimensional Riemannian manifold, $\mathcal{U} \subset \mathcal{M}$ be an open set and $\varphi : \mathcal{U} \to \mathcal{M}$ be a C^1-map. Suppose $\mathcal{K}$ and $\widetilde{\mathcal{K}}$ are compact sets satisfying the relation $\mathcal{K} \subset \varphi^j(\mathcal{K}) \subset \widetilde{\mathcal{K}} \subset \mathcal{U}$ for any $j = 1, 2, \ldots$. Suppose that for some numbers $a > 0$ and $d \in (0, n]$ of the form $d = d_0 + s$ with $d_0 \in \{0, \ldots, n-1\}$ and $s \in (0, 1]$ the following conditions are satisfied:*

(1) $\omega_{d,\widetilde{\mathcal{K}}}(\varphi) < \frac{1}{a}$;

(2) There are numbers l with $\omega_{d,\widetilde{\mathcal{K}}}(\varphi) < l < \frac{1}{a}$ and $j_0 \in \mathbb{N}$ such that for any natural number $j > j_0$ there exist a set $\mathcal{K}_j \subset \mathcal{K}$ and a number $\varepsilon_j > 0$ such that

$$\mu_H(\varphi^j(\mathcal{K}_j), d, \varepsilon) = \mu_H(\varphi^j(\mathcal{K}), d, \varepsilon),$$
$$\mu_H\left(\mathcal{K}_j, d, (d_0+1)^{-\frac{1}{2}} l^{-\frac{j}{d}} \varepsilon\right) \le a^j \mu_H(\mathcal{K}, d, \varepsilon)$$

holds for any $\varepsilon \in (0, \varepsilon_j]$.

Then $\dim_H \mathcal{K} \le d$.

Proof Like in the proof of Theorem 10.1 we have $\omega_{d,\mathcal{K}}(\varphi^j) \le \omega^j_{d,\widetilde{\mathcal{K}}}(\varphi) < l^j$ for any $j \in \mathbb{N}$. For any $\delta > 0$ we find an integer number $j_\delta > j_0$ so that the relation $2^{d_0}(d_0+1)^{\frac{d}{2}}(al)^j < \delta$ will be true for any $j > j_\delta$. Now let $j > j_\delta$ be fixed and consider $0 < \varepsilon \le \min\{\varepsilon_j, (d_0+1)^{-\frac{1}{2}} l^{-\frac{j}{d}} \varepsilon_0(l^j)\}$, where ε_0 is defined in Lemma 8.2, Chap. 8. Then condition (2) and Lemma 8.2, Chap. 8 result in the following inequalities:

$$\begin{aligned}\mu_H(\mathcal{K}, d, \varepsilon) &\le \mu_H(\varphi^j(\mathcal{K}), d, \varepsilon) = \mu_H(\varphi^j(\mathcal{K}_j), d, \varepsilon) \\ &\le 2^{d_0}(d_0+1)^{\frac{d}{2}} l^j \mu_H(\mathcal{K}_j, d, (d_0+1)^{-\frac{1}{2}} l^{-\frac{j}{d}} \varepsilon) \\ &\le 2^{d_0}(d_0+1)^{\frac{d}{2}} (al)^j \mu_H(\mathcal{K}, d, \varepsilon) \le \delta \mu_H(\mathcal{K}, d, \varepsilon).\end{aligned}$$

Since the number δ can be chosen arbitrarily small and $\mu_H(\mathcal{K}, d, \varepsilon)$ is finite, this means $\mu_H(\mathcal{K}, d, \varepsilon) = 0$ for any $\varepsilon \in \left(0, \min\{\varepsilon_j, (d_0+1)^{-\frac{1}{2}} l^{-\frac{j}{d}} \varepsilon_0(l^j)\}\right)$ and therefore, $\mu_H(\mathcal{K}, d) = 0$. Hence we get $\dim_H \mathcal{K} \le d$. □

Using now a Lyapunov-type function we get a corollary of this theorem analogous to Corollary 10.2.

Corollary 10.3 *Let* $(\mathcal{M}, g)$, $\mathcal{U}$, $\mathcal{K}$, $\widetilde{\mathcal{K}}$ *and* φ *be defined as in Theorem 10.2, and let* $\varkappa : \widetilde{\mathcal{K}} \to \mathbb{R}_+$ *be a continuous function, such that for some numbers* $a > 0$ *and* $d \in (0, n]$, $d = d_0 + s$ *with* $d_0 \in \{0, \dots, n-1\}$ *and* $s \in (0, 1]$ *the following conditions are satisfied:*

(1) $\sup_{u \in \widetilde{\mathcal{K}}} \left(\frac{\varkappa(\varphi(u))}{\varkappa(u)} \omega_d(d_u\varphi)\right) < \frac{1}{a}$;

(2) There are numbers l *with* $\sup_{u \in \widetilde{\mathcal{K}}} \left(\frac{\varkappa(\varphi(u))}{\varkappa(u)} \omega_d(d_u\varphi)\right) < l < \frac{1}{a}$ *and* $j_0 \in \mathbb{N}$ *such that for any natural number* $j > j_0$ *there exist a set* $\mathcal{K}_j \subset \mathcal{K}$ *and a number* $\varepsilon_j > 0$ *with*

$$\mu_H(\varphi^j(\mathcal{K}_j), d, \varepsilon) = \mu_H(\varphi^j(\mathcal{K}), d, \varepsilon),$$
$$\text{and} \quad \mu_H\left(\mathcal{K}_j, d, (d_0+1)^{-\frac{1}{2}} \left(l^j \frac{\sup_{u \in \widetilde{\mathcal{K}}} \kappa(u)}{\inf_{u \in \widetilde{\mathcal{K}}} \kappa(u)}\right)^{-\frac{1}{d}} \varepsilon\right) \le a^j \mu_H(\mathcal{K}, d, \varepsilon)$$

for any $\varepsilon \in (0, \varepsilon_j]$.

Then $\dim_H \mathcal{K} \le d$.

Example 10.2 (Example 10.1 cont'd) For the modified horseshoe map described before, in the two-dimensional case the first part of condition (2) of Theorem 10.2 is satisfied for arbitrary numbers $\varepsilon > 0$ and $d \in [1, 2]$. Furthermore, we can show the existence of a number l with $\omega_{d,\mathcal{K}}(\varphi) < l < \frac{1}{a}$ satisfying $(d_0 + 1)^{-\frac{1}{2}} l^{-\frac{j}{d}} \geq \gamma^j$. Together with the inequality stated above this yields the second part of condition (2). Thus we get $\dim_H \mathcal{K} \leq d$ for any number $d > \frac{\ln \alpha - \ln 4\beta_2}{\ln \alpha + \ln \gamma}$. In the limit this yields $\dim_H \mathcal{K} \leq \frac{\ln \alpha - \ln 4\beta_2}{\ln \alpha + \ln \gamma}$. For the parameters $\alpha = \frac{1}{3}$, $\beta_1 = 3$, $\beta_2 = 5$ we get $\dim_H \mathcal{K} \leq 1.863$.

If $\varphi(\mathcal{K} \setminus \mathcal{K}_1) \subset \mathcal{K}_1$ holds then the dimension estimate can be improved by means of Corollary 10.3. Using an appropriate Lyapunov-type function the condition (1) of Theorem 10.2 can be replaced by condition (1) of Corollary 10.3. Since here the singular value function is constant on $\mathcal{K}_1$ and $\mathcal{K} \setminus \mathcal{K}_1$, respectively, the simplest type of Lyapunov function is of the same kind, i. e. $\varkappa(u) = 1$ for $u \in \mathcal{K}_1$ and $\varkappa(u) = c > 0$ for $u \in \mathcal{K} \setminus \mathcal{K}_1$. Since the distance between the sets $\mathcal{K}_1$ and $\mathcal{K} \setminus \mathcal{K}_1$ is positive, such a function is continuous on $\widetilde{\mathcal{K}} = \mathcal{K}$. The constant c has to be chosen in such a way that $\sup_{u \in \mathcal{K}} \left(\frac{\varkappa(\varphi(u))}{\varkappa(u)} \omega_d(d_u \varphi) \right)$ becomes minimal. Because of

$$\frac{\varkappa(\varphi(u))}{\varkappa(u)} \omega_d(d_u \varphi) = \begin{cases} \frac{1}{c} \beta_2 \alpha^{d-1}, & \text{for } u \in \mathcal{K} \setminus \mathcal{K}_1, \\ c \beta_1 \alpha^{d-1}, & \text{for } u \in \mathcal{K}_1, \varphi(u) \in \mathcal{K} \setminus \mathcal{K}_1, \\ \beta_1 \alpha^{d-1}, & \text{for } u \in \mathcal{K}_1, \varphi(u) \in \mathcal{K}_1, \end{cases}$$

we have to choose c such that $\frac{1}{c} \beta_2 = c \beta_1$, i.e. $c = \sqrt{\frac{\beta_2}{\beta_1}}$. Thus we get the Lyapunov-type function

$$\varkappa(u) = \begin{cases} 1, & \text{for } u \in \mathcal{K}_1, \\ \sqrt{\frac{\beta_2}{\beta_1}}, & \text{for } u \in \mathcal{K} \setminus \mathcal{K}_1. \end{cases}$$

with this function $\varkappa$ for $d \in [1, 2]$ we get

$$\sup_{u \in \mathcal{K}} \left(\frac{\varkappa(\varphi(u))}{\varkappa(u)} \omega_d(d_u \varphi) \right) = \sqrt{\beta_1 \beta_2} \alpha^{d-1},$$

which is less than $\omega_{d,\mathcal{K}}(\varphi) = \beta_2 \alpha^{d-1}$. For any number $d > \frac{\ln \alpha - \ln 4\sqrt{\beta_1 \beta_2}}{\ln \alpha + \ln \gamma}$ the conditions of Corollary 10.3 are satisfied, and we get the improved estimate $\dim_H \mathcal{K} \leq \frac{\ln \alpha - \ln 4\sqrt{\beta_1 \beta_2}}{\ln \alpha + \ln \gamma}$. For the parameters $\alpha = \frac{1}{3}$, $\beta_1 = 3$, $\beta_2 = 5$ this means $\dim_H \mathcal{K} \leq 1.747$

Remark 10.1 In Example 10.2 we would have got the same improved result if we had changed the standard metric on $\mathbb{R}^2$ by multiplying the metric tensor with the Lyapunov-type function $\varkappa$.

Since condition (2) of Theorem 10.2 is not easy to check, especially if the map is not piecewise linear, we now give some stronger conditions which can be checked more easily.

Corollary 10.4 *Let $(\mathcal{M}, g)$ be a smooth n-dimensional Riemannian manifold, $\mathcal{U} \subset \mathcal{M}$ an open set, $\varphi : \mathcal{U} \to \mathcal{M}$ a C^1-map and $\mathcal{K} \subset \mathcal{U}$ a compact φ-invariant set. Suppose that for some numbers $a > 0$ and $d \in (0, n]$ of the form $d = d_0 + s$ with $d_0 \in \{0, \dots, n-1\}$ and $s \in (0, 1]$ the following conditions are satisfied:*

(1) $\omega_{d,\mathcal{K}}(\varphi) < \frac{1}{a}$.
(2) There is a number $j_0 \in \mathbb{N}$ such that for any natural number $j \geq j_0$ there exist a set $\mathcal{K}_j \subset \mathcal{K}$ with $\varphi^j(\mathcal{K}_j) = \mathcal{K}$, a natural number N_j, a number l_0 and C^1-maps $f_{i,j} : \mathcal{U} \to \mathcal{M}$ $(i = 1, \dots, N_j)$ with

$$\mathcal{K}_j = \bigcup_{i=1}^{N_j} f_{i,j}(\mathcal{K}), \quad \max_{i=1,\dots,N_j} \omega_{d,\mathcal{K}}(f_{i,j}) < l_0^j \quad \textit{and} \quad N_j \leq 2^{-d_0}(d_0+1)^{-\frac{d}{2}} a^j l_0^{-j}.$$

Then $\dim_H \mathcal{K} \leq d$.

Proof Using Lemma 8.2, Chap. 8 for any $j \in \mathbb{N}$ with $j > j_0$ there exists a number ε_j such that

$$\begin{aligned}\mu_H\left(\mathcal{K}_j, d, \sqrt{d_0+1} l_0^{\frac{j}{d}} \varepsilon\right) &\leq \sum_{i=1}^{N_j} 2^{d_0}(d_0+1)^{\frac{d}{2}} l_0^j \mu_H(\mathcal{K}, d, \varepsilon) \\ &= N_j 2^{d_0}(d_0+1)^{\frac{d}{2}} l_0^j \mu_H(\mathcal{K}, d, \varepsilon) \\ &\leq a^j \mu_H(\mathcal{K}, d, \varepsilon)\end{aligned}$$

holds for any $\varepsilon \in (0, \varepsilon_j]$. Because of $N_j \geq 1$ and of condition (2) we have $2^{-d_0}(d_0 + 1)^{-\frac{d}{2}} a^j l_0^{-j} \geq 1$ and therefore, $2^{d_0}(d_0+1)^{\frac{d}{2}} l_0^j \omega_{d,\mathcal{K}}^j(\varphi) < 1$ for any $j \geq j_0$. This means $l_0 \omega_{d,\mathcal{K}}(\varphi) < 1$, and because of condition (1), there are numbers $l \in \mathbb{R}$ and $j_0 \in \mathbb{N}$ such that $\omega_{d,\mathcal{K}}(\varphi) < l < \frac{1}{a}$ and $(l_0 l)^{\frac{j}{d}} < \frac{1}{d_0+1}$ for any $j > j_0$ are satisfied. For these numbers j and all $\varepsilon \in (0, \varepsilon_j]$ we have

$$\mu_H\left(\mathcal{K}_j, d, (d_0+1)^{-\frac{1}{2}} l^{-\frac{j}{d}} \varepsilon\right) \leq \mu_H\left(\mathcal{K}_j, d, \sqrt{d_0+1} l_0^{\frac{j}{d}} \varepsilon\right) \leq a^j \mu_H(\mathcal{K}, d, \varepsilon).$$

Applying Theorem 10.2 we get $\dim_H \mathcal{K} \leq d$. □

Example 10.3 (Example 10.2 cont'd) Since in our example of the modified horseshoe map the set $\mathcal{K}_j$ consists of 4^j linear copies of $\mathcal{K}$ we define $f_{i,j}$ to be the linear map of $\mathcal{K}$ onto the ith piece of $\mathcal{K}_j$, $i = 1, \dots, 4^j$. Then for $N_j = 4^j$ and $a > 4l_0 > 4\gamma^d$ condition (2) of Corollary 10.4 is satisfied. Condition (1) results in $a\beta_2\alpha^{d-1} < 1$, and the limit for $a \to 4\gamma^d$ yields $\dim_H(\mathcal{K}) \leq \frac{1}{2}\left(\frac{\ln 20}{\ln 3} + 1\right) \approx 1.863$. In this way we get the same result as before without a Lyapunov-type function, but we could reach it with less expense.

10.1.2 Fractal Dimension Estimates

The first theorem in this subsection provides an upper bound for the fractal dimension of a negatively invariant set if no information about the "degree of non-injectivity" is known.

Theorem 10.3 *Let $(\mathcal{M}, g)$ be a smooth n-dimensional Riemannian manifold, $\mathcal{U} \subset \mathcal{M}$ be an open set and $\varphi : \mathcal{U} \to \mathcal{M}$ be a C^1-map. Suppose $\mathcal{K} \subset \mathcal{U}$ is a compact set satisfying the relation $\mathcal{K} \subset \varphi(\mathcal{K}) \subset \mathcal{U}$. Assume that*

$$0 < \alpha_{\mathcal{K}}(\varphi) := \min_{u \in \mathcal{K}} \alpha_n(d_u\varphi) < n^{-\frac{1}{2}} \tag{10.2}$$

and there exists a number $d \in (0, n]$ such that

$$\omega_{n,\mathcal{K}}(\varphi)\alpha_{\mathcal{K}}^{d-n}(\varphi) \leq 8^{-n} n^{-\frac{d}{2}}. \tag{10.3}$$

Then $\dim_F \mathcal{K} \leq d$.

Proof Let $\eta \in (0, \alpha_{\mathcal{K}}(\varphi))$ be an arbitrary number and $r_1 > 0$ be so small that there exists an open set $\mathcal{V} \subset \mathcal{M}$ containing $\mathcal{K}$ which itself lies inside a compact subset of $\mathcal{U}$ such that

$$|\tau_{\varphi(v)}^{\varphi(u)} d_v\varphi\tau_u^v - d_u\varphi| \leq \eta \tag{10.4}$$

for any $u, v \in \mathcal{V}$ with $\rho(u, v) \leq r_1$ is satisfied, where $|\cdot|$ here denotes the operator norm. By $\rho(\cdot, \cdot)$ we mean the geodesic distance between the points of $\mathcal{M}$ and by τ_u^v we denote the isometry between the tangent spaces $T_u\mathcal{M}$ and $T_v\mathcal{M}$ defined by parallel transport.

Let $\exp_u : T_u\mathcal{M} \to \mathcal{M}$ denote the exponential map at an arbitrary point $u \in \mathcal{M}$. Since $\exp_u$ is a smooth map satisfying $|d_{O_u} \exp_u| = 1$ for any point $u \in \mathcal{M}$ we find a number $r_u > 0$ such that $|d_v \exp_u| \leq 2$ for any $v \in \mathcal{B}_{r_u}(O_u)$, where O_u denotes the origin of the tangent space $T_u\mathcal{M}$. Since $\mathcal{V}$ is contained in a compact set there is a number $r_2 > 0$ such that $|d_v \exp_u| \leq 2$ is satisfied for any $u \in \mathcal{V}$ and any $v \in \mathcal{B}_{r_2}(O_u)$. Furthermore there is a number $\alpha' > 0$ such that $\alpha_1(d_u\varphi) < \alpha'$ is satisfied for any $u \in \mathcal{V}$.

Now we can find a number $r_0 \leq \min\{r_1, \frac{r_2}{2+\alpha'+\eta}\}$ such that any ball $\mathcal{B}_{r_0}(u)$ containing points of $\mathcal{K}$ is entirely contained in $\mathcal{V}$. Let $r \in (0, r_0)$ be fixed. Since $\mathcal{K}$ is compact there is a finite number of points $u_j \in \mathcal{V}$, $j = 1, \ldots, N_r(\mathcal{K})$, such that $\mathcal{K} = \bigcup_{j=1}^{N_r(\mathcal{K})} \mathcal{B}_r(u_j) \cap \mathcal{K}$ and therefore,

$$\varphi(\mathcal{K}) = \bigcup_{j=1}^{N_r(\mathcal{K})} \varphi(\mathcal{B}_r(u_j) \cap \mathcal{K})$$

is satisfied. The Taylor formula for the differentiable map φ guarantees the relation

$$|\exp^{-1}_{\varphi(u_j)} \varphi(\upsilon) - d_{u_j}\varphi(\exp^{-1}_{u_j}(\upsilon))| \\ \leq \sup_{w \in \mathcal{B}_r(u_j)} |\tau^{\varphi(u_j)}_{\varphi(w)} d_w \varphi \tau^w_{u_j} - d_{u_j}\varphi| \cdot |\exp^{-1}_{u_j}(w)| \tag{10.5}$$

for every $\upsilon \in \mathcal{B}_r(u_j)$. Thus, using (10.4) and (10.5), the image of every ball $\mathcal{B}_r(u_j)$ under φ satisfies the inclusion

$$\varphi(\mathcal{B}_r(u_j)) \subset \exp_{\varphi(u_j)}(d_{u_j}\varphi(\mathcal{B}_r(O_{u_j})) + \mathcal{B}_{\eta r}(O_{\varphi(u_j)})).$$

Since $\mathcal{E}_j := d_{u_j}\varphi(\mathcal{B}_1(O_{\varphi(u_j)})))$ is an ellipsoid in $T_{\varphi(u_j)}\mathcal{M}$ we get for this $\mathcal{E}_j$ and $\mathcal{E}'_j = \left(1 + \frac{\eta}{\alpha_n(\mathcal{E}_j)}\right)\mathcal{E}_j$ with Lemma 7.4, Chap. 7

$$\varphi(\mathcal{B}_r(u_j)) \subset \exp_{\varphi(u_j)}\left(r(\mathcal{E}_j + \mathcal{B}_\eta(O_{\varphi(u_j)}))\right) \subset \exp_{\varphi(u_j)}\left(r\mathcal{E}'_j\right).$$

with

$$\alpha := \sqrt{n}\alpha_{\mathcal{K}}(\varphi) \tag{10.6}$$

we have

$$N_{\alpha r}(\varphi(\mathcal{K})) \leq N_r(\mathcal{K}) \max_{j=1,\dots,N_r(\mathcal{K})} N_{\alpha r}(\exp_{\varphi(u_j)}(r\mathcal{E}'_j))$$

and therefore,

$$\mu_F(\varphi(\mathcal{K}), d, \alpha r) \leq \left(\alpha^d \max_{j=1,\dots,N_r(\mathcal{K})} N_{\alpha r}(\exp_{\varphi(u_j)}(r\mathcal{E}'_j))\right)\mu_F(\mathcal{K}, d, r). \tag{10.7}$$

Every ball $\mathcal{B}_{\alpha r}(\upsilon)$, $\upsilon \in \mathcal{M}$ containing points of $\exp_{\varphi(u_j)}(r\mathcal{E}'_j)$ is contained in the ball $\mathcal{B}_{(2+\alpha_1(\mathcal{E}'_j))r}(u_j) \subset \mathcal{B}_{r_2}(u_j)$, and so we have $\mathcal{B}_{\alpha r}(\upsilon) \supset \exp_{\varphi(u_j)}(\mathcal{B}_{\frac{1}{2}\alpha r}(\exp^{-1}_{\varphi(u_j)}\upsilon))$. This means

$$N_{\alpha r}(\exp_{\varphi(u_j)}(r\mathcal{E}'_j)) \leq N_{\frac{1}{2}\alpha r}(r\mathcal{E}'_j).$$

Since $\alpha_{\mathcal{K}}(\varphi) \leq \alpha_n(d_{u_j}\varphi) = \alpha_n(\mathcal{E}_j) \leq \alpha_n(\mathcal{E}'_j)$ is satisfied, Lemma 7.5, Chap. 7 yields

$$N_{\frac{1}{2}\alpha r}(r\mathcal{E}'_j) \leq \frac{2^n\omega_n(r\mathcal{E}'_j)}{(\frac{1}{2}r\alpha_{\mathcal{K}}(\varphi))^n} = \frac{4^n\omega_n(\mathcal{E}'_j)}{\alpha^n_{\mathcal{K}}(\varphi)} \leq \frac{4^n\left(1 + \frac{\eta}{\alpha_{\mathcal{K}}(\varphi)}\right)^n \omega_n(\mathcal{E}_j)}{\alpha^n_{\mathcal{K}}(\varphi)} \leq \frac{8^n\omega_n(d_{u_j}\varphi)}{\alpha^n_{\mathcal{K}}(\varphi)}.$$

Using (10.6), (10.7) and the assumption (10.3) we get

$$\mu_F(\mathcal{K}, d, \alpha r) \leq \mu_F(\varphi(\mathcal{K}), d, \alpha r) \leq \alpha^d \frac{8^n\omega_{n,\mathcal{K}}(\varphi)}{\alpha^n_{\mathcal{K}}(\varphi)} \mu_F(\mathcal{K}, d, r) \\ = n^{\frac{d}{2}} 8^n \omega_{n,\mathcal{K}}(\varphi)\alpha^{d-n}_{\mathcal{K}}(\varphi)\mu_F(\mathcal{K}, d, r) < \mu_F(\mathcal{K}, d, r).$$

Because of (10.3) we have $\alpha < 1$. Therefore, for any $\varepsilon \in (0, r_0)$ we can find a number $l \in \mathbb{N}_0$ such that $\alpha^{l+1} r_0 \leq \varepsilon < \alpha^l r_0$ is satisfied. Finally we get

$$\mu_F(\mathcal{K}, d, \varepsilon) < \mu_F(\mathcal{K}, d, \alpha^{-l}\varepsilon) < N_{\alpha^{-l}\varepsilon}(\mathcal{K}) r_0^d \leq N_{\alpha r_0}(\mathcal{K}) r_0^d \leq \alpha^{-d} \mu_F(\mathcal{K}, d, \alpha r_0),$$

which yields $\mu_F(\mathcal{K}, d) < \infty$ and thus $\dim_F \mathcal{K} \leq d$. □

Corollary 10.5 *Let $(\mathcal{M}, g)$, $\mathcal{U}$ and φ be as in Theorem 10.3, $\mathcal{K}$ and $\widetilde{\mathcal{K}}$ be compact sets satisfying the relation $\mathcal{K} \subset \varphi^j(\mathcal{K}) \subset \widetilde{\mathcal{K}} \subset \mathcal{U}$ and $\alpha_{\widetilde{\mathcal{K}}}(\varphi) > 0$. Suppose that there exists a number $d \in (0, n]$ with*

$$\omega_{n,\widetilde{\mathcal{K}}}(\varphi) \alpha_{\widetilde{\mathcal{K}}}^{d-n}(\varphi) < 1. \tag{10.8}$$

Then $\dim_F \mathcal{K} \leq d$.

Proof From $\mathcal{K} \subset \varphi^j(\mathcal{K}) \subset \widetilde{\mathcal{K}} \subset \mathcal{U}$ we get $\mathcal{K} \subset \varphi^i(\mathcal{K}) \subset \mathcal{U}$ for any $i \in \mathbb{N}$. The iterates of φ satisfy the relations

$$\omega_{n,\mathcal{K}}(\varphi^i) \leq \omega^i_{n,\widetilde{\mathcal{K}}}(\varphi), \quad \alpha_{\mathcal{K}}(\varphi^i) \geq \alpha^i_{\widetilde{\mathcal{K}}}(\varphi) \tag{10.9}$$

and therefore,

$$\omega_{n,\mathcal{K}}(\varphi^i) \alpha_{\mathcal{K}}^{d-n}(\varphi^i) \leq \left(\omega_{n,\widetilde{\mathcal{K}}}(\varphi) \alpha_{\widetilde{\mathcal{K}}}^{d-n}(\varphi)\right)^i.$$

Furthermore we have from the definition

$$\omega_{n,\mathcal{K}}(\varphi^i) \alpha_{\mathcal{K}}^{d-n}(\varphi^i) \geq \alpha_{\mathcal{K}}^n(\varphi^i) \alpha_{\mathcal{K}}^{d-n}(\varphi^i) = \alpha_{\mathcal{K}}^d(\varphi^i). \tag{10.10}$$

By using (10.8), (10.9) and (10.10) without loss of generality we can assume

$$\omega_{n,\mathcal{K}}(\varphi) \alpha_{\mathcal{K}}^{d-n}(\varphi) \leq 8^{-n} n^{-\frac{d}{2}} \text{ and } \alpha_{\mathcal{K}}(\varphi) < n^{-\frac{1}{2}}.$$

In the opposite case consider the map φ^i with sufficiently large i. With Theorem 10.3 we get $\dim_F \mathcal{K} \leq d$. □

Corollary 10.6 *Let $(\mathcal{M}, g)$, $\mathcal{U}$, $\mathcal{K}$ and φ be defined as in Theorem 10.3 and $\varkappa : \mathcal{U} \to \mathbb{R}_+$ be a continuous function. Suppose that the following conditions are satisfied:*

(1) $\omega_n(d_u\varphi) = const \neq 0 \ \forall u \in \mathcal{K}$.
(2) There exists a number $s \in (0, 1]$ such that $\frac{\varkappa(\varphi(u))}{\varkappa(u)} \omega_{n-1+s}(d_u\varphi) < 1 \ \forall u \in \mathcal{K}$. Then $\dim_F \mathcal{K} \leq n - 1 + s$.

Proof According to condition (2) there exists a positive number $\kappa < 1$ with

$$\frac{\varkappa(\varphi(u))}{\varkappa(u)} \omega_{n-1+s}(d_u\varphi) \leq \kappa$$

for any $u \in \mathcal{K}$. Therefore, by the chain rule we have

$$\begin{aligned}
&\omega_{n-1+s,\mathcal{K}}(\varphi^i) \\
&\leq \max_{u\in\mathcal{K}} \frac{\varkappa(u)}{\varkappa(\varphi^i(u))} \frac{\varkappa(\varphi^i(u))}{\varkappa(\varphi^{i-1}(u))} \omega_{n-1+s}(d_{\varphi^{i-1}(u)}\varphi) \ldots \frac{\varkappa(\varphi(u))}{\varkappa(u)} \omega_{n-1+s}(d_u\varphi) \\
&\leq \frac{\max_{u\in\mathcal{K}} \varkappa(u)}{\min_{u\in\mathcal{K}} \varkappa(u)} \kappa^i .
\end{aligned}$$

Furthermore the relation $\omega_{n-1+s,\mathcal{K}}(\varphi^i) \geq \alpha_{\mathcal{K}}^{n-1+s}(\varphi^i)$ holds. Therefore, without loss of generality we can assume that $\omega_{n-1+s,\mathcal{K}}(\varphi) < 8^{-n} n^{\frac{n-1+s}{2}}$ and $\alpha_{\mathcal{K}}(\varphi) < n^{-\frac{1}{2}}$ is satisfied. In the opposite case consider φ^i with sufficiently large i. We take $u_0 \in \mathcal{K}$ such that $\alpha_n(d_{u_0}\varphi) = \alpha_{\mathcal{K}}(\varphi)$. Resulting from condition (1) we obtain

$$\omega_{n,\mathcal{K}}(\varphi)\alpha_{\mathcal{K}}^{s-1}(\varphi) = \omega_n(d_{u_0}\varphi)\alpha_n^{s-1}(d_{u_0}\varphi) = \omega_{n-1+s}(d_{u_0}\varphi) < 8^{-n} n^{\frac{n-1+s}{2}} .$$

with Theorem 10.3 we get $\dim_F \mathcal{K} \leq n - 1 + s$. □

Remark 10.2 Conditions analogous to (1) of Corollary 10.6 are considered in [6] for invertible maps as the Hénon system. In contrast to our results the fractal dimension estimates in [6] are given in terms of Lyapunov exponents and without use of a Lyapunov-type function $\varkappa$.

Now we want to include the "degree of non-injectivity" in the method of estimating the fractal dimension developed in Theorem 10.3.

Theorem 10.4 *Let $(\mathcal{M}, g)$ be a smooth n-dimensional Riemannian manifold, $\mathcal{U} \subset \mathcal{M}$ be an open set and $\varphi : \mathcal{U} \to \mathcal{M}$ be a C^1-map. Suppose $\mathcal{K} \subset \mathcal{U}$ is a compact set satisfying the relation $\mathcal{K} \subset \varphi(\mathcal{K}) \subset \mathcal{U}$. Suppose $\alpha_{\mathcal{K}}(\varphi) > 0$, and let $a, b > 0$ be numbers such that the following conditions are satisfied:*

(1) There exists a number $d \in (0, n]$ with

$$\begin{aligned}
\alpha_{\mathcal{K}}(\varphi) &< a^{-\frac{1}{n}} b^{\frac{d-n}{n}} n^{-\frac{1}{2}}, \\
\omega_{n,\mathcal{K}}(\varphi)\alpha_{\mathcal{K}}^{d-n}(\varphi) &\leq a^{-\frac{d}{n}} b^{\frac{d}{n}(d-n)} 8^{-n} n^{-\frac{d}{2}};
\end{aligned}$$

(2) For any $j \in \mathbb{N}$ there are a compact set $\mathcal{K}_j \subset \mathcal{K}$ and a number $\varepsilon_j > 0$ such that

$$\begin{aligned}
\mu_F(\varphi^j(\mathcal{K}_j), d, \varepsilon) &= \mu_F(\varphi^j(\mathcal{K}), d, \varepsilon), \\
\mu_F(\mathcal{K}_j, d, b^j\varepsilon) &\leq a^j \mu_F(\mathcal{K}, d, \varepsilon)
\end{aligned}$$

for any $\varepsilon \in (0, \varepsilon_j]$ are satisfied. Then $\dim_F \mathcal{K} \leq d$.

Proof Analogous to the proof of Theorem 10.3 let $\eta \in (0, \alpha_{\mathcal{K}}(\varphi))$ be an arbitrary number. Let $r_1, r_2 > 0$ be so small that there exists an open set $\mathcal{V} \subset \mathcal{M}$ containing $\mathcal{K}$ and that $\mathcal{V}$ is a contained in a compact subset of $\mathcal{U}$ such that the inequalities $|\tau_{\varphi(v)}^{\varphi(u)} d_v\varphi\tau_u^v - d_u\varphi| \leq \eta$ for any $u, v \in \mathcal{V}$ with $\rho(u, v) \leq r_1$ and $|d_v \exp_u| \leq 2$ for any $u \in \mathcal{V}$ and any $v \in \mathcal{B}_{r_2}(O_u)$ are satisfied. With α' defined in the proof of Theorem 10.3 we can find a number $r_0 \le \min\{r_1, \frac{r_2}{2+\alpha'+\eta}, \varepsilon_1\}$ such that any ball $\mathcal{B}_{r_0}(u)$ containing points of $\mathcal{K}$ is entirely contained in $\mathcal{V}$. Let $r \in (0, r_0)$ be fixed. Since $\mathcal{K}_1$ is compact there is a finite number of points $u_j \in \mathcal{V}$, $j = 1, \ldots, N_r(\mathcal{K}_1)$, such that $\mathcal{K}_1 = \bigcup_{j=1}^{N_r(\mathcal{K}_1)} \mathcal{B}_r(u_j) \cap \mathcal{K}_1$ and therefore,

$$\varphi(\mathcal{K}_1) = \bigcup_{j=1}^{N_r(\mathcal{K}_1)} \varphi(\mathcal{B}_r(u_j) \cap \mathcal{K}_1)$$

is satisfied. Using the Taylor formula we get that the image of every ball $\mathcal{B}_r(u_j)$ under φ satisfies the inclusion

$$\varphi(\mathcal{B}_r(u_j)) \subset \exp_{\varphi(u_j)}(d_{u_j}\varphi(\mathcal{B}_r(O_{u_j})) + \mathcal{B}_{\eta r}(O_{\varphi(u_j)})).$$

with $\mathcal{E}_j := d_{u_j}\varphi(\mathcal{B}_1(O_{\varphi(u_j)}))$ and $\mathcal{E}_j' = \left(1 + \frac{\eta}{\alpha_n(\mathcal{E}_j)}\right)\mathcal{E}_j$ we get by means of Lemma 7.4, Chap. 7

$$\varphi(\mathcal{B}_r(u_j)) \subset \exp_{\varphi(u_j)}\left(r(\mathcal{E}_j + \mathcal{B}_\eta(O_{\varphi(u_j)}))\right) \subset \exp_{\varphi(u_j)}\left(r\mathcal{E}_j'\right).$$

with $\alpha := a^{\frac{1}{n}} b^{\frac{n-d}{n}} \sqrt{n}\alpha_{\mathcal{K}}(\varphi)$ we obtain

$$N_{\alpha r}(\varphi(\mathcal{K}_1)) \le N_{br}(\mathcal{K}_1) \max_{j=1,\ldots,N_r(\mathcal{K}_1)} N_{\alpha r}(\exp_{\varphi(u_j)}(br\mathcal{E}_j'))$$

and therefore,

$$\mu_F(\varphi(\mathcal{K}_1), d, \alpha r) \le \left(\alpha^d b^{-d} \max_{j=1,\ldots,N_r(\mathcal{K}_1)} N_{\alpha r}(\exp_{\varphi(u_j)}(br\mathcal{E}_j'))\right)\mu_F(\mathcal{K}, d, r).$$

Analogous to the proof of Theorem 10.3 we get $N_{\alpha r}(\exp_{\varphi(u_j)}(br\mathcal{E}_j')) \le N_{\frac{1}{2}\alpha r}(br\mathcal{E}_j')$.

Since $\alpha_{\mathcal{K}}(\varphi) \le \alpha_n(d_{u_j}\varphi) = \alpha_n(\mathcal{E}_j) \le \alpha_n(\mathcal{E}_j')$ is satisfied, Lemma 7.5, Chap. 7 yields

$$\begin{aligned} N_{\frac{1}{2}\alpha r}(br\mathcal{E}_j') &\le \frac{2^n\omega_n(br\mathcal{E}_j')}{(\frac{1}{2}a^{\frac{1}{n}}b^{\frac{n-d}{n}}r\alpha_{\mathcal{K}}(\varphi))^n} = \frac{4^n b^d \omega_n(\mathcal{E}_j')}{a\alpha_{\mathcal{K}}^n(\varphi)} \le \frac{4^n b^d \left(1 + \frac{\eta}{\alpha_{\mathcal{K}}(\varphi)}\right)^n \omega_n(\mathcal{E}_j)}{a\alpha_{\mathcal{K}}^n(\varphi)} \\ &\le \frac{8^n b^d \omega_n(d_{u_j}\varphi)}{a\alpha_{\mathcal{K}}^n(\varphi)}. \end{aligned}$$

Thus we have

$$\begin{aligned}\mu_F(\mathcal{K}, d, \alpha r) &\le \mu_F(\varphi(\mathcal{K}), d, \alpha r) = \mu_F(\varphi(\mathcal{K}_1), d, \alpha r)\\ &\le \alpha^d b^{-d} \frac{8^n b^d \omega_n(d_{u_j}\varphi)}{a\alpha^n_{\mathcal{K}}(\varphi)} \mu_F(\mathcal{K}_1, d, br)\\ &\le a^{\frac{d}{n}} b^{\frac{d}{n}(n-d)} n^{\frac{d}{2}} \frac{8^n \omega_n(d_{u_j}\varphi)}{a\alpha^n_{\mathcal{K}}(\varphi)} \mu_F(\mathcal{K}_1, d, br)\\ &\le \mu_F(\mathcal{K}, d, r).\end{aligned}$$

Since $\alpha < 1$, analogous to the end of the proof of Theorem 10.3, this yields $\dim_F \mathcal{K} \le d$. □

Corollary 10.7 *Let $(\mathcal{M}, g)$, $\mathcal{U}$ and φ be as in Theorem 10.4, $\mathcal{K}$ and $\widetilde{\mathcal{K}}$ be compact sets satisfying the relation $\mathcal{K} \subset \varphi^j(\mathcal{K}) \subset \widetilde{\mathcal{K}} \subset \mathcal{U}$ and $\alpha_{\widetilde{\mathcal{K}}}(\varphi) > 0$. Let condition (2) of Theorem 10.4 be satisfied and assume that there exists a number $d \in (0, n]$ with*

$$\omega_{n,\widetilde{\mathcal{K}}}(\varphi)\alpha^{\frac{d-n}{\widetilde{\mathcal{K}}}}(\varphi) < a^{-\frac{d}{n}} b^{\frac{d}{n}(d-n)}. \tag{10.11}$$

Then $\dim_F \mathcal{K} \le d$.

Proof From $\mathcal{K} \subset \varphi^j(\mathcal{K}) \subset \widetilde{\mathcal{K}} \subset \mathcal{U}$ we get $\mathcal{K} \subset \varphi^i(\mathcal{K}) \subset \mathcal{U}$ for any $i \in \mathbb{N}$. The iterates of φ satisfy the relations

$$\omega_{n,\mathcal{K}}(\varphi^i) \le \omega^i_{n,\widetilde{\mathcal{K}}}(\varphi), \quad \alpha_{\mathcal{K}}(\varphi^i) \ge \alpha^i_{\widetilde{\mathcal{K}}}(\varphi),$$

and therefore,

$$a^{\frac{di}{n}} b^{\frac{di}{n}(n-d)} \omega_{n,\mathcal{K}}(\varphi^i)\alpha^{d-n}_{\mathcal{K}}(\varphi^i) \le \left(a^{\frac{d}{n}} b^{\frac{d}{n}(n-d)} \omega_{n,\widetilde{\mathcal{K}}}(\varphi)\alpha^{d-n}_{\widetilde{\mathcal{K}}}(\varphi)\right)^i.$$

Furthermore,

$$a^{\frac{d}{n}} b^{\frac{d}{n}(n-d)} \omega_{n,\mathcal{K}}(\varphi^i)\alpha^{d-n}_{\mathcal{K}}(\varphi^i) \ge a^{\frac{d}{n}} b^{\frac{d}{n}(n-d)} \alpha^n_{\mathcal{K}}(\varphi^i)\alpha^{d-n}_{\mathcal{K}}(\varphi^i) = \left(a^{\frac{1}{n}} b^{\frac{n-d}{n}} \alpha_{\mathcal{K}}(\varphi^i)\right)^d$$

holds. Thus without loss of generality we can assume

$$a^{\frac{d}{n}} b^{\frac{d}{n}(n-d)} \omega_{n,\mathcal{K}}(\varphi)\alpha^{d-n}_{\mathcal{K}}(\varphi) \le 8^{-n} n^{-\frac{d}{2}} \text{ and } a^{\frac{1}{n}} b^{\frac{n-d}{n}} \alpha_{\mathcal{K}}(\varphi) < n^{-\frac{1}{2}}.$$

Otherwise consider the map φ^i with sufficiently large i and substitute a by a^i and b by b^i. With Theorem 10.4 we get $\dim_F \mathcal{K} \le d$. □

Example 10.4 Let us again consider the modified horseshoe map in two dimensions. For the sets $\mathcal{K}_j$ defined before we have $N_\varepsilon(\mathcal{K}_j) \le 4^j N_{\frac{\varepsilon}{\gamma^j}}(\mathcal{K})$ for sufficiently small $\varepsilon > 0$, i.e. with $a = 4\gamma^d$ and $b = \gamma$ condition (2) of Theorem 10.4 is satisfied. Then condition (10.11) results in

$$4^{\frac{d}{2}} \gamma^d \beta_2 \alpha^{d-1} < 1, \tag{10.12}$$

which is equivalent to $d > \frac{\ln\alpha - \ln\beta_2}{\ln 2 + \ln\alpha + \ln\gamma}$. Corollary 10.7 can be applied for any such d and shows that $\dim_F \mathcal{K} \leq \frac{\ln\alpha - \ln\beta_2}{\ln 2 + \ln\alpha + \ln\gamma}$. For the parameters $\alpha = \frac{1}{3}$, $\beta_1 = 3$, $\beta_2 = 5$ we get $\dim_F \mathcal{K} \leq 1.800$.

By changing the metric with the Lyapunov-type function κ used in Example 10.2 we alter the form of the balls covering $\mathcal{K} \setminus \mathcal{K}_1$. However, again we have $N_{\gamma^j\varepsilon}(\mathcal{K}_j) \leq 4^j N_\varepsilon(\mathcal{K})$, i.e. condition (2) holds with $a = 4\gamma^d$ and $b = \gamma$. Condition (10.11) now results in $4^{\frac{d}{2}}\gamma^d\sqrt{\beta_1\beta_2}\alpha^{d-1} < 1$, which means $\dim_F \mathcal{K} \leq \frac{\ln\alpha - \frac{1}{2}\ln\beta_1\beta_2}{\ln 2 + \ln\alpha + \ln\gamma}$. For $\alpha = \frac{1}{3}$, $\beta_1 = 3$, $\beta_2 = 5$ we get $\dim_F \mathcal{K} \leq 1.631$.

Remark 10.3 For two-dimensional horseshoe maps the upper bound for the fractal dimension of an invariant set obtained by Corollary 10.7 is always smaller than the bound for the Hausdorff dimension by Theorem 10.2, because for $d < 2$ condition (10.12) is weaker than condition (1) of this theorem. For $d > 2$ this relation is reversed, i.e. for the considered horseshoe maps in more than two dimensions the estimates of Subsect. 10.1.2 really will be useful.

10.2 Dimension Estimates for Piecewise C^1-Maps

10.2.1 Decomposition of Invariant Sets of Piecewise Smooth Maps

One possibility to handle dimension estimates for non-differentiable maps on an n-dimensional manifold is to suppose that the set of non-differentiable points is a finite union of submanifolds having topological dimension less than the dimension of the manifold and to consider only orbits being without contact to the set of non-differentiability (see [2]). In contrast to this in the present subsection we consider a class of piecewise smooth maps whose preimage sets of non-differentiability points for various iterates are bounded. The maps under consideration are supposed to be differentiable on the elements of a partition of the given set such that the preimages of these elements under the iterated map are controllable in a certain sense. For some large classes of maps, our Hausdorff dimension bounds agree with the dimension of their invariant sets prevalently.

In concrete physical or technical systems the considered maps are often not only non-smooth but besides this even non-injective, i.e. they show a "many to one" behavior (see for instance [4, 5, 11]). For uniformly non-injective maps it is possible as it was shown in Sect. 10.1 to include into the dimension estimates some information about the "degree of non-injectivity". We apply the approach of [5], presented in Sect. 10.1, to our class of piecewise C^1-smooth maps.

In general, the singular value function computed for the differential of the basic map does not satisfy the contraction condition on all parts of the invariant set. Nevertheless in such a situation it may be possible to get a contraction for the outer Hausdorff measure if higher iterates of the given map are considered. For this purpose we divide the phase space into "good" and "bad" parts in order to compensate the growth of the outer measure under one iteration in bad parts by the decrease of such a measure in good parts of the phase space. By investigating the asymptotic behavior of the system we have to guarantee that any orbit stays sufficiently long in good parts and not too long in bad parts of the phase space.

The section is organized as follows. The definition of the considered class of piecewise smooth maps is given in Subsect. 10.2.2. Subsection 10.2.3 is concerned with the proof of the Douady-Oesterlé formula from [33] for piecewise smooth maps. The degree of non-injectivity is introduced in the Douady-Oesterlé estimate in Subsect. 10.2.4. In Subsect. 10.2.5, which is based on [36], some information on the statistical long time behavior of the orbits is considered in the Douady-Oesterlé formula.

10.2.2 A Class of Piecewise C^1-Maps

Let $(\mathcal{M}, g)$ be an n-dimensional smooth C^∞-Riemannian manifold and let $\rho(\cdot,\cdot)$ denote the geodesic distance between two arbitrary points of $\mathcal{M}$.

We now characterize a class of piecewise C^1-smooth maps $\varphi : \mathcal{U} \subset \mathcal{M} \to \mathcal{M}$, for which the Douady-Oesterlé estimate will be proved. Roughly speaking, these maps possess the property that the non-differentiability set in $\mathcal{U}$ have controllable preimage sets. It will be shown that some well-known classes of piecewise differentiable maps have this property.

Formally the definition is as follows. We say that a map $\varphi : \mathcal{U} \subset \mathcal{M} \to \mathcal{M}$ *satisfies the hypothesis* **(H)** if there exists a finite or infinite index set $I \subset \{1, 2, \ldots\}$ and a partition $\mathcal{U} = \bigcup_{i\in I} \mathcal{U}_i$ of subsets $\mathcal{U}_i \subset \mathcal{U}$ having $\mathcal{U}_i \cap \mathcal{U}_j = \emptyset$ for $i \neq j$ such that the following conditions **(H0)**–**(H2)** hold:

(H0) The restriction $\varphi|_{\mathrm{int}\,\mathcal{U}_i}$ is C^1 for every $i \in I$.

(H1) For any $m \in \mathbb{N}$ the set $\mathcal{U}$ can be decomposed into connected subsets $\mathcal{U}_i^{(m)}$, i.e. $\mathrm{int}\,\mathcal{U}_i^{(m)} \neq \emptyset$ is connected and $\mathcal{U}_i^{(m)} \subset \overline{\mathrm{int}\,\mathcal{U}_i^{(m)}}$, so that the map φ^m is C^1 on the set $\mathrm{int}\,\mathcal{U}_i^{(m)}$ and φ^m can be extended to a C^1-map on any $\overline{\mathcal{U}_i^{(m)}}$.

(H2) There is a natural number $\widetilde{k}$ with the property that for arbitrary $m \in \mathbb{N}$ there exists a real number $\widetilde{\varepsilon}(m) > 0$ such that for any $\varepsilon \in (0, \widetilde{\varepsilon}(m)]$ and any $u \in \mathcal{M}$ with $\mathrm{dist}(u, \mathcal{U}) \le \widetilde{\varepsilon}(m)$ the ball $\mathcal{B}(u, \varepsilon)$ can be decomposed into at most $\widetilde{k}$ connected subsets $\mathcal{U}_i^{(m)} \cap \mathcal{B}(u, \varepsilon)$.

The following two examples show that certain well-known classes of maps satisfy the condition **(H)**.

Example 10.5 (*Baker's map*) Consider the square $\mathcal{U} = (0, 1] \times [0, 1] \subset \mathbb{R}^2$ and the map $\varphi : \mathcal{U} \to \mathcal{U}$ defined by

$$\varphi(x, y) = \begin{cases} (2x, \lambda y), & \text{for } 0 < x \le \frac{1}{2},\ 0 \le y \le 1, \\ (2x - 1, \lambda y + \frac{1}{2}), & \text{for } \frac{1}{2} < x \le 1,\ 0 \le y \le 1 \end{cases}$$

with $\lambda \in (0, \frac{1}{2})$ as a parameter (see [10]). In order to show that the condition **(H)** is satisfied we choose the sets $\mathcal{U}_1 = \left(0, \frac{1}{2}\right] \times [0, 1]$, $\mathcal{U}_2 = \left(\frac{1}{2}, 1\right] \times [0, 1]$ and $I = \{1, 2\}$. Then it is easy to see that **(H)** is satisfied with $\widetilde{k} = 2$, $\widetilde{\varepsilon}(m) = \left(\frac{1}{2}\right)^{m+1}$ and

$\mathcal{U}_i^{(m)} = \left(\left(\frac{1}{2}\right)^m (i-1), \left(\frac{1}{2}\right)^m i\right] \times [0, 1]$ for $i = 1, \ldots, 2^m$, and every $m \in \mathbb{N}$.

Example 10.6 Consider the set $\mathcal{U} = (0, 1] \times [0, 1] \subset \mathbb{R}^2$ and the map $\varphi : \mathcal{U} \to \mathcal{U}$ defined by

$$\varphi(x, y) = \begin{cases} (l_1 x, r_1 y + a_1), & \text{for } 0 < x \le \frac{1}{l_1},\ 0 \le y \le 1, \\ \left(l_i\left(x - \sum_{k=1}^{i-1} \frac{1}{l_k}\right), r_i y + a_i\right), & \text{for } \sum_{k=1}^{i-1} \frac{1}{l_k} < x \le \sum_{k=1}^{i} \frac{1}{l_k},\ i = 2, \ldots, r, \\ & 0 \le y \le 1. \end{cases}$$

Here it is assumed that $r \ge 2$ is a fixed natural number, l_i, r_i and a_i are reals with $l_i > 1$, $0 \le a_i < 1$, $0 < r_i \le 1 - a_i$ $(i = 1, 2, \ldots, r)$ and $\sum_{i=1}^r \frac{1}{l_i} = 1$. Suppose further that $a_i + r_i < a_{i+1}$ for $i = 1, 2, \ldots, r-1$. We choose the sets $\mathcal{U}_1 = \left(0, \frac{1}{l_1}\right] \times [0, 1]$ and $\mathcal{U}_i = \left(\sum_{k=1}^{i-1} \frac{1}{l_k}, \sum_{k=1}^{i} \frac{1}{l_k}\right] \times [0, 1]$ $(i = 2, 3, \ldots, r)$ and the index set $I = \{1, 2, \ldots, r\}$. Obviously the condition **(H)** is satisfied with $\widetilde{k} = 2$ and $\widetilde{\varepsilon}(m) = \frac{1}{2}(\max_{i=1,\ldots,r} l_i)^{-m}$ for every $m \in \mathbb{N}$.

Since the maps from these examples are not everywhere differentiable the known methods from [5, 9, 20, 29, 38], presented in Chaps. 5 and 7, can not directly be applied to get dimension estimates of the invariant sets. We will come back to these examples later in Subsect. 10.2.3.

We present now an example which illustrates that **(H0)** and **(H1)** may be true although **(H2)** is not satisfied.

Example 10.7 (*Belykh map*) Consider the map $\widehat{\varphi} : \mathbb{R}^2 \to \mathbb{R}^2$ given by

$$\widehat{\varphi}(\theta, \eta) = (\theta + \delta_1 \eta - \delta_1 \delta_2 (F(\theta) - \delta_3), \eta (1 - \delta_4 \delta_1) - \delta_1 \delta_5 (F(\theta) - \delta_3)). \tag{10.13}$$

We suppose that in (10.13) the parameters $\delta_2, \delta_5, \delta_4$ are non-negative, δ_1 is positive and δ_3 is a real. Suppose that $F : \mathbb{R} \to \mathbb{R}$ is a 2π-periodic function. Note that the map (10.13) describes certain discrete systems of phase synchronization [4].

We now assume that $\delta_5 > 0$ and F is of the special type $F(\theta) = 1 - \frac{\theta}{\pi}$ (θ mod 2π). In this case the given map (10.13) can be transformed by $x = \frac{1}{2\pi}(\theta - (1 - \delta_3)\pi - \frac{\delta_1 \eta}{a})$, $y = \frac{1}{2\pi}(\theta - (1 - \delta_3)\pi + \frac{\delta_1 \eta}{b})$ with $\lambda = (1 - \delta_4 \delta_1) - a$, $\mu = (1 - \delta_4 \delta_1) + b$, $c = \frac{1 - \delta_3}{2}$, $a > 0$, $b > 0$ into the form

$$\varphi(x, y) = (\lambda x, \mu y) \quad \left(\left(\frac{ax + by}{a + b} + c\right) \bmod 1\right). \tag{10.14}$$

Note that $\mu - \lambda = a + b > 0$ and, under the condition $(\frac{\delta_1\delta_2}{\pi} + 1)(1 - \delta_4\delta_1) - \frac{\delta_1^2 s}{\pi} \neq 0$, also $\lambda\mu \neq 0$.

The map (10.14) can be considered on the two-dimensional torus T^2, defined by the equivalence relation

$$(x, y) \sim (x', y') \Longleftrightarrow x' = x + k + \frac{l}{\lambda}, \quad y' = y + k + \frac{l}{\mu}, \quad (k, l \in \mathbb{Z}).$$

Denote by $\pi : \mathbb{R}^2 \to T^2$ the canonical projection. Then the discontinuity set $\{k2\pi, k \in \mathbb{Z}\}$ of F transforms into the discontinuity set $\pi(\mathcal{G})$ of φ on T^2, where

$$\mathcal{G} := \Big\{(x, y) \in \mathbb{R}^2 : y = \frac{-c(a + b) - ax}{b}\Big\}.$$

In the following we suppose that the torus is represented as $T^2 = \pi(\mathcal{Q})$, where

$$\mathcal{Q} := \Big\{(x, y) \in \mathbb{R}^2 : 0 \leq \frac{ax + by}{a + b} + c < 1,\ 0 \leq \frac{a\lambda x + b\mu y}{a + b} + c < 1\Big\}.$$

Using this representation the discontinuity set of φ^m can be written as union of line segments

$$\pi\Big(\Big\{(x, y) \in \mathcal{Q} : \exists j \in \{1, \ldots, m\} \text{ with } \frac{a\lambda^{j-1}x + b\mu^{j-1}y}{a + b} + c \in \mathbb{Z}\Big\}\Big).$$

It follows that the sets $\mathcal{U}_i^{(m)}$ which have to be considered in condition **(H1)** posses bounderies consisting of discontinuity points of φ^m. Since φ^m here is linear, the C^1-extension of this function is possible. Thus, for the considered map (10.14) the conditions **(H0)** and **(H1)** are satisfied. In order to verify **(H2)** we distinguish two cases in parameter space.

Case 1 $c = 0$ and $\lambda < -\mu < 0$ or $\mu > \lambda > 0$.

Obviously, $\pi((0, 0)) \in T^2$ is a fixed point of φ. The family of discontinuity line segments for φ^m through $\pi((0, 0))$ is given by the projection of these segments in $\mathcal{Q}$

$$\frac{a\lambda^{j-1}x + b\mu^{j-1}y}{a + b} + c = 0, \quad \text{i.e.,} \quad y = -\frac{a}{b}\Big(\frac{\lambda}{\mu}\Big)^{j-1} x.$$

For sufficiently small $x > 0$ the projection of these points lies in T^2.

For $\mu = 2, \lambda = 1, a = \frac{2}{3}, b = \frac{1}{3}$ and $j = 1, 2, 3, 4$ these discontinuity sets are shown in Fig. 10.2a).

As a consequence the number of preimage sets $\mathcal{U}_i^{(m)}$ near $\pi((0, 0))$ which we have to consider in **(H1)** is proportional to m and so it is impossible to find a number $\widetilde{k}$ such that **(H2)** is satisfied.

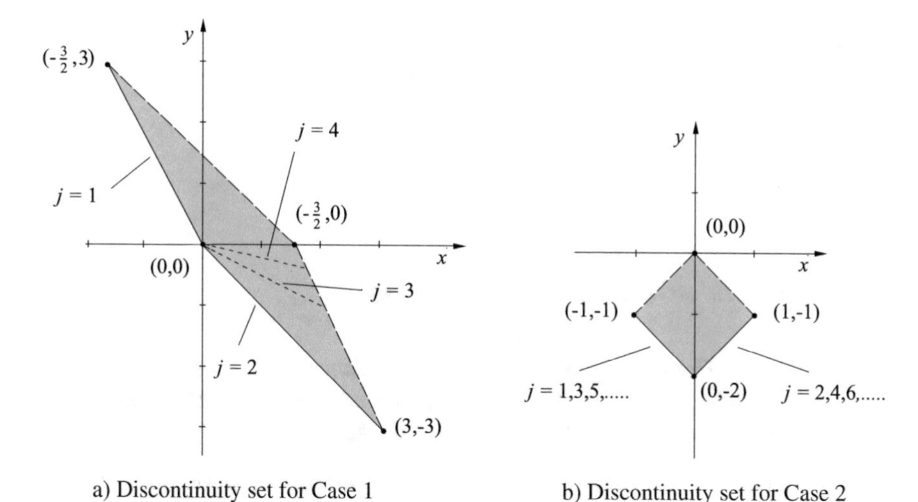

Fig. 10.2 Non-differentiability sets for the Belykh map

Case 2 $c \in \mathbb{R}$ arbitrary and $\mu = -\lambda (= \frac{a+b}{2} > 0)$. The discontinuity sets of φ^m are given by the projections of the line segments in Q written as

$$\exists k_j \in \mathbb{Z} : \frac{a\lambda^{j-1}x + b\mu^{j-1}y}{a+b} + c = k_j, \quad \text{i.e.,} \quad y = (-1)^j \, \frac{a}{b}x + \frac{2^{j-1}(k_j - c)}{b(a+b)^{j-2}}.$$

It follows that all these segments are parallel to the line segments $y = \frac{a}{b}x$ resp. $y = -\frac{a}{b}x$ in Q. Thus, for any $m \in \mathbb{N}$ there exist only finitely many such parallels and, consequently, they may have only finitely many (transversal) intersections. All these intersection points have a positive distance to neighboring parallels. Thus, in order to satisfy (**H2**) it is sufficient to take $\widetilde{k} = 4$ and to require that for any $m \in \mathbb{N}$ the number $\widetilde{\varepsilon}(m)$ is so small that $2\widetilde{\varepsilon}(m)$ is not greater than the minimum of the considered distances. For $\mu = 1$, $\lambda = -1$, $a = b = c = 1$ and various j Case 2 is illustrated in Fig. 10.2b).

Suppose now that $(\mathcal{M}, g)$ is an n-dimensional smooth Riemannian manifold and $\mathcal{U} \subset \mathcal{M}$ is a subset. For a map $\varphi : \mathcal{U} \to \mathcal{M}$ satisfying (**H**), a number $d \in [0, n]$ and an arbitrary $u \in \mathcal{U}$ we denote by $d_u\varphi : T_u\mathcal{M} \to T_{\varphi(u)}\mathcal{M}$ the tangent map of the C^1-extension of φ and define for any bounded set $\mathcal{K} \subset \mathcal{U}$ the function

$$\omega_{d,\mathcal{K}}(\varphi) := \sup_{u \in \mathcal{K}} \omega_d(d_u\varphi).$$

We formulate the following lemma which will be used in the next subsection and which can be proved similarly as the corresponding theorems in Chaps. 5 and 8.

Lemma 10.1 *Let $\varphi : \mathcal{U} \subset \mathcal{M} \to \mathcal{M}$ be a map satisfying* **(H)**. *Furthermore, let $\mathcal{K} \subset \mathcal{U}$ be a bounded and φ-invariant set (i.e. $\varphi(\mathcal{K}) = \mathcal{K}$) and $d \in [0, n]$ an arbitrary number. Then the following two statements are true:*

(a) $\omega_d(d_u\varphi^2) \le \omega_d(d_{\varphi(u)}\varphi) \cdot \omega_d(d_u\varphi)$ *for all $u \in \mathcal{K}$.*
(b) $\omega_{d,\mathcal{K}}(\varphi^m) \le (\omega_{d,\mathcal{K}}(\varphi))^m$ *for all $m \in \mathbb{N}$.*

10.2.3 Douady-Oesterlé-Type Estimates

In Chaps. 5 and 8 Douady-Oesterlé-type Hausdorff dimension estimates for invariant sets of C^1-smooth maps are derived. In this section we generalize some of these results for the class of piecewise smooth maps satisfying **(H)**.

Theorem 10.5 *Suppose that $\varphi : \mathcal{U} \subset \mathcal{M} \to \mathcal{M}$ satisfies* **(H)**. *Let $\mathcal{K} \subset \mathcal{U}$ be a bounded and φ-invariant set (i.e. $\varphi(\mathcal{K}) = \mathcal{K}$) and suppose that for some number $d \in (0, n]$ the inequality $\omega_{d,\mathcal{K}}(\varphi) < 1$ is satisfied. Then $\dim_H \mathcal{K} \le d$.*

Corollary 10.8 *Suppose that $\varphi : \mathcal{U} \subset \mathcal{M} \to \mathcal{M}$ satisfies* **(H)**. *Let $\mathcal{K} \subset \mathcal{U}$ be bounded and φ-invariant and suppose that there exists a number $d \in (0, n)$ with $\omega_{d,\mathcal{K}}(\varphi) = 1$ and $\omega_{\widetilde{d},\mathcal{K}}(\varphi) < 1$ for all $\widetilde{d} \in (d, n]$. Then $\dim_H \mathcal{K} \le d$.*

Proof By Theorem 10.5 we have $\dim_H \mathcal{K} \le \widetilde{d}$ for all $\widetilde{d} > d$. For $\widetilde{d} \to d + 0$ we get $\dim_H \mathcal{K} \le d$. □

Corollary 10.9 *Suppose that $\varphi : \mathcal{U} \subset \mathcal{M} \to \mathcal{M}$ satisfies* **(H)**. *Let $\mathcal{K} \subset \mathcal{U}$ be a compact and φ-invariant set and suppose that for a certain number $d \in (0, n]$ and some continuous function $\varkappa : \mathcal{K} \to \mathbb{R}_+ = \{x \in \mathbb{R} : x > 0\}$ the condition*

$$\sup_{u \in \mathcal{K}} \left(\frac{\varkappa(\varphi(u))}{\varkappa(u)} \, \omega_d(d_u\varphi) \right) < 1$$

is satisfied. Then $\dim_H \mathcal{K} \le d$.

Proof The proof is based on a technique considering $\varkappa$ as Lyapunov-type function as it is done in Chaps. 5 and 8. We set $\kappa := \sup_{u \in \mathcal{K}} \left(\frac{\varkappa(\varphi(u))}{\varkappa(u)} \omega_d(d_u\varphi) \right)$. By assumption the value κ is less than one. Using the chain rule and applying Lemma 10.1 we have

$$\begin{aligned} \omega_d(d_u\varphi^j) &\le \omega_d(d_{\varphi^{j-1}(u)}\varphi) \cdot \ldots \cdot \omega_d(d_u\varphi) \le \kappa \frac{\varkappa(\varphi^{j-1}(u))}{\varkappa(\varphi^j(u))} \cdot \ldots \cdot \kappa \frac{\varkappa(u)}{\varkappa(\varphi(u))} \\ &= \kappa^j \frac{\varkappa(u)}{\varkappa(\varphi^j(u))} \le \kappa^j \frac{\sup_{u \in \mathcal{K}} \varkappa(u)}{\inf_{u \in \mathcal{K}} \varkappa(u)} \end{aligned}$$

for all $j \in \mathbb{N}$ and for all $u \in \mathcal{K}$. Thus, there exists an integer j such that $\sup_{u \in \mathcal{K}} \omega_d(d_u\varphi^j) < 1$ holds. Now we can apply Theorem 10.5 to the map φ^j and we obtain $\dim_H \mathcal{K} \le d$. □

Before proving Theorem 10.5 we formulate a technical lemma, which generalizes a result in Chap. 8 for the case of bounded sets and piecewise C^1-maps on manifolds.

Lemma 10.2 *Suppose that $\varphi : \mathcal{U} \subset \mathcal{M} \to \mathcal{M}$ satisfies* **(H)**, *$d \in (0, n]$ is a number written as $d = d_0 + s$ with $d_0 \in \{0, 1, \ldots, n-1\}$ and $s \in (0, 1]$, C and λ are constants defined by $C = 2^{d_0}(d_0+1)^{d/2}$ and $\lambda = 2\sqrt{d_0+1}$. Suppose that for a bounded and φ-invariant set $\mathcal{K} \subset \mathcal{U}$ and a natural number m the inequality $\omega_{d,\mathcal{K}}(\varphi^m) \le k$ is satisfied. Then for every $l > k$ there exists a number $\varepsilon_0 > 0$ such that for all $\varepsilon \in (0, \varepsilon_0]$ and the constant $\widetilde{k}$ from* **(H)** *the inequality*

$$\mu_H(\varphi^m(\mathcal{K}), d, \lambda l^{1/d}\varepsilon) \le \widetilde{k}\, C l \mu_H(\mathcal{K}, d, \varepsilon) \tag{10.15}$$

holds.

Proof Let $\widetilde{\varepsilon}(m)$ be the number from hypothesis **(H)** and take $\varepsilon_1 < \widetilde{\varepsilon}(m)$ such that for the set $\mathcal{V} := \mathcal{U} \cap \bigcup_{u \in \mathcal{K}} \mathcal{B}(u, \varepsilon_1)$ the relation $k' := \sup_{u \in \mathcal{V}} \omega_d(d_u\varphi^m) < l$ holds. We choose numbers $\delta > 0$ and $\eta > 0$ such that $k' < \delta^d$, $\sup_{u \in \mathcal{V}} |d_u\varphi^m| \le \delta$ and the equality

$$\left[1 + \left(\frac{\delta^{d_0}}{k'}\right)^{1/s} \eta\right]^d k' = l$$

is satisfied. We take $\varepsilon_0 < \frac{1}{2}\varepsilon_1$ so small that

$$\left| \tau_{\varphi^m(v)}^{\varphi^m(u)} d_v\varphi^m \tau_u^v - d_u\varphi^m \right| \le \eta$$

is satisfied for all $u, v \in \mathcal{V}$ with $\rho(u, v) \le 2\varepsilon_0$. Here τ is the isometric map (see Appendix A).

For a fixed number $\varepsilon \le \varepsilon_0$ we consider a finite cover of $\mathcal{K}$ with balls $\{\mathcal{B}(u_j, \widetilde{r}_j)\}$ of radius $\widetilde{r}_j \le \varepsilon$, where each ball contains at least one point from $\mathcal{K}$.

Let j be fixed. Consider the decomposition

$$\mathcal{B}(u_j, \widetilde{r}_j) \cap \mathcal{K} \subset \bigcup_{i=1,\ldots,q_j} \mathcal{B}_i(v_{j,i}, r_j)$$

with $q_j \le \widetilde{k}$, $r_j \le 2\varepsilon$, $v_{j,i} \in \mathcal{K}$, $\mathcal{B}_i(v_{j,i}, r_j) = \mathcal{B}(v_{j,i}, r_j) \cap \mathcal{U}_s^{(m)}$ for a certain s such that the right-hand side is connected, and such that φ^m is on any such set extentiable to a C^1-map. Taylor's formula is now applied to the differentiable map φ^m along a continuous curve, which connects the points u and $v_{j,i}$ in $\mathcal{B}_i(v_{j,i}, r_j)$ due to **(H1)** and yields

$$\begin{aligned} &|\exp^{-1}_{\varphi^m(v_{j,i})} \varphi^m(u) - d_{v_{j,i}}\varphi^m(\exp^{-1}_{v_{j,i}}(u))| \\ &\qquad \le \sup_{w \in \mathcal{B}_i(v_{j,i}, r_j)} \left| \tau_{\varphi^m(w)}^{\varphi^m(v_{j,i})} d_w\varphi^m \tau_{v_{j,i}}^w - d_{v_{j,i}}\varphi^m \right| |\exp^{-1}_{v_{j,i}}(w)|. \end{aligned}$$

Thus, we have the inclusion

$$\varphi^m\Big(\mathcal{B}_i(\upsilon_{j,i}, r_j) \cap \mathcal{K}\Big) \subset \exp_{\varphi^m(\upsilon_{j,i})}\Big(d_{\upsilon_{j,i}}\varphi^m(\mathcal{B}(\mathcal{O}_{\upsilon_{j,i}}, r_j)) + \mathcal{B}(\mathcal{O}_{\varphi^m(\upsilon_{j,i})}, \eta r_j)\Big)$$

for any part $\mathcal{B}_i(\upsilon_{j,i}, r_j)$.

Using hypothesis **(H2)** and following the method in Chap. 8 we get finally the inequality (10.15). □

Proof of Theorem 10.5 From $\omega_{d,\mathcal{K}}(\varphi) < 1$ and the statement (b) of Lemma 10.1 it follows that $\omega_{d,\mathcal{K}}(\varphi^m)$ becomes arbitrary small for sufficiently large m. Therefore, the number l from Lemma 10.2 can be made so small that the inequalities $\lambda l^{\frac{1}{d}} < 1$ and $\widetilde{k}Cl < 1$ are satisfied, where λ and C are defined as in Lemma 10.2. Using $\mathcal{K} = \varphi(\mathcal{K}) = \ldots = \varphi^m(\mathcal{K})$, $\mu_H(\mathcal{K}, d, \varepsilon) < +\infty$ and Lemma 10.2 we obtain together with (10.15)

$$\mu_H(\mathcal{K}, d, \varepsilon) \le \mu_H(\mathcal{K}, d, \lambda l^{\frac{1}{d}}\varepsilon) = \mu_H(\varphi^m(\mathcal{K}), d, \lambda l^{\frac{1}{d}}\varepsilon) \le \widetilde{k}Cl\mu_H(\mathcal{K}, d, \varepsilon)$$

for all $\varepsilon \in (0, \varepsilon_0]$. Because of $\widetilde{k}Cl < 1$ as a consequence we get $\mu_H(\mathcal{K}, d, \varepsilon) = 0$ for all $\varepsilon \in (0, \varepsilon_0]$. This implies $\mu_H(\mathcal{K}, d) = 0$, and therefore we get $\dim_H \mathcal{K} \le d$.□

Now we apply the results of Theorem 10.5 to some examples. In all these cases the Hausdorff dimension estimates of the considered invariant sets, given on the basis of Corollary 10.5, are sharp.

Example 10.8 (Example 10.5 cont'd) We consider again the baker's map φ described in Example 10.5. An invariant set $\mathcal{K}$ of φ is given by $\mathcal{K} = \bigcap_{j=1}^{\infty} \varphi^j(\mathcal{U})$. The singular values of the linearization $d_{(x,y)}\varphi$ are $\alpha_1 = 2$ and $\alpha_2 = \lambda$ independently of $(x, y) \in \mathcal{K}$. Therefore, we have $\omega_{1+s,\mathcal{K}}(\varphi) < 1$ for all $s \in \big(-\frac{\ln 2}{\ln \lambda}, 1\big]$. With Corollary 10.8 we get the estimate $\dim_H \mathcal{K} \le 1 - \frac{\ln 2}{\ln \lambda}$. This is a sharp estimation because of the well-known fact that $\dim_H \mathcal{K} = 1 - \frac{\ln 2}{\ln \lambda}$ (see for example [10]).

Example 10.9 (Example 10.6 cont'd) For the map φ defined in Example 10.6 a φ-invariant set is, as in Example 10.8, given by $\mathcal{K} = \bigcap_{k=1}^{\infty} \varphi^k(\mathcal{U})$.

Suppose that the number $\widetilde{s} \in (0, 1]$ is determined by the condition $\sum_{i=1}^{r} r_i^{\widetilde{s}} = 1$, and furthermore $l_i = r_i^{-\widetilde{s}}$ for $i = 1, \ldots, r$ is satisfied. For all $i = 1, 2, \ldots, r$ and $s > \widetilde{s}$ we have $\omega_{1+s,\mathcal{K}\cap\mathcal{U}_i}(\varphi) = l_i r_i^s < l_i r_i^{\widetilde{s}} = 1$. It follows that $\omega_{1+s,\mathcal{K}}(\varphi) < 1$, and by Corollary 10.8 we get $\dim_H \mathcal{K} \le 1 + \widetilde{s}$.

Note that our invariant set may be represented in the form $\mathcal{K} = \mathcal{A} \times \mathcal{B}$, where $\mathcal{A} = (0, 1]$ and $\mathcal{B}$ is a modified Cantor set. Since $\mathcal{A}$ and $\mathcal{B}$ are Borel sets we have by Proposition 3.25, Chap. 3 that $\dim_H \mathcal{A} + \dim_H \mathcal{B} \le \dim_H(\mathcal{A} \times \mathcal{B})$. We pay our attention to the Hausdorff dimension of the set $\mathcal{B}$. With the estimate above and $\dim_H \mathcal{A} = 1$ we obtain $\dim_H \mathcal{B} \le \widetilde{s}$. Resulting from [14] or [10] (Theorem 9.3) we have $\dim_H \mathcal{B} = \widetilde{s}$. So our estimate is sharp.

For the special case $r = 2$, $l_1 = l_2 = 2$, $r_1 = r_2 = \frac{1}{3}$, $a_1 = 0$ and $a_2 = \frac{2}{3}$ we see that $\mathcal{B}$ is the standard Cantor set. Our estimate gives $\dim_H \mathcal{B} \le \widetilde{s} = \frac{\ln 2}{\ln 3}$ which coincides with the well-known value $\dim_H \mathcal{B} = \frac{\ln 2}{\ln 3}$.

Example 10.10 (*Sierpiński gasket*) Consider the map

$$\varphi(x,y,z)=\begin{cases}\left(3x,\,\frac{1}{2}y,\,\frac{1}{2}z\right), & \text{for } 0<x\le\frac{1}{3},\ (y,z)\in\mathcal{G},\\ \left(3x-1,\,\frac{1}{2}y+\frac{1}{2},\,\frac{1}{2}z\right), & \text{for } \frac{1}{3}<x\le\frac{2}{3},\ (y,z)\in\mathcal{G},\\ \left(3x-2,\,\frac{1}{2}y+\frac{1}{4},\,\frac{1}{2}z+\frac{\sqrt{3}}{4}\right), & \text{for } \frac{2}{3}<x\le 1,\ (y,z)\in\mathcal{G},\end{cases}$$

where $\mathcal{G}=\mathcal{G}_1\cup\mathcal{G}_2\subset\mathbb{R}^2$ and the sets $\mathcal{G}_i$ are given by

$$\mathcal{G}_1=\left\{(y,z):\ 0\le y\le\tfrac{1}{2},\ 0\le z\le\sqrt{3}y\right\}$$
$$\text{and } \mathcal{G}_2=\left\{(y,z):\ \tfrac{1}{2}\le y\le 1,\ 0\le z\le\sqrt{3}-\sqrt{3}y\right\}.$$

Analogously to Example 10.9 an invariant set $\mathcal{K}$ of this map can be represented as $\mathcal{K}=\mathcal{A}\times\mathcal{B}$, in the given situation with $\mathcal{A}=(0,1]$ and $\mathcal{B}$ the well-known Sierpiński gasket (see [10], Example 9.4). It can be shown that Corollary 10.8 gives the estimate $\dim_H\mathcal{B}\le\frac{\ln 3}{\ln 2}$. As to be seen in [10] this estimation is also sharp.

Remark 10.4 As it was noted above the Belykh map from Example 10.7 does not satisfy the condition **(H)** in general and thus a direct application of Theorem 10.5 for many parameters is not possible. For $\mu=|\lambda|=\frac{a+b}{2}<1$, however, Theorem 10.5 is applicable and for a compact invariant set $\mathcal{K}$ we have the inequality $\omega_{d,\mathcal{K}}(\varphi)=(\frac{a+b}{2})^d<1$ for $d\in(0,2]$. It follows that $\dim_H\mathcal{K}=0$. If in other parameter cases we extend the function φ both on $\mathcal{U}_1$ and on $\mathcal{U}_2$ to C^1-maps φ_1 and φ_2, respectively, then we have to consider at most two C^1-maps for any ball, which is an element of the cover of an invariant set $\mathcal{K}$. On the base in Chap. 8 this gives for $0<\lambda\mu<\frac{1}{2}$ the Hausdorff dimension estimate $\dim_H\mathcal{K}\le\frac{\ln\lambda-\ln\mu}{\ln\lambda+\ln 2}$. A different approach to the dimension investigation of the Belykh family is given in Sect. 10.2 (see also [35]).

10.2.4 *Consideration of the Degree of Non-injectivity*

In order to use the Douady-Oesterlé-type conditions of the previous subsection for dimension estimates it is necessary that the contraction condition of the singular value function $\omega_{d,\mathcal{K}}(\varphi)<1$ is satisfied. This may result in strong restrictions for the parameters as it happens for instance in Example 10.8, where $\lambda\in(0,\frac{1}{2})$ is required. The question raises how we can get a dimension estimate for the parameters $\lambda>\frac{1}{2}$, i. e. for the case $\omega_{d,\mathcal{K}}(\varphi)>1$.

It turns out that in certain cases, where the contraction condition of the singular value function is not fulfilled the exploration of an additional non-injectivity condition allows to get a contraction for the outer Hausdorff measures. Using the approach of Sect. 10.1 we investigate piecewise C^1-maps on n-dimensional smooth manifolds having the following property. Suppose that for the given d-dimensional Hausdorff outer measure $\mu_H(\cdot, d, \varepsilon)$ at level ε there exists a number $0 < a < 1$ and a family $\{\mathcal{K}_j\}_{j \geq j_0}$ of subsets of the φ-invariant set $\mathcal{K}$ such that for all $j \geq j_0$ this outer measure of $\mathcal{K}_j$ is at most a^j times the outer measure of $\mathcal{K}$ and the outer measures of the sets $\varphi^j(\mathcal{K}_j)$ and $\varphi^j(\mathcal{K})$ are equal. A map with such a property can be considered as piecewise expansive with respect to the Hausdorff measure on $\mathcal{K}$ and the factor a^{-1} is the expansion parameter which describes the "degree of non-injectivity".

The next two theorems are borrowed from [36].

Theorem 10.6 *Let $(\mathcal{M}, g)$ be a smooth n-dimensional Riemannian manifold. Suppose that $\varphi : \mathcal{U} \subset \mathcal{M} \to \mathcal{M}$ satisfies* **(H)**. *Let $\mathcal{K} \subset \mathcal{U}$ be a bounded and φ-invariant set. Suppose that for some numbers $a > 0$ and $d \in (0, n]$ of the form $d = d_0 + s$ with $d_0 \in \{0, \ldots, n-1\}$ and $s \in (0, 1]$ the following conditions are satisfied:*

(a) $\omega_{d,\mathcal{K}}(\varphi) < \frac{1}{a}$;

(b) *There are numbers l with $\omega_{d,\mathcal{K}}(\varphi) < l < \frac{1}{a}$ and $m_0 \in \mathbb{N}$ such that for every natural number $m > m_0$ there exist a set $\mathcal{K}_m \subset \mathcal{K}$ and a number $\varepsilon_m > 0$ such that*

$$\mu_H(\varphi^m(\mathcal{K}_m), d, \varepsilon) = \mu_H(\varphi^m(\mathcal{K}), d, \varepsilon),$$

$$\mu_H(\mathcal{K}_m, d, 2^{-1}(d_0+1)^{-\frac{1}{2}} l^{-\frac{m}{d}} \varepsilon) \leq a^m \mu_H(\mathcal{K}, d, \varepsilon)$$

hold for all $\varepsilon \in (0, \varepsilon_m]$.

Then $\dim_H \mathcal{K} \leq d$.

Proof The proof uses methods of Sect. 10.1 and Lemma 10.2. By Lemma 10.1 we have $\omega_{d,\mathcal{K}}(\varphi^m) \leq \omega^m_{d,\mathcal{K}}(\varphi) < l^m$ for all $m \in \mathbb{N}$. Thus, for every $\delta > 0$ we find an integer $m_\delta > m_0$ such that $\widetilde{k} 2^{d_0}(d_0+1)^{\frac{d}{2}}(al)^m < \delta$ will be true for all $m > m_\delta$. Let $m > m_\delta$ be fixed. Using the invariance of the set $\mathcal{K}$ and condition (b) we obtain

$$\mu_H(\mathcal{K}, d, \varepsilon) = \mu_H(\varphi^m(\mathcal{K}), d, \varepsilon) = \mu_H(\varphi^m(\mathcal{K}_m), d, \varepsilon) \tag{10.16}$$

for all $\varepsilon \in (0, \varepsilon_m]$. By applying the method of the proof of Lemma 10.2 we can show that there exists a number $\varepsilon_0 > 0$ such that

$$\mu_H(\varphi^m(\mathcal{K}_m), d, \varepsilon) \leq \widetilde{k}\, 2^{d_0}(d_0+1)^{\frac{d}{2}} l^m \mu_H(\mathcal{K}_m, d, 2^{-1}(d_0+1)^{-\frac{1}{2}} l^{-\frac{m}{d}} \varepsilon) \tag{10.17}$$

holds for every $\varepsilon \leq 2(d_0+1)^{\frac{1}{2}} l^{\frac{m}{d}} \varepsilon_0$. If $\varepsilon \in \big(0, \min\{\varepsilon_m, 2(d_0+1)^{\frac{1}{2}} l^{\frac{m}{d}} \varepsilon_0\}\big]$ then the condition (b) and the relations (10.16) and (10.17) imply now

$$\mu_H(\mathcal{K}, d, \varepsilon) \leq \widetilde{k}\, 2^{d_0}(d_0+1)^{\frac{d}{2}} (la)^m \mu_H(\mathcal{K}, d, \varepsilon) < \delta \mu_H(\mathcal{K}, d, \varepsilon),$$

where the number $\delta > 0$ can be chosen arbitrarily small. Since $\mu_H(\mathcal{K}, d, \varepsilon)$ is finite we conclude that $\mu_H(\mathcal{K}, d, \varepsilon) = 0$ holds for all $\varepsilon \in \left(0, \min\{\varepsilon_m, 2(d_0+1)^{\frac{1}{2}} l^{\frac{m}{d}} \varepsilon_0\}\right]$. Hence we get $\mu_H(\mathcal{K}, d) = 0$ and $\dim_H \mathcal{K} \le d$. □

Example 10.11 Consider the set $\mathcal{U} = [0, 1] \times [-1, 1) \subset \mathbb{R}^2$ and the map $\varphi : \mathcal{U} \to \mathcal{U}$ defined by

$$\varphi(x, y) = \begin{cases} (2x, \lambda y), & \text{for } 0 \le x \le \frac{1}{2}, -1 \le y < 1, \\ (2 - 2x, \lambda y), & \text{for } \frac{1}{2} < x \le 1, -1 \le y < 1, \end{cases}$$

where $\lambda \in (0, 1)$ is a parameter.

In order to satisfy the condition **(H)** we choose $\mathcal{U}_1 = \left[0, \frac{1}{2}\right] \times [-1, 1)$, $\mathcal{U}_2 = \left(\frac{1}{2}, 1\right] \times [-1, 1)$ and the index set $I = \{1, 2\}$. The condition **(H)** is satisfied if we choose $\widetilde{\varepsilon}(m) = \left(\frac{1}{2}\right)^{m+1}$ for $m \in \mathbb{N}$ and $\widetilde{k} = 2$. It is easy to see that $\mathcal{U} \supset \varphi(\mathcal{U}) \supset \ldots \supset \varphi^m(\mathcal{U})$ holds for every $m \in \mathbb{N}$, so we can define a φ-invariant bounded set $\mathcal{K} = \bigcap_{m=1}^{\infty} \varphi^m(\mathcal{U}) = [0, 1] \times \{0\}$. The singular values of $d_u\varphi^m$ are $\alpha_1(d_u\varphi^m) = 2^m$ and $\alpha_2(d_u\varphi^m) = \lambda^m$.

Define now $\mathcal{K}_1 := \mathcal{K} \cap \left[0, \frac{1}{2}\right] \times \{0\}$ and $\mathcal{K}_m := \mathcal{K}_{m-1} \cap \varphi^{-1}(\mathcal{K}_{m-1})$ $(m = 2, 3, \ldots)$. For arbitrary $m \in \mathbb{N}$ we have $\varphi^m(\mathcal{K}_m) = \mathcal{K} = \varphi^m(\mathcal{K})$. Thus, for all $\varepsilon > 0$ and $d \in (1, 2]$ the relation $\mu_H(\varphi^m(\mathcal{K}_m), d, \varepsilon) = \mu_H(\varphi^m(\mathcal{K}), d, \varepsilon)$ holds. If the set $\mathcal{K}$ is covered by balls of radii smaller than 1 then the set $\mathcal{K}_m$ can be covered by linear copies of sets of the covering for $\mathcal{K}_{m-1}$ scaled by the factor $\frac{1}{2}$. Thus, we get

$$\mu_H(\mathcal{K}_m, d, \varepsilon) \le \frac{1}{2^d} \mu_H(\mathcal{K}_{m-1}, d, 2\varepsilon)$$

for arbitrary $m \in \mathbb{N}$, $d \in (1, 2]$ and $\varepsilon < 1$. For $\varepsilon_m < 1$ it follows that $\mu_H(\mathcal{K}_m, d, 2^{-m}\varepsilon) \le 2^{-dm} \mu_H(\mathcal{K}, d, \varepsilon)$ holds for all $\varepsilon \in (0, \varepsilon_m]$. For $a = 2^{-d}$ and $l = 2$ we have $\omega_{d,\mathcal{K}}(\varphi) = 2\lambda^{d-1} < l < \frac{1}{a}$. Further, there holds $2^{-1-\frac{1}{2}} l^{-\frac{m}{d}} \ge 2^{-m}$ for sufficiently large $m \in \mathbb{N}$. Because of the property of the outer measure we conclude that

$$\mu_H(\mathcal{K}_m, d, 2^{-1-\frac{1}{2}} l^{-\frac{m}{d}} \varepsilon) \le a^m \mu_H(\mathcal{K}, d, \varepsilon)$$

for all $\varepsilon \in (0, \varepsilon_m]$ and $m \ge m_0 \ge \frac{3}{2} \frac{d}{d-1}$. So, all conditions of Theorem 10.6 are satisfied and we obtain $\dim_H \mathcal{K} \le d$ for arbitrary $d \in (1, 2]$. Applying Corollary 10.8 this yields $\dim_H \mathcal{K} \le 1$, which is a sharp estimate.

10.2.5 *Introduction of Long Time Behavior Information*

In this subsection we investigate maps φ for which the singular value function of the tangent map satisfies a contraction condition on a subset of the φ-invariant set $\mathcal{K}$ only. Using some information about the long time behavior of the system we derive dimension estimates.

Consider again a map $\varphi : \mathcal{U} \subset \mathcal{M} \to \mathcal{M}$, which possesses a bounded invariant set $\mathcal{K} \subset \mathcal{U}$, and satisfies the condition **(H)**. Suppose that there exists a partition of $\mathcal{K}$ of the form $\mathcal{K} = \left(\bigcup_{i=1}^{l} \mathcal{K}_i\right) \cup \left(\bigcup_{i=1}^{h} \mathcal{K}_i'\right)$ such that $\omega_{d,\mathcal{K}_i}(\varphi) < 1$, $i = 1, \ldots, l$, and $\omega_{d,\mathcal{K}_i'}(\varphi) \geq 1$, $i = 1, \ldots, h$. For $i = 1, \ldots, l$ we define the numbers (# denotes the cardinality of a set)

$$\mathbb{P}_i = \liminf_{m\to\infty} \inf_{u\in\mathcal{K}} \frac{1}{m} \#\{k \mid 0 \leq k \leq m-1, \varphi^k(u) \in \mathcal{K}_i\}$$

and for $i = 1, \ldots, h$ the numbers

$$\overline{\mathbb{P}}_i = \limsup_{m\to\infty} \sup_{u\in\mathcal{K}} \frac{1}{m} \#\{k \mid 0 \leq k \leq m-1, \varphi^k(u) \in \mathcal{K}_i'\}.$$

Note that in the particular case that $\{\varphi^m\}$ possesses an invariant ergodic probability measure μ on $\mathcal{U}$, for an arbitrary measurable set $\mathcal{A} \subset \mathcal{U}$ and for μ-almost every $u \in \mathcal{U}$ there holds the relation

$$\lim_{m\to\infty} \frac{1}{m} \sum_{k=0}^{m-1} \chi_{\mathcal{A}}(\varphi^k(u)) = \lim_{m\to\infty} \frac{1}{m} \#\{k \mid 0 \leq k \leq m-1, \varphi^k(u) \in \mathcal{A}\} = \mu(\mathcal{A}),$$

where $\chi_{\mathcal{A}}$ denotes the characteristic function of $\mathcal{A}$. The following theorem generalizes a result of [36] to the case of piecewise smooth maps on manifolds.

Theorem 10.7 *Let $(\mathcal{M}, g)$ be a smooth n-dimensional Riemannian manifold. Suppose that $\varphi : \mathcal{U} \subset \mathcal{M} \to \mathcal{M}$ satisfies the condition* **(H)** *and let $\mathcal{K} \subset \mathcal{U}$ be a bounded and φ-invariant set. Suppose that $\{\mathbb{P}_i\}_{i=1}^{l}$ and $\{\mathbb{P}_i'\}_{i=1}^{h}$ are the numbers defined above for φ with respect to a given partition $\mathcal{K} = (\bigcup_{i=1}^{l} \mathcal{K}_i) \cup (\bigcup_{i=1}^{h} \mathcal{K}_i')$. Let $\overline{\mathbb{P}}_i$ and $\overline{\mathbb{P}}_i'$ be numbers with $\overline{\mathbb{P}}_i \leq \mathbb{P}_i$, $i = 1, \ldots, l$, and $\overline{\mathbb{P}}_i' \geq \mathbb{P}_i'$, $i = 1, \ldots, h$, such that*

$$\left(\prod_{i=1}^{l} \omega_{d,\mathcal{K}_i}(\varphi)^{\overline{\mathbb{P}}_i}\right) \left(\prod_{i=1}^{h} \omega_{d,\mathcal{K}_i'}(\varphi)^{\overline{\mathbb{P}}_i'}\right) < 1 \tag{10.18}$$

is satisfied. Then $\dim_H \mathcal{K} \leq d$.

Proof Without loss of generality we suppose that the inequality (10.18) is satisfied with numbers $\overline{\mathbb{P}}_i < \mathbb{P}_i, i = 1, \ldots, l$, and $\overline{\mathbb{P}}_i' > \mathbb{P}_i, i = 1, \ldots, h$. Let m be sufficiently large such that

$$\overline{\mathbb{P}}_i < \frac{1}{m} \inf_{u\in\mathcal{K}} \#\{k \mid 0 \leq k \leq m-1, \varphi^k(u) \in \mathcal{K}_i\}, \quad i = 1, \ldots, l$$

and

$$\overline{\mathbb{P}}_i' > \frac{1}{m} \sup_{u\in\mathcal{K}} \#\{k \mid 0 \leq k \leq m-1, \varphi^k(u) \in \mathcal{K}_i'\}, \quad i = 1, \ldots, h$$

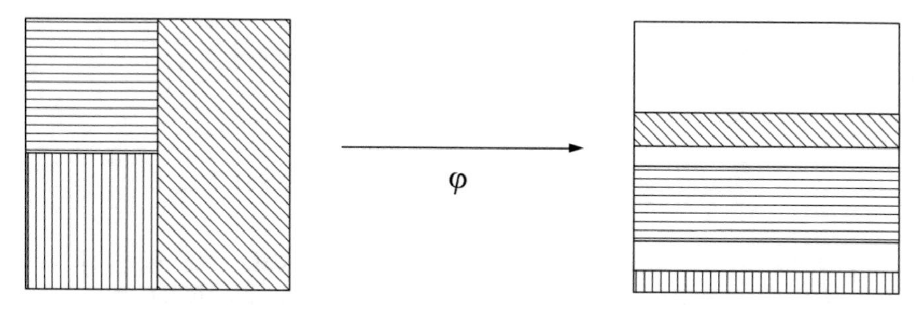

Fig. 10.3 Modified baker's map

are satisfied. It follows that an arbitrary orbit of length m starting in $u \in \mathcal{K}$ passes the set $\mathcal{K}_i$ more frequently than $\overline{\mathbb{P}}_i m$-times ($i = 1, \ldots, l$) and the set $\mathcal{K}'_i$ less than $\overline{\mathbb{P}}'_i m$-times ($i = 1, \ldots, h$). For the tangent map $d_u\varphi^m$ in an arbitrary point $u \in \mathcal{K}$ the chain rule

$$d_u\varphi^m = d_{\varphi^{m-1}(u)}\varphi \circ \ldots \circ d_u\varphi$$

holds. Using Lemma 10.1 we get the inequalities

$$\begin{aligned}
\omega_{d,\mathcal{K}}(\varphi^m) &\leq \sup_{u\in\mathcal{K}} \left[\omega_d(d_{\varphi^{m-1}(u)}\varphi)\omega_d(d_{\varphi^{m-2}(u)}\varphi)\ldots\omega_d(d_u\varphi)\right] \\
&\leq \left(\prod_{i=1}^{l} \omega_{d,\mathcal{K}_i}^{m\overline{\mathbb{P}}_i}(\varphi)\right)\left(\prod_{i=1}^{h} \omega_{d,\mathcal{K}'_i}^{m\overline{\mathbb{P}}'_i}(\varphi)\right) \\
&= \left[\left(\prod_{i=1}^{l} \omega_{d,\mathcal{K}_i}^{\overline{\mathbb{P}}_i}(\varphi)\right)\left(\prod_{i=1}^{h} \omega_{d,\mathcal{K}'_i}^{\overline{\mathbb{P}}'_i}(\varphi)\right)\right]^m < 1.
\end{aligned}$$

With respect to Theorem 10.5 we conclude that $\dim_H \mathcal{K} \leq d$. □

Example 10.12 (*Modified baker's map*) Let be $\mathcal{U} = (0, 1] \times [0, 1]$ and consider the map $\varphi : \mathcal{U} \to \mathcal{U}$ defined by

$$\varphi(x, y) = \begin{cases} (2x, \lambda_2 y), & \text{for } 0 < x \leq \frac{1}{2}, \frac{1}{2} \leq y \leq 1, \\ (2x, \lambda_1 y), & \text{for } 0 < x \leq \frac{1}{2}, 0 \leq y < \frac{1}{2}, \\ (2x - 1, \lambda_1 y + \frac{1}{2}), & \text{for } \frac{1}{2} < x \leq 1, 0 \leq y \leq 1, \end{cases}$$

where λ_1, λ_2 are parameters with $0 < \lambda_1 < \lambda_2 < \frac{1}{2}$ (see Fig. 10.3).

The singular values of the tangent map are $\alpha_1(d_u\varphi) = 2$ and $\alpha_2(d_u\varphi) \in \{\lambda_1, \lambda_2\}$. Based on Theorem 10.5 we get $\dim_H \mathcal{K} \leq 1 + \frac{\ln 2}{|\ln \lambda_2|}$.

We want to improve this estimate by using Theorem 10.7. For this purpose we consider the sets $\mathcal{U}'_1 = (0, \frac{1}{2}] \times [\frac{1}{2}, 1]$ and $\mathcal{U}_1 = \mathcal{U} \setminus \mathcal{U}'_1$. From $\varphi(\mathcal{U}'_1) \subset \mathcal{U}_1$ it follows that every point $u \in \mathcal{U}'_1$ satisfies $\varphi(u) \notin \mathcal{U}'_1$. So we conclude that

$$\limsup_{m\to\infty} \sup_{u\in\mathcal{U}_1'} \frac{1}{m}\#\{k \mid 0 \le k \le m-1, \varphi^k(u) \in \mathcal{U}_1'\} \le \frac{1}{2}$$

and

$$\liminf_{m\to\infty} \inf_{u\in\mathcal{U}_1} \frac{1}{m}\#\{k \mid 0 \le k \le m-1, \varphi^k(u) \in \mathcal{U}_1\} \ge \frac{1}{2}$$

are satisfied. For the invariant set $\mathcal{K} = \bigcap_{k=1}^{\infty} \varphi^k(\mathcal{U})$ we define $\mathcal{K}_1 = \mathcal{U}_1 \cap \mathcal{K}$ and $\mathcal{K}_1' = \mathcal{U}_1' \cap \mathcal{K}$. Note that $\omega_{d,\mathcal{K}_1}(\varphi) = 2\lambda_1^s$ and $\omega_{d,\mathcal{K}_1'}(\varphi) = 2\lambda_2^s$ hold for any $d = 1 + s$ with $s \in [0, 1]$. If we put $\overline{\mathbb{P}}_1 = \frac{1}{2} \le \mathbb{P}_1$ and $\overline{\mathbb{P}}_1' = \frac{1}{2} \ge \mathbb{P}_1'$, then the condition of Theorem 10.7 is fulfilled if $(2\lambda_1^s)^{\frac{1}{2}}(2\lambda_2^s)^{\frac{1}{2}} < 1$ holds. So we get the estimate $\dim_H \mathcal{K} \le 1 + \frac{\ln 2}{|\ln \sqrt{\lambda_1\lambda_2}|}$.

Remark 10.5 There have been done numerical investigations in order to approximate the essential dynamics of a given system (e.g. [8]). It can be expected that these methods can be applied to get estimates of the values $\overline{\mathbb{P}}_i$, $\overline{\mathbb{P}}_i'$ from Theorem 10.7 numerically.

10.2.6 Estimation of the Hausdorff Dimension for Invariant Sets of Piecewise Smooth Vector Fields

The results of this subsection are due to Noack [28] and Schmidt [36]. Suppose that on the n-dimensional manifold $\mathcal{M}$ of smoothness C^m $(m \ge 3)$ there is given a vector field $f : \mathcal{M} \to T\mathcal{M}$ of smoothness $C^r (1 \le r < m)$. Let us consider the differential equation

$$\dot{u} = f(u) \tag{10.19}$$

with the global flow $\varphi : \mathbb{R} \times \mathcal{M} \to \mathcal{M}$. Denote by $\Omega_r^k(\mathcal{M})$ the vector space of C^r-smooth k-forms on $\mathcal{M}$. Suppose that $\beta \in \Omega_l^k(\mathcal{M})$ is an arbitrary C^l-smooth k-form. The *Lie derivative* of β with respect to f at a point $u \in \mathcal{M}$ is given by [1]

$$L_f\beta(u) = \frac{d}{dt}_{|t=0} (\varphi^t)^*\beta(u) \tag{10.20}$$

where $(\varphi^t)^*\beta$ is the pullback of β. This pullback satisfies the *variational equation*

$$\frac{d}{dt}(\varphi^t)^*\beta = (\varphi^t)^* L_f\beta \tag{10.21}$$

of (10.19) with respect to the k-form β.

Assume now that $\beta \in \Omega_r^k(\mathcal{M})$ and $\mathcal{M}_k \subset \mathcal{M}$ is a k-dimensional submanifold of $\mathcal{M}$. We suppose that there is a *normalization* $\mathcal{N}_k$, i.e., a smooth map $\mathcal{N}_k : u \in \mathcal{M}_k \mapsto T_u\mathcal{M}_k \oplus \mathcal{N}_u$ with $T_u\mathcal{M}_k \oplus \mathcal{N}_u = T_u\mathcal{M}$ (see [28]). Then for arbitrary $u \in \mathcal{M}$ and

$t \in \mathbb{R}$ the pullback of $\beta_{|\mathcal{M}_k}$ is a k-form on $\mathcal{M}_k$ with

$$L_f \beta_{|\mathcal{M}_k}(u) = \operatorname{div}_{\beta, \mathcal{N}_k} f(u) \beta_{|\mathcal{M}_k}(u). \tag{10.22}$$

We call $\operatorname{div}_{\beta, \mathcal{N}_k} f(u)$ *divergence of f with respect to β and $\mathcal{N}_k$ at the point u*.

Suppose now that μ is a volume form on $\mathcal{M}$ and $\mu_{|\mathcal{N}_k}$ is the associated k-form on $\mathcal{M}_k$. Define the numbers

$$\lambda^{(k)}(u) := \sup_{\mathcal{N}_k} \operatorname{div}_{\mu, \mathcal{N}_k} f(u) \tag{10.23}$$

for $k = 1, 2, \ldots, n$, where $u \in \mathcal{M}_k$ and the supremum is taken over all normalizations $\mathcal{N}_k$ of $\mathcal{M}_k$.

For an arbitrary $d \in (0, n]$ written in the form $d = k + s$ with $k \in [0, \ldots, n-1]$ and $s \in [0, 1]$ we introduce the function

$$\kappa_d(u) := (1-s)\lambda^{(k)}(u) + s\lambda^{(k+1)}(u) \tag{10.24}$$

where $u \in \mathcal{M}_k$. For the singular value function $\omega_d(d_u\varphi^t)$ at an arbitrary point $u \in \mathcal{M}_k$ and for an arbitrary $t \geq 0$ we have (compare with (8.23), Chap. 8) the inequality

$$\omega_d(d_u\varphi^t) \leq \exp\left\{ \int_0^t \kappa_d(\varphi^\tau(u)) d\tau \right\}. \tag{10.25}$$

Remark 10.6 Suppose that μ is the volume form on $\mathcal{M}$ generated by the Riemannian metric. Then for arbitrary $u \in \mathcal{M}$, $k \in \{1, \ldots, n\}$ and a k-dimensional submanifold $\mathcal{M}_k \subset \mathcal{M}$ we have

$$\lambda^{(k)}(u) = \lambda_1(u) + \cdots + \lambda_k(u)$$

where $\lambda_1(u) \geq \cdots \geq \lambda_n(u)$ are the eigenvalues of the symmetrized covariant derivative $\frac{1}{2}(\nabla f^*(u) + \nabla f(u))$.

Suppose that $(\mathcal{M}, g)$ is an n-dimensional Riemannian C^m-manifold ($m > 3$) and let $f_k : \mathcal{M} \to T\mathcal{M}$ be C^r-smooth ($0 < r < m$) vector fields for $k \in I \subset \mathbb{N}$. Suppose also that for any $k \in I$ there exists the global flow to the differential equation

$$\dot{u} = f_k(u). \tag{10.26}$$

Denote this flow to (10.26) by $\{\varphi_k^{(\cdot)}(\cdot)\}$. Let us assume that there is a partition of the manifold into subsets $\mathcal{M}_i$ and there are C^1 functions $F_i : \mathcal{M} \to \mathbb{R}$ satisfying for all $i \in I$ the following conditions:

(E1) The sets $\mathcal{M}_i$ are mutually disjoint and connected;

(E2) If $u \in \mathcal{M}_j$ then there exists a $\tau = \tau(u) > 0$ such that $\varphi_j^t(u) \in \mathcal{M}_j$ for all $t \in [0, \tau]$;

(E3) For any $i \in I$ we have $F_i(u) = 0$ if and only if $u \in \partial\mathcal{M}_i \backslash \mathcal{M}_i$.

Let us introduce a sequence of maps $\xi_i : \mathcal{M} \to \mathcal{M}$ and a sequence of functions $t_i : \mathcal{M} \to \mathbb{R}_+ (i \in \mathbb{N}_0)$ satisfying $0 = t_0(u) < t_1(u) < \ldots$ for all $u \in \mathcal{M}$, using the formulas

$$\begin{aligned} \xi_i(u) &= \varphi^{t_i(u)}_{m_{i-1}(u)}(u), \\ t_{i+1}(u) &= t_1(\xi_i(u)) \end{aligned} \tag{10.27}$$

and suppose that $t_1(u)$ is the time such that $\varphi^t_k(u) \in \mathcal{M}_k$ for some $k \in I$ and all $t \in [0, t_1(u))$ and $\varphi^{t_1(u)}_k(u) \in \partial\mathcal{M}_k \backslash \mathcal{M}_k$.

Let us denote $\mathcal{T}(u) := \{t_j(u) | j \in \mathbb{N}_0\}$ for arbitrary $u \in \mathcal{M}$. Define the vector field $f : \mathcal{M} \to T\mathcal{M}$ by

$$f_{|\mathcal{M}_i} = f_i \quad \text{for all} \quad i \in I.$$

Then the differential equation

$$\dot{u} = f(u) \tag{10.28}$$

is called *piecewise smooth* and the points $\xi_i(u)$ are the *switching points* of the vector field (10.28) with respect to $u \in \mathcal{M}$. Let us also introduce the *coding functions* $m_i : \mathcal{M} \to \mathbb{R}$ which define for any $u \in \mathcal{M}$ and $i \in \mathbb{N}_0$ the index $j \in \mathbb{N}$ such that $\xi_i(u) \in \mathcal{M}_j$.

The solution of (10.28) is given by a map $\varphi : \mathbb{R}_+ \times \mathcal{M} \to \mathcal{M}$ which is defined by

$$\varphi^t(u) = \varphi^t_{m_0(u)}(t - t_0(u), \xi_k(u)) \quad \text{for} \quad t \in (t_k(u), t_{k+1}(u)]. \tag{10.29}$$

One can show that $\varphi^{(\cdot)}(\cdot)$ is a semiflow having the property $\varphi^{t+s}(u) = \varphi^t(\varphi^s(u))$ for all $t, s \in \mathbb{R}_+$ and $u \in \mathcal{M}$. Let us now introduce the sets $\Omega_0 = \mathcal{D}_0 = \mathcal{M}$ and

$$\Omega_{k+1} = \Omega_k \cap \left\{u \in \mathcal{M} \mid L_{f_{m_{k+1}(u)}} F_{m_k(u)}(\xi_{k+1}(u)) = \dot{F}_{m_k(u)}(\xi_{k+1}(u)) \neq 0\right\}, \tag{10.30}$$

$$\begin{aligned} \mathcal{D}_{k+1} = \mathcal{D}_k \cap \{u \in \mathcal{M} \mid \exists l \in \mathbb{N}_0 : \forall t \in [t_k(u), t_{k+1}(u)] \\ \exists \varepsilon > 0 : \xi_1(\varphi^t(\mathcal{B}(u,\varepsilon)) \cap \mathcal{M}_{m_k(u)}) \subset \mathcal{M}_l\right\}, \\ \mathcal{Q}_k = \Omega_k \cap \mathcal{D}_k \quad \text{for all} \quad k \in \mathbb{N}, \end{aligned}$$

$$\Omega := \bigcap_{k\in\mathbb{N}} \Omega_k, \qquad \mathcal{D} = \bigcap_{k\in\mathbb{N}} \mathcal{D}_k. \tag{10.31}$$

The next result follows directly from the Formula (10.30).

Lemma 10.3 *The sets* $\Omega := \bigcap_{k\in\mathbb{N}_0} \Omega_k, \mathcal{D} = \bigcap_{k\in\mathbb{N}_0} \mathcal{D}_k,$ *and, consequently,* $\mathcal{Q} = \Omega \cap \mathcal{D}$ *are positively invariant w.r.t. the semiflow* $\varphi^{(\cdot)}(\cdot)$ *to Eq. (10.28).*

The next lemmas are generalizations of similar results for piecewise smooth systems in $\mathbb{R}^n$ [12, 36].

Lemma 10.4 *Suppose $\bigcup_i \mathcal{M}_i$ is a partition of $\mathcal{M}$ and F_i are C^1-functions satisfying* **(E1)–(E3)**. *Suppose also that on $\mathcal{M}$ there is given a piecewise smooth vector field (10.28). Then for any $j \in \mathbb{N}$ and $k \in I$ with $\mathcal{M}_k \cap Q_j \neq \emptyset$ the function $t_j : \mathcal{M} \to \mathbb{R}_+$ and the map $\xi_j : \mathcal{M} \to \mathcal{M}$ from (10.27) is continuously differentiable on the set $\mathcal{M}_k \cap Q_j \neq \emptyset$.*

In addition to this for any $j \in \mathbb{N}_0, u_0 \in \mathcal{M}_k \cap \mathcal{Q}_{j+1}$ and $t \in (t_j(u_0), t_{j+1}(u_0))$ there exists a $\delta = \delta(t) > 0$ such that for all $u \in \mathcal{B}(u_0, \delta) \cap \mathcal{M}_k \cap \mathcal{Q}_{j+1}$ and $\varphi^t(\cdot)$ is continuously differentiable on $\mathcal{B}(u_0, \delta) \cap \mathcal{M}_k \cap \mathcal{Q}_{j+1}$.

Proof The proof can be done using the implicit function theorem. □

It follows from Lemma 10.4 that the map $\varphi^t(\cdot)$ is piecewise C^1 on $\mathcal{Q}$. As a consequence the differential $d_u\varphi^t$ has a jump at the transition moments $t_i(u), i \in \mathbb{N}$. Assume that the jump at time $t_j(u)$ can be described by a linear operator $S_j(u)$ which is defined by the property

$$\lim_{t \to t_j(u)+0} d_u\varphi^t = S_j(u) \circ \Big(\lim_{t \to t_j(u)-0} d_u\varphi^t \Big). \tag{10.32}$$

Let us define for arbitrary $j \in \mathbb{N}_0$ and $u \in \mathcal{Q}_j$ the transition operator $S_j(u) : T_{\xi_j(u)}\mathcal{M} \to T_{\xi_j(u)}\mathcal{M}$ with initial point $u \in \mathcal{M}$ by

$$S_j(u)\upsilon = \frac{d_{\xi_j(u)} F_{m_{j-1}(u)}\upsilon}{L_{f_{m_{j-1}(u)}} F_{m_{j-1}(u)}(\xi_j(u))} (f_{m_j(u)}(\xi_j(u)) - f_{m_{j-1}(u)}(\xi_j(u))) + \mathrm{id}_{d_{\xi_j(u)}\mathcal{M}} \tag{10.33}$$

for all $\upsilon \in T_{\xi_j(u)}\mathcal{M}$. By assumption the dominator in (10.33) is different from zero for $u \in \mathcal{Q}_j$.

Lemma 10.5 *Suppose that there are given a piecewise smooth vector field $f : \mathcal{M} \to T\mathcal{M}$ on the partition $\mathcal{M} = \bigcup_{i \in I} \mathcal{M}_i$ and the C^1-functions $F_i : \mathcal{M} \to \mathbb{R}$ satisfying the properties* **(E1)–(E3)**. *Then it holds:*

(1) The transition operator (10.33) $S_j(u) : T_{\xi_j(u)}\mathcal{M} \to T_{\xi_j(u)}\mathcal{M}$ has for arbitrary $j \in \mathbb{N}$ and $u \in \mathcal{Q}_j$ the property (10.32).
(2) Define for all $t \in \mathbb{R}_+$ and $u \in \mathcal{Q}$ the linear operator $Y(t, u) : T_u\mathcal{M} \to T_{\varphi^t(u)}\mathcal{M}$ as $Y(t, u) = d_{u\varphi^t}$ for $t \in T(u)$ and all $k \in \mathbb{N}$ by

$$Y(t_k(u), u) = S_k(u) \circ \Big(\lim_{t \to t_k(u)-0} Y(t, u) \Big).$$

Then $Y(t, u)$ has the form $Y(t, u) = d_u\varphi^t_{m_0(u)}$ for $t \in [0, t_1(u))$ and

$$Y(t, u) = Y_{m_j(u)}(t - t_j(u), \xi_j(u)) \circ S_j(u) \circ \Big(\lim_{\tau \to t_j(u)-0} Y(\tau, u) \Big) \tag{10.34}$$

for all $j \in \mathbb{N}$ and $t \in [t_j(u), t_{j+1}(u))$ with $Y_k(t,u) := d_u\varphi_k^t$. Moreover the map $Y(\cdot, u)$ is the normed for $t = 0$ fundamental solution of the variational equation

$$\frac{Dy}{dt} = \nabla f(\varphi^t(u))y, \tag{10.35}$$

given for all $t \in \mathbb{R}_+ \backslash \mathcal{T}(u)$.

Proof The first part of the assertion can be shown as in [16] for the linear space. Let us fix arbitrary $j \in \mathbb{N}_0 \backslash \{0\}$ and $u \in \mathcal{Q}_j$. For brevity we put $i := m_{j-1}(u)$ and $k := m_j(u)$. By Eq. 10.29 we have for $\varphi^{(\cdot)}(\cdot)$ with arbitrary $\upsilon \in T_u\mathcal{M}$

$$\begin{aligned} d_u\varphi^t\upsilon &= d_u(\varphi_k^{t-t_j(u)}(\xi_j(u)))\upsilon \\ &= -f_k(\varphi_k^{t-t_j(u)}(u))d_ut_j\upsilon + Y_k(t-t_j(u), \xi_j(u))d_u\xi_j\upsilon, \end{aligned} \tag{10.36}$$

where $d_ut_j : T_u\mathcal{M} \to \mathbb{R}$ and $d_u\xi_j : T_u\mathcal{M} \to T_{\xi_j(u)}\mathcal{M}$ are the differentials of the function t_j and the map ξ_j. Since $\varphi_i^0(p) = p$ and $d_\xi\varphi_i^0 = \mathrm{id}_{T_p\mathcal{M}}$ for all $p \in \mathcal{M}$ and $l \in I$ we have

$$\lim_{t \to t_j+0} d_u\varphi^t\upsilon = -f_k(\xi_j)d_ut_j\upsilon + d_u\xi_j\upsilon. \tag{10.37}$$

By definition we have $\xi_j(u) = \varphi^{t_j(u)}(u)$ and consequently

$$d_u\xi_j\upsilon = f_i(\xi_j(u))d_ut_j\upsilon + \lim_{t \to t_j-0} d_u\varphi^t\upsilon. \tag{10.38}$$

Let us consider the auxiliary map $\Xi_{i,j}^t(u) := \varphi_i^{t-t_{j-1}(u)}(\xi_{j-1}(u))$ and the auxiliary function $h_j^t(u) := F_i(\Xi_{i,j}(t,u))$ on $\mathcal{M}$. In particular we have $\Xi_{i,j}^{t_j(u)}(u) = \xi_j(u)$. Furthermore we receive for arbitrary $\upsilon \in T_u\mathcal{M}$

$$d_uh_j^t\upsilon = d_{\Xi_{i,j}^t(u)}F_i(d_u\Xi_{i,j}^t\upsilon)$$

and since $u \in \mathcal{Q}_j$ we have $\dot{h}_j^{t_j(u)}(u) = L_{f_i}F_i(\xi_j(u)) \neq 0$. By assumption **(E3)** we have the relation $h_j^{t_j(u)}(u) = 0$ for all $u \in \mathcal{Q}_j$. From this by differentiation we get

$$L_{f_i}F_i(\Xi_{i,j}^t(u))d_ut_j\upsilon + d_uh_j^t\upsilon = 0$$

for all $\upsilon \in T_u\mathcal{M}$. Consequently we get

$$d_ut_j\upsilon = -\frac{d_uh_j^{t_j(u)}\upsilon}{L_{f_i}F_i(\xi_j(u))} = -\frac{d_{\xi_j(u)}F_i(d_u\Xi_{i,j}^t\upsilon)}{L_{f_i}F_i(\xi_j(u))}. \tag{10.39}$$

If we put (10.38) and (10.39) into (10.37), we get for arbitrary $\upsilon \in T_u\mathcal{M}$

$$\lim_{t\to t_j+0} d_u\varphi^t\upsilon =(f_i(\xi_j(u)) - f_k(\xi_j(u))d_u t_j\upsilon + \lim_{t\to t_j-0} d_u\varphi^t\upsilon \tag{10.40}$$

$$=\frac{f_i(\xi_j(u)) - f_k(\xi_j(u))}{L_{f_i}F_i(\xi_j(u))} d_{\xi_j(u)}F_i(d_u\Xi^t_{i,j}\upsilon) + \lim_{t\to t_j-0} d_u\varphi^t\upsilon \tag{10.41}$$

and with $\lim_{t\to t_j-0} d_u\varphi^t\upsilon = d_u\Xi^t_{i,j}\upsilon$ we receive the assertion (1).

Let us show now that the operator defined in (2) has the form (10.34). Note that from the definition it follows immediately that $Y(t,u) = d_u\varphi^t_{m_0(u)}$ for the $t \in [0, t_1(u))$. Suppose now that $j \in \mathbb{N}_0\backslash\{0\}$ and $t \in (t_{j-1}(u), t_j(u))$ are arbitrary. Using the definition of $Y(t,u)$ and the fact that $f_k(\varphi^{(\cdot)}(u))$ is the solution of the variational equation $\frac{Dy}{dt} = \nabla f_k(\varphi^t(u))y$, we get with (10.36), (10.38) and (10.40) for arbitrary $\upsilon \in T_u\mathcal{M}$

$$\begin{aligned} Y(t,u)\upsilon = d_u\varphi^t\upsilon &= Y_k(t - t_j(u), \xi_j(u))[-f_k(\xi_j(u))d_u t_j\upsilon + d_u\xi_j\upsilon] \\ &= Y_k(t - t_j(u), \xi_j(u))[f_i(\xi_j(u)) - f_k(\xi_j(u))]d_u t_j\upsilon + \lim_{t\to t_j-0} d_u\varphi^t\upsilon \\ &= Y_k(t - t_j(u), \xi_j(u)) \lim_{t\to t_j+0} d_u\varphi^t\upsilon \end{aligned}$$

where again we have put $i = m_{j-1}(u)$ and $k = m_j(u)$.

From this and assertion (1) of the theorem it follows that (10.34) is true. It remains to show that any curve $t \mapsto y(t,u) = Y(t,u)y_0$ with $y(0,u) = y_0$ is a solution of the variational equation (10.35). Let an arbitrary $t \in (t_j(u), t_{j+1}(u))$ be fixed. Since $Y_{m_j(u)}(\cdot, u)$ is the normed for $t = 0$ fundamental operator of the variational equation $\frac{Dy}{dt} = \nabla f_{m_j(u)}y$, we get with the presentation (10.34) in coordinates of a chart x around $\varphi^t(u)$ with y^i, f^i, Y^i_k, S^i_k and $\widetilde{Y}^i_k$ as coordinates for y, $f_{m_j(u)}$, $Y_{m_j(u)}$, S_j resp. $\lim_{t-t_j-0} Y(t,u)$,

$$\begin{aligned} \frac{Dy^i}{dt} &= \qquad \dot{y}^i + \Gamma^i_{lm}f^l y^m = \dot{Y}^i_s y^s_0 + \Gamma^i_{lm}f^l y^m \\ &= \dot{Y}^i_s S^s_r \widetilde{Y}^r_p y^p_0 + \Gamma^i_{lm}f^l y^m \\ &= \frac{\partial f^i}{\partial x^k} Y^k_s S^s_r \widetilde{Y}^r_p y^p_0 + \Gamma^i_{lm}f^l y^m \\ &= \nabla_k f^i yk. \end{aligned}$$

But this is exactly the assertion in local coordinates of the chart x. □

In the next theorem we use together the approach from [36], developed for vector fields in $\mathbb{R}^n$, and a result from [28]. Let us consider for $t > 0$ the sets

$$\mathcal{U}_{i,k} := \{u \in \mathcal{M}_k \cap \mathcal{Q} \mid t \in (t_i(u), t_{i+1}(u))\} \tag{10.42}$$

and write them for brevity as $\{\mathcal{U}_j\}$ including in this family only non zero sets $\mathcal{U}_{j,k}$. Suppose that the family $\{\mathcal{U}_j\}$ has the following properties:

(U1) There exists an $k_0 \in \mathbb{N}_0$ such that for arbitrary $j \in \mathbb{N}_0$ there is an $\varepsilon_0(j) > 0$ with the property that for all $\varepsilon \in (0, \varepsilon_0]$ and all $u \in \mathcal{M}$ the number of j-tupel $(i_1, \ldots, i_j)$ satisfying $\mathcal{B}(u, \varepsilon) \cap \varphi^{-j+1}(\mathcal{U}_{i_j}) \cap \varphi^{-j+2}(\mathcal{U}_{i_{j-1}}) \cap \ldots \cap \mathcal{U}_{i_1} \neq \emptyset$ is not larger than k_0.

(U2) For any $j \in \mathbb{N}$ there exists an $\varepsilon_1(j) > 0$ such that for all $(i_1, \ldots, i_j)$ arbitrary two points $u, p \in \varphi^{-j+1}(\mathcal{U}_{i_j}) \cap \varphi^{-j+2}(\mathcal{U}_{i_j}) \cap \ldots \cap \mathcal{U}_{i_j}$ with $\rho(p, u) < \varepsilon_1$ can be connected by a continuous curve $c_{p,u}$ with $c_{p,u} \backslash \{p, u\} \subset \mathcal{B}(p, \rho(p, u)) \cap \varphi^{-j+1}(\mathcal{U}_{i_j}) \cap \varphi^{-j+2}(\mathcal{U}_{i_j}) \cap \ldots \cap \mathcal{U}_{i_j}$.

Theorem 10.8 *Suppose that there are given the piecewise smooth differential equation (10.28) on the partition $\bigcup_{i \in I} \mathcal{M}_i$ of $\mathcal{M}$ and C^1-smooth functions $F_i : \mathcal{M} \to \mathbb{R}$ such that the assumptions* **(E1)–(E3)** *are satisfied.*

Suppose further that $\mathcal{K} \subset \mathcal{Q} \subset \mathcal{M}$ is a compact set which is φ^t invariant for some $t > 0$. We set $N(t, u) := \max\{j \in \mathbb{N} | t_j(u) < t\}$. Assume that the sets $\mathcal{U}_i$, defined in (10.38), satisfy the properties **(U1)** *and* **(U2)** *and the inequality*

$$\int\limits_{[0,\tau]\backslash \mathcal{T}(u)} \kappa_d(\varphi^s(u))ds + \sum_{j=1}^{N(u)} \ln \omega_d(S_j(u)) < 0 \tag{10.43}$$

for all $u \in \mathcal{K}$. Then $\dim_H \mathcal{K} \leq d$.

Proof The first steps are similarly to the proof in [36]. Assume that $u \in \mathcal{K}$ is arbitrary. Let us introduce the abbreviation $N := N(t, u)$, $m_i := m_i(u)$, $t_i := t_i(u)$ and $\xi_i := \xi_i(u)$. From the representation (10.34) of the for $t = 0$ normed fundamental operator $Y(t, u) : T_u\mathcal{M} \to T_{\varphi^t(u)}\mathcal{M}$ we get

$$\begin{aligned} Y(t, u) = {} & Y_{m_N}(t - t_N, \xi_N) \circ S_N(u) \circ Y_{m_{N-1}}(t_N - t_{N-1}, \xi_{N-1}) \circ \cdots \\ & \circ Y_{m_1}(t_2 - t_1, \xi_1) \circ S_1(u) \circ Y_{m_0}(t_1, u). \end{aligned} \tag{10.44}$$

Using the generalization of Horn's lemma (Lemma 2.4, Chap. 2) with the abbreviation $Y_i := Y_{m_i}(t_{i+1} - t_i, \xi_i)$ and $S_i := S_i(u)$ and $Y_N^t := Y_{m_N}(t - t_N, \xi_i)$, we derive

$$\begin{aligned} \omega_d\Big(Y(t, u)\Big) & = \omega_d(Y_N^t \circ S_N \circ \cdots \circ Y_1 \circ S_1 \circ Y_0) \\ & \leq \omega_d(Y_N^t) \cdot \ldots \cdot \omega_d(Y_0)\omega_d(S_N) \cdot \ldots \cdot \omega_d(S_1). \end{aligned} \tag{10.45}$$

Since the linear operator $Y_{m_j}(\cdot, \xi_j)$ for any $j \in \{0, \ldots, N\}$ is the for $t = 0$ normed fundamental operator of the variational equation $\frac{Dy}{dt} = \nabla f_{m_j}(\varphi^t(u))y$ to the vector field f_{m_j}, we can use the Liouville formula (10.25) and get

$$\omega_d(Y_j) \leq \exp\left\{\int_{t_j}^{t_{j+1}} \kappa_d(\varphi^\tau(u))d\tau\right\} \tag{10.46}$$

for all $j \in \{1, \ldots, N - 1\}$ and

$$\omega_d(Y_N^t) \leq \exp\left\{\int_{t_j}^{t} \kappa_d(\varphi^\tau(u))d\tau\right\}. \tag{10.47}$$

From assumption (10.43) of the theorem and (10.45), (10.46) and (10.47) it follows that

$$\omega_d(Y(t,u)) < 1. \tag{10.48}$$

Now we consider the family of sets $\mathcal{U}_i$. Because of the properties of the semiflow $\varphi^{(\cdot)}(\cdot)$ and Lemma 10.4 the map $\varphi^t(\cdot)$ is on every set $\text{int}(\mathcal{U}_i)$ a C^1-map and extendable to a C^1-map on $\overline{\mathcal{U}_i}$. Assumptions **(U1)** and **(U2)** and the inequality (10.48) allow us to applicate Theorem 10.5. This directly gives the estimate $\dim_H K \leq d$. □

Remark 10.7 Suppose that the assumptions of Theorem 10.8 are satisfied. Additionally it is assumed that the vector field f is continuous. Then by Formula (10.33) the transition operator S_j is given as $S_j(u) = \text{id}_{T_{\xi_j(u)}\mathcal{M}}$ and the relation (10.32) has the form

$$\lim_{t \to t_j(u)+0} d_u\varphi^t = \lim_{t \to t_j(u)-0} d_u\varphi^t$$

for all $j \in \mathbb{N}$. This means that $d_u\varphi^{t_j(u)}$ exists for all $j \in \mathbb{N}$. Condition (10.43) simplifies to

$$\int_{[0,t]\setminus\mathcal{T}(u)} \kappa_d(\varphi^\tau(u))d\tau < 0$$

for all $u \in \mathcal{K}$.

10.3 Dimension Estimates for Maps with Special Singularity Sets

In this section, which is based on the results of [27], a Douady-Oesterlé-type estimate for another class of piecewise smooth dynamical systems is presented than in the previous section. A special assumption on the Hausdorff measure of the singularity set and its preimages is needed (see Theorem 10.9). In fact this as assumption is easy to check for some interesting classes of systems and especially the theorem is applicable in situation where the results from Sect. 10.2 can not be applied. The general results will be applied to three classes of systems illustrating these facts. Estimates of the Hausdorff dimension of invariant sets for the Belykh systems, for the Lozi systems and for a class of piecewise affine solenoid-like systems will be derived.

10.3.1 Definitions and Results

Let $\mathcal{M}$ be a C^∞-Riemannian manifold and $\mathcal{U} \subset \mathcal{M}$. We consider now a special class of piecewise smooth maps on $\mathcal{U}$.

Definition 10.1 We say that a map $\varphi : \mathcal{U} \to \mathcal{M}$ *fullfills condition* **(PC)** if for all $m \in \mathbb{N}$ there is a partition $\{\mathcal{U}_1^m, \ldots \mathcal{U}_{i(m)}^m\}$ of $\mathcal{U}$ with connected Borel sets that have compact closure such that $\varphi_k^m := \varphi^m|_{\mathcal{U}_k^m}$ is a C^1-map and is C^1-extendable to some open neighborhood of $\overline{\mathcal{U}_m^k}$ for all $k = 1, 2, \ldots, i(m)$.

Remark 10.8 If we would not suppose that the sets $\mathcal{U}_k^m$ are Borel and have compact closure but would suppose that they are connected then condition **(PC)** would be equivalent to condition **(H0)** and **(H1)** of Sect. 10.2

Given $\varphi : \mathcal{U} \to \mathcal{M}$ satisfying **(PC)** and $u \in \mathcal{U}$ we define the singular value function of φ by $\omega_d(\varphi, u) := \omega_d(d_u\varphi)$ where $d_u\varphi : T_u\mathcal{M} \to T_{\varphi(u)}\mathcal{M}$ is the tangent map of the C^1-extension of φ.

Let us state the main result [27].

Theorem 10.9 *Suppose $\varphi : \mathcal{U} \to \mathcal{M}$ satisfying condition* **(PC)**. *Let $\mathcal{K}$ be a compact φ-invariant, i.e. $\varphi(\mathcal{K}) = \mathcal{K}$, subset of $\mathcal{U}$. If we have a number $d \in (0, n]$ such that*

$$\sup_{x\in\mathcal{K}} \omega_d(\varphi, x) < 1 \quad \textit{and} \quad \forall m \in \mathbb{N} \quad \forall k \in \{1, \ldots, i(m)\} :$$

$$\mu_H\Big(\mathcal{K} \cap (\overline{\mathcal{U}_k^m} \setminus \mathcal{U}_k^m), d\Big) = 0$$

then $\dim_H \mathcal{K} \le d$ *holds.*

Remark 10.9 We compare this result with Theorem 10.5 of Sect. 10.2 about the Hausdorff dimension estimates for invariant sets piecewise smooth maps. We replaced condition **(H2)** of this section by the assumption that the intersection of $\mathcal{K}$ with the singularity set and its preimages has zero d-dimensional Hausdorff measure. We will see in Subsect. 10.3.3 that Theorem 10.9 is applicable in situations were Theorem 10.5 of Sect. 10.2 is not applicable.

Remark 10.10 In some situations it is useful to replace the map φ by a power of φ in order to get better dimension estimates by Theorem 10.9.

In the following corollary of Theorem 10.9 we introduce a Lyapunov function $\varkappa$ into the dimension estimate.

Corollary 10.10 *Suppose $\varphi : \mathcal{U} \to \mathcal{M}$ satisfying condition* **(PC)**. *Let $\mathcal{K}$ be a compact φ-invariant, i.e. $\varphi(\mathcal{K}) = \mathcal{K}$, subset of $\mathcal{U}$. If we have a number $d \in (0, n]$ and a continuous function $\varkappa : \mathcal{K} \to \mathbb{R}^+ := \{u \in \mathbb{R} | u > 0\}$ such that*

$$\sup_{u\in\mathcal{K}}\Big(\frac{\varkappa(\varphi(u))}{\varkappa(u)}\omega_d(\varphi,u)\Big)<1 \quad and \quad \forall m\in\mathbb{N} \ \ \forall k\in\{1,\dots,i(m)\}:$$

$$\mu_H\Big(\mathcal{K}\cap(\overline{\mathcal{U}_k^m}\setminus\mathcal{U}_k^m),d\Big)=0$$

then $\dim_H\mathcal{K}\le d$ *holds.*

Proof We get Corollary 10.10 from Theorem 10.9 by the same arguments that were used in the proof of Corollary 10.9 in Sect. 10.3. □

10.3.2 Proof of the Main Result

The following lemma is essential for the proof of Theorem 10.9:

Lemma 10.6 *Suppose that* $\varphi:\mathcal{U}\to\mathcal{M}$ *satisfies condition* **(PC)**. *Furthermore suppose that for* $d\in(0,n]$, $m\in\mathbb{N}$ *and for a compact* φ*-invariant set* $\mathcal{K}$ *the inequality* $\sup_{u\in\mathcal{K}}\omega_d(\varphi^m,u)<\delta$ *holds. Then there exists an* ε_0 *such that for all* $\varepsilon\in(0,\varepsilon_0]$ *we have*

$$\mu_H\Big(\varphi^m(\mathcal{K}),d,c(d)\delta^{1/d_\varepsilon}\Big)\le C(d)\delta\Big(\mu_H(\mathcal{K},d,\varepsilon)+\sum_{k=1}^{i(m)}\mu_H(\mathcal{K}\cap(\overline{\mathcal{U}_k^m}\setminus\mathcal{U}_k^m),d,\varepsilon)\Big)$$

where $c(d)=2\sqrt{2\lfloor d\rfloor+1}$ *and* $C(d)=2^{\lfloor d\rfloor}(\lfloor d\rfloor+1)^d$.

Proof Let $\widetilde{\varphi}_k^m$ be the C^1-extension of φ_k^m to some open neighborhood of $\overline{\mathcal{U}_k^m}$. We have

$$\varphi^m(\mathcal{K})=\bigcup_{k=1}^{i(m)}\varphi_k^m(\mathcal{K}\cap\mathcal{U}_k^m)\subset\bigcup_{k=1}^{i(m)}\widetilde{\varphi}_k^m(\mathcal{K}\cap\overline{\mathcal{U}_k^m}).$$

Hence

$$\mu_H\Big(\varphi^m(\mathcal{K}),d,c(d)\delta^{1/d_\varepsilon}\Big)\le\mu_H\left(\bigcup_{k=1}^{i(m)}\widetilde{\varphi}_k^m(\mathcal{K}\cap\overline{\mathcal{U}_k^m}),d,c(d)\delta^{1/d_\varepsilon}\right)$$

$$\le\sum_{k=1}^{i(m)}\mu_H\Big(\widetilde{\varphi}_k^m(\mathcal{K}\cap\overline{\mathcal{U}_k^m}),d,c(d)\delta^{1/d_\varepsilon}\Big).$$

Fix $k\in\{1,\dots,i(m)\}$. Since $\widetilde{\varphi}_k^m$ is C^1 on some open neighborhood of the compact set $\mathcal{K}\cap\overline{\mathcal{U}_k^m}$ and

$$\sup_{u\in\mathcal{K}\cap\overline{\mathcal{U}_k^m}}\omega_d(\widetilde{\varphi}_k^m,u)<\delta$$

We get from Lemma 8.2, Chap. 8 that there exits $\varepsilon_0(k)$ such that for all $\varepsilon \in (0, \varepsilon_0(k)]$

$$\mu_H\left(\widetilde{\varphi}_k^m(\mathcal{K} \cap \overline{\mathcal{U}_k^m}), d, c(d)\delta^{1/d_\varepsilon}\right) \leq C(d)\delta\mu_H(\mathcal{K} \cap \overline{\mathcal{U}_k^m}, d, \varepsilon).$$

Now let $\varepsilon_0 := \min\{\varepsilon_0(k) | k = 1, \ldots, i(m)\}$ and $\varepsilon \in (0, \varepsilon_0]$. We get

$$\mu_H\left(\varphi^m(\mathcal{K}), d, c(d)\delta^{1/d_\varepsilon}\right) \leq C(d)\delta \sum_{k=1}^{i(m)} \mu_H(\mathcal{K} \cap \overline{\mathcal{U}_k^m}, d, \varepsilon)$$

and, using the fact that $\mathcal{K} \cap \overline{\mathcal{U}_k^m} \subset (\mathcal{K} \cap \mathcal{U}_k^m) \cup (\overline{\mathcal{U}_k^m} \setminus \mathcal{U}_k^m)$,

$$\begin{aligned}\mu_H\left(\varphi^m(\mathcal{K}), d, c(d)\delta^{1/d_\varepsilon}\right) &\leq C(d)\delta \sum_{k=1}^{i(m)} \mu_H(\mathcal{K} \cap \mathcal{U}_k^m, d, \varepsilon) + \mu_H\left(\mathcal{K} \cap (\overline{\mathcal{U}_k^m} \setminus \mathcal{U}_k^m), d, \varepsilon\right) \\ &= C(d)\delta\left(\mu_H(\mathcal{K}, d, \varepsilon) + \sum_{k=1}^{i(m)} \mu_H(\mathcal{K} \cap (\overline{\mathcal{U}_k^m} \setminus \mathcal{U}_k^m), d, \varepsilon)\right).\end{aligned}$$

To get the last equality we used the fact that $\mathcal{U}_k^m \cap \mathcal{K}$ are disjoint Borel sets and $\mu_H(\cdot, d, \varepsilon)$ is a Borel measure (see [31]) and thus especially additive. □

We need one another simple lemma that is about the singular value function.

Lemma 10.7 *Suppose $\varphi : \mathcal{U} \to \mathcal{M}$ satisfying condition* **(PC)**. *Let $\mathcal{K}$ be a compact φ-invariant, i.e. $\varphi(\mathcal{K}) = \mathcal{K}$ subset of $\mathcal{U}$ and let $d \in (0, n]$. Then we have*

$$\sup_{u \in \mathcal{K}} \omega_d(\varphi^m, u) \leq (\sup_{u \in \mathcal{K}} \omega_d(\varphi, u))^m.$$

Proof This lemma follows immediately from Proposition 7.14 of Sect. 7. The proof can be done in exactly the same way as in the usual situation of C^1 maps (see the corresponding statements in Chap. 5. □

Proof of Theorem 10.9 . We know from Lemma 10.7 that if $\sup_{u \in \mathcal{K}} \omega_d(\varphi, u) < 1$ holds we have

$$\lim_{m \to \infty} \sup_{u, \in \mathcal{K}} \omega_d(\varphi^m, u) = 0.$$

Hence under first assumptions of our theorem there exist $m \in \mathbb{N}$ and $\delta \geq 0$ such that $C(d)\delta < 1$, $c(d)\delta^{1/d} < 1$ and $\sup_{u \in \mathcal{K}} \omega_d(\varphi^m, u) < \delta$. Using $c(d)\delta^{1/d} < 1$ and the invariance of $\mathcal{K}$ we have

$$\mu_H(\mathcal{K}, d, \varepsilon) \leq \mu_H\left(\mathcal{K}, d, c(d)\delta^{1/d_\varepsilon}\right) = \mu_H\left((\varphi^m(\mathcal{K}), d, c(d)\delta^{1/d_\varepsilon}\right).$$

Thus we get from Lemma 10.6 that there exits an ε_0 such that for all $\varepsilon \in (0, \varepsilon_0]$

$$\mu_H(\mathcal{K}, d, \varepsilon) \leq C(d)\delta\Big(\mu_H(\mathcal{K}, d, \varepsilon) + \sum_{k=1}^{i(m)} \mu_H(\mathcal{K} \cap (\overline{\mathcal{U}_k^m} \setminus \mathcal{U}_k^m), d, \varepsilon)\Big)$$

and, using $C(d)\delta < 1$,

$$\mu_H(\mathcal{K}, d, \varepsilon) \leq \frac{C(d)\delta}{1 - C(d)\delta} \sum_{k=1}^{i(m)} \mu_H\Big(\mathcal{K} \cap (\overline{\mathcal{U}_k^m} \setminus \mathcal{U}_k^m), d, \varepsilon\Big).$$

Using the second assumption of the theorem we see that the expression on the right hand tends to zero for $\varepsilon \to 0$. Hence we have $\mu_H(\mathcal{K}, d) = 0$ and consequently $\dim_H \mathcal{K} \leq d$. □

10.3.3 Applications

The Belykh Systems. We consider the class of Belykh systems given by the piecewise affine transformations

$$\varphi_\delta : [-1, 1]^2 \to [-1, 1]^2 \quad \text{with}$$
$$\varphi_\delta(x, y) = \begin{cases} (\delta_1 x + (1 - \delta_1), \delta_2 y + (1 - \delta_2)), & \text{for} \quad y \geq \delta_3 x, \\ (\delta_1 x - (1 - \delta_1), \delta_2 y - (1 - \delta_2)), & \text{for} \quad y < \delta_3 x, \end{cases}$$

where $\delta = (\delta_1, \delta_2, \delta_3)$ is a parameter with $\delta_1 \in (0, 1)$, $\delta_3 \in (-1, 1)$ and $\delta_2 \in (1, 2/(|\delta_3| + 1)])$ (see Fig. 10.4). The original Belykh map was introduced in [3]. This version of the Belykh map is due to by Pesin [30] who studied ergodic properties of this map. Dimensional theoretical properties of the Belykh attractor were studied by Schmeling [35].

We want to apply Theorem 10.9 to these systems.

Estimate 10.1 *Let $\delta_1 \in (0, 1)$, $\delta_3 \in (-1, 1)$ and $\delta_2 \in (1, 2/(|\delta_3| + 1)])$ be given and let $\mathcal{K}$ be a compact set which is invariant under φ_δ. Then*

$$\dim_H \mathcal{K} \leq 1 - \log \delta_2 / \log \delta_1.$$

Proof Let $d = (1 - \log \delta_2 / \log \delta_1) + \varepsilon$ where $\varepsilon > 0$. Note that $d > 1$. The singular values of $d_u \varphi_\delta$ are constant and given by $\delta_2 > \delta_1 > 0$. Hence $\omega_d(\varphi_\delta, (x, y)) = \delta_2 \delta_1^{d-1} = \delta_1^\varepsilon < 1$. Obviously we can choose the partitions $(\{\mathcal{U}_1^m, \ldots, \mathcal{U}_{i(m)}^m\})$ in a way such that the partition elements have one dimensional boundary and hence

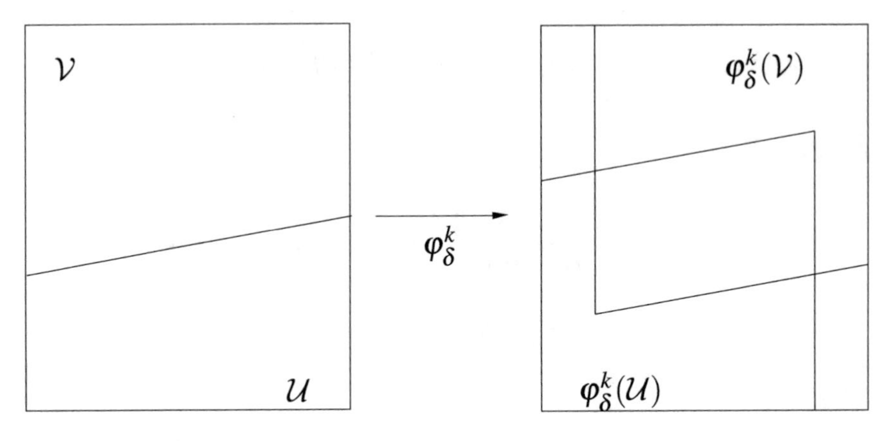

Fig. 10.4 The Belykh maps

$\mu_H(\mathcal{U}_k^m \setminus \mathcal{U}_k^m, d) = 0$. By Theorem 10.9 we get now $\dim_H \mathcal{K} \leq d$. Since $\varepsilon > 0$ was arbitrary our claim is proved. □

Remark 10.11

(a) Estimate 10.1 only gives some information if $\delta_2 < \delta_1^{-1}$. If this is not the case we have the trivial estimate by dimension two.
(b) Schmeling [35] showed that for almost all parameter values (with some technical restrictions) the Hausdorff dimension of the Belykh attractor is given by $\min\{2, 1 - \log \delta_2 / \log \delta_1\}$. Thus if $\delta_2 < \delta_1^{-1}$ the estimate obtained by Theorem 10.9 is at least generically sharp.
(c) In Sect. 10.2 it is remarked that we could not apply Theorem 10.5 to the Belykh systems in general. Thus we see that there are situations where Theorem 10.9 is more appropriate.

The Lozi Systems. Now for $b \in (0, 1)$ and $a \in (0, 2(1-b))$ we consider the class of Lozi systems (see [30]) given by the transformations

$$\varphi_{a,b} : [-1/(1-b), 1/(1-b)]^2 \to [-1/(1-b), 1/(1+b)]^2$$

$$\varphi_{a,b}(x, y) = (1 + by - a|x|, x).$$

The Lozi map (see Fig. 10.5) was introduced by Lozi [24]. Ergodic properties of the map were studied in [7] and estimates of the Hausdorff dimension were given by Ishii [15].

By Theorem 10.9 we get the following dimension estimate.

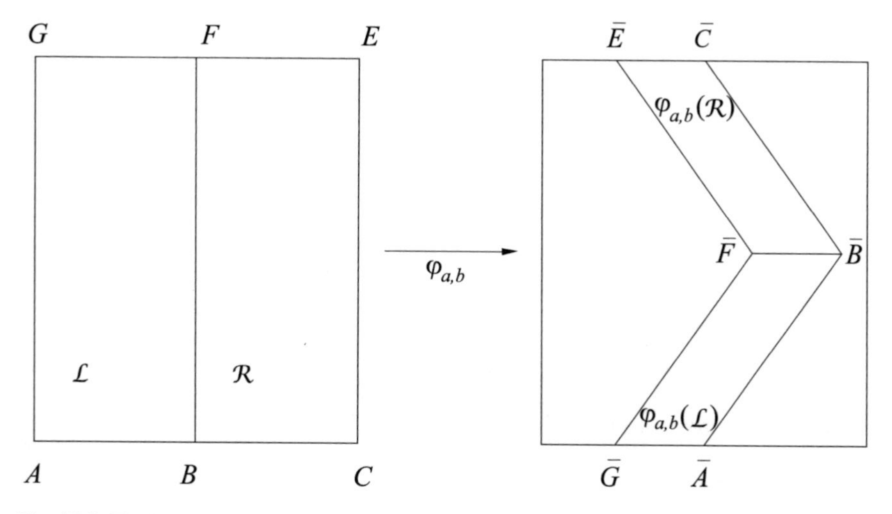

Fig. 10.5 The Lozi maps

Estimate 10.2 *Let $b \in (0, 1)$ and $a \in (0, 2(1 - b))$ and let $\mathcal{K}$ be a compact subset of the set $[-1/(1 - b), 1/(1 - b)]^2$ which is invariant under the Lozi map $\varphi_{a,b}$. Set*

$$\beta_1 = 1/2(\sqrt{(a^2 + b^2 + 1)^2 - 4b^2} + a^2 + b^2 + 1) \quad \text{and}$$
$$\beta_2 = 1/2(-\sqrt{(a^2 + b^2 + 1)^2 - 4b^2} + a^2 + b^2 + 1.$$

Furthermore assume that $\beta_2 < 1$. Then we have

$$\dim_H \mathcal{K} \leq 1 - \log \beta_1 / \log \beta_2.$$

Proof Let $d = 1 - \log \beta_1 \log \beta_2 + \varepsilon$ where $\varepsilon > 0$. Note that for all $(x, y) \in [-1/(1 - b), 1/(1 - b)]^2$ we have

$$d_{(x,y)}\varphi_{a,b} = \begin{pmatrix} a^2 + 1 & -ab \\ -ab & b^2 \end{pmatrix} \quad \text{if } x \geq 0,$$
$$d_{(x,y)}\varphi_{a,b} = \begin{pmatrix} a^2 + 1 & ab \\ ab & b^2 \end{pmatrix} \quad \text{if } x < 0.$$

A simple calculation shows that the singular values are constant and given by $\alpha_1 = \sqrt{\beta_1}$ and $\alpha_2 = \sqrt{\beta_2}$. We have $\omega_d(\varphi_{a,b}, (x, y)) = \alpha_1 \alpha_2^{d-1} = \alpha_2^{\varepsilon} < 1$. Furthermore the singularity set of the system $([-1/(1 - b), 1/(1 - b)]^2, \varphi_{a,b})$ is given by $\mathcal{S} = [-1, 1] \times \{0\}$ and since $\varphi_{a,b}$ is just a affine map on $[-1/(1 - b), 0]^2$ and $[0, 1/(1 - b)]^2$ we see that $\bigcup_{i=0}^{m} \varphi_{a,b}^{-i}(\mathcal{S})$ consists of a finite number of line segments. Hence we can choose the $\mathcal{U}_k^m$ as domains bounded by a polygon. The boundary of these sets is thus one dimensional and since $d > 1$ we get $\mu_H(\overline{\mathcal{U}_k^m} \setminus \mathcal{U}_k^m, d) = 0$. We

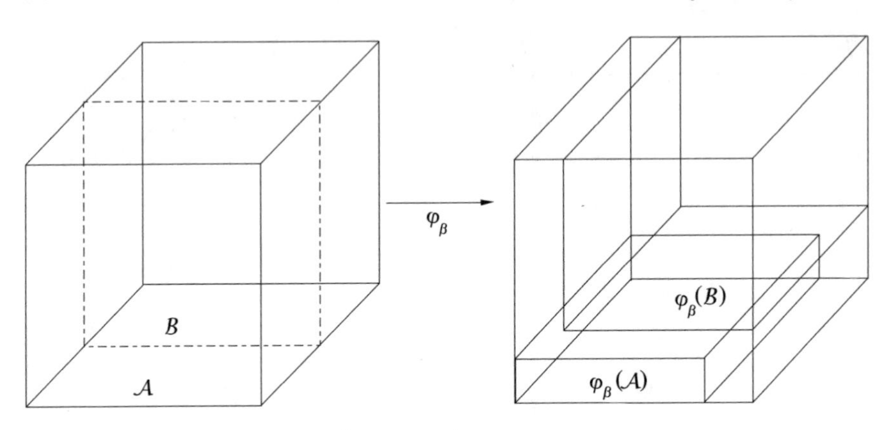

Fig. 10.6 The action of $\varphi_\beta : [-1, 1]^3 \mapsto [-1, 1]^3$

are thus in a situation where Theorem 10.9 applies and since $\varepsilon > 0$ was arbitrary our claim is proved. □

Remark 10.12 If we set $b = 1.7$ and $a = 0.1$ we get $\dim_H \mathcal{K} \leq 1.18761294\ldots$ which is better than the upper estimate by $1.247848\ldots$ found in [15]. In [15] a lower bound on the Hausdorff dimension is given by $1.16669\ldots$. In fact we do not know if our upper bound is sharp.

Piecewise Affine Solenoid Like Systems. Consider the following class of three dimensional piecewise affine maps

$$\varphi_\beta : [-1, 1]^3 \mapsto [-1, 1]^3$$
$$\varphi_\beta(x, y, z) = \begin{cases} (\beta_1 x + (1 - \beta_1), 2y - 1, \beta_3 z + (1 - \beta_3)) \text{ for } & y \geq 0 \\ (\beta_2 x - (1 - \beta_2), 2y + 1, \beta_4 z - (1 - \beta_4)) \text{ for } & y < 0 \end{cases}$$

where $\beta = (\beta_1, \beta_2, \beta_3, \beta_4) \in \mathcal{P} := \{(u, v, s, t) | u, v, s, \quad t \in (0, 1), s + t \geq 1, u + v < 1, u > v, s > t\}$ (see Fig. 10.6).

Again by Theorem 10.9 we can get a dimension estimate.

Estimate 10.3 *Let $\beta = (\beta_1, \beta_2, \beta_3, \beta_4) \in \mathcal{P}$ and $\mathcal{K}$ be a compact φ_β invariant set. Then*

$$\dim_H \mathcal{K} \leq 2 - \frac{\log(2\beta_1)}{\log(\beta_3)}.$$

Proof Let $d = 2 - \frac{\log(2\beta_1)}{\log(\beta_3)} + \varepsilon$ where $\varepsilon > 0$. The singular values at a point (x, y, z) are given by $2 \geq \beta_1 \geq \beta_3$ if $y \geq 0$ and $2 \geq \beta_1 \geq \beta_3$ if $y < 0$. Thus $\omega_d(\varphi_\beta, (x, y, z)) = 2\beta_1 \beta_3^{d-2}$ if $y \geq 0$ and $\omega_d(\varphi_\beta, (x, y, z)) = 2\beta_2 \beta_4^{d-2}$ if $y < 0$. But anyway we have $\omega_d(\varphi_\beta, (x, y, z)) < 1$. It is easy to see that one can choose the sets $\mathcal{U}_i^m$ with two dimensional boundary and since $d > 2$ we get $\mu_H(\mathcal{U}_k^m \setminus \mathcal{U}_k^m, d) = 0$. □

Remark 10.13 It follows from [26] that the Hausdorff dimension of the attractor of the map φ_β is always bounded from above by the solution x of $\beta_1\beta_3^{x-2} + \beta_2\beta_4^{x-2} = 1$ and that in some parts of the parameter space this upper bound is at least generically sharp (in the sense of Lebesgue measure). This example shows that the estimates obtained by Theorem 10.9 can not be expected to be sharp in general. The problem is that the singular value function is not constant here and Douady-Oesterlé-type estimates take the worst contraction rates of the system everywhere into consideration. If we consider the symmetric case $\beta_1 = \beta_2$ and $\beta_3 = \beta_4$ the singular value function is constant and the estimate in Theorem 10.9 is generically sharp (see again [26]).

10.4 Lower Dimension Estimates

10.4.1 Frequency-Domain Conditions for Lower Topological Dimension Bounds of Global $\mathcal{B}$-Attractors

In this subsection we derive a lower topological dimension estimate of a global $\mathcal{B}$-attractor which is based on the results of [18]. Clearly (see Proposition 3.20, Chap. 3) that such a bound is also a lower bound for the Hausdorff and fractal dimensions.

Suppose that a discrete-time system in $\mathbb{R}^n$ is given by

$$u_{t+1} = Au_t + b\,\phi\left((c, u_t)\right), \tag{10.49}$$

where A is an $n \times n$ matrix, b and c are n-vectors, $\phi : \mathbb{R} \to \mathbb{R}$ is a continuous piecewise linear function having only a countable set of discontinuities of the first derivative $\mathcal{Z} := \{\sigma_j \mid j = 1, 2, \ldots\}$. We suppose that all points of $\mathcal{Z}$ are isolated in a strong sense, i.e. there exists a number $\tau > 0$ such that

$$|\,\sigma_i - \sigma_j\,| \geq \tau, \quad \forall \sigma_i, \sigma_j \in \mathcal{Z}, i \neq j.$$

Let us also assume that

$$\det(A + bc^*\phi'(\sigma)) \neq 0, \quad \forall \sigma \in \mathbb{R}\backslash\mathcal{Z}, \tag{10.50}$$

and that there exist two real numbers $\kappa_1 < 0$ and $\kappa_2 > 0$ such that

$$(\phi'(\sigma) - \kappa_1)\,(\phi'(\sigma) - \kappa_2) \geq 0, \quad \forall \sigma \in \mathbb{R}\backslash\mathcal{Z}. \tag{10.51}$$

Introduce the transfer function of the linear part of (10.49) given for $z \in \mathbb{C}$ with $\det(A - zI) \neq 0$ by

$$W(z) := c^*(A - zI)^{-1}b.$$

Denote the discrete-time dynamical system, generated through (10.49) by $\{\varphi^t\}_{t\in\mathbb{N}_0}$. We assume that this dynamical system has a global minimal $\mathcal{B}$-attractor $\mathcal{A}_{\mathbb{R}^n,\min}$. The next theorem [18] gives a lower estimate for the topological dimension of this attractor.

Theorem 10.10 *Suppose that there is a number $\theta > 2$ such that the following conditions are satisfied:*

(1) The matrix $\frac{1}{\theta}(A + \kappa_1 bc^*)$ has m eigenvalues $(1 \le m \le n)$ outside the unit circle around the origin and $n - m$ eigenvalues inside this circle;

(2) $\mathrm{Re}\Big[\big(1 + \kappa_1 \overline{W(\theta z)}\big)\big(1 + \kappa_2 W(\theta z)\big)\Big] < 0, \quad \forall z \in \mathbb{C}, |z| = 1.$

Then $\dim_T \mathcal{A}_{\mathbb{R}^n,\min} \ge m$.

Proof According to conditions (1) and (2) of the theorem we can use the Kalman-Szegö theorem (Theorem 2.10, Chap. 2) to conclude that there exists a real symmetric $n \times n$ matrix P such that

$$\frac{1}{\theta^2}(P(Au + b\,\xi), Au + b\,\xi) - (Pu, u) \\ + \Big(\xi - \kappa_2(c, u)\Big)\Big(\xi - \kappa_1(c, u)\Big) < 0, \quad \forall u \in \mathbb{R}^n, \forall \xi \in \mathbb{R}, \quad |u| + |\xi| \ne 0. \tag{10.52}$$

Here $(\cdot, \cdot)$ denotes the scalar product in $\mathbb{R}^n$. If we put in (10.52) $\xi = \kappa_1(c, u)$, $u \in \mathbb{R}^n$, we get the inequality

$$\frac{1}{\theta^2}\Big(P(A + \kappa_1 bc^*)u,\ (A + \kappa_1 bc^*)u\Big) - (Pu, u) < 0, \quad \forall u \in \mathbb{R}^n, u \ne 0.$$

From this, condition (1) of the theorem and Lemma 2.8, Chap. 2, it follows that the matrix P has m negative and $n - m$ positive eigenvalues.

W.l.o.g. let us assume that the matrix P and a vector $u \in \mathbb{R}^n$ can be written as

$$P = \begin{pmatrix} -I_m & 0 \\ 0 & I_{n-m} \end{pmatrix}, \quad u = \begin{pmatrix} x \\ y \end{pmatrix},$$

where I_r denotes the $r \times r$ unit matrix and $x \in \mathbb{R}^m$, $y \in \mathbb{R}^{n-m}$. Introduce the quadratic form $V(u) := (Pu, u)$, $u \in \mathbb{R}^n$, which now can be written as $V(u) = |x|^2 - |y|^2$. Consider with $\delta > 0$ at a point $u_0 \in \mathbb{R}^n$ the closed ball $\mathcal{B}_\delta(u_0) := \{u \in \mathbb{R}^n \,\Big|\, |u - u_0| \le \delta\}$, where

$$\delta < \frac{\tau}{2|c|}. \tag{10.53}$$

Write u_0 as $u_0 = \binom{x_0}{y_0}$ with $x_0 \in \mathbb{R}^m$, $y_0 \in \mathbb{R}^{n-m}$, and consider the linear subspace

$$\mathbb{L}_m^0 := \{u = \tbinom{x}{y} \in \mathbb{R}^n \mid y = y_0\}.$$

It follows from the inequality (10.53) that at most one hyperplane of the type $\{u \mid (c, u) = \sigma_j, \sigma_j \in \mathcal{Z}\}$ can intersect $\mathcal{B}_\delta(u_0)$. Then there exists a point $u_1 \in \mathcal{B}_\delta(u_0) \cap \mathbb{L}_m^0$ such that the ball $\mathcal{B}_{\delta/2}(u_1)$ is included in $\mathcal{B}_\delta(u_0)$ and $\mathcal{B}_{\delta/2}(u_1)$ does not intersect the hyperplane $\{u \mid (c, u) = \sigma_j\}$. It follows from (10.51) to (10.52) that for $\in \mathcal{B}_{\delta/2}(u_1)$ we have the inequality

$$V(\varphi^1(v) - \varphi^1(u_1)) \leq \theta^2 V(v - u_1).$$

From this, the inequality $\theta > 2$ and the special representation of V it follows that

$$\mathcal{B}_\delta(\varphi^1(u_1)) \cap \mathbb{L}_m^1 \subset \varphi^1(\mathcal{B}_{\delta/2}(u_1) \cap \mathbb{L}_m^0),$$

where $\mathbb{L}_m^1$ is the linear set spanned by the elements of the set $\varphi^1(\mathcal{B}_{\delta/2}(u_1) \cap \mathbb{L}_m^0)$. It is easy to see that according to the piecewise linearity of ϕ and the choice of $\mathcal{B}_{\delta/2}(u_1)$ the set $\mathbb{L}_m^1$ is a linear m-dimensional subspace of $\mathbb{R}^n$. Note that the ball $\mathcal{B}_\delta(\varphi^1(u_1))$ according to (10.53) can intersect the hyperplane $\{u \mid (c, u) = \sigma_j\}$ at most for one $\sigma_j \in \mathcal{Z}$. The preimage of the intersection of the hyperplane with $\mathcal{B}_\delta(\varphi^1(u_1))$ is also part of some hyperplane. This follows from the fact that the inverse map $\varphi^{-1} := (\varphi^1)^{-1}$ exists as a linear and regular map on the set $\mathcal{B}_\delta(\varphi^1(u_1))$. It is evident that there exists a vector $u_2 \in \mathcal{B}_{\delta/2}(u_1) \cap \mathbb{L}_m^0$ such that $\mathcal{B}_{\delta/4}(u_2) \subset \mathcal{B}_{\delta/2}(u_1)$ and the ball $\mathcal{B}_{\delta/4}(u_2)$ does not have intersections with the set

$$\varphi^{-1}\Big(\{u \mid (c, u) = \sigma_j\} \cap \mathcal{B}_\delta(\varphi^1(u_1))\Big).$$

Again from (10.51) and (10.52) it is follows that, for each $v \in \mathcal{B}_{\delta/4}(u_2)$, we have the inequality

$$V(\varphi^2(v) - \varphi^2(u_2)) \leq \theta^4 V(v - u_2).$$

From this, the inequality $\theta > 2$ and from the special structure of V it follows that

$$\mathcal{B}_\delta(\varphi^2(u_2)) \cap \mathbb{L}_m^2 \subset \varphi^2(\mathcal{B}_{\delta/4}(u_2) \cap \mathbb{L}_m^0),$$

where $\mathbb{L}_m^2$ is the linear set spanned by the elements of the set $\varphi^2(\mathcal{B}_{\delta/4}(u_2) \cap \mathbb{L}_m^0)$. If we continue this procedure we get sequences of points $\{u_k\}_{k=0}^\infty$ and of linear m-dimensional sets $\{\mathbb{L}_m^k\}_{k=0}^\infty$ such that

$$\mathcal{B}_\delta(\varphi^k(u_k)) \cap \mathbb{L}_m^k \subset \varphi^k(\mathcal{B}_\delta(u_0)).$$

Since the sequence $\{\varphi^k(u_k)\}_{k=0}^\infty$ belongs for all sufficiently large k to an ε-neighborhood of the bounded global $\mathcal{B}$-attractor $\mathcal{A}_{\mathbb{R}^n,\min}$, this sequence is bounded. Thus we can choose a subsequence $k_i \to \infty$ such that $\varphi^{k_i}(u_{k_i}) \to \widetilde{u}$ as $i \to \infty$, where $\widetilde{u} \in \mathbb{R}^n$ is some point. But then we also find a subsequence, which we denote again by $\{k_i\}$, such that the sets $\mathcal{B}_{2\delta}(\widetilde{u}) \cap \mathbb{L}_m^{k_i}$ converge in the following sense: there exists a linear m-dimensional subspace $\widetilde{L}_m$ such that for each $v \in \mathcal{B}_{2\delta}(\widetilde{u}) \cap \widetilde{\mathbb{L}}_m$ we have

$\mathrm{dist}(\upsilon, \mathcal{B}_{2\delta}(\widetilde{u}) \cap \mathbb{L}_m^{k_i}) \to 0$ as $i \to \infty$. Here $\mathrm{dist}(\upsilon, \mathcal{Z})$ denotes the distance between a point $\upsilon \in \mathbb{R}^n$ and a set $\mathcal{Z} \subset \mathbb{R}^n$. But this implies that $\mathcal{B}_{\delta/2}(\widetilde{u}) \cap \mathbb{L}_m \subset \mathcal{A}_{\mathbb{R}^n,\min}$. From Proposition 3.3, Chap. 3, it follows that

$$\dim_T (\mathcal{B}_{\delta/2}(\widetilde{u}) \cap \widetilde{\mathbb{L}}_m) = m \leq \dim_T \mathcal{A}_{\mathbb{R}^n,\min}.$$

□

It is easy to generalize Theorem 10.10 to the situation where system (10.49) defines a discrete-time dynamical system on the flat cylinder. Suppose, for this, there exists a vector $\Delta \in \mathbb{R}^n$, $\Delta \neq 0$, such that

$$\varphi^t(u + k\Delta) = \varphi^t(u) + k\Delta, \quad t = 0, 1, 2, \ldots, \quad k = 0, 1, 2, \ldots, \quad u \in \mathbb{R}^n. \tag{10.54}$$

If this property is given, system (10.49) can be considered on the flat cylinder $\mathbb{R}^n/\mathbb{G}$ with $\mathbb{G} := \{k\Delta \mid k \in \mathbb{Z}\}$. W.l.o.g. we can assume that system (10.49) is given with

$$A = \begin{pmatrix} A_0 & 0 \\ c_0^* & 1 \end{pmatrix}, \quad b = \begin{pmatrix} b_0 \\ q_0 \end{pmatrix} \tag{10.55}$$

and a function ϕ as above with the additional property

$$\phi(\sigma + 2\pi) = \phi(\sigma), \quad \forall \sigma \in \mathbb{R}. \tag{10.56}$$

Here A_0 is an $(n-1) \times (n-1)$ matrix having all eigenvalues inside the unit disc, c_0 and b_0 are $(n-1)$-vectors and q_o is a real number.

The transfer function W for the linear part of (10.49), (10.55), and (10.56) is given by

$$W(z) = \frac{1}{1-z} \left[c_0^*(A_0 - zI)^{-1} b_0 - q_0 \right]. \tag{10.57}$$

Let us assume that system (10.49), (10.55), and (10.56), considered as dynamical system on the flat cylinder $\mathbb{R}^n/\mathbb{G}$ with $\mathbb{G} = \{k\Delta \mid k \in \mathbb{Z}\}$, $\Delta^* = (0, 0, \ldots, 2\pi)$, has a global minimal $\mathcal{B}$-attractor $\mathcal{A}_{\mathbb{R}^n/\mathbb{G},\min}$.

Theorem 10.11 *Suppose that there exists a $\theta > 2$ such that the transfer function W given by (10.57) satisfies the conditions (1) and (2) of Theorem 10.10.*

Then $\dim_T \mathcal{A}_{\mathbb{R}^n/\mathbb{G},\min} \geq m$.

Example 10.13 Consider a system (10.49), (10.55), and (10.56), with $n = 2$ and the transfer function (10.57) given by

$$W(z) = \frac{\beta_2 z}{(z-1)(z-\beta_1)}, \quad z \in \mathbb{C}, \ z \neq 1, \ z \neq \beta_1, \tag{10.58}$$

where β_1 and β_2 are real numbers satisfying $0 < |\beta_1| < 1$, $\beta_2 > 0$. Suppose that ϕ is a continuous piecewise linear 2π-periodic function satisfying (10.50) with $-\kappa_1 = \kappa_2 =: \kappa$.

Note that such a discrete-time dynamical system describes, for example, periodically kicked rotators [37] or systems of phase synchronization [22].

It follows from the stability of A_0 and the boundedness of ϕ that the system given by (10.58) is dissipative on the cylinder $\mathbb{R}^n/\mathbb{G}$ (see the proof of Proposition 1.6, Chap. 1). According to Proposition 1.6, Chap. 1, there exists a minimal global $\mathcal{B}$-attractor $\mathcal{A}_{\mathbb{R}^n/\mathbb{G},\min}$.

Let us check the conditions of Theorem 10.11. Assume that $\theta > 0$ is a number. Then the eigenvalues of $\frac{1}{\theta}(A + \kappa_1 bc^*)$ are the zeros of the polynomial $(\theta z - 1)(\theta z - \beta_1) - \kappa\beta_2\theta z$. Thus the assumptions (1) and (2) of Theorem 10.11 are satisfied if

$$(\kappa\beta_2 + \beta_1 + 1) + \sqrt{(\kappa\beta_2 + \beta_1 + 1)^2 - 4\beta_1} > 2\theta$$

and

$$\kappa\beta_2 > \frac{(1+\theta)(\beta_1 + \theta)}{\theta}.$$

It follows that in this case $\dim_T \mathcal{A}_{\mathbb{R}^n/\mathbb{G},\min} \geq 1$.

10.4.2 *Lower Estimates of the Hausdorff Dimension of Global $\mathcal{B}$-Attractors*

Suppose in this subsection that $\{\varphi^t\}_{t\in\mathbb{T}}$ is a dynamical system on the open set $\mathcal{D}$ of the Banach space $(\mathbb{E}, |\cdot|)$ and $\mathcal{A}_{\mathbb{E}} \subset \mathcal{D}$ is a compact subset.

The next theorem and the corollary are proved in [23]. For definitions see Sect. B.5, Appendix B.

Theorem 10.12 *Suppose:*

(a) $\mathcal{A}_{\mathbb{E}}$ *is a global* $\mathcal{B}$*-attractor of* $\{\varphi^t\}_{t\in\mathbb{T}}$;

(b) There is a bounded sequence $\{\Phi_k\}$ *of parameterized* $(m+1)$*-surfaces in* $\mathcal{D}$ *and a sequence* $\{t_k\}$ *in* $\mathbb{T}$, $t_k \to +\infty$, *such that the parameterized m-boundaries of* $(\varphi^{t_k} \circ \Phi_k)$, $k = 1, 2, \ldots$, *are simple and* δ*-linked for some* $\delta > 0$.

Then $\dim_H \mathcal{A}_{\mathbb{E}} \geq m + 1$.

Proof Let $\{\mathcal{B}_{r_i}\}$ be an open $\frac{\varepsilon}{2}$-cover of $\mathcal{A}_{\mathbb{E}}$ by balls of radius $r_i \leq \varepsilon/2$. Since $\mathcal{A}_{\mathbb{E}}$ attracts bounded sets, $\varphi^t(\bigcup_i \Phi_i(\overline{\mathcal{U}}_i)) \subset \bigcup_i \mathcal{B}_{r_i}$ for t sufficiently large, where $\overline{\mathcal{U}}_i$ is the domain of Φ_i. In particular, $(\varphi^{t_k} \circ \Phi_k)(\overline{\mathcal{U}}_k) \subset \bigcup_i \mathcal{B}_{r_i}$ when k is large and therefore

$$0 < c_m^{-1}\delta^{m+1} \leq \sum_i (2r_i)^{m+1} = 2^{m+1} \sum_i r_i^{m+1}$$

from Proposition B.3, Appendix B. Thus $\mu_H(\mathcal{A}_{\mathbb{E}}, m+1) > 0$. It follows from this and the property **(P4)**, Subsect. 3.2.1, Chap. 3, that $\dim_H \mathcal{A}_{\mathbb{E}} \geq m+1$. □

Corollary 10.11 *Suppose:*

(a) $\mathcal{A}_{\mathbb{E}}$ *is a global* $\mathcal{B}$*-attractor of* $\{\varphi^t\}_{t\in\mathbb{T}}$*;*
(b) $\mathcal{C} \subset \mathcal{A}_{\mathbb{E}}$*, where* $\mathcal{C}$ *is the trace of an ordinary* δ*-linked parameterized* m*-boundary of an parameterized* $(m+1)$*-surface in* $\mathcal{D}$*;*
(c) φ^t *is one-to-one on* $\mathcal{C}$ *and* $\varphi^t(\mathcal{C}) = \mathcal{C}$ *for some* $t > 0$*.*

Then $\dim_H \mathcal{A}_{\mathbb{E}} \geq m+1$.

Proof Let $\Phi : \overline{\mathcal{U}} \to \mathcal{D}$ be the parameterized $(m+1)$-surface in $\mathcal{D}$ such that $\Phi(\partial\mathcal{U}) = \mathcal{C} \subset \mathcal{A}_{\mathbb{E}}$. Let $\Phi_k := \Phi, k = 1, 2, \ldots$. Since $\Phi(\overline{\mathcal{U}})$ is bounded, $\{\Phi_k\}$ is a bounded sequence of parameterized $(m+1)$-surfaces in $\mathcal{D}$. Since φ^t is one-to-one on $\mathcal{C}$, if $t_k := kt$, Proposition B.2, Appendix B implies that the boundaries $\partial(\varphi^{kt} \circ \Phi_k) = \varphi^{kt} \circ \Phi_{|\partial\mathcal{U}}, k = 1, 2, \ldots,$ are δ-linked. Thus all conditions of Theorem 10.12 are satisfied. □

The estimation technique of M. Y. Li and J. S. Muldowney, represented in Theorem 10.12, is connected with the evolution of functionals and δ-linked boundaries under the dynamical system. It is naturally to use currents for such estimations. Currents are generalized surfaces. They are obtained by viewing an m-dimensional oriented surface as defining a continuous linear functional on the space of differential m-forms with compact support. In Sect. B.6, Appendix B we have briefly sketched the ideas from geometric measure theory needed for our presentation. The following are due to [32].

Let us assume that a C^∞-smooth dynamical system $(\{\varphi^t\}_{t\in\mathbb{T}}, \mathbb{R}^n, |\cdot|)$ is given.

Theorem 10.13 *Suppose that the following conditions are satisfied:*

(i) The set $\mathcal{A} \equiv \mathcal{A}_{\mathbb{R}^n}$ *is the global* $\mathcal{B}$*-attractor of the dynamical system* $\{\varphi^t\}_{t\in\mathbb{T}}$*;*
(ii) There is a sequence $\{T_k\}_{k=1}^\infty$ *of* $(m+1)$*-dimensional real flat chains such that* $\bigcup_{k=1}^\infty \operatorname{supp} T_k$ *is bounded, and there exists a sequence of times* $\{t_k\}_{k=1}^\infty$, $t_k \in \mathbb{T}$*, such that* $\varphi_*^{t_k} T_k \nrightarrow 0$ *as* $k \to \infty$*.*

Then $\dim_H \mathcal{A} \geq m+1$.

In order to prove this theorem we need the following lemma.

Lemma 10.8 *Let* T *be a* k*-dimensional real flat chain* $T \neq 0$*. Then* $\dim_H(\operatorname{supp} T) \geq m+1$.

Proof Suppose that $\dim_H(\operatorname{supp} T) < m+1$. Then the property **(P4)**, Subsect. 3.2.1, Chap. 3, implies that $\mu_H(\operatorname{supp} T, m+1) = 0$. From Theorem B.5, Appendix B it follows that $T = 0$. But this contradicts our assumption. □

Proof *(of Theorem* 10.13*)* Let $\varepsilon > 0$ and $\delta > 0$ be arbitrary numbers and $\{\mathcal{B}_i\}_{i\geq 1}$ be a countable cover of $\mathcal{A}$ by balls of radius $r_i \leq \delta$ such that

$$\sum_{i\geq 1} r_i^{m+1} \leq \mu_H(\mathcal{A}, m+1, \delta) + \varepsilon. \tag{10.59}$$

Since $\mathcal{A}$ attracts bounded sets there exists a $t_0 \in \mathbb{T}$ such that for all $t \geq t_0$, $t \in \mathbb{T}$, we have

$$\varphi^t\Big(\bigcup_{k\geq 1} \operatorname{supp} T_k\Big) \subset \bigcup_{i\geq 1} \mathcal{B}_i.$$

In particular,

$$\varphi^{t_k}(\operatorname{supp} T_k) \subset \bigcup_{i\geq 1} \mathcal{B}_i, \tag{10.60}$$

when k is large enough.

Since (see Sect. B.6, Appendix B) $\operatorname{supp}(\varphi_*^{t_k} T_k) \subset \varphi^{t_k}(\operatorname{supp} T_k)$ for $k = 1, 2, \ldots,$ we get from (10.60) that

$$\operatorname{supp}(\varphi_*^{t_k} T_k) \subset \bigcup_{i\geq 1} \mathcal{B}_i\,, \tag{10.61}$$

when k is large enough.

From Lemma 10.8 it follows that there exists a $\delta' > 0$ and a subsequence $\{t_{k_i}\}_{i=1}^{\infty}$ of $\{t_k\}_{k=1}^{\infty}$ such that (with δ from (10.59))

$$0 < \delta' \leq \mu_H(\operatorname{supp} \varphi_*^{t_{k_i}} T_{k_i}, m+1, \delta). \tag{10.62}$$

Now it follows from (10.59), (10.61), and (10.62) that

$$\delta' \leq \mu_H(\operatorname{supp} \varphi_*^{t_{k_i}} T_{k_i}, m+1, \delta) \leq \mu_H\Big(\bigcup_{i\geq 1} \mathcal{B}_i, m+1, \delta\Big) \leq \mu_H(\mathcal{A}, m+1, \delta) + \varepsilon.$$

But this implies $\mu_H(\mathcal{A}, m+1) > 0$. From **(P4)**, Chap. 3, it follows that $\dim_H \mathcal{A} \geq m+1$. □

Remark 10.14 The Koch curve $\mathcal{K}(\phi_1, \phi_2)$ of Subsect. 3.2.3, Chap. 3, supports an 1-dimensional integral flat chain T with $\operatorname{supp} T \subset \mathcal{K}(\phi_1, \phi_2)$. The construction of such T for $\mathcal{K}(\phi_1, \phi_2)$, and, more general, for arbitrary self-similar sets $\mathcal{K}(\phi_1, \ldots, \phi_m)$, is considered in [14]. In particular such integral flat chains can be used, in order to verify the conditions of Theorem 10.13.

Remark 10.15 Various types of other functionals for the Hausdorff dimension estimation are used by Leonov [19] and by Leonov and Florynskii [21]. These functionals are called Hausdorff-Lebesgue functionals [19] and, more general, Hausdorff functionals [21]. In a number of dimension estimations for attractors of dynamical systems the use of such functionals seems to be more efficient than the direct use of the outer Hausdorff measures.

References

1. Abraham, R., Marsden, J.E., Ratiu, T.: Manifolds, Tensor-Analysis, and Applications. Springer, New York (1988)
2. Afraimovich, V.S.: On the Lyapunov dimension of invariant sets in a model of active medium. In: Methods of Qualitative Theory of Differential Equations, pp. 19–29. Gorki State University, Gorki (1986) (Russian)
3. Belykh, V.N.: Qualitative Methods of the Theory of Nonlinear Oscillations in Finite Dimensional Systems. Gorki University Press, Gorki (1980) (Russian)
4. Belykh, V.N.: Models of discrete systems of phase synchronization. In: Shakhgil'dyan, V.V., Belyustina, L.N. (eds.) Systems of Phase Synchronization, pp. 161–176. Radio i Svyaz', Moscow (1982) (Russian)
5. Boichenko, V.A., Leonov, G.A., Franz, A., Reitmann, V.: Hausdorff and fractal dimension estimates of invariant sets of non-injective maps. Zeitschrift für Analysis und ihre Anwendungen (ZAA) **17**(1), 207–223 (1998)
6. Chen, Z.-M.: A note on Kaplan-Yorke-type estimates on the fractal dimension of chaotic attractors. Chaos Solitons Fractals **3**(5), 575–582 (1993)
7. Collet, P., Levy, Y.: Ergodic properties of the Lozi mappings. Comm. Math. Phys. **93**, 461–481 (1984)
8. Dellnitz, M., Junge, O.: On the approximation of complicated dynamical behavior. SIAM J. Num. Anal. **36**(2) (1999)
9. Douady, A., Oesterlé, J.: Dimension de Hausdorff des attracteurs. C. R. Acad. Sci. Paris, Ser. A **290**, 1135–1138 (1980)
10. Falconer, K.J.: Fractal Geometry: Mathematical Foundations and Applications. Wiley, Chichester (1990)
11. Franz, A.: Hausdorff dimension estimates for invariant sets with an equivariant tangent bundle splitting. Nonlinearity **11**, 1063–1074 (1998)
12. Giesl, P.: Necessary condition for the basin of attraction of a periodic orbit in non-smooth periodic systems. Discrete Contin. Dynam. Syst. **18**(2/3), 355–373 (2007)
13. Heineken, W.: Fractal dimension estimates for invariant sets of vector fields. Diploma thesis, University of Technology Dresden (1997)
14. Hutchinson, J.E.: Fractals and self-similarity. Ind. Univ. Math. J. **30**, 713–747 (1981)
15. Ishii, J.: Towars a kneading theory for the Lozi mappings. II: Monotonicity of topological entropy and Hausdorff dimension of attractors. Comm. Math. Phys. **190**, 375–394 (1997)
16. Kunze, M., Michaeli, B.: On the rigorous applicability of Oseledec's ergodic theorem to obtain Lyapunov exponents for non-smooth dynamical systems. In: Proceedings of the 2nd Marrakesh International Conference in Differential Equations (1995)
17. Ledrappier, F.: Some relations between dimension and Lyapunov exponents. Commun. Math. Phys. **81**, 229–238 (1981)
18. Leonov, G.A.: On lower dimension estimates of attractors for discrete systems. Vestn. S. Peterburg Gos. Univ. Ser. 1, Matematika, **4**, 45–48 (1998) (Russian); English transl. Vestn. St. Petersburg Univ. Math., **31**(4), 45–48 (1998)

19. Leonov, G.A.: Hausdorff-Lebesgue dimension of attractors. Int. J. Bifurcation and Chaos **27**(10) (2017)
20. Leonov, G.A., Boichenko, V.A.: Lyapunov's direct method in the estimation of the Hausdorff dimension of attractors. Acta Appl. Math. **26**, 1–60 (1992)
21. Leonov, G.A., Florynskii, A.A.: On estimations of generalized Hausdorff dimension. Vestn. St. Petersburg Univ. Math., T. 6 **64**(4), 534–543 (2019) (Russian)
22. Leonov, G.A., Reitmann, V., Smirnova, V.B.: Non-local Methods for Pendulum-like Feedback Systems. Teubner-Texte zur Mathematik, Bd. 132, B. G. Teubner Stuttgart-Leipzig (1992)
23. Li, M.Y., Muldowney, J.S.: Lower bounds for the Hausdorff dimension of attractors. J. Dynam. Diff. Equ. **7**(3), 457–469 (1995)
24. Lozi, R.: In attracteur étrange du type Hénon. J. Phys., Paris **39**, 69–77 (1978)
25. Mirle, A.: Hausdorff dimension estimates for invariant sets of k-1-maps. DFG-Schwerpunktprogramm "Dynamik: Analysis, effiziente Simulation und Ergodentheorie". Preprint 25 (1995)
26. Neunhäuserer, J.: Properties of some overlapping self-similar and some self-affine measures. Acta Mathematica Hungariaca **93**, 1–2 (2001)
27. Neunhäuserer, J.: A Douady-Oesterlé type estimate for the Hausdorff dimension of invariant sets of piecewise smooth maps. University of Technology Dresden, Preprint (2000)
28. Noack, A.: Dimension and entropy estimates and stability investigations for nonlinear systems on manifolds. Doctoral Thesis, University of Technology Dresden (1998) (German)
29. Noack, A., Reitmann, V.: Hausdorff dimension estimates for invariant sets of time-dependent vector fields. Zeitschrift für Analysis und ihre Anwendungen (ZAA) **15**(2), 457–473 (1996)
30. Pesin, Y.B.: Dynamical systems with generalised hyperbolic attractors: hyperbolic, ergodic and topological properties. Ergod. Theory Dyn. Syst. **12**, 123–151 (1992)
31. Pesin, Y.B.: Dimension Theory in Dynamical Systems: Contemporary Views and Applications. Chicago Lectures in Mathematics. The University of Chicago Press, Chicago and London (1997)
32. Reitmann, V.: Dimension estimates for invariant sets of dynamical systems. In: Fiedler, B. (ed.) Ergodic Theory, Analysis, and Efficient Simulation of Dynamical Systems, pp. 585–615. Springer, New York and Berlin (2001)
33. Reitmann, V., Schnabel, U.: Hausdorff dimension estimates for invariant sets of piecewise smooth maps. ZAMM **80**(9), 623–632 (2000)
34. Reitmann, V., Zyryanov, D.: The global attractor of a multivalued dynamical system generated by a two-phase heating problem. In: Abstracts, 12th AIMS International Conference on Dynamical Systems, Differential Equations and Applications, Taipei, Taiwan, 414 (2018)
35. Schmeling, J.: A dimension formula for endomorphisms—the Belykh family. Ergodic Theory Dyn. Syst. **18**, 1283–1309 (1998)
36. Schmidt, G.: Dimension estimates for invariant sets of differential equations with non-smooth right part and of locally expanding dynamical systems. Diploma Thesis, University of Technology Dresden (1996)
37. Schuster, H.G.: Deterministic Chaos. Physik-Verlag, Weinheim (1984)
38. Temam, R.: Infinite-Dimensional Dynamical Systems in Mechanics and Physics. Springer, New York and Berlin (1988)
39. Thieullen, P.: Entropy and the Hausdorff dimension for infinite-dimensional dynamical systems. J. Dynam. Diff. Equ. **4**(1), 127–159 (1992)

Appendix A
Basic Facts from Manifold Theory

A.1 Definition of a Differentiable Manifold

In this section we shall repeat some well-known facts and basic definitions on dynamical systems on finite-dimensional manifolds. Suppose $\mathcal{M}$ is an arbitrary set. An *n-dimensional chart* on $\mathcal{M}$ is a bijection $x : \mathcal{D}(x) \subset \mathcal{M} \to \mathcal{R}(x) \subset \mathbb{R}^n$, where $\mathcal{R}(x)$ is open in $\mathbb{R}^n$.

An *n-dimensional atlas of class* C^k $(k \geq 0)$ on $\mathcal{M}$ is a set $\mathbb{A}$ of n-dimensional charts such that:

(AT1) $\bigcup_{x \in \mathbb{A}} \mathcal{D}(x) = \mathcal{M}$;

(AT2) $x(\mathcal{D}(x) \cap \mathcal{D}(y))$ is open in $\mathbb{R}^n$ for arbitrary $x, y \in \mathbb{A}$;

(AT3) The map $y \circ x^{-1} : x(\mathcal{D}(x) \cap \mathcal{D}(y)) \to y(\mathcal{D}(x) \cap \mathcal{D}(y))$ is of class C^k for each $x, y \in \mathbb{A}$.

Suppose $\mathbb{A}$ is an n-dimensional atlas of class C^k on $\mathcal{M}$ and x is an arbitrary n-dimensional chart on $\mathcal{M}$. This chart is C^k*-compatible* with $\mathbb{A}$ if $\mathbb{A} \cup \{x\}$ is also an n-dimensional C^k-atlas on $\mathcal{M}$. An n-dimensional atlas of class C^k is called *maximal* if any C^k-compatible n-dimensional chart on $\mathcal{M}$ belongs to $\mathbb{A}$. Denote this (unique) maximal atlas by $\mathbb{A}_{\max}$. A pair $(\mathcal{M}, \mathbb{A}_{\max})$, where $\mathcal{M}$ is a set and $\mathbb{A}_{\max}$ is the maximal n-dimensional C^k-atlas on $\mathcal{M}$, is called *n-dimensional* C^k*-manifold*. The family of sets $\mathfrak{S} := \{\mathcal{D}(x) \subset \mathcal{M} | x \in \mathbb{A}_{\max}\}$ can be considered as the basis for a topology. The topology on $\mathcal{M}$ which is generated by $\mathfrak{S}$ is the *canonical topology* $\mathfrak{T}_{\text{can}}$. In the sequel, we assume that any n-dimensional C^k-manifold is Hausdorff, i.e. any two distinct points in $\mathcal{M}$ have disjunct neighborhoods.

Note that any n-dimensional C^k-manifold is *locally compact*, i.e. each point in $\mathcal{M}$ has a compact neighborhood. It follows that any manifold is *regular*, i.e. each point has, together with an open neighborhood, also a closed neighborhood. Any n-dimensional C^k-manifold is *locally connected*, i.e. each neighborhood of a point contains a connected neighborhood. An open set $\mathcal{U}$ of an n-dimensional C^k-manifold

N. Kuznetsov and V. Reitmann, *Attractor Dimension Estimates for Dynamical Systems: Theory and Computation*, Emergence, Complexity and Computation 38,
https://doi.org/10.1007/978-3-030-50987-3

$\mathcal{M}$ considered with a topology which is induced from the canonical topology of $\mathcal{M}$, is an n-dimensional C^k-manifold.

If $\mathcal{M}$ and $\mathcal{N}$ are n- and m-dimensional C^k-manifolds with the atlas $\mathbb{A}^{\mathcal{M}}$ and $\mathbb{A}^{\mathcal{N}}$, respectively, then the cartesian product $\mathcal{M} \times \mathcal{N}$, associated with the atlas

$\mathbb{A} = \{x \times y : \mathcal{D}(x) \times \mathcal{D}(y) \to \mathcal{R}(x) \times \mathcal{R}(y), x \in \mathbb{A}^{\mathcal{M}}, y \in \mathbb{A}^{\mathcal{N}}\}$, is an $(n+m)$-dimensional C^k-manifold, which is called a *product manifold*.

Example A.1 (a) The space $\mathbb{R}^n$ can be considered as an n-dimensional C^∞-manifold. A C^∞-atlas for $\mathbb{R}^n$ is $\mathbb{A} = \{\text{id}\}$ with id: $\mathbb{R}^n \to \mathbb{R}^n$ being the identical map. The maximal C^∞-atlas contains as charts all C^∞-diffeomorphisms

$$x : \mathcal{D}(x) \subset \mathbb{R}^n \to \mathcal{R}(x) \subset \mathbb{R}^n \quad \text{with } \mathcal{D}(x) \text{ and } \mathcal{R}(x) \text{ open.}$$

(b) Suppose $\Gamma = \left\{ \sum_{\varepsilon=1}^{m} k_i e_i, k_i \in \mathbb{Z} \right\}$ is a discrete subgroup of $\mathbb{R}^n$, $e_1, \ldots, e_m$ with $m \leq n$ are elements of the canonical basis. Consider the *canonical projection* $\pi : \mathbb{R}^n \to \mathbb{R}^n/\Gamma$ defined as $\upsilon \in \mathbb{R}^n \mapsto [\upsilon] = \upsilon + \Gamma$.

An n-dimensional atlas for $\mathbb{R}^n/\Gamma$ is given by $\mathbb{A} = \{\pi^{-1}_{|\pi(\mathcal{U})} | \pi(\mathcal{U}) \to \mathcal{U}$, $\mathcal{U} \subset \mathbb{R}^n$ open, $\pi : \mathcal{U} \to \mathbb{R}^n/\Gamma$ is injective$\}$. Then the n-dimensional C^∞-manifold $(\mathbb{R}^n/\Gamma, \mathbb{A}_{\max})$ is called a (flat) cylinder. If $m = n$ the (flat) cylinder is called (flat) torus.

The next theorem is called *Brouwer's theorem on the invariance of domain* [3, 6].

Theorem A.1 *Let $\mathcal{S}$ be an arbitrary subset of the n-dimensional Euclidean space $\mathbb{E}^n$ and ϕ a homeomorphism of $\mathcal{S}$ on another subset $\phi(\mathcal{S})$ of $\mathbb{E}^n$. Then if u is an interior point of $\mathcal{S}$ (with respect to $\mathbb{E}^n$), $\phi(u)$ is an interior point of $\phi(\mathcal{S})$ (with respect to $\mathbb{E}^n$). In particular, if $\mathcal{S}$ and $\mathcal{S}'$ are homeomorphic subsets of $\mathbb{E}^n$ and $\mathcal{S}$ is open, then $\mathcal{S}'$ is open.*

Remark A.1 (a) Theorem A.1 remains true if the n-dimensional Euclidean space $\mathbb{E}^n$ is replaced by an arbitrary n-dimensional C^k-manifold $\mathcal{M}$: For every point of $\mathcal{S}$ there is a neighborhood in $\mathcal{M}$ which is homeomorphic to $\mathbb{E}^n$.

(b) As it is mentioned in [6] Theorem A.1 includes the *Theorem on invariance of dimension of Euclidean spaces*, i.e. $\mathbb{E}^n$ and $\mathbb{E}^m$ are homeomorphic if and only if $n = m$.

The following classification theorem for smooth one-dimensional connected manifolds can be found in [9].

Theorem A.2 *Any connected C^∞-smooth one-dimensional manifold is diffeomorphic either to S^1 or to some interval of $\mathbb{R}$.*

Let us note that an interval of $\mathbb{R}$ is any connected subset of $\mathbb{R}$ different from a point. An interval can be finite or infinite, closed, open or semi-open. Since any interval is diffeomorphic either to $[0, 1]$, to $(0, 1]$ or $(0, 1)$, the above theorem says that there are exactly four different types of connected one-dimensional C^∞-manifolds.

The definition of a *manifold with boundary* $\mathcal{M}$ is similar to the definition of a manifold without boundary. However there are now two kinds of charts. Let $\mathbb{R}^n_+$ denote the region $\mathbb{R}^n_+ = \{(x^1, \ldots, x^n) \in \mathbb{R}^n \,\big|\, x^1 \geq 0\}$. In some charts $x : \mathcal{D}(x) \subset \mathcal{M} \to \mathcal{R}(x)$, the domain $\mathcal{D}(x)$ is mapped onto a certain open subset of $\mathbb{R}^n$, in some other charts $y : \mathcal{D}(y) \subset \mathcal{M} \to \mathcal{R}(y)$ the domain $\mathcal{D}(y)$ is mapped onto a certain (relative) open subset of $\mathbb{R}^n_+$. As before, the chart's domains $\{\mathcal{D}(x)\}$ cover $\mathcal{M}$, and two different charts define differentiable transition functions. The boundary $\partial\mathcal{M}$ of $\mathcal{M}$ is by definition the set of all points of $\mathcal{M}$ whose images under charts lie on the boundary of $\mathbb{R}^n_+$ defined by $x^1 = 0$. It is easy to see that $\partial\mathcal{M}$ is an $(n-1)$-dimensional manifold of the same class as $\mathcal{M}$.

A.2 Tangent Space, Tangent Bundle and Differential

Suppose that $\mathcal{M}$ is an n-dimensional C^k-manifold. If $p \in \mathcal{M}$ is a point, x and y are two arbitrary charts around p and $\xi, \eta \in \mathbb{R}^n$, then we introduce the equivalence relation $(p, x, \xi) \sim (p, y, \eta) \Leftrightarrow \eta = (y \circ x^{-1})'(x(p))\, \xi$. The equivalence class

$$[p, x, \xi] := \{(p, y, \eta) | (p, y, \eta) \sim (p, x, \xi)\}$$

is called *tangent vector* at p. The *tangent space* of $\mathcal{M}$ at $p \in \mathcal{M}$ is the set $T_p\mathcal{M}$ of all equivalence classes $[p, x, \xi]$ such that $p \in \mathcal{D}(x)$ and connected with a vector space structure given by

(1) $[p, x, \xi] + [p, x, \eta] := [p, x, \xi + \eta]\,,\ \forall \xi, \eta \in \mathbb{R}^n\,;$
(2) $\lambda[p, x, \xi] := [p, x, \lambda\, \xi]\,,\ \forall \lambda \in \mathbb{R}, \forall \xi \in \mathbb{R}^n\,.$

It can be shown that this definition is correct, i.e. does not depend on the chart x.

The *tangent bundle* $T\mathcal{M}$ of $\mathcal{M}$ is defined by $T\mathcal{M} := \bigcup_{p \in \mathcal{M}} T_p\mathcal{M}$ and the *natural projection* π is given by $\pi : T\mathcal{M} \to \mathcal{M}$ with $[p, x, \xi] \mapsto p$.

It can be shown that $T\mathcal{M}$ can be considered as Hausdorff $2\,n$-dimensional C^{k-1}-manifold.

Suppose that $\mathcal{M}$ and $\mathcal{N}$ are n-dimensional C^k-manifolds. The map $\phi : \mathcal{M} \to \mathcal{N}$ is said to be *C^r-differentiable* $(1 \leq r \leq k)$ *at* $p \in \mathcal{M}$ if there are charts x around p and y around $\phi(p)$ such that the map $y \circ \phi \circ x^{-1}$ is C^r-differentiable in $x(p)$. It is easy to see that the definition does not depend on the charts x and y. The map $\phi : \mathcal{M} \to \mathcal{N}$ is called C^r-differentiable if ϕ is C^r-differentiable at any point of $\mathcal{M}$, and is called a *C^r-diffeomorphism* if ϕ is bijective, ϕ is C^r-differentiable on $\mathcal{M}$ and ϕ^{-1} is C^r-differentiable on $\mathcal{N}$.

Suppose that $\phi : \mathcal{M} \to \mathcal{N}$ is of class C^1. The *differential* of ϕ at $p \in \mathcal{M}$ is the linear map $d_p\phi : T_p\mathcal{M} \to T_{\phi(p)}\mathcal{N}$ given by

$$d_p\phi\left([p, x, \xi]\right) = \left[\phi(p), y, (y \circ \phi \circ x^{-1})'(x(p))\xi\right],$$

where x is a chart at $p \in \mathcal{M}$, y is a chart at $\phi(p)$. One can easily show that this definition is independent of x and y.

The *rank* of the differential $d\phi$ at p is defined by rank $(d_p\phi)$ $= \text{rank}(y \circ \phi \circ x^{-1})'(x(p))$, where x and y are arbitrary charts around p and $\phi(p)$, respectively.

If $\mathcal{N} = \mathbb{R}^n$ we can write $T\mathbb{R}^n \cong \mathbb{R}^n \times \mathbb{R}^n$ and $d\phi : T\mathcal{M} \to \mathbb{R}^n \times \mathbb{R}^n$ is defined by

$$d\phi\,([p, x, \xi]) := (\phi\,(p), (\phi \circ x^{-1})'(x(p))\,\xi)\ .$$

In particular one can show that, if x is a chart on $\mathcal{M}$ then $dx([p, x, \xi]) = (x(p), (x \circ x^{-1})'(x(p))\xi) = (x(p), \xi)$ is a chart and $\{dx | x \in \mathcal{A}\}$ is a $2n$-dimensional C^{k-1}-atlas on $T\mathcal{M}$.

The subset $\mathcal{Z}$ of the n-dimensional C^k-manifold $\mathcal{M}$ is said to be a *submanifold* if there is a natural $m < n$ such that any point $p \in \mathcal{Z}$ belongs to a domain $\mathcal{D}(x)$ with

$$\begin{aligned} x(\mathcal{D}(x) \cap \mathcal{Z}) &= \{(x^1, \ldots, x^n) \in \mathcal{R}(x) | x^{m+1} = \cdots = x^n = 0\} \\ &= \mathcal{R}(x) \cap \mathbb{R}^m \times \{0\}\ . \end{aligned}$$

The C^r-differentiable map $\phi : \mathcal{M} \to \mathcal{N}$ is at $p \in \mathcal{M}$ called *regular* if rank $(d_p\phi) = \min(n, m)$. If ϕ is at any point regular, the map is called C^r*-submersion*, if $n \geq m$, C^r*-immersion* if $n \leq m$, and C^r*-embedding*, if $n \leq m$ and ϕ maps homeomorphly the manifold $\mathcal{M}$ on $\phi(\mathcal{M})$.

Whitney's embedding theorem [16] states the following. If $\mathcal{M}$ is a compact C^r-manifold of dimension n, then there exists a C^r-embedding $\phi : \mathcal{M} \to \mathbb{R}^{2n+1}$. Let $\mathcal{M}$ be a C^r-manifold of dimension n. We say that the set $\mathcal{Z} \subset \mathcal{M}$ has the Lebesgue measure zero if there exists a sequence of charts $x_i : \mathcal{D}(x_i) \to \mathcal{R}(x_i)$, $i = 1, 2, \ldots$, such that $\mathcal{Z} \subset \bigcup_{i=1}^{\infty} \mathcal{D}(x_i)$ and $\mu_L(x_i(\mathcal{D}(x_i) \cap \mathcal{Z})) = 0$ for every $i = 1, 2, \ldots$. (Here $\mu_L(\cdot)$ denotes the Lebesgue measure in $\mathbb{R}^n$.)

Suppose that $\mathcal{M}$ and $\mathcal{N}$ are C^r-manifolds of dimension n and m, respectively and $\phi : \mathcal{M} \to \mathcal{N}$ is a C^s-map, $s \leq r$. A point $q \in \mathcal{N}$ is called a *regular value* of ϕ if the map $d_p\phi : T_p\mathcal{M} \to T_q\mathcal{N}$ is surjective for any $p \in \phi^{-1}(q)$. The point $q \in \mathcal{N}$ is called a *critical value* of ϕ if q is not regular. The point $p \in \mathcal{M}$ is called a *critical point* of ϕ if there exists a critical value $q \in \mathcal{N}$ of ϕ such that $p \in \phi^{-1}(q)$.

Let us state now Sard's theorem [13], the proof of which can be found in many books.

Theorem A.3 *Let $\mathcal{M}$ and $\mathcal{N}$ be C^r-manifolds, $r \geq 1$, of dimension n and m, respectively, $\phi : \mathcal{M} \to \mathcal{N}$ a C^s-map,* $\max\{0, n - m\} < s \leq r$. *Then the set of critical values of ϕ has the Lebesgue measure zero.*

A.3 Tensor Products, Exterior Products and Tensor Fields

Suppose that $\mathcal{M}$ is an n-dimensional C^r-manifold, $p \in \mathcal{M}$ is an arbitrary point, x a chart around p. By this chart $x : \mathcal{D}(x) \to \mathcal{R}(x)$, we define an associated *isomorphism* $\Theta_{p,x} : T_p\mathcal{M} \to \mathbb{R}^n$ given by $\Theta_{p,x}([p, x, \xi]) := \xi \in \mathbb{R}^n$. We call ξ the *representant* of the tangent vector in the chart x. If $e_1, \dots, e_n$ is the canonical basis in $\mathbb{R}^n$, we define by $b_i := \Theta_{p,x}^{-1}(e_i),\ i = 1, \dots, n$, a basis in $T_p\mathcal{M}$. Let $C^\infty(\mathcal{M}, \mathbb{R})$ be the linear space of C^∞-functions over $\mathcal{M}$. Then every tangent vector $[p, x, \xi] \in T_p\mathcal{M}$ can be identified with the following map:

$$\partial_{[p,x,\xi]} : C^\infty(\mathcal{M}, \mathbb{R}) \to T_p\mathcal{M}\ ,\ \phi \mapsto d_p\phi\,([p, x, \xi]) = [\phi\,(p), \mathrm{id}, (\phi \circ x^{-1})'(x(p))\xi]\,,$$

i.e. with the directional derivative of ϕ in direction $[p, x, \xi]$. This means that $\partial_i(p) \equiv b_i$ is the canonical basis of $T_p\mathcal{M}$.

In order to define a canonical basis for the cotangential space $T_p^*\mathcal{M} \equiv (T_p\mathcal{M})^*$ we introduce the projection $\pi_2 : \mathbb{R}^n \times \mathbb{R}^n \to \mathbb{R}^n$ by $\pi_2(u, v) = v$. The dual basis $\Theta^i \equiv dx^i$ of $T_p^*\mathcal{M}$ is defined by $\Theta^i = e^i \circ \pi_2 \circ dx \in T_p^*\mathcal{M}$ and acts as $\Theta^i([p, x, \xi]) = e^i(\pi_2(d_p x([p, x, \xi]))) = e^i(\pi_2(x(p)\xi)) = e^i(\xi) = \xi_i \in \mathbb{R}$.
Since $\Theta^i(\partial_j) = \Theta^i(d_p x^{-1}(x(p), e_j)) = e^i(\pi_2(d_p x(d_p x^{-1}(x(p), e_j)))) = e^i(\pi_2(x(p), e_j)) = e^i(e_j) = \delta^i_j$ we see that $\Theta^i(p)$ is indeed the dual basis to $\partial_j(p)$.
For arbitrary numbers $k, h \in \mathbb{N}_0$ and $p \in \mathcal{M}$ we introduce the sets

$$(T_p\mathcal{M})^k_h := \underbrace{T_p\mathcal{M} \otimes \cdots \otimes T_p\mathcal{M}}_{k-\text{times}} \otimes \underbrace{T_p^*\mathcal{M} \otimes \cdots \otimes T_p^*\mathcal{M}}_{h-\text{times}}$$
$$\equiv (\overset{k}{\otimes}\, T_p\mathcal{M}) \otimes (\overset{h}{\otimes}\, T_p^*\mathcal{M})$$

and

$$T^k_h\mathcal{M} = \bigcup_{p \in \mathcal{M}} (T_p\mathcal{M})^k_h\ .$$

One can show again that if $\mathcal{M}$ is an n-dimensional Hausdorff C^r-manifold then $T^k_h\mathcal{M}$ has the canonical structure of an $n + n^{h+k}$-dimensional Hausdorff C^{r-1}-manifold. Analogously one shows that $\bigwedge^k T^*\mathcal{M} = \bigcup_{p\in\mathcal{M}} \bigwedge^k T_p^*\mathcal{M}$ is a smooth manifold. Denote by $\pi^k_h : T^k_h\mathcal{M} \to \mathcal{M}$ the projection operator. A C^m-tensor field of the type (k, h) on $\mathcal{M}$ is a C^m-section of the bundle $T^k_h\mathcal{M}$, i.e. a C^m-map $S : \mathcal{M} \to T^k_h\mathcal{M}$ with $\pi^k_h \circ S = \mathrm{id}_{\mathcal{M}}$. The tensor field of type $(1, 0)$ is also called a (*contravariant*) *vector field*. Tensor fields of the type $(0, h)$ we call h-times *covariant tensor fields.* A C^m-smooth *differential form* β of degree h (or an h-form of smoothness C^m) on $\mathcal{M}$ is a C^m-section of the bundle $\bigwedge^k(T^*\mathcal{M})$, i.e. a C^m-map $\beta : \mathcal{M} \to \bigwedge^k(T^*\mathcal{M})$. If β is a k-form and $\widetilde{\beta}$ is an l-form on $\mathcal{M}$ the *wedge product* $\beta \wedge \widetilde{\beta}$ is a $k + l$-form on $\mathcal{M}$ defined by $(\beta \wedge \widetilde{\beta})_p = \beta_p \wedge \widetilde{\beta}_p,\ \forall\, p \in \mathcal{M}$. The *exterior derivative* d for k-forms is defined as follows. If β is a k-form of class C^r, $r \geq 2$, on $\mathcal{M}$ then $d\beta$ is a $(k + 1)$-form of class C^{r-1} such that the following conditions are satisfied:

(i) If β is a 0-form, i.e. $\beta = \phi$ a C^r-function on $\mathcal{M}$ then $d\beta$ is the differential;
(ii) If $\beta, \widetilde{\beta}$ are k-forms of class C^r then $d(\beta + \widetilde{\beta}) = d\beta + d\widetilde{\beta}$;
(iii) If β and $\widetilde{\beta}$ are k-resp. l-forms of smoothness C^r on $\mathcal{M}$ then
$d(\beta \wedge \widetilde{\beta}) = d\beta \wedge \widetilde{\beta} + (-1)^k \beta \wedge d\widetilde{\beta}$;
(iv) $d(d\beta) = 0$ for any k-form β of class C^r .

A k-form on $\mathcal{M}$ is called *closed* if $d\beta = 0$ and exact if $\beta = d\widetilde{\beta}$ for some $(k-1)$-form $\widetilde{\beta}$. Suppose that $\mathcal{M}$ and $\mathcal{N}$ are C^k-smooth n-resp. m-dimensional manifolds, $\phi : \mathcal{M} \to \mathcal{N}$ is a C^r-map $(r \leq k)$ and β is a k-form $(k \geq 1)$ on $\mathcal{N}$. The *pullback* of β is the k-form $\phi^* \beta$ on $\mathcal{M}$ defined by

$$(\phi^* \beta)_p(\upsilon_1, \ldots, \upsilon_k) = \beta_{\phi(p)}(d_p\phi\, \upsilon_1, \ldots, d_p\phi\, \upsilon_k) ,$$
$$\forall\, p \in \mathcal{M},\ \forall\, \upsilon_1, \ldots, \upsilon_k \in T_p\mathcal{M} .$$

If $\phi : \mathcal{M} \to \mathcal{N}$ is a C^r-diffeomorphism, the k-form $\phi_* \beta$ on $\mathcal{N}$ defined by

$$(\phi_* \beta)_{\phi(p)}(d_p\phi\, \upsilon_1, \ldots, d_p\phi\, \upsilon_p) = \beta(\upsilon_1, \ldots, \upsilon_k) ,$$
$$\forall\, p \in \mathcal{M},\ \forall\, \upsilon_1, \ldots, \upsilon_k \in T_p\mathcal{M} .$$

is called *push-forward* of β. Given a C^r-smooth k-form β and a vector field F on $\mathcal{M}$, the *interior product* of β and F is a $(k-1)$-form which we denote by $\beta \lrcorner F$ and which is defined by $(\beta \lrcorner F)_p(\upsilon_1, \ldots, \upsilon_{k-1}) = \beta_p(F(p), \upsilon_1, \ldots, \upsilon_{k-1}),\ \forall\, p \in \mathcal{M},\ \forall\, \upsilon_1, \ldots, \upsilon_k \in T_p\mathcal{M}$. The *Lie derivative* of β *in direction* F is the k-form $L_p\beta$ given by $L_F\beta = d(\beta \lrcorner F) + d\beta \lrcorner F$. Note that if β is a 0-form, i.e. a function, then $L_F\beta = d\beta \lrcorner F$.

A.4 Riemannian Manifolds

A *Riemannian manifold* is a connected n-dimensional C^r-manifold equipped with a 2-covariant C^r-smooth tensor field g (the *Riemannian metric*) with the following properties.

(i) g is symmetric;
(ii) For any $p \in \mathcal{M}$ the bilinear form $g_{|_p}$ is non-degenerate, i.e. from $g_p(\upsilon, w) = 0,\ \forall\, \upsilon \in T_p\mathcal{M}$, it follows that $w = 0$.

The Riemannian manifold $(\mathcal{M}, g)$ is called *proper* if $g_p(\upsilon, \upsilon) > 0,\ \forall\, p \in \mathcal{M}$, $\forall\, \upsilon \in T_p\mathcal{M},\ \upsilon \neq 0$. In other case $(\mathcal{M}, g)$ is called *pseudo-Riemannian*.

Remark A.2 (a) A Riemannian metric of class $C^r (1 \leq r \leq k-1)$ can be defined on an n-dimensional C^k-manifold if at any point $p \in \mathcal{M}$ and any chart x around p there is given a positive definite (symmetric) $n \times n$ matrix $G_x(p)$ with the following properties:

(1) The map $G_x(\cdot) : \mathcal{D}(x) \to M_n(\mathbb{R})$ is C^r ;
(2) $\left[(y \circ x^{-1})'(x(p))\right]^T G_y(p) \left[(y \circ x^{-1})'(x(p))\right] = G_x(p)$ for any two charts x and y around p.

(b) Let us write the metric tensor in the canonical basis. In the dual basis $\{\Theta^i\}$ of $T_p^*\mathcal{M}$ the tensor g at the point p can be written as

$$g_p(\upsilon, w) = g_{ij}\Theta^i\Theta^j(\upsilon, w) = g_{ij}\upsilon^i w^j,$$
$$\text{where} \quad \upsilon = \upsilon^i \partial_i\,, \; w = w^i \partial_j \quad \text{and} \quad g_{ij} = g_{|p}(\partial_i, \partial_j)\,.$$

If $c : [a, b] \to \mathcal{M}$ is a continuous curve on the Riemannian n-dimensional C^k-manifold $(\mathcal{M}, g)$ with $c_{|(a,b)} \in C^1$ then $\ell(c) := \int_a^b \|\dot{c}\,(t)\|dt$ is the *length* of c. A *piecewise* C^1*-curve* on $\mathcal{M}$ is a continuous map $c\,(\cdot) : [a, b]$ for which there exists a finite number of points $a = t_1 < t_2 < \cdots < t_m = b$ such that $c_{|(t_i,t_{i+1})}(i = 1, \ldots, m-1)$ is C^1. The *length* of this piecewise C^1-curve c is

$$\ell(c) := \sum_{i=1}^{m-1} \ell\,(c_{|(t_i,t_{i+1})}).$$

Denote for arbitrary points $p, q \in \mathcal{M}$ by C_p^q the set of all piecewise C^1-curves from p to q. One shows that for any such points $C_p^q \neq \emptyset$. The *geodesic distance* on $\mathcal{M}$ is a function $\rho : \mathcal{M} \times \mathcal{M} \to \mathbb{R}$ defined by $\rho\,(p, q) = \inf_{c \in C_p^q} \ell(c)$. As an important property of ρ, it follows that ρ is a metric on $\mathcal{M}$. The topology $\mathfrak{T}_\rho$, generated by ρ coincides with the canonical topology $\mathfrak{T}_{\text{can}}$.

The set $\mathcal{U} \subset \mathcal{M}$ is called *Lebesgue measurable* if for any $x \in \mathbb{A}_{\max}$ the set $x(\mathcal{U} \cap \mathcal{D}(x)) \subset \mathbb{R}^n$ is Lebesgue measurable. The function $f : \mathcal{U} \subset \mathcal{M} \to \mathbb{R}$ is said to be *measurable* if $\mathcal{U}$ is measurable and for any $x \in \mathbb{A}_{\max}$ the function $f \circ x^{-1}$ is measurable on $x(\mathcal{U} \cap \mathcal{D}(x))$.

If $x : \mathcal{D}(x) \to \mathcal{R}(x)$ is a chart and g_{ij} is the metric tensor in this chart then the n-form

$$\mu = \sqrt{\det(g_{ij})}\, dx^1 \wedge \cdots \wedge dx^n$$

is the *canonical volume form* on $\mathcal{M}$.

Suppose $\mathcal{U} \subset \mathcal{D}(x)$ is measurable. The function $f : \mathcal{U} \to \mathbb{R}$ is *integrable* on $\mathcal{U}$ w.r.t. μ if $(f \circ x^{-1})(x^i)\sqrt{\det(g_{ij})}$ is integrable on $x(\mathcal{U})$.

Per definition we put

$$\int_{\mathcal{U}} f d\mu = \int_{x(\mathcal{U})} f(x^{-1}(x^i))\sqrt{\det(g_{ij})}\, dx^1 \ldots dx^n\,.$$

This definition is correct, i.e. independent on the chart x. Assume that $\phi : \mathcal{U} \subset \mathcal{M} \to \phi(\mathcal{U}) \subset \mathcal{M}$ is a diffeomorphism, and $\phi : \mathcal{U} \to \mathbb{R}_+$ a differentiable map. Then

$$\int_{\phi(\mathcal{U})} \phi\mu = \int_{\mathcal{U}} \phi^*(\phi\mu) \qquad \text{(change of variables in the integral).}$$

A.5 Covariant Derivative

Consider an n-dimensional Riemannian manifold $(\mathcal{M}, g)$ of the class C^r with $r \geq 2$. Suppose that $p \in \mathcal{M}$ is an arbitrary point, x is a chart around p and $g_{ij}(p)$ is the metric tensor in this chart. Denote for any i, j, m from $\{1, \ldots, n\}$ the partial derivative of the metric tensor by

$$g_{ij,m} := \frac{\partial (g_{ij} \circ x^{-1})(x(p))}{\partial x^m} .$$

The n^3 functions $\Gamma_{ij}^k : \mathcal{D}(x) \to \mathbb{R}$ defined by $\Gamma_{ij}^k := \frac{1}{2}\, g^{ks}[-g_{ij,s} + g_{js,i} + g_{si,j}]$ are the *Christoffel symbols of the second kind* on $\mathcal{M}$, computed with respect to the chart x.

Suppose that $x : \mathcal{D}(x) \to \mathcal{R}(x)$ is an arbitrary chart, Γ_{ij}^k are the Christoffel symbols of the second kind in this chart, $\{\partial_i(p)\}$ is the canonial basis of $T_p\mathcal{M}$ for $p \in \mathcal{D}(x)$, F is a C^s-vector field $(1 \leq s \leq r-1)$ on $\mathcal{M}$ written as $F(p) = f^i \partial_i(p)$, $\forall\, p \in \mathcal{D}(x)$, and $\upsilon \in T_p\mathcal{M}$ is an arbitrary vector given as $\upsilon = \upsilon^j \partial_j(p)$. Then $\nabla_i f^k := \frac{\partial f^k}{\partial x^i} + \Gamma_{ij}^k f^j$ is called the *covariant derivative* of $f^k(x^1, \ldots, x^n)$ with respect to x^i and $\nabla_\upsilon F(p) := \nabla_i f^k \partial_k(p)\upsilon^i$ is called the *covariant derivative* of F *in the direction* υ.

The linear operator $\nabla F(p) : T_p\mathcal{M} \to T_p\mathcal{M}$ defined by $\upsilon \in T_p\mathcal{M} \mapsto \nabla_\upsilon F(p)$ is the *covariant derivative* of F at p. In a chart $x : \mathcal{D}(x) \to \mathcal{R}(x)$ around p this linear operator is given by $(\upsilon^i) \mapsto \left[\frac{\partial f^k}{\partial x^i} + \Gamma_{ij}^k f^j\right]\upsilon^i$, where $\upsilon = \upsilon^i \partial_i(p)$ and $F(p) = f^i(x(p))\partial_j(p)$. If $\beta : \mathcal{M} \to \mathbb{R}$ is a C^1-smooth function (0-form), the *gradient* of β is the vector field grad β defined by

$$\langle \operatorname{grad} \beta(p), \upsilon \rangle_{T_p\mathcal{M}} = d_p\beta(\upsilon), \ \forall\, p \in M, \ \forall\, \upsilon \in T_p\mathcal{M}.$$

In a chart x around p, the canonical basis $\{\partial_i(p)\}$ of $T_p\mathcal{M}$ and with the metric tensor $g_{li}(p) = \langle \partial_l, \partial_i \rangle_{T_p\mathcal{M}}$, we can write for $\upsilon = \upsilon^i \partial_i(p) \langle \operatorname{grad} \beta(p), \upsilon \rangle_{T_p\mathcal{M}} = \langle a^l \partial_l, \upsilon^i \partial_i \rangle_{T_p\mathcal{M}} = g_{li} a^l \upsilon^i = \frac{\partial \beta}{\partial x^j} \upsilon^j$. It follows that $g_{lj} a^l = \frac{\partial \beta}{\partial x^j}$ and $a^s = g^{sj} \frac{\partial \beta}{\partial x^j}$. This means that in the chart x the gradient vector field grad β is given as grad $\beta(p) = g^{sj} \frac{\partial \beta}{\partial x^j} \partial_s(p)$.

A.6 Vector Fields

Suppose that $\mathcal{M}$ and $\mathcal{N}$ are C^k-manifolds of dimension n and m, respectively. The map $\phi : \mathcal{M} \to \mathcal{N}$ satisfies on $\mathcal{M}$ a *local Lipschitz condition* if in any chart x of $\mathcal{M}$ around p and any chart y of $\mathcal{N}$ the map

$$y \circ \phi \circ x^{-1} : x(\mathcal{D}(x) \cap \phi^{-1}(\mathcal{D}(y)) \to \mathcal{R}(y)$$

satisfies the usual local Lipschitz condition in $\mathbb{R}^n$. Assume that $F : \mathcal{M} \to T\mathcal{M}$ is a vector field, $x : \mathcal{D}(x) \to \mathcal{R}(x)$ is a chart around p. Then $F(p)$ can be written as $F(p) = [p, x, f(x(p))]$ where $f(x(p)) = (\pi_2 \circ dx \circ F \circ x^{-1})(x(p))$ we call $f = (f^1, \ldots, f^n)$ the *representant* of F in the chart x. In the canonical basis $\{\partial_j(p)\}$ of $T_p\mathcal{M}$ the vector field has the form

$$F(p) = f^i(x(p))[p, x, e_i] = f^i(x(p))\partial_i(p).$$

The C^1-curve $\varphi : (a, b) \to M$ with $0 \in (a, b)$ is called an *integral curve* of the vector field $F : \mathcal{M} \to T\mathcal{M}$ with initial condition p at $t = 0$ if

$$\dot{\varphi}(t) = F(\varphi(t)) \quad \text{for all} \quad t \in (a, b)$$

and $\varphi(0) = p$. The function $\varphi(\cdot)$ (or $\varphi(\cdot, p)$) is also called the *solution* of the differential equation with $\varphi(0) = p$. In a chart $x : \mathcal{D}(x) \to \mathcal{R}(x)$, the curve $\varphi : (a, b) \to \mathcal{D}(x)$ with $\varphi(0) = p$ is an integral curve of F iff

$$\dot{\varphi}(t) = f^i((x \circ \varphi)(t))\, \partial_i\, (\varphi(t))\ , \ \text{i.e. iff}\ \sigma = (\sigma^1, \ldots, \sigma^n) := x \circ \varphi$$

is the solution of the ODE in an open set $\mathcal{D}(x) \subset \mathbb{R}^n$, i.e.

$$\dot{\sigma}^j(t)\, \partial_j\, (\varphi(t)) = f^i(\sigma(t))\, \partial_i\, (\varphi(t))$$

or

$$\dot{\sigma}^j(t) = f^j(\sigma^1(t), \ldots, \sigma^n(t))\ , \quad j = 1, \ldots, n\ , \quad \text{in} \quad \mathcal{D}(x)\ .$$

The Picard-Lindelöf theorem for vector fields together with the theorem on uniqueness in the large, say that if $\mathcal{M}$ is an n-dimensional C^k-manifold ($k \geq 2$) and $F : \mathcal{M} \to T\mathcal{M}$ is a locally Lipschitzian vector field, then for every $p \in \mathcal{M}$ there exists on some open interval $\mathcal{J} \ni 0$ an integral curve φ of F satisfying $\varphi(0) = p$. Moreover, if $\varphi_1 : \mathcal{J}_1 \to \mathcal{M}$ and $\varphi_2 : \mathcal{J}_2 \to \mathcal{M}$ are two integral curves of F defined on open intervals $\mathcal{J}_1$ and $\mathcal{J}_2$, then $\varphi_1 = \varphi_2$ on $\mathcal{J}_1 \cap \mathcal{J}_2$. The above theorems imply that the union of all integral curves φ with $\varphi(0) = p$ of F defined on open intervals is the *maximal integral curve* $\varphi(\cdot, p)$ of F defined on the *maximal existence interval* (a_p, b_p) with $-\infty \leq a_p < 0 < b_p \leq +\infty$. As for ODE's in $\mathbb{R}^n$ the set $\mathcal{D} = \{(t, p) \in \mathbb{R} \times \mathcal{M} | a_p < t < b_p\}$ is open in $\mathbb{R} \times \mathcal{M}$ and the *maximal flow*

$\varphi : \mathcal{D} \to \mathcal{M}$, $(t, p) \mapsto \varphi(t, p)$ is continuous. If $\mathcal{D} = \mathbb{R} \times \mathcal{M}$ the local flow is called, for short, *flow*.

A vector field F is *complete* if F generates a flow. It can be shown that for any C^2-vector field there exists a C^1-smooth function $\psi : \mathcal{M} \to \mathbb{R}_+$ such that the vector field ψF is complete. (Note that ψF has the same integral curves as F, but different parametrizations.) A C^1-vector field on a compact manifold is complete.

Suppose that μ is an arbitrary volume form on $\mathcal{M}$. The *divergence* of a smooth vector field F w.r.t. μ is the scalar valued function $\mathrm{div}_\mu F$ defined by $L_F\mu = (\mathrm{div}_\mu F)\mu$. If μ is the canonical volume form $\mu = \sqrt{\det(g_{ij})}\, dx^1 \wedge \cdots \wedge dx^n$ then we have $\mathrm{div}_\mu F = \nabla_i f^i$ where $F = f^i \partial_i$ in a chart x. Let F be a C^r-vector field on the n-dimensional C^k-manifold $\mathcal{M}$ $(k \geq 2, 1 \leq r \leq k-1)$, let $(\{\varphi^t\}_{t\in\mathbb{R}}, \mathcal{M}, \rho)$ be the flow of F, $\rho : \mathcal{M} \to \mathbb{R}_+$ a smooth function and μ a volume form on $\mathcal{M}$.

Then we have, for any Lebesgue measurable set, $\mathcal{B} \subset \mathcal{M}$ and arbitrary $t \in \mathbb{R}$

$$\frac{d}{dt}\int_{\varphi^t(\mathcal{B})} \rho\mu = \int_{\varphi^t(\mathcal{B})} \mathrm{div}_\mu(\rho F)\mu \qquad \text{(Liouville's theorem)}.$$

Let us consider the flow $\{\varphi^t\}_{t\in\mathbb{R}}$ of F on the compact manifold $\mathcal{M}$ which *preserves* the volume form μ on $\mathcal{M}$, i.e. $(\varphi^t)^*\mu = \mu,\ \forall t \in \mathbb{R}$. Here $(\varphi^t)^*\mu$ denotes the pull-back of μ introduced above as

$$(\varphi^t)^*\mu_{|p}(\upsilon_1, \ldots, \upsilon_n) = \mu_{|\varphi^t(p)}(d_p\varphi^t\upsilon_1, \ldots, d_p\varphi^t\upsilon_n),\ \forall p \in \mathcal{M},\ \forall \upsilon_1, \ldots, \upsilon_n \in T_p\mathcal{M}\,.$$

Then for any measurable set $\mathcal{S} \subset \mathcal{M}$ with $\int_{\mathcal{S}} \mu > 0$ and any $T \geq 0$ there exists a time $t \geq T$ such that $\mathcal{S} \cap \varphi^t(\mathcal{S}) \neq \emptyset$ (Poincaré's theorem).

Analogously one says that $\{\varphi^t\}_{t\in\mathbb{R}}$ *preserves* a k-form β, if $(\varphi^t)^*\beta = \beta, t \in \mathbb{R}$, i.e. if $L_F\beta \equiv 0$. If β is a 0-form, i.e. β is a scalar valued function, preserving means that

$$(\varphi^t)^*\beta_{|p} = \beta(\varphi^t(p)) = \beta(p),\ \forall t \in \mathbb{R},\ \forall p \in \mathcal{M}\,.$$

In this case β is a *first integral* of the flow $\{\varphi^t\}_{t\in\mathbb{R}}$. The vector field F is called *conservative* with respect to the volume form μ if $\mathrm{div}_\mu F \equiv 0$ on $\mathcal{M}$. A *symplectic manifold* is a pair $(\mathcal{M}, \omega)$, where $\mathcal{M}$ is a $2n$-dimensional C^k-manifold and ω is a smooth closed non-degenerate two-form, i.e. $d\omega = 0$ and $\omega_{|p}(\upsilon_1, \upsilon_2) = 0 \quad \forall \upsilon_1 \in T_p\mathcal{M}$ at a point p and for a vector $\upsilon_2 \in T_p\mathcal{M}$ implies that $\upsilon_2 = 0$.

If $p \in \mathcal{M}$ is fixed the relation $\upsilon \in T_p\mathcal{M} \mapsto \omega_{|p}(\cdot, \upsilon)$ defines a 1-form on $T_p\mathcal{M}$. Since ω is non-degenerate, this relation is a linear bijection denoted by i. If $J := i^{-1}$ and β is an arbitrary 1-form on $\mathcal{M}$, the term $J\beta$ is a vector field on $\mathcal{M}$. For any smooth function $H : \mathcal{M} \to \mathbb{R}$ the vector field $X_H := JdH$ is called *Hamiltonian* and H is the associated *Hamiltonian*.

On a symplectic manifold $(\mathcal{M}, \omega)$ for arbitrary smooth functions $f, g : \mathcal{M} \to \mathbb{R}$ the *Poisson bracket* between f and g is $\{f, g\} := \omega(X_f, X_g) = df(X_g)$. We say that f and g *are in involution* if $\{f, g\} \equiv 0$. A smooth function $f : \mathcal{M} \to \mathbb{R}$ is a first integral of X_H on $\mathcal{M}$ iff $\{f, H\} = 0$ on $\mathcal{M}$.

A.7 Spaces of Vector Fields and Maps

Suppose that $\mathcal{M}$ is a compact n-dimensional C^r-manifold and $\mathrm{Diff}^1(\mathcal{M})$ denotes the set of all C^1-diffeomorphisms on $\mathcal{M}$. Since $\mathcal{M}$ is compact we can choose a finite set of charts $x_i : \mathcal{D}(x_i) \to \mathcal{R}(x_i),\ i = 1, \dots, m$, such that $\bigcup_{i=1}^m \mathcal{D}(x_i) = \mathcal{M}$. For arbitrary $\varphi, \psi \in \mathrm{Diff}^1(\mathcal{M})$ and $i, j, k \in \{1, \dots, m\}$ we introduce the maps $\varphi_{ij} = x_j \circ \varphi \circ x_i^{-1} : \mathcal{R}(x_i) \cap \varphi^{-1}(\mathcal{D}(x_j)) \to \mathbb{R}^n$ and $\psi_{ik} = x_k \circ \psi \circ x_i^{-1} : \mathcal{R}(x_i) \cap \psi^{-1}(\mathcal{D}(x_k)) \to \mathbb{R}^n$. Let $\mathcal{D}_{ijk} := \mathcal{R}(x_i) \cap \varphi^{-1}(\mathcal{D}(x_j)) \cap \varphi^{-1}(\mathcal{D}(x_k))$ and define the value

$$d_1(\varphi, \psi, x_i, x_j, x_k) := \max_{\xi \in \overline{\mathcal{D}_{ijk}}} |\varphi_{ij}(\xi) - \psi_{ik}(\xi)| + \max_{\xi \in \overline{\mathcal{D}_{ijk}}} \|D\varphi_{ij}(\xi) - D\psi_{ik}(\xi)\|$$

where $|\cdot|$ and $\|\cdot\|$ denote the Euclidean norm in $\mathbb{R}^n$ and the operator norm in $M_n(\mathbb{R})$, computed w.r.t. $|\cdot|$, respectively.

Then $d_1(\varphi, \psi) := \max_{i,j,k=1,\dots,m} d_1(\varphi, \psi, x_i, x_j, x_k)$ defines a metric in $\mathrm{Diff}^1(\mathcal{M})$. Equipped with this metric $\mathrm{Diff}^1(\mathcal{M})$ is a complete metric space. Suppose that $\mathcal{G} \subset \mathbb{R}^n$ is a domain with compact closure. Introduce the space $\mathrm{Diff}^1(\mathcal{G})$ as space of equivalence classes for C^1-diffeomorphisms $\varphi, \psi : \mathbb{R}^n \to \mathbb{R}^n$, i.e. $\varphi \sim \psi \Leftrightarrow \varphi(x) = \psi(x)\ \forall x \in \mathcal{G}$. For $\varphi, \psi \in \mathrm{Diff}(\mathcal{G})$ we define $d_0(\varphi, \psi) = \max_{x \in \overline{\mathcal{G}}} |\varphi(x) - \psi(x)|$ and

$$d_1(\varphi, \psi) = d_0(\varphi, \psi) + \max_{x \in \overline{\mathcal{G}}} \|D\varphi(x) - D\psi(x)\|\ .$$

Denote by $\mathrm{Diff}^i(\mathcal{G})$, $i = 0, 1$, the complete metric space derived from $\mathrm{Diff}(\mathcal{G})$ with metric d_0 resp. d_1. Let $\mathrm{Diff}_+(\mathcal{G})$ denote the set of all $\varphi \in \mathrm{Diff}(\mathcal{G})$ with $\varphi(\overline{\mathcal{G}}) \subset \mathcal{G}$. The set $\mathrm{Diff}_+(\mathcal{G})$ is open in $\mathrm{Diff}^i(\mathcal{G})$, $i = 0, 1$. Denote $\mathrm{Diff}^i_+(\mathcal{G}) := \mathrm{Diff}^i(\mathcal{G}) \cap \mathrm{Diff}_+(\mathcal{G})$.

Let us now consider C^1-vector fields on $\mathcal{M}$. Suppose again that $x_i : \mathcal{D}(x_i) \to \mathcal{R}(x_i)$, $i = 1, \dots, m$, is a finite set of charts for the compact manifold $\mathcal{M}$ with $\mathcal{M} = \bigcup_{i=1}^m \mathcal{D}(x_i)$. Let $F, G : \mathcal{M} \to T\mathcal{M}$ be two C^1-smooth vector fields and denote by f_i and g_i the realizations of F and G in the charts x_i. Introduce the value

$$d_1(F, G) := \max_{\xi \in \overline{\mathcal{R}(x_i)}} |g_i(\xi) - f_i(\xi)| + \max_{\xi \in \overline{\mathcal{R}(x_i)}} \|Dg_i(\xi) - Df_i(\xi)\|\ .$$

Then d_1 is a metric. Denote the metric space of all C^1-vector fields on $\mathcal{M}$ with this metric by $X^1(\mathcal{M})$.

Assume that $\mathcal{G} \subset \mathcal{U} \subset \mathbb{R}^n$ are open sets, $\overline{\mathcal{G}}$ is compact and $\overline{\mathcal{G}} \subset \mathcal{U}$. Assume also that $\partial\, \mathcal{G}$ is a smooth $(n-1)$ dimensional submanifold of $\mathbb{R}^n$. Denote by $X^1(\mathcal{G})$ the set of all equivalence classes of C^1-vector fields $f, g : \mathcal{U} \to \mathbb{R}^n$ w.r.t. the equivalence relation $f \sim g \Leftrightarrow f(x) = g(x),\ \forall x \in \mathcal{G}$. Any class of $X^1(\mathcal{G})$ is identified with some vector field $f : \mathcal{U} \to \mathbb{R}^n$. Now we define for any $f, g \in X^1(\mathcal{G})$ the value

$$d_1(f, g) := \max_{x \in \overline{\mathcal{G}}} |f(x) - g(x)| + \sup_{x \in \overline{\mathcal{G}}} \|Df(x) - Dg(x)\|.$$

Then d_1 is a metric and $X^1(\mathcal{G})$ with this metric is complete. Denote by $X^1_+(\mathcal{G}) \subset X^1(\mathcal{G})$ the set of all vector fields f satisfying $f(p) \notin T_p(\partial\,\mathcal{G}), \forall\, p \in \partial\,\mathcal{G}$, and $\varphi^t(p) \in \mathcal{G}$, $\forall p \in \partial\,\mathcal{G}$, $\forall t > 0$, sufficiently small ($\varphi^t(\cdot)$ is the flow of f). $X^1_+(\mathcal{G})$ is an open subset of $X^1(\mathcal{G})$.

The point $p \in \mathcal{M}$ is called a *wandering point* of the dynamical $\{\varphi^t\}_{t\in\mathbb{R}}$ on $(\mathcal{M}, g)$ if there exists a neighborhood $\mathcal{U}$ of p and a number $t_0 > 0$ such that

$$\bigcup_{|t|>t_0} \varphi^t(\mathcal{U}) \cap \mathcal{U} = \emptyset\,.$$

The set of all non-wandering points of $\{\varphi^t\}_{t\in\mathbb{R}}$ generated by the vector field F is denoted by $\mathcal{NW}(F)$.

One can show that if $\gamma(p)$ is a bounded orbit of the dynamical system $\{\varphi^t\}_{t\in\mathbb{R}}$ then $\omega(p) \cup \alpha(p) \subset \mathcal{NW}(F)$. This implies that if the dynamical system has a solution $\varphi^{(\cdot)}(p)$ such that $\gamma_+(p)$ or $\gamma_-(p)$ is bounded, then $\mathcal{NW}(F)$ is non-empty. It is also well-known that any non-wandering set $\mathcal{NW}(F)$ is closed and invariant.

Pugh's closing lemma [12] plays a crucial role in our global stability investigation:

Theorem A.4 *Let $\mathcal{X} \in \{X^1(\mathcal{M}), X^1_+(\mathcal{G})\}$, $F \in \mathcal{X}$, and $p \in \mathcal{NW}(F)$, $F(p) \neq 0$. Then for an arbitrary neighborhood $\mathcal{U}$ of F in $\mathcal{X}$ there is a vector field $G \in \mathcal{U}$ having a periodic orbit passing through p.*

Suppose that $\mathcal{M}$ and $\mathcal{N}$ are n-dimensional C^k-manifolds, $\varphi : \mathcal{M} \to \mathcal{M}$ and $\psi : \mathcal{N} \to \mathcal{N}$ are maps. These maps are called *topologically conjugated* (C^r-conjugated) if there exists a homeomorphism $h : \mathcal{M} \to \mathcal{N}$ (a C^r-diffeomorphism $h : \mathcal{M} \to \mathcal{N}$) such that $\varphi = h^{-1} \circ \psi \circ h$.

Suppose now that $F : \mathcal{M} \to T\mathcal{M}$ and $G : \mathcal{N} \to T\mathcal{N}$ are vector fields, $\{\varphi^t\}_{t\in\mathbb{R}}$ and $\{\psi^t\}_{t\in\mathbb{R}}$ are the associated flows. The vector fields F and G (or their flows) are called *C^r-equivalent* with $0 \le r \le k$ (*topologically equivalent* for $r = 0$) if there exists a C^r-diffeomorphism $h : \mathcal{M} \to \mathcal{N}$ (a homeomorphism for $r = 0$) which transforms the orbits of $\{\varphi^t\}_{t\in\mathbb{R}}$ into orbits of $\{\psi^t\}_{t\in\mathbb{R}}$ preserving the orientation of the orbits. If $\{\varphi^t\}_{t\in\mathbb{R}}$ and $\{\psi^t\}_{t\in\mathbb{R}}$ are C^r-equivalent (topologically equivalent) over $h : \mathcal{M} \to \mathcal{N}$ and if $h(\varphi^t(p)) = \psi^t(h(p))$, $\forall p \in \mathcal{M}$, $\forall t \in \mathbb{R}$, then $\{\varphi^t\}_{t\in\mathbb{R}}$ and $\{\psi^t\}_{t\in\mathbb{R}}$ are called *C^r-conjugated (topologically conjugated)*. Suppose that $\varphi \in \mathrm{Diff}^r(\mathcal{M})$ and $e \in \{$ topologically conjugated, C^r-conjugated$\}$ is an equivalence relation as defined above.

The diffeomorphism φ is called *e-stable* if there exists a neighborhood $\mathcal{U}$ of φ in $\mathrm{Diff}^r(\mathcal{M})$ such that any $\psi \in \mathcal{U}$ is e-equivalent to φ.

Suppose now that $F : \mathcal{M} \to T\mathcal{M}$ is a C^r-vector field with associated flow $\{\varphi^t\}_{t\in\mathbb{R}}$ and $\widetilde{e} \in \{$ topologically equivalent, C^r-equivalent, topologically conjugated, C^r-conjugated$\}$.

The vector field F (respectively, the flow $\{\varphi^t\}_{t\in\mathbb{R}}$) is said to be $\widetilde{e}$-*stable* if there exists a neighborhood $\mathcal{U}$ of F in $X^r(\mathcal{M})$ such that any $G \in \mathcal{U}$ is $\widetilde{e}$-equivalent to F. In case if e or $\widetilde{e}$ means "topologically equivalent" the e-or $\widetilde{e}$-stability is called *structural* stability. Recall that a topological space X has the Baire property, or is a *Baire space*, if every countable intersection of open dense sets in X is itself dense in X. Baire's theorem says that any complete metric space has the Baire property. As noted above, the metric spaces $\text{Diff}^0(\mathcal{M})$, $\text{Diff}^1(\mathcal{M})$, $\text{Diff}^0(\mathcal{G})$, $\text{Diff}^1(\mathcal{G})$, $X^0(\mathcal{G})$, $X^1(\mathcal{G})$, $X^0(\mathcal{M})$ and $X^1(\mathcal{M})$ are complete. It follows that they are Baire spaces. We say that in these spaces there is given a *generic property*, if such a property is satisfied for a set which contains a countable intersection of open and dense sets, then such a set is called *residual*.

A.8 Parallel Transport, Geodesics and Exponential Map

Suppose that $(\mathcal{M}, g)$ is a Riemannian n-dimensional C^k-manifold and $c : \mathcal{J} \to \mathcal{M}$ is a C^1-curve. A continuous map $X : \mathcal{J} \to T\mathcal{M}$ given by $t \in \mathcal{J} \mapsto X(t) \in T_{c(t)}\mathcal{M}$ is called a *vector field along c*.

The vector field $X : \mathcal{J} \to T\mathcal{M}$ along c is said to be C^r-*smooth* if for any chart $x \in \mathcal{A}$, the vector field has a representation $X(t) = \xi^i(t)\partial_i(c(t))$ with C^r-smooth functions $\xi^i(\cdot)$.

Let $F : \mathcal{M} \to T\mathcal{M}$ be a smooth vector field and $p \in \mathcal{M}$ be arbitrary. The covariant derivative $\nabla F(p) : T_p\mathcal{M} \to T_p\mathcal{M}$ in a chart x is given by

$$\nabla_v F(p) = \Big[\frac{\partial f^k}{\partial x^i} a^i + a^i f^j(p) \Gamma^k_{ij}(p)\Big]\partial_k(p) ,$$

where $F = f^i\partial_i$ and $v = a^i\partial_i$. Since the vector field X can be written in this chart as $x^i := (x \circ c)^i$, we get with $\xi^k(t) := f^k(c(t))$ for the vector $v = \dot{c}(t)$

$$\nabla_{\dot{c}} X(t) = [\dot{\xi}^k(t) + \dot{x}^i(t)\, \xi^j(t)\, \Gamma^k_{ij}(c\,(t))]\, \partial_k\,(c(t)),\ t \in \mathcal{J} \quad \text{s.t.} \quad c\,(t) \in \mathcal{D}(x).$$

The vector field $X : \mathcal{J} \to T\mathcal{M}$ is *parallel* along c if $\nabla_{\dot{c}} X(t) = 0$ on $\mathcal{J}$. Locally in a chart x this means that

$$\dot{\xi}^k(t) + \dot{x}^i(t)\, \xi^j(t)\, \Gamma^k_{ij}(c\,(t)) = 0\,.$$

Using the existence and uniqueness theorem for linear differential equations, one can show that for a given smooth curve $c : \mathcal{J} \to \mathcal{M}$, arbitrary $t_0 \in \mathcal{J}$ and arbitrary $v \in T_{c(t_0)}\mathcal{M}$, there exists exactly one vector field $X_v(t)$ which is parallel along c and for which $X_v(t_0) = v$ holds.

Suppose $c : \mathcal{J} \to \mathcal{M}$ is a given smooth curve, $t_0 < t_1$ are values in $\mathcal{J}$. The map $\tau(c)|_{t_0}^{t_1} : T_{c(t_0)}\mathcal{M} \to T_{c(t_1)}\mathcal{M}$ which is defined by $v \in T_{c(t_0)}\mathcal{M} \mapsto X_v(t_1) \in T_{c(t_1)}\mathcal{M}$, where $X_v(\cdot)$ is the unique vector field parallel c with $X_v(t_0) = v$, is called *parallel transport* along c from $c(t_0)$ into $c(t_1)$. One can show that the map $\tau(c)|_{t_0}^{t_1}$ is linear and is an isomorphism.

The smooth curve $c : \mathcal{J} \to \mathcal{M}$ on the Riemannian C^k-manifold $(\mathcal{M}, g)$ is called a *geodesic* if the vector field $\dot{c}$ is parallel along c, i.e. if $\nabla_{\dot{c}}\, \dot{c}\,(t) = 0$ on $\mathcal{J}$. In a chart $x : \mathcal{D}(x) \to \mathcal{R}(x)$ we have with $x^k(t) := (x \circ c)^k(t)$ and $X(t) = \dot{x}^i(t)\, \partial_i(c(t))$ the representation $\nabla_{\dot{c}}\, \dot{c}\,(t) = \left[\ddot{x}^k(t) + \dot{x}^i(t)\, \dot{x}^j(t)\ \Gamma^k_{ij}(c(t))\right] \partial_k(c(t))$. It follows that c is a geodesic iff in any chart x $\ddot{x}^k + \Gamma^k_{ij} \dot{x}^i \dot{x}^j = 0$. From the local existence and uniqueness theorem for ODE's, it follows that for arbitrary $p \in \mathcal{M}$ and $v \in T_p\mathcal{M}$, there exists an $\varepsilon > 0$ and a unique geodesic $c : (-\varepsilon, \varepsilon) \to \mathcal{M}$ with $c(0) = p$ and $\dot{c}(0) = v$.

The parallel transport remains the scalar product in the following sense: Suppose $c : \mathcal{J} \to \mathcal{M}$ is a smooth curve, $t_0 \in \mathcal{J}$ is arbitrary and $X, Y : \mathcal{J} \to \mathcal{M}$ are two vector fields which are parallel along c. Then $g(X(t), Y(t)) = g(X(t_0), Y(t_0))$, $\forall t \in \mathcal{J}$. Suppose $(\mathcal{M}, g)$ is a Riemannian n-dimensional C^k-manifold ($k \geq 3$). One can show that for any $p \in \mathcal{M}$ and any $v \in T_p\mathcal{M}$ there exists a unique geodesic $\varphi(\cdot, p, v)$, defined for $|t| < \varepsilon$ and satisfying $\varphi(0, p, v) = p, \dot{\varphi}(0, p, v) = v$.

The map $(t, p, v) \mapsto \varphi(t, p, v)$ is C^{k-2}-differentiable for $v \in T_p\mathcal{M}$ with sufficiently small $|v|$.

The *exponential map* $v \mapsto \exp_p v := \varphi(1, p, v)$ is C^{k-2} in a neighborhood of $0 \in T_p\mathcal{M}$. If $\mathcal{V}$ is a sufficiently small open neighborhood of 0 then the map $\exp_p : \mathcal{V} \to \exp_p \mathcal{V}$ is a C^{k-2}-diffeomorphism.

Suppose that $p \in \mathcal{M}$ is arbitrary and $\varepsilon > 0$ is so small that $\exp_p$ is a C^{k-2}-diffeomorphism on $\mathcal{B}_\varepsilon(0_p) \subset T_p\mathcal{M}$. Then for any $v \in \mathcal{B}_\varepsilon(0_p)$, the curve $t \mapsto c(t) := \exp_p(tv), t \in [0, 1]$, defines the geodesic c on $\mathcal{M}$ with $c(0) = p$ and $\dot{c}(0) = v$.

A.9 Curvature and Torsion

Suppose $(\mathcal{M}, g)$ is a Riemannian $C^k (k \geq 2)$-manifold of dimension n and let F and H be C^r-vector fields ($2 \leq r \leq k-1$) on $\mathcal{M}$. The *Lie bracket* of F and H is that C^{r-1}-vector field $[F, H]$ on $\mathcal{M}$, whose components in a chart $x : \mathcal{D}(x) \to \mathcal{R}(x)$ are defined as

$$[F, H]_{|\mathcal{D}(x)} = [f^i \partial_i h^j - h^i \partial_i f^j]\, \partial_j \ ,$$

$$\text{where} \quad F_{|\mathcal{D}(x)} = f^i \partial_i \quad \text{and} \quad H_{|\mathcal{D}(x)} = h^j \partial_j \ .$$

One can show that this definition is correct, i.e. does not depend on the chart x.

Let F, G and H be C^r-vector fields on the Riemannian C^k-manifold $(\mathcal{M}, g)$ of dimension n ($k \geq 3, 2 \leq r \leq k-1$), let λ and μ be numbers and $\rho, \gamma : \mathcal{M} \to \mathbb{R}$ be C^r-functions. Then we have:

1. $[F, G] = -[G, F]$; $[F, F] = 0$; 2. $[\lambda F + \mu G, H] = \lambda [F, H] + \mu$ $[G, H]$;

3. $d_{[F,G]}\rho = d_F(d_G\rho) - d_G(d_F\rho)$; 4. $[\rho F, G] = \rho\,[F, G] - (d_G\rho)F$;
5. $\nabla_{\rho F+\gamma G}H = \rho\,\nabla_F H + \gamma\nabla_G H$; 6. $\nabla_F(G+H) = \nabla_F G + \nabla_F H$;
7. $\nabla_F(\rho\, G) = (d_F\rho)G + \rho\,\nabla_F G$.

Suppose that for the above manifold $(\mathcal{M}, g)$ and vector fields F, G, H we have $k \geq 4$ and $3 \leq r \leq k-1$. The *curvature tensor field* is the C^{r-2}-smooth tensor field R of type $(1, 3)$ given by

$$R\,(F, G)H = \nabla_F\nabla_G H - \nabla_G\nabla_F H - \nabla_{[F,G]}H\ ,$$

the *torsion tensor field* is the C^{r-1}-smooth tensor field of type $(1, 2)$ given by

$$T(F, G) = \nabla_F G - \nabla_G F - [F, G].$$

By Bianchi's first identity we have

$$R\,(F, G)H + R\,(G, H)F + R\,(H, F)G = 0\,.$$

Suppose $p \in \mathcal{M}$ and $f, g, h \in T_p\mathcal{M}$ with $f = F(p)$, $g = G(p)$ and $h = H(p)$ are arbitrary. Then $R_p(f, g)h := R(F, G)H_{|p}$ is the *curvature tensor* at p computed in f, g, h. Analogously the *torsion tensor* T is defined. The components of the curvature tensor and the torsion tensor in a chart $x : \mathcal{D}(x) \to \mathcal{R}(x)$ and the associated Christoffel symbols are

$$R^l_{kij} = \partial_i\Gamma^l_{jk} - \partial_j\Gamma^l_{ik} + \Gamma^m_{jk}\,\Gamma^l_{im} - \Gamma^m_{ik}\,\Gamma^l_{jm}$$

and $T^k_{ij} = \Gamma^k_{ij} - \Gamma^k_{ji}$. Other curvature type tensors are the *Riemannian curvature* tensor

$$R_{ijkh} := g_{ir}R^r_{jkh} = \partial_k\Gamma_{jk,i} - \partial_h\Gamma_{jk,i} + \Gamma^r_{jh}\Gamma_{ih,r} - \Gamma^r_{jh}\Gamma_{ik,r}\ ,$$

where g_{ij} denotes the metric tensor in the chart x, the *Ricci tensor*

$$R_{ij} := R^k_{ijk} = \frac{\partial^2 \log\sqrt{|\det g_{ij}|}}{\partial x^i\partial x^j} - \frac{\partial}{\partial x^k}\Gamma^k_{ij} - \Gamma^h_{ij}\frac{\partial}{\partial x^h}\log\sqrt{|\det g_{ij}|} + \Gamma^k_{hj}\Gamma^h_{ik}$$

and the *scalar curvature*

$$R := R^i_i = g^{ik}R_{ik} = g^{ik}R^j_{ikj}.$$

The Theorema egregium by Gauss states that for a 2-dimensional elementary surface $\mathcal{S} \subset \mathbb{R}^3$ with induced metric tensor g_{ij} the Riemannian curvature $K(p)$ at a point $p \in \mathcal{S}$ is given by

$$K(p) = \frac{R_{1212_{|p}}}{\det\,(g_{ij})_{|p}}\ .$$

Suppose that$(\mathcal{M}, g)$ is a Riemannian manifold of dimension 2, R_{1212} is a component of the curvature tensor and g_{ij} is the metric tensor in a chart x. Then $K(p) := \frac{R_{1212|p}}{\det(g_{ij})_{|p}}$ is called the *Gaussian curvature* of $\mathcal{M}$ at p. Let $(\mathcal{M}, g)$ be a Riemannian C^k-manifold of dimension $n \geq 2$ and let $p \in \mathcal{M}$ be arbitrary. If $\mathbb{L}$ is an arbitrary 2-dimensional subspace of $T_p\mathcal{M}$ spanned by the vectors u and υ, the number

$$K(\mathbb{L}) := \frac{g(R(u, \upsilon)\upsilon, u)}{g(u, u)\, g(\upsilon, \upsilon) - g(u, \upsilon)^2}$$

is called the *section curvature* at p with respect to $\mathbb{L}$, i.e. this definition does not depend on the choice of u and υ in $\mathbb{L}$.

For a Riemannian C^k-manifold $(\mathcal{M}, g)$ of dimension 2 we have

$$K(p) = K(T_p\mathcal{M}) = \frac{1}{2}R(p)\ , \ \forall\, p \in \mathcal{M}.$$

A.10 Fiber Bundles and Distributions

A C^k-*fiber bundle* is given by a surjective submersion $\pi : \mathcal{P} \to \mathcal{M}$, where $\mathcal{P}$ and $\mathcal{M}$ are C^k-manifolds. The submersion is assumed to be *locally trivial*, i.e. there exists a manifold $\mathcal{N}$ such that, for each point $p \in \mathcal{M}$, there exists a neighborhood $\mathcal{U}$ of p and a C^k-diffeomorphism $h : \pi^{-1}(\mathcal{U}) \to \mathcal{U} \times \mathcal{N}$ with $\pi \circ h^{-1}_{|\mathcal{U}} = \mathrm{id}_{\mathcal{U}}$. We say that $\mathcal{P}$ is the *bundle space*, $\mathcal{M}$ is the *base space*, π is the *projection*, h is a *bundle chart* and $\mathcal{N}$ is the *typical fiber*. For any $p \in \mathcal{M}$ the set $\pi^{-1}(p)$ is the *fiber over* p.

A C^k-*vector bundle* is a C^k-fiber bundle whose fibers have a vector space structure. Suppose that $\mathcal{M}$ is a C^k-manifold $(k \geq 2)$ of dimension n and $T\mathcal{M} = \bigcup_{p\in\mathcal{M}} T_p\mathcal{M}$ is the tangent bundle, which has the canonical structure of a C^{k-1}-manifold of dimension $2n$. Then the natural projection $\pi : T\mathcal{M} \to \mathcal{M}$ defines a C^{k-1}-vector bundle. The typical fiber in this case is an n-dimensional vector space $\mathbb{V}$. Similarly, if $T^*\mathcal{M}$ denotes the cotangent space of $\mathcal{M}$, the projection $\pi : T^*\mathcal{M} \to \mathcal{M}$ is also a C^{k-1}-vector bundle.

A C^k-vector bundle $\pi : \mathcal{P} \to \mathcal{M}$ with an n-dimensional C^k-manifold $\mathcal{M}$ as base space and an m-dimensional vector space $\mathbb{V}$ as typical fiber has the canonical structure of an $(n + m)$-dimensional C^k-manifold. A C^r-*section*, $1 \leq r \leq k$, in a C^k-vector bundle $\pi : \mathcal{P} \to \mathcal{M}$ is a C^r-map $s : \mathcal{M} \to \mathcal{P}$ such that $\pi \circ s = \mathrm{id}_{\mathcal{M}}$. Suppose that $\mathcal{E}^r(\mathcal{P})$ is the set of all C^r-sections of $\pi : \mathcal{P} \to \mathcal{M}$. For the given vector bundle, the base manifold $\mathcal{M}$ can be realized as a submanifold of $\mathcal{P}$, identifying any point $p \in \mathcal{M}$ with the zero vector in $\pi^{-1}(p)$. We denote this submanifold by $\mathcal{Z}(\mathcal{P})$ and call it *zero section* of $\mathcal{P}$. A subset $\mathcal{P}' \subset \mathcal{P}$ of a vector bundle $\pi : \mathcal{P} \to \mathcal{M}$ is a *subbundle* if there exists a subspace $\mathbb{W} \subset \mathbb{V}$ and for any $b \in \mathcal{P}'$ a bundle chart $h : \pi^{-1}(\mathcal{U}) \to \mathcal{U} \times \mathbb{V}$ with $b \in \mathcal{U}$ and such that $h(\pi^{-1}(\mathcal{U}) \cap \mathcal{P}') = \mathcal{U} \times (\mathbb{W} \times \{0\})$.

Suppose that $l, m \in \mathbb{N}_0$ are arbitrary. Then $T_k^l(\mathcal{P}) := \bigcup_{p \in \mathcal{M}} T_k^l(\mathcal{P}_p)$ is the tensor bundle of type (l, m) on $\mathcal{M}$ with respect to $\mathcal{P}$. A *C^r-tensor field* of type (l, m) is a C^r-section of $T_k^l(\mathcal{P})$. A *bundle metric* g of smoothness $C^r (1 \le r \le k)$ on a C^k-vector bundle $\pi : \mathcal{P} \to \mathcal{M}$ is a C^r-smooth symmetric tensor field $g \in \mathcal{E}^r(T_2^0(\mathcal{P}))$ for which $g_{|p}(\cdot, \cdot)$ at any $p \in \mathcal{M}$ is a positive definite bilinear form on $\mathcal{P}_p$.

Suppose $\pi : \mathcal{P} \to \mathcal{M}$ is a C^{k+1}-vector bundle and consider the C^k-vector bundle $d\pi : T\mathcal{P} \to T_{\mathcal{M}}$. The *vertical subbundle* for the vector bundle $\pi : \mathcal{P} \to \mathcal{M}$ is the subbundle of $T\mathcal{P}$ defined by $VT\mathcal{P} = \ker(d\pi)$. A vector field on $\mathcal{P}$ is called *vertical* if it takes only values in $VT\mathcal{P}$. In a similar manner one can define the *horizontal* subbundle of $T^*\mathcal{P}$ which is that subbundle $HT^*\mathcal{P}$ of $T^*\mathcal{P}$ that annihilates $V\mathcal{P}$. A one-form on $\mathcal{P}$ will be called *horizontal* if it takes values in $HT^*\mathcal{P}$.

Denote by $\pi_{\mathcal{P}} : T\mathcal{P} \to \mathcal{P}$ the natural projection. A *connection* on $\mathcal{P}$ is a map $C : \mathcal{P} \times T\mathcal{M} \to T\mathcal{P}$ for which $(\pi_{\mathcal{P}}, d\pi) \circ C = \mathrm{id}_{\mathcal{P} \times T\mathcal{M}}$ and $C_p : \mathcal{P}_p \times T_p\mathcal{M} \to T\mathcal{P}$ is for any $p \in \mathcal{M}$ bilinear. A connection on the vector bundle defines the *horizontal* subbundle $HT\mathcal{P}$ of $T\mathcal{P}$ which is the image of $\mathcal{P} \times T\mathcal{M}$ under C and which is complementary to $VT\mathcal{P}$, i.e. $T\mathcal{P} = VT\mathcal{P} \oplus HT\mathcal{P}$. A C^1-curve $\gamma : \mathcal{J} \to \mathcal{P}$ is called *horizontal* if $\dot{\gamma}(t) \in HT_{\gamma(t)}\mathcal{P}$ for any $t \in \mathcal{J}$. Suppose $c : \mathcal{J} \to \mathcal{M}$ is C^1-curve in the base manifold $\mathcal{M}$. The *horizontal lift* of c is the horizontal C^1-curve $\widetilde{c} : \mathcal{J} \to \mathcal{P}$ with $\pi \circ \widetilde{c} = c$. For a C^1-curve $c : \mathcal{J} \to \mathcal{M}$, $t_0 \in \mathcal{J}$, and $\upsilon \in \pi^{-1}(c(t_0))$, there exists exactly one horizontal lift $\widetilde{c}_\upsilon : \mathcal{J} \to \mathcal{P}$ satisfying $\widetilde{c}_\upsilon(t_0) = \upsilon$. For a given C^1-map $c : \mathcal{J} \to \mathcal{M}$ and arbitrary times $t_0, t_1 \in \mathcal{J}$, $t_0 \le t_1$, we call the map $\tau_{c(t_0)}^{c(t_1)} : \mathcal{P}_{c(t_0)} \to \mathcal{P}_{c(t_1)}$ which associates to any vector $\upsilon \in \mathcal{P}_{c(t_0)}$ the vector $\widetilde{c}_\upsilon(t_1) \in \mathcal{P}_{c(t_1)}$ *parallel transport* of υ along c from $c(t_0)$ to $c(t_1)$. The connection is called *metric* if the parallel transport along curves is isometric with respect to the bundle metric.

Suppose that $s \in \mathcal{E}^r(\mathcal{P})$, $1 \le r \le k$, is a C^r-section of the C^k-vector bundle $\pi : \mathcal{P} \to \mathcal{M}$ and $c : \mathcal{J} \to \mathcal{M}$ is a C^1-curve. The *covariant derivative* of the section s at the time $t \in \mathcal{J}$ in direction $\dot{c}(t)$ is defined as

$$\nabla_{\dot{c}(t)} s(t) = \lim_{h \to 0+} \frac{\left[\tau_{c(t)}^{c(t+h)}\right]^{-1} s\,(c\,(t+h)) - s(c\,(t))}{h}.$$

Suppose $w \in T_p\mathcal{M}$ and $s \in \mathcal{E}^r(\mathcal{P})$, $1 \le r \le k$, is a section which is defined in a neighborhood of $p \in \mathcal{M}$. The *covariant derivative* of s *at the point p in direction w* is given by $\nabla_w s(p) = \nabla_{\dot{c}(0)} s(0)$, where $c : [-\varepsilon, \varepsilon] \to \mathcal{M}$ is an arbitrary C^1-curve with $c(0) = p$ and $\dot{c}(0) = w$. It can be shown that this definition does not depend on the choice of c if the initial conditions are satisfied.

The covariant derivative defines a map $\nabla : \mathcal{E}^r(\mathcal{P}) \times \mathcal{E}^r(T\mathcal{M}) \to \mathcal{E}^{r-1}(\mathcal{P})$.

The *absolute* derivative of a smooth section $X : \mathcal{J} \to \mathcal{P}$ along the C^1-curve $c : \mathcal{J} \to \mathcal{M}$ is given at $t \in \mathcal{J}$ as $\frac{DX(t)}{dt} = \nabla_{\dot{c}(t)} X(t)$. A smooth section $X : \mathcal{J} \to \mathcal{P}$ is called *parallel along the curve c* if $\nabla_{\dot{c}(t)} X(t) = 0$, $t \in \mathcal{J}$.

Suppose the connection is metric w.r.t. the metric g and $(\cdot, \cdot)$ is the induced scalar product in the bundle. Then we have for the absolute derivative of arbitrary C^1-sections $X_1, X_2 : \mathcal{J} \to \mathcal{P}$ along a C^1-curve $c : \mathcal{J} \to \mathcal{P}$ the formula $\frac{d}{dt}(X_1(t), X_2(t)) = \left(\frac{D}{dt} X_1, X_2\right) + (X_1, \frac{D}{dt} X_2)$. Let $\mathcal{M}$ be an n-dimensional C^k-manifold ($k \ge 3$).

A *distribution* $\mathcal{D}$ on $\mathcal{M}$ is a subbundle of $T\mathcal{M}$, i.e. the union over all $p \in \mathcal{M}$ of linear subspaces of $T_p\mathcal{M}$. The *rank* of $\mathcal{D}$ at p, i.e. of $\mathcal{D}(p)$, is the dimension of the subspaces in $\mathcal{D}(p)$. Given for $1 \leq r < k$ a family of C^r-vector fields $\mathcal{V} := \{F_1, \ldots, F_l\}$ on $\mathcal{M}$, we can define a *distribution of rank l and smoothness C^r* by $\mathcal{D}_{\mathcal{V}}(p) = \text{span}\{F_1(p), \ldots, F_l(p)\}$, $p \in \mathcal{M}$. The distribution $\mathcal{D}_{\mathcal{V}}$ is said to be *involutive* if for any pairs $F, G \in \mathcal{V}$ and any $p \in \mathcal{M}$ we have for their Lie bracket the inclusion $[F, G](p) \in \mathcal{D}(p)$. An *integral manifold* $\mathcal{Z}$ of $\mathcal{D}_{\mathcal{V}}$ is a differentiable submanifold of $\mathcal{M}$ such that $T_p\mathcal{Z} \subset \mathcal{D}(p)$ for all $p \in \mathcal{M}$. The distribution $\mathcal{D}_{\mathcal{V}}$ is said to be *integrable* if, for all $p \in \mathcal{M}$, there exists an integral manifold with the same dimension as the rank of $\mathcal{D}$. This submanifold of $\mathcal{M}$ is called the *maximal integral manifold* of $\mathcal{D}$. Frobenius' theorem states that involutivity and integrability of distributions are locally equivalent notions.

Suppose that $(\mathcal{M}, g)$ is a C^k-manifold ($k \geq 3$) of dimension $n \geq 2$ and $\mathcal{F} = \{\mathcal{L}_\alpha | \alpha \in A\}$ is a partition of $\mathcal{M}$ in disjunct and connected subsets which are called *leaves*. $\mathcal{F}$ is called a *C^k-foliation of $\mathcal{M}$ of dimension m* if for any point $p \in \mathcal{M}$ there exists a chart $x : \mathcal{D}(x) \to \mathcal{R}(x)$ of the manifold such that $\mathcal{R}(x) = \mathcal{U}_1 \times \mathcal{U}_2$, where $\mathcal{U}_1 \subset \mathbb{R}^m$ and $\mathcal{U}_2 \subset \mathbb{R}^{n-m}$ are open connected sets, and for any leaf $\mathcal{L} \in \mathcal{F}$ we have $\mathcal{D}(x) \cap \mathcal{L} = x^{-1}(\mathcal{U}_1 \times v_0)$ with some $v_0 \in \mathcal{U}_2$. The number $n - m$ is the *codimension* of the foliation. Any leaf $\mathcal{L}$ is a connected immersion of dimension m. A C^k-foliation $\mathcal{F}$ on $\mathcal{M}$ defines through $T\mathcal{F} = \bigcup_{\alpha \in A} \bigcup_{p \in \mathcal{L}_\alpha} T_p\mathcal{L}_\alpha$ a subbundle of $T\mathcal{M}$ which is called the *tangent bundle of the foliation* $\mathcal{F}$. With $T^\perp\mathcal{F}$ we denote the *normal bundle* of the foliation. Thus the tangent bundle of $\mathcal{M}$ can be written as $T\mathcal{M} = T\mathcal{F} \oplus T^\perp\mathcal{F}$.

Given a fiber bundle $\pi : \mathcal{E} \to \mathcal{M}$ with principal fiber $\mathcal{F}$ we can define a *product bundle* by taking the total space $\widetilde{\mathcal{E}} = \mathcal{M} \times \mathcal{F}$ with the regular projection $\pi_2 : \mathcal{M} \times \mathcal{F} \to \mathcal{F}$. The fiber $\pi_2^{-1}(p) = \{p\} \times \mathcal{F}$ has a unique natural homeomorphism which is $(p, w) \mapsto w$ for $(p, w) \in \mathcal{M} \times \mathcal{F}$. Let us construct two fiber bundles over S^1. Start with the product bundle $\mathcal{D}^{n-1} \times [0, 1]$ over $[0, 1]$ $(n \geq 2)$, and glue $\mathcal{D}^{n-1} \times \{0\}$ to $\mathcal{D}^{n-1} \times \{1\}$ by some homeomorphism. The result is a $\mathcal{D}^{n-1}$-bundle over S^1. One can show that there are two classes of such $\mathcal{D}^{n-1}$-bundles, one is orientable and one is not. The total space of the orientable $\mathcal{D}^{n-1}$-bundle over S^1 is the *fibered solid torus*, the total space of the non-orientable $\mathcal{D}^{n-1}$-bundle over S^1 is the *fibered Klein bottle*. The boundary of the fibered solid Klein bottle is the *regular Klein bottle*. The total space of the non-orientable $\mathcal{D}^1$-bundle over S^1 is the *Möbius band*. Similarly one can define the fibered Klein bottle as total space of the non-orientable $\mathbb{R}^{n-1}$-bundle over S^1. Note that the *trivial fibered solid n-torus* over S^1 is the set $\mathcal{D}^{n-1} \times S^1$ or $\mathbb{R}^{n-1} \times S^1$.

Appendix B
Miscellaneous Facts

B.1 Totally Ordered Sets

A set $\mathcal{X}$ is called *partially ordered* if for certain pairs x, y of its elements there is defined an *order relation* $x \le y$ satisfying the following conditions:

$(i)\ x \le x;\quad (ii)\ x \le y, y \le x \Rightarrow x = y;\quad (iii)\ x \le y, y \le z \Rightarrow x \le z.$

A partially ordered set $(\mathcal{X}, \le)$ is said to be *totally ordered* if either $x \le y$ or $y \le x$, for any $x, y \in \mathcal{X}$. Any bijective map f of a totally ordered set $(\mathcal{X}, \le)$ onto a totally ordered set $(\mathcal{Y}, \le)$ is called *similarity*, if $x \le x'$ in $\mathcal{X}$ implies $f(x) \le f(x')$, for any $x, x' \in \mathcal{X}$. Two totally ordered sets $(\mathcal{X}, \le)$ and $(\mathcal{Y}, \le)$ are *similar* or have the *same ordering number* if there exists a similarity which maps $\mathcal{X}$ onto $\mathcal{Y}$. The ordering numbers of infinite totally ordered sets are called *transfinite numbers*. Suppose $\mathcal{A}$ is a set of transfinite numbers. A transfinite number ξ such that $\xi = \lim_{x \in \mathcal{A}} x$ is called *limit transfinite number* (see [1]). All natural numbers and the number zero are called *transfinite numbers of the first class*. The transfinite numbers of countable totally ordered sets are called *transfinite numbers* of *the second class*.

The following *Baire-Hausdorff theorem* [1] holds.

Theorem B.1 *Suppose that $\mathcal{X}$ is a topological space with a countable base and there is a totally ordered decreasing system of closed sets of this space, ordered by all transfinite numbers of the first and second classes:*

$$\mathcal{S}_0 \supset \mathcal{S}_1 \supset \mathcal{S}_2 \supset \cdots \supset \mathcal{S}_\alpha \supset \cdots . \tag{B.1}$$

Then there exists a transfinite number α such that all sets of (B.1), beginning with some ordering number α, coincide:

$$\mathcal{S}_\alpha = \mathcal{S}_{\alpha+1} = \mathcal{S}_{\alpha+2} = \cdots .$$

N. Kuznetsov and V. Reitmann, *Attractor Dimension Estimates for Dynamical Systems: Theory and Computation*, Emergence, Complexity and Computation 38,
https://doi.org/10.1007/978-3-030-50987-3

B.2 Recurrence and Hyperbolicity in Dynamical Systems

Let $\{\varphi^t\}_{t\in\mathbb{T}}$ be a dynamical system with $\mathbb{T} \in \{\mathbb{R}, \mathbb{Z}\}$ on the n-dimensional Riemannian C^k-manifold $(\mathcal{M}, g)$. The orbit $\gamma(p)$ is called *positively recurrent* if $p \in \omega(p)$ and *negatively recurrent* if $p \in \alpha(p)$.

Kneser [7] showed in 1924 that a continuous flow on the regular Klein bottle $\mathcal{K}^2$ without equilibrium points has at least one periodic orbit. Later it was shown by Markley [10] that every positively or negatively recurrent orbit of a continuous flow on the regular Klein bottle $\mathcal{K}^2$ is periodic.

For any non-empty, closed, bounded and invariant set $\mathcal{K} \subset \mathcal{M}$ there exists at least one minimal set in $\mathcal{K}$ (Proposition 1.4, Chap. 1). It follows that if $\gamma_+(p)$ (resp. $\gamma_-(p)$) is bounded then $\omega(p)$ (resp. $\alpha(p)$) contains at least one minimal set. The next theorem (*Lemma of Schwartz*, [14]) is the Poincaré-Bendixson theorem for 2-dimensional manifolds.

Theorem B.2 *Suppose that $\mathcal{M}$ is a 2-dimensional manifold of class C^2, $\{\varphi^t\}_{t\in\mathbb{R}}$ is a C^2-smooth flow on $\mathcal{M}$ and $\mathcal{N} \subset \mathcal{M}$ is a non-empty compact minimal set of the flow. Then the following three cases are possible.*
(i) $\mathcal{N}$ is an equilibrium; (ii) $\mathcal{N}$ is a periodic orbit; (iii) $\mathcal{N} = \mathcal{M}$.

In the case (iii) $\mathcal{M}$ is compact and the flow doesn't have equilibrium points. One can show in this case that $\mathcal{M}$ is homeomorphic either with the torus or with the regular Klein bottle. But on the regular Klein bottle without equilibrium states for $\{\varphi^t\}_{t\in\mathbb{R}}$, this dynamical system always has, by Kneser's theorem a cycle. Thus this case is impossible.

Suppose that $\{\varphi^t\}_{t\in\mathbb{T}}$ is a C^1-smooth dynamical system with $\mathbb{T} \in \{\mathbb{N}_0, \mathbb{Z}, \mathbb{R}_+, \mathbb{R}\}$ on the open set $\mathcal{U}$ of the Riemannian n-dimensional C^k-manifold $(\mathcal{M}, g)(k \geq 3)$. Assume that $\mathcal{K} \subset \mathcal{U}$ is an arbitrary set and $\mathcal{E} \subset T_{\mathcal{K}}\mathcal{M}$ is a subbundle. The subbundle $\mathcal{E}$ is called *$d\varphi^t$-equivariant* if $\mathcal{K}$ is φ^t-invariant and $d_p\varphi^t\mathcal{E}_p = \mathcal{E}_{\varphi^t p}$ for any $p \in \mathcal{K}$ and any $t \in \mathbb{T}$. Let $\{\varphi^t\}_{t\in\mathbb{T}}$ be a C^1-smooth dynamical system with $\mathbb{T} \in \{\mathbb{R}, \mathbb{Z}\}$ on the n-dimensional Riemannian C^k-manifold $(\mathcal{M}, g)$. A compact set $\mathcal{K} \subset \mathcal{M}$ is called *partially hyperbolic* for $\{\varphi^t\}_{t\in\mathbb{T}}$ if $\mathcal{K}$ is φ^t-invariant and there exist numbers $C > 0, \lambda > 0$ and a splitting $T_{\mathcal{K}}\mathcal{M} = \mathcal{E}^s \oplus \mathcal{E}^u \oplus \mathcal{E}^c$ of the tangent bundle $T_K\mathcal{M}$ into the $d\varphi^t$-invariant subbundles $\mathcal{E}^s$, $\mathcal{E}^u$ and $\mathcal{E}^c$ such that

$$\|d\varphi^t_{|\mathcal{E}^s}\| \leq C\lambda^t \quad \text{and} \quad \|d\varphi^{-t}_{|\mathcal{E}^u}\| \leq C\lambda^t$$

are satisfied for all $t > 0, t \in \mathbb{T}$.

If $\mathcal{E}^c$ belongs to the zero-section, the set $\mathcal{K}$ is called a *hyperbolic set* of $\{\varphi^t\}_{t\in\mathbb{T}}$. The dynamical system $\{\varphi^t\}_{t\in\mathbb{T}}$ on $(\mathcal{M}, g)$ is called *partially hyperbolic* (resp. *hyperbolic*) if $\mathcal{M}$ is compact and is a partially hyperbolic (resp. hyperbolic) set of $\{\varphi^t\}_{t\in\mathbb{T}}$.

For a time-discrete hyperbolic system $\{\varphi^t\}_{t\in\mathbb{Z}}$, the C^1-diffeomorphism $\varphi^1 : \mathcal{M} \to \mathcal{M}$ is called *Anosov diffeomorphism.* A hyperbolic flow $\{\varphi^t\}_{t\in\mathbb{R}}$ generated by F is said to be an *Anosov flow* [2] if $\dim \mathcal{E}^c_p = 1$ and $F(p) \in \mathcal{E}^c_p$, $\forall\, p \in \mathcal{M}$.

A C^1-diffeomorphism $\varphi : \mathcal{M} \to \mathcal{M}$ is said to be an *Axiom A diffeomorphism* [15] and the associated dynamical system $\{\varphi^t\}_{t\in\mathbb{Z}}$ is an *Axiom A system* if the set $\mathcal{NW}(\varphi)$ of non-wandering points is hyperbolic and the set of periodic points of φ is dense in $\mathcal{NW}(\varphi)$.

A flow $\{\varphi^t\}_{t\in\mathbb{R}}$ is called *Axiom A system* if the non-wandering set $\mathcal{NW}(\varphi)$ is hyperbolic and can be written as $\mathcal{Z}_1 \cup \mathcal{Z}_2$, where $\mathcal{Z}_1$ and $\mathcal{Z}_2$ are disjunct compact invariant sets, $\mathcal{Z}_1$ contains a finite number of equilibrium points and in $\mathcal{Z}_2$ the periodic orbits of $\{\varphi^t\}_{t\in\mathbb{R}}$ are dense.

B.3 Degree Theory

Suppose γ is a simple curve in $\mathbb{R}^2$, i.e. a compact, without intersection and oriented curve, given by the continuous and injective parameterization $c : [a, b] \to \mathbb{R}^2$. Suppose also that f is a continuous vector field along γ that does not vanish along this curve. The winding number or *rotation of f along γ* is an integer number $w(f, \gamma)$ which shows how many times the closed continuous path $f : \gamma \to \mathbb{R}^2 \setminus \{0\}$ rotates in the mathematically positive direction around the origin in $\mathbb{R}^2$.

In the case when γ is an oriented C^1-regular curve, $f = (f^1, f^2)$ is a non-vanishing C^1-vector field along γ and α is a 1-form given on $\mathbb{R}^2 \setminus \{0\}$ by $\alpha = \frac{xdy-ydx}{x^2+y^2}$, the winding number can be computed by the formula

$$w(f, p) = \frac{1}{2\pi} \int\limits_{\gamma} \frac{f^1 df^2 - f^2 df^1}{|f|^2} . \tag{B.2}$$

This formula can be used to generalize the winding number to the n-dimensional space $\mathbb{R}^n$. Suppose for this that $\Omega \subset \mathbb{R}^n$ is a bounded domain with C^1-boundary $\partial\Omega$ and $f = (f^1, \ldots, f^n)$ is a non-vanishing C^1-vector field along $\partial\Omega$. Then the *winding number $w(f, \partial\Omega)$ of f* along $\partial\Omega$ is defined by the integral (*Kronecker's integral*)

$$w(f, \partial\Omega) = \frac{1}{\mathrm{vol}(S^{n-1})} \int\limits_{\partial\Omega} \sum_{i=1}^{n} (-1)^{i+1} \frac{f^i}{|f|^n} df^1 \wedge \cdots \wedge \widehat{df^i} \wedge \cdots \wedge df^n . \tag{B.3}$$

Here $\mathrm{vol}(S^{n-1})$ is the volume of the unit sphere S^{n-1} and $\widehat{df^i}$ means that this expression is missing in the exterior product.

It can be shown that if $\Omega \subset \mathbb{R}^n$ is a bounded domain with C^2-boundary $\partial\Omega$ and f is a C^1-vector field on $\overline{\Omega}$ with $0 \notin f(\partial\Omega)$ then

$$w(f, \partial\Omega) = \deg(f, \Omega, 0) . \tag{B.4}$$

If we denote $S := S^{n-1}$ and use a well-known formula for the volume element dS of S we can write Kronecker's integral (B.3) in the form

$$w(f, \partial\Omega) = \frac{\int_{\partial\Omega} f^* dS}{\int_S dS} \,. \tag{B.5}$$

This formula, which can also be used for n-dimensional oriented smooth manifolds, shows again, how many times the sphere S is covered by the image $f(\partial\Omega)$.

Under the consideration of formula (B.4) the relation (B.5) leads to the definition of the (global) degree of a map ϕ. Suppose that $\phi : \mathcal{M} \to \mathcal{N}$ is a smooth map between the n-dimensional smooth manifolds $\mathcal{M}$ and $\mathcal{N}$. Then the uniquely defined number $\deg\phi$ which satisfies the relation

$$\int_{\mathcal{M}} \phi^* \omega = \deg \phi \int_{\mathcal{N}} \omega \tag{B.6}$$

for any n-form on $\mathcal{N}$, is called the *global degree* of ϕ.

A measure of the non-injectivity of a given map is the multiplicity function. Suppose that $\mathcal{M}_1$ and $\mathcal{M}_2$ are two arbitrary sets and $\phi : \mathcal{M}_1 \to \mathcal{M}_2$ is a map. The *multiplicity function* $\mathcal{N}(\phi, \mathcal{K}, u)$ of ϕ with respect to a set $\mathcal{K} \subset \mathcal{M}_1$ at the point $u \in \mathcal{M}_2$, is the cardinality of the set $\{v \in \mathcal{K} | \phi(v) = u\}$. Suppose that $\phi : \mathcal{M} \to \mathcal{M}$ is a C^1-map on the orientable n-dimensional smooth Riemannian manifold $(\mathcal{M}, g)$. It can be shown that if the determinant $\det(d_u\phi)$ of the tangent map $d_u\phi$ is positive on $\mathcal{M}$ then the multiplicity function of ϕ with respect to $\mathcal{K}$ at u coincides with the local degree of ϕ at u.

In the following, we need the extension theorem of Dugundij [4].

Theorem B.3 *Suppose that $(\mathcal{X}, d)$ is a metric space, $(\mathcal{Y}, \|\cdot\|)$ is a Banach space, $\mathcal{N} \neq \emptyset$ is a closed subset of $\mathcal{X}$ and $T : \mathcal{N} \to \mathcal{Y}$ is a continuous map. Then there exists a continuous extension $\widetilde{T} : \mathcal{X} \to \mathrm{co}\, T(\mathcal{N})$ (convex hull).*

Consider a map $\phi : \overline{\mathcal{D}} \subset \mathcal{M} \to \mathcal{N}$ where $\mathcal{M}, \mathcal{N}$ are oriented differentiable n-dimensional manifolds and $\mathcal{D}$ is relatively compact in $\mathcal{M}$.

Suppose that $\phi \in C^0(\overline{\mathcal{D}}) \cap C^1(\mathcal{D})$. Let $\mathcal{Z}$ denote the critical points of ϕ on $\mathcal{D}$, i.e. set of points in $\mathcal{D}$ at which the Jacobian J_ϕ of ϕ vanishes. Let $q \in \phi(\overline{\mathcal{D}})$ be such that $q \in \phi(\partial\mathcal{D})$ and $\phi^{-1}(q) \cap \mathcal{Z} = \emptyset$. By the inverse function theorem the set $\phi^{-1}(q) \subset \overline{\mathcal{D}}$ is discrete and therefore finite since $\overline{\mathcal{D}}$ is compact. Because the manifolds are oriented, the sign of J_ϕ is defined at each point of $\phi^{-1}(q)$. The *degree* $\deg(\phi, \mathcal{D}, q)$ *of ϕ at q with respect to $\mathcal{D}$* is defined by

$$\deg(\phi, \mathcal{D}, q) = \sum_{p \in \phi^{-1}(q)} \operatorname{sign} J_\phi(p) \,.$$

For $q \notin \phi(\overline{\mathcal{D}})$ we define $\deg(\phi, \mathcal{D}, q) = 0$. An important theorem says that the degree of ϕ is constant in every connected component of $\mathcal{N} \backslash \phi(\partial\mathcal{D})$. In order to define the degree for any $\phi \in C^0(\overline{\mathcal{D}})$ we use the following theorem: The map $C^1(\mathcal{D}) \cap C^0(\overline{\mathcal{D}}) \to \mathbb{Z}$ given by $\phi \mapsto \deg(\phi, \mathcal{D}, q)$, $q \notin \phi(\partial\mathcal{D})$, is continuous with respect to the C^0-topology. This theorem allows us to define the degree for a continuous map ϕ at q with respect to $\mathcal{D}$ as

$$\deg(\phi, \mathcal{D}, q) = \lim_{n\to+\infty} \deg(\phi_n, \mathcal{D}, q),\ q \notin \phi(\partial\mathcal{D})$$

where $\{\phi_n\}$ is a sequence of C^1-maps which converges to ϕ in the C^0-topology.

Theorem B.4 *(Brouwer's fixed point theorem) Let $\mathcal{X}$ be a topological space which is homeomorphic to a bounded and convex set of $\mathbb{R}^n$. If T is a continuous map, which maps $\mathcal{X}$ into itself, then T has at least one fixed point in $\mathcal{X}$.*

B.4 Homology Theory

Suppose that in the Euclidean space $\mathbb{R}^m$ there are given $k+1$ points $p_0, p_1, \dots, p_k$ such that the vectors $p_i - p_0$, $i = 1, \dots, k$, are linearly independent. The convex hull of $\{p_0, \dots, p_k\}$, i.e. the set of all points $x = \sum_{i=0}^k \lambda_i p_i$ with $\sum_{i=0}^n \lambda_i = 1$, $\lambda_i \geq 0$, is called *k-simplex* and is denoted by $\sigma^k \equiv (p_0, p_1, \dots, p_k)$. The points $p_0, \dots, p_k$ are the *vertices of* σ^k . A *face* of a k-simplex σ^k is any simplex whose vertices are a subset of those of σ^k. The orientation of the k-simplex $\sigma^k \equiv (p_0, p_1, \dots, p_k)$, is by definition the orientation of the linear space with basis $p_1 - p_0, p_2 - p_0, \dots, p_k - p_0$. If σ^k is an oriented simplex, $-\sigma^k$ is the same set of points with opposite orientation. A *simplicial complex* $\mathcal{K}$ is a finite collection of simplices of $\mathbb{R}^m$ such that if $\sigma_1^k \in \mathcal{K}$ then so are all its faces, and if $\sigma_1^k, \sigma_2^l \in \mathcal{K}$ then $\sigma_1^k \cap \sigma_2^l$ is either a face of σ_1^k or is empty.

Let $\mathcal{K}$ be a simplicial complex and consider the set theoretic union $|\mathcal{K}| \subset \mathbb{R}^m$ of all simplices from $\mathcal{K}$. Introduce on $|\mathcal{K}|$ a topology that is the strongest of all topologies in which the embedding of each simplex into $|\mathcal{K}|$ is continuous. The set $|\mathcal{K}|$ is the associated *polyhedron* . The polyhedron $|\mathcal{K}|$ is said to be *triangulated* by the simplicial complex. A *triangulation* of a manifold $\mathcal{M}$ is a simplicial complex $\mathcal{K}$ together with a homeomorphism $h : |\mathcal{K}| \to \mathcal{M}$. A *k-chain* in the simplicial complex $\mathcal{K}$ with coefficients in an abelian group G is a formal sum $c_k = \sum g_i \sigma_i^k$, $g_i \in G$, σ_i^k a k-simplex in $\mathcal{K}$. The set $C_k(\mathcal{K}, G)$ of k-chains in $\mathcal{K}$ with coefficients in $(G, +)$ together with the addition

$$c_k + c'_k = \sum (g_i + g'_i)\sigma_i^k$$

form an abelian group. Let $\sigma^{k+1} \equiv (p_0, \dots, p_{k+1})$ be an oriented $k+1$ simplex. The *boundary* $\partial\sigma^{k+1}$ is the k-chain defined by

$$\partial\sigma^{k+1} = \sum_{i=0}^{k+1} (-1)^i (p_0, \dots, \widehat{p_i}, \dots, p_{k+1})$$

where $\widehat{\ }$ means that the symbol should be deleted. The boundary of the k-chain $\sum_i g_i \sigma_i^k$ is $\sum_i g_i \partial\sigma_i^k$. A direct computation shows that the maps

$$C_{k+1}(\mathcal{K}, G) \xrightarrow{\partial} C_k(\mathcal{K}, G) \xrightarrow{\partial} C_{k-1}(\mathcal{K}, G)$$

satisfy $\partial^2 = \partial \circ \partial = 0$. The space of *k-cycles* is given by

$$Z_k(\mathcal{K}, G) = \{c \in C_k(\mathcal{K}, G) | \partial c = 0\},$$

the space of *k-boundaries* is defined by

$$B_k(\mathcal{K}, G) = \{\partial c \,|\, c \in \mathcal{C}_{k+1}(\mathcal{K}G)\}.$$

Then the factor-group $H_k(\mathcal{K}, G) = Z_k(\mathcal{K}, G)/B_k(\mathcal{K}, G)$ defines the k-th *homology group* of $\mathcal{K}$ with coefficients from G. For various triangulations $h : |\mathcal{K}| \to \mathcal{M}$ of a given manifold, the associated homology groups $H_k(\mathcal{K}, G)$ are independent of the concrete triangulation. Thus we can introduce the k-th *homology group* $H_k(\mathcal{M}, G)$ of the manifold $\mathcal{M}$ with coefficients from G.

Let us now consider $G = \mathbb{Z}$ and let us introduce the abbreviations

$$\mathcal{B}_k(\mathcal{M}) \equiv \mathcal{B}_k(\mathcal{M}, \mathbb{Z}), \quad \mathcal{C}_k(\mathcal{M}) \equiv \mathcal{C}_k(\mathcal{M}, \mathbb{Z}) \quad \text{and} \quad H_k(\mathcal{M}) \equiv H_k(\mathcal{M}, \mathbb{Z}).$$

It can be shown that $H_k(\mathcal{M})$ is a finitely generated abelian group, i.e. there exists a finite number of elements $h_1, \dots, h_p$ such that any element in $H_k(\mathcal{M})$ may be written as $h = \sum_{i=1}^p a_i h_i$ with $a_i \in \mathbb{Z}$ and none of the elements $h_1, \dots, h_p$ can be written as such a sum of the remaining ones. The number p is the rank of the group. It is well-known that any group $H_k(\mathcal{M})$ has only generators of finite or infinite order and can be written as the direct sum of one-dimensional free subgroups and one-dimensional subgroups of finite order, i.e.

$$H_k(\mathcal{M}) = \underbrace{\mathbb{Z} \oplus \mathbb{Z} \oplus \cdots \oplus \mathbb{Z}}_{b_k(\mathcal{M})-\text{times}} \oplus F_1 \oplus \cdots \oplus F_{m_k(\mathcal{M})}.$$

The number $b_k(\mathcal{M})$ of free generators is the k-th *Betti-number*, the orders $r_1, \dots, r_{m_k(\mathcal{M})}$ of the remaining generators are the *torsion coefficients*. Let us recall that there are many similarities between chains and differential forms. A closed differential form ω on $\mathcal{M}$, i.e. such that $d\omega = 0$, is also called *cocycle*. An exact differential form ω, i.e. such that there exists another differential form Θ with $\omega = d\Theta$, is also called *coboundary*. Let $Z^k(\mathcal{M})$ denote the set of all cocycles with the natural structure of an additive group over $\mathbb{Z}$ and let $\mathcal{B}^k(\mathcal{M})$ denote the subgroup of $\mathcal{Z}^k(\mathcal{M})$ consisting of all coboundaries of degree k. The factor-group $H^k(\mathcal{M}) =$

$Z^k(\mathcal{M})/\mathcal{B}^k(\mathcal{M})$ is the *k-th cohomology group* of $\mathcal{M}$. By the de Rham theorem this group is isomorphic to the dual of $H_k(\mathcal{M})$.

For a given triangulation $h : |K| \to \mathcal{M}$ of the manifold $\mathcal{M}$ we put

$$\chi = \sum_{k=0}^{\dim \mathcal{M}} (-1)^k \text{ card } \{\sigma \mid \sigma \quad \text{is a } k\text{-dimensional simplex from } K\} .$$

One can show that χ does not depend on the triangulation of $\mathcal{M}$ and this defines a topological invariant $\chi(\mathcal{M})$ which is called *Euler characteristic* of $\mathcal{M}$.

The *Euler-Poincaré formula* says that on a compact orientable manifold $\mathcal{M}$

$$\chi(\mathcal{M}) = \sum_{k=0}^{\dim \mathcal{M}} (-1)^k b_k(\mathcal{M}) ,$$

where $b_k(\mathcal{M})$ are the k-th Betti-numbers of $\mathcal{M}$. The Euler characteristic $\chi(\mathcal{M})$ of a compact orientable manifold is also equal to the sum of the indicies of the (isolated) zeros of any smooth vector field f on $\mathcal{M}$ (*Poincaré Hopf theorem*).

B.5 Simple δ-Linked Parameterized m-Boundaries

Let us introduce the following concept [8].

Suppose $(\mathcal{X}, \|\cdot\|)$ is a Banach space and m is a non-negative integer. A *parameterized m-boundary* (resp. *parameterized* $(m+1)$*-surface*) in $\mathcal{D} \subset \mathcal{X}$ is a continuous function Φ whose domain of definition is $\partial\,\mathcal{U}$ (resp. $\overline{\mathcal{U}}$), the boundary (resp. closure) of a non-empty bounded open connected set $\mathcal{U} \subset \mathbb{R}^{m+1}$ and whose range is in $\mathcal{D}$.

A parameterized m-boundary Φ is the *parameterized boundary* of a parameterized $(m+1)$-surface Ψ, denoted $\Phi = \partial\,\Psi$, if $\Phi = \Psi_{|\partial\mathcal{U}}$, the restriction of Ψ to $\partial\,\mathcal{U}$. A parameterized m-boundary is *simple* if it is one-to-one on its domain. The *trace* of a parameterized m-boundary (resp. $(m+1)$-surface) Φ is the set $\Phi(\partial\,\mathcal{U})$ (resp. $\Phi(\overline{\mathcal{U}})$). The extension theorem of Dugundji (Theorem B.3) implies that any parameterized m-boundary $\Phi : \partial\,\mathcal{U} \to \mathcal{X}$ is the parameterized boundary of a parameterized $(m+1)$-surface in the convex hull $\text{co}\,\Phi(\partial\,\mathcal{U})$.

A sequence of parameterized $(m+1)$-surfaces $\Phi_k : \overline{\mathcal{U}}_k \to \mathcal{X}$ is *compact* (resp. *bounded*) if the closure of the set $\bigcup_k \Phi_k(\overline{\mathcal{U}}_k)$ is compact (resp. bounded). If $u_0 \in \mathbb{R}^{m+1}$, $\delta > 0$, let $\mathcal{N}_\delta(u_0) = \{u \in \mathbb{R}^{m+1} \big| \, |u - u_0| < \delta\}$, where $|\cdot|$ is the Euclidean norm in $\mathbb{R}^{m+1}$.

Let $\mathcal{N}_0, \mathcal{N}_-, \mathcal{N}_+$ denote the sets of points $u \in \mathcal{N}_1(0)$, $u = (u_1, \ldots, u_m, u_{m+1})$, for which $u_{m+1} = 0$, $u_{m+1} < 0$, $u_{m+1} > 0$, respectively.

A point $u \in \partial\,\mathcal{U}$ is an *ordinary point* of $\partial\,\mathcal{U}$ if there exists a neighborhood $\mathcal{W}$ of u_0 and a homeomorphism $h : \mathcal{N}_1(0) \to \overline{\mathcal{W}}$ such that $h(\mathcal{N}_0) = \mathcal{W} \cap \partial\,\mathcal{U}$, $h\,(\mathcal{N}_-) = \mathcal{W} \cap (\mathbb{R}^{m+1} \setminus \overline{\mathcal{U}})$ and $h(\mathcal{N}_+) = \mathcal{W} \cap \mathcal{U}$.

A parameterized m-boundary Φ has an *m-dimensional tangent* space $\mathcal{X}_1$ at $x_0 = \Phi(u_0)$ if u_0 is an ordinary point of $\partial\mathcal{U}$ (realized by a homeomorphism h), $\Phi_0 h_{|\mathcal{N}_0}$ is Frechét differentiable at 0 and the range $\mathcal{X}_1$ of the derivative $L = (\Phi_0 h_{|,\upsilon_0})(0)$ has dimension m. Let $\mathcal{X}_1$ be an m-dimensional subspace of $\mathcal{X}$. Then there exists a basis $\{e_1, \ldots, e_m\}$ of $\mathcal{X}_1$ (called the *Auerbach basis*), such that $\|e_i\| = 1$ and $1 = \|q_i\|_1 = \sup\{\|q_i x\| \,\big|\, x \in \mathcal{X}_1, \|x\| = 1\}$, where $q_i x$ is the ith coordinate of x referred to this basis, $i = 1, \ldots, m$.

The Hahn-Banach is used to extend the linear functionals q_i to $\mathcal{X}$ with $\|q_i\| = 1$ and to define the projections P_1, P_2 by $P_1 = \sum_{i=1}^m e_i q_i$, $P_2 = I - P_1$. It follows that $P_j^2 = P_j$ and $\mathcal{X} = \mathcal{X}_1 \oplus \mathcal{X}_2$, where $\mathcal{X}_j = P_j\mathcal{X}$, $j = 1, 2$. The projections satisfy $\|P_1\| \leq m$, $\|P_2\| \leq m + 1$. Define a norm $\|\cdot\|_1$ on $\mathcal{X}$ by

$$\|x\|_1 = (\|q_1 x\|^2 + \cdots + \|q_m x\|^2 + \|P_2 x\|^2)^{1/2}.$$

Then

$$(m+1)^{-1/2}\|x\| \leq \|x\|_1 \leq (m + (m+1)^2)^{1/2}\|x\|$$

so that $\|\cdot\|$ and $\|\cdot\|_1$ are equivalent norms and the equivalence is uniform with respect to the choice of $\mathcal{X}_1$. If $\delta > 0$ and $x_0 \in \mathcal{X}$, let $\mathcal{B}_\delta^1(x_0) = \{x \in \mathcal{X} \,\big|\, \|x - x_0\|_1 < \delta\}$. If $\gamma \geq \delta$, let

$$\mathcal{T}_{\gamma,\delta}^1(x_0) = \bigcup_x \{\mathcal{B}_\delta^1(x) \,\big|\, \|x - x_0\|_1 = \gamma,\ P_1(x - x_0) = 0\},$$

a *torus centered on x_0 with axis P_1*. A simple parameterized m-boundary $\Phi : \partial\mathcal{U} \to \mathcal{X}$ is called *δ-linked* (with some $\delta > 0$) if there exist $\gamma \geq \delta$, an ordinary point $u_0 \in \partial\mathcal{U}$, a parameterized m-boundary $\widehat{\Phi} : \partial\mathcal{U} \to \mathcal{X}$, $x_0 = \widehat{\Phi}(u_0)$ and an m-dimensional subspace $\mathcal{X}_1$ of $\mathcal{X}$ such that

(D1) $\widehat{\Phi}$ is one-to-one on $\widehat{\Phi}^{-1}(\mathrm{co}\mathcal{T}_{\gamma,\delta}^1(x_0))$;
(D2) $\widehat{\Phi}(\partial\mathcal{U}) \cap \mathrm{co}\mathcal{T}_{\gamma,\delta}^1(x_0) = (x_0 + \mathcal{X}_1) \cap \mathrm{co}\mathcal{T}_{\gamma,\delta}^1(x_0)$;
(D3) $\|\Phi(u) - \widehat{\Phi}(u)\|_1 < d_1(\widehat{\Phi}(u), \mathcal{T}_{\gamma,\delta}^1(x_0))$, if $u \in \partial\mathcal{U}$. (Here d_1 is the distance defined by $\|\cdot\|_1$.)

Remark B.1 (a) Note that the perturbation $\widehat{\Phi}$ of Φ satisfying **(D1)** – **(D3)** need not be a simple parameterized boundary.
(b) A simple parameterized m-boundary without points where a tangent space exists, may be δ-linked. The Koch curve (Sect. 3.2.3, Chap.3) is δ-linked.

Let us state the following three propositions, due to M. Y. Li and Muldowney [8].

Proposition B.1 *When $m > 0$ is an integer, a simple parameterized m-boundary Φ is δ-linked for some $\delta > 0$ if it has an m-dimensional tangent space $\mathcal{X}_1$ at some point $x_0 = \Phi(u_0)$, $u_0 \in \partial\mathcal{U}$.*

Proposition B.2 *Let Φ and Ψ be two ordinary parameterized m-boundaries which have the same domain and range. Then if Φ is δ-linked, so also is Ψ.*

If $|\cdot|$ is any norm on $\mathbb{R}^{m+1}$ or $\mathcal{X}$, and if $\mathcal{B}$ is a set in one of these spaces, let $|\mathcal{B}| = \sup\{|x-y| \,|\, x, y \in \mathcal{B}\}$ be the *diameter* of $\mathcal{B}$.

Proposition B.3 *Suppose $\Phi : \overline{\mathcal{U}} \to \mathcal{X}$ is a parameterized $(m+1)$-surface and that its parameterized m-boundary is simple and δ-linked. Then*

$$c_m^{-1}\delta^{m+1} \leq \sum_i |\mathcal{B}_i|^{m+1}$$

if $\{\mathcal{B}_i\}$ is any collection of sets such that $\Phi(\overline{\mathcal{U}}) \subset \bigcup_i \mathcal{B}_i$, where $c_m = 2^{m+1}[m + (m+1)^2]^{(m+1)/2}$.

B.6 Geometric Measure Theory

All of the following basic facts from geometric measure theory can be found in [5, 11]. Suppose $m \geq 0$ is a positive integer. A set $\mathcal{Z} \subset \mathbb{R}^n$ is *m-rectifiable* if $\mathcal{Z}$ is $\mu_H(\cdot, m) \equiv \mu_H^m(\cdot)$-measurable, $\mu_H(\mathcal{Z}, m) < \infty$, and there exist m-dimensional C^1-submanifolds $\{\mathcal{M}_i\}_{i=1}^{\infty}$ in $\mathbb{R}^n$ such that $\mu_H(\mathcal{Z} \setminus \bigcup_{i=1}^{\infty} \mathcal{M}_i, m) = 0$. For $\mu_H(\cdot, m)$ - a.a $x \in \mathcal{Z}$, the tangent spaces at x to distinct $\mathcal{M}_i$ containing x are equal. Let $T_x\mathcal{Z}$ be this tangent space where it exists. Both the standard inner product and the duality pairing for all spaces is denoted by $\langle \cdot, \cdot \rangle$.

The space of all C^∞-differential m-forms in $\mathbb{R}^n$ with compact support is $\mathcal{D}^m$. For $\beta \in \mathcal{D}^m$ we define

$$\|\beta\| := \sup\{\langle \beta, \xi \rangle \mid \xi \in \Lambda^m(\mathbb{R}^n),\ |\xi| = 1, \xi \text{ simple } m\text{-vector}\}.$$

The dual space is denoted by $\mathcal{D}_m$ and is called the space of *m-dimensional currents*. Suppose θ is a *multiplicity function* on $\mathcal{Z}$, i.e. an $\mu_H(\cdot, m)$-measurable function θ with domain $\mathcal{Z}$ and range a subset of the positive integers, such that $\int_{\mathcal{Z}} \theta d\mu_H^m < \infty$.

Suppose also that there is an *orientation* ξ, i.e. a μ_H^m-measurable function ξ with domain $\mathcal{Z}$ such that for μ_H^m - a.a. $x \in \mathcal{Z}$ the orientation $\xi(x)$ is one of the two simple m-vectors associated with $T_x\mathcal{Z}$, i.e. for μ_H^m - a.a. $x \in \mathcal{Z}$ we have $\xi(x) = \tau_1(x) \wedge \cdots \wedge \tau_m(x)$, where $\tau_1(x), \ldots, \tau_m(x)$ is an orthonormal basis in $T_x\mathcal{Z}$.

A linear operator on C^∞ m-forms β given by

$$T(\beta) := \int_{\mathcal{Z}} \theta(x) < \xi(x), \beta(x) > d\mu_H^m$$

is called *m-dimensional rectifiable current*. The set of all m-dimensional rectifiable currents forms an abelian group which is denoted by $\mathcal{R}_m$.

For each $T \in \mathcal{R}_m$, $m \geq 1$, we define the *boundary operator* ∂T given by Stokes formula

$$\partial T(\beta) = T(d\beta)\,,\ \beta \text{ an arbitrary } C^\infty\ (m-1)\text{-form}\,.$$

It is not necessarily true that $\partial T \in \mathcal{R}_{m-1}$. The abelian group of m-dimensional *integral currents* is given by

$$\mathbf{I}_m := \{T \in \mathcal{R}_m \mid \partial T \in \mathcal{R}_{m-1}\}, \ m = 1, 2, \ldots,$$
$$\mathbf{I}_0 := \mathcal{R}_0 .$$

We enlarge $\mathcal{R}_m$ to the abelian group of m-dimensional *integral flat chains*, or *m-chains*, defined by

$$\mathcal{F}_m := \{R + \partial S \mid R \in \mathcal{R}_m, S \in \mathcal{R}_{m+1}\} .$$

The operator ∂ is extendible to a group homomorphism $\partial : \mathcal{F}_m \to \mathcal{F}_{m-1}$ if $m \geq 1$.

For $T \in \mathcal{R}_m$ we define the *mass* of T by

$$\mathbf{M}(T) := \int_{\mathcal{Z}} \theta d\mu_H^m .$$

One can extend the definition of $\mathbf{M}$ to $\mathcal{D}_m$. For $T \in \mathcal{D}_m$ we define

$$\mathbf{M}(T) := \sup\{T(\beta) \mid \|\beta\| \leq 1 , \ \beta \in \mathcal{D}^m\}.$$

Then one has

$$\mathcal{R}_m = \mathcal{F}_m \cap \{T \mid \mathbf{M}(T) < \infty\} ,$$
$$\mathbf{I}_m = \mathcal{R}_m \cap \{T \mid \mathbf{M}(\partial T) < \infty\} .$$

One now defines the *integral flat "norm"* on $\mathcal{F}_m$ by

$$\mathcal{F}(T) := \inf\{\mathbf{M}(R) + \mathbf{M}(S) \mid T = R + \partial S, R \in \mathcal{R}_m, T \in \mathcal{R}_{m+1}\}$$

and the *integral flat metric* by $\mathcal{F}(T_1, T_2) := \mathcal{F}(T_1 - T_2)$.

If $T \in \mathcal{F}_m$, $m \geq 1$, and $\partial T = 0$ (or if $T \in \mathcal{F}_0$), we say that T is an m-dimensional *integral flat cycle* or *m-cycle*.

If $m \geq 1$, it follows by a cone construction that $T = \partial S$ for some $S \in \mathcal{F}_{m+1}$.

If $T \in \mathcal{F}_m$ and $T = \partial S$ for some $S \in \mathcal{F}_{m+1}$, we say T is an m-dimensional *integral flat boundary*, or *m-boundary*. Thus if $m \geq 1$, every m-cycle is an m-boundary. Let $\mathcal{E}_m$ denote the space of all currents $T \in \mathcal{D}_m$ with compact support.

If $T \in \mathcal{E}_m$ and $\phi : \mathbb{R}^n \to \mathbb{R}^n$ is a C^∞-map, then one defines the *push-forward* $\phi_* T \in \mathcal{D}_m$ by

$$(\phi_* T)(\beta) = T(\phi^* \beta) , \ \forall \beta \in \mathcal{D}^m .$$

Here $\phi^* \beta$ denotes the pullback of the k-form β (see Sect. A.3). In case T corresponds to some oriented manifold $\mathcal{Z}$, then $\phi_* T$ corresponds to the oriented image $\phi(\mathcal{Z})$. The push-forward $\phi_* T$ has the following important properties:

(a) $\phi_* \partial T = \partial \phi_* T$;
(b) $\operatorname{supp} \phi_* T \subset \phi(\operatorname{supp} T)$.

The *normal currents* are given by

$$N_m := \{T \in \mathcal{E}_m \mid \mathbf{M}(T) + \mathbf{M}(\partial T) < \infty\}.$$

For $T \in \mathcal{D}_m$ we define the *flat norm* $\mathbf{F}$ of currents by

$$\mathbf{F}(T) := \min\{\mathbf{M}(A) + \mathbf{M}(B) \mid T = A + \partial B, \\ A \in \mathcal{E}_m, B \in \mathcal{E}_{m+1}\}.$$

The set $\mathbf{F}_m :=$ $\mathbf{F}$-closure of N_m in $\mathcal{E}_m$ defines the *real flat chains*. It is shown in [5] that $\mathcal{F}_m \subset \mathbf{F}_m$.

As a corollary from a theorem in 4.1.20 [5] we have the following:

Theorem B.5 *If* $T \in \mathbf{F}_m(\mathbb{R}^n)$ *and* $\mu_H(\operatorname{supp} T, m) = 0$, *then* $T = 0$.

References

1. Alexandrov, P.S.: Introduction to Set Theory and General Topology. Nauka, Moscow (1977). (Russian)
2. Anosov, D.V.: Geodesic flows on closed Riemannian manifolds with negative curvature. Proc. Steklov Inst. Math. **90**, (1967). (Russian)
3. Brouwer, L.E.J.: Beweis der Invarianz der Dimensionszahl. Math. Ann. **70**, 161–165 (1911)
4. Dugundij, J.: An extension of Tietze's theorem. Pacific J. Math. **1**, 353–367 (1951)
5. Federer, H.: Geometric Measure Theory. Springer, New York (1969)
6. Hurewicz, W., Wallman, H.: Dimension Theory. Princeton University Press, Princeton (1948)
7. Kneser, H.: Reguläre Kurvenscharen auf Ringflächen. Math. Ann. **91**, 135–154 (1924)
8. Li, M.Y., Muldowney, J.S.: Lower bounds for the Hausdorff dimension of attractors. J. Dyn. Diff. Equ. **7**(3), 457–469 (1995)
9. Milnor, J.W.: Topology from the Differentiable Viewpoint. Virginia University Press, Charlottesville (1965)
10. Markley, N.G.: The Poincaré-Bendixson theorem for the Klein bottle. Trans. Amer. Math. Soc. **135**, 139–165 (1969)
11. Morgan, F.: Geometric Measure Theory. A Beginners's Guide. Academic Press INC, San Diego, CA (1988)
12. Pugh, C.C.: An improved closing lemma and a general density theorem. Amer. J. Math. **89**, 1010–1021 (1967)
13. Sard, A.: The measure of the critical values of differentiable maps. Bull. Amer. Math. Soc. **48**, 883–890 (1942)
14. Schwartz, A.J.: A generalization of the Poincaré-Bendixson theorem to closed two-dimensional manifolds. Amer. J. Math. **85**, 453–458 (1963)
15. Smale, S.: Differential dynamical systems. Bull. Amer. Math. Soc. **73**, 747–817 (1976)
16. Whitney, H.: Differentiable manifolds. Ann. Math., II. Ser. **37**, 645–680 (1936)

Index

N. Kuznetsov and V. Reitmann, *Attractor Dimension Estimates for Dynamical Systems: Theory and Computation*, Emergence, Complexity and Computation 38,
https://doi.org/10.1007/978-3-030-50987-3

T

Printed in the United States
by Baker & Taylor Publisher Services